THE INDUSTRIAL RAILWAYS
&
LOCOMOTIVES
OF COUNTY DURHAM

WITH A HISTORY OF THEIR OWNERS & SITES

PART 1

(excluding the National Coal Board and the fluorspar industry)

**COLIN E. MOUNTFORD
&
DAVE HOLROYDE**

INDUSTRIAL RAILWAY SOCIETY 2006

Published by the INDUSTRIAL RAILWAY SOCIETY
at 24, Dulverton Road, Melton Mowbray, Leicestershire. LE 13 0SF

© INDUSTRIAL RAILWAY SOCIETY 2006

ISBN 1 901556 38 7

Visit the Society at www.irsociety.co.uk

This book is one of a series of books about industrial railways and locomotives published by the Industrial Railway Society and covering parts of Britain and certain foreign countries. Details of the other titles currently available can be obtained by sending a stamped, self-addressed envelope to Mr.S.Geeson, Hon. Sales Officer, 24, Dulverton Road, Melton Mowbray, Leicestershire. LE13 0SF

British Library Cataloguing in-Publication Data. A catalogue record of this book is available from the British Library.

All rights reserved. No part of this publication may be reproduced, stored in a retrieval system or transmitted in any form or by any means without prior permission in writing from the Industrial Railway Society. Within the UK, exceptions are allowed in respect of any fair dealing for the purpose of private research or private study or criticism or review as permitted under the Copyright, Designs & Patents Act 1988.

Designed and printed by Hillprint Media, Prime House, Park 2000, Newton Aycliffe, Co.Durham, DL5 6AR, 01325 245 555.

Front cover:
Crane tank locomotives were a feature of Durham's shipbuilding and iron & steel industries. ROKER, RSHN 7006/1940, pauses at the shipyard of William Doxford & Sons Ltd, Sunderland, on 25th August 1969.

Back cover:
The magnificent company crest of The Lambton, Hetton & Joicey Collieries Ltd, the largest coal company in the world when it was formed in 1924.

Contents

	Page
Introduction, Acknowledgments and Explanatory Notes	1
Key Maps and list of System Maps	7
John Abbot & Co Ltd	31
William Barker & Co	38
The Cargo Fleet Iron Co Ltd	71
The Darlington Forge Ltd	116
The Easington Coal Co Ltd	151
The Felling Coal, Iron & Chemical Co Ltd	156
Gateshead Borough Council	164
The Hamsterley Colliery Co Ltd	166
Ibbotson & Co	202
Jarrow Chemical Co Ltd	208
Lafarge Cement UK	212
The Mainsforth Coal Co	240
Newall's Insulation Co Ltd	246
Ord & Maddison Ltd	267
Page Bank Brick Co Ltd	273
Raine Industries plc	321
Sir S. A. Sadler Ltd	337
Tarmac (Northern) Ltd	387
H. R. Vaughan & Co Ltd	395
The Wallsend & Hebburn Coal Co Ltd	398
Index 1 Alphabetical list of owners	442
Index 2 Locations	450
Index 3 Locomotives	457

Frontispiece : The six Spencer kilns at Fulwell & Southwick Quarries, Sunderland, owned by Sir Hedworth Williamson's Limeworks Ltd (copied from the September 1954 edition of the British Bulletin of Commerce). *One of the Hudswell Clarke 0-4-0STs is propelling wagons of limestone to charge the kilns, which were over 80 feet high. The kilns produced hydrated lime, which can be seen in the wagons at the loading bays.*

Introduction

Soon after its formation in 1949, the Birmingham Locomotive Club (Industrial Locomotive Information Section) began to publish small 'pocket books', bringing together the information then known on industrial locomotives in a particular county or area of Britain. The pocket book covering Durham was published in 1962. More information came to light, and in 1977 the Industrial Railway Society, as the BLC (ILIS) had become, published a much larger, 500-page 'Durham Handbook', for which Les Charlton and Colin Mountford were joint editors. This included not only locomotives but also details of the sites where they worked and the companies which owned them, together with summaries of the history and operation of the related private railways. Subsequent titles in the Handbook series continued these developments. As the interest in local history began to grow dramatically, so these books have gained a wider circulation. Now the amount of material available for Durham is so great that three volumes are needed! The first, *The Private Railways of County Durham* (ISBN 1-901556-29-8), was published as a 'stand alone' monograph in 2004. It took advantage of all the benefits of computerisation and digital technology, which had already been adopted for the 'Handbook' series itself, but also adopted a new page size. The present volume, the first half of a new edition of the Durham Handbook proper, comprises an alphabetical list of all the owners of industrial locations in Durham with the exception of the National Coal Board and the fluorspar industry, giving details of ownership and a summary of the history of each site, followed by information about their locomotives. The final volume, Part 2 of the new Durham Handbook, will comprise the NCB and the fluorspar industry, the work of and locomotives used by contractors, dealers and hirers and also locomotives now preserved. A wide range of maps, together with appropriate indexes, is also included. Readers should note that 'Durham' now observes the southern boundary established by changes in local government established in 1974, when the areas of Stockton, Hartlepool and Billingham were transferred to Cleveland. Since Durham's private railways are covered in detail in the monograph cited, only a summary of each system and its locomotive fleet have been included here.

Durham has been one of the most industrialised counties in Britain. All but the far south of the county fell within the 'Great Northern Coalfield', and with its rivers giving access to the sea and to London, it was the first major coalfield to be developed, the first of many 'firsts' in the county. As new pits were sunk at greater distances from the rivers, transporting the coal became a major issue, and early in the nineteenth century Durham saw the development of the steam locomotive and the first public railway. But the county's hilly terrain was unsuitable for the universal adoption of locomotives and so rope haulage, in its many forms, was widely adopted. As railways developed, so did heavy engineering. Local supplies of iron ore led to the growth of an iron industry, followed by steel once scientific discoveries made mass production possible. As the demand for coal increased, the first iron screw collier was built on the River Tyne, and shipbuilding and ship-repairing became a major industry along the rivers, adding further heavy engineering and foundries and creating even more demand for steel. Durham coal produced excellent coke, in demand all over Britain, and hundreds of Durham collieries built coke ovens, firstly of the beehive type, but latterly vertical 'batteries' recovering the by-products. Many collieries also produced refractory clay (known locally as seggar clay) and used it to make firebricks to line the ovens, while thick beds of ordinary clay stimulated the brick and sanitaryware industries. A chemical industry initially based on minerals extracted from sea water and later on imported ores developed on the River Tyne. All over the county were major deposits of limestone, not least in Weardale, so industry spread here too. And not just limestone: Weardale had long produced lead, followed by fluorspar, while Teesdale produced whinstone (for road setts), barytes and smaller quantities of other minerals. In 1913 alone the members of the Durham Coal Owners Association produced 41,748,687 tons of coal, together with 299,683 tons of clay, and manufactured 5,099,236 tons of coke – huge figures by any standards.

Colin Mountford first came to Durham as a university student in 1961, Dave Holroyde as a local government officer in 1973. If someone had said then that by the mid 1990s there would be no collieries left in Durham it would have seemed unbelievable. But not only no collieries; no coke ovens, no shipbuilding, no iron and steel works, almost no brickworks, almost no heavy engineering or quarrying, and very little of the infrastructure that served them. Durham has seen the loss of almost all of its old industrial base, and with a few exceptions, the new industries which were developed after 1960 have vanished too as manufacturing has sought cheaper labour costs abroad in 'the global economy'. With the beauty of its countryside re-emerging after the demise of its industries, albeit with fragments of its industrial heritage preserved, Durham in the twenty-first century now looks to improving its prosperity from tourism and the growth of new sources of employment.

This volume represents nearly a century of work by many people. Much of the pioneering early research was done by R.H. Inness (Darlington) and L.G. Charlton (Newcastle upon Tyne), by members of the Stephenson Locomotive Society's Newcastle branch and by visitors to the area before and after World War 2. As car travel developed from the 1950s, so people were able to visit many more places, seeking out records which have sadly since been lost. Russell Wear very kindly made available the considerable archive of the Industrial Locomotive Society. Similarly, the Industrial Railway Society's resources, its library and its huge archive of reports and observations by its members covering more than fifty years, have been used extensively. So many people, from all parts of the country, have kindly and willingly shared their particular specialist knowledge, and we are sincerely very grateful. Equally, since the 1960s a considerable quantity of material and photographs have come into public deposit. The Durham Record Office was the first to put all its catalogues on to the internet, making research into its records so much easier, and we would like to record special thanks to the Chief Archivist, Jennifer Gill, and all her staff for all their help and patience in handling hundreds of requests. The North of England Institute of Mining & Mechanical Engineers in Newcastle upon Tyne has a large and invaluable library archive, and reference has been made to material held by the Tyne & Wear Archives Service and Tyne & Wear Museums, both in Newcastle upon Tyne, at the Regional Resource Centre of Beamish, The North of England Open Air Museum, the Darlington Railway Centre & Museum and the Local Studies Centres in the central libraries of Newcastle upon Tyne, Gateshead, Sunderland, Darlington and Middlesbrough. Information about the companies has been sought from both the National Archives at Kew and Companies House at Cardiff. We are equally grateful to all the firms and private individuals that have willingly made their records available to us over the past forty years. Without all this help this book could not have been written, and we are sincerely grateful.

We should explain that because Durham's private railways were covered in The Private Railways of County Durham, published by the Industrial Railway Society in 2004, in this volume no photographs or maps of them are included, while descriptions are only thumbnail sketches, although full locomotive lists and footnotes are given. The National Coal Board and the fluorspar industry, together with details of contractors, dealers, repairers and hirers and preserved locomotives, will all be included in Part 2.

The Maps
Before the main section of the book, the reader will find 21 maps covering most of Durham in a grid. In the list of owners, each location has been given a map location reference letter and number, so that, for example, location J148 will be found on Map J. thirteen other maps are included in the text itself for specific locations where the system was particularly complicated. All the maps have again been superbly drawn by Roger Hateley, to whom we are especially grateful for his skill, commitment and patience.

Photographs
Photographs taken in industrial locations and available to us may be roughly divided into five main categories.

1. Vernacular – taken by ordinary people years ago, usually employees. Some were copied in the 1940s-1960s by enthusiasts; some remain in private hands; some are now in public deposit, not least at Beamish, The North of England Open Air Museum, which began collecting them when few other public bodies were willing to take them and now has by far the biggest public collection in the region.

2. Professional – taken by professional photographers a century or more ago. The scenes that professional photographers thought could be sold as commercial postcards to be sent through the post never cease to amaze. Many of them are now commercially traded by postcard dealers, and we are especially grateful to George Nairn of Chester-le-Street, one of the leading dealers in the country, for allowing us to use photographs from his collection.

3. Enthusiast – taken by those interested in industrial locations and their locomotives from the 1930s onwards. Many continue in the ownership of those who took them, or of their relatives, or have been purchased by private individuals, while some are now held by organisations like the Industrial Railway Society.

4. Company – taken by or for the companies themselves for their own purposes. Some of these collections have now been placed in public deposit.

5. Public Deposit – held by Beamish Museum, mentioned above, the Durham County Record Office, Tyne & Wear Archives, Tyne & Wear Museums, Newcastle Central Library and the Darlington Railway Centre & Museum, which also holds the photographs once owned by R.H. Inness and the great North Eastern Railway expert, Ken Hoole. The Tyne & Wear Archives Service holds the collection of the commercial photographer W. Parry & Son of South Shields, while Beamish houses the collection of Ward Philipson, the Newcastle upon Tyne photographers, amongst others.

All these sources made it possible to examine many hundreds of photographs, although of course these excluded any from the private railways and also any covered by the final volume, including the National Coal Board. We drew up three criteria – to have as wide and interesting a range of photographs as possible; to include examples from all of the North East's locomotive builders, big or small, and to include pictures of locomotives working in their context (for example, a shipyard); for not only have the locomotives themselves all gone, but many of the places where they worked have also been reclaimed and thus have vanished too. Every photograph was then scanned, or transferred from its disc, cropped and then repaired as necessary, using digital techniques, by our friend Malcolm Young. We were very conscious of the ethical issues involved in undertaking this work, and our prime aim was to do nothing that would alter the view the original photographer saw. We have therefore corrected tilt, repaired creases, scratches, hair marks, dust spots and similar marks, and we have altered the brightness and contrast where we thought it necessary; but the final result is still the view as the original photographer saw it. With so many photographs, this involved many hours of work, and we are very grateful indeed to Malcolm for all his skill and patience, again without which this book could not have been compiled. We were conscious of the slight short-comings involved in using 90g paper in *The Private Railways of County Durham*, and we hope that the use of 115g paper here will remedy these. We have taken the opportunity offered by the page format to make the captions as informative and helpful as possible. We have also tried to show the huge variety of locomotives that worked in Durham's industries, probably greater than anywhere else in the country.

Many of those controlling photographs in public deposit now insist that each photograph's reference number be included in the credits. Where photographs have come from private collections, we also thought it appropriate to include the original photographer's name, where known.

J.W. Armstrong Trust 50; 60.
Beamish, The North of England Open Air Museum 30 70001; 31 31229; 71 15467; 88 14836; 89 74887; 104 31385; 139 17696; 140 17678; 151 21352.
A.J. Booth 19; 21.
K. Buckle 167.
I.S. Carr 55; 121; 134; 136.
R.M. Casserley Collection/H.C. Casserley 94; 125.
D.G. Charlton 40.
D.G.C harlton Collection/C. Cairns 85.
D.G. Charlton Collection/L.G. Charlton 5; 6; 7; 28; 32; 33; 37; 38; 53; 61; 69; 70; 95; 96; 97; 111; 113; 119; 141; 142; 168; 169; 171; 172; 174.
D.G. Charlton Collection/J. Clewley 132; 164.
D.G. Charlton Collection/I.M. Gardiner 66.
D.G. Charlton Collection/E. Haigh 49.
D.G. Charlton Collection/J. Hayes 73; 165.
D.G. Charlton Collection/R.H. Inness 129.
D.G. Charlton Collection/J.M. Jarvis 26.
D.G. Charlton Collection/J.A. Peden 120.
D.G. Charlton Collection/B.D. Stoyel 87.
D.G. Charlton Collection/A. Wright 123.
D.G. Charlton Collection 9; 17; 34; 36; 52; 81; 93; 99; 112; 116; 145; 154; 155; 156; 163; 166.
June Crosby Collection 152.
Darlington Railway Centre & Museum, Ken Hoole Study Centre Collection 25; 41; 42; 64.
Darlington Railway Centre & Museum, Ken Hoole Study Centre/S.E. Teasdale 2.
R.R. Darsley 16; 115.
Durham County Council 57; 58; 78; 79; 80.
A.R. Etherington Collection 8.
F.W. Harman Collection 10; 143.
Dr.W.F. Heyes Collection 98.
Historical Model Railway Society 175.
D.W. Holroyde 20; 23; 65; 173.
D.W. Holroyde Collection/W. Greenfield 4.
Hunslet-Barclay Ltd 157.
J.M. Hutchings Collection 162.
Industrial Railway Society Collection/J. Faithfull 43; 159.
Industrial Railway Society Collection/C.H.A. Townley 114.
Industrial Railway Society Collection/B. Webb 18; 122.
P. Jackson Collection 39.
F. Jones 63.
F. Jones Collection 59.
Lambton Locomotives Trust 170.
K. Lane 14; 45; 68.
S.A. Leleux Front cover; 54.
C.E. Mountford 1; 24; 82; 148; 160; 161.
C.E. Mountford Collection/J.M. Jarvis 27; 126; 130; 131.
C.E. Mountford Collection/R.G. Jarvis 29; 47; 127; 128; 144.
C.E. Mountford Collection 3; 22; 51; 75; 84; 86; 102; 118; 133; 146; 147; 150; 158.

G. Nairn Collection 11; 12; 62; 67; 72; 74; 92.
Newcastle City Libraries 117.
North Eastern Railway Association Collection, housed at the Ken Hoole Study Centre, Darlington Railway Centre & Museum:
   R.H. Inness, Vol.4 44; 48; 76; 77; 103; 105; 106; 107; 138.
   R.H. Inness, Vol.5 13; 46; 91; 100; 137; 153.
Northumberland County Record Office, Berwick branch Back cover.
L. Pitcher 56.
Port of Sunderland Authority 108; 109.
Seaham Harbour Dock Company 135.
C. Shepherd 15.
M. Stead 110.
N.E. Stead Collection 35.
Sunderland Library & Arts Centre Frontispiece.
A. Thompson Collection 124.
Tyne & Wear Archives Service, W. Parry & Son Collection 90; 101; 149.
Tyne & Wear Museums 83.

Despite the nature of their operations (and fortunately before the days of Health & Safety restrictions), almost invariably the firms gave a friendly welcome to those who visited them. We hope their history, a record of a unique county, will be as equally interesting and enjoyable.

27, Glencoe Avenue, Cramlington, Northumberland NE23 6EH   Colin E. Mountford
127, Lindisfarne Road, Durham City, Co.Durham DH1 5YU   Dave Holroyde

1. The last locomotive working in industry in Durham – 4wDH TH 287V/1980, at T.J. Thomson & Son Ltd's scrapyard at Tyne Dock, on 9th July 2005.

Explanatory Notes

Gauge
The gauge of a railway system is given at the head of the list.

Map references
Each location mentioned in the text is given a six-figure Ordnance Survey grid reference. It will also be numbered and prefixed by a letter indicating the kep map on which the site appears (see Maps).

Locomotive numbers and names
If a number or name was removed during a locomotive's use at a site, it is shown in brackets. If the name or number was unofficial, it is shown in inverted commas.

Locomotive type
The Whyte system of wheel classification is used wherever possible, but where the driving wheels are not connected by outside rods but by chains or motors, they are shown as 4w, 6w, etc. If each axle is driven independently, this is shown by 2w-2w, etc. The following abbreviations are used:-

T	Fitted with side tanks, invariably fastened to the frame.
BT	Fitted with a tank to the rear of the cab, under the coal bunker.
Cr	Crane
CT	Crane Tank, a side tank locomotive equipped with load-lifting apparatus.
PT	Pannier Tank - fitted with tanks either side of the boiler, but supported from the boiler.
ST	Saddle Tank, where a single tank covers the top of the boiler. These were made in a variety of shapes, round and curved, including 'ogee' and 'box'.
IST	Inverted Saddle Tank, where the tank passed under the boiler.
WT	Well Tank – a tank located between the frames below the level of the boiler.
VB	Vertical boilered locomotive.
DM	Diesel locomotive with mechanical transmission.
DE	Diesel locomotive with electric transmission.
DH	Diesel locomotive with hydraulic transmission.
PM	Petrol locomotive with mechanical transmission.
BE	Battery-powered electric locomotive.
WE	Overhead wire-powered electric locomotive.
F	Fireless locomotive (operated by steam piped into a receiver from an external source).
CA	Compressed air locomotive.

Diesel or battery-electric locomotives that were flame-proofed for working underground are shown with the suffix F to the type, for example, 4wDMF.

Cylinder position

IC	Inside cylinders		OC	Outside cylinders
VC	Vertical cylinders		G	Geared transmission (used with IC, OC or VC)

Locomotive manufacturer
The abbreviations used to denote makers are given in the Locomotive Index at the end of the book.

Makers number and date
The first column shows the maker's works number, the second shows the date that appeared on the maker's plate, or the date the locomotive was built if this was not shown on the plate. It should be noted that the ex-works date in the Locomotive Index may be a different year from that recorded as the building date or that shown on the maker's plate.

Rebuilding details are denoted by the abbreviation 'reb', usually involving significant alterations to the locomotive.

Source of locomotive
'New' indicates that the locomotive was delivered new by the makers to a location. A bracketed letter indicates that a locomotive was transferred to this location from elsewhere, and details are given in the first of the two sets of footnotes, which are listed in the locomotives' chronological order of arrival. The date of arrival is given, where known.

Disposal of locomotive
A locomotive transferred to another location or sold to another owner is shown by a bracketed number, and details are given in the second set of footnotes. If a locomotive was broken up for scrap, the abbreviation 'Scr' is used. The date of departure or scrapping is given, where known. If the disposal of a

locomotive is not known, this is shown by the abbreviation 's/s' (scrapped or sold).

Many sales of locomotives were effected through dealers and contractors and details are given where known. If a locomotive went to a dealer's premises before re-sale, this is denoted by 'via'; if a sale was effected without this happening, this is denoted by 'per'.

Doubtful information

Information which is known to be of a doubtful nature is shown by the wording chosen, or else shown in brackets with a question mark, for example, (1910?).

2. The top of the Randolph Incline at Evenwood on 6th August 1966, owned by The Randolph Coal & Coke Co Ltd. Although the full wagons of coke descended the incline, because wagons of coal had to be brought in, the incline by this date was being worked by a stationary engine (see No.125). The 'short end' of the rope waits to be attached to the two wagons on the left.

Maps

In the alphabetical list of owners, each location entry is numbered in a sequence beginning at 1. This number is prefixed by a letter indicating the key map on which the site appears. Thus 'V149' indicates that the site will be found on Map V.

Key Maps

	County Durham, showing Arrangement of Key Maps
A	Ryton, Chopwell and Rowlands Gill
B	Blaydon and Swalwell
C	Gateshead
D	Felling and South Shields
E	Hebburn and Jarrow
F	Consett from 1948
G	Stanley, Birtley and Chester-le-Street
H	Washington and Sunderland
J	Houghton-le-Spring, Hetton-le-Hole and Seaham
K	Durham and the Deerness Valley
L	Easington, Shotton and Wingate
M	Tow Law, Crook and Willington
N	Spennymoor and Ferryhill
P	Woodland, Cockfield and Evenwood
Q	Bishop Auckland
R	Newton Aycliffe
S	Darlington
T	Lower Weardale
U	Frosterley
V	Upper Weardale
W	Upper Teesdale (transferred to County Durham in 1974)

System Maps

Fig.1	Collieries, ironworks and quarry owned by Bolckow, Vaughan & Co Ltd
Fig.2	Consett and Leadgate, 1857
Fig.3	The Consett area, 1885/1894
Fig.4	Burnhope Reservoir
Fig.5	Cornsay, Hamsteels and Malton Collieries
Fig.6	Hartlepool Dock & Railway and its associated area, 1846
Fig.7	South Moor Colliery, Annfield Plain
Fig.8	Pease & Partners Ltd, Crook
Fig.9	Blaydon Burn and Blaydon Main waggonways, 1855
Fig.10	South Medomsley Colliery
Fig.11	Lead mines worked by The Weardale Lead Co Ltd
Fig.12	Quarries and waggonways in the Southwick and Fulwell areas of Sunderland in 1855
Fig.13	Fulwell and Southwick Quarries in 1919 (simplified)

Notes

1. On a Key Map, where a rope incline was subsequently replaced by locomotive haulage the route is shown representing the latter.
2. On both sets of maps, while narrow gauge lines have been included as far as possible, rope haulage on them has been largely omitted, because (a) to identify the type of rope haulage on every narrow gauge line shown would be a monumental and probably imposssible task, and (b) because of the difficulty of distinguishing cartographically between main-and-tail, top endless and bottom endless rope haulage. However, examples of these have been included in the photographs.

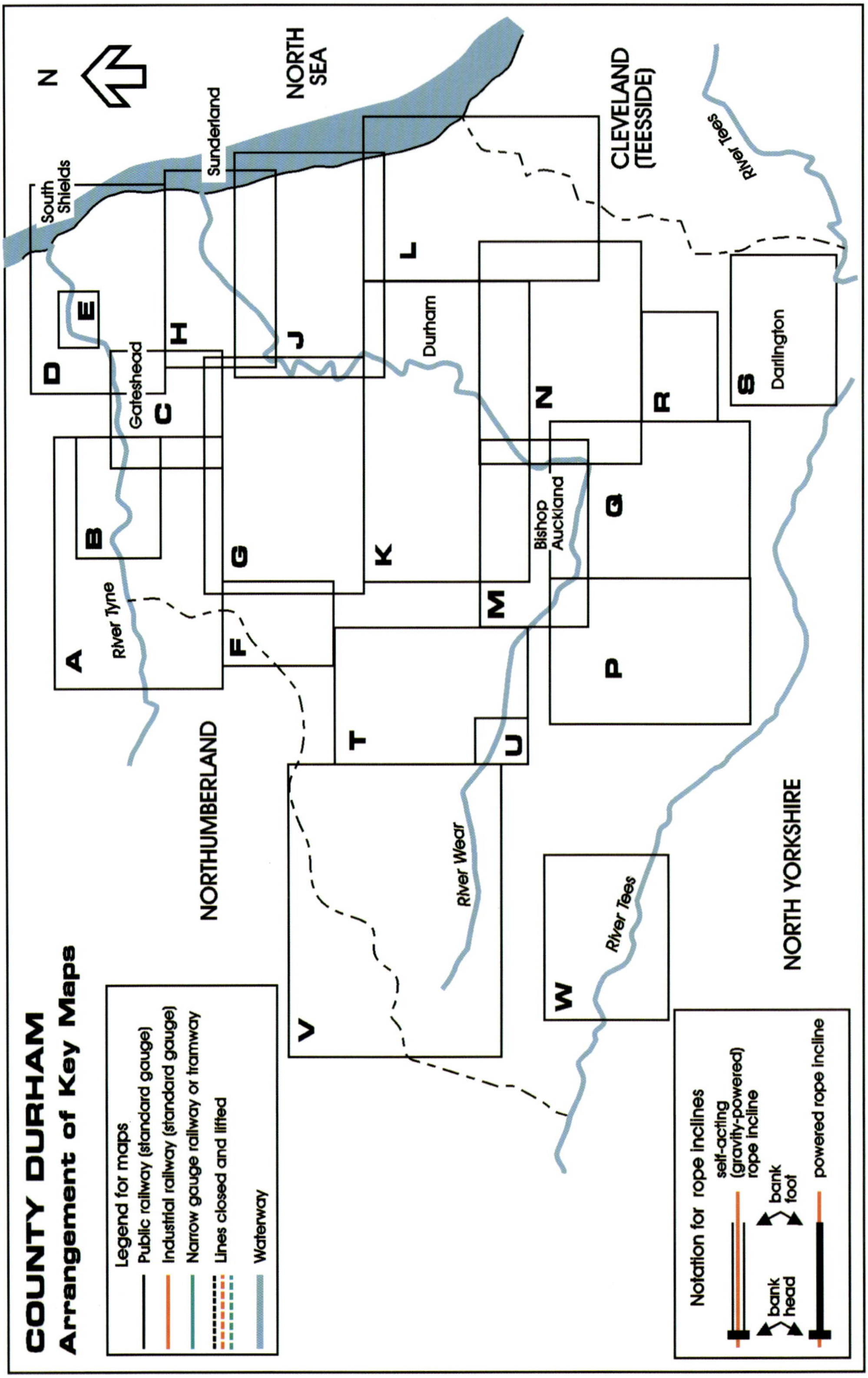
COUNTY DURHAM
Arrangement of Key Maps
Legend for maps
Public railway (standard gauge)
Industrial railway (standard gauge)
Narrow gauge railway or tramway
Lines closed and lifted
Waterway
Notation for rope inclines
self-acting (gravity-powered) rope incline
bank head
bank foot
powered rope incline
N
NORTH SEA
South Shields
Sunderland
Gateshead
Durham
Bishop Auckland
Darlington
CLEVELAND (TEESSIDE)
NORTHUMBERLAND
NORTH YORKSHIRE
River Tyne
River Wear
River Tees
River Tees
A
B
C
D
E
F
G
H
J
K
L
M
N
P
Q
R
S
T
U
V
W

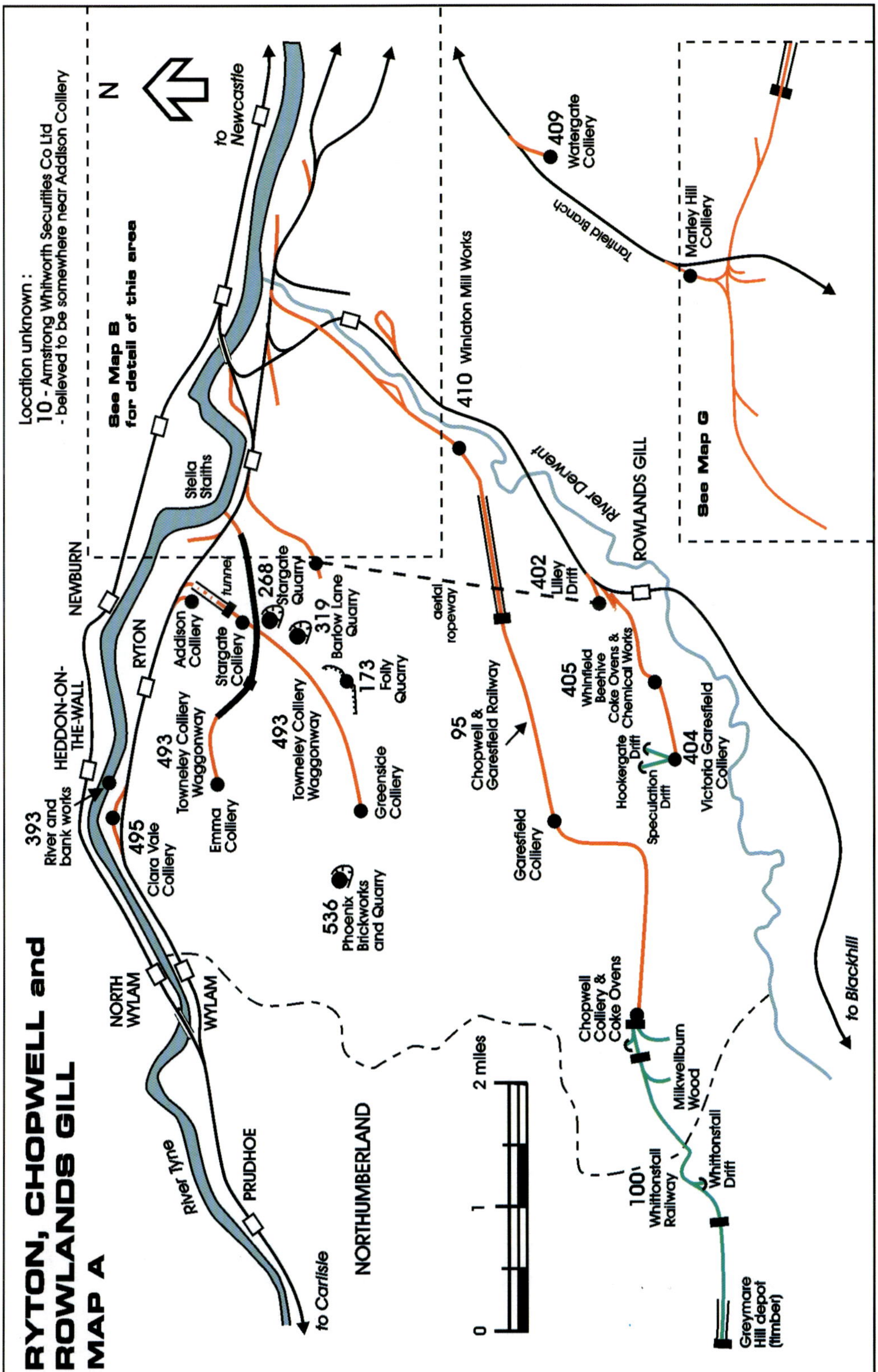

RYTON, CHOPWELL and ROWLANDS GILL
MAP A
Location unknown :
10 - Armstrong Whitworth Securities Co Ltd
- believed to be somewhere near Addison Colliery
N
See Map B
for detail of this area
See Map G
to Newcastle
to Carlisle
to Blackhill
River Tyne
River Derwent
NEWBURN
HEDDON-ON-THE-WALL
RYTON
NORTH WYLAM
WYLAM
PRUDHOE
NORTHUMBERLAND
ROWLANDS GILL
Stella Staiths
393 River and bank works
495 Clara Vale Colliery
Emma Colliery
493 Towneley Colliery Waggonway
493 Towneley Colliery Waggonway
Addison Colliery
Stargate Colliery
tunnel
268 Stargate Quarry
319 Barlow Lane Quarry
173 Folly Quarry
Greenside Colliery
536 Phoenix Brickworks and Quarry
aerial ropeway
95 Chopwell & Garesfield Railway
Garesfield Colliery
410 Winlaton Mill Works
402 Lilley Drift
405 Whinfield Beehive Coke Ovens & Chemical Works
Hookergate Drift
Speculation Drift
404 Victoria Garesfield Colliery
Chopwell Colliery & Coke Ovens
Milkwellburn Wood
100 Whittonstall Railway
Whittonstall Drift
Greymare Hill depot (timber)
409 Watergate Colliery
Tanfield Branch
Marley Hill Colliery
0
1
2 miles

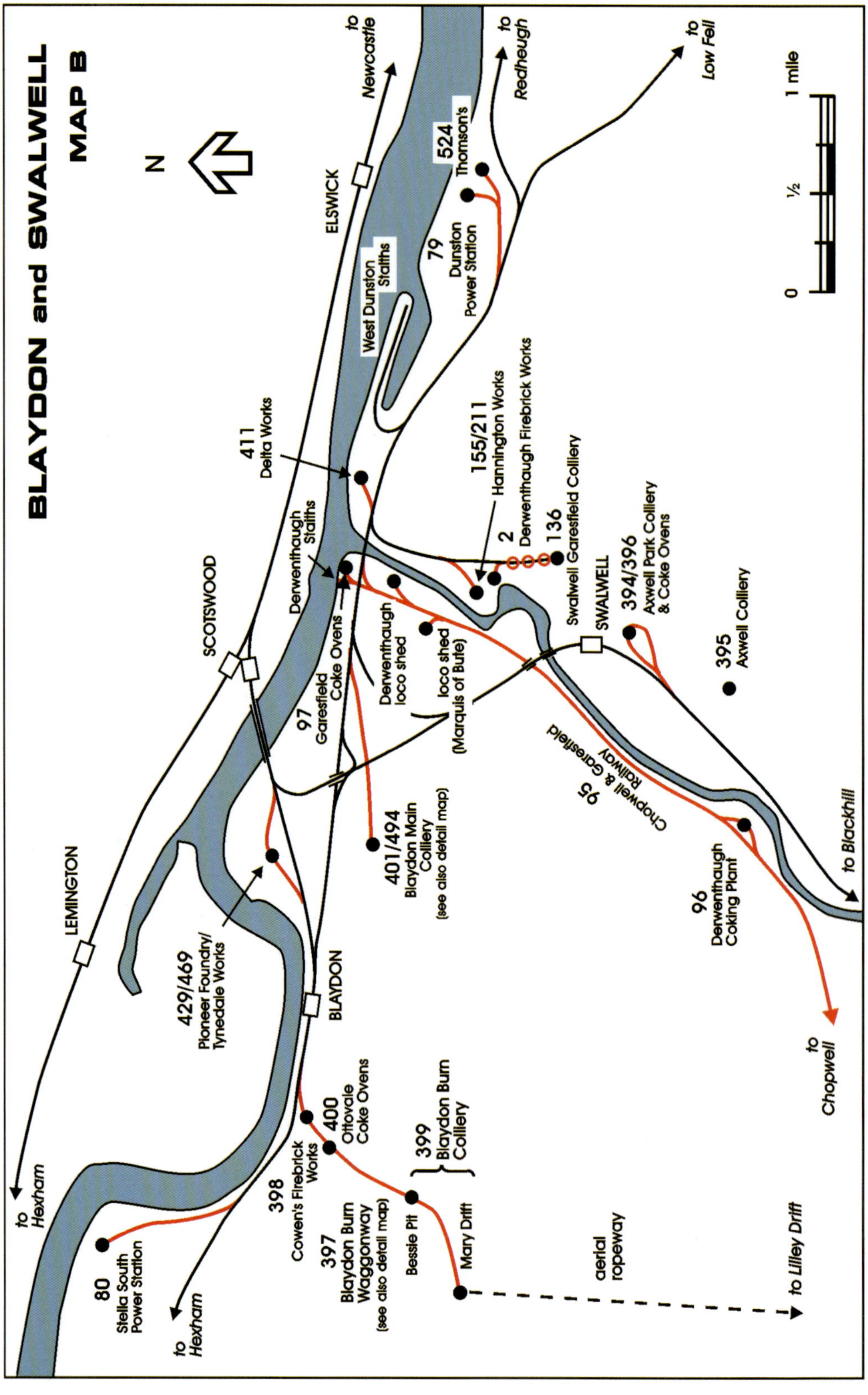
BLAYDON and SWALWELL
MAP B
N
to Newcastle
to Redheugh
to Low Fell
1 mile
½
0
ELSWICK
West Dunston Staiths
79
Dunston Power Station
524
Thomson's
411
Delta Works
155/211
Hannington Works
Derwenthaugh Firebrick Works
2
136
Swalwell Garesfield Colliery
Derwenthaugh Staiths
SCOTSWOOD
SWALWELL
394/396
Axwell Park Colliery & Coke Ovens
395
Axwell Colliery
97
Garesfield Coke Ovens
Derwenthaugh loco shed
loco shed (Marquis of Bute)
95
Chopwell & Garesfield Railway
401/494
Blaydon Main Colliery (see also detail map)
96
Derwenthaugh Coking Plant
to Blackhill
LEMINGTON
429/469
Pioneer Foundry/ Tynedale Works
BLAYDON
to Chopwell
400
Ottovale Coke Ovens
399
Blaydon Burn Colliery
398
Cowen's Firebrick Works
397
Blaydon Burn Waggonway (see also detail map)
Bessie Pit
Mary Drift
aerial ropeway
to Lilley Drift
to Hexham
to Hexham
80
Stella South Power Station

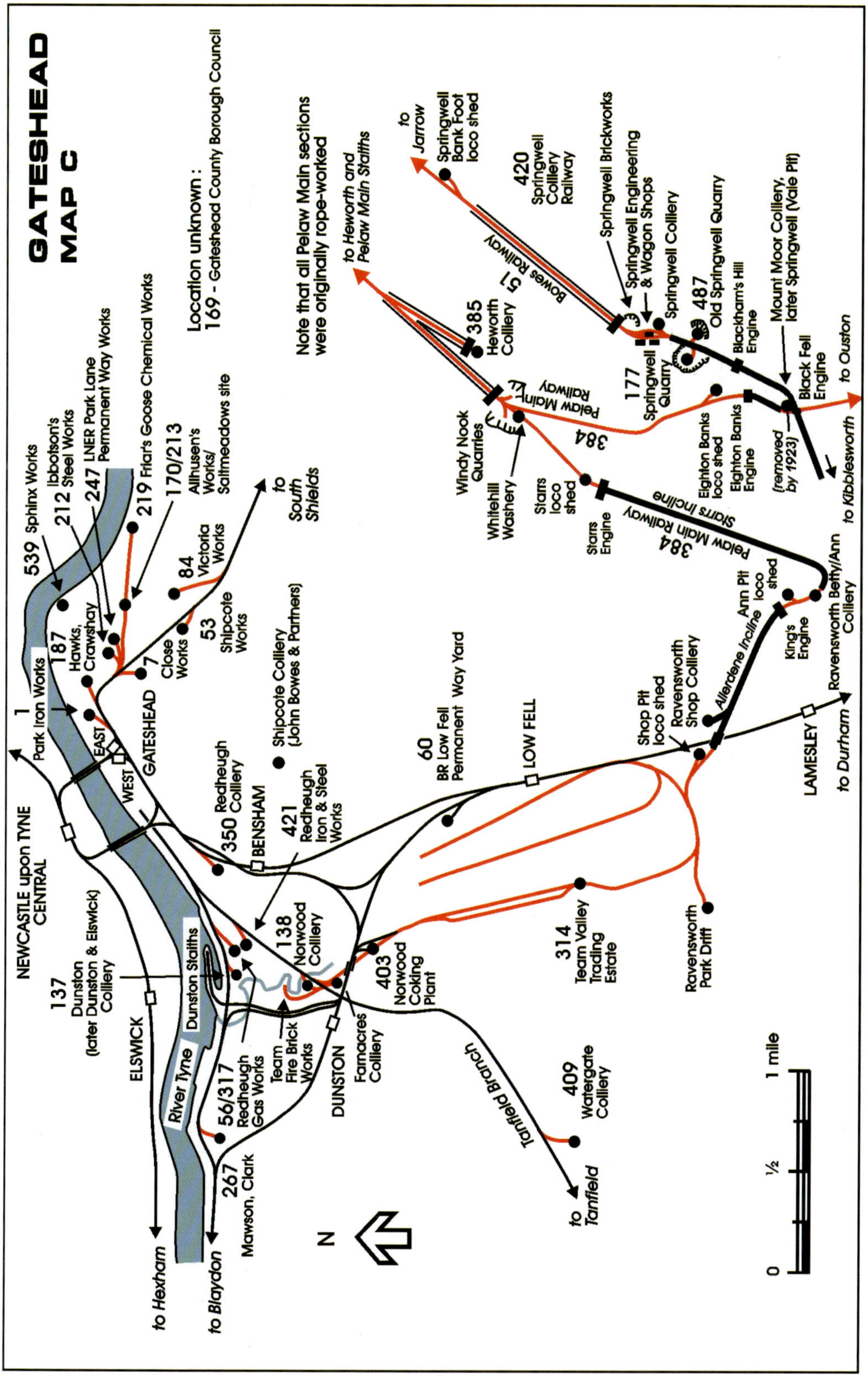
GATESHEAD
MAP C
Location unknown :
169 - Gateshead County Borough Council
Note that all Pelaw Main sections
were originally rope-worked
to Heworth and
Pelaw Main Staiths
to
Jarrow
Springwell
Bank Foot
loco shed
420
Springwell
Colliery
Railway
Bowes Railway
51
Springwell Brickworks
Springwell Engineering
& Wagon Shops
Springwell Colliery
487
Old Springwell Quarry
Blackham's Hill
Engine
Mount Moor Colliery,
later Springwell (Vale Pit)
Black Fell
Engine
to Ouston
to Kibblesworth
(removed
by 1923)
Eighton Banks
Engine
Eighton Banks
loco shed
177
Springwell
Quarry
385
Heworth
Colliery
Pelaw Main
Railway
384
Windy Nook
Quarries
Whitehill
Washery
Starrs
loco
shed
Starrs
Engine
Starrs Incline
Pelaw Main Railway
384
Ravensworth Betty/Ann
Colliery
Ann Pit
loco
shed
King's
Engine
Allerdene Incline
Ravensworth
Shop Colliery
Shop Pit
loco shed
LAMESLEY
to Durham
539 Sphinx Works
212 Ibbotson's
Steel Works
247 LNER Park Lane
Permanent Way Works
219 Friar's Goose Chemical Works
170/213
Allhusen's
Works/
Saltmeadows site
84
Victoria
Works
to
South
Shields
53
Shipcote
Works
7
Close
Works
187
Hawks,
Crawshay
1
Park Iron Works
EAST
WEST
GATESHEAD
Shipcote Colliery
(John Bowes & Partners)
60
BR Low Fell
Permanent Way Yard
LOW FELL
NEWCASTLE upon TYNE
CENTRAL
350
Redheugh
Colliery
BENSHAM
421
Redheugh
Iron & Steel
Works
138
Norwood
Colliery
403
Norwood
Coking
Plant
314
Team Valley
Trading
Estate
Ravensworth
Park Drift
137
Dunston
(later Dunston & Elswick)
Colliery
Dunston Staiths
ELSWICK
River Tyne
56/317
Redheugh
Gas Works
Team
Fire Brick
Works
DUNSTON
Farnacres
Colliery
Tanfield Branch
409
Watergate
Colliery
to
Tanfield
267
Mawson, Clark
to Hexham
to Blaydon
N
0
½
1 mile

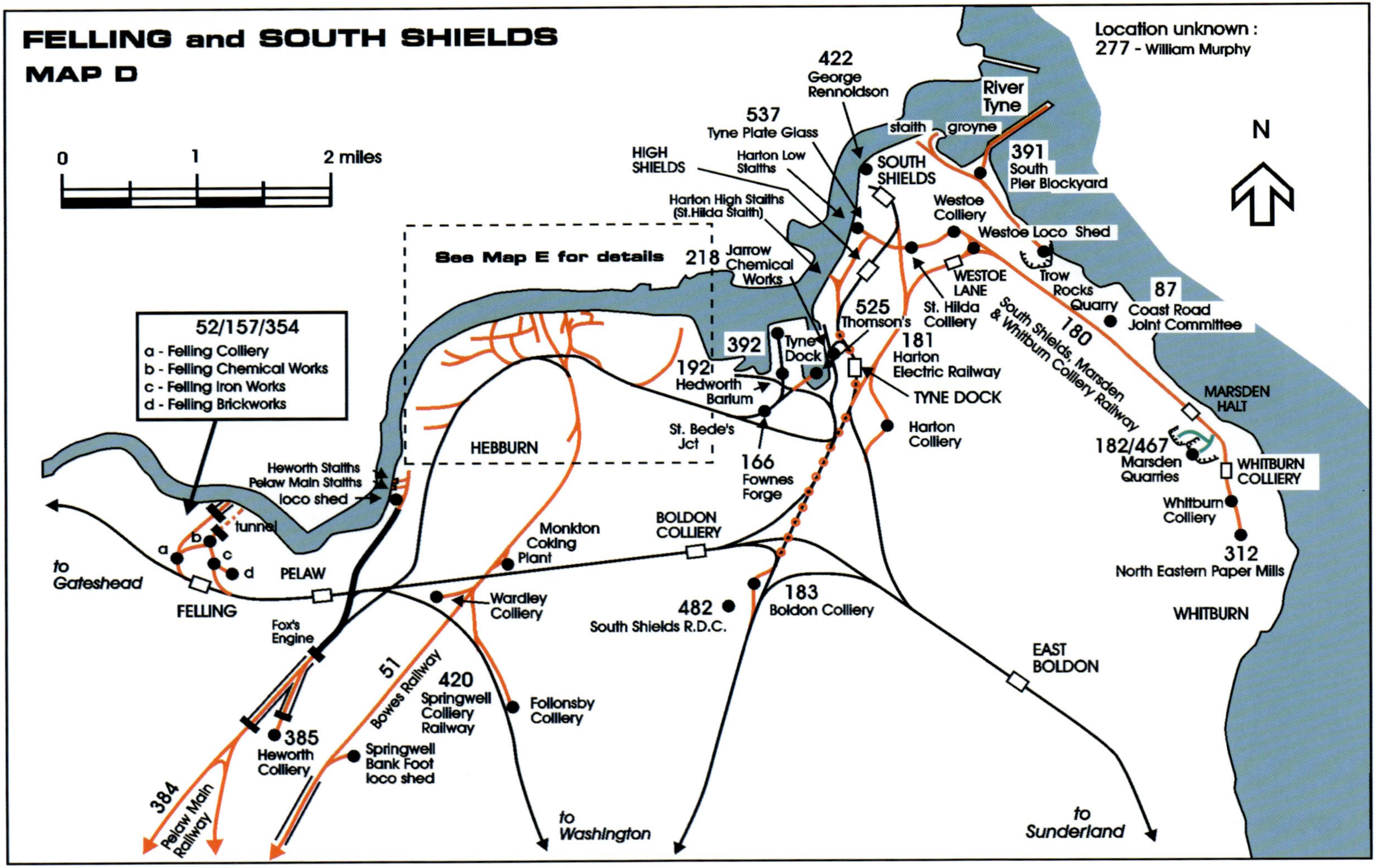
FELLING and SOUTH SHIELDS
MAP D
0
1
2 miles
Location unknown :
277 - William Murphy
N
See Map E for details
52/157/354
a - Felling Colliery
b - Felling Chemical Works
c - Felling Iron Works
d - Felling Brickworks
422
George Rennoldson
537
Tyne Plate Glass
HIGH SHIELDS
Harton Low Staiths
Harton High Staiths (St.Hilda Staith)
218 Jarrow Chemical Works
River Tyne
staith
groyne
SOUTH SHIELDS
391
South Pier Blockyard
Westoe Colliery
Westoe Loco Shed
WESTOE LANE
Trow Rocks Quarry
87
Coast Road Joint Committee
525
Thomson's
St. Hilda Colliery
180
South Shields, Marsden & Whitburn Colliery Railway
392
Tyne Dock
181
Harton Electric Railway
192
Hedworth Barium
TYNE DOCK
MARSDEN HALT
Harton Colliery
182/467
Marsden Quarries
WHITBURN COLLIERY
St. Bede's Jct
166
Fownes Forge
Whitburn Colliery
312
North Eastern Paper Mills
WHITBURN
HEBBURN
Heworth Staiths
Pelaw Main Staiths
loco shed
tunnel
Monkton Coking Plant
BOLDON COLLIERY
a
b
c
d
to Gateshead
PELAW
FELLING
Fox's Engine
Wardley Colliery
482
South Shields R.D.C.
183
Boldon Colliery
EAST BOLDON
51
Bowes Railway
420
Springwell Colliery Railway
Follonsby Colliery
385
Heworth Colliery
Springwell Bank Foot loco shed
384
Pelaw Main Railway
to Washington
to Sunderland

HEBBURN and JARROW
MAP E

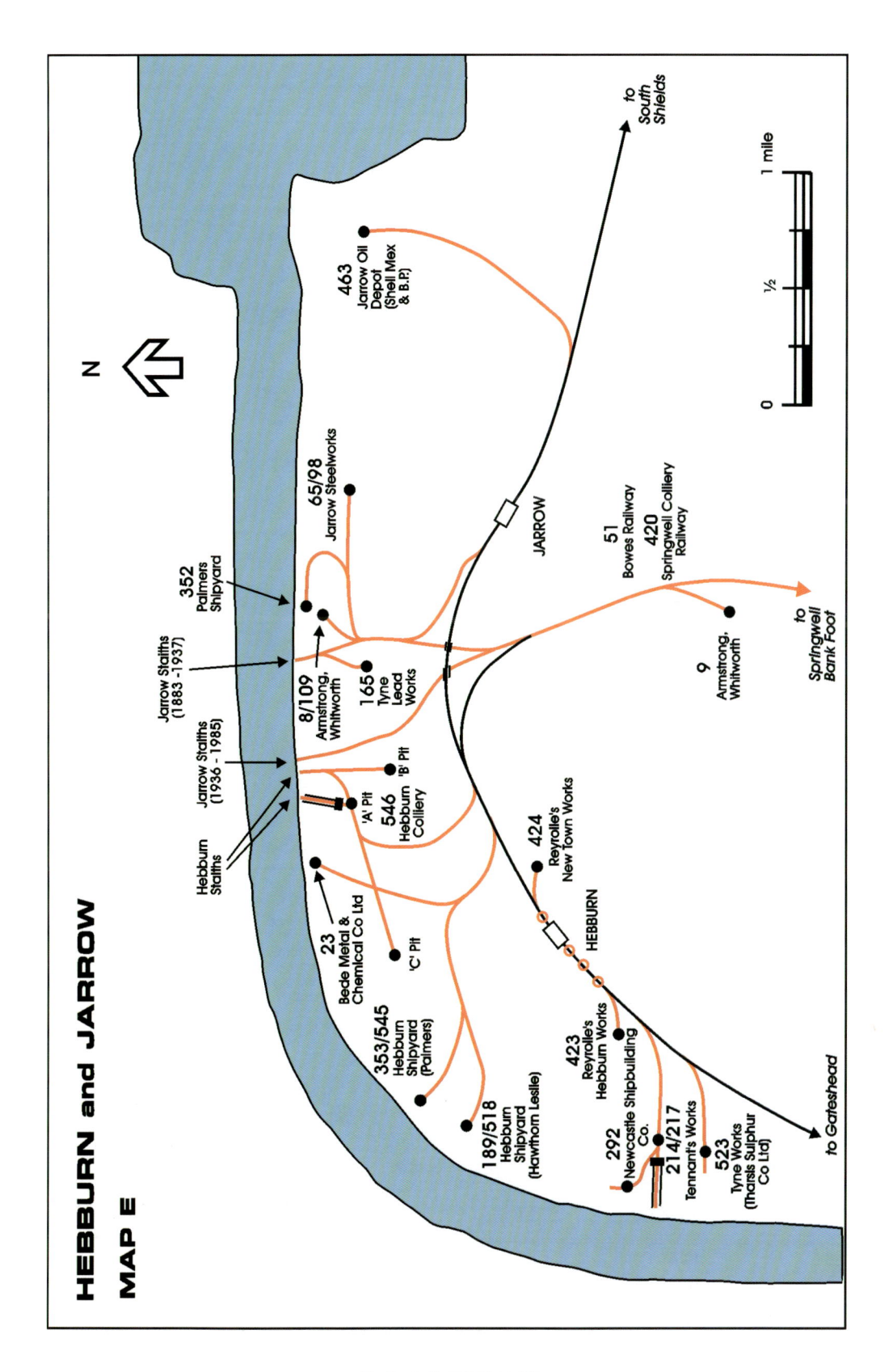

- 463 Jarrow Oil Depot (Shell Mex & B.P.)
- 65/98 Jarrow Steelworks
- 352 Palmers Shipyard
- Jarrow Staiths (1883-1937)
- Jarrow Staiths (1936-1985)
- 8/109 Armstrong, Whitworth
- 165 Tyne Lead Works
- Hebburn Staiths
- 546 Hebburn Colliery ('A' Pit, 'B' Pit, 'C' Pit)
- 23 Bede Metal & Chemical Co Ltd
- 353/545 Hebburn Shipyard (Palmers)
- 189/518 Hebburn Shipyard (Hawthorn Leslie)
- 424 Reyrolle's New Town Works
- 423 Reyrolle's Hebburn Works
- 292 Newcastle Shipbuilding Co.
- 214/217 Tennant's Works
- 523 Tyne Works (Marsts Sulphur Co Ltd)
- 51 Bowes Railway
- 420 Springwell Colliery Railway
- 9 Armstrong, Whitworth

JARROW
HEBBURN

to South Shields
to Springwell Bank Foot
to Gateshead

N

0 ½ 1 mile

Durham Part 1 Page 13

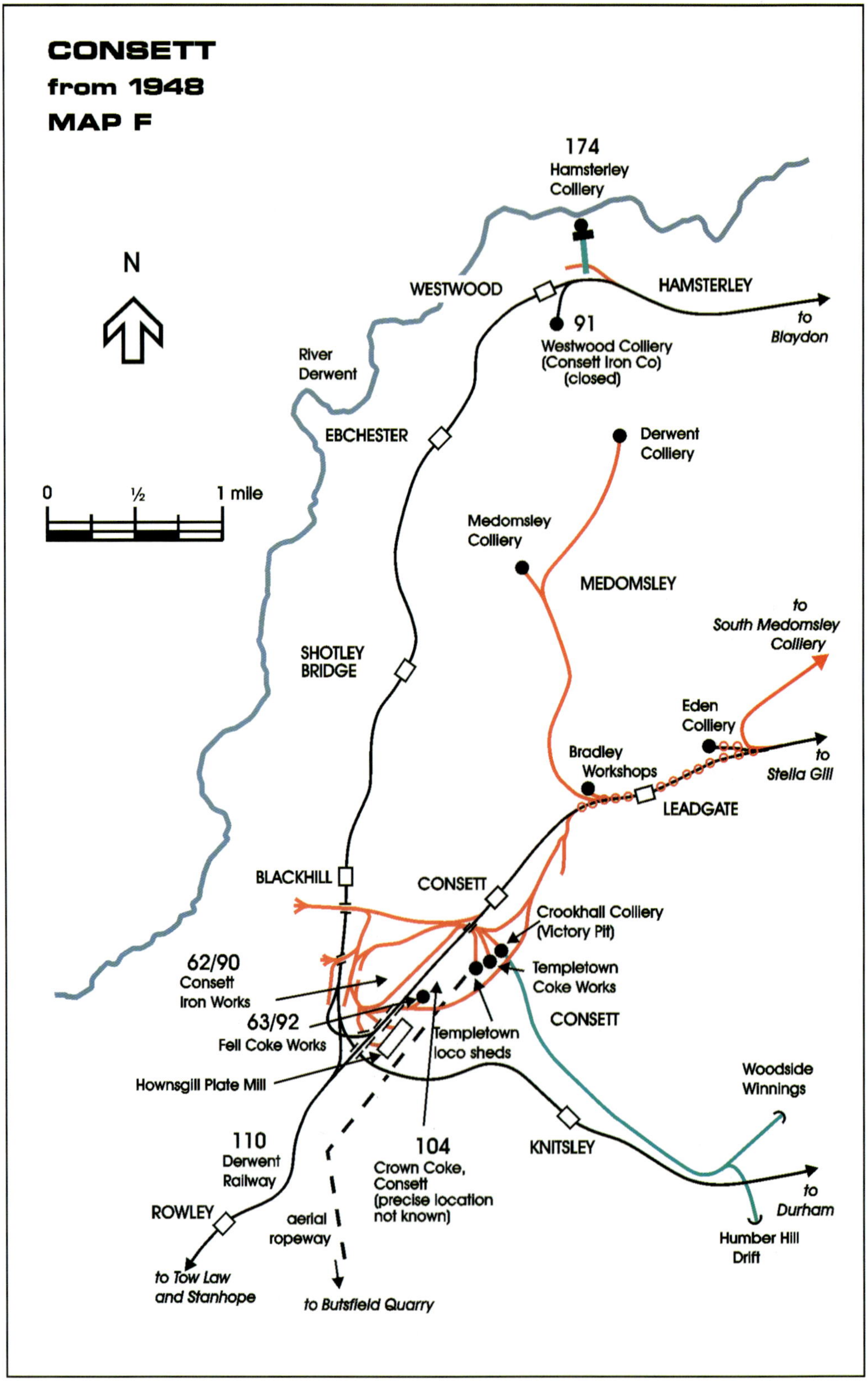
CONSETT
from 1948
MAP F
N
0
½
1 mile
174
Hamsterley
Colliery
WESTWOOD
HAMSTERLEY
to
Blaydon
91
Westwood Colliery
(Consett Iron Co)
(closed)
River
Derwent
EBCHESTER
Derwent
Colliery
Medomsley
Colliery
MEDOMSLEY
to
South Medomsley
Colliery
SHOTLEY
BRIDGE
Eden
Colliery
to
Stella Gill
Bradley
Workshops
LEADGATE
BLACKHILL
CONSETT
Crookhall Colliery
(Victory Pit)
Templetown
Coke Works
62/90
Consett
Iron Works
CONSETT
63/92
Fell Coke Works
Templetown
loco sheds
Hownsgill Plate Mill
Woodside
Winnings
110
Derwent
Railway
104
Crown Coke,
Consett
(precise location
not known)
KNITSLEY
to
Durham
ROWLEY
aerial
ropeway
Humber Hill
Drift
to Tow Law
and Stanhope
to Butsfield Quarry

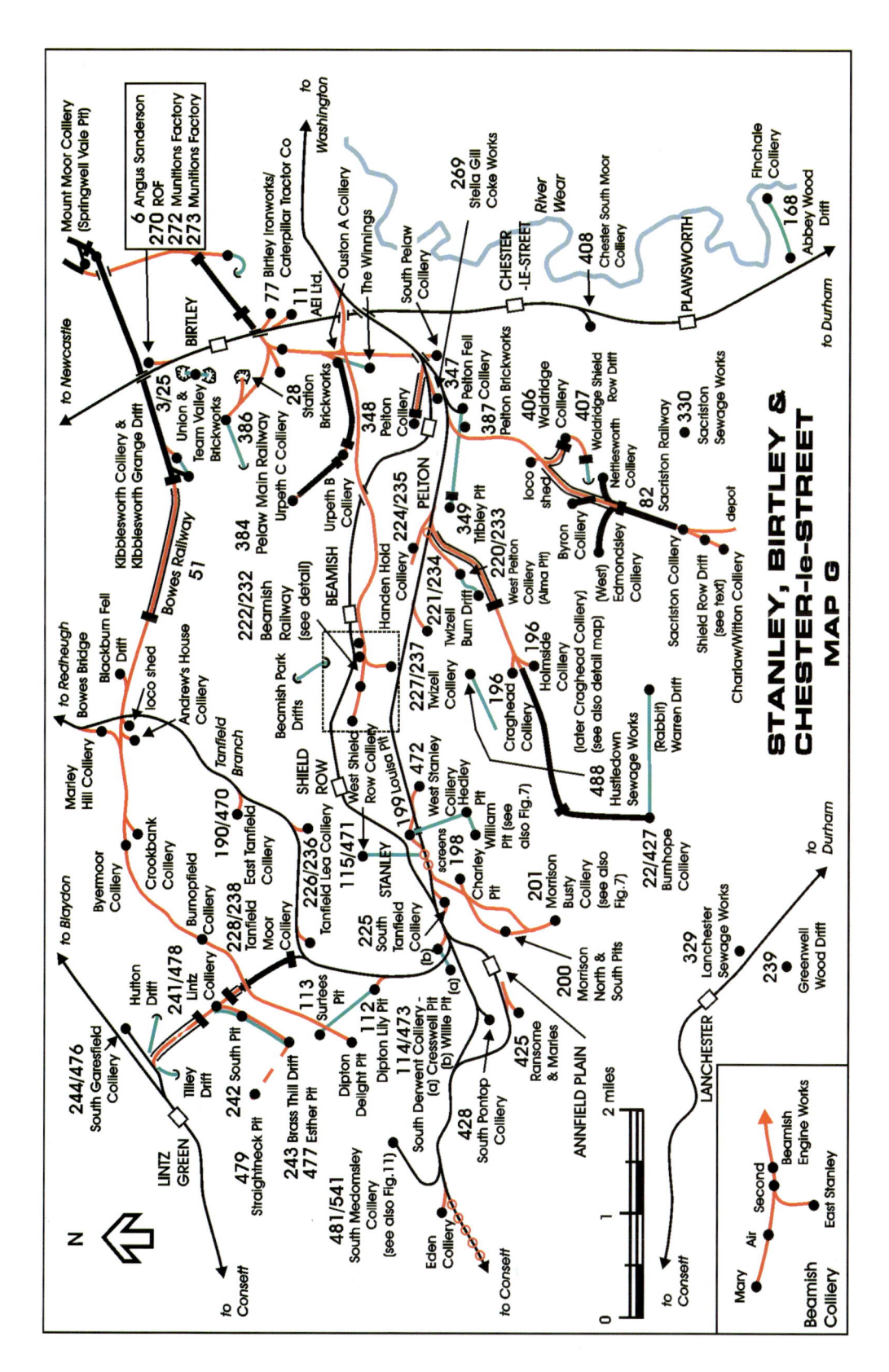

STANLEY, BIRTLEY & CHESTER-le-STREET
MAP G

Map showing collieries, railways, and industrial sites in the Stanley, Birtley and Chester-le-Street area of Durham.

Key locations and numbered sites shown on map:

- Mount Moor Colliery (Springwell Vale Pit)
- 6 Angus Sanderson
- 270 ROF
- 272 Munitions Factory
- 273 Munitions Factory
- BIRTLEY
- 77 Birtley Ironworks / Caterpillar Tractor Co
- 11 AEI Ltd.
- Ouston A Colliery
- The Winnings
- South Pelaw Colliery
- 269 Stella Gill Coke Works
- CHESTER-LE-STREET
- River Wear
- 408 Chester South Moor Colliery
- PLAWSWORTH
- 168 Finchale Colliery / Abbey Wood Drift
- 3/25 Union & Team Valley Brickworks
- Kibblesworth Colliery & Kibblesworth Grange Drift
- 386
- 43
- 28 Station Brickworks
- 348 Pelton Colliery
- Urpeth C Colliery
- 347 Pelton Fell
- 387 Pelton Brickworks
- 406 Waldridge Shield
- Waldridge Colliery
- 407 Waldridge Shield Row Drift
- 82 Sacriston Railway
- 330 Sacriston Sewage Works
- Bowes Railway
- Blackburn Fell Drift
- loco shed
- Andrew's House Colliery
- 51 Bowes Railway
- 384 Pelaw Main Railway
- 222/232 Beamish Railway (see detail)
- BEAMISH
- Urpeth B Colliery
- Handen Hold Colliery 224/235
- PELTON
- 349 Tribley Pit
- 220/233
- West Pelton Colliery (Alma Pit)
- Byron Colliery
- Edmondsley Colliery (West)
- Sacriston Colliery
- Shield Row Drift (see text)
- Charlaw/Witton Colliery
- Marley Hill Colliery
- Tanfield Branch
- 190/470 Crookbank Colliery
- East Tanfield Colliery
- Beamish Park Drifts
- 227/237 Twizell
- Twizell Colliery
- 221/234 Burn Drift
- 196 Holmside Colliery
- 196 Craghead Colliery [later Craghead Colliery] (see also detail map)
- 488 Hustledown Sewage Works
- (Rabbit) Warren Drift
- Byemoor Colliery
- Burnopfield Colliery
- 228/238 Tanfield Moor Colliery
- 226/236 Tanfield Lea Colliery
- 115/471
- STANLEY
- SHIELD ROW
- West Shield Row Colliery
- 472 Louisa Pit
- 199
- West Stanley Colliery
- Hedley Pit
- Charley William Pit (see also Fig.7)
- 201 Morrison Busty Colliery (see also Fig.7)
- 22/427 Burnhope Colliery
- to Durham
- 329 Lanchester Sewage Works
- LANCHESTER
- 239 Greenwell Wood Drift
- 244/476 South Garesfield Colliery
- Hutton Drift
- 241/478 Lintz Colliery
- 479 Straightneck Pit
- 242 South Pit
- 243 Brass Thill Drift
- 477 Esther Pit
- 113 Surtees Pit
- 112 Dipton Lily Pit
- 114/473 Dipton Delight Pit
- 225 South Tanfield Colliery
- (b) screws
- 198
- (a)
- 200 Morrison North & South Pits
- LINTZ GREEN
- Tilley Drift
- 481/541 South Medomsley Colliery (see also Fig.11)
- Eden Colliery
- South Derwent Colliery – (a) Cresswell Pit (b) Willie Pit
- 428 South Pontop Colliery
- 425 Ransome & Marles
- ANNFIELD PLAIN

Beamish Colliery (inset):
- Mary
- Air
- Second
- Beamish Engine Works
- East Stanley

Directions: to Newcastle, to Washington, to Durham, to Redheugh Bowes Bridge, to Blaydon, to Consett

Scale: 0 – 1 – 2 miles

Durham Part 1 Page 15

WASHINGTON and SUNDERLAND
MAP H

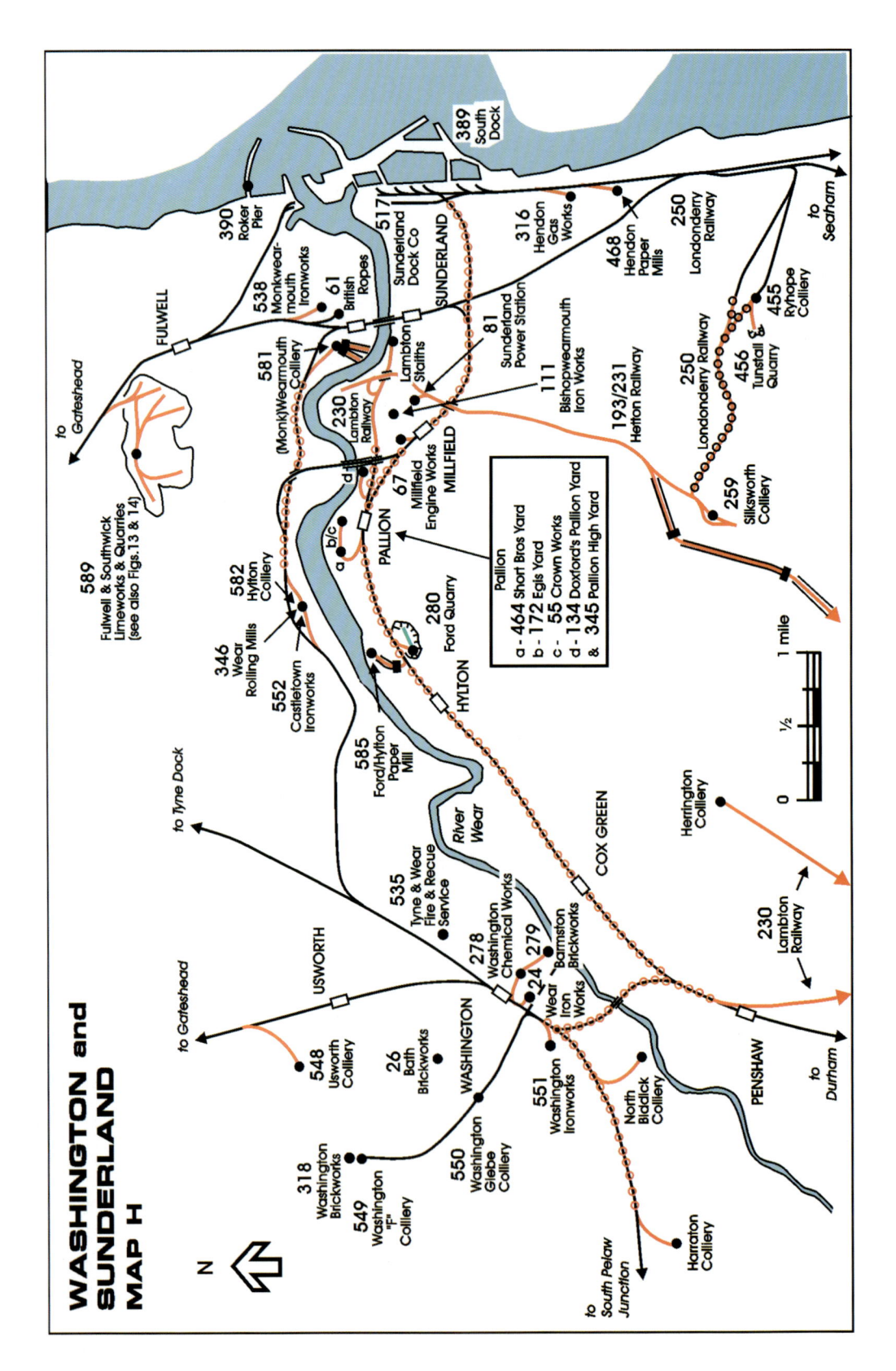

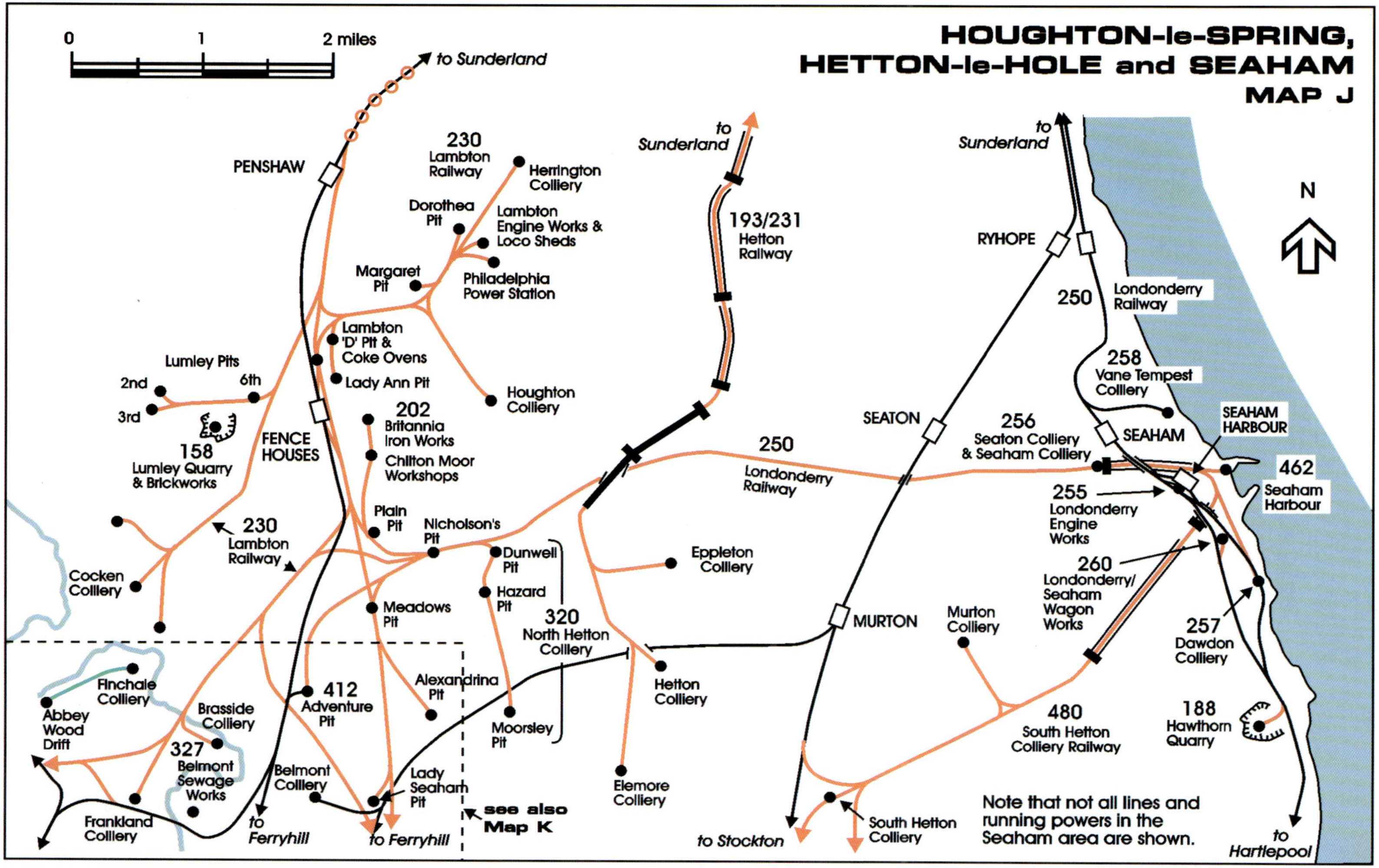
HOUGHTON-le-SPRING, HETTON-le-HOLE and SEAHAM
MAP J
0
1
2 miles
N
to Sunderland
PENSHAW
230
Lambton
Railway
Herrington
Colliery
Dorothea
Pit
Lambton
Engine Works &
Loco Sheds
Margaret
Pit
Philadelphia
Power Station
Lambton
'D' Pit &
Coke Ovens
Lady Ann Pit
Houghton
Colliery
Lumley Pits
2nd
3rd
6th
158
Lumley Quarry
& Brickworks
FENCE
HOUSES
202
Britannia
Iron Works
Chilton Moor
Workshops
Plain
Pit
Nicholson's
Pit
230
Lambton
Railway
Cocken
Colliery
Dunwell
Pit
Hazard
Pit
320
North Hetton
Colliery
Meadows
Pit
Alexandrina
Pit
412
Adventure
Pit
Moorsley
Pit
Finchale
Colliery
Abbey
Wood
Drift
Brasside
Colliery
327
Belmont
Sewage
Works
Belmont
Colliery
Lady
Seaham
Pit
Frankland
Colliery
to
Ferryhill
to Ferryhill
see also
Map K
to
Sunderland
193/231
Hetton
Railway
250
Londonderry
Railway
Eppleton
Colliery
Hetton
Colliery
Elemore
Colliery
to Stockton
to
Sunderland
RYHOPE
250
Londonderry
Railway
258
Vane Tempest
Colliery
SEATON
256
Seaton Colliery
& Seaham Colliery
SEAHAM
SEAHAM
HARBOUR
462
Seaham
Harbour
255
Londonderry
Engine
Works
260
Londonderry/
Seaham
Wagon
Works
MURTON
Murton
Colliery
257
Dawdon
Colliery
480
South Hetton
Colliery Railway
188
Hawthorn
Quarry
South Hetton
Colliery
Note that not all lines and
running powers in the
Seaham area are shown.
to
Hartlepool

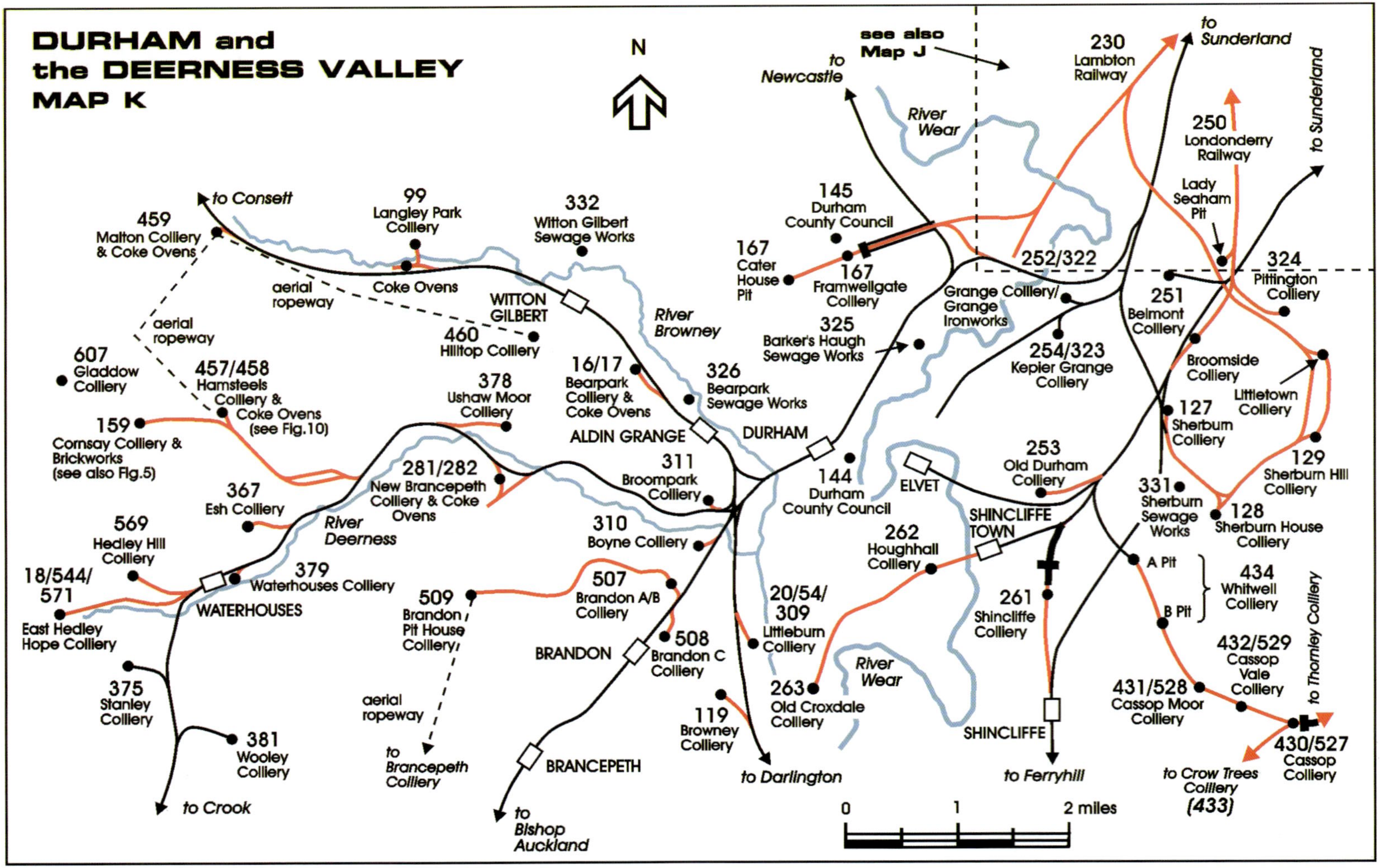
DURHAM and the DEERNESS VALLEY MAP K
N
see also Map J
to Newcastle
to Sunderland
230 Lambton Railway
250 Londonderry Railway
to Sunderland
Lady Seaham Pit
324 Pittington Collery
River Wear
145 Durham County Council
167 Cater House Pit
167 Framwellgate Colliery
252/322
Grange Colliery/ Grange Ironworks
251 Belmont Colliery
Broomside Colliery
Littletown Colliery
127 Sherburn Colliery
129 Sherburn Hill Colliery
325 Barker's Haugh Sewage Works
254/323 Kepier Grange Colliery
253 Old Durham Colliery
331 Sherburn Sewage Works
128 Sherburn House Colliery
to Consett
459 Malton Colliery & Coke Ovens
99 Langley Park Colliery
Coke Ovens
332 Witton Gilbert Sewage Works
aerial ropeway
WITTON GILBERT
River Browney
460 Hilltop Colliery
aerial ropeway
607 Gladdow Colliery
457/458 Hamsteels Colliery & Coke Ovens (see Fig.10)
378 Ushaw Moor Colliery
16/17 Bearpark Colliery & Coke Ovens
326 Bearpark Sewage Works
ALDIN GRANGE
DURHAM
159 Cornsay Colliery & Brickworks (see also Fig.5)
281/282 New Brancepeth Colliery & Coke Ovens
311 Broompark Colliery
144 Durham County Council
ELVET
367 Esh Colliery
River Deerness
310 Boyne Colliery
262 Houghhall Colliery
SHINCLIFFE TOWN
A Pit
B Pit
434 Whitwell Colliery
569 Hedley Hill Colliery
379 Waterhouses Colliery
18/544/ 571
East Hedley Hope Colliery
WATERHOUSES
509 Brandon Pit House Colliery
507 Brandon A/B Colliery
20/54/ 309 Littleburn Colliery
261 Shincliffe Colliery
to Thornley Colliery
432/529 Cassop Vale Colliery
508 Brandon C Colliery
BRANDON
River Wear
431/528 Cassop Moor Colliery
375 Stanley Colliery
aerial ropeway
263 Old Croxdale Colliery
119 Browney Colliery
SHINCLIFFE
381 Wooley Colliery
to Brancepeth Colliery
BRANCEPETH
to Darlington
to Ferryhill
to Crow Trees Colliery (433)
430/527 Cassop Colliery
to Crook
to Bishop Auckland
0
1
2 miles

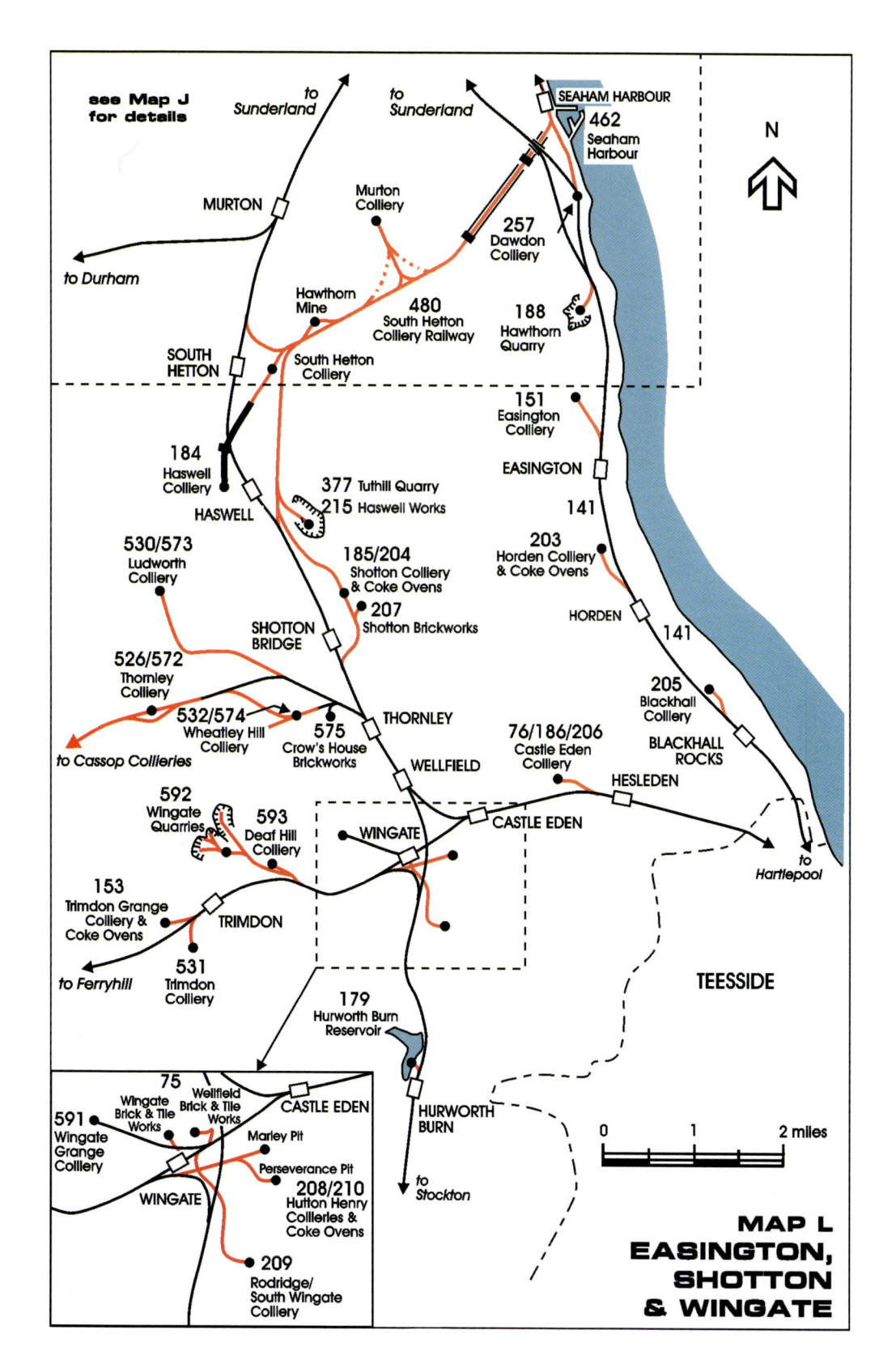

MAP L
EASINGTON, SHOTTON & WINGATE

see Map J for details

to Sunderland
to Sunderland
to Durham
to Cassop Collieries
to Ferryhill
to Stockton
to Hartlepool

- MURTON
- Murton Colliery
- SEAHAM HARBOUR — 462 Seaham Harbour
- 257 Dawdon Colliery
- Hawthorn Mine
- 480 South Hetton Colliery Railway
- 188 Hawthorn Quarry
- SOUTH HETTON
- South Hetton Colliery
- 151 Easington Colliery
- EASINGTON
- 184 Haswell Colliery
- HASWELL
- 377 Tuthill Quarry
- 215 Haswell Works
- 141
- 203 Horden Colliery & Coke Ovens
- 530/573 Ludworth Colliery
- 185/204 Shotton Colliery & Coke Ovens
- 207 Shotton Brickworks
- SHOTTON BRIDGE
- HORDEN
- 141
- 526/572 Thornley Colliery
- 532/574 Wheatley Hill Colliery
- 575 Crow's House Brickworks
- THORNLEY
- WELLFIELD
- 76/186/206 Castle Eden Colliery
- 205 Blackhall Colliery
- BLACKHALL ROCKS
- HESLEDEN
- 592 Wingate Quarries
- 593 Deaf Hill Colliery
- WINGATE
- CASTLE EDEN
- 153 Trimdon Grange Colliery & Coke Ovens
- TRIMDON
- 531 Trimdon Colliery
- 179 Hurworth Burn Reservoir
- HURWORTH BURN
- TEESSIDE

Inset (WINGATE):
- 75 Wellfield Brick & Tile Works
- Wingate Brick & Tile Works
- CASTLE EDEN
- 591 Wingate Grange Colliery
- Marley Pit
- Perseverance Pit
- WINGATE
- 208/210 Hutton Henry Collieries & Coke Ovens
- 209 Rodridge/South Wingate Colliery

N

0 1 2 miles

Durham Part 1 Page 19

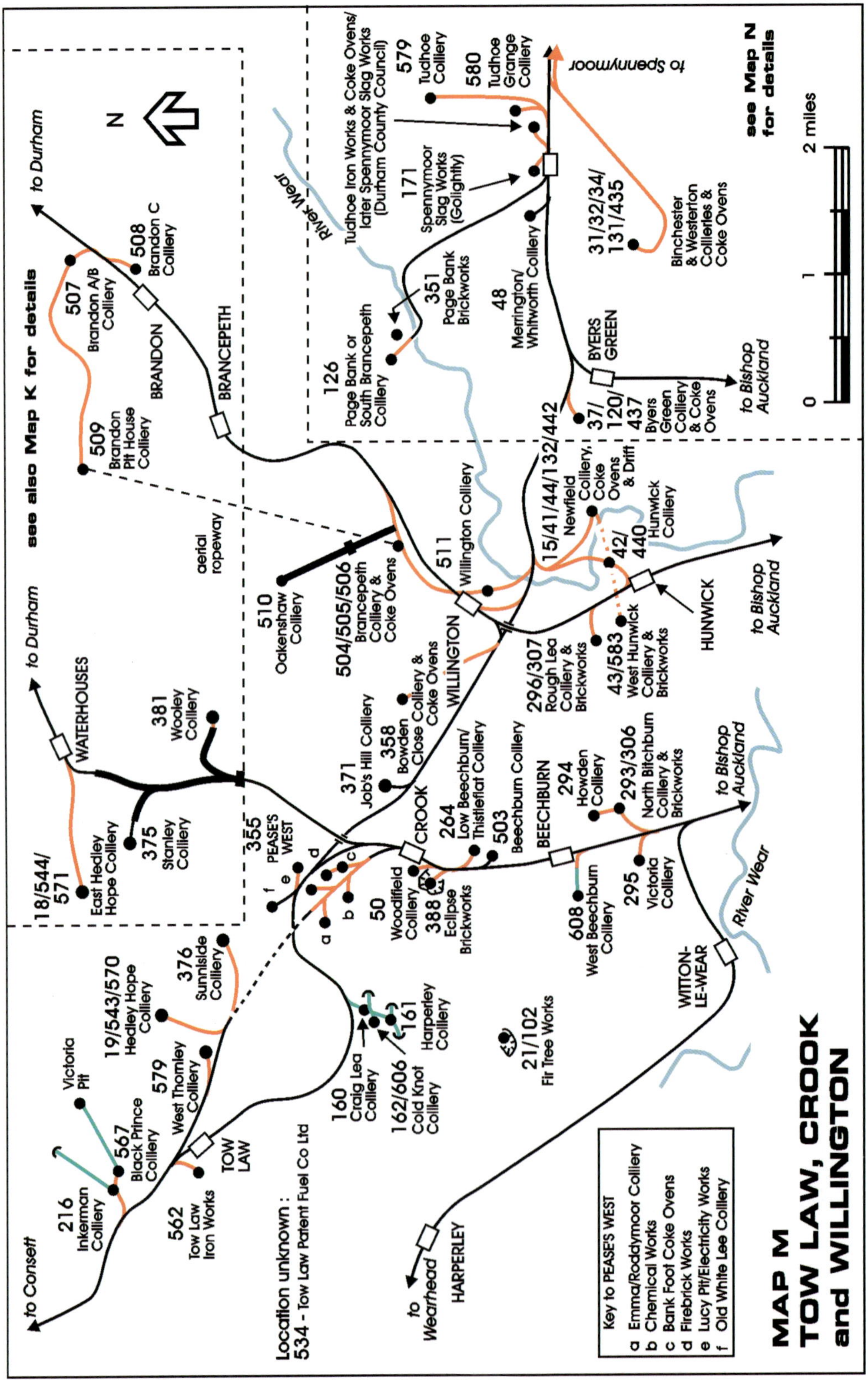
MAP M
TOW LAW, CROOK
and WILLINGTON
see also Map K for details
see Map N for details
N
to Durham
to Consett
to Wearhead
to Bishop Auckland
to Spennymoor
River Wear
0
1
2 miles
WATERHOUSES
BRANDON
BRANCEPETH
WILLINGTON
HUNWICK
BYERS GREEN
CROOK
PEASE'S WEST
BEECHBURN
WITTON-LE-WEAR
HARPERLEY
TOW LAW
aerial ropeway
18/544/571 East Hedley Hope Colliery
375 Stanley Colliery
381 Wooley Colliery
509 Brandon Pit House Colliery
507 Brandon A/B Colliery
508 Brandon C Colliery
510 Oakenshaw Colliery
504/505/506 Brancepeth Colliery & Coke Ovens
511 Willington Colliery
126 Page Bank or South Brancepeth Colliery
351 Page Bank Brickworks
48 Merrington/Whitworth Colliery
171 Spennymoor Slag Works (Golightly)
Tudhoe Iron Works & Coke Ovens/ later Spennymoor Slag Works (Durham County Council)
579 Tudhoe Colliery
580 Tudhoe Grange Colliery
31/32/34/131/435 Binchester & Westerton Collieries & Coke Ovens
37/120/437 Byers Green Colliery & Coke Ovens
15/41/44/132/442 Newfield Colliery, Coke Ovens & Drift
42/440 Hunwick Colliery
43/583 West Hunwick Colliery & Brickworks
296/307 Rough Lea Colliery & Brickworks
358 Bowden Close Colliery & Coke Ovens
371 Job's Hill Colliery
264 Low Beechburn/Thistleflat Colliery
503 Beechburn Colliery
294 Howden Colliery
293/306 North Bitchburn Colliery & Brickworks
295 Victoria Colliery
608 West Beechburn Colliery
355
50 Woodifield Colliery
388 Eclipse Brickworks
a
b
c
d
e
f
216 Inkerman Colliery
567 Black Prince Colliery
Victoria Pit
579 West Thornley Colliery
19/543/570 Hedley Hope Colliery
376 Sunniside Colliery
562 Tow Law Iron Works
160 Craig Lea Colliery
161 Harperley Colliery
162/606 Cold Knot Colliery
21/102 Fir Tree Works
Location unknown :
534 - Tow Law Patent Fuel Co Ltd
Key to PEASE'S WEST
a Emma/Roddymoor Colliery
b Chemical Works
c Bank Foot Coke Ovens
d Firebrick Works
e Lucy Pit/Electricity Works
f Old White Lee Colliery

MAP N SPENNYMOOR and FERRYHILL

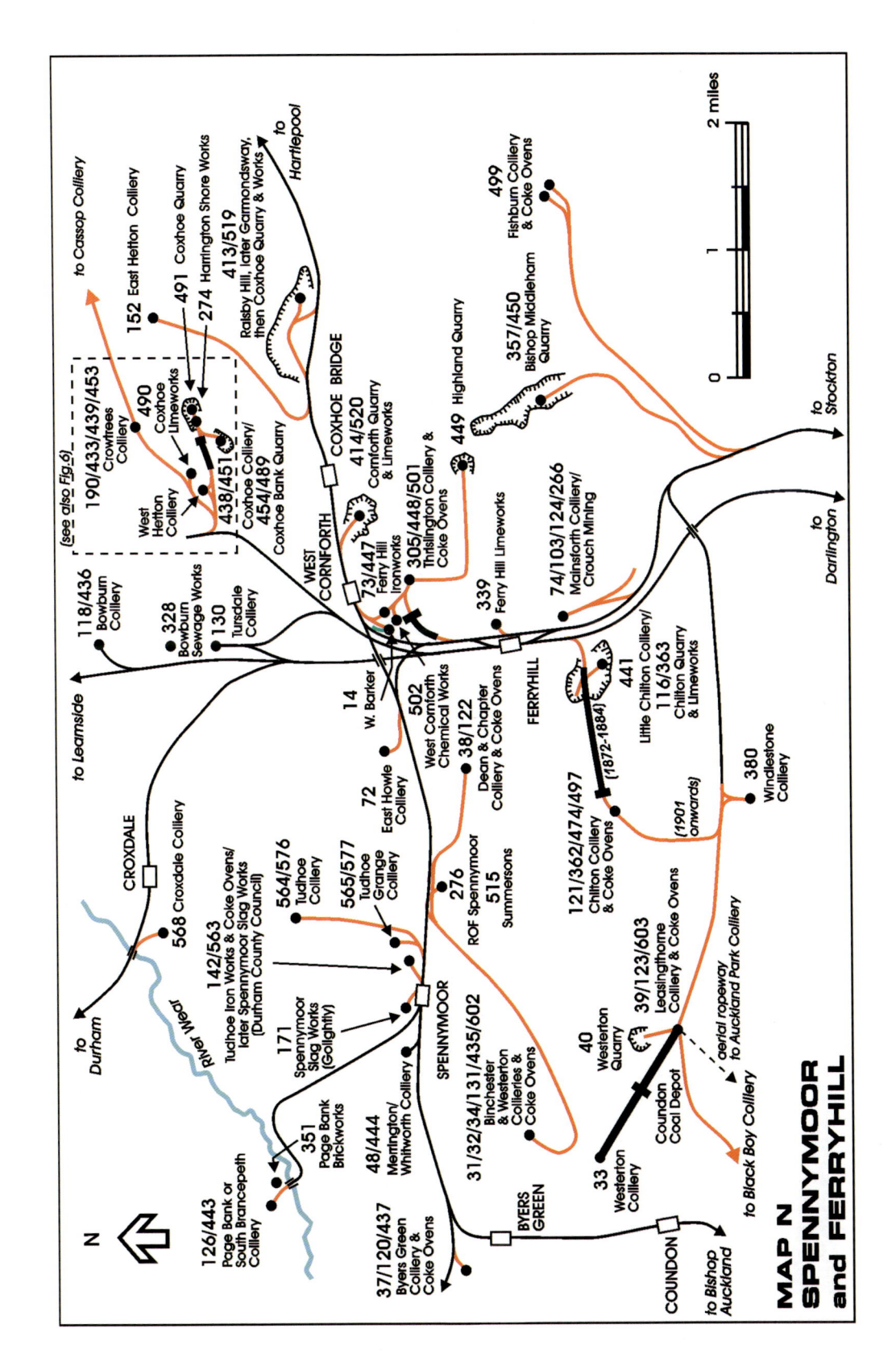

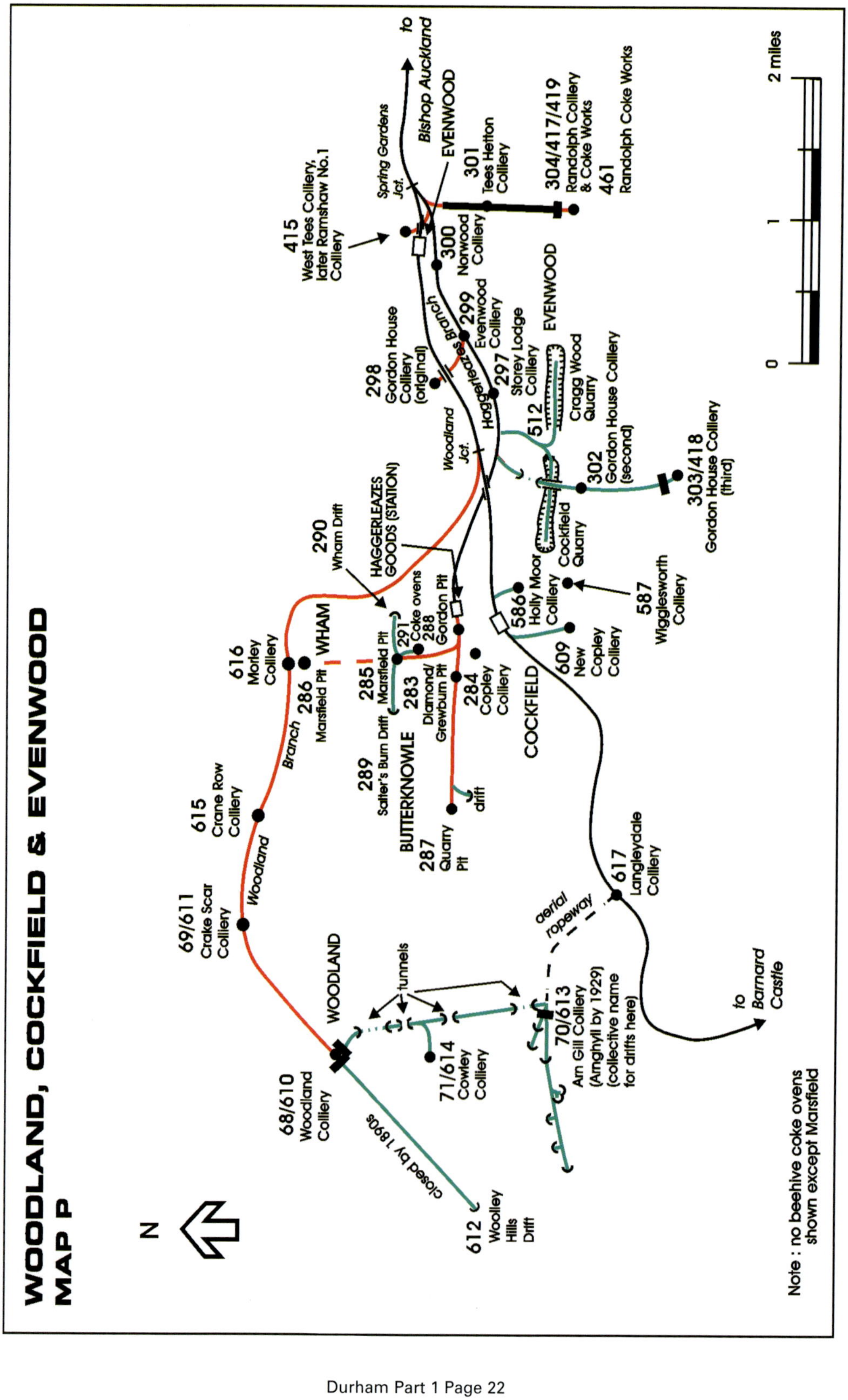

WOODLAND, COCKFIELD & EVENWOOD
MAP P
N
to Bishop Auckland
Spring Gardens Jct.
EVENWOOD
415 West Tees Colliery, later Ramshaw No.1 Colliery
300 Norwood Colliery
301 Tees Hetton Colliery
304/417/419 Randolph Colliery & Coke Works
461 Randolph Coke Works
298 Gordon House Colliery (original)
299 Evenwood Colliery
Haggerleazes Branch
297 Storey Lodge Colliery
EVENWOOD
512 Cragg Wood Quarry
Woodland Jct.
302 Gordon House Colliery (second)
303/418 Gordon House Colliery (third)
Cockfield Quarry
290 Wham Drift
HAGGERLEAZES GOODS (STATION)
291 Coke ovens
288 Gordon Pit
586 Holly Moor Colliery
587 Wigglesworth Colliery
609 New Copley Colliery
616 Morley Colliery
WHAM
286 Marsfield Pit
285 Marsfield Pit
283 Diamond/ Grewburn Pit
284 Copley Colliery
COCKFIELD
289 Salter's Burn Drift
BUTTERKNOWLE
Woodland Branch
615 Crane Row Colliery
287 Quarry Pit
drift
69/611 Crake Scar Colliery
617 Langleydale Colliery
aerial ropeway
WOODLAND
tunnels
68/610 Woodland Colliery
71/614 Cowley Colliery
70/613 Arn Gill Colliery (Arnghyll by 1929) (collective name for drifts here)
to Barnard Castle
closed by 1890s
612 Woolley Hills Drift
Note : no beehive coke ovens shown except Marsfield
0
1
2 miles

MAP G
BISHOP AUCKLAND

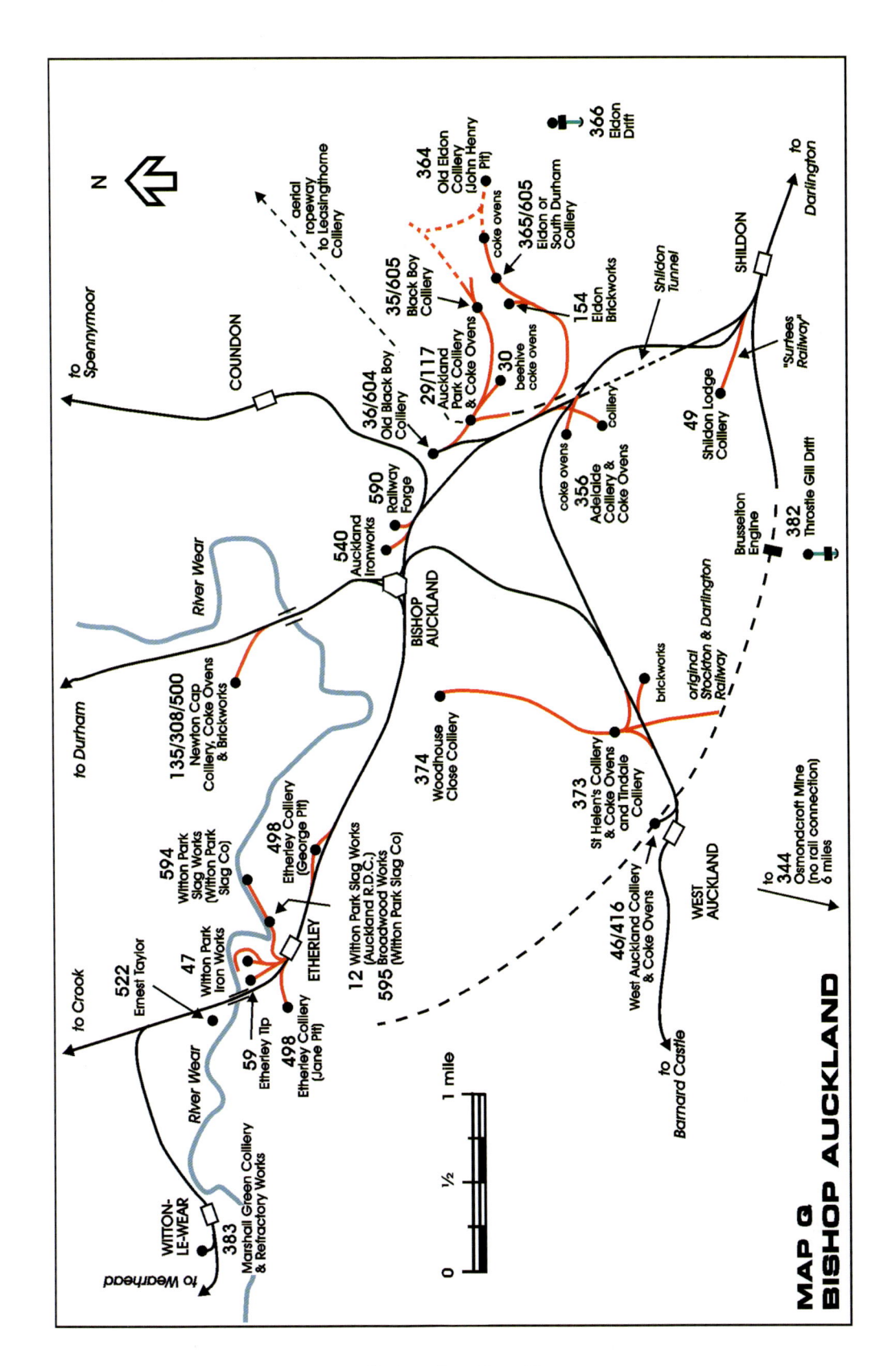

Durham Part 1 Page 23

MAP R
NEWTON AYCLIFFE

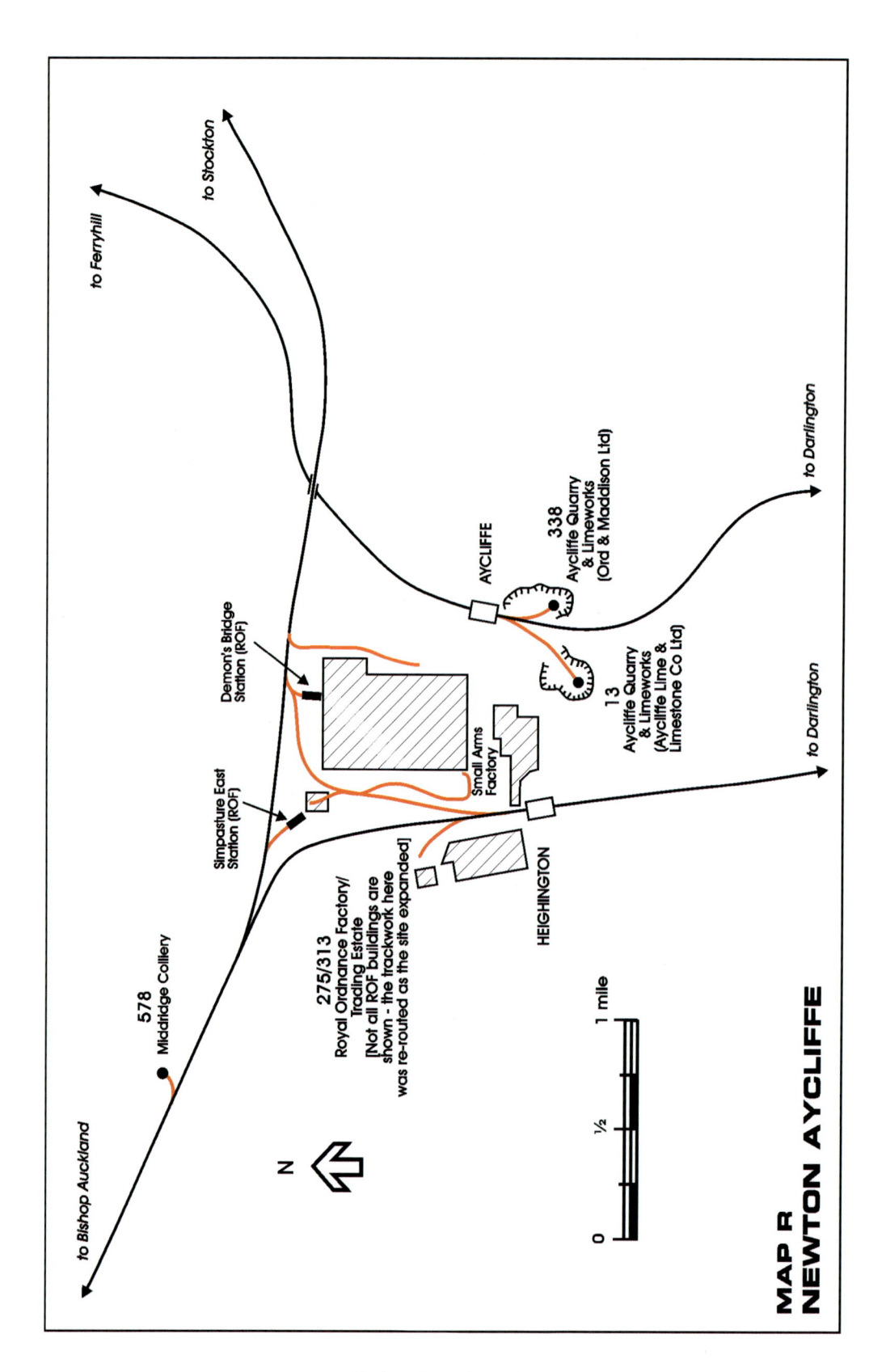

MAP 8 DARLINGTON

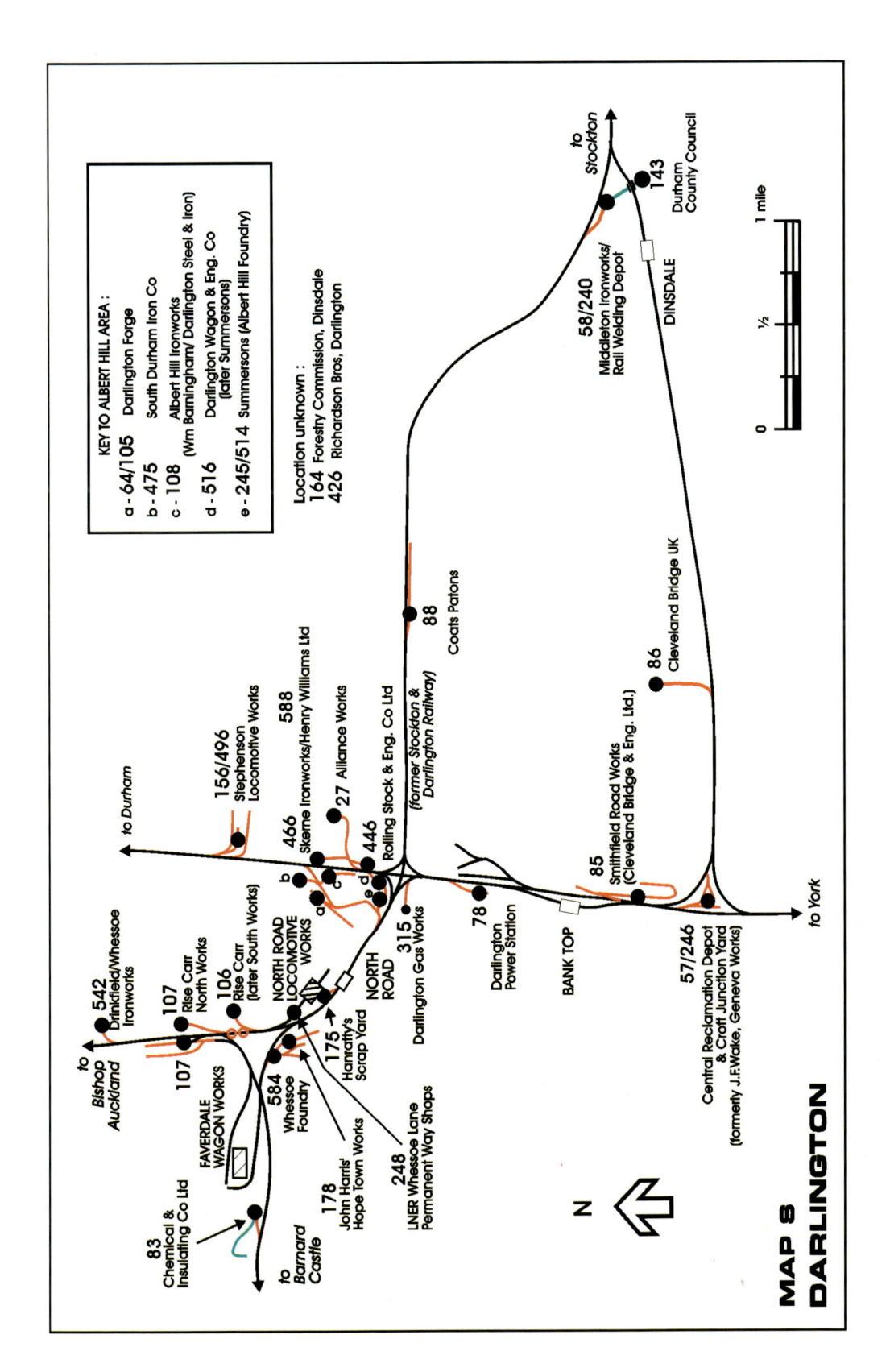

KEY TO ALBERT HILL AREA:

- a - 64/105 Darlington Forge
- b - 475 South Durham Iron Co
- c - 108 Albert Hill Ironworks (Wm Barningham/Darlington Steel & Iron)
- d - 516 Darlington Wagon & Eng. Co (later Summersons)
- e - 245/514 Summersons (Albert Hill Foundry)

Location unknown:
- 164 Forestry Commission, Dinsdale
- 426 Richardson Bros, Darlington

Scale: 0 – ½ – 1 mile

to Stockton
143 Durham County Council
58/240 Middleton Ironworks/Rail Welding Depot
DINSDALE
Coats Patons 88
86 Cleveland Bridge UK
85 Smithfield Road Works (Cleveland Bridge & Eng. Ltd.)
to York
57/246 Central Reclamation Depot & Croft Junction Yard (formerly J.F.Wake, Geneva Works)
BANK TOP
78 Darlington Power Station
315 Darlington Gas Works
NORTH ROAD
NORTH ROAD LOCOMOTIVE WORKS
446 Rolling Stock & Eng. Co Ltd (former Stockton & Darlington Railway)
466 Skerne Ironworks/Henry Williams Ltd
27 Alliance Works
588
156/496 Stephenson Locomotive Works
to Durham
106 Rise Carr (later South Works)
107 Rise Carr North Works
542 Drinkfield/Whessoe Ironworks
to Bishop Auckland
FAVERDALE WAGON WORKS
107
584 Whessoe Foundry
175 Hanratty's Scrap Yard
178 John Harris' Hope Town Works
248 LNER Whessoe Lane Permanent Way Shops
83 Chemical & Insulating Co Ltd
to Barnard Castle

N

LOWER WEARDALE
MAP T

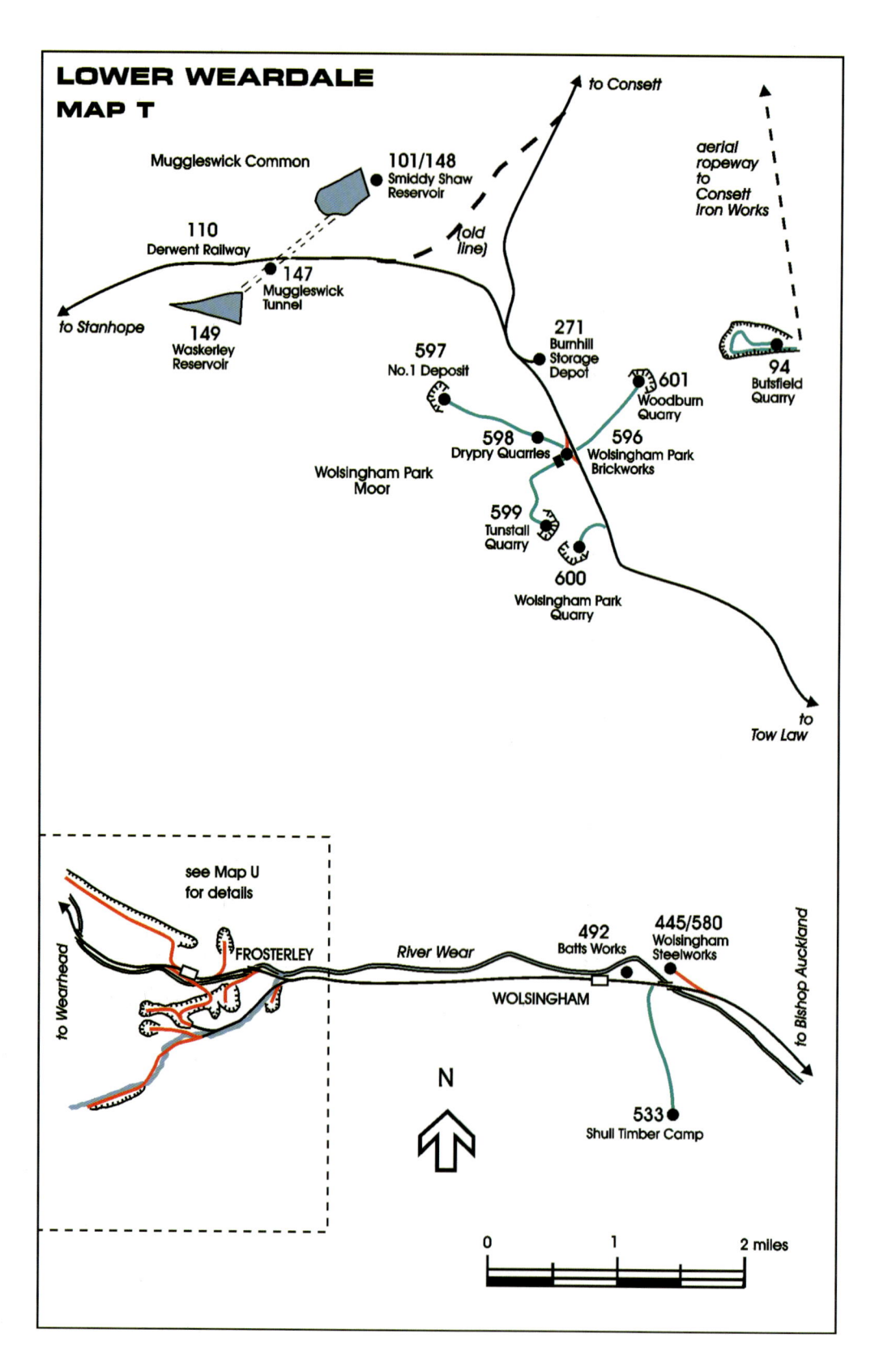

Muggleswick Common

to Consett

101/148 Smiddy Shaw Reservoir

aerial ropeway to Consett Iron Works

110 Derwent Railway

(old line)

to Stanhope

147 Muggleswick Tunnel

149 Waskerley Reservoir

271 Burnhill Storage Depot

597 No.1 Deposit

601 Woodburn Quarry

94 Butsfield Quarry

598 Drypry Quarries

596 Wolsingham Park Brickworks

Wolsingham Park Moor

599 Tunstall Quarry

600 Wolsingham Park Quarry

to Tow Law

see Map U for details

FROSTERLEY

River Wear

492 Batts Works

445/580 Wolsingham Steelworks

to Wearhead

WOLSINGHAM

to Bishop Auckland

N

533 Shull Timber Camp

0   1   2 miles

Durham Part 1 Page 26

FROSTERLEY
MAP U

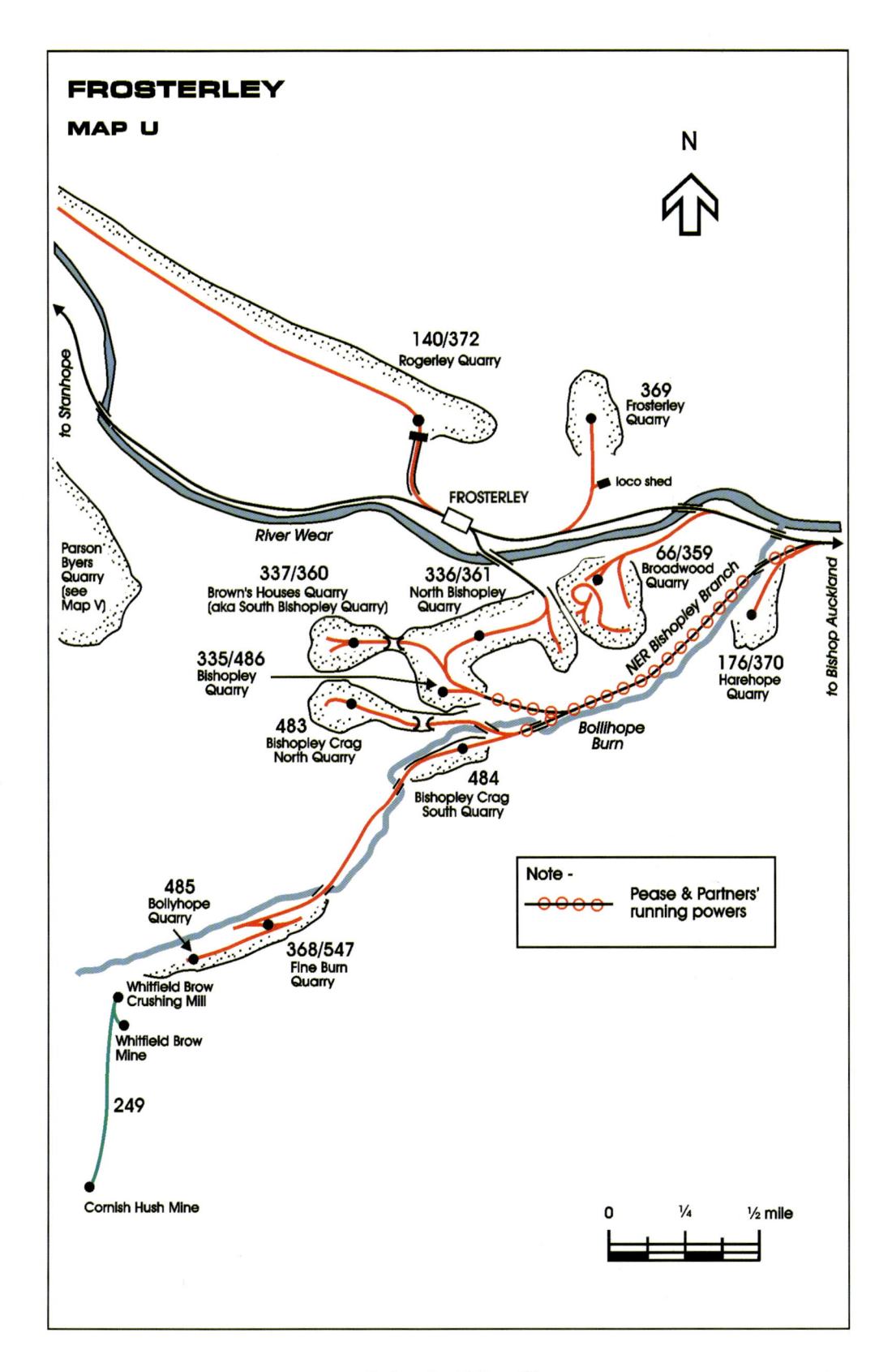

- to Stanhope
- 140/372 Rogerley Quarry
- 369 Frosterley Quarry
- loco shed
- FROSTERLEY
- River Wear
- Parson Byers Quarry (see Map V)
- 337/360 Brown's Houses Quarry (aka South Bishopley Quarry)
- 336/361 North Bishopley Quarry
- 66/359 Broadwood Quarry
- NER Bishopley Branch
- to Bishop Auckland
- 335/486 Bishopley Quarry
- 176/370 Harehope Quarry
- 483 Bishopley Crag North Quarry
- Bollihope Burn
- 484 Bishopley Crag South Quarry
- 485 Bollyhope Quarry
- 368/547 Fine Burn Quarry
- Whitfield Brow Crushing Mill
- Whitfield Brow Mine
- 249
- Cornish Hush Mine

Note - Pease & Partners' running powers

0   ¼   ½ mile

Durham Part 1 Page 27

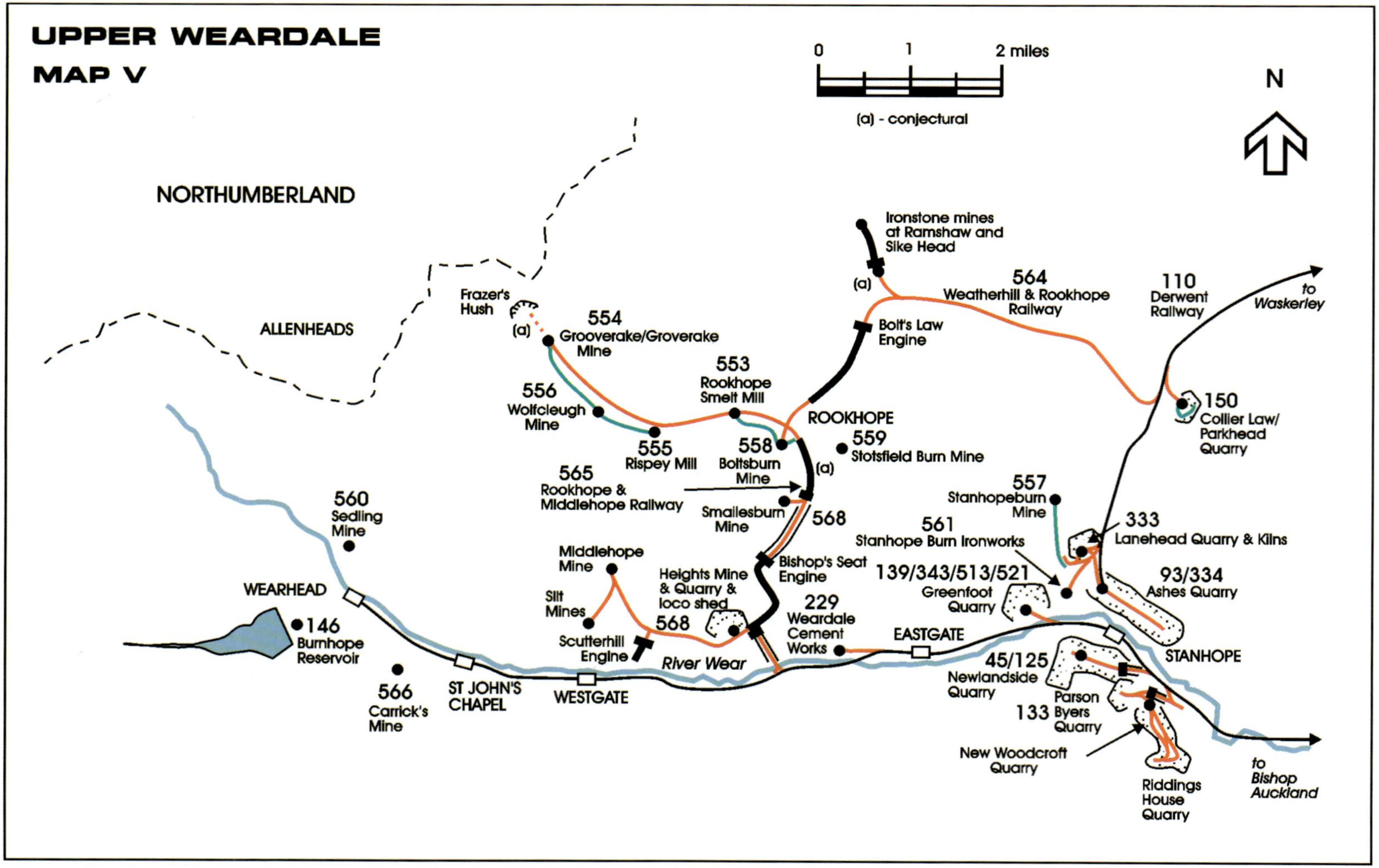
UPPER WEARDALE
MAP V
0
1
2 miles
(a) - conjectural
N
NORTHUMBERLAND
ALLENHEADS
Frazer's Hush
(a)
554 Grooverake/Groverake Mine
556 Wolfcleugh Mine
555 Rispey Mill
553 Rookhope Smelt Mill
558 Boltsburn Mine
565 Rookhope & Middlehope Railway
Smailesburn Mine
568
(a)
ROOKHOPE
559 Stotsfield Burn Mine
Ironstone mines at Ramshaw and Sike Head
(a)
Bolt's Law Engine
564 Weatherhill & Rookhope Railway
110 Derwent Railway
to Waskerley
150 Collier Law/ Parkhead Quarry
557 Stanhopeburn Mine
333 Lanehead Quarry & Kilns
561 Stanhope Burn Ironworks
139/343/513/521 Greenfoot Quarry
93/334 Ashes Quarry
Bishop's Seat Engine
229 Weardale Cement Works
Heights Mine & Quarry & loco shed
568
Middlehope Mine
Slit Mines
Scutterhill Engine
River Wear
560 Sedling Mine
WEARHEAD
146 Burnhope Reservoir
566 Carrick's Mine
ST JOHN'S CHAPEL
WESTGATE
EASTGATE
STANHOPE
45/125 Newlandside Quarry
133 Parson Byers Quarry
New Woodcroft Quarry
Riddings House Quarry
to Bishop Auckland

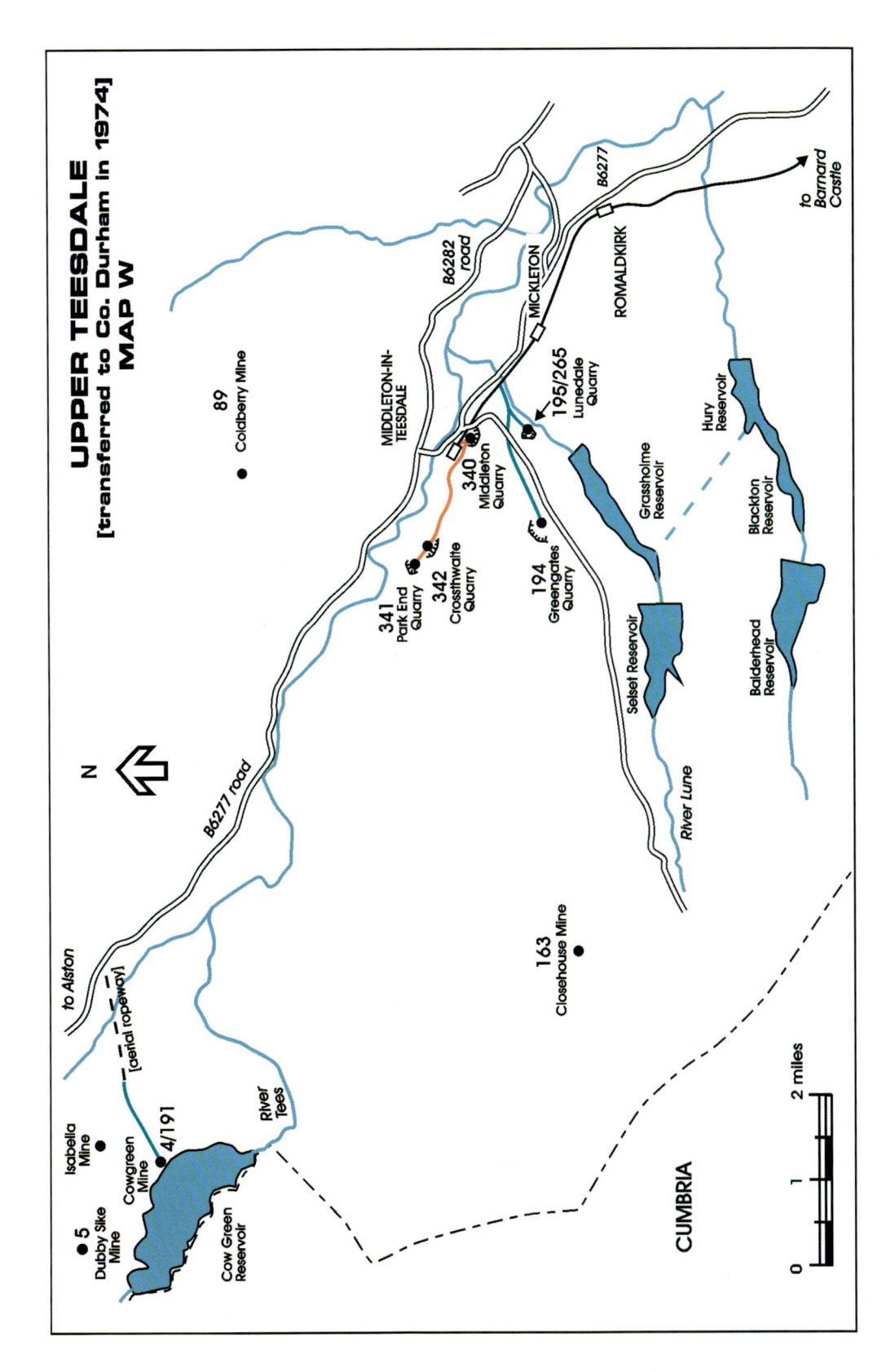

UPPER TEESDALE
[transferred to Co. Durham in 1974]
MAP W

3. A general view of the East and West rolling mills at the Consett Iron Works in 1892, taken from a booklet produced for a visit to the works that year by the Iron & Steel Institute. The 0-4-0ST on the right of the picture is believed to be the John Harris locomotive, B No 4.

Alphabetical List of Owners

(excluding the National Coal Board and the fluorspar industry)

JOHN ABBOT & CO LTD

John Abbot & Co until 1/7/1864
PARK IRON WORKS, Gateshead   C1
                                                                                                          NZ 256637

This firm was established before 1790, and its works was eventually served by sidings north of the NER Gateshead - South Shields line, ¼ mile east of Gateshead East Station. Shunting was probably done by horses before the arrival of first loco below. This was a sizeable works, with 33 puddling furnaces and four rolling mills by 1870. The company absorbed its neighbours, Hawks, Crawshay & Sons (which see), about August 1891. The NER Sidings map for this area, drawn in 1884 and revised in 1895, shows premises entitled '**Abbotts Team Rolling Mills** (closed)' (NZ 237626) served by the NER Redheugh branch just west of Redheugh Junction, on the southern bank of the River Tyne facing Dunston Staiths, and it would seem very likely that they were connected with this firm. The company went into liquidation in October 1909 and the works was dismantled by Thos W. Ward Ltd in 1910.

Gauge: 4ft 8½in

-	0-4-0ST	OC	BH	13	1866	New	s/s
ADAMSON	0-4-0ST	OC	AB	693	1891	New	(1)
HAWK	0-4-0ST	OC	BH	682	1883	(a)	(2)
ABBOT	0-4-0ST	OC	HL	2425	1899	New	(3)

The firm was reputed to have built a locomotive for its own use.

(a)   ex Hawks, Crawshay & Sons, Gateshead, with this firm's site, c8/1891.
(1)   a 12in cylinder loco, which this loco was, was offered for sale on 29/4/1910 in both the *Glasgow Herald* and *Contract Journal*; scrapped on site by Thos W. Ward Ltd, c/1910.
(2)   offered for sale 29/4/1910, (as above); to Thos W. Ward Ltd, Charlton Works, Sheffield, Yorkshire (WR), c/1910.
(3)   offered for sale 29/4/1910, (as above); to Skinningrove Iron Co Ltd, Carlin How, Yorkshire (NR), via Thos W. Ward Ltd, c/1910.

ADAMS PICT FURNACE BRICK CO LTD

Firebrick Works, Derwenthaugh   B2
G.H. Ramsey & Co until 1/8/1925
G.H. Ramsey & Son until /1880; originally **G.H. Ramsey**   NZ 203627

This firebrick works was begun in 1830 by George Ramsey (1790-1880), who developed a wide variety of industrial interests on Tyneside, including the **Blaydon Main Colliery** nearby (see The Stella Coal Co Ltd). In 1860 he patented the crushing of coal to a powder for coke making, a practice later widely adopted. In May 1847 the Newcastle & Carlisle Railway opened its short Swalwell Branch, which ran past the eastern side of the yard, but curiously it seems that some years passed before it was connected. After Ramsey's death the works continued to be operated under his name.

About 100 yards south of the brickworks lay **Swalwell Garesfield Colliery** (NZ 205625), opened in 1887 and later owned by Dunston Garesfield Collieries Ltd (which see). The NER extended its branch to serve it and gave the brickworks running powers between the colliery and the brickworks to bring up fireclay mined at the pit. It is not known whether the two were effectively under the same ownership at this time. By 1900 the brickworks was one of the largest on Tyneside.

In 1925 the yard was purchased by Adamsez & Co Ltd, the sanitaryware manufacturers at Scotswood in Newcastle upon Tyne, who set up this new company. In his book *Brickworks of the North East*, Davison gives the new company's name as Adams-Pict Firebrick Co. Swalwell Garesfield Colliery closed in 1940, causing the works to obtain clay and coal from elsewhere. Rail traffic eventually ceased, though the works continued until the early 1970s.

It would appear that the works had at least two locomotives, but the only details known are given below.

Gauge: 4ft 8½in

-	0-4-0ST	OC	BH	298	1875	(a)	(1)

(a) ex NER, 996; but NER records give two different versions of the sale of this loco. One gives it sold to G.H. Ramsey on 3/12/1891; the other gives it sold to Lingford, Gardiner & Co Ltd, Bishop Auckland, on 17/12/1891. Although it cannot now be proved which version is correct, the authors believe the former is more likely.

(1) to Dunston Garesfield Collieries Ltd, Swalwell Garesfield Colliery, Swalwell.

ALLISON, ENGLISH & CO LTD (registered in 1906)
UNION & TEAM VALLEY BRICKWORKS, Birtley   G3

The **UNION BRICKWORKS** (NZ 264556 approx.), north of Station Road in Birtley, was begun in 1879 by **Burleson, Todd & Co**, who operated it until 1897. In 1906 this new company opened the **TEAM VALLEY BRICKWORKS** (NZ 266558) on land to the east of the Union Works, which it re-started in 1908. The two works were served by a link to the Bewicke Main branch of the Pelaw Main Railway (see The Pelaw Main Collieries Ltd), which worked the company's traffic between the works and Birtley Station on the NER Newcastle – Durham line. The locomotive below was presumably acquired to work on the tramways between the two quarries and the works. The firm went out of business in 1925. The two yards were re-opened, probably in 1934, by **The Birtley Brick Co Ltd** (which see).

Gauge : 2ft 0in

-	4wPM	RH	114563	1923	(a)	(1)

(a) exhibited, apparently new (it is possible that the loco was built in 1921), by the makers at The Royal Show, Newcastle upon Tyne, 3-7/7/1923; sold after the Show to this firm.

(1) said to have passed, with the yard, to The Birtley Brick Co Ltd (which see), but unconfirmed.

ANGLO-AUSTRAL MINES LTD
subsidiary of **Imperial Smelting Corporation**, Avonmouth, Somerset
COW GREEN MINE, near Langdon Beck, near Middleton-in-Teesdale   W4
                                                                                NY 810305

Mining in this area of Teesdale began in 1853; between 1898 and 1920 it was carried on by **The Hedworth Barium Co Ltd** (which see). This firm commenced operations here about 1939.

The main mine was situated about eight miles north-west of Middleton-in-Teesdale, about 1500 feet above sea level. It consisted of a drift, originally called the **Low Level** but latterly the **Horse Level** because horses hauled the tubs, together with a shaft nearby known as **Wrentnall Shaft**. It was one of the world's chief sources of barytes, with about 300 tons of dressed ore per week being taken by road to Middleton-in-Teesdale Station to be sent for processing to Orr's Zinc White Ltd's works at Widnes in Lancashire.

In 1950 the firm re-opened **DUBBY SIKE MINE** (NY 795319) **(W5)**, which was actually a drift. This mine, first worked in the nineteenth century, lay about 1½ miles north-west of the main site. Working here ended in 1952.

Loco haulage underground to the Wrentnall Shaft was introduced in 1949 (one source says in 1950). The mine closed in April 1954. The site is now submerged under the Cow Green Reservoir.

References : *Out of the Pennines*, ed. B. Chambers, 1997, *Mine Alone*, J.R. Foster-Smith (mine manager here from 1948 -1954); *The Mines of Upper Teesdale*, R.A. Fairbairn, 2005.

Gauge : 1ft 6in

-	0-4-0BE	WR	4146	1949	New	(1)
-	0-4-0BE	WR	4147	1949	New	(1)
-	0-4-0BE	WR	4148	1949	New	(2)

(1) s/s; possibly to Gasswater Mine, Cronberry Moor, Ayrshire, c/1954.
(2) to Gasswater Mine, Cronberry Moor, Ayrshire, c/1954.

SIR WILLIAM ANGUS, SANDERSON & CO LTD (registered in 1910)
BIRTLEY MOTOR WORKS, Birtley G6
NZ 265564

This company purchased the former National Projectile Factory No.9 from the **Ministry of Munitions** (which see) at the end of 1919 and converted it to manufacture the luxury 14 h.p. Angus-Sanderson car. It was served by sidings east of the NER Newcastle-Durham line, ½ mile north of Birtley Station. However, the high demand produced a wait of at least a year, and in 1921 the company moved its production to Hendon in Middlesex.

The premises were purchased for dismantling by the Darlington dealer S.F. Hill, who advertised the two locomotives below on 10th March 1922. They were offered again in an auction for Hill by Wheatley Kirk on 17th May 1922, and Hill continued to offer No.10 for sale until 1923. The premises subsequently became a Government Training Centre, and then in 1936 they were taken over for the development of a **Royal Ordnance Factory** (which see).

Gauge : 4ft 8½in

No.8	0-4-0ST	OC	AE	1054	1874	(a)	(1)
No.10	0-4-0ST	OC	AE	1056	1875	(a)	(2)

(a) ex Ministry of Munitions, with site, c12/1919.

(1) offered for sale, 10/3/1922; offered for auction, 17/5/1922, without any motion; believed to have been sold at this auction for scrap.
(2) offered for sale at various dates between 10/3/1922 and 1/1923; s/s.

ARMSTRONG-WHITWORTH ROLLS LTD
subsidiary of **Davy-Ashmore Ltd** from 10/1968
Armstrong Whitworth (Metal Industries) Ltd until 10/1968
Sir W.G. Armstrong Whitworth & Co. (Ironfounders) Ltd until 15/6/1953
Sir W.G. Armstrong Whitworth & Co. (Engineers) Ltd until 10/4/1930
Sir W.G. Armstrong Whitworth & Co Ltd until 1/1/1929

In 1928 Sir W.G. Armstrong, Whitworth & Co Ltd began re-organising its activities. From 31st December 1928 its shipbuilding activities were transferred to Sir W.G. Armstrong, Whitworth & Co (Shipbuilders) Ltd and its engineering and ironfounding activities to Sir W.G. Armstrong, Whitworth & Co (Engineers) Ltd, though both of these new companies were not registered until 8th July 1929. The ironfounding was separated off to another new company, Sir W.G. Armstrong, Whitworth & Co (Ironfounders) Ltd, with effect from 31st December 1929, though this company was not registered until 10th April 1930. Meanwhile Sir W.G. Armstrong, Whitworth & Co Ltd itself was converted into a holding company, those above all being subsidiaries, with the new title of Armstrong Whitworth Securities Co Ltd, registered on 6th July 1929. See also the entry for the Close Works below.

CLOSE WORKS, Gateshead C7
NZ 258633

This works was taken over by Sir W.G. Armstrong, Whitworth & Co Ltd from C.A. Parsons & Co Ltd in 1915. It included a large part of the former locomotive works of Black, Hawthorn & Co Ltd/Chapman & Furneaux. In October 1904 most of these premises had been sold to Ernest Scott & Mountain Ltd, a firm of electrical engineers, who transferred their works here from Newcastle upon Tyne. They built a small number of electric locomotives here. This business had been taken over in March 1913 by C.A. Parsons & Co Ltd, also of Newcastle upon Tyne. The works, which retained the Black Hawthorn office block, was served by sidings west of the BR Gateshead - South Shields line, ¼ mile east of Gateshead Station. Its main products were iron castings, special pig iron and pneumatic tools. The demand for pneumatic tools grew steadily, until by 1950 the works produced nothing else and at a date unknown a new company, Sir W.G. Armstrong, Whitworth & Co (Pneumatic Tools) Ltd had been set up to run it. This firm was purchased by the Thor Power Tool Co of Illinois, USA, in February 1951, which in the summer of 1953 took all of the equipment and the workforce to a new factory in North Shields, Northumberland, with production at the Close Works ceasing on 31st July 1953. The empty works then reverted to Armstrong Whitworth (Metal Industries) Ltd, the new name for Sir W.G. Armstrong, Whitworth & Co (Ironfounders) Ltd, who re-equipped it for ironfounding and the production of steel rolls.

Shunting was done by rail crane for some years, and then rail traffic ceased. In July 1970 the works passed to The Davy Roll Co Ltd, which continues (2006) to operate it.

Gauge: 4ft 8½in

	-	0-4-0ST	OC	?	?	?	(a)	s/s	
No.2	(formerly 12)	0-4-0ST	OC	HL	2357	1896	(b)	Scr c/1955	
1799		0-4-0T	IC	Ghd	4	1897	(c)	(1)	
No.1		4wVBT	VCG	S	9558	1953	(d)	(2)	
185	DAVID PAYNE	0-4-0DM		JF	4110006	1950	(e)	(3)	

Two steam rail cranes also worked here.

(a) possibly built by Peckett; origin and date of arrival unknown.
(b) ex Elswick Works, Newcastle upon Tyne, c/1917.
(c) ex LNER, Tyne Dock loco shed, Tyne Dock, by 25/3/1937, hire; LNER Y7 class.
(d) S demonstration loco; ex trials at The Seaham Harbour Dock Co, Seaham Harbour, by 13/11/1953; purchased from S soon afterwards.
(e) ex Dorman, Long & Co Ltd, Dock Street Foundry, Middlesbrough, Yorkshire (NR), 5/1965.

(1) returned to LNER, Tyne Dock loco shed, Tyne Dock, at end of hire.
(2) to Jarrow Works, Jarrow, 2/1965.
(3) to Jarrow Works, Jarrow, 9/1965.

JARROW WORKS, Jarrow   E8 & E9
Armstrong Whitworth (Metal Industries) Ltd until 10/1968
Jarrow Metal Industries Ltd until 30/9/1960 (see also below)
subsidiary of **Armstrong Whitworth (Metal Industries) Ltd** from 15/6/1953
previously subsidiary of **Sir W.G. Armstrong, Whitworth & Co (Ironfounders) Ltd** (see above)

The first to help Jarrow after the demolition of Palmers' shipyard (see Palmers Shipbuilding & Iron Co Ltd) was Sir John Jarvis, High Sheriff of Surrey and a wealthy businessman and philanthropist. On part of the Palmers site he established **Jarrow Metal Industries Ltd** (NZ 321654) **(E8)** to manufacture steel and light alloy castings, while a mile away, on the site of the former Palmers slag heap, he set up **Jarrow Tube Works Ltd** (NZ 323645) **(E9)** to make steel tubes. The two works were linked by a branch, also previously operated by Palmers, which ran alongside the Bowes Railway of John Bowes & Partners Ltd, crossing it near the site of the southern works. It was operated by a Jarrow Metal Industries Ltd loco. Both works opened in May 1938. The tube works company was subsequently absorbed into Jarrow Metal Industries Ltd, which became a subsidiary of Sir W.G. Armstrong, Whitworth & Co (Ironfounders) Ltd, both probably between 1939 and 1941. Rail traffic ceased in 1969. In July 1970 the two sites, with the firm, were incorporated into the Davy Roll Co Ltd (which see).

4. *LNER Y7 class 983, at the Jarrow Works, on hire from Tyne Dock loco shed, on 24th May 1939.*

Gauge: 4ft 8½in

	ELSIE	0-4-0ST	OC	HL	3895	1937	New	(1)
983		0-4-0T	IC	Dar		1923	(a)	(2)
No.2	(formerly No.1)	0-4-0ST	OC	RSHN	7297	1945	New	(3)
No.1		4wVBT	VCG	S	9558	1953	(b)	(4)
185	DAVID PAYNE	0-4-0DM		JF	4110006	1950	(c)	(5)

(a) ex LNER, Tyne Dock loco shed, Tyne Dock, by 24/5/1939, hire; LNER Y7 class.
(b) ex Close Works, Gateshead, 2/1965.
(c) ex Close Works, Gateshead, 9/1965.

(1) to Vickers Armstrong Ltd, Openshaw, Lancashire, /1941.
(2) returned to LNER, Tyne Dock loco shed, Tyne Dock, at end of hire.
(3) partially scrapped in 6/1965, the frame being fitted with a diesel engine to make a furnace charging machine.
(4) the boiler was removed and the frame converted to a ladle carrier, /1967.
(5) to Davy Roll Co Ltd, with works, 7/1970, out of use.

A loco was occasionally borrowed from the adjacent works of The Consett Iron Co Ltd (which see), and on several occasions other LNER Y7 class 0-4-0T's were hired from Tyne Dock Shed, but the identities of the other locomotives and the dates are not known.

5. A number of companies in the North East bought the high steam pressure Sentinel 100 h.p. and 200 h.p. machines in the 1950s. 4wVBT S 9558/1953, a 100 h.p. locomotive, pauses at Jarrow on 7th April 1965.

ARMSTRONG WHITWORTH SECURITIES CO LTD
For the history of this company see above. It was put into liquidation on 16th September 1943, by which time the Armstrong Whitworth companies (see above) were owned by Vickers Ltd.

Unknown Location A10
On 17th December 1929 the auctioneers Wheatley Kirk held an auction of plant owned by this company at the "Addison Yard, Blaydon". Addison Colliery, owned by The Stella Coal Co Ltd (which see) lay about 1½ miles east of Blaydon, but a sale in a colliery yard seems unlikely. There were also two sets of sidings owned by the LNER near Blaydon known as Addison Yard, but one of these was not developed until the Second World War. The sale would seem to have been held in this area, but it has not been possible to determine its precise location.

The plant for sale included a four-coupled saddletank by KS and a six-wheeled saddletank by MW, both presumably standard gauge. It <u>may</u> be that the plant had some connection with the disposal of equipment acquired during the First World War by the Ministry of Munitions, which in 1923 set up the George Cohen & Armstrong Disposal Corporation to handle what remained. Armstrong Whitworth itself seems never to have owned a Kerr Stuart locomotive, and had only one Manning Wardle locomotive, at the Elswick Works in Newcastle upon Tyne.

ASSOCIATED ELECTRICAL INDUSTRIES LTD
BIRTLEY WORKS, Birtley G11
W.T. Henley's Telegraph Works Co Ltd until 1/1/1960 NZ 272547

This works replaced an older works elsewhere in Birtley that Henley's had taken over in 1945. Officially opened on 16th October 1950, it was served by sidings east of the BR Newcastle - Durham line, one mile south of the former Birtley Station. Rail traffic ceased in 1970, although the factory continues in production (2006) under different owners.

Gauge : 4ft 8½in

    -    4wDM   RH 265615 1948 New Scr 6/1970

AUCKLAND RURAL DISTRICT COUNCIL
WITTON PARK SLAG WORKS, Witton Park Q12
NZ 177304

The former Witton Park Ironworks of Bolckow, Vaughan & Co Ltd (which see), which had been closed in 1884, had produced a large amount of slag, which stretched away eastwards along the southern bank of the River Wear. The works was not demolished until 1896. Auckland Rural District Council, in whose area the works was situated, decided to work the slag for its own use and to sell it, as well as to remove it and to provide employment. The surviving information, both about its operations and the locomotives that worked here, is limited almost entirely to the council's minute books, now deposited at the Durham County Record Office (ref: DRO/UD/BA).

It would appear from a minute in 1915 that the council started work at Witton Park in 1891, leasing the slag heaps from Bolckow Vaughan and developing a small works at the western end of the site, near to the ironworks. This was served by a ¼ mile branch from Etherley Station on the NER line between Bishop Auckland and Crook. In 1898 an asphalt plant was added, and subsequently the manufacture of tarmacadam was begun. In December 1912 Bolckow Vaughan sold the slag heaps to Mr.T.B. Maughan, trading as The Witton Park Slag Co, which leased them back to the council in October 1913. However, on 27th February 1915 the council closed the works down, apparently because the NER could not supply suitable wagons, and it did not re-open until March 1918, initially with the council using a sub-contractor, though eventually it seems the council resumed full control. When The Witton Park Slag Co began processing slag on its own account is uncertain; it may have been as early as 1913, when the council agreed its charges for handling Slag Co traffic, which presumably means that the council's line had been extended to serve the Slag Co's site. Work at the council's works stopped in November 1920 due to shortage of coal, and this, plus a withdrawal by the NER of a supply of wagons suitable to convey tar, led to permanent closure in July 1921 and the lease being surrendered on 31st December 1921. The auction of surplus plant was agreed in May 1922. Meanwhile the heaps continued to be worked by The Witton Park Slag Co (which see), which had opened a new works in 1921.

It is impossible from the minutes to piece together a clear picture of the council's use of locomotives or to identify them, and it would seem best to record the information in chronological order. It is assumed that all references are to standard gauge locomotives.

25/6/1896	Mention of charges and hours worked by Bolckow Vaughan's locomotive [presumably on hire].
6/2/1900	A locomotive supplied by Wake & Hollis, contractors, of Darlington was being used. Previously horses had been employed for shunting.
6/11/1900	Wake & Hollis could no longer undertake the haulage with their locomotive.
8/1/1901	Wake & Hollis again handling the haulage with their locomotive.
14/10/1902	Agreed to purchase 10 wagons as the NER refused to supply its wagons.
3/5/1904	It was agreed that the council should purchase a small locomotive for £300. This arrived at the end of June 1904 and was subsequently approved after trials. On 4/10/1904 the council approved a payment of £300 to Lingford, Gardiner & Co Ltd of Bishop Auckland [presumably for the locomotive, though this is not stated].

In the *Colliery Guardian* for 7/9/1906 a R.R. Deans (background unknown) advertised for sale at Witton Park a 10in (cylinders) loco built by Black Hawthorn and the frame, boiler and fittings of a 12in (cylinders) loco, also built by Black Hawthorn, together with a considerable quantity of crushing plant. The owner is not stated, but there is no mention of any of this in the council's minutes, and one is left to conclude that this sale had nothing to do with the council. No other source of this equipment has come to light.

3/12/1907   Agreed to hire a locomotive while the council's own locomotive was under repair. The latter was working again in 1908.

7/10/1913   Under the arrangements made with The Witton Park Slag Co for a lease of slag deposits the council agreed the charge for its locomotive undertaking haulage for the Slag Co.

There is no further reference to any locomotive after this. However, George Alliez noted in his records that the first two locomotives below worked at Witton Park Slag Works, apparently before The Witton Park Slag Co set up its own works.

Gauge : 4ft 8½in

SPIDER	0-4-0ST	OC	GW	108	1861	(a)	s/s
EGYPT	0-6-0T	IC	JF	1539	1871	(b)	(1)
-	0-4-0ST?	?	?	?	?	(c)	(2)

(a)   ex Pease & Partners Ltd, Tees Ironworks, Cargo Fleet, Middlesbrough, Yorkshire (NR).
(b)   ex Pease & Partners Ltd, Broadwood Quarry, near Frosterley, c/1910, loan.
(c)   ex Lingford, Gardiner & Co Ltd, dealers, Bishop Auckland; hired during period 1910-1914 when EGYPT was under repair.

(1)   returned to Pease & Partners Ltd, Broadwood Quarry, near Frosterley, ex loan, c/1914.
(2)   returned to Lingford, Gardiner & Co Ltd, Bishop Auckland, ex hire.

AYCLIFFE LIME & LIMESTONE CO LTD (registered in 1920)
latterly a subsidiary of **Gjers, Mills & Co Ltd, Middlesbrough**, Yorkshire (NR)
AYCLIFFE QUARRY & LIMEWORKS, Aycliffe   R13
NZ 283221 approx.

This limestone quarry and limeworks was started by a local farmer, **George Chapman**, in the 1880s. By 1914 the business was being run by **Hanson, Brown & Co Ltd**. The quarry was served by sidings west of the NER Durham – Darlington line, immediately south of Aycliffe Station. Hanson, Brown & Co

6. HAZELS, CF 1189/1900, at Aycliffe on 27th August 1948. The only standard gauge well tank locomotive built by CF, she had another small tank added between the frames by her original owners, the glass manufacturers Pilkington Bros at St.Helens.

Ltd's siding agreements with the NER began in 1916. So far as is known, neither of these owners had their own locomotives, traffic being worked by the NER.

It would seem that the quarry and limeworks closed down about 1920, and that this firm then took over. Latterly it became a subsidiary company of Gjers, Mills & Co Ltd, the owners of Ayresome Iron Works in Middlesbrough. Rail traffic was shunted by road tractor from 1961, and the quarry and works were subsequently closed (Gjers, Mills & Co Ltd itself closed down on 3rd July 1965).

Gauge : 4ft 8½in

	HAZELS	0-4-0WT	OC	CF	1189	1900		
		reb 0-4-0WT + T	OC	Pilkington		?	(a)	Scr /1951
	-	2-2-0WT	IC	Aycliffe		?	(b)	Scr by 1/1949
	AYRESOME No.5	0-4-0ST	OC	MW	777	1881	(c)	Scr /1956
12	AYRESOME No.12	0-4-0ST	OC	MW	1903	1916	(d)	Scr 4/1962
	-	0-4-0ST	OC	P	1058	1906	(e)	(1)

(a)   ex Pilkington Bros Ltd, St.Helens, Lancashire, via H.W. Johnson, dealer, Rainford, Lancashire, /1923; this purchase would appear to arise from an advertisement in *Machinery Market* dated 1/6/1923, placed by J. Worth on behalf of the firm, seeking a four-coupled locomotive with cylinders approximately 14in x 20in.

(b)   built from a Foden steam lorry, Works No.8360, acquired from Davey, Paxman & Co Ltd, Colchester, Essex, /1937.

(c)   ex Gjers, Mills & Co Ltd, Middlesbrough, Yorkshire (NR), 6/1950.

(d)   ex Gjers, Mills & Co Ltd, Middlesbrough, Yorkshire (NR), 3/1956.

(e)   ex Gjers, Mills & Co Ltd, Middlesbrough, Yorkshire (NR), after 3/1957.

(1)   to Gjers, Mills & Co Ltd, Middlesbrough, Yorkshire (NR), /1962.

7. This extraordinary machine, also at Aycliffe on 27th August 1948, was built by the firm from a Foden steam lorry.

WILLIAM BARKER & CO
Slag Works, West Cornforth

N14
NZ 303338

By 1894 William Barker was producing crushed slag from the huge heaps left behind after the closure of the Ferry Hill Iron Works (see Carlton Iron Co Ltd). By 1912 it would seem that the firm was owned by Mr.T. Benjamin Maughan, who also owned The Witton Park Slag Co (which see). From 1926 the slag was being combined with tar from the adjacent West Cornforth Chemical Works of Dent, Sons & Co Ltd (see Henry Stobart & Co Ltd) to produce tarmacadam. Production is believed to have ended in

1938. Nearby was Thrislington Colliery (North Bitchburn Coal Co Ltd/Henry Stobart & Co Ltd), which up to 1915 was served by a short branch from West Cornforth Station on the NER line from Ferryhill-Hartlepool (formerly the Clarence Railway). The slag-crushing plant was served by a line, with sidings, from this link, which joined the NER Durham-Darlington line north of Ferryhill Station. However, in 1915 Thrislington Colliery added its own link to the NER Durham-Darlington line, eventually abandoning its link to West Cornforth Station. If Barker & Co was still dispatching by rail, then presumably this new link had to be used.

The reference below describes the wagons of slag from the heaps being worked by a petrol locomotive, presumably the one below. How the wagons were worked before its arrival is not known.

Reference : *A History of Thrislington, Cornforth & West Cornforth*, Robin Walton, 1991.

Gauge : 2ft 0in

| - | 4wPM | MR | 3869 | 1928 | (a) | s/s |

(a) ex MR, c11/1928 (originally built as MR 1730/1918, re-numbered 2187 in 1922 and 3869 in 1928).

J. BARTLETT

NEWFIELD DRIFT MINE, Newfield **M15**
NZ 209332

This mine, abandoned by Dorman Long (Steel) Ltd (which see) in December 1959, was re-opened by this new owner in August 1980, to work coal and fireclay from the Tilley 'P' seam. The locomotives listed below were used entirely underground, with rope haulage up the drift to the surface, though they were brought to the surface for battery charging. There was no main line rail connection. The mine was closed in January 1985 and abandoned in 1986.

Reference : *British Small Mines (North)*, A.J. Booth, Industrial Railway Society, 2000

Gauge : 2ft 6in

6/44	0-4-0BE	WR	6595	1962	(a)	(1)
6/53	0-4-0BE	WR	6704	1962	(a)	Scr c10/1983
6/44	0-4-0BE	WR	C6710	1963	(a)	(2)
6/41	0-4-0BE	WR	6133	1959	(b)	(1)
6/46	0-4-0BE	WR	6593	1962	(b)	(1)

(a) ex British Steel Corporation, Beckermet Mine, Cumbria, c12/1980 (by 29/5/1981).
(b) ex Treloar Bros, Haltwhistle, Northumberland, after 5/1982, by 19/7/1982; formerly British Steel Corporation, Beckermet Mine, Egremont, Cumbria.
(1) to Ayle Colliery Co Ltd, Alston, Cumbria, after 3/1985, by 7/1985.
(2) loaned to F. Shepherd, Flow Edge Colliery, Alston, Cumbria, after 3/1982; returned by 10/1983; to Ayle Colliery Co Ltd, Alston, Cumbria, after 3/1985, by 7/1985.

THE BEARPARK COAL & COKE CO LTD (registered 6/5/1872)
latterly a subsidiary of **Sir S.A. Sadler Ltd** (which see).
BEARPARK COLLIERY & COKE WORKS, Bearpark **K16**
NZ 243434

Mineral Statistics suggest that initially the colliery was called **Bearpark Brancepeth Colliery** and its owners were **The Bearpark Brancepeth Coal & Coke Co Ltd**; but the details of the company's registration are recorded as shown above, and if "Brancepeth" was used in the colliery's name it was very quickly dropped, probably because of confusion with the Brancepeth Colliery a few miles away owned by Strakers & Love. The sinking of the colliery began on 18th March 1872, with production starting in 1876. The colliery was served by sidings west of NER Consett - Durham line (Lanchester Valley branch), one mile north of Relly Mill Junction. Batteries of beehive coke ovens were also built. In 1884 the company constructed a battery of 50 Simon Carves waste heat coke ovens here, the second installation of by-product ovens in Britain after those built by Pease & Partners Ltd at Crook. They were closed in 1918, replaced by new Simon Carves ovens, 60 in 1916, 15 more in 1917 and a further five in 1920. The last beehive ovens closed in January 1921. In addition to coke and gas for industrial use, the by-product ovens also had facilities for the manufacture of tar, sulphate of ammonia, crude benzole and naphthalene. A **brickworks** making common bricks was begun adjacent to the colliery in 1887. In the twentieth century the company developed a number of drifts, the most important of which was **HOLLINSIDE DRIFT** (NZ 228448), some two miles to the north-west of the colliery and served by a rope-worked tramway.

8. *This is believed to be BEARPARK, FW 265/1875; the works plate shows 1875 but carries no maker's number.*

The colliery, coke works and brickworks were vested in the NCB Northern Division No.5 Area on 1st January 1947.

Gauge : 4ft 8½in

	BEARPARK	0-6-0ST	OC	FW	265	1875	(a)	(1)
No.1	WALKER	0-4-0T	OC	GW	231	1866	(b)	(2)
(No.25)		0-6-0ST	OC	CF	1155	1898	New	(3)
	BRISTOL	0-6-0ST	OC	FW	171	1873	(c)	Scr
	FLORENCE	0-6-0ST	OC	HC	880	1910	New	(4)
	NEWPORT	0-6-0ST	OC	FW	169	1872	(d)	Scr
	PRINCE	0-4-0ST	OC	Harris?	?		(e)	(5)
	"LITTLEBURN"	0-4-0ST	OC	KS	4143	1919	(f)	(4)
	MOSTYN	0-4-0ST	OC	MW?	?	?		
		reb		LG		1906	(g)	(4)

(a) some versions of the FW works list show this loco as New here, which may well be correct; others show it as New to Swan, Coates & Co, Cargo Fleet Ironworks, Middlesbrough, Yorkshire (NR).
(b) ex Swan, Coates & Co, Cargo Fleet Ironworks, Middlesbrough, Yorkshire (NR); see also the first paragraph below these locomotive footnotes.
(c) ex J.T. Firbank, contractor, Fishguard Harbour contract, Pembrokeshire, c/1902.
(d) this loco was offered for sale by Crown Coke Co Ltd, Consett, in 1/1912; date of arrival here said to be c/1915.
(e) ex Lingford, Gardiner & Co Ltd, dealers, Bishop Auckland, hire.
(f) ex Littleburn Colliery, Meadowfield.
(g) ex East Hedley Hope Colliery, near Tow Law, for repairs, after 16/6/1946.
(1) one source gives hired to The Furness Withy Shipbuilding Co Ltd, Haverton Hill (then in Co. Durham) and returned; scr c/1912.
(2) one version gives Scr c/1898, another gives Scr c/1915; 1898 would seem more likely.
(3) to East Hedley Hope Colliery, near Tow Law, c/1945.
(4) to NCB No. 5 Area, with colliery and coke works, 1/1/1947.
(5) returned to Lingford, Gardiner & Co Ltd, dealers, Bishop Auckland.

This company shared some directors with Swan, Coates & Co (see above), which also had an 0-6-0ST OC called BEARPARK, FW 245/1874 New. The versions of the FW works list disagree over which company it was new to.

The company is also said to have had a 0-6-0ST IC named VICTORIA on hire from Lingford, Gardiner & Co Ltd, Bishop Auckland, but this locomotive does not appear in a list of LG hire locomotives compiled by R.H. Inness. A locomotive of this name and description, HE 484/1889, did work in Durham, for Sir B. Samuelson & Co Ltd at Sherburn Hill Colliery.

The firm offered for sale a six-coupled saddle tank with 11inch cylinders in *Machinery Market* on 23/4/1920; its identity is not known.

BEARPARK COKE OVENS **K17**

The first of these beehive ovens began production in January 1877, the total rising to 431 by 1891. They consisted of rows, or "batteries", along the top of which ran a tramway for the tubs used to fill, or "charge", the ovens from above. In most places the tubs were pushed by hand, but some places hand-tramming was replaced by locomotive working, as here. Recently-discovered evidence shows that from 1894 the locomotives were also used to charge the by-product ovens, the only place in Britain where this is known to have happened. From 1894 locomotives were also used to fill the by-product ovens, the only place in Britain where this was done. These closed in 1918 and the last of the beehive ovens followed in January 1921.

9. This loco is either JF 2820 or 2821 of 1876, one of three locomotives used on the 3ft 0in gauge system which served the beehive coke ovens here. Note the features derived from traction engine practice.

Gauge : 3ft 0in

-	0-4-0WTG	OC	JF	2820	1876	New	(1)
-	0-4-0WTG	OC	JF	2821	1876	New	(1)
-	0-4-0WTG	OC	JF	5653	1888	New	(1)

(1) all three were still at work in 10/1899, and "small locos" were reported still in use in Durham Coal Owners Association Return 427 dated 27/4/1901; one was said to have "gone out of use in 1921" and possibly all of them did so; s/s.

EAST HEDLEY HOPE COLLIERY, near Tow Law **K18**

NZ 158404

This colliery was opened in 1875 by **Thomas Vaughan & Co** (which see) and after various closures and changes of owner, which included both Sir B. Samuelson & Co Ltd and Dorman, Long & Co Ltd (which see), it passed from the **Hedley Hope Coal Co Ltd**, together with **HEDLEY HOPE COLLIERY** (see below) to the control of The Bearpark Coal & Coke Co Ltd in 1936. East Hedley Hope Colliery was re-opened in June 1936, Hedley Hope about two months later. East Hedley Hope was served by a one mile long branch from the freight-only extension of the LNER Deerness Valley branch, and so far as is known there were no locomotives here before the introduction of the home-made machines below. East Hedley Hope Colliery was vested in NCB Northern Division No. 5 Area on 1st January 1947.

Gauge : 4ft 8½in

-	2-2-0PM	Bearpark	?	(a)	(1)
-	2-2-0PM	Bearpark	?	(a)	(2)

MOSTYN	0-4-0ST	OC	MW?	?	?		
	reb		LG		1906	(b)	(3)
(No.25)	0-6-0ST	OC	CF	1155	1898	(c)	(2)

(a) both machines were constructed from Daimler motor lorries fitted with flanged wheels at Bearpark Colliery, Bearpark, and then dispatched here.
(b) ex Randolph Coal Co Ltd, Randolph Colliery, Evenwood, 4/1945.
(c) ex Bearpark Colliery, Bearpark, c/1945.

(1) dismantled by 16/6/1946; scrapped (before take-over by NCB?).
(2) to NCB No. 5 Area, with colliery, 1/1/1947.
(3) to Bearpark Colliery, for repairs, after 16/6/1946.

HEDLEY HOPE COLLIERY, near Tow Law M19
NZ 135394

This colliery was opened in 1866, and was linked to the NER Sunniside branch by an NER stationary engine-worked incline about ½ mile long. It was eventually acquired by Sir B. Samuelson & Co Ltd, passing to Dorman, Long & Co Ltd in 1923, and was acquired with East Hedley Hope Colliery (see above) in 1936. At the change of ownership the colliery was closed and the incline had been lifted. It was re-opened about September 1936, without rail traffic, only to be closed again in January 1945. One source stated that a former NER loco was converted and used here to work a drift, presumably in some stationary engine form. When this was done, and how long it lasted, are both unknown.

LITTLEBURN COLLIERY, Meadowfield K20
NZ 255395

This colliery was formerly owned by **The North Brancepeth Coal Co Ltd** (which see). It was taken over by The Bearpark Coal & Coke Co Ltd and re-opened about October 1931; the shafts were no longer coal drawing and working was via a drift. It was served by sidings east of the LNER Durham - Darlington line, 2¾ miles south of Durham Station. It was closed again on 9th July 1935. In 1941 the mine was re-opened by **The Brancepeth Coal Co Ltd** (which see).

Gauge : 4ft 8½in

-	0-4-0ST	OC	BH	1096	1896	(a)	(1)
-	0-4-0ST	OC	KS	4143	1919	(a)	(2)

(a) ex The North Brancepeth Coal Co Ltd, with colliery, c10/1931 (assumed).

(1) remained here after closure of colliery on 9/7/1935; to The Brancepeth Coal Co Ltd, with colliery, /1941.
(2) to Bearpark Colliery & Coking Plant, Bearpark, c/1935 (by 30/10/1938).

In November 1936 the company opened **FIR TREE DRIFT**, near Crook (NZ 134344) **(M21)**, which was closed on 19th April 1940 and subsequently sold to J.Crossley & Sons Ltd (which see). On 25th February 1939 the company took over **BURNHOPE COLLIERY** at Burnhope (NZ 191482) **(G22)** from **Halmshaw & Partners**. An aerial flight was built to link Burnhope to Bearpark Colliery. The colliery was vested in NCB Northern Division No. 6 Area on 1st January 1947.

BEDE METAL & CHEMICAL CO LTD
Bede Metal Co until 16/5/1872
HEBBURN WORKS, Hebburn E23
NZ 299657

This firm was a member of the industrial empire of Sir Charles Mark Palmer (1822-1907) (see Palmers Shipbuilding & Iron Co Ltd) in its early days. The works was originally established in 1865 and sometime after 1872 it was linked by a ½ mile long branch to the NER Gateshead - South Shields line, ¾ mile east of Hebburn Station, a line that also served the shipyards of R. & W. Hawthorn, Leslie & Co Ltd and Robert Stephenson & Co Ltd, a yard later purchased by Palmers Shipbuilding & Iron Co Ltd. The company unloaded minerals at its own quay on the River Tyne, owning the Killingdal Mine near Trondheim in Norway, and latterly the Herrerias Mine (copper) in Spain, which it sold in 1912. At Hebburn it smelted copper and manufactured iron ore briquettes.

The company went into liquidation in September 1957, and in 1959 the site was cleared and used for an extension of the Palmers Hebburn shipyard, by then owned by Vickers Armstrongs Ltd.

Gauge: 4ft 8½in

B M & C No.1	0-4-0ST	OC	BH	316	1874	New	s/s by 8/1951
	reb		HL		1903		
No.2	0-4-0ST	OC	HL	2152	1890	(a)	
	reb		HL	6970	1914		(1)
BEDE No.3	0-4-0ST	OC	HL	3654	1927	New	s/s by /1956
-	0-4-0DE		AW	D24	1933	(b)	(2)
No.4	0-4-0DM		HC	D607	1938	New	s/s by 1/1957

(a) ex John Bowes & Partners Ltd, Felling Colliery, (c/1901?).
(b) ex AW, for demonstration, /1936.
(1) to Priestman Collieries Ltd, Norwood Coke Works, Dunston, c/1930 (precise date not known; may have gone to Norwood before Priestman Collieries Ltd took over the plant.
(2) returned to AW, after demonstration, /1936.

BELL BROTHERS LTD
Bell Brothers until 27/11/1873

The partnership of Bell Brothers was established in 1844 to take over the Wylam Ironworks in Northumberland. Ten years later they opened the Clarence Ironworks at Port Clarence, on the mouth of the River Tees near Middlesbrough, to produce pig iron and ships' plates and angles. The Wylam works was closed in 1864. The limited company was registered as shown to take over from 1st October 1872 the business of Bell Brothers, then run by Isaac Lowthian Bell (1816-1904) (later Sir Lowthian Bell), John Bell and Thomas Bell. From 1889 the company diversified into the manufacture of steel rails. It went into voluntary liquidation on 6th August 1895, with a new company of the same name registered the next day. This company was in turn liquidated and a new public company with the same name formed on 24th January 1899, now including Dorman family members as directors (see Dorman, Long & Co Ltd), who held half of the ordinary shares and who in 1902 bought out the Bell family holding to obtain full control, although the company continued to trade under its registered name until merged into Dorman, Long & Co Ltd on 2nd May 1923.

WEAR IRON WORKS, Washington   **H24**
Bell, Hawks & Co until 1867   NZ 319555

It may well be that the original owners were one of the Bell brothers in partnership with one of the Hawks from Hawks, Crawshay & Sons (which see). The construction of this works began in 1856, but its one furnace was not put into blast until 1859. It was situated immediately to the south of the Washington Chemical Works (see Newalls Insulation Ltd), and was served by sidings south of the NER Pontop & South Shields branch at Washington Station. It would seem that Bell, Hawks & Co failed in 1867 and Bell Brothers took over, although production was not re-started until 1870. Curiously, there would seem to be no link between Bell Brothers and the lease of Broomside Colliery from the Marquis of Londonderry in March 1867 by two other brothers named Bell, Charles William Bell and William Morrison Bell. The furnace was put out of blast in 1875 and there was no further production. In 1887 the premises were converted into the Washington Wire Rope Works owned by R.S. Newall & Co, the family which owned the adjacent Washingon Chemical Co.

It would seem very likely that one or more locomotives were used here, but no details are known.

For the collieries and quarry owned by Bell Brothers Ltd in Durham see the entries for Dorman, Long & Co Ltd and Dorman Long (Steel) Ltd.

THE BIRTLEY BRICK CO LTD
UNION and TEAM VALLEY BRICKWORKS, Birtley   **G25**

This company was one of the largest brick manufacturers at Birtley. In 1934 (Davison gives 1938) it re-opened the **UNION BRICKWORKS** (NZ 264556 approx.), which lay north of Station Road in Birtley and made common bricks, though its clay pit, accessed via a tunnel, lay to the south of Station Road. This yard, begun in 1879, had been worked from 1908 to 1925 by **Allison, English & Co Ltd** (which see). About the same time the company also took over the **TEAM VALLEY BRICKWORKS** (NZ 266558), a little to the north-east, originally opened in 1906 and also owned by Allison, English & Co Ltd. Although the Union works had originally been linked to the Pelaw Main Railway (see Pelaw Main Collieries Ltd), this link had been removed by the time that The Birtley Brick Co Ltd took over. Both works were linked to sidings along the LNER Newcastle upon Tyne-Stanley-Consett line, ¼ mile north of Birtley Station. It is believed that the locomotive below worked on the extensive tramway system linking the claypits north of Station Road with the two works. The Union Works is believed to have closed at the outbreak of the Second World War in 1939, and the locomotive is known to have stood derelict for a considerable time before its eventual sale. The works and quarry were subsequently re-opened, and operated until 1983, without any further locomotives.

Gauge : 2ft 0in

83	LANCHESTER	0-4-0T	OC	AE	2071	1933	(a)	(1)

(a) ex Durham County Water Board, Burnhope Reservoir, Wearhead, c2/1937.

(1) it is believed to be this locomotive which was offered for sale by J.G.R. Herbert of Newcastle upon Tyne in *Machinery Market* on 25/10/1940 and 28/11/1941; by 1946 it had been acquired by R.R. Dunn, dealer, Bishop Auckland, who offered it for sale in *Machinery Market* on 13/12/1946; by 1947 it had "passed" to H. Dunn Plant & Machinery Co Ltd, dealers (H. Dunn was the father of R.R. Dunn); it was sold to Dinorwic Slate Quarry Co Ltd, Llanberis, Caernarvonshire, 7/1948.

One source claims that the firm also owned 2ft 0in gauge 4wPM RH 114563/1923, which had been used by Allison, English & Co Ltd (which see), the previous owners of the yard. This source also says that The Birtley Brick Co Ltd had sold this loco by 1932, when it was being used at "The Scarborough Pleasure Park"; however, this date would appear to be before The Birtley Brick Co Ltd took over the yard.

The makers' records for the locomotive below state that it was purchased by "The Birtley Brick Co Ltd, New Washington". New Washington was a small settlement about one mile north of Washington village. Both the Ordnance Survey maps and Davison show only one brickworks here, the **(Washington) Bath Brick Works (H26)**, to the east of Spout Lane (NZ 312564). This is shown working on the 2nd edition O.S. map of 1896, but appears in an apparently disused state on the 3rd edition map (1921) and is clearly abandoned on the 4th edition map (1939). Neither a brickworks nor the Birtley Brick Co Ltd is listed here in Kelly's Directory for Durham in 1934, the year after the loco was built, and the company is not known to have had a brickworks at Washington. It has therefore not proved possible to explain the maker's reference.

Gauge : 2ft 0in

-	4wPM	HU	46851	1933	New	s/s

BLAKE BOILER, WAGON & ENGINEERING CO LTD (registered 23/10/1905)
(latterly a subsidiary of **Metropolitan-Cammell Carriage, Wagon & Finance Co Ltd**, Birmingham)
ALLIANCE WORKS, Albert Hill, Darlington   S27
                                                                                                NZ 299167

This works was opened in the mid-1850s as an ironworks by **Wilson Brothers & Co**. It was served by an extension of the short branch from Parkgate Junction, about one mile north of Darlington (Bank Top) Station on the NER Darlington-Durham line, which first served Skerne Ironworks (see Skerne Ironworks Co Ltd).

In the mid-1870s it was taken over by the **Darlington Railway Wagon Co**, founded in 1867, initially to repair railway wagons but latterly also to build them, and it was subsequently named the Alliance Works. On 21st June 1884 this firm became **Darlington Wagon & Engineering Co Ltd**. The company subsequently developed a wagon works at York Street (see Thomas Summerson & Sons Ltd), converting the Alliance works to manufacture bridges and wheels. About 1900 it was closed down, leaving the premises empty until Blake Boiler, Wagon & Engineering Co Ltd re-opened them in 1905. This firm closed down sometime between 1921 and 1925. After its demise the premises passed through the hands of several owners for different purposes, but without rail transport.

Almost nothing is known about locomotives that worked here. Blake Boiler, Wagon & Engineering Co Ltd owned the locomotive below, and also used a steam crane here.

Gauge : 4ft 8½in

	ALLIANCE	0-4-0T	OC	?	?	?	(a)	s/s

(a) built either by GW or HG; origin uncertain – possibly ex The Darlington Forge Ltd, Darlington; said to have been rebuilt by HL; here by 1920s.

10. ALLIANCE, unidentified, but almost certainly built by GW or HG, probably photographed in the 1920s.

BLYTHE & SONS (BIRTLEY) LTD
BIRTLEY STATION BRICK WORKS, Birtley
G28
NZ 265556 approx.

This was another family firm, founded in 1858, that exploited the layer of clay 90 feet thick overlying the coalfield in the Birtley area. Its works, for which the abbreviated **STATION BRICK WORKS** is also found, lay between the works of The Birtley Brick Co Ltd and the Bewicke Main branch of the Pelaw Main Railway, to which it was connected until 1932; thereafter it had no main line rail connection. Its quarry lay to the south of Station Road, and the monorail system, with the powered wagon shown below, is believed to have been used to bring clay from the quarry to the works. The system was removed sometime between 1968 and 1970. The firm and its quarry closed down in 1978.

Monorail

| | 2wPH | RM | 9792 | 1960 | New s/s c/1968-70 |

BOLCKOW, VAUGHAN & CO LTD
Bolckow, Vaughan & Co until 19/11/1864

This vast industrial empire was founded in 1840 when Henry Bolckow (1806-1878) entered into partnership with John Vaughan (1799-1868) to build a small ironworks at Middlesbrough, which opened in May 1841. Their fortune was made ten years later when Vaughan discovered the Cleveland ironstone deposits. The firm also led the way with the development of the Gilchrist-Thomas process, which made possible the large-scale manufacture of steel from Cleveland ironstone after 1879. Described in the late nineteenth century as the largest manufacturing firm in the world, its business initially fell into four sections - ironworks on Tees-side; collieries, usually in south-west Durham, and often with large numbers of beehive coke ovens, later by-product ovens; ironstone mines in Cleveland and Spain and limestone quarries in Weardale and North Yorkshire. In 1911 the firm rejected an approach from Dorman, Long & Co Ltd for a merger, and subsequently it began acquiring controlling interests in other companies, notably the Scottish bridge-builders, Redpath, Brown & Co Ltd. In the mid-1920s the company fell on hard times, and on 1st November 1929 it finally surrendered to Dorman, Long & Co Ltd.

Locomotives

What scant information survives about Bolckow Vaughan's locomotives and their movements comes from makers' lists and spares orders, from boiler reports seen at Newlandside Quarry and from workmen interviewed near the beginning of the Second World War in 1939. Inevitably this leaves conflicting and incomplete information, and the results below are believed to be the best available summary.

AUCKLAND PARK COLLIERY, Coundon Grange   Q29
The Black Boy Coal Co until 1/4/1872   NZ 227285

This colliery was developed on the site of the former Black Boy Colliery's Machine Pit by the Black Boy Coal Co, and opened in 1866. Its development meant a major re-organisation of the railway arrangements serving it and the nearby **BLACK BOY COLLIERY** (see below). The former Stockton & Darlington Black Boy branch was replaced by a new NER Black Boy branch, ½ mile long, from a junction one mile south of Bishop Auckland Station on the NER Bishop Auckland-Shildon line. The branch which formerly served the Black Boy Colliery's Gurney Pit was re-routed at its western end to join the new system, traffic for Black Boy having to reverse at Auckland Park. The locomotive(s) formerly at Black Boy Colliery were transferred to Auckland Park. Four are recorded in DCOA Return 102 of November 1876, though the **stationary engine** serving Black Boy Colliery is also listed; this had gone by 1890.

The Auckland Park coke ovens (see below) lay to the south-east of the colliery, while the Black Boy Brickworks lay to the south-west. To replace the beehive coke ovens (see below) the company built two batteries of 50 Semet-Solvay waste heat ovens, which began production in the quarter ending 31st December 1908. The colliery and coke works passed to Dorman, Long & Co Ltd on 1st November 1929.

Gauge : 4ft 8½in

8		0-4-0ST	OC	BH	427	1877	New	s/s
	VICKERS	0-6-0ST	OC	FW	249	1874	(a)	(1)
3		0-6-0ST	IC	MW	194	1866	(b)	Scr c/1904
101		0-6-0ST	IC	HL	2429	1899	New	Scr /1927
108	HECTOR	0-6-0ST	OC	HL	2613	1905	New	(2)
113	PLUTO	0-6-0ST	OC	HL	2655	1906	New	(3)
		?	?	?	?	?	(c)	s/s
	HARE	0-4-0ST	OC	GH		1908	(d)	(3)

(a) ex Cleveland Works, Middlesbrough, Yorkshire (NR)? This loco was also reported at Leasingthorne Colliery, near Coundon (see below), but whether she was there before or after coming here is not known.
(b) ex Black Boy Colliery; probably one or two other locos unknown also came from there.
(c) oral tradition remembered a loco coming from Lingford, Gardiner & Co Ltd, Bishop Auckland, dealers & repairers, "with large wheels; had number 1040".
(d) ex Newlandside Quarry, Stanhope, by 7/1929.

(1) possibly to Leasingthorne Colliery, near Coundon (see note (a) above); otherwise s/s.
(2) to Leasingthorne Colliery, near Coundon, and returned; to Dorman, Long & Co Ltd, with colliery and coke ovens, 1/11/1929.
(3) to Dorman, Long & Co Ltd, with colliery and coke ovens, 1/11/1929.

AUCKLAND PARK COKE OVENS (beehive)   Q30

These beehive ovens, 431 in all, lay to the south-east of the colliery in two long rows. The locomotives shunted the tubs along the top of the ovens, positioning them over the oven charging hole for coal to be discharged. Several Bolckow, Vaughan & Co Ltd collieries used locomotives on their coke ovens, but only one of them has been identified. The customer detail in the BH works list usually gives delivery to a Bolckow Vaughan steelworks, whereas certainly some of these orders must have been for the various coke ovens. The locomotives involved were BH 402/3 of 1876, 435/6 of 1877, together with 442 of 1877, which is shown as "for Binchester Colliery", 446/7 and 459 of 1878 (447 was not delivered until 1880), 494/5 of 1879, 557/8/9/561 of 1880 and 595/6/7/8/9 and 601 of 1881. All of these were built to 3ft 0in gauge.

Two coke oven locomotives were recorded at Auckland Park in January 1880. It would appear from Durham Coal Owners Association Return No.427 that Auckland Park North and South Pits [ovens] were each using a locomotive in April 1901. The ovens ceased production about April 1909, having been replaced by the Semet-Solvay by-product ovens mentioned earlier.

11. This magnificent photograph, taken at Auckland Park Colliery, shows the whole process of coke making using beehive coke ovens. A narrow gauge BH loco propels a train of conical steel tubs on to the tops of the ovens. The coal will be dropped through the charging hole, raked level and the door bricked up. The flow of air into the oven is controlled via holes in the bricks. After about 72 hours, the bricks were pulled down and the red hot coke was brought out on large flat rakes on to the 'bench', where it was sprayed with water. When cool it was then loaded, in quite large lumps, into the NER wagons.

However, when C.H.A. Townley visited Auckland Park in January 1949 he was told that the narrow gauge system had been 2ft 4in gauge, that there were three locomotives and that one had "lasted for another job until about 1925". It has not been possible to identify any of these locomotives.

Gauge : 2ft 4in (one report states 2ft 6in)

1	0-4-0ST	OC	BH?	?	?	(a)	(1)
2	0-4-0ST	IC	?	?	?	(b)	(1)
3	0-4-0ST	OC	BH?	?	?	(a)	(1)

(a) if these were BH locos, then they may well be some of those in the list above.
(b) identity, origin and date of arrival unknown; said to have been "internally geared"; another report describes it as a tender engine with inside cylinders.
(1) whichever locomotives were still here in 1909 ceased work with the closure of the beehive coke ovens, with the exception of one, which was retained for "another job" until c/1925.

BINCHESTER & WESTERTON COLLIERIES, near Westerton   **N31**
(see notes below)   NZ 242316
WESTERTON COLLIERY, near Westerton   **N32**
The Black Boy Coal Co until 1/4/1872; see also below   NZ 235308

The first **WESTERTON COLLIERY (N33)** was opened in June 1841 by **Nicholas Wood & Partners** (which see). This mine lay south west of Westerton village and was served by an extension of the branch serving Leasingthorne Colliery (see below) and its coal was shipped at Hartlepool. For the first ½ mile from the colliery the waggons were hauled up a single line incline to a stationary engine at NZ 244306, from where they descended a further ½ mile incline to Leasingthorne. One oral tradition claimed that this latter section was a self-acting incline, but as the 1st ed. O.S. map shows it as a single line this cannot be correct, and it may be that the engine house worked both inclines. By the time of his death in 1865, Wood and his partners owned Westerton, Leasingthorne and Black Boy Collieries, which in 1866 passed to **The Black Boy Coal Co**, and in 1872 were acquired by Bolckow, Vaughan & Co Ltd. A map of 1876 shows it as working and still linked to Leasingthorne, and DCOA Return 102 also records the stationary engine here in November 1876.

Meanwhile in 1871 (almost certainly) **BINCHESTER COLLIERY (N31)** was acquired by Bolckow Vaughan as one of the three collieries it took over from **The Hunwick Coal Co**, though it was not working at the time. This colliery had been opened to the north of Westerton village in 1855 by **J.Robson & Co** (which see). It was subsequently served by a 2¾ mile branch from Binchester Junction, one mile east of Spennymoor Station on the NER Bishop Auckland & Ferryhill branch, though whether this line dates from 1855 is not known. Beehive coke ovens were built here (see below) and also a limeworks. It became one of the collieries acquired by the **West Hartlepool Harbour & Railway Co** (which see), and was acquired by **The Hunwick & Newfield Coal Co** in 1865, which changed its name to **The Hunwick Coal Co** in 1866. Under Bolckow Vaughan the colliery was re-developed, with a new sinking beginning production in 1874.

Although what happened next is known, the dates are not. First, Westerton Colliery and its line to Leasingthorne were closed, but apparently not before two new shafts were sunk on the Binchester Colliery site, to different seams, and then named **WESTERTON COLLIERY (N32)**, one of the rare examples of two collieries with different names sharing the same surface. Binchester Colliery closed in January 1908. Westerton's beehive ovens closed about July 1910. The colliery followed in August 1924, though the site continued in use as a pumping station and was taken over by Dorman, Long & Co Ltd on 1st November 1929.

The operation of the original Westerton Colliery is described above, while no locomotives are known at Binchester Colliery before Bolckow Vaughan acquired it. BH 391 listed below was ordered for the new Binchester Colliery, presumably marking the beginning of the period when Bolckow Vaughan locos shunted what became the two collieries. The DCOA Returns records four locomotives here in March 1890, so there were clearly more here than those listed below.

Gauge : 4ft 8½in

	TINY	0-4-0ST	OC	FJ	31	1864	(a)	(1)
7		0-6-0ST	OC	BH	391	1877	New	s/s
107	ATLAS	0-6-0ST	OC	HL	2612	1905	(b)	(2)
	HENRY CORT	0-4-0ST	OC	BH	607	1881	(c)	(3)

(a) ex Black Boy Colliery, Coundon Grange (possibly by 11/1876).
(b) ex Leasingthorne Colliery, near Coundon.
(c) origin uncertain; one source gives ex Cleveland Works, Middlesbrough, Yorkshire (NR), another gives ex Dean & Chapter Colliery, Ferryhill; also whether transfer here was before or after the Dorman Long take-over is not known.

(1) to Newlandside Quarry, Stanhope, by 5/1908.
(2) to Leasingthorne Colliery, near Coundon.
(3) if here under BV ownership, then to Dorman, Long & Co Ltd, with colliery,1/11/1929.

It is possible that COMET 0-4-0ST OC BH 544/1880 was also used here.

BINCHESTER COKE OVENS   N34

These lay adjacent to the colliery, being constructed at the same time as the sinking of the (new) pit, and were brought into production in 1877. Only one locomotive is recorded in January 1880, and another source states that only one loco was used here; however DCOA Return 427 lists two locomotive drivers being employed daily, which might suggest that two locomotives were in use at that time. The ovens ceased production in the quarter ending December 1908.

Gauge : 3ft 0in

-	0-4-0ST	OC	BH	442	1877	New	s/s
-	0-4-0ST	OC	BH	?	?	(a)	s/s

(a) if two locomotives were used here; see list under Auckland Park Coke Ovens.

BLACK BOY COLLIERY, Coundon Gate   Q35
The Black Boy Coal Co until 1/4/1872; see also below

The history of this colliery is extremely complicated. The original Black Boy Colliery (NZ 230291)**(Q36)**, named after a nearby public house, was opened in 1827 and was served by the Black Boy Branch of the Stockton & Darlington Railway running north from Shildon Station. The colliery was at this time part owned by **Jonathan Backhouse**, the S&DR Treasurer. The Stockton & Darlington's Black Boy Branch, opened in July 1827, used a stationary engine at Shildon Bank Top to haul waggons up from the pit, from where a self-acting incline took them down to the junction near Denburn Beck.

By 1860 this colliery had been replaced by two new shafts, the Machine Pit (NZ 230282) and the

Gurney Pit (NZ 237282). These two shafts, half a mile apart, were now collectively called **Black Boy Colliery**, the old colliery now being called **Old Black Boy Colliery**. The Machine Pit was situated alongside the Black Boy Branch about ½ mile south of the old colliery and the extension beyond it to the old colliery was eventually lifted. Immediately south of the Machine Pit was a **brick & tile works** and from here a branch ran ½ mile east to the Gurney Pit. This branch was operated by a rope incline, which from DCOA Return 102 appears to have been operated by a **stationary engine**.

Backhouse died in October 1842, and the collieries passed to his family, who advertised Black Boy and Leasingthorne for sale on 28th January 1851 (see also West Auckland Colliery below). They passed next to **Nicholas Wood & Partners**, who already owned Westerton Colliery near the latter (see Binchester Colliery above). It would seem that Wood introduced locomotive working, apparently only on the line serving the Machine Pit. At some date after 1855, possibly initiated by Wood, a link 1¾ miles long was constructed between Leasingthorne Colliery and Black Boy Colliery's Gurney Pit, presumably to divert Black Boy coal to run eastwards, rather than via the Stockton & Darlington Railway.

Wood died in 1865, and in 1866 the three collieries passed to **The Black Boy Coal Co.**, who in the same year began the re-development of the Machine Pit as **AUCKLAND PARK COLLIERY**, leaving only the Gurney Pit as Black Boy Colliery. For the subsequent locomotive arrangements see the entry for Auckland Park Colliery.

The Gurney Pit lay only yards away from South Durham Colliery (see Pease & Partners Ltd), the two being separated merely by a small stream. The colliery was closed in December 1924.

Gauge : 4ft 8½in

| - | 0-4-0ST | OC | FJ | 31 | 1864 | New | (1) |
| BLACK BOY COAL CO 3 | 0-6-0ST | IC | MW | 194 | 1866 | New | (2) |

The name of MW 194 suggests that there was at least one more locomotive here.

(1)   to Binchester Colliery, near Westerton.
(2)   to Auckland Park Colliery, Coundon Grange.

BYERS GREEN COLLIERY & COKE OVENS, Byers Green   N37
West Hartlepool Harbour & Railway Co until /1865; see also below   NZ 223335

The Clarence Railway 'opened' its 5-mile long Byers Green Branch on 31st March 1837, but this had

12. This photo was taken by the commercial photographer Herbert Coates of Willington, whose negative number dates it to the late 1890s. The loco carries no identification at all, although the design would suggest it could be an early Black Hawthorn or a Joicey loco.

more to do with the imminent expiry of parliamentary powers than any immediate traffic. On 12th June 1840 the West Durham Railway opened its eastern section from an end-on junction with the Byers Green Branch, which by then had been brought up to operational condition. The sinking of the long-awaited colliery was begun in January 1840 and was completed in August 1841; but soon afterwards it was drowned out and it did not resume production until 12th March 1845. In May 1853 the Clarence Railway was purchased by the West Hartlepool Harbour & Railway Company, which subsequently acquired the colliery from the Trustees of J.Robson & Co (which see). This company's ownership of collieries, undertaken to safeguard its coal traffic, was subsequently declared illegal, and in 1865 the colliery was acquired by Bolckow Vaughan.

After the NER opened its line from Bishop Auckland to Burnhouse Junction, two miles west of Spennymoor, in 1885 and closed the former West Durham Railway's rope inclines west of Todhills in 1891, the colliery became served by what was now a one mile line between Burnhouse Junction and Todhills, ½ mile west of the junction. There were beehive coke ovens at the colliery, but there is no record of locomotives being used on them. They were closed in July 1913. In April 1916 50 Semet-Solvay waste heat by-product ovens began production here. The colliery, coke ovens and an associated **brickworks** were all closed in March 1926 at the beginning of the miners' strike. In 1927 the underground workings were combined with Newfield Colliery, where it is believed that all coal was then drawn, with no locomotives being used here after 1926.

The DCOA Returns give one locomotive here in April 1871, November 1876 and March 1890.

Gauge: 4ft 8½in

	-	0-4-0T	OC	FJ	47	1865	(a)	Scr by /1926
	CHEETHAM	?	?	?	?	1877?	(b)	(1)
	-	0-4-0ST	OC	?	?	?	(c)	s/s
	-	0-4-0ST	OC	BH		1877	(d)	(2)
118	BELMONT	0-6-0ST	OC	HL	2909	1911	(e)	(3)
105	KELVIN	0-4-0ST	OC	CF	1211	1901	(e)	(3)
	JUNO	0-4-0ST	OC	BH	606	1881	(e)	(4)
110	HERCULES	0-6-0ST	OC	HL	2654	1906	(f)	(3)
108	HECTOR	0-6-0ST	OC	HL	2613	1905	(f)	(2)

(a)   ex ?, (by 1878?); previously Robert Sharpe & Sons, contractors, North Devon Railway contract.
(b)   ex Cleveland Works, Middlesbrough, Yorkshire (NR), 9/1899.
(c)   identity, origin and date of arrival unknown; see photo no.12.
(d)   ex Lingford, Gardiner & Co Ltd, Bishop Auckland, following repairs; its previous location was presumably owned by Bolckow Vaughan; see note on this loco under Newfield Colliery.
(e)   ex Cleveland Works, Middlesbrough, Yorkshire (NR).
(f)   ex Leasingthorne Colliery, near Coundon.

(1)   boiler exploded, 19/3/1901, but not seriously; s/s.
(2)   to Leasingthorne Colliery, near Coundon.
(3)   to Newfield Colliery & Brickworks, Newfield.
(4)   to Darlington Rolling Mills Co Ltd, Rise Carr Rolling Mills, Darlington.

DEAN & CHAPTER COLLIERY & COKE OVENS, Ferryhill   **N38**
                                                                                           NZ 272331

The sinking of this colliery, with three shafts, began in 1902 and production commenced two years later. It was served by a one mile long branch from sidings at Binchester Junction on the NER Bishop Auckland & Ferryhill branch, 1¼ miles east of Spennymoor Station, the point where the 2¾ mile branch from Binchester and Westerton Collieries also joined the NER, and it is clear that it was possible to work between the two branches without passing over NER metals. A major coke ovens and by-product plant was developed here, the latter probably operated, at least for a time, by the **Deanbank Chemical Co Ltd** (a photograph of one of this company's tank wagons is known): Dean Bank was the name of the village built nearby to serve the site. The first coke ovens were 60 Coppee waste heat ovens, which began production about August 1905. 60 more of these were added during 1906. They were joined in June 1910 by 100 Semet-Solvay waste heat ovens. 60 of the Coppee ovens were closed down about March 1912, and to the remaining 60 Coppee ovens were added 40 more Semet-Solvay ovens, unusually on the same battery, which began production in the quarter ending 30th June 1913. The Coppee ovens were closed down in 1919 as a result of the miners' strike that year.

The colliery, the 140 Semet-Solvay ovens and the by-products plant all passed to Dorman, Long & Co Ltd on 1st November 1929.

Gauge : 4ft 8½in

105	KELVIN	0-4-0ST	OC	CF	1211	1901	New	(1)
	HENRY CORT	0-4-0ST	OC	BH	607	1881	(a)	(1)
106	ERIMUS	0-6-0ST	OC	HL	2595	1904	New	(2)
116	GEORGE V	0-6-0ST	OC	HL	2833	1910	New	(2)
148	TAURUS	0-4-0ST	OC	HL	3384	1919	New	(2)
No.26	JOHN EVANS	0-6-0ST	IC	P	629	1896	(a)	(2)
No.10		0-4-0ST	OC	BH	1095	1896	(b)	(2)

(a)   ex Cleveland Works, Middlesbrough, Yorkshire (NR).
(b)   ex Sir W.G. Armstrong, Whitworth & Co Ltd, Elswick Works, Newcastle upon Tyne.

(1)   to Cleveland Works, Middlesbrough, Yorkshire (NR).
(2)   to Dorman, Long & Co Ltd, with colliery and coke ovens, 1/11/1929.

LEASINGTHORNE COLLIERY & COKE OVENS, near Coundon   **N39**
The Black Boy Coal Co until 1/4/1872; see also below   NZ 252304

The lease of the coal here was let on 24th March 1833 to **Christopher Mason**, who also had a similar lease for the proposed Great Chilton Colliery nearby (see The South Durham Coal Co Ltd). To serve both proposed collieries the Clarence Railway, in which Mason was a shareholder, proposed to build a 5 miles long branch from Chilton Junction on its main line, 1¼ miles south of Ferry Hill Station. But Mason was forced to abandon the sinking of Great Chilton in 1835, and on 29th February 1836 he entered a sub-lease jointly with Messrs. King, Mease & Campion to sink Leasingthorne Colliery. The railway to the colliery was completed before the sinking, and in 1841 an extension from just east of Leasingthorne was constructed to serve Westerton Colliery (see the entry for Binchester Colliery above). Also in 1841 the partnership sold the colliery to **James Reid** of Newcastle upon Tyne.

Coal was reached at last on 30th July 1842. However, about 1845 Reid sold the colliery to **Andrew Spottiswoode**, Queen Victoria's printer in London. He did not keep it long, and then sold it to **Jonathan Backhouse & Co**, one of the Darlington Quakers. This company advertised the colliery on 28th January 1851, following which it passed, with Black Boy Colliery (see above), to **Nicholas Wood & Partners**, who already owned Westerton Colliery (see above). Wood & Partners hauled their own traffic over the Clarence Railway to Port Clarence, using at least one locomotive (see the entry for Nicholas Wood & Partners). Wood died in 1865, and in 1866 the three collieries passed to **The Black Boy Coal Co**, and were acquired by Bolckow, Vaughan & Co Ltd in 1872.

At first the colliery was shunted by horses, but then a **stationary engine** was installed at the top (west end) of the colliery yard to haul the waggons up for them to be let down under the screens by gravity. It is not known when locomotives replaced the stationary engine to shunt the colliery. At an unknown date, but certainly after 1855, a line about 1¾ miles long was built running south-west from Leasingthorne Colliery, through Coundon village, to connect to **Black Boy Colliery (Gurney Pit)**, diverting its coal eastwards instead of down the Stockton & Darlington Railway. A curve near Black Boy Colliery also linked this line to **Eldon Colliery (John Henry Pit)** (see the entry for Pease & Partners Ltd).

The DCOA Return 102 lists one locomotive at Leasingthorne Colliery in November 1876, but its identity is unknown. The extension to Westerton Colliery was still in use at this date, but is believed to have closed, with the colliery, soon afterwards, leaving just a ½ mile stub serving **WESTERTON QUARRY** (NZ 247310) **(N40)**, which presumably supplied limestone to the limeworks. The line to Black Boy was also closed and lifted by the mid 1890s, presumably as the working of Black Boy Colliery became closely linked with Auckland Park Colliery (see above), leaving another stub just over ½ mile long serving the Coundon coal depot.

In 1889 the company opened 60 Coppee non-recovery coke ovens here, the first Coppee ovens to be built in Britain. They were closed down in December 1910. 36 were re-started in September 1912, only to be closed finally in June 1913. Meanwhile 48 Otto-Hilgenstock waste heat ovens began production here in the quarter ending 30th September 1904, to be followed by a further 48 in the quarter ending 30th June 1905. All 96 were closed down at the start of the miners' national strike in March 1921 and did not re-open.

The colliery passed to Dorman, Long & Co Ltd on 1st November 1929.

Gauge : 4ft 8½in

	YORK	0-4-0ST	OC	BH	526	1880	New	s/s
14		0-6-0ST	IC	MW	1138	1889	New	Scr by 11/1929
No.102		0-4-0ST	OC	HL	2449	1900	New	(1)

No.104		0-6-0ST	IC	MW	1469	1900	(a)	(2)
	VICKERS	0-6-0ST	OC	FW	249	1874	(b)	(3)
107	ATLAS	0-6-0ST	OC	HL	2612	1905	New	(1)
110	HERCULES	0-6-0ST	OC	HL	2654	1906	New	(4)
120	LEEHOLME	0-4-0ST	OC	HL	2916	1912	New	(5)
108	HECTOR	0-6-0ST	OC	HL	2613	1905	(c)	(4)

(a) ex Cleveland Works, Middlesbrough, Yorkshire (NR)? Whether she arrived here in the chronological position shown is not known, though she is said to have been one of the earliest locomotives here.
(b) ex Cleveland Works, Middlesbrough, Yorkshire (NR)? This loco was also reported at Auckland Park Colliery & Coke Ovens, Coundon Grange (see above), but whether she went there before or after coming here is not known.
(c) ex Auckland Park Colliery, Coundon Grange.

(1) to Dorman, Long & Co Ltd, with colliery, 1/11/1929.
(2) to Cleveland Works, Middlesbrough, Yorkshire (NR)?
(3) possibly to Auckland Park Colliery & Coke Ovens, Coundon Grange; otherwise s/s.
(4) to Byers Green Colliery, and returned here; then to Auckland Park Colliery & Coke Ovens, Coundon Grange.
(5) sold to unknown purchaser, /1923.

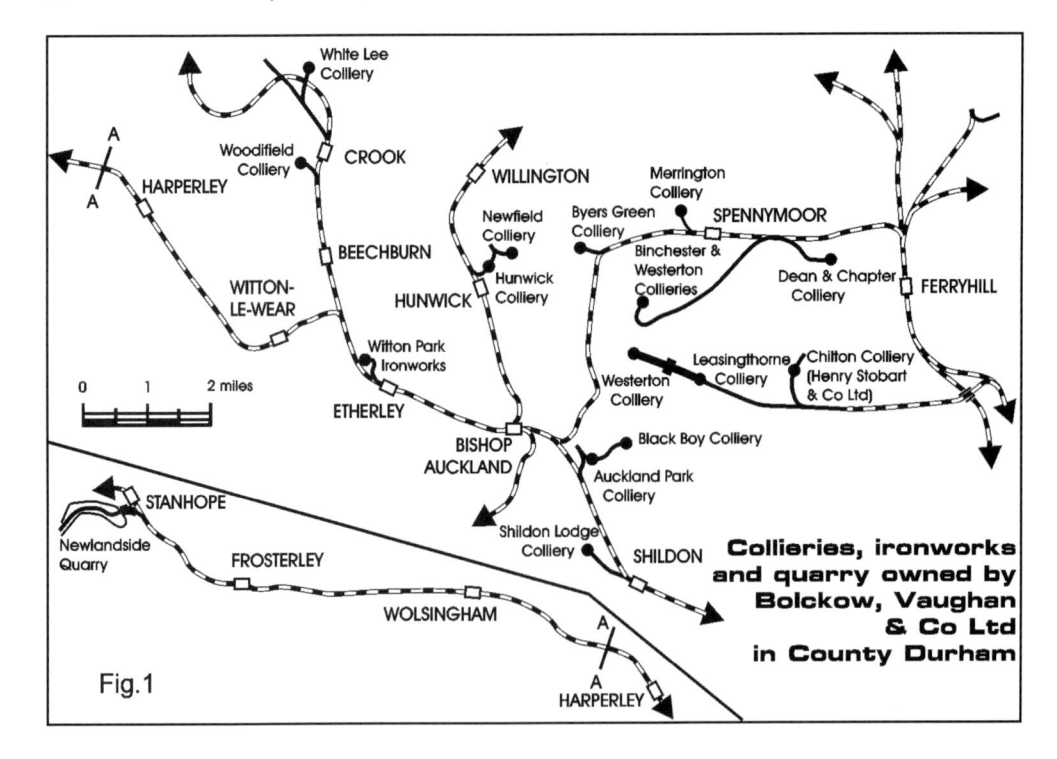

Fig.1 Collieries, ironworks and quarry owned by Bolckow, Vaughan & Co Ltd in County Durham

NEWFIELD COLLIERY, Newfield		**M41**
The Hunwick Coal Co until /1871; see also below		NZ 209332
HUNWICK COLLIERY, near Hunwick		**M42**
The Hunwick Coal Co until /1871; see also below		NZ 210328

NEWFIELD COLLIERY was opened in 1840 by **J. Robson & Partners**, subsequently **J. Robson & Co** (which see). It was originally served by a ½ mile branch from the foot of the Tod Hills Incline on the West Durham Railway, itself only partially opened at that date. **HUNWICK COLLIERY**, on the opposite bank of the River Wear and sunk by J. Robson & Co in 1844, was served by a ½ mile extension of this line via a timber trestle bridge across the River Wear, this extension being operated by a **stationary engine** at Hunwick. In the mid 1840s the firm hauled its own traffic over the Clarence Railway to Port Clarence. Like the other collieries owned by this company, Newfield and Hunwick eventually came under the illegal ownership of the **West Hartlepool Harbour & Railway Co** (which see). They were acquired in 1865 by **The Hunwick & Newfield Coal Co**, which shortened its title to **The Hunwick Coal Co** in 1866. Newfield was closed in 1867-68, but when Bolckow Vaughan purchased the collieries in 1871 for £70,000, Newfield was promptly re-opened.

In 1872 a ½ mile branch was built from Hunwick Colliery under the NER line to serve **West Hunwick Colliery** (NZ 195329) **(M43)**, being sunk by the Lackenby Iron Co. This line is said to have been rope-worked and to have used ancient 6-ton waggons. This branch did not last long; for the subsequent history of West Hunwick see West Hunwick Refractories Ltd.

By 1890 the NER had decided to close the almost entirely rope-worked section of the former West Durham Railway between Crook and Todhills Engine House, using nearby railways to handle the coal traffic. Hunwick Colliery lay adjacent to Hunwick Station on the NER Bishop Auckland-Durham line, and so a short link was put in here. The closure of the link to Todhills Bank Foot now meant a reverse to get out of Newfield, so that NER locos had to propel to the end of the reverse and then work very hard up to Hunwick Station. The West Durham section was closed in 1891.

By the 1890s a **firebrick works** had been opened at Newfield. Hunwick Colliery was abandoned in 1921, though it may have closed before this. Newfield Colliery was re-opened in 1921 after a period of closure. Its underground workings were combined with Byers Green Colliery in 1927. It passed to Dorman, Long & Co Ltd on 1st November 1929.

A loco was recorded here in February 1868, one each at Newfield and Hunwick in April 1871 and November 1876 and three at Newfield in March 1890 (with none at Hunwick, by then shunted presumably by the NER); but the only known information is that given below.

Gauge : 4ft 8½in

-		0-4-0T	OC	FJ	72	1867	(a)	(1)
No.1		0-4-0ST	OC	P	916	1901	(b)	(2)
No.2		0-4-0ST	OC	BH	*	1877	(c)	
		reb		B Vaughan		1893		(2)
105	KELVIN	0-4-0ST	OC	CF	1211	1911	(c)	(2)
118	BELMONT	0-6-0ST	OC	HL	2909	1911	(c)	(2)

*   in later years the worksplates were very worn. Various visitors between 1939 and 1951 reported that whilst 1877 was visible, their best guess was that there was a 6 in the works number.

(a)   New to Hunwick Coal Co (per FJ records).
(b)   ex West Auckland Colliery, West Auckland, /1921.
(c)   ex Byers Green Colliery, Byers Green.
(1)   to West Hunwick Colliery & Brickworks; as the date of sale is not known, the owners of West Hunwick Colliery at the time cannot be determined; see the entry for West Hunwick Refractories Ltd.
(2)   to Dorman, Long & Co Ltd, with Newfield Colliery and brickworks, 1/11/1929.

NEWFIELD COKE OVENS   M44

These beehive ovens were adjacent to the colliery, and were operated in the usual way. A locomotive was being used here in November 1876. The ovens ceased production in August 1913.

Gauge : 3ft 0in

-		0-4-0ST	OC	BH	?	?	(a)	(1)
No.1	NEWFIELD	0-4-0ST	OC	BH	459	1878	New?	(1)

There may also have been transfers unknown.

(a)   see list of locomotives under the entry for Auckland Park Colliery.
(1)   one locomotive was in daily use here in 4/1901 (DCOA Return 427); s/s.

NEWLANDSIDE QUARRY, Stanhope.   V45
Newlandside Mining Co until /1880   NY 995383 approx.

The quarrying of this large area of limestone was due to the opening of the Wear Valley Railway by the Stockton & Darlington Railway in April 1862, the first traffic to the quarry being run on 30th April. From 1865 Bolckow Vaughan was the quarry's main, and perhaps sole, customer. When this contract expired in 1880 the Newlandside Mining Co declined both to renew it and to continue operating the quarry, so Bolckow Vaughan was compelled to purchase it. The quarry became one of the largest in Weardale. It was connected by a **self-acting incline**, ¾ mile long, down to the NER Wear Valley Branch,

¼ mile east of Stanhope Station. It passed to Dorman, Long & Co Ltd, on 1st November 1929.

There may well have been other locomotives here besides those given below.

Gauge: 4ft 8½in

NEWLANDSIDE	0-4-0ST	OC	BH	365	1876	New	
	reb		Wake	2432	?		(1)
WITTON	0-4-0ST	OC	BH	367	1877	(a)	(1)
COMET	0-4-0ST	OC	BH	544	1880	(b)	s/s
TINY	0-4-0ST	OC	FJ	31	1864	(c)	s/s
HAVERTON	0-4-0ST	OC	AB	656	1890	(d)	(1)
HARE	0-4-0ST	OC	GH		1908	(e)	(2)

(a) ex either Witton Park Ironworks, Witton Park, or Cleveland Works, Middlesbrough, Yorkshire (NR).
(b) ex Cleveland Works, Middlesbrough, Yorkshire (NR), by 5/1908.
(c) here by 5/1908, possibly ex Cleveland Works, Middlesbrough, Yorkshire (NR) or possibly ex Binchester Colliery, near Westerton.
(d) ex South Bank Works, Middlesbrough, Yorkshire (NR), by /1913.
(e) ex Eston Mines, Yorkshire (NR), by 9/1914.

(1) to Dorman, Long & Co Ltd, with quarry, 1/11/1929.
(2) to Auckland Park Colliery, Coundon Grange, by 7/1929.

WEST AUCKLAND COLLIERY AND COKE OVENS, West Auckland   Q46
NZ 184267

This colliery was situated in the western area of the junction between the Stockton & Darlington Railway's line to Etherley Colliery and its Haggerleazes Branch, immediately east of West Auckland Station. It was sunk in 1837 and coal was won in 1838, almost certainly under the auspices of **Jonathan Backhouse**, the senior partner in the Darlington banking firm of Backhouse & Co, and also treasurer of the Stockton & Darlington Railway. Besides this colliery, he also owned Black Boy and Leasingthorne Collieries (see above) and White Lee and Woodifield Collieries (see below). Backhouse died in October 1842 and the collieries passed to his family, who offered them for sale in 1851. West Auckland Colliery had passed to Bolckow Vaughan by 1859.

60 Coppee patent non-byproduct ovens were opened here in January 1900. 24 were closed in 1912, followed by the remaining 36 in October 1913. The colliery was closed in September 1925. It was re-opened by **The Ramshaw Coal Co Ltd** (which see) in 1944, without locomotives, and passed to NCB Northern Division No. 4 Area on 1st January 1947.

The only knowledge of the first locomotive below is an order to Peckett for spares to be delivered here in 1917. It would seem likely that the colliery was normally shunted by the Stockton & Darlington Railway and then the NER.

Gauge : 4ft 8½in

-	0-4-0ST	OC	P	916	1901	(a)	(1)
DART	0-4-0(ST?) ?	?	?	?	?	(b)	s/s

(a) this loco was new to W.S. Laycock Ltd, Sheffield, Yorkshire (WR); whether Bolckow Vaughan bought it direct from this company or via some other firm or dealer is not known; it was here by 15/3/1917 (spares order to P).
(b) probably here following the First World War; no other details are known.

(1) to Newfield Colliery & Brickworks, Newfield, /1921.

WITTON PARK IRONWORKS, Witton Park   Q47
NZ 174306

This works was commissioned in October 1845, with the furnaces first being tapped on 14th February 1846. It was situated within a meander of the River Wear, and was served by ¼ mile branch from Etherley Station on what was then the Bishop Auckland & Weardale Railway (an extension of the Stockton & Darlington Railway), later the NER. The works initially produced pig iron, but later moved into the manufacture of railway rails, and by 1876 it had six blast furnaces. Slag from the furnaces was deposited eastwards from the works along the river bank. The works suffered short time closures in both 1877-1878 and 1882, and was closed again on 19th May 1884. This was initially intended to be temporary, but became permanent, as the demand for steel rails rose and was met from investment at Middlesbrough. The works appears to have stood unused until 1896 before being demolished. The slag heaps were subsequently worked by Auckland Rural District Council and The Witton Park Slag Co Ltd (both which see).

Given that the works had rail traffic from its opening, there may well have been locomotives before those given below.

Gauge : 4ft 8½in

VAUGHAN	0-4-0ST	OC	MW	416	1872	New	s/s
WITTON	0-4-0ST	OC	BH	367	1877	New	(1)
WEAR	0-4-0ST	OC	BH	368	1877	New	(2)
WHITWORTH	0-4-0ST	OC	BH	519	1879	New	(3)

(1) to Newlandside Quarry, Stanhope, or possibly to Cleveland Works, Middlesbrough, Yorkshire (NR).
(2) the most likely probability is that this loco was transferred to Byers Green Colliery, Byers Green, via repairs at Lingford, Gardiner & Co Ltd, Bishop Auckland (see the note for this loco under the entry for Newfield Colliery & Brickworks).
(3) to Cleveland Works, Middlesbrough, Yorkshire (NR) (?)

Note : It has been suggested that either BH 367 or BH 368 may be the locomotive offered for sale by R.R. Deans at Witton Park in 1906 (see Auckland Rural District Council), but there is no evidence to support this.

The company also owned (latterly) **MERRINGTON COLLIERY**, near Spennymoor (NZ 252338) **(N48)**, closed in July 1927; **SHILDON LODGE COLLIERY** (NZ 223264) **(Q49)** at Old Shildon, served by the one mile long "Surtees Railway" (named after the local landowner) and closed in May 1928; **WHITE LEE COLLIERY** (NZ 155375) (Map M), near Crook, which was sold in 1889 to Pease & Partners Ltd (which see) and **WOODIFIELD COLLIERY** (NZ 160354) **(M50)** also near Crook, which closed in 1911 (see The Woodifield Coal Co Ltd). White Lee and Woodifield Collieries were acquired from the Backhouse family (see above) in the 1850s. So far as is known none of these collieries had any Bolckow Vaughan locomotives. The firm also owned collieries in Yorkshire through subsidiary companies, notably The Upton Coal Co Ltd, which passed to Dorman, Long & Co Ltd.

JOHN BOWES & PARTNERS LTD
John Bowes, Esq, & Partners until 21/7/1886
The Marley Hill Coal Co until 11/1847
This firm became one of the largest coal owners in North-East England, with collieries in both Durham and Northumberland. In addition to its collieries, the firm operated its coke making for quite a time in the nineteenth century under the title of **The Marley Hill Coke Co**. In the twentieth century it set up **The Marley Hill Coking Co Ltd** and then **The Marley Hill Coke & Chemical Co Ltd** for the same purpose. The exact dates between which these companies operated are not known.

PONTOP & JARROW RAILWAY: BOWES RAILWAY from 1932   C/D/E/G51
Originally the partners owned only **MARLEY HILL COLLIERY** (NZ 206575), where a new sinking was begun on 8th January 1840 and completed on 28th June 1841. The colliery was linked to what was then the Brandling Junction Railway's Tanfield Branch (opened after rebuilding in November 1839) by a line just under ½ mile long, joining just north of the Branch's Bowes Bridge Engine House. The link was extended between 1842 and 1844 to serve **CROOKBANK COLLIERY** (NZ 187571) and **BURNOPFIELD COLLIERY** (NZ 173562), and then **ANDREWS HOUSE COLLIERY** (NZ 205573), all of which the firm purchased. Locomotives were introduced at Marley Hill in June 1847.

In North-East Durham lay the **Springwell Colliery Railway**, owned by Lord Ravensworth & Partners (which see). This had been opened in 1826 to take coal from **MOUNT MOOR COLLIERY** (NZ 279577) and **SPRINGWELL COLLIERY** (NZ 285589) to staiths at Jarrow, with locomotives being used for the final 4¾ miles to the River Tyne. The line had been extended in 1842 to serve **KIBBLESWORTH COLLIERY** (NZ 243562), owned by George Southern. By 1845, but possibly from 1843, a passenger service had been introduced between **Springwell Station** (NZ 312624) and Jarrow; this lasted until 1872.

John Bowes (1811-1885) was also one of the members of Lord Ravensworth & Partners, and in 1850 his own firm acquired the railway and the first two collieries, adding Kibblesworth Colliery in the following year. The connection between the two was constructed between 1853 and 1854, and at the same time the railway was extended to the newly-sunk **DIPTON COLLIERY** (NZ 158358). The line, fifteen miles long between Dipton and Jarrow, became fully operational in April 1855, having earlier been given the title of **PONTOP & JARROW RAILWAY**.

Between Dipton and Birkheads, at the top of the western side of the Team valley, a distance of 5¼ miles, traffic was worked by locomotives from **MARLEY HILL LOCO SHED** (NZ 207573), except for the ¾ mile long Hobson Bank, north of Burnopfield Colliery, which was worked by a stationary engine until sometime between 1890 and 1900. On this section Crookbank Colliery was replaced by **BYERMOOR COLLIERY** (NZ 187573), while nearby a short branch served **Crookgate Quarries** (NZ 184573). Andrews House Colliery closed in 1920, but about ¾ mile to the east **BLACKBURN FELL DRIFT** (NZ 214573) was opened in 1937.

The section between Birkheads and Springwell Bank Foot, at the Leam Lane near Wardley, a distance of six miles, was worked entirely by rope inclines. The first and last were self-acting, the others were worked by three stationary engines. About ½ mile below Kibblesworth Colliery **KIBBLESWORTH GRANGE DRIFT** (NZ 249564) was opened in 1914. It was closed in 1932, but with the sinking of a new shaft here the site was then used for the colliery screens, with the wagons beings shunted by an auxiliary hauler. At Black Fell, Mount Moor Colliery became **SPRINGWELL (VALE PIT)**. This was closed in 1931, to be following by Springwell Colliery itself in 1932. The latter's workshops were then developed to become **ENGINEERING SHOPS**, with a large coal bunker being converted into a **WAGON SHOP**.

On the final 4¾ mile section to Jarrow **WARDLEY COLLIERY** (NZ 306620), served by a ½ mile branch from Wardley, was acquired in 1868. In 1911 this was closed and replaced by **FOLLONSBY COLLIERY** (NZ 313608), served by a ¾ mile branch. Near the branch junction in the 1920s the company built the **WARDLEY DRY-CLEANING PLANT** (NZ 308622) and a major marshalling yard was also developed here. At Jarrow the original staiths were replaced in 1854-1855, with further replacements in 1883 and 1936, the last so far upstream that although they were still called the **JARROW STAITHS** (NZ 318657), they were actually in Hebburn; these last staiths were owned by the Tyne Improvement Commission.

Like many other major colliery companies, the firm got into serious difficulties in the early 1930s and in 1932 it was taken over by new directors, although the Strathmore family retained its involvement. The new owners re-named the system the **BOWES RAILWAY**.

The firm was famous for its coke production, all the western collieries sharing hundreds of beehive or rectangular ovens. These were phased out in favour of the **by-product recovery ovens** built at **MARLEY HILL** from 1908 onwards. These in turn were replaced in 1937 by a new plant at Wardley, called originally the **BOWES COKE WORKS** but soon re-named the **MONKTON COKE WORKS** (NZ 315626). In February 1939 Springwell and Follonsby Collieries were sold to The Washington Coal Co Ltd, although their traffic continued to be handled by the Railway.

The Bowes Railway, together with the collieries at Burnopfield, Byermoor, Marley Hill, Kibblesworth and Blackburn Fell Drift, Monkton Coke Works and its branch to Follonsby Colliery, passed to NCB Northern Division No. 6 Area on 1st January 1947.

References : *The Private Railways of County Durham*, Colin E. Mountford, Industrial Railway Society, 2004; *The Bowes Railway*, 2nd edition, Colin E. Mountford, Industrial Railway Society, 1976; a considerable Bowes Railway deposit is held by the Tyne & Wear Archives Service, Newcastle upon Tyne.

The loco sheds are coded as follows:

    MH   Marley Hill
    SBF   Springwell Bank Foot

Gauge : 4ft 8½in

	BOWES	2-4-0?	?	?	?	?	(a)	MH	(1)
	GIBSIDE	2-4-0	IC	RS	?	?	(a)	MH	(1)
	RAVENSWORTH	2-4-0?	?	?	?	?	(a)	MH	(1)
	STRATHMORE	?	?	?	?	?	(b)	SBF	s/s
No.2 (?)		0-6-0	OC	RWH	476	1846	(c)	MH	(2)

(a) the first locomotives to work at Marley Hill arrived in 6/1847, their identity and origin being unknown; whether any are linked to these locos is also not known.
(b) believed to have come from Lord Ravensworth & Partners, with the Springwell Colliery Railway, 1/1/1850, but not confirmed.
(c) ex North British Railway, No.29, 11/1855.

(1) said to have been scrapped in the 1870s.
(2) to Northumberland & Durham Coal Co (a subsidiary company of John Bowes, Esq., & Partners), Blackwall, London, by 1/1859.

No.1	"BULL"	0-4-0	VC	RS		1826	(a)	
	SBF-MH (via RS, repairs) 1/1851							(1)

No.	Name	Wheel	Cyl	Mkr	Wks	Year	Acq	Notes	Ref
1		0-6-2ST	IC	BH	937	1888	New		
		reb		HL		1901			
	SBF								(2)
No.2		0-4-0	VC	RS		1826	(a)		
	SBF								(3)
No.2	(later 2)	0-6-0	IC	RS	1516	1864	New		
		reb 0-6-0ST	IC	RS	2902	1898			
		reb		HL	8243	1915			
	SBF								(4)
No.3	STREATLAM	0-4-0	OC	RS	795	1851	New		
	SBF								(5)
No.3	(later 3)	0-6-2ST	IC	BH	938	1888	New		
		reb		HL	3045	1903			
	SBF-MH 30/11/1934-SBF 19/3/1942								(6)
No.4	MARLEY HILL	0-4-0	OC	RS	816	1851	New		
	MH								s/s c/1886
No.4	(later 4)	0-6-0ST	OC	BH	891	1887	New		
	MH-SBF 31/10/1907-MH 12/6/1908								(7)
4		0-4-0ST	OC	KS	4030	1919	(b)		
	SBF								(8)
No.5	DANIEL O'ROURKE	0-4-0ST	IC	Marley Hill		1854	New		
	MH								(9)
No.5		0-6-0ST	OC	RWH	1986	1884	(c)		
		reb		HL	3593	1913			
	SBF-MH by /1904-SBF 29/1/1932								(2)
No.5		0-6-0T	IC	Ghd	7	1897	(d)		
	SBF								(10)
No.6		0-6-0	IC	RS	1074	1856	New		
	? -MH by 1885								s/s c/1890
No.6		0-4-0T	OC	FJ	125	1874	(e)		
	MH								(11)
No.6	(later 6)	0-6-0ST	OC	HL	2515	1901	New		
		reb		HL	1779	1911			
		reb		LG		1930			
	MH-RSHN, repairs, 23/8/1942-MH 16/5/1945-SBF 11/6/1945								(12)
No.7		?		?	?	?	(f)		s/s by /1874
No.7	(later 7)	0-6-0ST	IC	BH	304	1874	New		
	SBF-MH by /1904								Scr /1933
	BOWES No.7	0-4-0ST	OC	CF	1203	1901			
		reb		HL	9270	1909	(g)		
	SBF- MH 4/12/1934-SBF 13/5/1938-MH 21/11/193-SBF 24/7/1941								(13)
No.8		?		?	?	?	(f)		s/s by /1882
No.8	(later 8)	0-6-0ST	IC	BH	692	1882	New		
		reb		HL	668	1910			
	MH								Scr c9/1934
No.8	(later BOWES No.8)	0-6-0ST	IC	SS	4594	1900	(h)		
	SBF								(14)
No.9	(later 9)	0-6-0ST	IC	?	?	?	(j)		
		reb		RS	109	1867			
		reb		RS	2821	1894			
		reb		HL	6830	1914			
		rep		Ridley Shaw		1927			
	MH-SBF ?/?-MH 31/10/1907-SBF 12/6/1908-MH 12/1/1926								
	-SBF 23/12/1927-MH 15/10/1929-SBF 29/4/1931								Scr 2/1935

No.9		0-6-0ST	IC	SS	4051	1894		
	reb	0-6-0PT	IC	Caerphilly		1930	(k)	
SBF								(15)
No.10	(HARTLEPOOL)	0-6-0	IC	TR	252	1854	New	
	reb	0-6-0ST	IC	BH		1877		
SBF								(16)
10		0-6-2ST	IC	BH	1071	1892	New	
	reb		HL			1906		
	reb		T.D.Ridley			1917		
SBF								Scr 10/1931
BOWES No.10		0-6-0ST	IC	NBA	16628	1905		
	reb	0-6-0PT	IC	Sdn		1924	(m)	
SBF								(12)
No.11(later 11)		0-6-0	IC	RS	1313	1860	New	
	reb	0-6-0ST	IC	RS		1875?		
SBF								Scr /1915
11		0-6-0ST	OC	HL	3103	1915	New	
MH-SBF 7/9/1945								(12)
No.12		0-6-0ST	IC	RS	1612	1864	New	
SBF-MH after /1872								Scr /1885
12		0-6-0ST	OC	HL	2719	1907	New	
	reb		RS			1932		
MH								(12)
No.13		0-6-0	IC	RS	1611	1864	New	
SBF?								Scr /1896
13		0-6-0ST	IC	HL	2545	1902	New	
SBF-MH 10/7/1931-SBF 1/11/1937								(12)
No.14		0-6-0ST	IC	RS	1800	1866	New	
?								(17)
14		0-6-2T	IC	CF	1158	1898	(n)	
MH-SBF 13/3/1914								Scr 9/1923
14		0-6-0ST	IC	HL	3569	1923	New	
SBF-MH 13/2/1942								(12)
15		0-6-0T	IC	HE	1506	1930	New	
MH								(12)
39		0-6-0ST	IC	RWH	1422	1867	(p)	
SBF								(18)
16		0-6-0ST	IC	VF	5288	1945	(q)	
SBF								(12)
17		0-6-0ST	IC	VF	5298	1945	(r)	
MH								(12)
"18"	75317	0-6-0ST	IC	VF	5307	1945	(s)	
SBF								(19)

(a) ex Lord Ravensworth & Partners, with the Springwell Colliery Railway, 1/1/1850.
(b) ex Felling Colliery, Felling, 1/4/1926.
(c) ex HL, 3/1886; built new in 1884 and carried RWH works plate.
(d) ex LNER, 1787, Gateshead, 11/8/1936, following accident to RWH 1986/1884; LNER J79 class; previously NER H2 class.
(e) ex Lord Dunsany & Partners Ltd, Pelton Colliery, Pelton Fell, c/1880 (if identification is correct).
(f) on 19/12/1854 the LNWR Southern Committee agreed the sale to Charles Mark Palmer, the firm's Managing Partner, of locos Nos.47 and 54. Both were 2-2-0 tender locos built by Bury, Curtis & Kennedy, No.47 in 1840 and No.54 in 1841. However, both locos seem to have survived on the LNWR until 1856, and it is not known whether either came to the Pontop & Jarrow Railway.
(g) ex Felling Colliery, Felling, date unknown, but first recorded repair at SBF was 23/2/1932; ran as FELLING No.1 until 11/1934.

(h)   ex GWR, 713, 7/1936, per R.H. Longbotham & Co Ltd, dealers, Northwood, Middlesex; arrived 13/8/1936; previously Barry Railway, 'F' class, 52.
(j)   identity uncertain; may have been built by Ralph Coulthard, Gateshead; its date of arrival is also uncertain – the Bowes Railway Locomotive Ledger gives 1860, R.H. Inness gave 1862.
(k)   ex GWR, 717, 11/1934, via R.H. Longbotham & Co Ltd, dealers, Northwood, Middlesex; arrived 26/11/1934; previously Barry Railway, 'F' class, 71.
(m) ex GWR, 725, 11/1934, via R.H. Longbotham & Co Ltd, dealers, Northwood, Middlesex; arrived 26/11/1934; previously Barry Railway, 'F' class, 127.
(n)   ex The Harton Coal Co Ltd, South Shields, Marsden & Whitburn Colliery Railway, No.7, /1912 (after 22/7/1912), per Robert Frazer & Sons Ltd, dealers, Hebburn.
(p)   hired from The Lambton, Hetton & Joicey Collieries Ltd, Lambton Railway, Philadelphia, 18/12/1937.
(q)   ex War Department, 75298, Longmoor Military Railway, Liss, Hampshire, 4/1946; arrived 5/1946.
(r)   ex War Department, 75308, Longmoor Military Railway, Liss, Hampshire, 4/1946; arrived 13/5/1946.
(s)   ex War Department, 75317, Longmoor Military Railway, Liss, Hampshire; oral tradition claimed this locomotive worked here, but written confirmation is lacking; if she was here, she arrived between 5/1946 and 8/1946, and was presumably on loan.

(1)   the *Transactions of the Institute of Mining Engineers*, Vol.14, 1912-1913, p.111, state that "the engines of that locomotive, which were vertical, could be seen until a few years ago driving lathes and other tools in the [Marley Hill] colliery workshops."
(2)   sold as scrap to Robert Frazer & Sons Ltd, Hebburn, 8/1936; scrapped on site, 9/1936.
(3)   to Killingworth Colliery, Killingworth, Northumberland, /1863.
(4)   sold as scrap to Robert Frazer & Sons Ltd, Hebburn, 2/1937; scrapped on site, 3/1937.
(5)   to Killingworth Colliery, Killingworth, Northumberland, c/1880.
(6)   to NCB No. 6 Area, with the Railway, 1/1/1947; out of use.
(7)   to Sir W.G. Armstrong, Whitworth & Co Ltd (at Lemington, Northumberland ?), /1917.
(8)   returned to Felling Colliery, Felling, 14/10/1927; to SBF, 6/12/1929; to W.G. Bagnall Ltd, Stafford, Staffordshire, for repairs, 21/9/1942; to MH, 8/9/1943; to NCB No. 6 Area, with the Railway, 1/1/1947.
(9)   scrapped c/1885, but the saddletank survived outside Springwell Bank Foot shed until about 1954, being used to hold the clay for making firebrick arches.
(10) sold as scrap to D. Sep.Bowran Ltd, Gateshead, 4/1946; scrapped, almost certainly at Follonsby Colliery, 10/1946.
(11) to Felling Colliery, Felling, c/1895.
(12) to NCB No. 6 Area, with the Railway, 1/1/1947.
(13) loaned to The Bedlington Coal Co Ltd, Northumberland, 18/11/1942; returned to SBF, 21/8/1944; sold as scrap to D. Sep.Bowran Ltd, Gateshead, 4/1946; scrapped, almost certainly at Follonsby Colliery, 10/1946.
(14) sold as scrap to D. Sep.Bowran Ltd, Gateshead, 4/1946; scrapped, almost certainly at Follonsby Colliery, 9/1946.
(15) loaned to The Harton Coal Co Ltd, Boldon Colliery, 20/2/1943; returned to SBF, 14/8/1943; to MH, 24/10/1945; to NCB No. 6 Area, with the Railway, 1/1/1947.
(16) to Killingworth Colliery, Killingworth, Northumberland, c/1866; returned to SBF; boiler exploded at Jarrow Staithes, 14/6/1882; to MH, c/1883; to Killingworth Colliery, Killingworth, Northumberland, c/1886.
(17) said to have been sold c/1880, possibly to a firm in Scotland.
(18) returned to The Lambton, Hetton & Joicey Collieries Ltd, Philadelphia, 28/3/1938.
(19) to The Weardale Steel, Coal & Coke Co Ltd, Thornley Colliery, Thornley, by 8/1946.

FELLING COLLIERY, Felling   D52
Sir George Elliot until 1/3/1883 (**George Elliot** until 15/5/1874)   NZ 275623

This colliery had a long and complicated history. The first colliery at Felling was Brandling Main Colliery, opened in 1779. This was closed on 19th January 1811 and replaced by a new shaft, the John Pit, which commenced production in May 1811. Exactly a year later it was the scene of the then worst explosion in mining history in which 91 people died, the disaster leading directly to the invention of the miners' safety lamp. At this time the colliery was owned by the brothers John and William Brandling, together with two others. They sold out in 1818, the Brandlings moving east to develop the very large Harton royalty (see The Harton Coal Co Ltd).

Originally the colliery's coal was taken by a waggonway about ¾ mile long to staiths on the River Tyne, a line which by the 1850s was worked by a **stationary engine**. Then when in 1839 the Brandling brothers opened their Brandling Junction Railway between Gateshead and South Shields, the colliery was linked to it via sidings at Felling Station.

Meanwhile in 1833 the **Felling Chemical Works** (NZ 279626) had been opened on the east side of the waggonway about ½ mile from the colliery, by Hugh Lee Pattinson & Co (which see). In 1854 the same company opened the **Felling Iron Works** (NZ 279623), with two blast furnaces, while to the south of the iron works **Felling Brick Works** (NZ 278622) was also opened. The history of these enterprises is very complicated - see the entries for H.L. Pattinson and for The Felling Coal, Iron & Chemical Co Ltd, which acquired the iron works in 1871. George Elliot is believed to have acquired the colliery in 1863.

The supplement to the *Mining Journal* dated 21st January 1871 describes the transport arrangements here as follows: "the wagon beam engine (2 drums), near the Felling Pit, hauls empty wagons from the Felling Shore, 1200 yards, and also laden wagons from the pit up to the siding on the NER by means of a return wheel; the wagons run down by their gravity the contrary direction". The drawings of part of the line which exist in the Bell Papers at Tyne & Wear Archives show a meetings, presumably indicating that two sets were run simultaneously.

13. The official works photo of FELLING No.1, CF 1203/1901, one of the last locomotives built by the firm. Note the characteristic CF design of chimney.

However, about this time a new link with the NER was established a few yards to the west, by-passing the colliery and its stationary engine. The DCOA Return 102 records the hauler in April 1871, but the entry for November 1876 does not, and instead lists one locomotive. This would suggest that in the early 1870s the line was rebuilt to its final condition, with loco haulage for most of its length and a short self-acting incline down the steep bank to the staiths.

Sir George Elliot sold the colliery to John Bowes, Esq., & Partners as from 1st March 1883. It was closed in April 1931, temporarily at first but then permanently, finally being abandoned in December 1933.

There were clearly one or more locomotives here before the first one listed below. The identity of the locomotive recorded in November 1876 is unknown, while the name of HL 2152 would suggest that there was a FELLING No.1 here when it arrived in 1889, which may well have lasted until the arrival of CF 1203 in 1901.

Gauge : 4ft 8½in

FELLING No.2	0-4-0ST	OC	HL	2152	1889	New	(1)
BEACONSFIELD	0-4-0ST	OC	N	?	?	(a)	s/s
ABBEY	0-4-0ST	OC	BH	317	1874	(b)	(2)

FELLING No.2	0-4-0T	OC	FJ	125	1874	(c)		
	reb		CF		1901		Scr /1917	
FELLING No.1	0-4-0ST	OC	CF	1203	1901	New		
	reb		HL	9270	1909			(3)
FELLING No.2	0-4-0ST	OC	KS	4030	1919	New		(4)

(a) origin and date of arrival unknown; LG is said to have had a hire loco built by N.
(b) ex BH, hire.
(c) ex Pontop & Jarrow Railway, Marley Hill Loco Shed, Marley Hill, c/1895.

(1) to The Bede Metal & Chemical Co Ltd, Hebburn, (c/1901 ?).
(2) returned to BH.
(3) to Bowes Railway, Springwell Bank Foot Loco Shed, Wardley, c3/1932.
(4) to Pontop & Jarrow Railway, Springwell Bank Foot Loco Shed, Wardley, 1/4/1926; returned, 14/12/1927; to Springwell Bank Foot Loco Shed, Wardley, 6/12/1929.

Additional note
John Bowes and his first wife Josephine were major collectors of a wide variety of art and antiques, and to house their collections for public display they began in 1869 the construction of The Bowes Museum at Barnard Castle. The completion of the Museum, which was not opened until 1892, seven years after Bowes' death, together with Bowes' will, had a major impact on the colliery business. The final collieries in Northumberland, Seaton Burn and Dinnington, were disposed of in 1899, by which time all the other collieries in Durham besides those above had also been sold. The Bowes Museum now has an international reputation.

D. SEP. BOWRAN LTD (latterly a subsidiary of **T.J. Thomson & Son Ltd**, which see)
SHIPCOTE WORKS, Gateshead　　　　　　　　　　　　　　　　　　　　　　　　　　**C53**
NZ 261629

This scrapyard was situated on the site of the former BR Park Lane Goods Depot, and was served by sidings south of the BR Gateshead-South Shields line, 3/4 mile east of Gateshead Station. The yard was shunted by two steam cranes before the arrival of the first locomotive below. It closed in March 1977.

Gauge : 4ft 8½in

CHURCHILL	0-4-0DM	RH	281270	1951	(a)	(1)
(No.5)	0-6-0DM	HC	D835	1954	(b)	Scr 10/1976
No.3 P.S.A. No.20	0-4-0DM	RH	327969	1954	(c)	(2)

(a) ex Hawthorn Leslie (Engineers) Ltd, St.Peter's Works, Newcastle upon Tyne, 4/1974.
(b) ex Central Electricity Generating Board, North Tees Power Station, Haverton Hill, Teesside, /1974.
(c) ex Clarke Chapman Ltd, Gateshead, 13/10/1976.

(1) to T.J. Thomson & Son Ltd, Tyne Depot, Dunston, c1/1977.
(2) to T.J. Thomson & Son Ltd, Tyne Depot, Dunston, 3/1977.

THE BRANCEPETH COAL CO LTD (formed by 1937)
LITTLEBURN COLLIERY, Meadowfield　　　　　　　　　　　　　　　　　　　　　　　**K54**
NZ 255395

This colliery had been closed in 1935 by The Bearpark Coal & Coke Co Ltd, but was re-opened in 1941 by this company, formed for the purpose by the Summerson family of Darlington (which see). The shafts had long ceased to draw coal, production being via a drift. The colliery was served by sidings east of the LNER Durham – Darlington line, 2¾ miles south of Durham Station. It was vested in NCB Northern Division No. 5 Area on 1st January 1947.

Gauge : 4ft 8½in

-	0-4-0ST	OC	BH	1096	1896	(a)	(1)

(a) presence here at this date not confirmed, but if here, then ex The Bearpark Coal & Coke Co Ltd, former owners of the colliery, /1941.

(1) to NCB No. 5 Area, with the colliery, 1/1/1947 (?).

BRITISH CRANE & EXCAVATOR CORPORATION LTD
CROWN WORKS, Pallion, Sunderland H55
Steels Engineering Products Ltd until 28/3/1964 NZ 375580
(both firms were subsidiaries of **Steel Group Ltd** (**Steel & Co Ltd** until 1/10/1965))

The firm of Steel & Co started in 1879 as a builder's merchants. In 1939 it took over Coles Engineering Co Ltd and the adjacent site of the former Egis Shipyard (see William Gray & Co Ltd) and set up the Crown Works to manufacture cranes, using the Coles brand name. The works was served by a branch (½ mile) from Pallion Station on the LNER Sunderland - Durham line. A loco from this works also shunted the adjacent shipyard of Short Brothers Ltd, and this has led to considerable confusion about the ownership of the locomotives at the two sites; the version below and the list in the entry for Short Brothers (which see) are believed to be correct; putting this another way, after the departure of LNER 1799 in 1946, Short Brothers shunted the traffic until an expansion of the works in the mid-1950s resulted in the construction here of its own locomotive, as shown. After Short Brothers Ltd closed in January 1964 the firm purchased 12 acres of that site for further expansion. Rail traffic ceased in 1967. The works continued, under different owners, until closure on 18th December 1998.

Gauge : 4ft 8½in

	-		0-4-0ST	OC	HL	2496	1901	(a)	(1)
	JUNO		0-4-0ST	OC	BH?	606?	1881?	(b)	s/s after c/1952
	-		0-4-0	OC?	Butterley		?		
		reb	0-4-0ST	OC	Butterley		1895	(c)	(2)
	BASIL		0-4-0ST	OC	I'Anson		1875	(d)	s/s
1799			0-4-0T	IC	Ghd	4	1897	(e)	(3)
	-		4wDE		Coles	16640	1956	(f)	(4)

(a) ex Ewesley Quarry Co Ltd, Ritton Whitehouse Quarry, near Netherwitton, Northumberland, c/1939.
(b) ex Gateshead County Borough, Saltmeadows Clearance Site, Gateshead, 9/1939. However, a photograph of an 0-4-0ST OC named JUNO taken here c/1952 shows a completely different locomotive from the JUNO photographed at Gateshead, other than the nameplate, which appears identical; if the nameplate was transferred to a different locomotive, nothing is known of its identity, origin and date of arrival.
(c) ex Butterley Co Ltd, Ripley, Derbyshire, B12C, c/1940; prior to the 1895 rebuilding the locomotive is believed to have been a 0-4-0 tender engine, probably with outside cylinders.
(d) ex The South Medomsley Colliery Co Ltd, South Medomsley Colliery, near Dipton, c/1945.
(e) ex LNER 1799, Tyne Dock loco shed, Tyne Dock, 2/1946, hire; LNER Y7 class.
(f) New; constructed by the company at the works here.

(1) to Sir Hedworth Williamson's Limeworks Ltd, Fulwell Quarries, Sunderland, /1940.
(2) to Ridley, Shaw & Co Ltd, dealers, Middlesbrough, Yorkshire (NR), by /1942; rebuilt by Ridley Shaw in 1942 and re-sold to Sir S.A. Sadler Ltd, Middlesbrough, Yorkshire (NR), /1942.
(3) returned to LNER, Tyne Dock loco shed, Tyne Dock, 7/1946, after hire.
(4) to Crane Machinery Services Ltd, Feltham, Middlesex, 11/1967.

The works also had two steam rail cranes and one diesel rail crane.

BRITISH GAS CORPORATION, NORTHERN DIVISION
REDHEUGH WORKS, Gateshead C56
Northern Gas Board until 1/1/1973 NZ 237625

This works was closed at the time of the Corporation's creation, with the locomotive listed below awaiting disposal. The very large site became part of a huge reclamation scheme undertaken prior to the Gateshead Garden Festival in 1990 (see Part 2) and was subsequently used for housing development.

Gauge : 4ft 8½in

-	4wDM	RH	476140	1963	(a)	(1)

(a) ex Northern Gas Board, with site, 1/1/1973.
(1) to North of England Open Air Museum, Marley Hill Loco Shed, for preservation, 3/1973.

BRITISH RAILWAYS BOARD

From the nationalisation of the railways in January 1948 stockyards, depots, etc., were shunted by locomotives in the capital list. However, from about 1955 these duties were taken over by service locomotives belonging to the Chief Civil Engineer, though they too were included in capital stock. This continued until about the spring of 1969, when the duties were resumed by normal locomotives in capital stock.

CENTRAL RECLAMATION DEPOT and CROFT JUNCTION STOREYARD, Darlington
S57
NZ 295130

The Reclamation depot, which was controlled by the Chief Civil Engineer, York, and the Croft Junction Storeyard, which came under the District Civil Engineer, were situated together, about ½ mile south of Darlington (Bank Top) Station, on the BR Darlington - York line, in the triangle formed by the main line, the Stockton - Darlington line and the Geneva Loop. The site had once been the Geneva Works of the dealer and repairer J. F. Wake (see Part 2). Both yards were subsequently closed.

Gauge: 4ft 8½in

87	4wDM	RH	463152	1961	New	(1)
82	4wDM	RH	425485	1958	(a)	(2)

(a) ex Dinsdale Rail Welding Depot, Dinsdale, 1/1968.

(1) to Darlington Motive Power Depot, Darlington, 4/1968.
(2) to Darlington Motive Power Depot, Darlington, by 14/9/1968.

DINSDALE RAIL WELDING DEPOT, Dinsdale
S58
NZ 348137

This depot was served by sidings from Oak Tree Signal Box on the BR Stockton - Darlington line, 3½ miles east of Darlington Bank Top Station. It was developed on the site of the former **Linthorpe-Dinsdale Smelting Co Ltd's Middleton Ironworks** (which see), which had subsequently been reclaimed by **Durham County Council** (which see).

Gauge : 4ft 8½in

82	4wDM	RH	425485	1958	New	(1)
No.89	0-6-0DM	HE	5664	1961	(a)	(2)

(a) ex Gateshead Motive Power Depot, Gateshead, 25/1/1964; formerly D2615.

(1) to Croft Junction Storeyard, Darlington, 1/1968.
(2) to Darlington Motive Power Depot, Darlington, 11/1967.

ETHERLEY TIP, Witton Park
Q59
NZ 172307 approx.

This tip for waste ballast was situated alongside the north side of the BR Wear Valley branch and was served by a ½ mile siding from the site of the former Etherley Station, two miles west of Bishop Auckland Station. It occupied the westernmost area of the site formerly occupied by the Witton Park Ironworks owned by Bolckow, Vaughan & Co Ltd (which see). The tip appears to have been open by 1950, and may have been set up before this. It would seem to have been closed in the late 1980s.

Gauge: 4ft 8½in

56	4wDM	RH	338424	1955	(a)	(1)

(a) ex Barnard Castle, /1963, following track recovery work on the Darlington-Barnard Castle line.
(1) to Darlington Motive Power Depot, Darlington, by 14/4/1968.

LOW FELL PERMANENT WAY STOCKYARD, Gateshead
C60
NZ 248608

This stockyard lay alongside the BR freight-only line from Low Fell to Dunston, about ½ mile north of this line's junction with the Newcastle-Darlington line at Low Fell. The yard continued in use until the early 1990s, after which it was abandoned.

Gauge : 4ft 8½in

83	4wDM	RH	432477	1959	New	(1)

(1) to Heaton Motive Power Depot, Newcastle upon Tyne, 4/1969.

BRITISH ROPES LTD (registered 6/2/1924)
Monkwearmouth, Sunderland

H61
NZ 398582

This works manufactured hemp, rope and binder twine. Situated on Fulwell Road, it was served by a link to the Portobello Goods Yard at Monkwearmouth, itself served by a link to the LNER South Shields-Sunderland line at Wearmouth Junction. It is not known what the locomotive below was used for, nor when rail traffic ceased. The factory closed about 1986.

Gauge : unknown

		4wE *	WR	859	1934	New	s/s by c/1960

* described in WR records as a "trolley rail" loco.

Two WR rail cranes were also delivered new here.

BRITISH STEEL CORPORATION
CONSETT WORKS, Consett
Teesside Division by 1/1/1977
General Steels Division, Northern Tubes Group until 29/3/1970
The Consett Iron Co Ltd until 1/7/1968

F62

For the earlier history of this works see Consett Iron Co Ltd.

This was a major steelworks, covered by NZ 0949/1049 and NZ 0950/1050. The large locomotive shed and repair facilities were at **Templetown** (NZ 109500). By the late 1960s the works was effectively at the end of the heavily-graded BR freight-only line from South Pelaw via Annfield Plain to Consett, entering the works at the Low Yard. All the works' raw materials had to come in this way - iron ore from Tyne Dock, and later from Teesside, together with coal from the Durham coalfield, and from August 1969 molten steel from the Cargo Fleet works at Middlesbrough. These heavy transport costs hastened the works' closure. Under a rationalisation scheme the Hownsgill Plate Mill (NZ 104496 approx.) closed in October 1979, but this did not save the works. Production ceased on 5th September 1980, with final closure on 12th September 1980, with 3,700 people made redundant, the largest-ever British figure for a single day up to that time. The demolition of the works was completed in the

14. The last locomotives to come to Consett were four 0-4-0DH Sentinels, transferred from the South Teesside Works. 61, S 10069/1961, passes the Consett Works offices with a slag ladle on 3rd July 1978.

autumn of 1983 and the area was subsequently landscaped or developed for housing, leaving it difficult to imagine such a huge complex had ever existed.

The Consett Works was also responsible for the locomotives at the Jarrow Works, with transfers as required. These are shown as a footnote to the main locomotive list and also under the Jarrow Works entry.

Note : Locomotives acquired by the North of England Open Air Museum for preservation were stored here between 1969 and 1973 prior to being transferred to a permanent home at Beamish (see Part 2).

Besides the locomotives below six diesel rail cranes also worked here.

Gauge : 4ft 8½in

	B No 13	0-4-0ST	OC	HL	3953	1938	(a)	Scr 5/1971

(a) ex The Consett Iron Co Ltd, with works, 1/7/1968; stored.

E No 1	2-4-0VBCr	OC	BH	897	1887	(a)	(1)
E No 6	0-4-0VBCr	OC	BH	1049	1892	(a)	(2)
E No 9	2-4-0VBCr	OC	RS	2854	1898	(a)	(3)
E No 10	0-4-0VBCr	OC	CF	1206	1901	(a)	(4)

(a) ex The Consett Iron Co Ltd, with works, 1/7/1968.
(1) to The North of England Open Air Museum, Beamish, for preservation, 22/3/1978.
(2) to J.A. Lister & Sons Ltd, Consett, for scrap, 6/1969.
(3) dismantled; boiler fitted to E No.1, /1973; remains scrapped, c12/1978.
(4) to J.A. Lister & Sons Ltd, Consett, for scrap, 9/1969.

1		0-6-0DM		HE	3504	1947	(a)	(1)
2		0-6-0DM		HE	3580	1949	(a)	(1)
3		0-4-0DM		HE	4010	1950	(a)	(1)
4		0-4-0DM		HE	4011	1950	(a)	Scr 5/1970
5		0-6-0DE		BBT	3020	1951	(a)	Scr 5/1970
6		0-6-0DE		BBT	3021	1951	(a)	(2)
7		0-4-0DM		HE	4431	1953	(a)	(3)
8		0-4-0DM		HE	4432	1953	(a)	(3)
9		0-6-0DM		Consett		1956	(a)	Scr 2/1971
10		0-6-0DM		Consett		1958	(a)	(4)
11		0-6-0DM		HE	4987	1956	(a)	Scr 2/1971
12		0-6-0DM		HE	4988	1957	(a)	Scr 2/1971
13		0-6-0DM		HE	4989	1957	(a)	Scr 2/1971
14		0-6-0DM *		HE	5173	1957	(a)	(5)
15		0-6-0DM *		HE	5174	1957	(a)	Scr 11/1973
16		0-6-0DM		HE	5175	1957	(a)	Scr 2/1971
17		0-6-0DM		HE	5375	1958	(a)	(6)
18		0-6-0DM		HE	5376	1958	(a)	Scr 5/1975
20		0-6-0DM		HE	5378	1958	(a) #	(7)
21		0-6-0DM		HE	5379	1958	(a) #	(7)
22		0-6-0DM		HE	5380	1958	(a) #	(7)
23		0-6-0DM		HE	5381	1958	(a)	(7)
24		0-4-0DM		HE	5384	1959	(a) #	(7)
25		0-4-0DM		HE	5385	1959	(a) #	(8)
28		0-6-0DH		HE	5392	1959	(a)	Scr 9/1970
29		0-6-0DH		HE	5393	1959	(a)	(9)
30		0-6-0DH		HE	5394	1959	(a)	(6)
-		0-6-0DH		HE	6663	1967	(b)	(10)
-		0-6-0DH		RR	10278	1968	(c)	(11)
17	(19 until 8/1973)	0-6-0DM		HE	5377	1958	(d)	(12)
26		0-4-0DM		HE	5386	1959	(e) #	(13)
31		0-6-0DH		RR	10285	1969	New	(14)
32		0-6-0DH		RR	10286	1969	New	(15)
33		0-6-0DH		RR	10287	1969	New	(15)
34		0-6-0DH		RR	10288	1969	New	(16)
35		4wDH		TH	221V	1970	New	(17)
36		4wDH		TH	222V	1970	New	(17)
37		4wDH		TH	223V	1970	New	(17)

38	4wDH	TH	224V	1970	New	(17)
39	0-6-0DH	RR	10289	1970	New	(15)
40	0-6-0DH	RR	10290	1970	New	(16)
41	0-6-0DH	S	10079	1961	(f)	Scr 10/1983
42	0-6-0DH	S	10082	1961	(g)	(17)
43	0-6-0DH	S	10088	1961	(f)	(18)
44	0-6-0DH	S	10081	1961	(f)	(21)
45	0-6-0DH	S	10080	1961	(f)	(21)
46	0-6-0DH	S	10071	1961	(f)	(20)
47	0-6-0DH	S	10050	1960	(h)	(20)
134	0-6-0DH	S	10153	1963	(j)	Scr 4/1978
60	0-4-0DH	S	10067	1961	(k)	(22)
61	0-4-0DH	S	10069	1961	(m)	(23)
62	0-4-0DH	S	10084	1961	(n)	(23)
63	0-4-0DH	S	10066	1961	(n)	(22)

* modified to work the coke car at Fell Coke Works, if required.

\# Transfers to/from Jarrow Works:
   20 (HE 5378/1958); to Jarrow, 3/1971; returned, 2/1976.
   21 (HE 5379/1958); to Jarrow, 12/1969; returned, 3/1971.
   22 (HE 5380/1958); to Jarrow, 19/8/1975; returned, 2/1976.
   24 (HE 5384/1959); to Jarrow, 9/2/1977; returned, 5/1977.
   25 (HE 5385/1959); to Jarrow, 9/1969; returned, 4/1973.
   26 (HE 5386/1959); to Jarrow, 4/1973; returned, 2/1977.

(a) ex The Consett Iron Co Ltd, with works, 1/7/1968.
(b) ex The Consett Iron Co Ltd, with works, 1/7/1968; on trial from HE.
(c) RR demonstration loco; ex Lindsey Oil Refinery Ltd, Immingham, Lincolnshire, 16/1/1969.
(d) ex Jarrow Works, Jarrow, 10/1969.
(e) ex Jarrow Works, Jarrow, 12/1969.
(f) ex Redbourn Works, Lincolnshire, 11/1974.
(g) ex Redbourn Works, Lincolnshire, 16/11/1974.
(h) ex South Teesside Works, Cleveland, 4/1/1977.
(j) ex South Teesside Works, Cleveland, 11/1977, for spares.
(k) ex South Teesside Works, Cleveland, 17/2/1978.
(m) ex South Teesside Works, Cleveland, 16/2/1978.
(n) ex South Teesside Works, Cleveland, 10/7/1978.

(1) scrapped by J.A. Lister & Sons Ltd, 3/1970.
(2) to Trostre Works, Llanelli, Carmarthenshire, 8/1971.
(3) to J.A. Lister & Sons Ltd, Consett, for scrap, 8/1972; scrapped 6/1974.
(4) to Tyne & Wear County Council, Monkwearmouth Station Museum, Sunderland, for preservation, 1/1976.
(5) to J.A.Lister & Sons Ltd, Consett, for scrap, 9/1978; scrapped 10/1978.
(6) to J.A.Lister & Sons Ltd, Consett, for scrap, 12/1972; scrapped after 22/6/1973.
(7) to J.A.Lister & Sons Ltd, Consett, for scrap, 9/1978; scrapped after 11/3/1979.
(8) sold for scrap, 5/1975.
(9) sold for scrap, 11/1974.
(10) returned to HE by 8/1968.
(11) to NCB Northumberland Area, Bates Colliery, Blyth, Northumberland, 4/1969; returned to Consett Works, 6/1969; to Ravenscraig Works, Motherwell, Lanarkshire, Scotland, 7/1969.
(12) to Jarrow Works, Jarrow, 2/1976.
(13) to Jarrow Works, Jarrow, 5/1977.
(14) to Ravenscraig Works, Motherwell, Lanarkshire, Scotland, after 15/2/1993, by 24/4/1984.
(15) to Ravenscraig Works, Motherwell, Lanarkshire, Scotland, 5/1981.
(16) to GKN (South Wales) Ltd, Tremorfa Works, Cardiff, South Glamorgan, 28/4/1981.
(17) to Hartlepool Works, Hartlepool, Cleveland, 20/5/1981.
(18) to GKN (South Wales) Ltd, Tremorfa Works, Cardiff, South Glamorgan, 6/5/1981.
(19) to Ravenscraig Works, Motherwell, Lanarkshire, Scotland, 6/1981.
(20) scrapped after 19/4/1983.
(21) to Skinningrove Works, Carlin How, Cleveland, 10/7/1981.

(22) to Jarrow Works, Jarrow, 12/1980.
(23) to Jarrow Works, Jarrow, 18/12/1979.

15. On 20th September 1980, a week after the closure of the Consett works, the loco drivers came in for one last shift to bring out all of the locomotives for a visit by the Festiniog Railway Society. Five of them, 43, S 10088/1961, 41, S 10079/1961, 32, RR 10286/1969, 38, TH 224V/1970 and 39, RR 10289/1970, are lined up for inspection.

FELL COKE OVENS, Consett
F63
NZ 098499

These coke ovens, first opened in 1924, were situated within the Consett Works. They supplied coke for the blast furnaces as well as producing a range of by-products from the coal. The 54 Woodall-Duckham Becker regenerative combination ovens were closed down in 1968-69, being replaced by 52 Gibbons Wilputte regenerative combination ovens; 17 of these had been opened in 1953, followed by 20 more in December 1955. Of these 89 ovens, 37 were shut down in 1975-76, with the remainder lasting till the closure of the works, the last coke being pushed on 13th September 1980.

The locomotives below were used on the bench side of the ovens. The coke car was positioned alongside the oven to be pushed (emptied), and when full it was propelled into the quenching tower for water to be sprayed on it. When this process was complete the car was hauled back and its doors opened for the coke to fall out on to the bench for removal. An oven was pushed every 19-27 minutes, depending on the type of coke being produced. One locomotive was in use, with one spare, though locomotives from the normal stock could be used in an emergency.

Gauge : 4ft 8½in

			4wWE	WSO	529	1924		
				Goodman	3576	1924	(a)	Scr /1977
	-		4wWE	GB	2368	1952	(a)	
		reb	0-4-0WE	Consett		1972		Scr /1981
	3		4wWE	GB	420306	1972	New	(1)

(a) ex The Consett Iron Co Ltd, with works, 1/7/1968.
(1) to Hartlepool Works, Hartlepool, Cleveland, 5/1981.

DARLINGTON FORGE WORKS, Darlington   **S64**
The Darlington Forge Ltd (subsidiary of **English Steel Corporation Ltd**)   NZ 295157
until 1/7/1968

This works, served by sidings north of the BR Bishop Auckland - Darlington line immediately north of Albert Hill Junction, had been closed in February 1967, but survived to be included in the British Steel Corporation on 1st July 1968. The locomotives were occasionally used in salvage work before disposal. One steam and three diesel rail cranes were also used here.

Gauge : 4ft 8½in

		0-4-0ST	OC	RSHD	7013	1940	(a)	Scr /1970
No.46		0-4-0DM		HC	D1159	1961	(a)	(1)
No.48		0-4-0DM		HC	D1161	1961	(a)	(1)

(a)   ex The Darlington Forge Ltd, with site, 1/7/1968.

(1)   to Llanelly Steel Co Ltd, Carmarthenshire, c6/1971, via Howard & Pepperell Ltd, Sheffield, Yorkshire (WR).

JARROW WORKS, Jarrow   **E65**
BSC Sections   NZ 324655
Teesside Division from 1/1/1977
General Steels Division, Northern Tubes Group until 29/3/1970
The Consett Iron Co Ltd until 1/7/1968

This works was served by a ½ mile branch from the BR Gateshead-South Shields line at Jarrow Station. When this route was taken over by the Tyne & Wear Passenger Transport Executive for incorporation into the new Metro system BR retained running powers to the works, though for some years before closure there was little or no BR traffic. Besides the tube works there was also a **basic refractory brick works**. This was latterly run by BSC's Refractory Division and was closed in 1985. The works was closed on 25th July 1986 and the site was cleared.

16. 0-6-0DM 17, HE 5377/1958, 0-4-0DH 61, S 10069/1961 and 0-4-0DM 26, HE 5386/1959, alongside the loading platform in the works on 24th December 1979. The two Hunslets were examples from the initial dieselisation of the Consett Works with Hunslet locomotives in the 1950s.

Until 1980 the locomotives here were the responsibility of the Consett Works, which changed them as required; after this date this responsibility passed to the Scunthorpe Works. There were also, at various times, one steam and three diesel rail cranes here.

Gauge : 4ft 8½in

19	(17 from 8/1973)	0-6-0DM	HE	5377	1958	(a)	(1)
26		0-4-0DM	HE	5386	1959	(a)	(2)
25		0-4-0DM	HE	5385	1959	(b)	(3)
21		0-6-0DM	HE	5379	1958	(c)	(4)
20		0-6-0DM	HE	5378	1958	(d)	(5)
22		0-6-0DM	HE	5380	1958	(e)	(5)
24		0-4-0DM	HE	5384	1959	(f)	(6)
61		0-4-0DH	S	10069	1959	(g)	(7)
62		0-4-0DH	S	10084	1959	(g)	Scr 1/1984
60		0-4-0DH	S	10067	1959	(h)	(7)
63		0-4-0DH	S	10066	1959	(h)	Scr 1/1984
42		0-6-0DE	YE	2766	1960	(j)	(8)
53		0-6-0DE	YE	2793	1961	(k)	(9)

(a) ex The Consett Iron Co Ltd, with works, 1/7/1968.
(b) ex Consett Works, Consett, 9/1969.
(c) ex Consett Works, Consett, 12/1969.
(d) ex Consett Works, Consett, 3/1971.
(e) ex Consett Works, Consett, 19/8/1975.
(f) ex Consett Works, Consett, 9/2/1977.
(g) ex Consett Works, Consett, 18/12/1979.
(h) ex Consett Works, Consett, 12/1980.
(j) ex Scunthorpe Works, Humberside, 6/1983.
(k) ex Scunthorpe Works, Humberside, by 3/5/1985.

(1) to Consett Works, Consett, 10/1969; to Jarrow Works, 2/1976; to J.A. Lister & Sons Ltd, Consett, for scrap, 6/1980; scrapped after 13/6/1981.
(2) to Consett Works, Consett, 12/1969; to Jarrow Works, 4/1973; to Consett Works, Consett, 2/1977; to Jarrow Works, 5/1977; to J.A. Lister & Sons Ltd, Consett, for scrap, 6/1980; scrapped after 13/6/1981.
(3) to Consett Works, Consett, 4/1973.
(4) to Consett Works, Consett, 3/1971.
(5) to Consett Works, Consett, 2/1976.
(6) to Consett Works, Consett, 5/1977.
(7) believed sold for scrap during /1987.
(8) to Scunthorpe Works, Humberside, by 11/3/1985.
(9) to Scunthorpe Works, Humberside, 10/1/1987.

BROADWOOD LIMESTONE CO
BROADWOOD QUARRY, near Frosterley U66
NZ 033366 approx.

This large limestone quarry was owned for many years by Pease & Partners Ltd (which see). In 1926 it was taken over by The Witton Park Slag Co (which see), which set up the Broadwood Limestone Company as a trading name to operate it. This company operated both as limeburners, for which kilns were constructed, and as producers of road materials, for which a tarmacadam plant was built. Initially a narrow gauge system brought the limestone from the quarry to the processing area, which lay near to the LNER Wear Valley branch and was served by a ¼ mile branch one mile east of Frosterley Station. The sidings agreement with the LNER was signed on 5th February 1926.

By the early 1950s the narrow gauge system had been replaced by dumper trucks, and subsequently the standard gauge shunting was done by a tractor. The tarmac plant closed in January 1952. Quarrying ceased in 1969, though the lime kilns continued to be used until March 1972. Part of the quarry was then leased to Swiss Aluminium Mining (UK) Ltd for the construction of a plant to process fluorspar (see the Fluorspar section in Part 2). The Broadwood Limestone Co continues (2006) to own the quarry, and sub-lets the quarrying of stone, which resumed in 2001.

Note : references to the locomotives below usually, though not always, refer to The Witton Park Slag Co Ltd, as the parent company, rather than to Broadwood Limestone Co as the trading company.

Gauge : 4ft 8½in

(No.1?)		0-6-0T	IC	JF	1539	1871	(a)	(1)
(No.2?)		4wPM		MR	2096	1922	(b)	s/s after 10/1949
-		4wPM		MR	1951	1920	(c)	s/s

(a)   ex Durham County Council, reclamation scheme at Tudhoe Ironworks, Spennymoor.
(b)   ex Golightly Tar & Stone Co, Spennymoor.
(c)   ex The Witton Park Slag Co Ltd, Witton Park Slag Works, Witton Park, after 9/1928, by /1930.

(1)   two standard gauge steam locomotives were offered for sale by The Witton Park Slag Co in 4/1927; as the firm only owned two steam locomotives at this date, the other being at Witton Park Slag Works, it is assumed that JF 1539 must have been one of those offered; s/s.

Note: the last order for spares from MR was delivered here on 11/3/1954, but whether for a standard or narrow gauge MR loco is not known.

Gauge : 2ft 0in

-	0-4-0ST	OC	KS	?	?	(a)	s/s by 15/8/1931
JIMMIE	0-4-0ST	OC	KS	4246	1922	(b)	Scr by /1939
-	4wPM		MR	3833	1926	(c)	s/s
-	4wPM		MR	4577	1930	(d)	s/s

(a)   this locomotive was almost certainly one of KS 4003/1918 or KS 4004/1918; both of these had been sold to Watts, Hardy & Co (1920) Ltd, dealers, Newcastle upon Tyne, by the Coast Road Joint Committee (which see) in 5/1927.
(b)   ex Watts, Hardy & Co (1920) Ltd, dealers, Newcastle upon Tyne, /1927; previously Coast Road Joint Committee until either 5/1927 or 6/1927.
(c)   ex The Witton Park Slag Co Ltd, Witton Park Slag Works, Witton Park.
(d)   ex The Witton Park Slag Co Ltd, Witton Park Slag Works, Witton Park, by 2/2/1940.

J. BURLINSON & CO
MILLFIELD ENGINE WORKS, Millfield, Sunderland   H67
                                                                                    NZ 384572

This firm had begun an engineering business in Bedford Street, Sunderland (in the town centre) in the mid-1820s, which thirty years later needed to expand. Then on 20th December 1852 what subsequently became the NER opened its Pensher (latterly Penshaw) Branch from Pensher to Sunderland's South Dock. Soon afterwards, and certainly by 1856, the company had moved to this new works north of and alongside the line, immediately west of Millfield Station. It was a major business, for besides general engineering, including locomotive repairs, it manufactured boilers and cranes.

In 1862 W.D. Burlinson died, and on 21st July 1862 G. Hardcastle auctioned the plant (the press advertisement gives the owners as Burlison), which included one tank locomotive, presumably standard gauge. However, the works continued in operation after this for a considerable time, first owned by Close, Burlinson & Co and then by Close & Co Ltd.

MARQUIS OF BUTE
GARESFIELD WAGGONWAY

One of the most important eighteenth century waggonways on Tyneside was the Main Way, running from Pontop to Derwenthaugh on the River Tyne; it was certainly operating by 1710. Having undergone various alterations over the years, by the early nineteenth century what was left was called the "Garesfield Waggonway". By this time the line served a number of pits near the hamlet of High Spen, together with **Winlaton Ironworks** (NZ 186604) at Winlaton Mill, the works begun by Ambrose Crowley in 1691, once famous for its swords and at one time the largest ironworks in Europe. From 1766 the chief owner of the royalty was the 4th Earl of Bute (1744-1814), raised to the marquisate in 1796.

A report in the Northumberland Record Office (NRO 3410/East/1) dated May 1819 described the line and recommended improvements, though there is no evidence that any of these were implemented. The line was worked entirely by horses, using "dandy carts" for them to ride in when sets descended to Winlaton Mill by gravity. Oral tradition claimed that wooden rails were still in place on some sections in the 1840s, which if correct would almost certainly make the line the last timber-built line in use in the North-East.

In 1837 the Marquis opened the **Bute Pit**, which subsequently became called **GARESFIELD COLLIERY** (NZ 139598) (sometimes called **Spen Colliery**) at the hamlet of High Spen, and the old waggonway was extended to serve it. Although horses continued to be used to shunt at the colliery, three stationary engines were erected to work the most of the remainder of the line. The first was the **Barlowfield** or **Spen High Engine** (NZ 153598), about ¼ mile east of the colliery. This worked rope inclines on either side of it, but had only one drum, so that the same rope had to be used for all the operations. The loaded wagons were hauled up to the engine house, where the rope was dropped off and the wagons ran forward by momentum. The rope was then attached to the rear for the 1¼ miles' descent to Low Thornley. Here the rope was attached to empties and the same process followed to work them through to the pit. The **Low Thornley** or **Spen Low Engine** (NZ 174598) worked the mile-long incline down to Winlaton Mill (NZ 185607), again a single line bank with sets worked alternately. The 1½ miles, between Winlaton Mill and Swalwell Bridge (NZ 197623), was worked by horses, four being used for each set of waggons. Over the next ½ mile the waggons were worked by the **Derwenthaugh Engine** (NZ 203629) using a main-and-tail system, with a return wheel at Swalwell Bridge. The final 220 yards was again worked by horses, bringing the full waggons to the Garesfield Staith (NZ 204634) and beehive coke ovens nearby. Originally the Garesfield line crossed what became the NER Redheugh branch on the level; later a link was put in to join the NER just west of its bridge over the River Derwent.

The three stationary engines survived until at least November 1876, as they are listed in DCOA Return 102 (with no locomotives recorded). However, the NER Sidings map of 1894 shows only the Spen High and Low Engines, and the existence of only two engines in the line's final years is confirmed by the reference below. Clearly the Derwenthaugh Engine had ceased to work; given that it would appear to have been working in 1876, it might well have been the last in Durham to operate standard gauge main-and-tail rope haulage. No locomotive is known before 1885, and there is clearly a strong assumption that its arrival coincided with the end of rope and horse haulage between Winlaton Mill and the Tyne; indeed, the 2nd edition of the O.S. maps suggests that the former Derwenthaugh Engine House building had become a **loco shed**. The oral tradition that the locomotive worked at Garesfield Colliery is disproved by the minutes of the directors of The Consett Iron Co Ltd, which describe the colliery as being shunted by a **stationary engine** (presumably different from the others above) when they took it over in 1890.

In July 1889 the colliery (and a brickworks, opened in 1875 immediately to the south of the pit), together with its railway and Garesfield Staith, with its associated beehive coke ovens, were sold for £140,000 to The Consett Iron Co Ltd, which took possession on 1st January 1890. The new owners then extended the railway to serve their new Chopwell Colliery and rebuilt the section between Garesfield and Winlaton Mill, the line then being known as the Chopwell & Garesfield Railway.

For a more detailed description of the waggonway see *The Private Railways of County Durham*, Colin E. Mountford, Industrial Railway Society, 2004.

Gauge: 4ft 8½in

GARESFIELD	0-4-0ST	OC	BH	854	1885	New	(1)
BURLEY	0-6-0ST	OC	?	?	?	(a)	(2)

(a) ex ?, contractor, Leeds, Yorkshire (WR), on hire, supposedly while BH 854 was under repair.
(1) to The Consett Iron Co Ltd, with colliery and railway, 1/7/1889.
(2) returned to hirer.

THE CARGO FLEET IRON CO LTD

This company was registered on 29th January 1883, though its origins went back to 1865. It owned the Cargo Fleet Ironworks in Middlesbrough, which by 1900 was an obsolete works on an outstanding site, including river frontage. In 1904 the firm was acquired by Sir Christopher Furness (1852-1912), using The Weardale Steel, Coal & Coke Co Ltd; for Sir Christopher's industrial empire see the entry for this latter company. It became a subsidiary of The Weardale Steel, Coal & Coke Co Ltd from 24th January 1905, and its works was completely rebuilt to manufacture angles, joists and rails. The Cargo Fleet company subsequently acquired control of The South Durham Steel & Iron Co Ltd of Hartlepool, with which it was merged in 1928, although it retained its trading title until 3rd October 1953.

WOODLAND COLLIERY, Woodland   **P68**
The Woodland Collieries Co Ltd until /1914;   NZ 066266
latterly a subsidiary of **The Cargo Fleet Iron Co Ltd**

In 1914 this company absorbed its erstwhile subsidiary company, The Woodland Collieries Co Ltd (which see), which owned collieries on the extreme south-western edge of the Durham coalfield. This company had gone into voluntary liquidation in November 1911, then to be taken over by The Cargo Fleet Iron Co Ltd, which allowed the company to continue trading until it took over direct control in 1914.

WOODLAND COLLIERY lay at the end of the five mile long and privately-owned Woodland Branch from Woodland Junction on the NER Barnard Castle - Bishop Auckland line, 1¼ miles north-east of

Cockfield Fell Station. A mile east of Woodland Colliery lay **CRAKE SCAR COLLIERY** (NZ 082276) **(P69)**. After 1911 this colliery received heavy investment, with both a new coal washing plant and new coke ovens, the latter resulting in the closure of the old coke ovens at Woodland by 1915. However, coke production had ceased by 1917, perhaps because of shortage of men caused by the First World War. The new owners also revived mining to the south of Woodland at **ARNGILL COLLIERY** (NZ 072243) **(P70)**, served by a narrow gauge main-and-tail rope system 1¼ miles long between Woodland and Arngill, worked by a stationary engine at the latter. By this date **COWLEY COLLIERY** (NZ 067255) **(P71)**, originally served by a short branch from the "main line", was merged with Arngill Colliery, but how many of the drifts west of Arngill Colliery were still being worked (see the entry for The Woodland Collieries Ltd) is unknown.

From Woodland Colliery itself the full standard gauge wagons travelled downhill by gravity, a practice begun in the 1890s (see The Woodland Collieries Co Ltd), either just to Crake Scar, where coal was needed for the coking plant, or the full five miles to Woodland Junction, with a locomotive following behind to push when required and the loco fireman acting as a brakesman to pin down or release wagon brakes. The usual number of wagons in a set was supposed to be 24, but oral tradition reported that a lot more were run on occasion, 66 being the known maximum, and sometimes very quickly! On arrival at the junction the locomotive would haul back the empties.

Arngill Colliery was closed in 1914, probably towards the end of that year. Crake Scar Colliery was closed in 1920 and Woodland Colliery followed in March 1921, together with the Woodland Branch. Woodland Colliery was dismantled in 1923. However, mining in the area continued, carried on by various small companies. By 1928 Arnghyll (the new spelling) and Cowley Collieries were owned by the Arnghyll & Cowley Colliery Co, although not being worked. When they were re-opened, coal was dispatched from the Arnghyll site by an aerial flight to sidings alongside the LNER Barnard Castle – Bishop Auckland line, 2½ miles south-west of Cockfield Station. Ten years later the owners were the West Pits Colliery Co, but both collieries were closed in 1939. On 1st January 1947, still closed, they passed to NCB Northern Division No. 4 Area.

Gauge : 4ft 8½in

	NELSON		0-6-0ST	IC	K	1786	1871		
		reb			LG		?	(a)	(1)
	ELEANOR		0-6-0ST	OC	AB	694	1891	(a)	Scr /1923
	-		0-6-0	IC	Dar		1876		
		reb	0-6-0ST	IC	Dar		1891	(a)	Scr c/1923
	GEORGE		0-6-0ST	IC	?	?	?	(b)	Scr c/1923
No.1			0-4-0ST	OC	HL	2412	1899	(c)	(2)
	"TOBY"		0-4-0ST	OC	Joicey		1870	(d)	(3)

(a) ex The Woodland Collieries Co Ltd, with collieries, /1914.
(b) ex ?, South Wales, /1916; possibly built by Peckett.
(c) ex The Weardale Steel, Coal & Coke Co Ltd, Tudhoe, c/1918.
(d) ex Cargo Fleet Works, Cargo Fleet, Yorkshire (NR), c/1920.

(1) to J.F. Wake, dealer, Darlington, by 6/1918; then to Cardiff Corporation, construction of Llwynon reservoir, Glamorgan, 9/1919.
(2) assisted in dismantling work after closure; to Irchester Iron Co Ltd, Irchester Quarries, Northamptonshire, c/1926 (this company too was a subsidiary of The Cargo Fleet Iron Co Ltd).
(3) to Cargo Fleet Works, Cargo Fleet, Yorkshire (NR), c/1923.

THE CARLTON IRON CO LTD

The North of England Industrial Iron & Coal Co Ltd until 1/5/1877; company change of name.

This company operated on the southern edge of the Durham coalfield. It established in 1854 the Carlton Iron Works at Stillington, a village which in April 1983 was part of an area transferred to the county of Cleveland. The company went into voluntary liquidation on 25th June 1914 and a new company with the same name was registered on 7th July 1914. In 1919 it was purchased by Dorman, Long & Co Ltd, which absorbed it on 2nd May 1923.

EAST HOWLE COLLIERY, near Ferryhill **N72**
East Howle Coal Co until 1874; see also below   NZ 291339

This colliery was opened in 1867 by **Willis & Co**, and had passed to the **East Howle Coal & Firebrick Co** by 1869 and the **East Howle Coal Co** by 1870. It was served by a ¼ mile branch north of the NER Ferryhill - Bishop Auckland line, ½ mile west of Coxhoe Junction. The colliery also had a **brickworks**.

The beehive coke ovens were served by a further ¼ mile extension of the line. On 23rd March 1905 a huge fire destroyed most of the buildings on the surface, and the colliery never resumed production, being replaced by Mainsforth Colliery (which see).

Two locomotives, almost certainly standard gauge, were recorded here in March 1890, but their identities are unknown.

FERRY HILL IRON WORKS, West Cornforth   N73
NZ 304335

This ironworks had been opened in 1859 and was operated by the **Rosedale & Ferry Hill Iron Co Ltd** (which see) until 1877. The works, which had eight blast furnaces, stood idle until being taken over by this company, probably in 1883. It was close to East Howle Colliery above, and was served by a link to the NER Spennymoor-Hartlepool line (formerly the Clarence Railway) at West Cornforth Station.

Blast Furnace Statistics give the owner from 1883 as John Rogerson and state that no furnaces worked under his ownership; however, local sources suggest that The Carlton Iron Co Ltd did re-commence production, though briefly, and that the works was closed in 1890. It would appear to have been demolished about 1895, by which time the slag heaps were being worked by William Barker (see William Barker & Co), while in 1899 another part of the site was used for the construction of a by-product coking plant for the adjacent Thrislington Colliery (see North Bitchburn Coal Co Ltd).

Assuming that the works did re-commence production, clearly locomotives would have been needed, but no details are known.

MAINSFORTH COLLIERY, Ferryhill Station   N74
NZ 307316

This colliery was begun in 1873 by the **Mainsforth Coal Co** (which see), but was closed down soon after coal was reached in 1877. Site work began again in 1900, and the colliery commenced production in November 1905, to replace East Howle Colliery (see above), which had suffered a serious fire. It was served by a ½ mile branch from the NER Ferryhill - Stockton line, one mile south of Ferryhill Station. It was taken over by Dorman, Long & Co Ltd (which see) on 2nd May 1923.

Which locomotives worked here before 1923 is uncertain, but it is believed that those listed below did so at some period.

Gauge: 4ft 8½in

-		0-4-0ST	OC	Hopper		?	(a)	s/s
5		0-4-0ST	OC	MW	60	1862	(b)	(1)
6	LOTTIE	0-4-0ST	OC	HL	2110	1888	(b)	
		reb				1906		(2)
-		0-4-0ST?	?	?	?	?	(c)	(3)

(a) source and date of arrival unknown.
(b) ex Carlton Iron Works, Stillington (see Cleveland & North Yorkshire Handbook).
(c) origin and identity unknown.

(1) to Carlton Iron Works, Stillington (?) (see Cleveland & North Yorkshire Handbook).
(2) to Dorman, Long & Co Ltd, with the colliery, 2/5/1923.
(3) to Bell Bros Ltd, Port Clarence Works, near Billingham, 4/1918 (see Cleveland & North Yorkshire Handbook).

G. & W.H. CARTER
WINGATE BRICK & TILE WORKS, Wingate   L75
WELLFIELD BRICK & TILE WORKS, Wingate   L75

These two works lay adjacent to each other. The Wellfield works (NZ 407369) was served by a siding from the NER Wingate Colliery branch; the Wingate works (NZ 408370) was served by sidings from the NER Ferryhill-Hartlepool line (the former Clarence Railway), immediately west of Wingate Colliery Junction. Production from this site had begun by 1879, and from the Wellfield works by the mid 1890s. The Wingate works was by far the larger, with extensive sidings and a tramway to bring clay from the pits to the works. By 1900 it is clear that the two works were regarded as one and were under the sole ownership of this firm.

Local tradition claimed there was both a "tank loco" here and a former Hartlepool Tramways locomotive, geared with two horizontal cylinders. The Hartlepool system was 3ft 6in gauge and closed down in 1888. Whether the Hartlepool locomotive was used on the tramway or to shunt the standard gauge sidings is unknown. The site, latterly known solely as the **Wingate Brick & Tile Works**, was closed between 1910 and 1914, the firm concentrating production at its Greatham Works near Billingham.

THE CASTLE EDEN COAL CO LTD
The Castle Eden Coal Co until 6/10/1882

CASTLE EDEN COLLIERY, Hesleden L76
The Haswell, Shotton & Easington Coal & Coke Co Ltd until between 1878 and 1881 NZ 437381
possibly **The Haswell & Shotton Coal Co Ltd** before the above
The Haswell Coal Co until 22/8/1865

This sinking of this colliery, situated on the extreme south-eastern edge of the Durham coalfield, began on 28th September 1840. It began shipping coal at Hartlepool via the Hartlepool Dock & Railway (which see) in April 1842. It would appear to be one of the collieries that operated its own trains over the HD&R using its own locomotives, for after the York & Newcastle Railway took over the HD&R in October 1846 and ended the operation of private trains, the "Castle Eden Coal Company" received £4072 in compensation, a sum sufficiently high to suggest that one or more locomotives must have been included. If this was the case, no details of the locomotives are known.

The colliery was served by sidings immediately west of Hesleden Station and was worked by the NER until the arrival of the locomotive below. However, for a period up to about May 1881 the company had the use of a NER loco and carriage to run workmen's trains between Hesleden Station and Shotton Bridge Station, three miles to the north. It may be that this working began after the closure of Shotton Colliery (which see) in November 1877. After 1881 the company provided the locomotive, but with the NER still providing the coach, five trips each way being run per day. It is not known how long this arrangement lasted.

From the early 1890s the colliery suffered from serious water problems, coal working having to be completely stopped to try to reduce the flooding. It was closed in January 1894, the equipment being dismantled and auctioned in April 1894. In 1900 it was acquired by The Horden Collieries Ltd (which see).

Gauge: 4ft 8½in

| | | 0-4-0ST | OC | BH | 613 | 1881 | (a) | (1) |

There may have been at least one further locomotive; the company advertised for one in the *Colliery Guardian* of 7th March 1884.

(a)   ex West Hartlepool Steel & Iron Co Ltd, West Hartlepool (see Cleveland & North Yorkshire Handbook).

(1)   to be auctioned by A.T. & E.A. Crow, 17-20/4/1894; to The Hutton Henry Coal Co Ltd, Hutton Henry Colliery, Station Town.

CATERPILLAR (U.K) LTD (registered 27/7/1950)
BIRTLEY WORKS, Birtley G77
The Birtley Co Ltd until 8/1956 NZ 272552

This works was reconstructed from the premises of the former Birtley Iron Co's Birtley Ironworks (see The Pelaw Main Collieries Ltd), and for many years it was served by sidings from the Pelaw Main Railway at "Birtley Tail", the Railway working the Birtley Company's traffic to and from Birtley Station on the LNER Newcastle - Durham line. The gas works owned by the company had closed by 1939 and iron foundry work may also have ceased by then.

With the closure, albeit temporarily, of this section of the Pelaw Main Railway in 1940 because of the Second World War, the company was obliged to purchase its own locomotive to shunt its traffic. Having undertaken fabrication work during the War, the company signed an agreement in 1947 with the Caterpillar Tractor Co of the USA to manufacture scrapers and bulldozers. Caterpillar purchased the premises in 1956.

After the change of ownership the site was completely re-developed, with new buildings and a new rail link ¼ mile south of Birtley Station. Rail traffic ceased in July 1966 and the works was closed in June 1984, though it was subsequently reopened by Komatsu UK Ltd. In 2004 this firm refurbished the siding to test road/rail vehicles manufactured by the firm.

Gauge: 4ft 8½in

| | MC 156 | 0-4-0DM | JF | 22900 | 1941 | New | (1) |

(1) to J. Parker, Slaggyford Station, Northumberland, for preservation, 8/1974.

CENTRAL ELECTRICITY GENERATING BOARD
Central Electricity Authority until 1/1/1958
British Electricity Authority until 1/4/1955

When the electricity industry was nationalised on 1st April 1948 the British Electricity Authority was created to be responsible for the generating and distribution of electricity. By the time that the industry was privatised again in the early 1990s there were no power, or generating, stations left in Co. Durham.

DARLINGTON POWER STATION S78
Darlington Corporation Electricity Department until 1/4/1948 NZ 294147

This station was opened in December 1900, and was served by sidings west of the NER Durham-Darlington line, ½ mile north of Darlington (Bank Top) Station. It was shunted by the NER/LNER before the arrival of the first locomotive listed below. Rail traffic was replaced by road transport about 1964 and the station was subsequently closed and demolished in 1978.

Gauge : 4ft 8½in

	JOHN HINKS	0-4-0ST	OC	RSHN	7066	1942	New	(1)
	-	0-6-0ST	IC	HC	1674	1937	(a)	(2)
(No.11)		0-4-0ST	OC	HL	3641	1926	(b)	Scr /1969

(a) ex Sir Robert McAlpine & Sons Ltd, Dunston depot, No.82, 9/1950 (by 25/9/1950); intended for Dunston Power Station, but diverted here when RSHN 7066/1942 needed repairs.
(b) ex Carville Power Station, Newcastle upon Tyne, 11/1960.
(1) to RSHD, for repairs, by 25/9/1950, and returned; s/s c/1962.
(2) to Carville Power Station, Newcastle upon Tyne, 9/1952, possibly via Dunston Power Station, Dunston.

Shortly after the North-Eastern Electric Supply Co Ltd was formed in 1932, it introduced a numbering scheme for its locomotives, based on their age, at its North Tees (Haverton Hill), Dunston and Carville (Newcastle upon Tyne) Power Stations. This scheme was continued for some time after nationalisation, and was extended to the new Stations, Stella South in Co. Durham and Stella North and Blyth in Northumberland, before being abandoned. It would appear that the Dunston locomotives carried the same number throughout their existence, but what numbers, if any, were carried previously by other locomotives is unknown.

DUNSTON POWER STATION, Dunston B79
North-Eastern Electric Supply Co Ltd until 1/4/1948 NZ 238626
The Newcastle upon Tyne Electric Supply Co Ltd until 30/9/1932 (founded in 1889)

The original station, later known as Dunston A Station, was opened in July 1910 and was served by sidings from Dunston West Junction on the NER Redheugh branch. An interesting point is that one of its boiler units was driven by gas from Norwood Coke Works nearby, owned latterly by The Priestman Collieries Ltd (which see). In January 1933 a new station, known as Dunston B Station, was opened to the west, though it was not brought up to its full 320 megawatt capacity until after the Second World War. This reduced the original station to standby status; after the war its site was cleared and redeveloped for ship-breaking yards. The overhead electric system was disused by October 1956. Rail traffic ceased about 1974, when the station was reduced to standby status. It was closed in 1981, though not demolished until some years afterwards.

Gauge : 4ft 8½in

No.3		4wWE		W/DK*		1908	New	s/s c/1964
No.4		4wWE		W/DK*		1910	New	(1)
	STRETFORD	0-6-0ST	IC	HC	299	1888	(a)	(2)
No.11		0-4-0ST	OC	HL	3641	1926	New	(3)
No.8		0-4-0ST	OC	HL	3772	1930	New	Scr 8/1968
-		0-4-0DE		AW	D10	1932	(b)	(4)
No.14		0-4-0DE		AW	D21	1933	(c)	(5)
No.5		4wBE		W/DK?*		1920	(d)	Scr /1953

Durham Part 1 Page 75

No.13	0-4-0ST	OC	HL	3732	1928	(e)	(6)
No.15	0-4-0ST	OC	RSHN	7063	1942	New	(7)
No.69	0-6-0ST	IC	HC	1609	1934	(f)	(8)
No.17	0-4-0ST	OC	RSHN	7679	1951	New	Scr 8/1968
-	4wVBT	VCG	S	9558	1953	(g)	(9)
No.25	4wVBT	VCG	S	9597	1955	(h)	(10)
No.12	0-4-0ST	OC	HL	3651	1926	(j)	(11)
No.30	0-4-0DE		RH	412707	1957	(k)	(12)
No.55	0-4-0DE		RH	381751	1955	(m)	(13)

* it is believed that these locomotives were ordered from The British Westinghouse Electric & Manufacturing Co Ltd of Manchester, who sub-contracted the actual construction to Dick, Kerr & Co Ltd, with Westinghouse supplying the motors. No.3 carried only a Westinghouse makers' plate, and it is possible that the others were similar.

(a) ex Richard Siddall & Co Ltd, Stackshead Quarry, Bacup, Lancashire, /1916; this loco appears not to have been included in the company's numbering scheme.
(b) ex AW, Newcastle upon Tyne, for demonstration.
(c) on demonstration from AW, before purchase in 2/1934.
(d) ex North Tees Power Station, Haverton Hill, (see Cleveland & North Yorkshire Handbook)); here by 23/9/1943.
(e) ex Carville Power Station, Newcastle upon Tyne, c/1936.
(f) ex Sir Robert McAlpine & Sons Ltd, Dunston depot, hire, /1947 (after 1/8/1947).
(g) S demonstration loco; ex S for trials by 19/10/1953.
(h) ex Stella North Power Station, Newburn, Northumberland.
(j) ex North Tees Power Station, Haverton Hill, c/1962 (see Cleveland & North Yorkshire Handbook).
(k) ex Port of Tyne Authority, Albert Edward Dock, North Shields, Northumberland, 11/12/1971.
(m) ex Port of Tyne Authority, Tyne Dock, Tyne Dock, 12/1971.

(1) to Clayton & Davie, Dunston, for scrap, /1952.
(2) the minutes of the company's Power Station Committee for 20/5/1926 record that while Dunston Power Station was being run with voluntary labour during the General Strike, 'the steam engine collided with loaded coal trucks and is so severely damaged that it is not worth repairing. The locomotive when we first purchased it was then 26 years old...' This would seem very likely to refer to this loco. To replace it the Committee ordered HL 3641, with the remains of HC 299 being sent for scrap.
(3) to Carville Power Station, Newcastle upon Tyne, c/1936.
(4) returned to AW, Newcastle upon Tyne, after demonstration.
(5) to Hexham Rolling Stock Group, for preservation, 8/1974, but stored on site; to National Railway Museum, York, 23/10/1977.
(6) to J. Parker, Slaggyford Station, Northumberland, for preservation, 12/1973.
(7) to North Yorkshire Moors Railway Preservation Society Ltd, Goathland, Yorkshire (NR), for preservation, 1/6/1972.
(8) returned to Sir Robert McAlpine & Sons Ltd, Hayes Depot, Hayes, Middlesex, 5/9/1950.
(9) returned to S after demonstration; then sent from S to The Seaham Harbour Dock Co, Seaham Harbour, for trials, c10/1953.
(10) to Stella North Power Station, Newburn, Northumberland.
(11) to North Tees Power Station, Haverton Hill, c6/1963 (see Cleveland & North Yorkshire Handbook).
(12) to Stella South Power Station, Blaydon, 11/1981.
(13) to T.J. Thomson & Son Ltd, Dunston, 11/1981.

STELLA SOUTH POWER STATION, Blaydon

B80
NZ 174644

This station was opened in 1953, and was served by sidings north of the BR Newcastle-Carlisle line, ½ mile east of Blaydon Station. Rail traffic was suspended at the beginning of the miners' strike in April 1984, the station by then being only operated as required. It was officially closed about April 1986 and demolished in 1995.

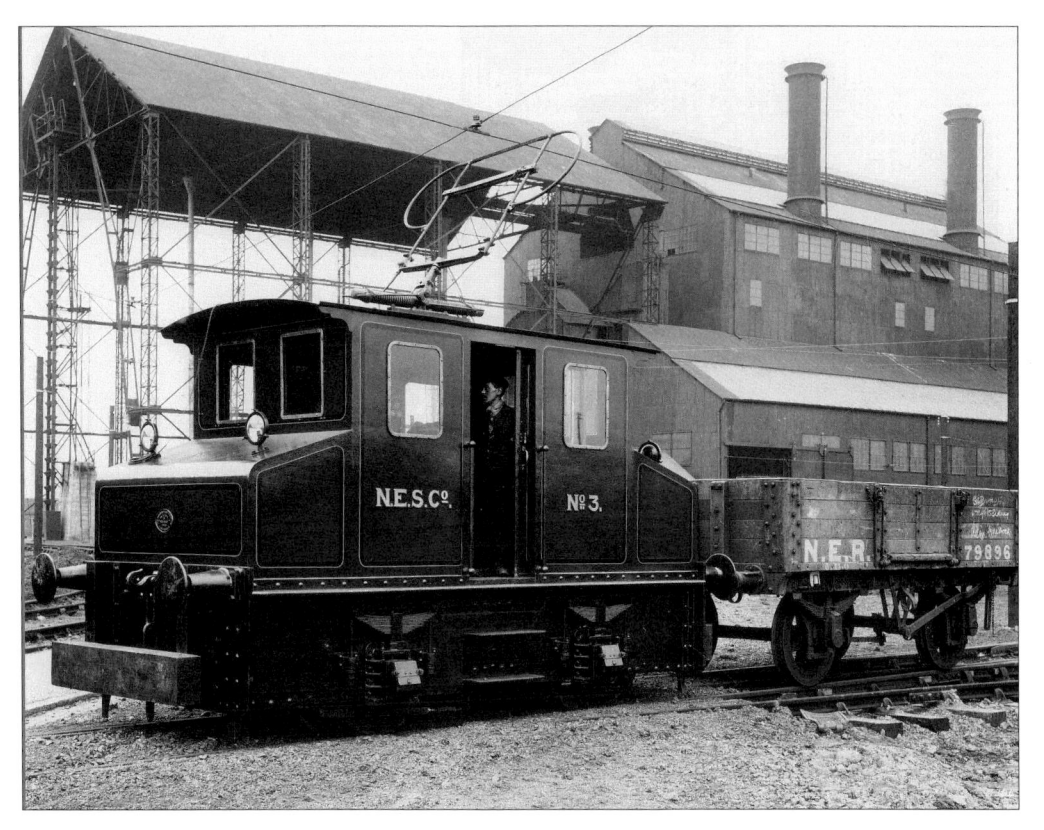

17. No.3, pictured here probably not long after it was delivered new to Dunston Power Station in 1908, owned then by The Newcastle upon Tyne Electric Supply Co Ltd.

18. No.8, HL 3772/1930, a standard Hawthorn Leslie design with 14in x 22in cylinders and 3ft 6in wheels, while behind stands No.17, RSHN 7679/1951, the same design 21 years later, taken at Dunston on 18th July 1968.

Gauge : 4ft 8½in

"No.18"	0-4-0ST	OC	HL	3090	1914	(a)	(1)
19	0-6-0ST	IC	HC	1674	1937	(b)	(2)
No.20	0-4-0ST	OC	RSHN	7743	1953	New	Scr 6/1971
21	0-4-0ST	OC	RSHN	7796	1954	New	(3)
No.22	0-4-0ST	OC	RSHN	7744	1953	(c)	(4)
No.12	0-4-0ST	OC	HL	3651	1926	(d)	(5)
(No.29)	0-4-0DM		JF	4240013	1962	(e)	Scr 4/1982
T.I.C. No.31	0-4-0DE		RH	412714	1957	(f)	(6)
(1) (formerly No.1)	0-6-0DH		JF	4240020	1964	(g)	(6)
(THORPE MARSH) No.2	0-6-0DH		JF	4240015	1962	(h)	(7)
No.30	0-4-0DE		RH	412707	1957	(j)	(6)

(a) ex G.E. Simm (Machinery) Ltd, Sheffield, Yorkshire (WR), dealer, 12/1952; previously Thos Hall & Sons (Llansamlet) Ltd, Llansamlet, Glamorgan; arrived in 2/1953.
(b) ex Carville Power Station, Newcastle upon Tyne, 3/1953.
(c) ex Carville Power Station, Newcastle upon Tyne, 3/1956.
(d) ex North Tees Power Station, Haverton Hill, by 5/1966 (see Cleveland & North Yorkshire Handbook).
(e) ex Blyth Power Station, Blyth, Northumberland, 5/1968.
(f) ex Port of Tyne Authority, Albert Edward Dock, North Shields, Northumberland, 2/1970.
(g) ex Drax Power Station, near Selby, Yorkshire (NR), 8/1972.
(h) ex Thorpe Marsh Power Station, Doncaster, South Yorkshire, 7/1976.
(j) ex Dunston Power Station, Dunston, 11/1981.

(1) to Blyth Power Station, Blyth, Northumberland, 11/1955.
(2) to Blyth Power Station, Blyth, Northumberland, /1957.
(3) to North of England Open-Air Museum, Marley Hill Loco Shed store, Marley Hill, 5/1973 (property of Stephenson & Hawthorn Locomotive Trust, for preservation).
(4) to Hams Hall Power Station, Warwickshire, for repairs, c9/1959; returned by 4/1960; to Thos W. Ward Ltd, Sheffield, Yorkshire (WR), for scrap, 12/1968.
(5) returned to North Tees Power Station, Haverton Hill (see Cleveland & North Yorkshire Handbook).
(6) to C.F. Booth (Rotherham) Ltd, Rotherham, South Yorkshire, for scrap, 5/1987.
(7) to Hartlepool Nuclear Power Station, Hartlepool, Cleveland, 23/2/1987.

19. 0-6-0DH JF 42400015/1962 at Stella South Power Station on 27th June 1979, one of three JF locomotives that were used here.

SUNDERLAND POWER STATION, Holsgrove Street, Sunderland   **H81**
Sunderland Corporation Electricity Department until 1/4/1948   NZ 389574

This station was opened in March 1895 and was served by a ½ mile branch from the Hetton Railway (see The Lambton, Hetton & Joicey Collieries Ltd and NCB) from a junction south of the Railway's Farringdon Row Tunnel. Just north of the junction were sidings where traffic was exchanged, the station's locomotive working between here and the power station itself. There was no main line rail connection. When the station's own locomotive was unavailable, traffic was worked by a locomotive from the Railway, latterly from its Lambton Staiths loco shed. Rail traffic was replaced by road transport about 1959, following the end of coal working over the Railway from the Hetton collieries. The station was closed in November 1976 and the whole area redeveloped.

Gauge : 4ft 8½in

-	4wWE	BTH *	1780	1901	New	Scr c/1953
-	4wWE	EE	1214	1943		
		Bg	3054	1943	New	Scr /1963

    *   the mechanical equipment was supplied by the General Electric Company of America.

Gauge : 2ft 0in

-	4wPM	FH	?	?	(a)	(1)
-	4wPM	L	20882	1943	New	(2)

(a)   ex Thos W. Ward Ltd, Sheffield, Yorkshire (WR), hire, 8/1942; Buckton Quarry Co Ltd, Stalybridge, Cheshire, until 8/1941.

(1)   returned to Thos W. Ward Ltd, Sheffield, Yorkshire (WR), by 3/1943.
(2)   derelict in the Power House by 6/1960; scrapped by 8/1965.

THE CHARLAW & SACRISTON COLLIERIES CO LTD (registered 12/8/1890)

This firm, an amalgamation of some smaller concerns, owned the
SACRISTON RAILWAY   G82
In its final form this ran from **WITTON COLLIERY** (NZ 233477) to a junction with the LNER Pontop & South Shields branch at Stella Gill, a distance of 4¼ miles.

The system began with the Waldridge Waggonway, opened between **WALDRIDGE 'A' COLLIERY** (NZ 233474) and the Ouston Waggonway in 1831, but shortened to join the newly-opened Stanhope & Tyne Railway at Stella Gill in November 1834. A ½ mile **self-acting incline** took waggons down to the Cong Burn, from where horses were used. Meanwhile, 3¾ miles to the south, the sinking of **CHARLAW COLLIERY** (NZ 233477) and **SACRISTON COLLIERY** (NZ 237479) was begun in 1833. The two collieries were linked to the Waldridge line with traffic beginning on 29th August 1839, although with the two sections remaining under separate ownership. From Charlaw to Sacriston was horse-worked, but from here the **Sacriston Engine** (NZ 238488) hauled waggons up for ½ mile to the top of Daisy Hill, from where a **self-acting incline** took fulls down to a junction at the Cong Burn. By 1857 the line north of the Cong Burn had been converted to rope haulage under the control of the **Waldridge Engine** (NZ 252505), although horses continued to be used to shunt waggons at the junction here.

In the middle of the nineteenth century other collieries were sunk and joined to the Railway. The first, in 1845, was **NETTLESWORTH COLLIERY** (NZ 244491), which was served by a branch from the Sacriston Engine, which subsequently also served a branch westwards to **WEST EDMONDSLEY**, later **EDMONDSLEY COLLIERY** (NZ 234491). Alongside the self-acting incline down from the Engine was **BYRON COLLIERY** (NZ 243498), which was worked by using a unique auxiliary cylinder alongside the Engine house. Last came **WITTON COLLIERY**, about 180 yards south of Charlaw Colliery, where sinking began in 1859. A **brickworks** was opened at Sacriston in the following year. Charlaw closed in 1884, Byron Colliery followed soon afterwards and Nettlesworth closed in 1894.

The ownership of these collieries was extremely complicated, but in 1890 the various small firms around Sacriston combined to form The Charlaw & Sacriston Collieries Co Ltd. The new firm built a beehive coke works at Sacriston in 1891, which lasted until 1939, and also introduced locomotives on its section. The **locomotive shed** (NZ 234476) was situated about fifty yards north of Witton Colliery. The owners of Waldridge Colliery also amalgamated into a larger company called The Priestman Collieries Ltd (which see) in 1899, having two years earlier also replaced the horses and the Waldridge Engine by locomotives, the site of the engine house being used for the locomotive shed for this section.

Both companies opened drift mines in their royalties, but with the closure of Waldridge D Colliery in 1925 and Waldridge Shield Row Drift the following year, Priestman Collieries abandoned working in this area and about November or December 1926 sold their section of the Railway to Charlaw & Sacriston Collieries. What was now East Edmondsley Colliery was abandoned in March 1929, whilst in 1942 the drifts forming Sacriston Shield Row Drift, begun in 1897, were merged into Witton Colliery.

For the locomotives used up to 1926 at Waldridge Colliery and Waldridge Bank Foot loco shed, see the entry for The Priestman Collieries Ltd. The Railway, together with Sacriston and Witton Collieries and Sacriston Brickworks, was vested in NCB Northern Division No.5 Area on 1st January 1947.

Reference : *The Private Railways of County Durham*, Colin E. Mountford, Industrial Railway Society, 2004.

SACRISTON COLLIERY

Gauge : 4ft 8½in

	-	0-4-0ST	OC	MW	455	1874	(a)	(1)
	CHARLAW	0-4-0ST	OC	BH	1037	1891	New	(2)
1	SACRISTON	0-4-0ST	OC	CF	1210	1901	New	(3)
2	CHARLAW	0-4-0ST	OC	P	1180	1912	New	(4)

(a) ex P. & H. Hodgson, contractors, Workington, Cumberland.

(1) to Benton & Woodwiss, contractors.
(2) to W. Whitwell & Co Ltd, Thornaby, Yorkshire (NR), /1912.
(3) loaned to Ministry of Works (location unknown) during 1914-1918 War, and returned; to NCB No. 5 Area, with colliery, 1/1/1947.
(4) to NCB No. 5 Area, with colliery, 1/1/1947.

BANK FOOT LOCO SHED, Waldridge
The Priestman Collieries Ltd until November or December 1926

Gauge : 4ft 8½in

MARGARET	0-6-0ST	IC	AB	1005	1904		
		reb	AB	8833	1924	(a)	(1)
CECIL	0-6-0T	IC	HC	1524	1924	(a)	(1)

(a) ex The Priestman Collieries Ltd, c11-12/1926.

(1) to NCB No. 5 Area, with railway, 1/1/1947.

HENRY CHAYTOR

This gentleman, who lived at Witton Castle, Witton Park, appears to have entered the coal trade in 1874, when he purchased both Ushaw Moor Colliery, near Durham, and Witton Park Colliery at Witton Park. He also acquired Carterthorne Colliery, near Evenwood. The latter two both closed in 1875, although Carterthorne was re-opened by 1888 under new owners.

One source acquired information in 1949 that a pit near Witton Park owned by Chaytor had once been linked to the NER east of Etherley Station, on the line between Bishop Auckland and Crook (Map Q), by a zig-zagged branch line, and that there had been a locomotive there, which had been broken up perhaps between 1900 and 1914. However, nothing has been found to substantiate this information. The main mining in this area was controlled by Henry Stobart & Co Ltd (which see).

Ushaw Moor Colliery continued to be operated by Chaytor until 1893, when it was acquired by Pease & Partners Ltd (which see).

THE CHEMICAL & INSULATING CO LTD (registered in 1927)
Faverdale, Darlington
S83
NZ 272165

This works was opened in 1928, and was served by sidings north of the LNER Darlington-Barnard Castle line, 1½ miles west of Darlington (North Road) Station. With the closure of this line on 18th June 1962, the works was sited at the end of the surviving section. It was unusual in employing locomotives on three different gauges. Standard gauge rail traffic ceased about the end of January 1980. The works was closed on 28th August 1994 and was demolished in 1996.

20. The Chemical & Insulating Co Ltd unusually used locomotives on three gauges. Its standard gauge 4wWE, built by GEC for the new factory in 1928 and seen here dealing with incoming coal wagons on 21st May 1975, handled rail traffic until it ended in 1980.

21. Working entirely inside the autoclave shop, 2ft 0in gauge 4wBE GB 2848/1957 pushes trolleys of blocks into the ovens on 8th May 1970.

22. The 1ft 8in gauge system carried waste from the factory on to what became a huge heap. The first two 'locomotives' were built by the firm itself from Morris cars, this one photographed in 1935.

23. Later the company had a fleet of four-wheeled RH diesel locomotives. Here No.1, RH 375360/1955, hauls three tubs of waste up on the heap on 12th May 1976.

Gauge : 4ft 8½in

| | - | 4wWE | GEC | | 1928 | New | (1) |

A diesel locomotive was hired from BR Darlington for five days in October 1970.

(1) to Darlington Railway Preservation Society, Hopetown Goods Depot, for preservation, 3/1982.

Gauge : 2ft 0in

This very short system was used inside a building for the charging and discharging of the autoclaves of the indurating plant.

| | - | 4wBE | GB | | 2848 | 1957 | New | (1) |

(1) to R. Stewart, c/o Skinningrove Motors, Skinningrove, Cleveland, for preservation, by 20/9/1994.

Gauge : 1ft 8in

This system was used for handling materials between sections of the processing plant and also for the disposing of waste material. By 1986 the section that was still in use was quite short and locomotive haulage was replaced by a tractor and chain.

	-	4wPM	Chem & Ins	c1932	(a)	Scr c7/1967
	-	4wPM	Chem & Ins	c1935	(b)	Scr c7/1967
No.1		4wDM	RH	375360	1955	New	(1)
(No.2)		4wDM	RH	402428	1956	New	(2)
No.3		4wDM	RH	476124	1962	New	(1)
No.4	MOSELEY	4wDM	RH	354013	1953	(c)	(1)

(a) built by company from Morris Cowley car.
(b) built by company from Morris Oxford car.
(c) ex Cheadle Moseley Grammar School, Stockport, Greater Manchester, 2/1977.
(1) to Darlington Railway Preservation Society, Hopetown Goods Depot, Darlington, for preservation, 8/1988.
(2) dismantled for spares for other locos, 6/1977; remains scrapped about 1982.

CLARKE CHAPMAN LTD

VICTORIA WORKS, Gateshead   **C84**
Clarke Chapman-John Thompson Ltd until 11/6/1974   NZ 263630
Clarke, Chapman & Co Ltd until 27/7/1970
Clarke, Chapman & Co until 14/6/1893; see also below

This engineering business was begun in 1862 by **Clarke, Watson & Gurney** in a works on South Shore Road in Gateshead. In 1874 the firm was joined by Abel Chapman to become **Clarke, Chapman & Co** and moved into the new Victoria Works. This was served by sidings north of the NER Gateshead-South Shields line, 1¼ miles east of Gateshead East Station. Chapman, with another Clarke, Chapman director, John Furneaux, acquired the locomotive business of Black, Hawthorn & Co Ltd in 1896, running it as Chapman & Furneaux.

The works became well-known for its winches and similar equipment. There was no rail traffic between 1969 and 1973 and it finally ceased about 1976. Besides the locomotives there were also three steam cranes and one diesel crane here at various times. The works continues in operation, under the same title.

Gauge : 4ft 8½in

	-	0-4-0ST	OC	HL	2249	1892	New	Scr 10/1955
	(FOWNES)	0-4-0CT	OC	HL	2499	1901	(a)	Scr 5/1962
	-	4wDM		RH	221642	1944	(b)	(1)
No.20		0-4-0DM		RH	327969	1954	(c)	(2)

(a) ex The Fownes Forge & Engineering Co Ltd, Tyne Dock.
(b) ex Sevenoaks Brick Works Ltd, Sevenoaks, Kent, c/1953; another source gives ex Standard Brick & Sand Co Ltd, Holmethorpe Sand Pits, Redhill, Surrey, c/1953.
(c) ex Port of Sunderland Authority, via Northern Supply Co, Sunderland, 6/1973.
(1) to Leslie Sanderson Ltd, dealer, Birtley, 6/1969.
(2) to D. Sep. Bowran Ltd, Gateshead, for scrap, 13/10/1976.

THE CLEVELAND BRIDGE & ENGINEERING CO LTD (registered 19/3/1877)

The share capital of the company was acquired by Cementation Co Ltd in 1967. This latter company was acquired by Trafalgar House Investments Ltd in 1969. Through both of these changes the original company name was left unaltered.

SMITHFIELD ROAD WORKS, Darlington　　　　　　　　　　　　　　　　　　　　S85
NZ 295136

This works was opened in 1877 by some former employees from Skerne Ironworks (see Skerne Ironworks Co Ltd). It was served by sidings immediately south of Bank Top Station, Darlington, on the NER line from Darlington to York. The original firm went into liquidation in 1883, and a new company with the same name took over. It became one of the world's famous bridge-building companies. A new works having been built at Yarm Road in Darlington, this old works was closed on 24th December 1981; for the Yarm Road Works see the next entry.

There were 31 steam cranes and three diesel rail cranes here, most of which were frequently away working on contracts.

Gauge : 4ft 8½in

	ADAM	0-4-0ST	OC	P	1430	1916	(a)	(1)
	-	0-4-0ST	OC	BLW	45285	1917	(b)	(2)
	NAIROBI	0-6-0ST	OC	WB	2169	1922	(c)	(3)
	-	4wDH		TH/S	111C	1961	(d)	(4)
185	DAVID PAYNE	0-4-0DM		JF	4110006	1950	(e)	(5)

There may have been at least one other locomotive before the first one listed here.

(a) ex Ministry of Munitions, Chilwell, Nottinghamshire, via Thos W. Ward Ltd, by 12/1919.
(b) ex Willys-Overland-Crossley Ltd, Levenshulme, Lancashire, via Thos W. Ward Ltd, Grays, Essex, /1927.
(c) ex contract by company for the Great Western Railway at the goods yard at High Wycombe Station, High Wycombe, Buckinghamshire, c /1939; previously Pauling & Co, Swansea, Glamorgan.
(d) new to firm, but rebuild by TH of unidentified 4wVBT Sentinel locomotive.
(e) ex Britannia Bridge contract, Anglesey, after 8/1975, by 3/1976, but never used here.
(1) to Cambrian Railways Society Ltd, Oswestry, Shropshire, for preservation, 3/12/1977.
(2) to Thos W. Ward Ltd, Brightside, Sheffield, Yorkshire (WR), /1930; re-sold to Preston Corporation, Preston Gas Works, Preston, Lancashire.
(3) sold for scrap, c4/1961.
(4) to Yarm Road Works, Darlington, 10/5/1982.
(5) to Darlington Railway Preservation Society, for preservation, /1982; then moved to Skipper Trucks, Yarm Road, Darlington; to DRPS, Hopetown Depot, c/1983.

The company acquired numerous locomotives for work on contract sites; those that are known to have been here are listed in the Contractors' Section (Part 2).

CLEVELAND BRIDGE UK LTD
Kvaerner Cleveland Bridge Ltd until 31/5/2000
Cleveland Structural Engineering Ltd until 1/11/1996
The Cleveland Bridge & Engineering Co Ltd until 1/11/1990; see also below

In 1969 the whole share issue of The Cleveland Bridge & Engineering Co Ltd (see the previous entry) had been acquired by Trafalgar House Investments Ltd, although the firm's trading name was left unaltered. In 1982 what was now Trafalgar House plc acquired Redpath Dorman Long Ltd, and on 1st November 1990 Trafalgar House amalgamated Redpath Dorman Long Ltd with The Cleveland Bridge & Engineering Co Ltd to form Cleveland Structural Engineering Ltd. In 1996 Trafalgar House plc was acquired by Kvaerner plc, a Norwegian-based company, which then changed this subsidiary company's name to the above. At the end of May 2000 Kvaerner plc sold the company to the recently-formed Cleveland Group plc, which then set up Cleveland Bridge UK Ltd to operate the Darlington works.

YARM ROAD WORKS, Darlington　　　　　　　　　　　　　　　　　　　　　　　S86
NZ 320137 approx.

From 1877 the firm had occupied a site east of the East Coast main line south of Darlington (Bank Top)

Station, but in 1981 a large new works was developed on a green field site at Yarm Road, to the south east of the town. The old works was closed on 24th December 1981 and the new works opened on 3rd January 1982. This was served by a ¾ mile branch from Network Rail's Darlington-Middlesbrough line, 1½ miles east of Croft Junction on the East Coast main line. Rail traffic ceased about 1993, but the works continues in production.

Gauge : 4ft 8½in

| - | 4wDH | TH/S | 111C | 1961 | (a) | (1) |

(a) ex Smithfield Road Works, Darlington, 10/5/1982; rebuilt by TH of unidentified 4wVBT Sentinel loco.

(1) to Foxfield Steam Railway, Blythe Bridge, Staffordshire, for preservation, 10/11/2001.

COAST ROAD JOINT COMMITTEE
(responsible to South Shields Borough Council and South Shields Rural District Council)

South Shields – Marsden – Whitburn Road   D87

This Committee was formally constituted on 13th May 1924 to superintend the construction of a new coast road from South Shields to Marsden (2½ miles), undertaken with Government support to reduce local unemployment. Work began on 23rd June 1924. The construction of the road involved the realignment of the South Shields, Marsden & Whitburn Colliery Railway (see The Harton Coal Co Ltd) for 1050 yards near Marsden. The work was closed down on 11th September 1926 through lack of materials, but was then unable to re-start due to financial problems, not least the increasing cost of the Lighthouse bridge carrying the railway over the road south of Marsden. Eventually the Ministry of Transport agreed to fund the outstanding cost and an extension to Whitburn village (½ mile). Work recommenced in November 1928 and the road was officially opened on 2nd November 1929. It is now part of the A183.

References : Tyne & Wear Archives Service, Newcastle upon Tyne; Minutes of South Shields Road special committee, 1923-1932, Ref No. T179/444-445. See also : *The Industrial Locomotive*, Vol.4, Nos. 38-40, 1985/1986 (NOTE : some of the details in this article are superseded by the reference above).

Gauge : 2ft 0in

-	4wPM		MR	429	1917	(a)	(1)
-	?		?	?	?	(b)	(2)
-	4wPM		HU		1924	(c)	(1)
-	0-4-0ST	OC	KS	4003	1918	(d)	(3)
-	0-4-0ST	OC	KS	4004	1918	(e)	(3)
-	0-4-0ST	OC	KS	4246	1922	(f)	(3)

(a) ex N.E. Potts, East Boldon, 6/1924; this loco was "free on rail at Ravensworth Colliery", which has yet to be explained; New to Ministry of Munitions, France.

(b) ex Cleadon Quarry, Cleadon, South Shields, c8/1924. A committee report dated 14/10/1924 states "the locomotive used at Cleadon Quarry had been used temporarily to get the material away". Cleadon Quarry had been owned from 1918 by South Shields Borough Council, but it was disused and it is not known to have been worked or to have used a locomotive during the council's ownership. The locomotive may be the locomotive used on a new road being constructed at Boldon Colliery village by South Shields Rural District Council (which see) in 1921.

(c) New from James Teasdale & Co, motor agents and engineers, South Shields, 12/1924.

(d) ex Watts, Hardy & Co (1920) Ltd, Newcastle upon Tyne (acting as dealers), 2/1925; it was lying at Sheffield, Yorkshire (WR), when purchased (Committee minutes); originally Air Ministry, Henlow, Bedfordshire; curiously, this loco was advertised for sale in *Engineering Trader* in both 4/1925 and 5/1925, without an owner being given.

(e) ex Watts, Hardy & Co (1920) Ltd, Newcastle upon Tyne (acting as dealers), 5/1925; originally Air Ministry, Henlow, Bedfordshire; this loco was advertised for sale in *Engineering Trader* in both 4/1925 and 5/1925, without an owner being given.

(f) ex Watts, Hardy & Co (1920) Ltd, Newcastle upon Tyne (acting as dealers), 10/1925; orig. R.H. Neal, contractor, on a contract at Barkingside, Ilford, Essex.

(1) one of these, probably MR 429/1917, was sold to Watts, Hardy & Co (1920) Ltd, Newcastle upon Tyne, in exchange for KS 4004/1918, 5/1925; no disposal is recorded for the other locomotive.

(2) returned to Cleadon Quarry, Cleadon, South Shields, c9/1924 (see note (b) above).

(3) one of these was sold to Watts, Hardy & Co (1920) Ltd in 5/1927 and a second one seems to have been sold to the same firm either later the same month or in 6/1927. One of these was KS 4246/1922, which later in 1927 had arrived at The Witton Park Slag Co, Broadwood Quarry, Frosterley (which see). The final loco was sold to White & Co, Sunderland (believed to be dealers), in 11/1929.

COATS PATONS LTD
J & P Coats, Patons & Baldwins Ltd until 26/7/1967
Patons & Baldwins Ltd (registered 16/4/1920) until 30/12/1960
DARLINGTON WORKS, Darlington   **S88**
                                                                    NZ 305153

The construction of this very large works, the last example of Darlington's once famous woollen industry and covering 140 acres, was begun in February 1946 on a green field site, south of what had been the original Stockton & Darlington Railway, two miles east of Darlington (North Road) Station, but latterly at the end of a freight branch, 1¾ miles east of Haughton Lane Junction, after the closure of the remainder. The works was opened in December 1947.

By the 1960s parts of the premises were being sub-let to other firms as production was scaled down. Rail traffic ceased in 1973, by which time production was limited only to knitting yarn. This was transferred to Alloa in Scotland in 1980, leaving latterly only warehousing and distribution facilities.

24. The last large woollen mill in Darlington was built by Patons & Baldwins Ltd in 1947. To handle its shunting it purchased a fireless locomotive. The receiver was charged with pressurised steam from the factory about every eight hours. WB 2898/1948, painted light blue, stands outside its shed on 26th June 1967.

Gauge : 4ft 8½in

	PATONS	0-4-0F	OC	WB	2898	1948	New	(1)
	-	0-4-0D		?	?	?	(a)	(2)

(a) ex RSHD, on hire, while WB 2898 was at RSHD for repairs, /1959; this may well have been 0-4-0DM RSHN 7869/1956, which RSH was using as a hire loco about this period.

(1) to Darlington (North Road) Railway Museum for preservation, 27/10/1979.
(2) returned to Robert Stephenson & Hawthorns Ltd, Darlington, /1959.

COLDBERRY LEAD CO LTD
COLDBERRY LEAD MINE, near Middleton-in-Teesdale

W89
NY 940291

This lead mine, worked by the London Lead Company between 1813 and 1903, lay about four miles north of Middleton-in-Teesdale. It comprised of various levels and had no main line rail connection. After other owners, this firm reopened it in 1952. About 200 yards of track connected the North Level with the ore dressing mill and the locos also worked in underground from the surface. The mine was closed in 1956.

Gauge : 2ft 0in

| - | 0-4-0DM | HE | 4675 | 1954 | New | s/s |
| - | 0-4-0DM | HE | 4979 | 1955 | New | (1) |

(1) to The Weardale Lead Co Ltd, Stotsfield Burn Mine, Rookhope, c/1957.

THE CONSETT IRON CO LTD
Derwent & Consett Iron Co Ltd until 4/4/1864 (not wound up till 23/6/1864)
Derwent Iron Co until 20/7/1858

CONSETT AREA F90

The Consett Iron Co Ltd was formed in 1864 to take over the business of the Derwent & Consett Iron Co Ltd, itself formed to take over the business of the Derwent Iron Co. (which see). This company had been formed towards the end of 1839 and had greatly expanded iron-making at Consett, but had failed when the Northumberland & Durham District Bank failed in 1858, to which it owed some £1 million. The operation had been kept going through the support of the North Eastern and Stockton & Darlington Railways.

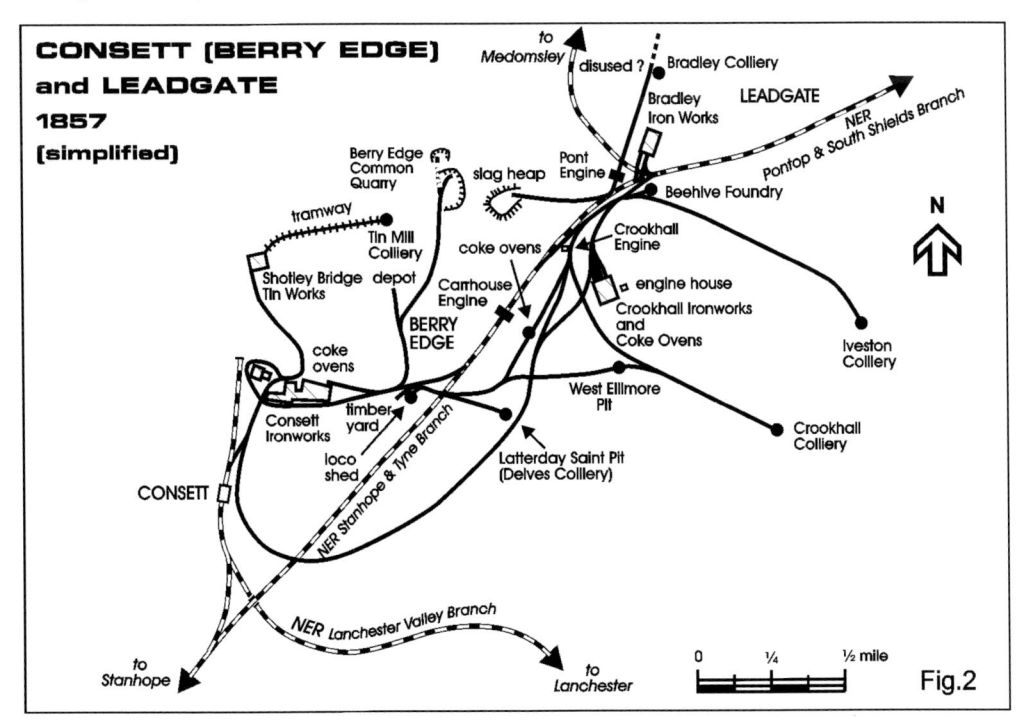

Fig.2 CONSETT (BERRY EDGE) and LEADGATE 1857 (simplified)

The new company, containing some of the partners who had been managing the business since 1858, finally took possession on 15th August 1864, though the company's existence was backdated to 4th April 1864. At this date the firm owned fourteen blast furnaces (although only six were working), seven at Consett and seven at Crookhall, together with 99 puddling furnaces, also at Consett. These works were served by various connections to the North Eastern Railway, the Stockton & Darlington Railway having amalgamated with it in July 1863. The company also owned the **BISHOPWEARMOUTH IRON WORKS** at Sunderland (NZ 388569), served by both the Hetton and Lambton Railways, and via them the NER, where there were a further 31 puddling furnaces. The company sold the Bishopwearmouth works in 1865 in order to concentrate on Consett.

To the north of the Consett Works lay the property of the **Shotley Bridge Iron Co**, which included the **Shotley Bridge Tin Works** and the adjacent **Tin Mill Colliery**, together with the **Bradley Iron Works** at Leadgate. About November 1866 the firm went into voluntary liquidation, and the assets were acquired by the Consett Iron Co Ltd, although documentary evidence shows this did not happen until 1st December 1872, rather than in 1866, as previously believed. These premises were rail-served and may well have had their own locomotives, possibly passing to the new owners, but no details survive. The manufacture of tin plates was replaced by iron plates, although the site was known as the 'Tin Mill' for many years afterwards. Bradley Iron Works, like the Crookhall Works nearby, was subsequently closed, the site of the former being re-used for workshops.

In 1871 the company opened its first colliery not directly linked to the Consett complex when Westwood Colliery near Hamsterley began production. By 1872 all of the original blast furnaces at Consett had been pulled down and replaced with six much larger ones, to which a seventh was added in 1880.

From 1851 the works had begun bringing iron ore from Cleveland, but in 1872 the company joined with the Dowlais Iron Co Ltd in South Wales, Herr Krupp of Essen in Germany and Messrs. Ybarra in Spain to set up the Orcanera Iron Co Ltd, which acquired large hematite mines at Bilbao in Spain and from which the works obtained its supplies for many years. Eventually the company's furnaces, mills, collieries, coke ovens, by-product plants, quarries and brickworks lay all around Consett, especially to the south and the east.

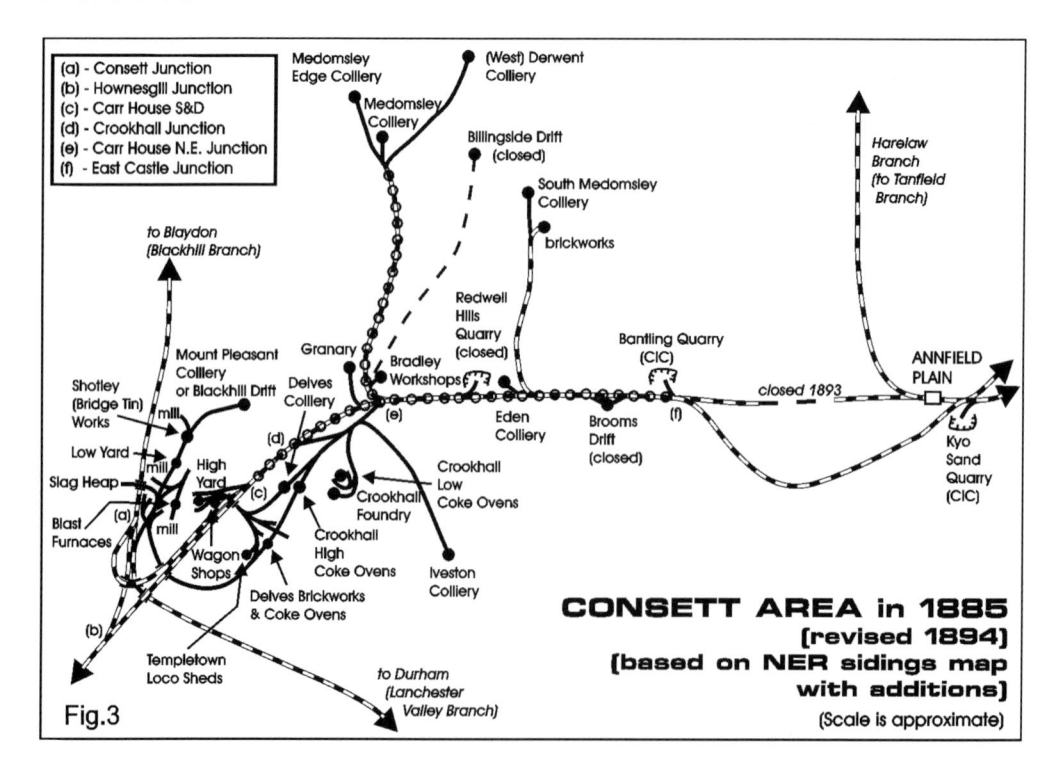

Fig.3 CONSETT AREA in 1885 (revised 1894) (based on NER sidings map with additions) (Scale is approximate)

Initially the company concentrated on iron rails and plates; but this market collapsed in the 1870s and the company was forced to concentrate on iron ship plates. The company began steel-making, using the Siemens process, in 1883. In July 1889 the company purchased a large royalty from the Marquis of Bute (which see) that included Garesfield Colliery and its railway to Derwenthaugh on the River Tyne, and subsequently built the Chopwell & Garesfield Railway and the Whittonstall Railway to serve the new collieries it developed in this area. It also built staiths on the River Tyne at Derwenthaugh and a large coking plant at Winlaton Mill, opened in the spring of 1929. Its agreements to obtain limestone from Stanhope having come to an end, the company was compelled to purchase its own quarry, taking over the **ASHES QUARRY** at Stanhope (NZ 995400 approx.) from the North Eastern Railway in October 1900.

25. *Of the very early locomotives at the Consett Iron Works, only two photographs are known, of tender engines 2 and 4. No details of No.2 have survived.*

26. *About 1872 the company divided its locomotives into five classes, lettered A to E, although the 'C' class did not last long. The 'A' class were all six-coupled, 'long-boilered' engines, with all their wheels in front of the firebox. The first four, built by Kitsons at Leeds, were all saddletanks. A No 3, K 1845/1872, shunts on 5th July 1936, possibly near Medomsley Colliery.*

To manufacture coke the company built hundreds of beehive ovens in various locations around Consett, as well as at Langley Park, Chopwell and Garesfield. Its first by-products coking plant was opened at Templetown in May 1906, which was followed by a plant at Langley Park in 1915. After the First World War the company decided on the construction of a second by-product works at Consett, the **FELL COKE WORKS** (NZ 104497), which was the most innovative single development in U.K. coke making in the twentieth century. Constructed by the Coppee Company in 1922-23, these 60 slot ovens were the first to be constructed lined with high-heat silica bricks. Because bricks of the required specification were unavailable in Britain, the company also established the **TEMPLETOWN BRICKWORKS** to manufacture them, linking it by aerial ropeway to **BUTSFIELD QUARRY** (NZ 096445), where ganister for the bricks was quarried. For further details of these locations, see below. In 1940 the company opened a rolling mill at Jarrow, operated initially through a subsidiary company, **The New Jarrow Steel Co Ltd**.

Eventually "The Works" at Consett covered over 650 acres, including not only the blast furnaces and cokeworks but also steel melting shops, plate, slabbing, blooming and billet mills, a foundry, rolling mills, fabrication shops and brickworks, as well as major workshop facilities, all served by an extensive railway system, and employing over 6,000 people. Besides its steel products, the company was a major seller of coal, coke and bricks, including silica products.

The collieries that supplied the Consett Works itself were called by the company the "Home Collieries" and the remainder the "Sale Collieries". The collieries that were still working were vested in NCB Northern Division Nos. 5 and 6 Areas on 1st January 1947, although the company's Templetown Sheds continued to house the NCB locomotives at Consett until about the middle of 1950. Under the Iron & Steel Act of 1949 the company was scheduled to be vested into the British Iron & Steel Corporation on 16th February 1951, though this only lasted until 15th July 1953, when the Conservative government de-nationalised the industry; the Consett company returned to private ownership in December 1955.

Between 1950 and the early 1960s the company embarked on two huge development programmes, including a new plant for handling iron ore imported through Tyne Dock from Sweden, Africa and America; new coke ovens; a large new steel-making plant, the first in the world to use both the Kaldo and L.D. Oxygen processes; a new power station, a new £14 million Plate Mill at Hownsgill, commissioned in September 1960, and the complete replacement of all of the steam locomotives. All this increased annual production to one million tons of steel ingots and ½m tons of steel plate, with the works employing over 8,000 people.

Under the second government nationalisation scheme, the company was vested into the **British Steel Corporation** on 1st July 1968 (which see).

The early development and working of the company's railway operations is both complex and obscure. About 1844 the Pontop & South Shields Railway allegedly sold Medomsley Colliery (see below) to the Derwent Iron Co, which also took over the working of the traffic on its 1½ mile branch. It would seem that the branch was subsequently 'sold back', but the working retained, and this "divided ownership" lasted until 1924, when the LNER sold the branch to the company. In August 1861 the NER was asked to build a 1¼ mile branch to D.Baker & Co's new Pontop Hall Colliery, but once again the Iron Company worked the traffic, the NER paying £1000 per year for this from 1863. With the opening of the Ann Pit at Pontop Hall in 1864, the colliery became called South Medomsley Colliery and the branch the South Medomsley Colliery Branch. The Consett Iron Company continued to operate the traffic until 1884, when it was handed back to the NER. Page 118 of *Railways of Consett and North West Durham* by G.Whittle, published in 1971, states "The Consett Iron Co (Derwent & Consett Iron Co before 1864) appears to have operated some of the mineral traffic on the Stanhope & Tyne line west of the Annfield inclines up to the NER/S&D amalgamation of 1863 and even thereafter into the 1880s", though the basis for this statement is not known. Certainly the company possessed extensive running powers over the NER/LNER in the Consett area. The NER Sidings Maps of 1894 records these as extending eastwards from Carr House N.E.Junction at Leadgate for 1¾ miles eastwards along the Pontop & South Shields line as far as Bantling Quarry at East Castle Junction, where the then new NER Avoiding Line re-joined the original route. In later years these powers extended only as far as Eden Colliery (see below).

Around Consett itself an accurate and clear story is also very difficult to achieve. The same location may have been served by different lines at different times; lines were altered to make way for new development; in some cases the location is known but not its history, in others the reverse, while the frequent changes of nomenclature add to the confusion. It would seem that in the early days the "system" was operated partly by locomotives and partly by stationary engines and that some of the latter survived for many years, but again details are sketchy. A detailed plan of the works included in a document (D/Co/107A) dated 30th August 1859 shows a "stationary engine" at Knitsley level crossing near the then centre of the works, but what its purpose was is not known. However, by 1915 it would appear that all of the company's works, collieries, etc., around Consett were operated by its own locomotives except **WESTWOOD COLLIERY** (NZ 113554) **(F91)**. This was opened in 1871 and worked by the NER; it closed in 1941.

Besides small quarries and brickworks, the system around Consett, known latterly as **BANK TOP HOME RAILWAYS**, served the following locations, either directly or via running powers over the NER:

27. Thereafter new 'A' class locomotives were built as pannier tanks. They were a powerful design, which was continued long after it had become obsolete, in the interests of standardisation. When a order for a new locomotive was placed, the builders were simply sent the drawings. A No 13 was built by the small firm of The New Lowca Engineering Co Ltd in Whitehaven, Cumberland, in 1908, and is seen here at Consett on 5th July 1936. The unusual device on top of the boiler was water-softening apparatus. Note the long pole lying on the running plate, which was used for 'parallel shunting' – propelling wagons on the next road to the locomotive.

28. Incredibly, perhaps, an order for three 'A' class locomotives was placed with Robert Stephenson & Hawthorns Ltd in 1941. One of those, A No.1, RSHN 7027/1941, pauses on 5th June 1955, displaying the lime green livery with red lining then recently adopted. Eleven of the class passed to the National Coal Board in 1947, and happily, one of this group, A No.5, is preserved and in working order.

BANTLING QUARRY, East Castle (NZ 151519). This was a sandstone quarry situated north of the NER Pontop & South Shields line at East Castle Junction, and was worked via running powers (see above).

BERRY EDGE COMMON QUARRY, near Leadgate (NZ 110515). This was one of a number of quarries in this area. It was being worked by the 1850s and was served by a ¾ mile branch from just east of Consett Iron Works. Almost certainly it closed in the mid nineteenth century.

BILLINGSIDE DRIFT, near Leadgate. There were several coal workings north-west of the Billingside Plantation, but this drift was probably situated at NZ 129534, at the end of the **Pont Burn Railway**. This ran for just over a mile from a junction west of Leadgate with the NER Pontop & South Shields branch, close to the junction of the Medomsley Branch. Near the junction lay the **Pont Engine** (NZ 117516), but how this operated is not known. The DCOA returns of April 1871 and November 1876 list this drift as having one locomotive, but whether it was stationed here or came daily from Consett is not known. The drift closed on 23rd January 1879.

BRADLEY COLLIERY, near Leadgate (NZ 122522). This was also served by the Pont Burn Railway, about ¼ mile north of the junction with the NER. Its closure date is not known, but was probably before 1875.

BRADLEY IRON WORKS, Leadgate (NZ 121518). This works was opened in 1854 by E.Richardson & Co and had four furnaces. It was served by sidings north of the NER Pontop & South Shields branch at Leadgate. The works appears to have been shut down in 1871. By the 1890s the site had been redeveloped into the **BRADLEY WORKSHOPS**, which passed to NCB Northern Division No.6 Area in 1947. They were worked by the running powers above.

BROOM'S DRIFT, near East Castle (NZ 143520). This was served by sidings alongside the NER Pontop & South Shields branch, 1½ miles east of Leadgate. Like Billingside, the DCOA returns for April 1871 and November 1876 list one locomotive here. It was also worked by the running powers from Carr House N.E.Junction. The drift closed on 28th November 1877.

CONSETT IRON WORKS. This grew into a vast complex, mostly covered by NZ 0949/1049 and NZ 0950/1050. Included latterly were the **Fell Coke Works**, the **Templetown Coke Works** (see below) and the **Templetown Refractory Works**. The original works was served by a ½ mile branch from the Stanhope & Tyne Railway (later NER Pontop & South Shields Branch) south-west of Berry Edge just west of the Carrhouse Engine, but subsequently by the NER Lanchester Valley Branch (opened in September 1862) and the NER Derwent Valley Branch (opened in December 1867).

On the NER Sidings map of 1894 traffic with the Iron Works is listed as being exchanged at Carr House Junction N.E. (the junction with the Medomsley branch), Crook Hall Junction, Carr House S & D (also known as Delves Lane), all of these being on the Pontop & South Shields line, and at Consett Junction (the end-on junction of the Blackhill and Lanchester Valley lines. Latterly most traffic was exchanged here, via the very large Low Yard.

CROOKHALL COLLIERY, near Consett. The original Crookhall Colliery was sited at NZ 127503 and was served by the Crookhall Railway, a 1¼ mile branch from the NER Pontop & South Shields branch between Berry Edge and Leadgate. This had the **Crookhall Engine** (NZ 116513) near the junction with the NER, but how this operated is unknown. There was also a link near the colliery westwards to Consett Iron Works, with at the junction **WEST ELLIMORE PIT** (NZ 120508). In the second half of the nineteenth century the **Crookhall Foundry** and **Crookhall Low Coke Ovens** were built nearby. **Crookhall High Coke Ovens** were situated near Delves Colliery (see below). The closure dates of all of these are not known. Later the name **CROOKHALL COLLIERY** was given to a new colliery, the **VICTORY PIT** (NZ 113503), sunk in the early 1920s on the site of the former Delves Colliery (which see). From here a narrow gauge rope-worked line two miles long was built to serve **HUMBER HILL DRIFT** (NZ 140470) and **WOODSIDE WINNINGS** (NZ 138490), the latter being opened in October 1928. These three were all vested in NCB Northern Division No. 6 Area on 1st January 1947.

CROOKHALL IRON WORKS, near Consett (NZ 119509). This was opened in 1845, and by 1847 there were seven furnaces here. It was also served by a link to the Pontop & South Shields branch between Berry Edge and Leadgate. There were two sets of beehive coke ovens here. The ironworks appears to have ceased production in 1871.

DELVES COLLIERY, near Consett (NZ 113503). This colliery was sunk in 1847 (possibly termed **Delfts Colliery** originally) and was served by a short branch from the main complex. From the 1850s it was known as the **Latterday Saint Pit**. It also had coke ovens and a brickworks, which was opened in 1875. The colliery was closed in September 1913. However, the site was later re-opened as **Crookhall Colliery (Victory Pit)** (see above).

DERWENT COLLIERY, Medomsley (NZ 122545). This colliery, also called West Derwent Colliery, was opened in 1856 and was served by a ¾ mile extension of the Medomsley Branch (see above). The

29. The 'B' class comprised four-coupled saddletanks, almost all of them the standard Black Hawthorn and Hawthorn Leslie designs, albeit with different sizes of cylinders and wheels. B No.9, BH 552/1880, with 12in x 19in cylinders and 3ft 2in wheels, stands at Consett in July 1935.

30. B No 31, HL 3023/1913, stands outside the loco sheds at Templetown. She had 13in x 19in cylinders and 3ft 4½in wheels – note the single crosshead slide – and carries the livery of black with red and white lining used in the late 1940s.

O.S. map of this period entitles this extension the Derwent Railway and also names it Derwent Bank, but it is not known whether this means the extension was originally a rope-worked incline. Latterly locomotives were used. The name Derwent Colliery subsequently included the **HUNTER PIT**, (NZ 123548), sunk in 1889; its output was raised via the Derwent Pit shaft from early in 1911. The colliery was vested in NCB Northern Division No. 6 Area on 1st January 1947.

EDEN COLLIERY, Leadgate (NZ 135521). This was opened in 1850 and was served by sidings north of the NER Pontop & South Shields Branch near the junction of the South Medomsley branch (see below), one mile east of Leadgate. It was worked via running powers over the NER. The DCOA returns for April 1871 and November 1876 record one locomotive here, though whether this was a permanent allocation or the loco came daily from Consett is not known. By the 1930s three drifts to the south-east sent coal via a ¾ mile tramway and a gantry over the LNER to the screens. It was vested in NCB Northern Division No. 6 Area on 1st January 1947.

FELL COKE WORKS, Consett (NZ 104497). A set of 60 by-product coke ovens, consisting of Coppee ovens to an American design – Wilputte-Coppee regenerative - was opened here in March 1924. As noted above, they were the most innovative single development in the British coke industry in the twentieth century, incorporating the first use of high heat-resistant silica bricks. They adopted many American features, including the quenching of the coke in a remote tower. They were planned to carbonise 6000 tons of coal per week. This battery was eventually closed down about 1955. Before then 54 Woodall-Duckham Becker under-jet combination ovens had been built nearby, probably opening in May 1948, replacing those at Templetown (see below), closed in the following year. Tar, sulphate of ammonia and rectified benzole were retrieved as by-products. 17 Gibbons Wilputte ovens were started in August 1953, followed by 20 more in 1955, the first coke being pushed on 23rd December 1955.

IVESTON COLLIERY, Iveston (NZ 131509). This colliery was won in June 1839 by Black, Reay & Co and was later taken over by the Iron Company, who connected it to the main complex by a ¾ mile branch. The DCOA Return 102 lists one locomotive here. It was closed on 23rd January 1892. Later mining in this area was undertaken by The Lanchester & Iveston Coal Co Ltd (which see).

KYO QUARRY, Annfield Plain (NZ 174519). This was a short-lived sand quarry about ½ mile east of Annfield Plain Station served by sidings from the NER Pontop & South Shields Branch. It is believed that the NER worked the traffic.

MEDOMSLEY COLLIERY, Medomsley (NZ 115537). The early history of this colliery is not clear. Whittle (see above) says that this colliery, then owned by the Stanhope & Tyne Railway, provided the first coal shipped at South Shields on 10th September 1834, the Railway's opening day. Latimer's Local Records 1832-1857 give the opening date as 30th December 1839 and the owners as the Derwent Iron Co; but as this predates the founding of the iron works, this may be suspect. Whittle goes on to say that the colliery was sold by the Pontop & South Shields Railway to the Derwent Iron Co "about 1844". However, it was leased to John Bowes, Esq. & Partners between July 1850 and April 1852, so it may be that the Derwent Iron Co did not acquire it until this latter date.

It was served by the 1½ mile Medomsley Branch from a junction at Leadgate, and appears initially to have been rope-worked by the Derwent Engine near the colliery. Once the colliery was owned by the Iron Company it worked the traffic, though the branch continued to be owned by (latterly) the NER. The DCOA Returns list a locomotive here in April 1871. By the 1890s there were also coke ovens and a **brickworks** here. The NER Sidings Map of 1894 shows the branch also serving Medomsley Edge Colliery, ¼ mile to the south-west, a separately-owned and short-lived landsale colliery. In 1924 the LNER sold the branch to the Consett Iron Co Ltd. The colliery was vested in NCB Northern Division No.6 Area on 1st January 1947.

REDWELL HILLS QUARRY, near Leadgate (NZ 131520). This was a short-lived quarry, served by the running powers from Carr House N.E. Junction. It had closed by 1894.

SHOTLEY BRIDGE TIN WORKS, Blackhill (NZ 099511) (note that this was situated some way from the village of Shotley Bridge). This was owned by the **Shotley Bridge Iron Co** (Shotley Bridge Tin Co and E.Richardson & Co are also found) and it was served by a ¾ mile branch from the Consett Iron Works. For many years it was believed that the Consett Iron Co Ltd took over the works in 1866, but surviving documents show that in fact the Consett company did not purchase the works until 1st December 1872. After the take-over this works was re-constructed to manufacture iron plates and became an integral part of the main complex, but the name "Tin Mill" remained in use for many years.

TEMPLETOWN COKE OVENS, Consett (NZ 109500). For many years the company operated a large number of beehive coke ovens, both at locations around Consett and at all of its collieries. The move into by-product ovens began at Templetown, Consett, on 9th May 1906 with the opening of 55 Otto-Hilgenstock waste heat ovens. 25 more began production in July 1912 and a further 30 followed in

31. The tunnel which had to be driven between the two sections of Ashes Quarry at Stanhope was very restricted, and when a new locomotive was needed, Hawthorn Leslie modified its standard design with 14in x 22 cylinders and 3ft 6in wheels, fitting side tanks and a chimney and cab of reduced height.

32. In 1901 Andrew Barclay had built B No 24 to a Hawthorn Leslie design, but with a new locomotive required urgently at the beginning of the Second World War, there was no time for this, and Andrew Barclay supplied one of their standard 0-4-0STs with 16in x 24in cylinders and 3ft 8in wheels. B No 12, AB 2078/1939, also in the lime green livery with red lining, pauses during some heavy shunting at Consett on 5th June 1955.

January 1913. The number of beehives gradually dwindled, the closure of those at Delves Brickworks (Consett) and Garesfield Colliery in April 1921 leaving only those at Chopwell, which survived until 1940. Adjacent to the new Templetown ovens lay the **Templetown Tar Works**, where sulphate of ammonia and crude benzole were also manufactured, the latter being processed by a plant initially operated by The Newcastle Benzole Co Ltd. The two works were closed in October 1949.

TEMPLETOWN REFRACTORY WORKS, Consett (NZ 112502). This works was opened in 1922 to manufacture refractory bricks from the ganister sent from Butsfield Quarry (which see) by aerial ropeway. It lay to the north of Templetown Coke Works and to the east of Templetown Loco Sheds. It developed into the largest refractory works in North East England, selling to owners of coke ovens all over the North East. When Butsfield Quarry was closed ganister was obtained from Harthope Quarry near St. John's Chapel in Weardale.

TIN MILL COLLIERY, Blackhill (NZ 104515). This colliery was owned by the Shotley Bridge Iron Co until 1872 (see above), and seems to have been called Consett Colliery originally. At first it was linked to the Tin Works above by a tramway; possibly its coal tubs were sent down to be tippled at the works. Later it was linked to the main Consett system by a ¾ mile branch. In later years the colliery had various names, including **Mount Pleasant Colliery**, **Blackhill Colliery** and latterly **Blackhill Drift**. It was closed on 10th September 1910.

The company also worked:

SOUTH MEDOMSLEY COLLIERY, near Dipton (NZ 144531). This was originally called Pontop Hall Colliery and was opened in 1862 by D. Baker & Co. It was served by a 1¼ mile branch which the NER built but the Iron Company operated. The junction was one mile north-east of the later Leadgate Station. In 1864 the Ann Pit was sunk nearby, the colliery's name was changed to South Medomsley Colliery and thus the branch became known as the South Medomsley Colliery Branch. In 1884 the NER took over the working, the colliery owners (now the Owners of South Medomsley Colliery) providing their own locomotive to shunt at the pit. For later details see the entry for The South Medomsley Colliery Co Ltd.

Away from Consett, the company also owned Ashes Quarry, Stanhope; Butsfield Quarry, near Consett; the Chopwell & Garesfield Railway, serving Chopwell and Garesfield Collieries and their coke ovens, Derwenthaugh Coke Works and Derwenthaugh Staiths; Langley Park Colliery, Langley Park; the Jarrow Works, Jarrow and the Whittonstall Railway, serving the Whittonstall Drifts. Details of these places will be found below the general locomotive lists.

The Locomotive Department
It would appear that well before the end of the nineteenth century the company had established a Locomotive Department at Consett. This developed the large **TEMPLETOWN LOCO SHEDS** (NZ 110500) – an "Engine Shed" at about the same site is shown on a map of 1858. There was also a small shed on the western side of the **LOW YARD** (NZ 097497), which appears to have been used solely to house cranes and locomotive cranes, whilst a third "Engine Shed" (NZ 099503) is shown on the 1939 O.S.map just south of the **PLATE MILLS**. The Low Yard shed was still being used in 1950, but how long these latter two sheds survived is not known. The Department also had a dedicated and well-equipped **Locomotive Repair Shop** near the main sheds. The 1892 booklet describes the building then as 180ft x 42ft, housing a fitting shop, machine shop and smiths' shop, together with stores and offices. After 1920 locomotives were no longer sent to local firms for repairs, and it is clear that the workshops were amongst the most extensive in the area. Boiler repairs were undertaken separately from locomotive repairs, and spare boilers added to the pool, so that boilers could be interchanged, and a repaired boiler was quickly fitted when a locomotive came in for repairs. From 1925 the company even began constructing its own locomotive boilers, ten having been made by 1941. Sometimes new boilers were held in stock for years; a boiler made for D10 in 1918 was not fitted until 1935, a boiler made for D No 2 in 1921 was not fitted until 1936. This policy of standardisation was undoubtedly the reason why the construction of the 'A' class long-boilered pannier tanks, a long out-dated design, continued until 1941. Similarly, AB 895/1901 was not a Andrew Barclay design, but a Hawthorn Leslie design built to Consett instructions; B No 12, ordered from Barclay in 1939, was very much the exception to the rule. With all of this went both a centralised system of locomotive organisation - locomotives working away from Consett and needing major overhaul were simply exchanged with a replacement from Consett – and a centralised system of recording; all day-to-day running repairs, even those carried out away from Consett, had to sent to the Department for entering into the Department's Locomotive Repair Registers, which were meticulously maintained. Seven of these, covering the majority of steam repairs for the first half of the twentieth century, survive. They are not currently (2006) available to the public, but have been carefully consulted and are the basis on which the transfers and disposals up to 1954 are given below, together with information taken from records kept at Derwenthaugh Loco Shed. However, it is unclear whether the electric locomotives at

33. In 1955 the Templetown workshops built a fireless locomotive for use at the coke ovens. The frames, running gear and cab came from B No 14, itself a rebuild from HL 3906/1937 and RSHN 7022/1941, but the new locomotive incorporated the wheels from 'E' class rail crane BH 1051/1892. She was virtually completed when photographed here, on 7th September 1955. But the drivers were said to be scared of it, and she was scrapped in 1962.

Chopwell ever visited Consett, and those from Butsfield Quarry certainly did not. When the Jarrow Works first opened its locomotives were kept separate from Consett, but after the New Jarrow Steel Co Ltd was merged into its parent company its locomotives too were integrated into the Consett system.

After the collieries were taken over by the National Coal Board in 1947 the company continued to service and repair the NCB-owned locomotives at Consett until the NCB opened its new shed at Leadgate about the summer of 1950.

The Department was well capable of constructing new locomotives, both steam and diesel, and of undertaking major experiments, such as oil-firing, and its fleet of locomotives was perhaps the best maintained and organised in Co. Durham.

The company also had an extensive fleet of private owner wagons, and to service them **WAGON SHOPS** (NZ 106504) were established west of Knitsley Road Level Crossing within the main Iron Works complex.

In the locomotive lists below the following abbreviations have been used for locations where locomotives were shedded. The details of locations not near Consett have been given after the loco lists. Short-term transfers involving Derwenthaugh and Garesfield have not been included. Similarly, where Consett records listed a locomotive as "rebuilt" at Consett this has only been shown if the locomotive underwent physical change and has not been shown if only an identical new boiler was fitted.

AP	Axwell Park; transfers here, rather than to Derwenthaugh, are specifically named in the Locomotive Repair Registers, and it would appear they are associated with the Derwenthaugh Coke Works, which was situated here
AQ	Ashes Quarry, Stanhope (see below)
Ch	Chopwell Colliery, Chopwell
C	Consett Iron Works (Templetown Sheds and Workshops)
D	Derwenthaugh Loco Shed, Swalwell
G	Garesfield Colliery, High Spen
J	Jarrow Works (see below)
LP	Langley Park Colliery, Langley Park (see below)

Early locomotives

R.H. Inness stated that the first two locomotives at Consett were of the return-flue design, presumably tender engines with outside cylinders, possibly purchased second-hand. For many years it was thought that four locomotives ordered from Thomas Richardson at Hartlepool worked at Consett. These were TR 208, 209, 210 (delivered April 1852) and 232 (delivered June 1853). More recent research has shown that certainly the first three were used on the Derwent Iron Company's tramway to its ironstone workings at Upleatham in the North Riding of Yorkshire; they were 2ft 1in gauge tender locomotives, the first genuinely narrow gauge locomotives ever built. Whether TR 232 was also ordered for Upleatham is not known. It would seem almost certain that there must have been more early locomotives than those listed below, especially given the existence of two photographs, unfortunately of poor quality, showing tender engines numbered 2 and 4. One source says that three early 0-4-0STs were built by John Harris of Darlington (see next section). Another source says that there were ten locomotives at Consett in 1870. The numbering of BH 191/192 when delivered in 1871 (see below) suggests that a numbering scheme was in use then. The first list gives a summary of some of these early locomotives.

Gauge : 4ft 8½in

2		0-6-0	OC	?	?	?	(a)	C	s/s
4		0-4-2	IC	RWH?	?	?	(b)	C	s/s
-		2-4-0T	OC	RS	1085	1857	New	C	
	reb	2-4-0	OC	Consett		?			s/s
DERWENT		0-4-0ST	OC	MW	112	1864	(c)	C	-
BEN		0-4-0T	OC	FJ	87	1871	New	C	s/s

(a) origin and identity unknown; photograph in collection of Beamish Regional Resource Centre.
(b) origin and identity unknown; photograph in private collection, with loco attributed to RWH.
(c) The MW Engine Book gives this loco as New to The Consett Iron Co Ltd; one version says that the order was cancelled and instead the locomotive was sold to Pease, Hutchinson & Co, Skerne Iron Works, Darlington; another version says that it was delivered to Consett and then returned to MW, who re-sold it to Pease, Hutchinson & Co c/1871.

About 1872 the company decided to divide its locomotives into three classes, using a letter for each class. The 'A' class consisted of six-coupled engines and was developed into a group of long-boilered pannier tanks with interchangeable parts; whichever firm received the order for one or more new locomotives was sent the original Kitson drawings to build from. The 'B' class consisted of four-coupled saddletanks, later developed into three basic types - those with 12in x 19in (later 13in x 19in) cylinders, those with 14in x 22in cylinders and those with 16in x 24in cylinders. Almost certainly up to six 0-4-0ST acquired before 1872 were included in B Nos.4-9 below, together with B No.10 (see its footnote). Three of these are said to have been built by John Harris of Darlington. A booklet prepared by The Consett Iron Co Ltd for visitors from the Iron & Steel Institute in 1892 describes the locomotives and cranes in some detail, and using this it is possible to work out that the John Harris loco then surviving was B No 4. The 'C' class was made up of four very small four-coupled saddletanks, with 9in x 16in cylinders and 2ft 10in driving wheels. Two of these were the former Nos. 12 and 13, BH 191/192, mentioned above. These were used "about the blast furnaces" (1892 booklet), but were not developed. Subsequently the company added the 'D' class, which was originally intended for crane tanks but in which some four-wheeled cranes were later included, and the 'E' class, which consisted entirely of vertical-boilered cranes. These were amongst the first of this type in the country, and were designed by the company's Chief Engineer, John Roe, with the detail being worked out by Black, Hawthorn & Co Ltd, and were intended for use in the steel melting shops; the 1892 booklet gives a detailed description of those built up to that time. The standard locomotive livery in the first half of the twentieth century was black with red lining.

All locomotives were new to Consett Iron Works unless stated otherwise in the lists below. Locomotive transfers to and from Consett are meticulously recorded in the Locomotive Registers, but these must be interpreted with some care. Firstly, whilst records of transfers going back to the 1880s from volumes now lost have been copied up into later volumes, it may be that there were other transfers before 1914 that have not survived. Secondly, they do not in general record transfers between different locations on the Chopwell & Garesfield Railway. Such information as is known about these from the 1940s onwards came from repair books kept by the locomotive foreman at Derwenthaugh, and inspected in the 1960s. Thirdly, there has to be a suspicion that sometimes the Registers regard a location here as synonymous with the Railway itself, so that, for example, a locomotive sent to "Derwenthaugh" might actually have been sent to Chopwell once it arrived at Derwenthaugh. Finally, short-term transfers to Consett were nearly always for major overhaul in the Locomotive Repair Shop, followed by a short period of running-in; only transfers of longer than 9-12 months should be regarded as a transfer which involved working at Consett.

It was clearly the company's intention to replace the old A Nos 1 and 2 with the new A Nos 1 and 2 delivered in 1941; in the event, presumably because of the demands of the Second World War, the old A Nos 1 and 2 continued working alongside the new A Nos 1 and 2 until 1944 and 1946 respectively.

Gauge : 4ft 8½in

A No 1		0-6-0ST	IC	K	1843	1872	New	
			reb	K		1886		
	C-HC (repairs) ?/? –C 28/10/08							Scr 9/1944
A No 1		0-6-0PT	IC	RSHN	7027	1941	New	
	C							Scr c6/1960
A No 2		0-6-0ST	IC	K	1844	1872	New	
			reb	K		1888		
			reb	HL	7008	1907		
	C-G ?/?-C 22/7/20-D 31/5/29-C 18/9/29							Scr /1946
A No 2		0-6-0PT	IC	RSHN	7028	1941	New	
	C-D 27/6/41-C 12/46							(1)
A No 3		0-6-0ST	IC	K	1845	1872	New	
			reb	BH		1889		
	C							Scr 4/1938
A No 3		0-6-0PT	IC	HL	3951	1938	New	
	C-D 18/12/46							(2)
A No 4		0-6-0ST	IC	K	1998	1874	New	
			reb	HL		1892		
	C-HL(repairs) ?/? –C 10/10/92-G 14/11/01-C /05-D 16/3/06							
	-C 9/7/07-HL ?/?-C 6/12/20							Scr 5/1938
A No 4		0-6-0PT	IC	HL	3952	1938	New	
	C							(3)
A No 5		0-6-0PT	IC	K	2509	1883	New	
			reb	HC		1900		
	C-D 6/12/22-C 25/10/28-D 22/2/29-C 29/5/35-D 3/6/36-C 10/10/40							
	-Ch 2/12/41-G ?/?-D 3/3/44-G 22/3/44-D 21/6/45-G 22/9/45-Ch by 12/46							(4)
A No 6		0-6-0PT	IC	K	2510	1883	New	
			reb	RS	2915	1899		
	C-RS(repairs) ?/? –C 28/2/99-D 6/19-C 1/10/23-D 9/1/24							
	-Ch ?/?-C 12/9/29-Ch 28/2/30-C 8/11/34-Ch 4/1/35-C 8/11/40-D 3/1/41							
	-C 2/4/43-D 23/5/44-C 19/7/46							(4)
A No 7		0-6-0PT	IC	K	3905	1899	New	
	D**--C ?/?-D 14/11/19-G ?/?-C 17/1/24-D 28/5/24							
	-C 13/10/27-D 27/1/28-C 31/5/29-D 12/7/29-C 1/4/32							
	-D 11/5/32-C 30/5/34-D 4/9/45							(4)
A No 8		0-6-0PT	IC	K	3906	1899	New	
			reb	HL	7749	1915		
	D**--C 23/11/99							Scr 3/1950
A No 9		0-6-0ST	IC	SS	2260	1872	(a)	
			reb	AB	7952	1904		
	D-C-D 2/19							(5)
A No 9		0-6-0PT	IC	HL	3891	1936	New	
	C-D 19/7/46-Ch by 12/46							(4)
A No 10		0-6-0PT	IC	K	4051	1901	New	
	D##-C?/? -G 15/3/07-C 7/11/08--G 8/10-C 9/11							
	-D 25/10/11-C 11/1/27-D 13/10/27-C 25/9/30-D 21/11/30							
	-C 21/12/33-D 26/6/34-C 28/10/37-D 19/1/38-C 20/6/41							
	-D 19/11/45-Ch ?/?-C 25/2/46-Ch 1/3/46-D 23/7/46							(4)

A No 11		0-6-0PT	IC	HL	2641	1906	New	
C-D 14/1/23-C 25/5/23-D 24/9/23-C 8/3/27-D 19/12/33-C 13/8/36								
-D 28/10/37-C 11/10/38-D 25/11/38-C 9/1/45-D 16/5/45								(4)
A No.12		0-6-0PT	IC	HC	809	1907	New	
D-C 3/11/11 -G 29/3/12-C 16/1/13-D 6/14-C 16/1/22-D 16/3/22-C 14/3/29								
-D 18/7/29-C 17/1/35-D 29/5/35-G 10/5/40-D 12/7/40-G 15/7/40-C 16/11/45								Scr c6/1960
A No 13		0-6-0PT	IC	NLE	249	1908	New	
G-C ?/?-D 30/4/14-C 7/6/14-D16/2/20-C 7/6/24								(4)
A No 14		0-6-0PT	IC	HL	3080	1914	New	
C-D 25/9/30-C 5/12/30-D 14/3/31-C 11/5/32-D 11/10/38-C 26/5/44-D 9/1/45								(4)
A No15		0-6-0PT	IC	K	5179	1917	New	
C-HL(repairs) ?/?-C 1/3/20								(4)
A No 16		0-6-0PT	IC	HC	1448	1921	New	
C								(4)
A No 17		0-6-0PT	IC	HC	1449	1921	New	
C-D 8/6/21-C 18/1/23-D 25/5/23-C 12/7/29-D 18/9/29-C 16/3/31-D 28/9/32								
-C 3/6/36-D 13/8/36-C 11/38-D 2/6/39-C 23/4/45-D 4/9/45								(6)
A No 18		0-6-0PT	IC	HL	3905	1937	New	
C								(4)
A No 19		0-6-0PT	IC	RSHN	7029	1941	New	
C								(1)
-		?	?	?	?	?	(b)	
D								(7)

**	ordered for Derwenthaugh, but according to the Locomotive Registers it was new to Garesfield; this could be a generic entry for the Railway itself.
##	ordered for Derwenthaugh, but according to the Locomotive Registers it was new to Consett.
(a)	the minutes of The Consett Iron Co Ltd's directors' meeting dated 13/12/1898 record that a second-hand locomotive with 17in cylinders had been purchased for Derwenthaugh; SS 2260/1872 had 17in cylinders and almost certainly the minute refers to it, yet allegedly it was not sold by the Brecon & Merthyr Railway to the Bute Works Supply Co, Cardiff, Glamorgan, until 2/1899, being previously Brecon & Merthyr Railway, 29; it would appear from the Consett minutes that the locomotive needed minor repairs before delivery, which would suggest that it did not arrive until January or February 1899.
(b)	hired from Hudswell, Clarke & Co Ltd, Leeds, by 13/12/1898; almost certainly a six-coupled engine; no further details are known.
(1)	sold for scrap, 3/1962.
(2)	loco worked from Derwenthaugh loco shed for NCB No. 6 Area until returned to Consett, 10/1947; Scr c6/1960.
(3)	converted to oil burning, /1951; then removed, replaced and finally removed, 4/1953; to NCB Durham No. 6 Area, Leadgate Shed, 11/1957.
(4)	to NCB No. 6 Area, 1/1/1947.
(5)	to A. Bainbridge, Thornaby-on-Tees, Yorkshire (NR), for scrap, 10/1923.
(6)	loco worked from Derwenthaugh loco shed for NCB No. 6 Area until returned to Consett, 7/1947; Scr c6/1960.
(7)	presumably returned to Hudswell, Clarke & Co Ltd, Leeds, at end of hire period, probably early in 1899.

B No 1		0-4-0ST	OC	BH	289	1873	New	
C-BH(repairs) ?/?-C 9/9/91								(1)
B No 1		0-4-0ST	OC	HL	3390	1919	New	
C								Scr 7/1958
B No 2		0-4-0ST	OC	BH	326	1874	New	
			reb	HL		1893		
			reb	HL		1914		
C-HL(rebuild) ?/?-C 11/5/93-HL(repairs) ?/?-C 18/1/15-AQ 3/2/15								
-C 4/1/28-D 20/1/28-C 20/3/30-AQ 7/5/30-C by 5/49								Scr 7/1950

Loco	Type	Cyls	Builder	Works No	Year	Acquired	Notes	Disposal
B No 3	0-4-0ST	OC	BH	327	1874	New		
	reb		BH		1889		(2)	
C-BH(repairs) ?/?-C 16/5/89								
B No 3	0-4-0T	OC	HL	3495	1920	New		Scr /1952
C-AQ 4/1/21-C 14/4/44								
B No 4	0-4-0ST	OC	Harris			New?	(3)	
B No 4	0-4-0ST	OC	CF	1205	1901	New		Scr 7/1950
C-LP 2/12/05-C 22/6/09								
B No 5	?	?	?	?	?	?	(4)	
B No 5	0-4-0ST	OC	RS	2654	1888	New		
C							(5)	
B No 5	0-4-0ST	OC	HL	3473	1920	New		
	reb		Consett		1952			
C								Scr 5/1959
B No 6	?	?	?	?	?	?	(4)	
B No 6	0-4-0ST	OC	HL	2235	1892	New		
C							(6)	
B No 7	?	?	?	?	?	?	(4)	
B No 7	0-4-0ST	OC	RS	2655	1888	New		
C							(5)	
B No 7	0-4-0ST	OC	HL	3474	1920	New		
C-J 10/2/44-C 20/12/45							(7)	
B No 8	?	?	?	?	?	(a)	(4)	
B No 8	0-4-0ST	OC	BH	551	1880	New		
C							(8)	
B No 8	0-4-0ST	OC	HL	3475	1920	New		Scr 4/1959
C								
B No 9	?	?	?	?	?	(a)	(4)	
B No 9	0-4-0ST	OC	BH	552	1880	New		Scr 7/1950
C-AQ 22/10/17-C 25/3/20-AQ 8/21-C 31/3/33-AQ 31/3/44-C 5/47								
B No 10	0-4-0ST	OC	BH	328	1875	(b)		
	reb		RS	2811	1893			
C-RS(repairs) ?/? –C 29/12/1893-AQ 14/11/16-C 29/10/17							(9)	
B No 10	0-4-0ST	OC	HL	3476	1920	New		
C-D 23/3/37-C 29/6/43							(10)	
B No 11	0-4-0ST	OC	BH	553	1880	New		
C							(11)	
B No 11	0-4-0ST	OC	HL	3391	1919	New		
C-D by 25/6/1932-C ?/?/-D 4/9/41							(7)	
B No 12	0-4-0ST	OC	BH	698	1882	New		Scr 4/1936
C-AQ /05-C 14/3/19-AQ 23/6/19-C 4/1/28-D (for G) 11/10/28-C 6/3/30								
B No 12	0-4-0ST	OC	AB	2078	1939	New		Scr 6/1958
C								
B No 13	0-4-0ST	OC	HL	2176	1890	New		
	reb		HL	7324	1914			
C							(12)	
B No 13	0-4-0ST	OC	HL	3953	1938	New		
C-J c/53-C 7/56-J 5/57							(13)	
B No 14	0-4-0ST	OC	HL	3906	1937	New		
	reb		Consett		1951	(c)		
C							(15)	
B No 14	0-4-0F	OC	Consett		1955	(d)		Scr 3/1962
C								

Durham Part 1 Page 101

B No 15		0-4-0ST	OC	RS	2724	1890	New	
	C							(16)
B No 15		0-4-0ST	OC	HL	3873	1936	New	
	C							(17)
B No 16		0-4-0ST	OC	RS	2725	1890	New	
			reb	HL	5809	1920		

C-HL(repairs) ?/?-C 8/12/20-LP 10/12/20-C 2/10/25-LP by 24/2/26
-C 15/9/28-LP 19/1/29-C 20/6/30-LP 12/7/30-C 7/1/32-LP 11/3/32
-C 13/9/33-LP 18/10/33-C 23/2/35-LP 21/3/35-C 28/10/36-LP 3/12/36
-C 20/9/38-LP 28/10/38-C 5/6/40-LP 2/1/41   (18)

B No 17		0-4-0ST	OC	HL	2236	1892	New	
	C							Scr /1928
B No 17		0-4-0ST	OC	HL	3753	1930	New	

C-D 25/9/31-C 21/3/34-D 17/4/34-C 12/2/37-D 2/4/37
-C 3/5/39-D 23/8/39-C 24/4/40-D 7/4/41-C 15/9/43   Scr 8/1958

B No 18		0-4-0ST	OC	BH	854	1885	(e)	

G-C /04-G 15/12/04-AQ ?/?-C c/19-AQ 19/3/20-C 31/3/33   Scr /1942

B No 19		0-4-0ST	OC	BH	1113	1895	New	

G-C /04-G 15/4/04-C 30/11/06   (19)

B No 19		0-4-0ST	OC	HL	3752	1930	New	

D-C 17/3/34-D 30/5/34-C 2/4/37-D 30/6/37-C 5/2/41-D 22/4/41   (7)

B No 20		0-4-0ST	OC	RS	2852	1897	New	
	C							Scr /1929
B No 20		0-4-0ST	OC	HL	3745	1929	New	

C-D 3/5/39-C 23/8/39   (20)

B No 21		0-4-0ST	OC	HL	2377	1897	New	

G-C 13/6/12-LP 3/9/12-C 23/10/19-LP 15/12/28-C 19/1/29   (7)

B No 22		0-4-0ST	OC	CF	1163	1898	New	
			reb	HL	4077	1919		

C**-HL(repairs) ?/?-C 15/10/19-LP 23/10/19-C 29/3/24-LP 21/6/24
-C 30/4/30-LP 20/6/30-C 21/11/31-LP 7/1/32-C 23/6/33-LP 13/9/33
-C 17/7/34-LP 28/8/34-C 18/10/35-LP 13/11/35-C 3/12/36-LP 29/1/37
-C 2/12/37-LP 30/12/37-C 24/1/40-LP 5/6/40-C 2/1/41-D by 12/46   (7)

B No 23		0-4-0ST	OC	HL	2404	1899	New	
	C							(21)
B No 23		0-4-0ST	OC	HL	3744	1929	New	
	C							(22)
B No 24		0-4-0ST	OC	AB	895	1901	New	

C-LP 2/12/37-C 30/12/37   Scr 4/1959

B No 25		0-4-0ST	OC	HC	702	1904	New	

C-LP 15/9/15-C by 2/23-LP 29/3/24-C 21/6/24-D 18/3/32-C 28/9/32
-D 15/2/33-C 3/7/34-LP 17/7/34-C 28/8/34-D 4/9/34-C 10/10/39
-J 2/48-C 2/49   Scr 6/1950

B No 26		0-4-0ST	OC	HL	2639	1905	New	

C-G by 16/4/15-C after 23/8/19-D 2/12/20-C 2/12/21-D 20/2/29
-C 16/10/29-D 4/2/30-C 11/12/35-D 3/2/37-C 2/12/42-D 30/6/43   (7)

B No 27		0-4-0ST	OC	HL	2640	1905	New	

C-LP 18/6/09-C 4/4/19-LP 27/5/19-C 10/12/20-D 13/10/29-C 1/11/38   Scr 4/1959

B No 28		0-4-0ST	OC	HL	3003	1913	New	

C-LP 30/4/30-C 12/7/30-LP 21/11/31-C 11/3/32   (23)

B No 29		0-4-0ST	OC	HL	3004	1913	New	

C-LP 5/10/33-C 18/10/33-LP 28/2/35-C 21/3/35
-Ch 26/6/41-C 18/8/43-J 17/11/53-C by 6/4/55   Scr 6/1958

B No 30		0-4-0ST	OC	HL	3022	1913	New	
	C-D 28/2/14-G by 1/1/15-D ?/?-C 19/8/21-D 20/2/23-C 4/9/34							
	-LP 18/10/35-C 13/11/35-D 11/12/35-C 19/1/38-D 18/3/38							
	-C 24/5/40-D 26/9/40-C 4/9/45-J 18/12/45-C 4/48-J 2/49							(24)
B No 31		0-4-0ST	OC	HL	3023	1913	New	
	C-LP 24/9/25-C 16/1/26-LP 23/1/33-C 12/10/33-LP 20/9/38							
	-C 28/10/38-D 1/11/38-J 6/7/43-C 9/3/45-J 7/53							(25)
B No 32		0-4-0ST	OC	HL	3251	1917	New	
	C-G 6/3/30-C 4/9/41-LP 20/3/42-C 28/9/43							Scr 6/1952
B No 33		0-4-0ST	OC	HL	3252	1917	New	
	C-LP 28/10/36-C 29/1/37-LP 24/1/40-J 23/6/42-C 5/8/43							Scr 8/1955
B No 34		0-4-0ST	OC	HL	3253	1917	New	
	C							(26)
B No 35		0-4-0ST	OC	HL	3254	1917	New	
	C-LP 8/9/43							(18)
B No 36		0-4-0ST	OC	HL	3471	1921	New	
	C-D 29/8/21-C 6/5/41-D 2/4/42-C 12/46							Scr 8/1958
B No 37		0-4-0ST	OC	HL	3472	1921	New	
	C-D 23/4/45-C 16/5/45							Scr 9/1959
B No 38		0-4-0ST	OC	HL	3496	1921	New	
	C-D 23/11/21-C 3/3/27-J 7/53-C by 11/55-J by 28/11/56-C by 8/8/1957							(27)
B No 39		0-4-0ST	OC	HL	3497	1921	New	
	C							(28)
B No 40		0-4-0ST	OC	RSHD	7011	1940	New	
	C							Scr 7/1959
B No 41		0-4-0ST	OC	RSHD	7016	1940	New	
	C							(17)
B No 42		0-4-0ST	OC	RSHD	7022	1941	New	
	C							(29)
	(JARROW No.1)	0-4-0ST	OC	AB	2091	1940	New	
	C-J 23/3/41-C 15/2/44-AB (repairs) 10/3/44-C 27/1/45-J 16/2/45-C by 11/53							(30)
B 42		0-4-0ST	OC	Consett		1954	(f)	
	C-J 8/54							(31)
	VULPES	0-4-0ST	OC	YE	480	1891	(g)	
	C							(32)
	PARK	0-6-0T	IC	HC	1250	1916	(h)	
	C							(33)

** according to the records of the directors' meetings, this loco was ordered for Langley Park Colliery; but the Locomotive Registers give it as new to Consett, while oral tradition claimed it was the first B class locomotive at Derwenthaugh.

(a) R.H. Inness claimed that three 'B' class locomotives were built by John Harris of Darlington, and that these were B Nos.8-10; however, B No 4 is known to have been a Harris locomotive, and B No 10 may well have been one too; which 'B' class number the third was, if there were only three, is unknown.

(b) "boiler made by Consett Iron Co in 1870 and Engine Built up of parts of an old one. Engine remodelled by Messrs. Black, Hawthorn & Co of Gateshead in 1875" (Consett Locomotive Register); R.H. Inness claimed the original loco was built by John Harris of Darlington.

(c) rebuilt with frames, tanks and wheels from B No 42.

(d) built from B No 14 (Consett 1951) and wheels from E No 7 between 7/9/1954 and 16/9/1955.

(e) ex Marquis of Bute, with Garesfield Colliery and its railway, 1/7/1889.

(f) built from parts of B No 14 (Consett 1955) and JARROW No.1.

(g) ex Hudswell, Clarke & Co Ltd, Leeds, on hire, 1/1906.

(h) ex Hudswell, Clarke & Co Ltd, Leeds, almost certainly on hire, 20/4/1936.

(1) sold, 24/10/1919.
(2) sold, /1919.
(3) still in existence in 1892; presumably s/s c/1901.
(4) assumed to have existed, but latterly no records survived at Consett; probably one of early locomotives listed above.
(5) to Thos W. Ward Ltd, Tinsley Works, Sheffield, Yorkshire (WR), 8/7/1920.
(6) sold - for scrap? - 6/1926.
(7) to NCB No. 6 Area, 1/1/1947.
(8) to Sir Hedworth Williamson's Limeworks Ltd, Fulwell, Sunderland, 14/10/1919.
(9) sold, /1919; "loaded on NER trolley, 24/10/1919".
(10) to The Seaham Harbour Dock Co, Seaham, 2/1960; delivered, 17/3/1960.
(11) no repairs after 23/8/1917; sold, /1919; "loaded on to NER trolley, 31/10/1919".
(12) to The Steetley Lime and Basic Co Ltd, Coxhoe, 22/6/1923.
(13) converted to oil-burning, 1954-1957; used as stationary boiler at Jarrow Works from 1961; to Consett Works by 3/1967; to British Steel Corporation, with works, 1/7/1968.
(14) to J. Tait & Partners, dealers, Middlesbrough, Yorkshire (NR); re-sold to The Wingate Limestone Co Ltd, Wingate Quarry, near Trimdon Colliery, /1920.
(15) taken out of traffic, /1954 and parts used to build 0-4-0F B No 14 and B 42.
(16) to The Steetley Lime & Basic Co Ltd, Coxhoe, /1922.
(17) sold to The Seaham Harbour Dock Co, Seaham, 2/1960; delivered, 28/4/1960.
(18) to NCB No. 5 Area, with Langley Park Colliery, 1/1/1947.
(19) to The Darlington Forge Ltd, Darlington, after 27/3/1928.
(20) to Raine & Co Ltd, Blaydon, 8/1957; delivered, 1/10/1957.
(21) no repairs after 22/2/1920; "sold – dismantled and loaded into trucks, 7/1/1921"; almost certainly it was this locomotive which was acquired 'from the Middlesbrough area' by Poynton Collieries Ltd, Cheshire, in 1925.
(22) to The Seaham Harbour Dock Co, Seaham, 11/2/1960.
(23) to Darlington Rolling Mills Co Ltd, Darlington, loan, 26/11/1944-21/12/1944; to NCB No.6 Area, 1/1/1947.
(24) scrapped at Jarrow Works, 6/1953.
(25) scrapped at Jarrow Works, 11/1959.
(26) scrapped on site by Geo. Cohen & Sons Ltd, 6/1950.
(27) converted to oil burning, 4/1953; removed (at Jarrow), c9/1953; to Raine & Co Ltd, Blaydon, loan, /1958, and returned; sold to The Seaham Harbour Dock Co, Seaham, 2/1960; delivered, 28/4/1960.
(28) to RSH, Newcastle upon Tyne, for repair, /1952, but found to be beyond repair and scrapped there, c6/1952.
(29) taken out of traffic, 5/1951, and parts used in building 0-4-0F B No 14, with remaining parts being scrapped.
(30) taken out of traffic, /1954, and parts used in building B 42.
(31) built with oil-burning equipment fitted; this was removed, /1956; scrapped at Jarrow Works, 3/1960.
(32) sold, almost certainly from Consett, to Tees Furnace Co Ltd, Lackenby Ironworks, near Grangetown, Yorkshire (NR), per A.M. Terry, dealer, Newcastle upon Tyne, 16/6/1906.
(33) returned to HC, Leeds, Yorkshire (WR), 25/9/1936, presumably at end of hire period.

All of the 'C' class worked at Consett.

C No 1	(formerly No.12)	0-4-0ST	OC	BH	191	1871	New	(1)
C No 2	(formerly No.13)	0-4-0ST	OC	BH	192	1871	New	(1)
C No 3	(formerly 3C)	0-4-0ST	OC	BH	247	1872	New	(1)
C No 4	(formerly 4C)	0-4-0ST	OC	BH	248	1872	New	(1)

(1) all scrapped in early 1900s.

All of the 'D' class worked solely at Consett unless otherwise shown.

D No 1		0-4-0CT	OC	D	1758*	1883	New	Scr 6/1927
D No 1		0-4-0CT	OC	AB	2111	1941	New	Scr 8/1954
D No 2		0-4-0CT	OC	D	2063	1884	New	
			reb	Consett		1922		(1)
D No 3 "THE COFFEE POT"		0-4-0VBCr	OC	BH	831	1885	New	Scr 1/1940
D3		4wVBCr		J.Booth		?	(a)	Scr 7/1950

34. For crane tank locomotives the company went to Dubs & Co in Glasgow, whose design was very different from the later Hawthorn Leslie design (see photo 71); here the jib had to be turned by hand. D No 4, D 2365/1888, being oiled outside the Templetown sheds, worked until the end of the Second World War.

D No 4		0-4-0CT	OC	D	2365	1888	New	(2)
D No 5		0-4-0CT	OC	D	2366	1888	New	Scr /1927
D No 6		4wVBCr		TSR	3469	1887	New	(3)
D No 6		4wVBCr		TSR	13654	1941	New	(4)
D No 7		4wVBCr		TSR	4121	1891	New	(5)
D No 7		4wVBCr		TSR	13653	1941	New	(4)
D No 8		4wVBCr		TSR		1891	New	Scr 3/1929
D No 8		4wVBCr		Priestman		?	(b)	(4)
D No 9		4wVBCr		TSR	4731	1895	New	Scr 3/1929
D9		4wVBCr		Priestman		?	(c)	(6)
D10		4wVBCr		TSR	5784	1900	New	
	C-D 16/3/00-C ?/?-D 27/9/21-C 20/11/43							(1)
D11		4wVBCr		TSR	5986	1901	New	(1)
D12		4wVBCr		TSR	6199	1902	New	
	C-J 1/5/46							(7)
D13		4wVBCr		TSR	8162	1913	New	Scr 8/1953
D14		0-4-0CT	OC	NBQ	21522	1917	New	Scr 10/1952
D15		4wVBCr		TSR	9298	1919	New	(8)
D16		0-4-0CT	OC	AB	1665	1920	New	
	reb	0-4-0T	OC	Jarrow		c1952		
	C-J c /52							(9)
D17		0-4-0CT	OC	AB	1715	1920	New	Scr 2/1955
D No 18		4wVBCr		TSR	9586	1920	New	
	D-C 11/10/21							Scr 3/1956
D No 19	KATIE	4wVBCr		TSR	9634	1921	New	Scr 6/1954
D20		4wVBCr		J.Booth		?	(d)	Scr c/1962

* the Consett Locomotive Repair Registers give the works number as 1708.

Note : how many others in this class carried DX numberplates (i.e., with the No omitted) besides those above is not known.

(a) origin unknown, 2/7/1940.
(b) origin unknown, 14/10/1940.
(c) origin unknown, 19/8/1940.
(d) ex Head, Wrightson & Co Ltd, Stockton, 8/1957 (see Cleveland & North Yorkshire Handbook).

(1) scrapped on site by G. Cohen, Sons & Co Ltd, 7/1950.
(2) not repaired after 1945; scrapped, /1951.
(3) sold, 29/6/1932.
(4) to Thos W. Ward Ltd, 5 /1952.
(5) "crane broken up and boiler sent to Low Yard for heating new Fitting Shop; boiler prepared for examination, 19/8/1925" (Consett Locomotive Repair Registers).
(6) to H.D. Ward Ltd, Wolsingham, 5/1952, for use as stationary boiler.
(7) scrapped at Jarrow Works, /1957.
(8) to Thos W. Ward Ltd, 7/1952.
(9) scrapped at Jarrow Works, 1/1954, and boiler returned to Consett.

35. By the mid-1880s the company wanted some powerful rail cranes, and the Chief Engineer, J.P.Roe, worked with Black Hawthorn to produce this completely new design. E No 1, BH 897/1887, could lift 12 tons. In 1978, still at Consett, she was donated to Beamish, The North of England Open Air Museum, for preservation.

All of the 'E' class worked at Consett unless otherwise shown.

E No 1		2-4-0VBCr OC	BH	897	1887	New	(1)
E No 2		2-4-0VBCr OC	BH	898	1887	New	(2)
E No 3		0-4-0VBCr OC	BH	931	1888	New	Scr /1927
E No 4		0-4-0VBCr OC	CoS	1749	1892	New	(3)
E No 5		0-4-0VBCr OC	BH	1048	1892	New	Scr /1927
E No 6		0-4-0VBCr OC	BH	1049	1892	New	(1)
E No 7		0-4-0VBCr OC	BH	1051	1892	New	
	C-AP 25/9/28-C 13/2/29						(4)
E No 8		0-4-0VBCr OC	RS	2853	1897	New	Scr 6/1954
E No 9		2-4-0VBCr OC	RS	2854	1898	New	(1)
E No 10		0-4-0VBCr OC	CF	1206	1901	New	(1)
E No 11		0-4-0VBCr OC	CC	7519	1907	New	(5)
E No 12		0-4-0VBCr OC	CC	7520	1907	New	(6)
E No 13	ROSE	0-4-0VBCr OC	HL	2984	1913	New	
	C-D 18/10/32						(7)

E No 14		0-4-0VBCr	OC	CoS	4101	1920	New	
	C-AP 19/6/28-C 19/1/33-AP 30/6/33-D ?/?-C 17/10/33							
	-AP 24/7/34-D ?/?-C 23/8/34							Scr c6/1961
E No 15		4wVBCr	VC	MFlm	3358	1921	New	(8)
E No 16		4wVBCr	VC	MFlm	3397	1921	New	(9)
E No 17		4wVBCr	VC	MFlm	3397	1921	New	(10)

* According to the Locomotive Repair Registers all three Marshall Fleming cranes carried the maker's number 3358.

(1) converted to oil firing by /1963; to British Steel Corporation, with works, 1/7/1968.
(2) withdrawn 8/1955 and used for spares; finally scrapped by 2/1964.
(3) no repairs after 29/4/1926; sent to Scrap Yard to be cut up, 2/5/1932.
(4) scrapped 9/1954; wheels used in the rebuilding of B No 14.
(5) no repairs recorded after 12/8/1950; scrapped, /1953.
(6) withdrawn from traffic by 11/1956 to provide spares for other rail cranes; remains scrapped, 10/1962.
(7) to NCB No. 6 Area, 1/1/1947.
(8) to C.A. Parsons & Co Ltd, Newcastle upon Tyne, 8/1925.
(9) exploded, 11/1959; Scrapped, c/1962.
(10) the jib was removed on 7/3/1934 and the boiler taken off on 17/1/1936; there are no further entries in the Locomotive Repair Registers; parts used to repair E No 16, remains scrapped.

36. The company was one of the first steel companies in the country to introduce diesel locomotives into its locomotive fleet. No.1, HE 3504/1947, incorporating a 186/204hp Gardner engine, underwent extensive trials. This view, taken from a Hunslet advertisement in the October 1950 edition of The Locomotive, shows her hauling one of the slag ladles, probably soon after being delivered to Consett.

All the diesel locomotives worked solely at Consett unless otherwise shown.

1		0-6-0DM	HE	3504	1947	New	
	C-HE (repairs), 8/6/1948 -C 14/7/1948						(1)
2		0-6-0DM	HE	3580	1949	New	(1)
No.3	later 3	0-4-0DM	HE	4010	1950	New	(1)
No.4	later 4	0-4-0DM	HE	4011	1950	New	(1)
5		0-6-0DE	BBT	3020	1951	New	(1)

6		0-6-0DE	BBT	3021	1951	New	(1)
7		0-4-0DM	HE	4431	1953	New	(1)
8		0-4-0DM	HE	4432	1953	New	(1)
9		0-6-0DM	Consett		1956	New	(1)
10		0-6-0DM	Consett		1958	New	(1)
11		0-6-0DM	HE	4987	1956	New	(1)
12		0-6-0DM	HE	4988	1957	New	(1)
13		0-6-0DM	HE	4989	1957	New	
	C-J 1/67-C by 7/68						(1)
14		0-6-0DM	HE	5173	1957	New	(1)
15		0-6-0DM	HE	5174	1957	New	(1)
16		0-6-0DM	HE	5175	1957	New	
	C-J 9/61-C by 9/63						(1)
17		0-6-0DM	HE	5375	1958	New	(1)
18		0-6-0DM	HE	5376	1958	New	(1)
19		0-6-0DM	HE	5377	1958	New	
	C-J 4/68						(2)
20		0-6-0DM	HE	5378	1958	New	(1)
21		0-6-0DM	HE	5379	1958	New	
	C-J 3/59-C 10/60-J 1/62-C 4/68						(1)
22		0-6-0DM	HE	5380	1958	New	(1)
23		0-6-0DM	HE	5381	1958	New	(1)
24		0-4-0DM	HE	5384	1959	New	(1)
25		0-4-0DM	HE	5385	1959	New	(1)
26		0-4-0DM	HE	5386	1959	New	
	C-J 3/64						(2)
27		0-4-0DM	HE	5387	1959	New	
	J-C 4/61-J 5/61						(3)
28		0-6-0DH	HE	5392	1959	New	(1)
29		0-6-0DH	HE	5393	1959	New	(1)
30		0-6-0DH	HE	5394	1959	New	(1)
-		0-6-0DH	HE	6663	1967	(a)	(4)

(a) ex HE, on trial, by 7/12/1967.
(1) to British Steel Corporation, at Consett Works, 1/7/1968.
(2) to British Steel Corporation, at Jarrow Works, 1/7/1968.
(3) to Raine & Co Ltd, Blaydon, loan, 1/1965; returned to Consett, 2/1965; sold to Raine & Co Ltd, Blaydon, 2/1965.
(4) to British Steel Corporation, at Consett Works, 1/7/1968, still on trial.

There was also a diesel rail crane, un-numbered, built by Coles in 1950, New. This had gone by 1959, to be replaced by three diesel electric rail cranes, viz., a 6-ton crane built by Coles, numbered 1, a 25 ton crane also built by Coles, numbered 2 and a 18 ton crane, built by Jones, numbered 3. All were acquired New in 1958-59.

FELL COKE WORKS, Consett F92
Gauge : 4ft 8½in

-		4wWE	WSO	529	1924	
			Goodman	3576	1924	New	(1)
-		0-4-0WE	GB	2368	1952	New	(1)

(1) to British Steel Corporation, with works, 1/7/1968.

The powered wagons below were also supplied to The Consett Iron Co Ltd, but where they worked is not known.

Monorail

-		2wGasH	RM	14262	1965	New	(1)
-		2w-H	RM	14798	1966	(a)	s/s

(a) New, but supplied without an engine.
(1) returned to RM, c4/1966; re-sold to W.A. Dawson Ltd, (location not known), c10/1966.

37. The company also purchased two 355hp diesel electric locomotives from Brush Bagnall Traction. One of these much more powerful locomotives, 6, BBT 3021/1951, was shunting one of the company's vans on 5th June 1955. But the company chose the Hunslet designs, and also appointed Mr. George Cowell from Hunslet as its Locomotive Superintendant.

38. The fireless locomotive completed, the Templetown workshops turned to building two diesels of its own, designed by Mr. Cowell but based on the Hunslet design. The first, No.9, built at Consett in 1956, is seen on 7th August 1957. She too has been preserved.

39. The Wilputte-Coppee ovens at the Fell Coke Works, opened in 1924, were the single most innovative development in the British coke industry in the 20th century. When the coke was ready, the doors at each side of an oven were removed and a ram pushed the coke out into a steel car. This was then propelled by an electric locomotive under a quenching tower, before being hauled back for the coke to be discharged down a bench on to a conveyor, as seen here. The Wellman Smith Owen Engineering Corporation Ltd were the chief contractors for the Works, but the electric locomotive was sub-contracted to Goodman Bros of Chicago in the USA. WSO 529/Goodman 3576/1924 is seen here when new.

The company also used locomotives at the following locations away from Consett:

ASHES QUARRY, Stanhope   **V93**
North Eastern Railway/Ord & Maddison Ltd until 1/10/1900   NY 995400 approx.

For many years the company did not own limestone quarries, as under an agreement dating back to the early years of the Derwent Iron Co the NER, as it became, agreed to supply limestone, and worked its own quarries in order to do so. Latterly Consett's requirements came from this quarry, probably begun in the 1870s, which was served by a separate bank foot of the Crawleyside Incline on the NER Stanhope & Tyne branch. The quarry also had a private-owned link to Lanehead Quarry, where the lime kilns had continued to operate after the closure of the quarry itself about 1890. Although owned by the NER, both Lanehead and Ashes were operated by Ord & Maddison Ltd, whose headquarters were in Darlington.

When The Consett Iron Co Ltd took over from the NER on 1st October 1900 there was a locomotive at the quarry. It was immediately declared "unsafe" and a 'B' class locomotive was sent to replace it. The company objected to its inclusion in the valuation of the quarry equipment, and it is not known whether the NER removed it or whether the company cut it up.

It would seem that the Consett company also took over the lime kilns at Lanehead Quarry, whose output was dispatched by a separate bank foot on the Crawleyside Incline. The kilns were probably closed about the beginning of the First World War.

At Ashes, named after a nearby house, a large new area of limestone lay further east of the existing workings. This the company leased, but it was unable to acquire the two intervening fields to link the old workings to this new area. So about 1920 a narrow tunnel had to be driven to provide this link, the new loco B No 3 (HL 3495/1920) being specially designed to pass through this. Eventually the quarry stretched eastwards for about a mile. To make locomotive working easier the old loco shed near the level crossing over the road into Stanhope was supplemented by a new shed ½ mile into the quarry.

Some time after the First World War Consett began purchasing limestone from The Steetley Lime Co Ltd at Coxhoe. Ashes Quarry ceased production in 1944, after which the company also obtained limestone from the Durham & Yorkshire Whinstone Co Ltd's Frosterley Quarry.

Gauge : 4ft 8½in

The Minutes of the meeting of directors of The Consett Iron Co Ltd for 2/10/1899, the day after the company had taken possession of the quarry, record that a "four wheeled coupled locomotive" had been sent here already. It was almost certainly a 'B' class engine, because a replacement 'B' class engine to replace the engine sent to the quarry was ordered in the same month. No transfers are recorded in the Locomotive Registers before 1915.

Locomotives recorded here are: B No 2, B No 9, B No 3 (HL 3495), B No 12 (BH 698) and B No 18 (BH 854). For full details see the main list above. There may have been other locomotives here not recorded.

BUTSFIELD QUARRY, near Consett   T94
NZ 096445

With the design of blast furnaces involving ever-higher temperatures the bricks lining the furnaces had to be capable of withstanding these temperatures. This led to the manufacture of refractory bricks utilising ganister, from which silica could be obtained. A large deposit of suitable ganister was found some 3¾ miles south-west of Consett, at West Butsfield. Quarrying here certainly began in the latter half of the nineteenth century, for in March 1898 the company renewed its lease. At this time the quarry was worked both manually and by horses, which were also used to pull the carts to Rowley Station on the NER Consett to Tow Law line, where it was transferred into wagons for the two mile journey to Consett.

This lease expired in October 1918, and it may be at this date that the actual working of the quarry was taken over by G.F. Butterfield, a haulage contractor from Tow Law. Initially he continued to use horses on a 2ft gauge system between the quarry face and the loading point for the steam lorries now being used. However, in 1922 the company opened at Consett the Templetown Refractory Works, one of the largest of its type in the country, and clearly the transport system between the quarry face at Butsfield and Consett was inadequate for the increased output now required. So a new 3ft 0in gauge system

40. The ganister for the high heat-resistant bricks needed to line the new ovens came from Butsfield Quarry, 3¾ miles from Consett, in which was developed an extensive 3ft gauge system. WB 2084/1919 awaits scrap inside the loco shed in 1951 after the system was replaced by road transport.

was introduced within the quarry, worked from 1924 by locomotive haulage, with an aerial ropeway about four miles long to carry the ganister to the Templetown Works. It would seem that at first the first locomotive below was owned by G.L. Butterfield, for the spares orders for it up to 1929 were ordered by him; thereafter they were ordered by the company, although the locomotives are not included in the Consett Locomotive Registers and do not seem to have visited Consett for repairs. Locomotive haulage was abandoned in favour of road transport in 1945. The quarry was closed in 1953, the Consett company then obtaining its ganister from other sources (see above).

Gauge : 3ft 0in

CTS 1	0-4-0ST	OC	WB	2058	1917	(a)		Scr by 6/1951
-	0-4-0ST	OC	WB	2084	1919	(b)		Scr 10/1951
15	0-4-0ST	OC	HC	485	1897	(c)		(1)
FYLDE	0-4-0T	OC	P	1671	1924	(d)		(1)

(a) ex William Patterson & Co, Newcastle upon Tyne, via repairs at Forth Engine & Motor Works, Newcastle upon Tyne, 1924 (CIC information); formerly Board of Trade, Timber Supplies Dept, and used in Slaley Forest, Northumberland.
(b) ex Dorman, Long & Co Ltd, Burley Ironstone Quarries, Rutland, c 5/1929, via repairs at WB.
(c) ex C.D. Phillips, Newport, Monmouthshire, dealer, /1935; formerly Guest, Keen & Nettlefolds Ltd, Cyfarthfa Ironworks, Merthyr Tydfil, Glamorgan.
(d) ex Davies, Middleton & Davies Ltd, Caerphilly, Glamorgan, /1943.

(1) scrapped on site between 6/1951 and 10/1951.

CHOPWELL & GARESFIELD RAILWAY A/B95

This railway was developed from the Garesfield Waggonway, purchased from the 3rd Marquis of Bute (which see) in July 1889. To serve the company's new colliery at **CHOPWELL** (NZ 118584), developed between 1894 and 1909, the waggonway was extended by 2½ miles between 1892 and 1894. Then from 1895 onwards the majority of the section between **GARESFIELD COLLIERY** (NZ 139598) at High Spen and Winlaton Mill was replaced by a new route incorporating the penultimate self-acting incline to be built in Durham, the 1¾ mile Garesfield Incline, designed to handle 3,000 tons of coal per ten-hour shift. The line then travelled northwards to the River Tyne to Derwenthaugh, near Swalwell, where a new staith was opened in 1898, rebuilt and enlarged in 1913. This new railway, 7¼ miles long, was finally opened in July 1899, and is entitled the Chopwell & Garesfield Railway on Ordnance Survey maps, although Garesfield Railway is also found.

The Railway had **loco sheds** at **Chopwell Colliery, Garesfield Colliery** and at **Derwenthaugh**, near Swalwell (NZ 204630), where there were repair facilities for both locomotives and wagons, although all major locomotive repairs were undertaken at Consett. At Chopwell an electricity power station was built, together with the last major batteries of beehive coke ovens in Durham. An electrically-driven car, very probably similar to the one used at Axwell Park Colliery (see p.315), was used to charge the ovens. The old Garesfield Staith and the Garesfield beehive coke ovens nearby were replaced by new staiths in 1898, which were in turn replaced in 1913. The company's last major investment on this system was **DERWENTHAUGH COKE WORKS** (NZ 193615), opened about ½ mile south of Winlaton Mill in the spring of 1929. These incorporated 56, later increased to 66, Otto twin regenerative ovens, the first of this so-called 'hair pin' design in Britain, with the by-products being tar, sulphate of ammonia and rectified benzole. To these were added in 1936 a battery of 20 Simon-Carves Otto twin regenerative ovens. Semaphore signalling, with signal boxes near both ends of the coke works, was introduced on this section of the line, possibly at the same time as the coke works was opened, in order to avoid conflict between coke works traffic and 'main line' working. The beehive ovens at Chopwell closed in 1940.

On 1st January 1947 the Railway, together with Chopwell Colliery and its power station, Garesfield Colliery and its **brickworks**, latterly producing only common bricks, Derwenthaugh Coke Works and Derwenthaugh Staiths, passed to NCB Northern Division No.6 Area.

Transfers of locomotives, locomotive cranes and rail cranes to sheds on the Railway were controlled by the company's Locomotive Department, and are shown in the main locomotive list above.

For a longer and more detailed description of the Railway and its history see *The Private Railways of County Durham*, Colin E. Mountford, Industrial Railway Society, 2004.

DERWENTHAUGH COKE OVENS, Axwell Park B96

Note : this locomotive entry is an addition to the description of these ovens given in chapter 3 of *The Private Railways of County Durham*. It was the first coke car locomotive actually built by the firm.

Gauge : 4ft 8½in

-	4wWE	WSO	1252	1928	New	(1)

(1) to NCB No. 6 Area, with the ovens, 1/1/1947.

GARESFIELD COKE OVENS, Derwenthaugh B97
(Chopwell & Garesfield Railway) NZ 204631

These beehive coke ovens were situated on the southern bank of the River Tyne at Derwenthaugh, and should not be confused with the ovens with the same name at Garesfield Colliery. Oral tradition claimed that the locomotive below was used here. The ovens were cleared in 1898 to make way for new staiths.

Gauge : 4ft 8½in

BETTY	0-4-0ST	OC	?	?	?	(a) (1)

(a) believed to have been ex John Jackson, contractor for the extension of the Garesfield Waggonway from Garesfield to Chopwell, which was completed late in 1893 or early in 1894; said to have been a "four-wheeled type" with a "launch-type boiler"; identity unknown; not taken into The Consett Iron Co Ltd's locomotive stock.

(1) said to have been used as a pumping engine; s/s.

JARROW WORKS, Jarrow E98
New Jarrow Steel Co Ltd (formed 6/7/1938) until 6/1948 NZ 324655
(this was a subsidiary of **The Consett Iron Co Ltd** from 6/7/1938 to 6/1948)

This rolling mill, initially called the **CONSETT JARROW ROLLING MILLS**, was opened on 15th May 1940, on the site of part of the former Palmers Shipbuilding & Iron Co Ltd's shipyard (which see). It was served by a ½ mile branch from the LNER Gateshead-South Shields line at Jarrow Station. In 1952 the company began basic refractory brick making here. The works passed to the British Steel Corporation on 1st July 1968.

Gauge : 4ft 8½in

Locomotives which worked here, in order of arrival, were (a) steam locomotives: JARROW No.1, B No 33, B No 31, B No 7 (HL 3474), B No 30, B No 25, B No 38, B No 29, B No 13 (HL 3953) and B 42; (b) crane tanks: D No 12, D No 16, and (c) diesel locomotives: 21, 27, 16, 26, 13 and 19. For full details see the main list above.

19 (HE 5375/1958) and 26 (HE 5386/1959) passed to British Steel Corporation on 1/7/1968.

Besides the locomotives above, one steam and one diesel rail crane were used here up to 1968.

LANGLEY PARK COLLIERY & COKE OVENS, Langley Park K99
  NZ 211456

The sinking of this colliery was begun in November 1873 and the Busty seam was reached on 24th April 1875. To serve it the NER built in 1875 a ¾ mile branch from immediately west of Witton Gilbert Station on the NER Lanchester Valley branch (Durham-Consett) with a reverse before crossing the River Browney via a bridge to reach the colliery. At the same time two batteries of beehive coke ovens were built on the south side of the river north of the first section of the branch, coke production beginning by the end of 1876. At the same time the colliery began the first of quite a number of drifts developed higher up the valley side to work the seams that outcropped there. All of these were served by 2ft 0in gauge lines operated by main-and-tail rope haulage. In the mid-1890s the colliery was reconstructed, and it may be from this time that coal was taken still in its tubs by an endless rope system ¾ mile long, much of it on an elevated wooden gantry, to screens near the ovens, where some was crushed before direct transfer into the ovens. From possibly about the beginning of the twentieth century, an electrically-operated hopper vehicle was used to charge the ovens. Further re-modelling took place in 1913-1915. The beehive ovens were closed in October 1913, the screens were demolished and new screens built at the colliery; in their place four batteries of fifteen German Otto by-product ovens were completed in May 1915 (NZ 210453). A further fifteen had been built by 1917.

The DCOA Returns list no locomotive here in March 1890, while the first-known transfer listed in the Consett locomotive ledgers was dated 1905. It may therefore well be that the colliery and coke ovens were shunted by the NER until then. It is not known whether the NER or LNER sold the branch to the Consett Iron Co Ltd in a similar way to the Medomsley branch (see above). The loco shed lay at the eastern end of the coke ovens area.

The colliery and coke ovens passed to NCB Northern Division No. 5 Area on 1st January 1947; at this date three seams were worked via the shafts and three, the Five Quarter, the Main Coal and the Hutton, by drifts on the valley side served by main-and-tail rope haulage; the system had previously also served the Harvey Drift, closed in October 1942.

Reference : *Langley Park Colliery, Centenary*, 1875-1975, edited by Colin E. Mountford and published by the colliery in 1975 to mark its centenary and closure.

Gauge : 4ft 8½in

Locomotives recorded here, in order of known arrival, were: B No 4 (CF 1205), B No 27, B No 16, B No 31, B No 21, B No 28, B No 29, B No 25, B No 33, B No 22, B No 24, B No 35 and B No 32. According to the records of directors' meetings, B No 22 (AB 895/1901) was ordered new for Langley Park, but the Locomotive Registers show it as new to Consett.

B No 16 (RS 2725/1890) and B No 35 (HL 3254/1917) passed to NCB No. 5 Area on 1/1/1947.

WHITTONSTALL RAILWAY   A100

Having largely completed the major investment at Chopwell (see above), the company turned next to the royalty immediately to the west, at Whittonstall in Northumberland, where **WHITTONSTALL DRIFT** (NZ 087572) was completed early in 1908. To serve this a 2ft 2in gauge line about two miles long was built from Chopwell No.2 Pit using electric locomotives. This was the first electrically-operated railway to be built in Durham, and so the company was very much leading the field; indeed, it remained the only electric narrow gauge line ever built in the county. It was opened on 11th September 1908, the locomotives, or "cars", being provided with a shed at Chopwell. Miners were carried to and from the drift in empty tubs. In 1916 the line was extended beyond Whittonstall to handle sawn timber from the forests there, and branches were built at Chopwell for the same purpose. All of these were rope-worked, and had closed by 1921.

The locomotives were not the success hoped for, and in February 1913 main-and-tail rope haulage operated by the **Ravenside Engine** (NZ 113586) replaced them on the section between Whittonstall and the Engine, leaving them handling only the shunting over the remaining ½ mile to the screens at Chopwell No.2 Pit. Their work was further cut back in 1923 and they were dispensed with altogether about 1929, after which all railway working was handled by main-and tail rope haulage. The drift and railway were closed in 1940 because of shortage of men caused by the Second World War. They passed, still closed, to NCB Northern Division No.6 Area on 1st January 1947.

For a longer and more detailed description of the Railway and its history see *The Private Railways of County Durham*, Colin E. Mountford, Industrial Railway Society, 2004.

Gauge : 2ft 2in

WHITTONSTALL No.1	0-4-4-0WE Siemens	450	1908	New	Scr c/1930
WHITTONSTALL No.2	0-4-4-0WE Siemens	454	1909	New	Scr c/1930
WHITTONSTALL No.3	0-4-4-0WE Hanomag	5968	1910		
	Siemens	460	1910	(a)	Scr c/1930

(a)   New; built by Hanomag, with the electrical parts supplied by Siemens.

CONSETT WATERWORKS CO

SMIDDY SHAW RESERVOIR, near Waskerley   T101
                                                                                       NZ 043463

This company was created by Act of Parliament in 1860, and constructed a number of reservoirs south-west of Consett. The biggest of these was **SMIDDY SHAW** near Waskerley, which was constructed between 1869 and 1877. It covered 65 acres and held 225 million gallons. It is believed that the site was served by a link to the NER Stanhope & Tyne branch near Waskerley. In April 1870 the company advertised for two standard gauge locomotives not over 20 tons in weight. It is believed that the locomotives below worked here during the construction period.

The company subsequently amalgamated with the Weardale Water Company to form the Weardale & Consett Water Company, which on 31st December 1920 was taken over by the Durham County Water Board (which see). The reservoir continues in use.

Gauge : 4ft 8½in

-	0-6-0	OC	Kitching	1840	(a)	(1)
25 DERWENT	0-6-0	OC	Kitching	1845	(b)	(2)

(a) ex NER, which had ordered its scrapping in 1867; formerly Stockton & Darlington Railway, No.26, PILOT.
(b) hired from Joseph Pease & Partners, Pease's West, Crook, 1870.

(1) borrowed back by the NER and paraded at Stockton & Darlington Railway 50th anniversary celebrations in 1875 as No.10 AUCKLAND, ostensibly to illustrate locomotives built by Timothy Hackworth; said to have been returned to the company afterwards; s/s.
(2) returned to Joseph Pease & Partners, Pease's West, Crook, 1872.

There were said to have been another three former Stockton & Darlington Railway locomotives here, two built by Hackworth and one by Alfred Kitching.

41. *A fake! Once Stockton & Darlington Railway No.26, PILOT, built by Kitching in 1840, the NER sold her to the Consett Waterworks Company for the construction of the Smiddy Shaw reservoir near Waskerley – only to borrow her back in 1875 for the celebrations to mark the 50th anniversary of the Stockton & Darlington, and for her to be altered to look like No.10, a Hackworth locomotive with two tenders! This photograph is believed to have been taken at North Road Works, Darlington, after the alterations were completed.*

CROSSLEY SANITARY PIPES LTD
originally **J. Crossley & Sons Ltd**, the firm subsequently becoming a holding company and setting up the firm above as a subsidiary
FIR TREE WORKS, Fir Tree, near Crook M102
NZ 134344

This works was developed by the firm in 1946 on the site of the former Fir Tree Drift owned by Bearpark Coal & Coke Co Ltd to replace the firm's large works at Commondale in the North Riding of Yorkshire. It produced both sanitaryware, including pipes, and bricks. It had no rail connection, the locomotives below being used only within the site.

The works was taken over by The North Bitchburn Fireclay Co Ltd in the early 1960s. Whether either of the first two locomotives below was in use then is unknown. Pipe making ceased about 1969 and the works closed with the end of brickmaking in 1977.

Gauge : 2ft 0in

-	4wPM	Crossley	?	(a)	Scr
-	4wDM	RH	247174	1947	New	Scr
-	4wDM	HE	3496	1947	(b)	(1)

(a) built by the firm using an engine from a Morris car.
(b) ex NCB No. 5 Area, Hole-in-the-Wall Colliery, Crook, on loan, 4/1955.

(1) returned to NCB No. 5 Area, Hole-in-the-Wall Colliery, Crook.

CROUCH MINING LTD
former MAINSFORTH COLLIERY YARD, Ferryhill Station
N103
NZ 307316

When Crouch Mining Ltd's contract to operate the Widdrington Disposal Point in Northumberland ceased in the summer of 1995, the company brought its locomotives here for temporary storage, in the open. On 11th July 1995 the firm's parent company, Crouch Group plc, went into liquidation, but this company survived. There was no main line rail connection.

Gauge : 4ft 8½in

MP342	0-6-0DH	EEV	D924	1966	(a)	(1)
-	0-6-0DH	EEV	D1201	1967	(a)	(2)
MP201	0-6-0DH	EEV	D1202	1967	(a)	(3)
MP202	0-6-0DH	EEV	D1230	1969	(a)	(4)
-	0-6-0DH	EEV	3994	1970	(a)	(5)

(a) ex Widdrington Disposal Point, Widdrington, Northumberland, c8/1995 (after 3/7/1995, by 15/10/1995).

(1) to Yorkshire Engine Co, Long Marston, Warwickshire, 16/2/2000, where scrapped by Stratford Car Breakers, Stratford-on-Avon, Warwickshire, by 18/5/2002.

(2) to Yorkshire Engine Co, Long Marston, Warwickshire, 19/2/2000; rebuilt as YEC L182/2000, and then hired to Ford Motor Co Ltd, Bridgend, West Glamorgan.

(3) to Yorkshire Engine Co, Long Marston, Warwickshire, 3/2/2000; re-sold to A.Briddon, South Yorkshire, 1/2002.

(4) to Yorkshire Engine Co, Long Marston, Warwickshire, 7/2/2000; rebuilt as YEC L179/2000 and re-sold to Tibbett & Britten Group, Neasden, Greater London, 11/12/2000.

(5) to Yorkshire Engine Co, Long Marston, Warwickshire, 3/2/2000; rebuilt as YEC L180/2000, and re-sold to Burrows Brothers (Sales) Ltd, Oxcroft Disposal Point, Stanfree, near Clowne, Derbyshire, 2/8/2000.

CROWN COKE CO LTD (registered 9/1907)
Templetown, Consett
F104

This company owned a plant, the exact site of which is not known, which washed coke waste. It is said to have been linked to the Consett Iron Co Ltd's system in the Templetown area, from which a locomotive was hired on occasions. It appears to have closed before the First World War.

Gauge : 4ft 8½in

NEWPORT (?)	0-6-0ST	OC	FW	169	1872	(a)	(1)

(a) ex Sir B. Samuelson & Co Ltd, Newport Works, Middlesbrough, Yorkshire (NR).

(1) advertised for sale, 1/1912; to The Bearpark Coal & Coke Co Ltd, Bearpark, c/1915.

THE DARLINGTON FORGE LTD
(subsidiary of **English Steel Corporation** from 11/1933)
The Darlington Forge Co Ltd until 6/6/1919
The Darlington Forge Co until 25/9/1873
DARLINGTON FORGE WORKS, Darlington
S105
NZ 295157 (initially - see below)

The date when this works opened is uncertain, but it was not before 1855. It is described at first as little more than a "heavy blacksmith's shop", but this gradually developed into a major forging business. The works was served by a short link to the Stockton & Darlington/NER Darlington-Bishop Auckland line from Albert Hill Junction to the South Durham Iron Co (which see), which lay adjacent to the north-east. In its early days two of the partners in the company were John Cowans and E.P. Sheldon of the engineering and crane construction company Cowans, Sheldon & Co of Carlisle in Cumberland.

One source alleges that in 1872 the company purchased part of the nearby works of the Darlington Iron Co, but this is inaccurate. However, the South Durham Iron Co (which see) went into voluntary liquidation in June 1877, and in 1886 Darlington Forge purchased its site for its own expansion. East of the former South Durham works lay the premises of Darlington Steel & Iron Co Ltd (which see). This works closed in 1894 and in May 1898 its site too was sold to Darlington Forge, the sale being completed in March 1899. This expanded the Forge site to 30 acres.

42. WOODBANK, BH 387/1876, although fitted with new frames supplied by BH as 1019/1890. The 20-ton NER coal wagon she is shunting dwarfs her!

43. After the firm became a subsidiary of the English Steel Corporation, ESC locomotives were sent to Darlington; No.9, HC 1688/1937, on 24th August 1956.

At first the firm operated mainly in the railway market, but it later became famous for its marine castings. The company was re-structured in 1919, and in the same year the foundry in what had been the Darlington Steel & Iron Co Ltd's area of the works was sold to its southern neighbours, Thomas Summerson & Sons Ltd. During the late 1920s the company hit hard times, and the works closed in 1932. However, in the following year the English Steel Corporation took over the company from the receivers and early in 1936 re-opened the works under the old title. The works eventually ceased production in February 1967, but survived to be merged into the British Steel Corporation on 1st July 1968 (which see).

There may well have been one or more locomotives here in the early years of the firm. There was certainly at least one more, because it exploded in July 1873, with such force that only the frame plates survived. There were also 11 steam rail cranes here at various times.

Gauge : 4ft 8½in

		PELAW	0-6-0T	OC	BH	?	?	(a)	(1)
9/40		WOODBANK	0-4-0ST	OC	BH	387	1876	New	
			reb		Dar Forge		1890	(b)	Scr 3/1960
		BLACK PRINCE	0-6-0ST	OC	BH	354	1875	(c)	(2)
		-	0-6-0WT	OC	Barningham	?		(d)	s/s by 3/1928
		ALLIANCE	0-4-0T	OC	GW/HG	?	?	(e)	(3)
		UNITY	0-4-0ST	OC	HL	2890	1911	New	Scr 2/1963
		CONCORD	0-4-0ST	OC	HL	3300	1917	New	Scr 2/1960
		ADVANCE	0-4-0ST	OC	BH	1113	1895	(f)	(4)
9/5007		(KING GEORGE VI)	0-4-0ST	OC	RSHD	7013	1940	New	(5)
		-	0-4-0ST	OC	HL	2839	1910	(g)	(6)
		TOM BARRON	0-4-0ST	OC	AB	1287	1914	(h)	(7)
		E.S.C. No.1	0-4-0ST	OC	HC	1199	1916	(j)	Scr c/1963
		E.S.C. No.9	0-4-0ST	OC	HC	1688	1937		
			reb		ESC		1954	(k)	Scr 1/1960
No.46			0-4-0DM		HC	D1159	1959	New	(5)
No.48			0-4-0DM		HC	D1161	1959	New	(5)

(a) if the identity of its disposal (see (1) below) is correct, then this loco was one of the locomotives advertised in the sale of plant owned by J.C. Tone, contractor for the NER Pelaw to Jarrow branch, in *Colliery Guardian*, 29/5/1872.
(b) rebuilt with new frame plates supplied by BH as BH 1019/1890.
(c) included with the premises of South Durham Iron Co when Darlington Forge acquired them in 1886.
(d) included with the premises of Darlington Steel & Iron Co Ltd when Darlington Forge acquired them in 3/1899.
(e) origin unknown; here by 27/3/1928; may be the GW/HG locomotive from Blake Boiler, Wagon & Engineering Co Ltd, Darlington (which see).
(f) ex The Consett Iron Co Ltd, Consett, after 27/3/1928.
(g) ex Thos W. Ward Ltd, Sheffield, Yorkshire (WR), 9/1942, hire.
(h) loaned from Darlington Corporation, Darlington Gas Works, sometime between 1939 and 1945.
(j) ex English Steel Corporation, River Don Works, Sheffield, Yorkshire (WR), 7/1954.
(k) ex English Steel Corporation, River Don Works, Sheffield, Yorkshire (WR), 5/1956.
(1) this locomotive was offered for sale in *The Engineer* of 13/12/1878, where it is described as a "six coupled tank locomotive with outside cylinders 14in by 20in and 3ft 6in wheels". The most likely candidate would seem to be BH 60/1868, which had been for sale in 1872 (see (a) above). If the identity is correct, then the loco went to The Birtley Iron Co, Birtley.
(2) loaned to The Newcastle upon Tyne & Gateshead Gas Company, Redheugh Gas Works, Gateshead, and returned; to Skinningrove Iron Co Ltd, Carlin How, Yorkshire (NR), 1/1891.
(3) an 0-4-0T OC named ALLIANCE, later at Blake Boiler, Wagon & Engineering Co Ltd, Darlington, is believed to have come there from The Darlington Forge Ltd.
(4) said to have been scrapped "in the early 1930s", but here on 12/3/1935; s/s.
(5) loaned to Darlington Corporation, Darlington Gas Works, and returned, both dates unknown; to British Steel Corporation, with works, 1/7/1968.
(6) to Tees Side Bridge & Engineering Works Ltd, Cargo Fleet, Yorkshire (NR), 3/1944 (still owned by Thos W. Ward Ltd, Sheffield, Yorkshire (WR).
(7) returned to Darlington Corporation, Darlington Gas Works, c/1945.

DARLINGTON & SIMPSON ROLLING MILLS LTD

Darlington Rolling Mills Co Ltd until 31/3/1945 (see also below)
Sir Theodore Fry & Co Ltd until 19/7/1909 (liquidation)
Fry, Ianson & Co Ltd until 17/4/1894 (name change of existing company) *
Fry, I'Anson & Co until 1/10/1890
originally **Charles I'Anson & Co Ltd**, possibly until 26/4/1887
* in the official registration of the company on 1/10/1890 the family name of I'Anson is spelled Ianson in fact, there were no I'Anson shareholders in the company.

RISE CARR IRON WORKS, later ROLLING MILLS, Darlington　　　　　　　　　　S106/S107
South Works, NZ 276166 **(S106)** and **North Works**, NZ 275170 and 274170 (two sites) **(S107)**

The original works, Charles I'Anson's second in the area (see Whessoe Products Ltd), opened during the summer of 1865 for the production of puddled iron. The works was named after a nearby farm. Both I'Anson (1809-1884) and Fry were Darlington Quakers, Fry being a member of the Bristol chocolate family. The I'Anson locomotives in the list below were almost certainly built at Charles I'Anson's Whessoe Works (see Whessoe Products Ltd) rather than here, as previously thought. This works was sold to Fry, probably in 1887, after the death of Charles I'Anson in 1884 and the retirement of his son James (1845-1898) in the following year. After the change of ownership the Rise Carr Works expanded its production into various rolled steel items.

In 1920 half of the shares were acquired by Bolckow, Vaughan & Co Ltd of Middlesbrough, who took over full control in 1922. On 24th December 1927 the production of iron from the puddling furnaces ceased. In 1928 Bolckow Vaughan entered into a partnership with the Crittall Manufacturing Co Ltd of Braintree in Essex, who manufactured metal window frames for council houses. On 1st November 1929 Bolckow Vaughan's share passed to Dorman, Long & Co Ltd on the two companies' amalgamation. On 7th June 1935 Dorman Long and Crittall acquired F.R. Simpson & Co Ltd of Wednesbury in Staffordshire, setting up a new holding company, Darlington & Simpson Rolling Mills Ltd, to combine the businesses, with Darlington Rolling Mills Co Ltd as a subsidiary to this firm. On 31st March 1945 this structure was abolished and the holding company, still jointly owned, became the operating company.

By the 1930s the firm was operating on what it regarded as two separate sites, both alongside the LNER Darlington-Bishop Auckland line; the South Works, which was the original premises, lay to the east of the line ½ mile north of Darlington (North Road) Station, and the North Works, on both sides of the line ½ mile further north, which the company shunted by using running powers between them.

44. A rare photograph of a locomotive built by Charles I'Anson & Co of Darlington, probably at the firm's Whessoe Works. RISE CARR dates from 1875.

Under the Iron & Steel Act of 1949 the company was scheduled to be vested into the British Iron & Steel Corporation on 16th February 1951, though this only lasted until 15th July 1953, when the Conservative government de-nationalised the industry.

Rail traffic to the North Works ceased in the autumn of 1978 and to the South Works about 1985. The company subsequently became part of British Steel plc, rolling special steel sections, and the works continues (2006) in production.

45. 103 years later the Rise Carr Works was being shunted by a modern diesel locomotive : No.1, TH 129V/1963, taken on 1st July 1978.

Gauge : 4ft 8½in

	-	0-4-0ST	OC?	I'Anson		?	New	(1)
	RISE CARR	0-4-0ST	OC	I'Anson		1875	New	(2)
	-	0-4-0ST	OC	HL	?	?	(a)	
		reb	HL	9242	1900			(3)
No.1		0-4-0ST	OC	HL	2039	1885	(b)	Scr
No.2		0-4-0ST	OC	AB?	?	?		
		reb	AB	9070	c1899	(c)	s/s	
No.3		0-4-0ST	OC	BH	?	?		
		reb	LG		?	(d)	s/s	
	JUNO	0-4-0ST	OC	BH	606	1881	(e)	(3)
	WYLLIE	0-4-0ST	OC	BH	524	1880	(f)	(4)
	PRINCE	0-4-0ST	OC	Harris?	?	?	(g)	s/s
No.1		0-4-0ST	OC	HL	3237	1917	(h)	(5)
No.2		0-4-0ST	OC	RSHN	7073	1943	New	(6)
	B No 28	0-4-0ST	OC	HL	3003	1913	(j)	(7)
No.3		0-4-0ST	OC	RSHN	7160	1945	New	(5)
No.4		0-4-0ST	OC	RSHN	7660	1950	New	(8)
	-	4wDH		TH	118C	1962	(k)	(9)
No.1		4wDH		TH	129V	1963	New	(10)
No.2		4wDH		TH	131V	1963	(m)	(10)

(a) origin and date of arrival unknown.
(b) ex R. & W. Hawthorn, Leslie & Co Ltd, Hebburn Shipyard, Hebburn, /1913.
(c) origin uncertain; owned by J. Best, contractor, until at least /1912.

(d) ex Lingford, Gardiner & Co Ltd, dealers, Bishop Auckland.
(e) ex Bolckow, Vaughan & Co Ltd, Cleveland Works, Middlesbrough, Yorkshire (NR).
(f) ex Bolckow, Vaughan & Co Ltd, hire; believed to have come from Byers Green Colliery, Byers Green.
(g) ex Bolckow, Vaughan & Co Ltd, Cleveland Works, Middlesbrough, Yorkshire (NR); may be the LG hire loco named PRINCE.
(h) ex Thos W. Ward Ltd, after repair at HL, /1934; previously Palmers Shipbuilding & Iron Co Ltd, Jarrow.
(j) ex The Consett Iron Co Ltd, Consett, 26/11/1944, loan.
(k) ex TH on demonstration, c2/1963.
(m) New ex TH, 29/11/1963, after demonstration at Firth Brown Ltd, Sheffield, Yorkshire (WR) and at East Midlands Gas Board, Carr House Works, Rotherham, Yorkshire (WR).

(1) to R.W. Crosthwaite Ltd, Thornaby, Yorkshire (NR).
(2) to Joseph Love & Son, Malton Colliery, near Lanchester.
(3) to Gateshead County Borough Council, Saltmeadows Clearance Site, Gateshead, via The Rolling Stock Co Ltd, Darlington, acting as dealers, 1/1936.
(4) returned to Bolckow, Vaughan & Co Ltd, Cleveland Works, Middlesbrough, Yorkshire (NR).
(5) to T.J. Thomson & Son Ltd, Stockton, for scrap, /1963.
(6) to ?, for scrap, /1963.
(7) returned to The Consett Iron Co Ltd, Consett, 21/12/1944.
(8) to W. Arnott, Young & Co Ltd, Darlington, for scrap, 5/1965.
(9) returned to TH after demonstration, 5/3/1963.
(10) to North Yorkshire Moors Railway, Grosmont, North Yorkshire, for preservation, 26/2/1986.

An 0-4-0ST OC with 10in x 18in cylinders was offered for sale in *Machinery Market*, 5/3/1903, and *Contract Journal*, 18/3/1903.

DARLINGTON STEEL & IRON CO LTD
Darlington Iron Co Ltd until 15/6/1882
Darlington Iron Co until 7/11/1872
William Barningham until /1864
ALBERT HILL IRONWORKS, Darlington   S108
                                                                            NZ 296158

William Barningham, a Yorkshireman, having set up a very successful factory in Manchester to produce wrought iron railway rails, sought to do the same in the north-east, and in 1858, at the age of only 33, he built this works, which began production in 1859. It was served by a reverse extension of the branch from Albert Hill on the Stockton & Darlington Railway's Darlington-Bishop Auckland line which served the Darlington Forge Co and the South Durham Iron Co (both which see), Barningham's works lying immediately to the east of the latter. Subsequently a second works, the **SPRINGFIELD WORKS** (NZ 296160), was built immediately north of the Albert Hill Works to produce wrought iron from puddled pig iron.

In 1864 Barningham set up the Darlington Iron Co, remaining a major shareholder until his death in 1882. The source which alleges that he sold the Albert Hill works to The Darlington Forge Co in 1872 is clearly inaccurate. There was a court order to wind the company up in 1881, but it survived until Barningham's death and was then re-organised under the chairmanship of Hugh Bell of Bell Bros Ltd, the Middlesbrough ironmasters and colliery owners.

Facing the need to replace the production of wrought iron by steel the new company installed two Bessemer steel furnaces in the Albert Hill Works, and by 1885 the Springfield Works was derelict. But by the early 1890s it was proving more expensive to manufacture steel at Darlington compared with works at Middlesbrough, and this, coupled with a slump in trade, led to the works being closed in September 1894, this marking the end of iron and steel production in the town. The company went into liquidation in May 1896, and dismantling began in October 1897. The (combined) site was sold to the neighbouring firm of The Darlington Forge Co Ltd in May 1898, though the sale was not finally completed until March 1899.

Almost nothing is certain about the locomotives that were used here. One source claims that in 1867 three 0-4-0ST's were built at Barningham's Pendleton Works in Manchester (the Manchester Steel & Railway Plant Co) for this works. Another source claims that Barningham built four 0-6-0WT locomotives which worked here. Equally, one or more of the Barningham locomotives reputed to have worked here is said to have been built here also. How many locomotives the firm built for its own use is uncertain; other versions give a different number and different types. See also the footnotes below.

46. William Barningham is said to have built a number of locomotives at his Manchester factory for his Darlington works, and this is the only known photograph of one of them, after it was rebuilt from a 0-4-0ST to a 0-6-0ST – although the front end looks very similar to the 0-4-0STs built by John Harris, also in Darlington.

In January 1866 experiments were carried out here with a locomotive to take coal, iron ore and limestone up a gradient of about 1 in 14 to the blast furnaces. Its identity, whether it was adopted permanently and its disposal are all unknown.

Gauge : 4ft 8½in

-		0-4-0ST	OC	?	?	(a)
	reb	0-6-0ST	OC	Barningham c1870	#	(1)
-		0-4-0ST	OC	?	?	(a) (2)
-		0-4-0ST	OC	?	?	(a) (2)
-		0-6-0WT	OC	?	?	(a) (3)

\#   A photograph of this locomotive as a 0-6-0ST, included here, suggests a strong similarity with 0-4-0ST's built by John Harris at his Hope Town Foundry in Darlington.

(a)   origin, builder and date of arrival all unknown; see notes above and below.

(1)   one 0-6-0WT loco, allegedly built by Barningham, is said to have been sold to North Cleveland Ironstone Co, Yorkshire (NR).

(2)   two 0-4-0ST locos, allegedly built by Barningham, are said to have been sold either to Sir B. Samuelson & Co, Slapewath Mines, Slapewath, Yorkshire (NR), or to Aysdalegate Mine, Slapewath, Yorkshire (NR), which was owned by W. Barningham (possibly about 1876).

(3)   another 0-6-0WT, allegedly built by Barningham, is said to have passed with the premises when they were acquired by The Darlington Forge Co Ltd in March 1899.

A locomotive was hired for a very short period to Kirkleatham Ironstone Co, Kirkleatham, Yorkshire (NR) about January or February 1876.

THE DAVY ROLL CO LTD (registered 17/1/1920; latterly a subsidiary of **Davy-Ashmore Ltd**)
JARROW WORKS, Jarrow   E109
Armstrong-Whitworth Rolls Ltd (which see) until 7/1970   NZ 321654

This works was served by sidings north of the BR Gateshead-South Shields line, ½ mile west of Jarrow Station. However, rail traffic had ceased in 1969, before this company took over, and the locomotive was awaiting disposal. The works ceased production in April 1978 and was demolished in 1978-79.

Gauge : 4ft 8½in

| 185 | DAVID PAYNE | 0-4-0DM | JF 4110006 | 1950 | (a) | (1) |

(a) ex Armstrong-Whitworth Rolls Ltd, with site, 7/1970.

(1) to Leslie Sanderson Ltd, dealer, Birtley, 12/1970; re-sold to The Cleveland Bridge & Engineering Co Ltd, Darlington, for its contract on the Britannia Bridge, Anglesey, by 5/1972.

The works also owned a mechanised ladle carrier built as a 2-2wDM using the frames from 4wVBT VCG S 9558/1953, previously used at the works. This was last seen on 11/8/1979 and had gone by 24/12/1979.

The firm also acquired the former **CLOSE WORKS** in Gateshead owned by Armstrong-Whitworth Rolls Ltd (which see), which it continues (2006) to operate.

THE DERWENT IRON CO

The discovery that coal measures in west Durham contained bands of ironstone led to the creation of the Derwent Iron Company and its construction of blast furnaces at Carr House near Berry Edge (later Consett) in 1840. The works lay close to the Stanhope & Tyne Railway to bring in limestone from Stanhope and take away the finished products. The Stanhope & Tyne Railway, 32 miles from Stanhope to South Shields, had been opened in 1834; but despite the opening of collieries along it, it was in continual financial difficulty because of the high wayleave charges for crossing land, as it had been built without an Act of Parliament. As bands of ironstone had been found associated with some coal seams in North-West Durham, towards the end of 1839 four men formed a partnership to build an ironworks at Carrhouse, near Berry Edge, later called Consett, linking it to the Stanhope & Tyne Railway. Soon afterwards the partnership was offered the **RIDSDALE** (sometimes called **REDESDALE**) **IRON WORKS** (NY 909844), near West Woodburn in Northumberland and begun by the Chesterhope Iron Company in 1838. This had one blast furnace and used locally available coal and iron ore to produce top quality cast iron. The exact sequence is not entirely clear, but in September 1840 a new partnership of five men was drawn up, and the partnership changed its title to The Derwent Iron Company, apparently in 1841.

However, at the end of 1840 the Stanhope & Tyne Railway company collapsed and in February 1841 it was dissolved, with the section from the foot of the Carr House East Incline at Berry Edge to South Shields being taken over by a new company, the Pontop & South Shields Railway (P & SSR). With the section west of Carr House now closed, the Berry Edge ironworks was cut off from its source of limestone. With the P&SSR unwilling to take over the former Stanhope & Tyne line west of Carr House, the Derwent Iron Company was compelled to purchase this section and the quarry at Stanhope in order to safeguard its supply of limestone. The exact date of the purchase is unknown, but Tomlinson attributes it to early in 1842. The new owners called their line the **DERWENT RAILWAY (V110/T110)**.

Meanwhile the Iron Company had purchased in 1841 the **BISHOPWEARMOUTH IRONWORKS** (NZ 388569) **(H111)** at Sunderland, which had been started by 1828 and was served by the Lambton Railway (which see). The iron produced at the Ridsdale Works was sent to the Bishopwearmouth Works for refining, a tortuous journey, given that there was as yet no railway in north Northumberland.

From the foot of the Pontop & South Shields Railway's Carr House East Incline, to the east of the Berry Edge works, a branch ran northwards for 1½ miles to serve Medomsley Colliery. Information about the ownership of the colliery in the early 1840s is uncertain (see the entry for the Consett Iron Co Ltd), but certainly from the mid 1840's the branch was operated by the Iron Company, although owned by the P&SSR. It was worked by the **Derwent Engine** near the colliery.

Then in December 1843 the Iron Company, in association with the Brandling Junction Railway (BJR), re-laid the Tanfield Moor branch, which ran from the foot of the Annfield East Incline northwards to Harelaw to join the BJR's Tanfield Branch. This meant all the Iron Company's traffic could now go via the BJR to Gateshead rather than on the P&SSR to South Shields, albeit travelling over part of the P&SSR to achieve this.

Meanwhile further west the Iron Company was already looking to link up with the Stockton & Darlington (S&D) Railway's Bishop Auckland & Weardale Railway, which was due to reach Crook in 1843. The S&D agreed to purchase the Derwent Railway by acquiring the wayleaves and to build a link to it at Waskerley Park; but the former was not completed until New Year's Day, 1845, whereupon the S&D re-named the system the Weardale & Derwent Junction Railway. The link was finally opened on 16th May 1845.

By 1847 the company had invested £160,000 in the Ridsdale Works, which now had three blast furnaces, and then expanded it further to produce malleable iron. But an attempt to sink a new coal pit proved the financial last straw for this works, which closed during 1849 and was never re-opened (see *The High Level Bridge and Newcastle Central Station*, John Addyman and Bill Fawcett, Appendix 2 (a), Newcastle,1999).

In January 1851 the Derwent Iron Company began obtaining its ironstone from Bolckow & Vaughan's mine at Eston in North Yorkshire, which it supplemented a few months later by obtaining its own lease for ironstone at Upleatham. This was linked to the Middlesbrough & Redcar Railway by a narrow gauge tramway between four and five miles long. This was opened in April 1852, and it is now known that TR 208, 209 and 210 were built for this line and not for Consett. They were 2ft 1in gauge tender locomotives, the first genuine narrow gauge locomotives ever built. The Upleatham mines passed to J. & J.W. Pease in July 1857.

TR 247/1854 is recorded as built for 'Mounsey & Co'. The Mounsey family was another of the Quaker families living at Darlington, and they were prominent in Durham industry in both the 19th and 20th centuries. One possibility is that TR 254 came to Consett, another that it went to the Earl of Durham, with 'Mounsey & Co' acting as agents.

By the mid 1850s various branches from the system had been built, both at Consett and at Leadgate; for these, and for a map showing them (Fig. 2), see the entry for The Consett Iron Co Ltd.

In the middle 1850s the Iron Company was involved in another new railway scheme. To give the ironworks a better outlet to Cumberland, the Company promoted the Stockton & Darlington & Newcastle & Carlisle Union Railway, which obtained its Act in July 1856. This was for a line 8 miles long to run between Cold Rowley, west of Hownes Gill on what was now the Stockton & Darlington Railway, and Stocksfield on the Newcastle & Carlisle Railway. About a mile of this line had been built, when in November 1857 the Northumberland & Durham District Bank failed, in turn causing the collapse of the Iron Company, which owed the Bank £1 million. The Managing Director of the Bank, Jonathan Richardson, was also a leading partner in the Derwent Iron Co. By now the ironworks at Consett was so big - there were 18 blast furnaces alone - that it could not be allowed to close, and control passed to the Court of Chancery. Operations were temporarily vested in a new company set up jointly by the Stockton & Darlington and North Eastern Railways, until in 1864 **The Consett Iron Co Ltd** (which see) was created and took over the Consett and Bishopwearmouth works.

Reference : *The Private Railways of County Durham*, Colin E. Mountford, Industrial Railway Society, 2004.

ROBERT DICKINSON & PARTNERS
LILY COLLIERY (NZ 159535) and **SURTEES PIT** (NZ 155546)   G112/G113

This company owned various collieries in West Durham in the latter part of the nineteenth century. Amongst these was the **LILY COLLIERY** at Dipton. This lay only ¼ mile to the east of Dipton Colliery (Delight Pit), owned by John Bowes & Partners Ltd (which see). Situations where one firm worked the upper seams on a royalty and another firm worked the lower seams are sometimes found, despite the obvious difficulties. The Lily Pit was connected by a tramway, probably narrow gauge and rope worked, to the **SURTEES PIT**. The output from the two pits was dispatched via what was now the NER Harelaw Branch (see the reference to the Tanfield Moor branch under the entry for the Derwent Iron Co), which passed immediately to the east. The two pits seem to have closed in the 1890s, the Lily Colliery by 1895 (perhaps earlier) and the Surtees Pit, sometimes called the Collierley Pit, perhaps after this.

The firm is not known to have owned any locomotives; but on 3rd and 4th October 1900 A.T. & E.A. Crow held an auction on behalf of the Marquis of Bute, who owned the royalty and was dismantling the pits. This included two six-wheeled tank locomotives, with 10in and 12in cylinders. Their identity and fate is not known.

The company also owned **SOUTH DERWENT COLLIERY (G114)** at Annfield Plain. This comprised two pits, the Willie Pit (NZ 169516), which was sunk by Robert Dickinson & Co and began production on 14th October 1872, added to subsequently by the Cresswell Pit (NZ 164514), and **WEST SHIELD ROW COLLIERY** (NZ 193530) **(G115)** near Annfield Plain. Both of these also used surface narrow gauge tramways and both passed to The South Derwent Coal Co Ltd (which see).

F.W. DOBSON & CO LTD (registered 16/7/1921)
CHILTON QUARRY & LIMEWORKS, near Ferryhill Station N116
Pease & Partners Ltd (which see) until c/1945 NZ 302315

This business was developed on the site of Little Chilton Colliery, closed in 1865 and formerly owned by the West Hartlepool Harbour & Railway (which see). A limestone quarry and limeworks had been opened by 1894 and was served by a ½ mile branch from the NER Durham-Darlington line, ¼ mile south of Ferryhill Station. Besides limestone, dolomite, clay, sand and gravel were also produced. Rail traffic had ceased by September 1968, and operations were abandoned in December 1968.

Gauge : 4ft 8½in

24		0-4-0ST	OC	HL	2453	1900	(a)	
		reb		Tees Iron Wks		1935		(1)
	HAWTHORN	0-4-0ST	OC	RSHN	7308	1946	(b)	(2)
	F.W. DOBSON							
(formerly JAMES BLUMER)		4wDM		RH	236362	1946	(c)	Scr 11/1969

(a) ex Pease & Partners Ltd, with quarry, c/1945.
(b) ex Hawthorn Limestone Co Ltd, Hawthorn Quarry, near Seaham Harbour, by 26/3/1949.
(c) ex Cornforth Limestone Co Ltd, Cornforth, by 29/4/1950.
(1) to Hawthorn Limestone Co Ltd, Hawthorn Quarry, near Seaham Harbour, by 28/5/1949; returned c/1955; to T.J. Thomson & Son Ltd, Stockton, for scrap, 9/1955.
(2) to Hawthorn Limestone Co Ltd, Hawthorn Quarry, near Seaham Harbour.

DORMAN, LONG & CO LTD

This company was founded in 1876 as a private company by Arthur Dorman and Albert de Lande Long to manufacture iron bars and angles in Middlesbrough for the shipbuilding industry, initially at West Marsh Ironworks and from 1879 at the Britannia Ironworks, which it purchased from Bernhard Samuelson, the owner of the Britannia Iron Works Co Ltd. It was registered as a limited company on 2nd November 1889. Open hearth steel furnaces were added in the 1880s, and by the end of the nineteenth century the firm was also operating wire and rolling mills. From this point it began to acquire controlling interests in a wide variety of companies in Middlesbrough and South Durham, beginning with Bell Brothers Ltd when the latter became a public company in 1899, initially by acquiring half of the ordinary shares and the remaining half in 1902; for the full history of this company see the entry for Bell Brothers Ltd. This company thus became the first of what Dorman Long called its "Allied Companies" - Bell Brothers Ltd, Sir B. Samuelson & Co Ltd (acquired in 1917), The North Eastern Steel Co Ltd and The Carlton Iron Co Ltd (acquired in 1919), all of which it merged into the parent company on 2nd May 1923. This was followed by the take-over of the failing firm of Bolckow, Vaughan & Co Ltd on 1st November 1929, making it then the largest industrial concern in North-East England. Unfortunately, the acquisition of Bolckow Vaughan dragged the company down financially. Negotiations for a merger with either The Consett Iron Co Ltd or The South Durham Steel & Iron Co Ltd of Hartlepool both collapsed, and in 1933 a receiver was appointed. As with the similar case of Pease & Partners Ltd, a settlement was eventually agreed in the High Court, and the firm was able to survive. In 1954 it reorganised its company structure into manufacturing units (see the next entry). Its operations on Teesside, in North Yorkshire and in the part of County Durham taken into the new county of Teesside in 1974 will be found in the Cleveland & North Yorkshire Handbook; the company also controlled Upton Colliery in the West Riding of Yorkshire and the Burley ironstone quarry in Rutland.

All the collieries in Co. Durham except Newfield passed to the National Coal Board on 1st January 1947; for Newfield Colliery & Brickworks and Parson Byers Quarry, Stanhope, see the entry for Dorman Long (Steel) Ltd.

Locomotive history
Dorman Long maintained no central workshop nor any centralised repair and transfer system in the same way as The Consett Iron Co Ltd did. As a result, information about locomotives and their movements is almost entirely dependent on makers' lists and orders for spares, on the researches of individuals about the beginning of the Second World War, looking at boiler reports and interviewing workmen, and such old records as survived into the 1960s at collieries formerly owned by the firm. However, the firm did publish in 1929, unfortunately just before the take-over of Bolckow, Vaughan & Co Ltd, three volumes entitled *Schedule of Properties*, in which details of the locomotives at all of their locations are given, and this at least gives a point in time at which official information is available. Nevertheless, the information overall is conflicting and incomplete, and the lists below are believed to be the best available in the circumstances.

AUCKLAND PARK COLLIERY & COKE WORKS, Coundon Grange   Q117
Bolckow, Vaughan & Co Ltd until 1/11/1929 (which see)   NZ 227285

This colliery, opened in 1866, was served by a ½ mile branch from a junction on the NER Bishop Auckland-Shildon line one mile south of Bishop Auckland Station. There were also 100 Semet-Solvay waste heat coke ovens here, but these were closed down in May 1930. In the 1930s an aerial ropeway was built to link this colliery to Leasingthorne Colliery, possibly to carry coal to the coke ovens there. Auckland Park ceased production in June 1943 but remained in use for pumping, and locomotives continued to be sent to the workshops here for repairs. The colliery passed to NCB Northern Division No. 4 Area on 1st January 1947.

There may have been more locomotives here than those listed below.

Gauge : 4ft 8½in

108	HECTOR	0-6-0ST	OC	HL	2613	1905	(a)	(1)
113	PLUTO	0-6-0ST	OC	HL	2655	1906	(a)	(2)
	HARE	0-4-0ST	OC	GH		1908	(a)	(3)
	ACTIVE	0-4-0ST	OC	RS	3075	1901		
		reb		Ellis		1927		
		reb		R. Shaw	16	1931	(b)	(4)
	HENRY CORT	0-4-0ST	OC	BH	607	1881	(c)	(4)

(a) ex Bolckow, Vaughan & Co Ltd, with colliery and coke works, 1/11/1929.
(b) ex Mainsforth Colliery, Ferryhill Station, c/1935 (after 8/5/1934).
(c) ex Westerton Colliery, near Westerton, c/1935.

(1) to Dean & Chapter Colliery, Ferryhill, c/1934 (by 8/5/1934).
(2) to Chilton Colliery, Chilton Buildings, c/1935 (after 8/5/1934).
(3) to Newlandside Quarry, Stanhope, by 8/1931.
(4) to NCB No. 4 Area, with colliery, 1/1/1947.

BOWBURN COLLIERY, Bowburn   N118
Bell Brothers Ltd until 2/5/1923   NZ 304379

This was the second colliery to bear this name. The original Bowburn Colliery (NZ 318436) (see Fig. 6) was sunk in the 1840s and was linked to the end of the line at Coxhoe of the Clarence Railway's proposed Durham City branch. It was abandoned in 1857.

The sinking of this new colliery was begun in July 1906, with coal production commencing on 28th August 1908. Its purpose was to reduce the cost of underground haulage at Tursdale Colliery (which see), which had previously worked coal in this area. To act as the new colliery's man-riding shaft, the firm re-opened the former Heugh Hall Colliery shaft (NZ 323379) (see Fig.6), closed in the 1880s, until the Bowburn up-cast shaft was sunk about 1911. The colliery lay east of the LNER Leamside-Ferryhill line 1½ miles north of Tursdale Junction.

When the colliery was first opened its rail traffic was shunted by a steam crane (Coles 11978/1908 New), which was subsequently joined by the chain-driven locomotive described below; the first conventional locomotive does not appear to have arrived until 1928. The colliery was merged with Tursdale Colliery after the latter ceased coal-winding in June 1931, and was vested in NCB Northern Division No. 4 Area on 1st January 1947.

Gauge : 4ft 8½in

	-	VB		?		1916	(a)	s/s after 6/1929
	HUGO	0-4-0ST	OC	MW	1517	1900	(b)	(1)
	HENRIETTA	0-4-0ST	OC	MW			(c)	s/s
	COLONEL BELL	0-4-0ST	OC	MW	1697	1906	(d)	(2)
	MARY BELL	0-4-0ST	OC	MW	1422	1899	(e)	(3)
No.30		0-4-0ST	OC	P	669	1897		
		reb		Ridley Shaw		1930	(f)	(4)
113	PLUTO	0-6-0ST	OC	HL	2655	1906	(g)	(5)
120	CARLTON No.1	0-6-0ST	OC	HL	2732	1907	(h)	(5)
(153)	(PATRIOT)	0-6-0T	IC	MW	153	1865		
		reb		B Vaughan		1920	(j)	(5)

(a) in Dorman Long's *Schedule of Properties*, published in 1929, the only locomotive at Bowburn is described as a "Chain Driven locomotive, made by Bell Bros Ltd in 1916, with vertical boiler (Newton Chambers 1907)". One suggestion has been that it was constructed incorporating parts from one of the machines manufactured by Newton, Chambers & Co Ltd of Sheffield to rake coke out of beehive coke ovens, which ran on rails on the ovens' bench side; however, Newton Chambers are not known to have sold any of these after 1897.

(b) ex Browney Colliery, Meadowfield, /1928.
(c) origin and date of arrival unknown; a locomotive of this description was reported at the colliery to have been here.
(d) ex Tursdale Colliery, near Cornforth, by 8/1932.
(e) ex Tursdale Colliery, near Cornforth, by 5/1933.
(f) ex Tursdale Colliery, near Cornforth, /1933.
(g) ex Sherburn Hill Colliery, Sherburn Hill, 7/4/1937.
(h) ex Chilton Colliery, Chilton Buildings, 20/5/1942.
(j) ex Sherburn Hill Colliery, Sherburn Hill, 2/1945.

(1) to Parson Byers Quarry, Stanhope, by 4/1929.
(2) to Browney Colliery, Meadowfield, by 2/1936; returned by 11/1936; to NCB No. 4 Area, with colliery, 1/1/1947.
(3) to Mainsforth Colliery, Ferryhill Station, /1933; ex Mainsforth Colliery, 5/1945; to NCB No. 4 Area, with colliery, 1/1/1947.
(4) to Tursdale Colliery, near Cornforth, /1933.
(5) to NCB No. 4 Area, with colliery, 1/1/1947.

No.2 0-4-0ST BH 992/1890, No.26 0-6-0ST P629/1896, ACTIVE 0-4-0ST RS 3075/1901 and an unidentified 0-6-0ST IC are also said to have worked here.

BROWNEY COLLIERY, near Meadowfield   K119
Bell Brothers Ltd until 2/5/1923   NZ 250388

This colliery was opened in 1873 and was served by sidings west of the LNER Durham-Darlington line, 1¾ miles south of Relly Mill Junction. A locomotive, probably four-coupled and second-hand, was purchased in the quarter ending March 1873. A description in the *Colliery Guardian* of 22/11/1893 gives one locomotive here. Another locomotive, again possibly second-hand, was purchased in April 1901. No further information is known about either loco. Whether the Hopkins Gilkes locomotive arrived here before MW 1328 is also unknown.

There were beehive coke ovens here, the last of which closed in July 1922. The colliery closed in July 1938 after partial flooding underground.

Gauge : 4ft 8½in

-	0-4-0ST	OC	HG	?	?	(a)	Scr
ELSA	0-4-0ST	OC	MW	1328	1898	New	(1)
SIR LOWTHIAN	0-4-0ST	OC	MW	1658	1905	New	Scr /1938
OLIVIA	0-4-0ST	OC	HL	3426	1919	(b)	(2)
HUGO	0-4-0ST	OC	MW	1517	1900	(c)	(3)
COLONEL BELL	0-4-0ST	OC	MW	1697	1906	(d)	(4)

(a) ex Port Clarence Ironworks, Port Clarence (see Cleveland & North Yorkshire Handbook).
(b) ex Parson Byers Quarry, Stanhope, after 2/1921, by 11/1921.
(c) ex Parson Byers Quarry, Stanhope, after 2/1924 (one source says ex Tursdale Colliery, near Cornforth).
(d) ex Bowburn Colliery, Bowburn, by 2/1936.

(1) to Parson Byers Quarry, Stanhope, by 10/1918.
(2) to Parson Byers Quarry, Stanhope, after 11/1921, by 9/1922.
(3) to Bowburn Colliery, Bowburn, /1928.
(4) to Bowburn Colliery, Bowburn, by 11/1936.

BYERS GREEN COLLIERY & COKE OVENS, Byers Green   N120
Bolckow, Vaughan & Co Ltd until 1/11/1929 (which see)   NZ 223335

This colliery and coke works, the latter comprising 50 Semet-Solvay waste heat ovens, was served by what had become the one mile long LNER Todhills Branch from the Spennymoor-Bishop Auckland line from Burnhouse Junction, two miles west of Spennymoor Station. Both the colliery and the coke works had been closed in March 1926, but both were re-opened, if briefly, under Dorman Long's ownership, the coke ovens closing in January 1930 and the colliery, now combined with Newfield Colliery (see below), in July 1931. It is not known whether locomotives were used here again during this brief period of operation.

CHILTON COLLIERY & COKE OVENS, near Chilton Buildings N121
NZ 278307

This colliery, whose history went back to 1835, was formerly owned by Pease & Partners Ltd (which see), which had closed it in 1930. It was subsequently acquired by Dorman Long & Co Ltd, and re-commenced production about August 1934. There was also a coke works, with 52 Simplex waste heat ovens, and a by-products plant. These too were re-opened in 1934, but only briefly, closing again in 1938 or 1939. The two sites were served by a branch (¾ mile) from the LNER Chilton branch. Its workshops also repaired locomotives for Leasingthorne Colliery (see below), and there was probably more interchange of locomotives between the two collieries than is recorded here. The colliery passed to NCB Northern Division No. 4 Area on 1st January 1947.

Gauge : 4ft 8½in

No.1	"CLARENCE"	0-4-0ST	OC	BH	985	1890		
		reb	AB	2847	1904	(a)	(1)
	CARLTON No.7	0-6-0ST	OC	HL	2607	1905	(b)	(2)
120	CARLTON No.1	0-6-0ST	OC	HL	2732	1907	(c)	(3)
113	PLUTO	0-6-0ST	OC	HL	2655	1906	(d)	(4)
No.4		0-6-0ST	OC	HC	1335	1918	(e)	(5)
108	HECTOR	0-6-0ST	OC	HL	2613	1905	(f)	(6)
107	ATLAS	0-6-0ST	OC	HL	2612	1905	(g)	(7)
110	HERCULES	0-6-0ST	OC	HL	2654	1906	(h)	(8)

(a) ex Port Clarence Distillation Works, Port Clarence, 20/1/1934 (see Cleveland & North Yorkshire Handbook).
(b) ex Carlton Iron Works, Stillington, /1934 (see Cleveland & North Yorkshire Handbook).
(c) ex Carlton Iron Works, Stillington, 8/1934 (see Cleveland & North Yorkshire Handbook).
(d) ex Auckland Park Colliery, Coundon Grange, c/1935 (after 8/5/1934).
(e) ex Sherburn Hill Colliery, Sherburn Hill.
(f) ex Auckland Park Colliery, Coundon Grange, c/1935 (by 11/1937).
(g) ex Leasingthorne Colliery, near Coundon, by 22/3/1940.
(h) ex Dean & Chapter Colliery, Ferryhill.

(1) to Mainsforth Colliery, Ferryhill Station.
(2) to Dean & Chapter Colliery, Ferryhill (may have worked at Leasingthorne Colliery before this transfer, and have gone to Dean & Chapter Colliery from Leasingthorne Colliery; returned to Chilton Colliery by 22/3/1940; to Dean & Chapter Colliery, Ferryhill, by /1943.
(3) to Bowburn Colliery, Bowburn, 20/5/1942.
(4) to Sherburn Hill Colliery, Sherburn Hill.
(5) to Sherburn Hill Colliery, Sherburn Hill, by 22/3/1940.
(6) to Leasingthorne Colliery, near Coundon, by 22/3/1940; ex Leasingthorne Colliery for repairs by 12/5/1946; to NCB No. 4 Area, with colliery, 1/1/1947.
(7) to NCB No. 4 Area, with colliery, 1/1/1947
(8) to Leasingthorne Colliery, near Coundon, c6/1946.

DEAN & CHAPTER COLLIERY & COKE OVENS, Ferryhill N122
Bolckow, Vaughan & Co Ltd until 1/11/1929 (which see) NZ 272331

This colliery, opened in 1904, was served by a ½ mile branch from sidings at Binchester Junction on the LNER Bishop Auckland – Ferryhill line, 1¼ miles east of Spennymoor Station. From the same sidings a 2¾ mile branch ran south-westwards to serve **WESTERTON COLLIERY** (NZ 242316) (see below) near Westerton. This was closed when Dorman Long took over. It was re-opened in 1930, only to be closed again in September 1931; but a shaft was retained as **WESTERTON PUMPING STATION**, and its branch remained open to serve it, a Dean & Chapter locomotive working down as required. The coke works and by-product plant at Dean & Chapter included 140 Semet-Solvay ovens; of these, 40 were closed in July 1930, 75 more in October 1930 and the remaining 25 early in 1931. The colliery, together with Westerton Pumping Station, was vested in NCB Northern Division No. 4 Area on 1st January 1947.

Gauge : 4ft 8½in

No.10		0-4-0ST	OC	BH	1095	1896	(a)	(1)
No.26	JOHN EVANS	0-6-0ST	IC	P	629	1896	(a)	(2)
106	ERIMUS	0-6-0ST	OC	HL	2595	1904	(a)	(3)
116	GEORGE V	0-6-0ST	OC	HL	2833	1910	(a)	(3)

148	TAURUS	0-4-0ST	OC	HL	3384	1919	(a)	(4)
107	ATLAS	0-6-0ST	OC	HL	2612	1905	(b)	(5)
No.14		0-4-0ST	OC	HL	3248	1917	(c)	(6)
	CARLTON No.7	0-6-0ST	OC	HL	2607	1905	(d)	(7)
108	HECTOR	0-6-0ST	OC	HL	2613	1905	(e)	(8)
110	HERCULES	0-6-0ST	OC	HL	2654	1906	(f)	(9)
	ANGELA	0-4-0ST	OC	AE	1793	1918	(g)	(3)

(a) ex Bolckow, Vaughan & Co Ltd, with colliery, 1/11/1929.
(b) ex Leasingthorne Colliery, near Coundon, c/1930.
(c) ex Port Clarence Works, Port Clarence, 16/9/1932 (see Cleveland & North Yorkshire Handbook).
(d) ex Chilton Colliery, near Chilton Buildings, or possibly ex Leasingthorne Colliery, near Coundon (see these entries).
(e) ex Auckland Park Colliery, Coundon Grange, c/1934 (by 8/5/1934).
(f) ex Newfield Colliery & Brickworks, Newfield, by 22/3/1940.
(g) ex Burley Ironstone Quarries, Rutland, c/1946.

(1) to Mainsforth Colliery, Ferryhill Station.
(2) to Sherburn Hill Colliery, Sherburn Hill, c/1931.
(3) to NCB No. 4 Area, with colliery, 1/1/1947.
(4) to Mainsforth Colliery, Ferryhill Station, after 11/1941, by 12/5/1946.
(5) to Leasingthorne Colliery, near Coundon, c/1931.
(6) to Warrenby Works, Middlesbrough, Yorkshire (NR), after 22/3/1940, before 1/1/1947.
(7) to Chilton Colliery, near Chilton Buildings, by 22/3/1940, and returned; to Sherburn Hill Colliery, Sherburn Hill, /1943, and returned; to NCB No. 4 Area, with colliery, 1/1/1947.
(8) to Leasingthorne Colliery, Coundon, between 1935 and 11/1937.
(9) to Chilton Colliery, near Chilton Buildings.

LEASINGTHORNE COLLIERY, near Coundon N123
Bolckow, Vaughan & Co Ltd until 1/11/1929 (which see) NZ 252304

This colliery was opened in 1842 and was served by a two mile private extension of the LNER Chilton Branch. After Dorman Long took over Chilton Colliery (see above) Leasingthorne locomotives needing repair were sent along the branch to the workshops at Chilton Colliery for this work, Chilton sending a locomotive as replacement, but few precise details are known. In the 1930s an aerial ropeway was built to link the colliery to Auckland Park Colliery, and the colliery was at least administratively, and perhaps underground also, linked to Dean & Chapter Colliery. Adjacent to the colliery was **Leasingthorne Brickworks**, which manufactured firebricks. The colliery and its brickworks were vested in NCB Northern Division No. 4 Area on 1st January 1947.

Gauge : 4ft 8½in

No.102		0-4-0ST	OC	HL	2449	1900	(a)	Scr by 11/1937
107	ATLAS	0-6-0ST	OC	HL	2612	1905	(a)	(1)
	NEWLANDSIDE	0-4-0ST	OC	BH	365	1876		
		reb		Wake	2432	?	(b)	(2)
108	HECTOR	0-6-0ST	OC	HL	2613	1905	(c)	(3)
153	PATRIOT	0-6-0T	IC	MW	153	1865		
		reb		B Vaughan		1920	(d)	(4)
110	HERCULES	0-6-0ST	OC	HL	2654	1906	(e)	(5)

(a) ex Bolckow, Vaughan & Co Ltd, with colliery, 1/11/1929.
(b) ex Newlandside Quarry, Stanhope, date unknown; may have been before the take-over by Dorman, Long & Co Ltd.
(c) ex Dean & Chapter Colliery, Ferryhill, between 1935 and 11/1937.
(d) ex Cleveland Works, Middlesbrough, Yorkshire (NR).
(e) ex Chilton Colliery, near Chilton Buildings, c6/1946.

(1) to Dean & Chapter Colliery, Ferryhill, c/1930; returned c/1931; to Chilton Colliery, near Chilton Buildings, by 22/3/1940.
(2) returned to Newlandside Quarry, Stanhope.
(3) to Chilton Colliery, Chilton Buildings, by 11/1937; returned by 22/3/1940; to Chilton Colliery, near Chilton Buildings, for repairs, c6/1946.
(4) to Sherburn Hill Colliery, Sherburn Hill, by 24/3/1940.
(5) to NCB No. 4 Area, with colliery, 1/1/1947.

MAINSFORTH COLLIERY, Ferryhill Station N124
The Carlton Iron Co Ltd until 2/5/1923 (which see) NZ 307316

This colliery, opened in 1905, was served by a ½ mile branch from the LNER Ferryhill - Stockton line, one mile south of Ferryhill Station. It was vested in NCB Northern Division No. 4 Area on 1st January 1947.

Gauge: 4ft 8½in

6	LOTTIE	0-4-0ST	OC	HL	2110	1888			
			reb	?		1906	(a)	(1)	
	ACTIVE	0-4-0ST	OC	RS	3075	1901			
			reb	Ellis		1927	(b)		
			reb	R.Shaw	16	1931		(2)	
	CHARLIE	0-4-0ST	OC	HC	1402	1922	(c)	(3)	
No.104		0-6-0ST	IC	MW	1469	1900	(d)	(4)	
No.10		0-4-0ST	OC	BH	1095	1896	(e)	(5)	
	MARY BELL	0-4-0ST	OC	MW	1422	1899	(f)	(6)	
No.1	"CLARENCE"	0-4-0ST	OC	BH	985	1890			
			reb	AB	2847	1904	(g)	(7)	
148	TAURUS	0-4-0ST	OC	HL	3384	1919	(h)	(3)	
No.30		0-4-0ST	OC	P	669	1897			
			reb	Ridley Shaw		1930	(j)	(3)	

(a) ex The Carlton Iron Co Ltd, with colliery, 2/5/1923.
(b) ex The Priestman Collieries Ltd, Blaydon Burn Colliery, near Blaydon, via J.W. Ellis & Co Ltd, Swalwell, /1927.
(c) ex Tursdale Colliery, near Cornforth, by mid 1929.
(d) ex Cleveland Works, Middlesbrough, Yorkshire (NR), after mid 1929.
(e) ex Dean & Chapter Colliery, Ferryhill, after 1/11/1929.
(f) ex Page Bank Colliery, Page Bank, /1931.
(g) ex Chilton Colliery, near Chilton Buildings.
(h) ex Dean & Chapter Colliery, Ferryhill, after 11/1941, by 12/5/1946.
(j) ex Sherburn Hill Colliery, Sherburn Hill.

(1) here in mid 1929, according to *Schedule of Properties*; scrapped, after 5/6/1932.
(2) to Auckland Park Colliery, Coundon Grange, after 8/5/1934.
(3) to NCB No. 4 Area, with colliery, 1/1/1947.
(4) scrapped c/1943, although parts were still to be seen on 12/5/1946.
(5) to Sherburn Hill Colliery, Sherburn Hill, /1931.
(6) to Tursdale Colliery, near Cornforth, /1933; returned from Bowburn Colliery, Bowburn, /1933; to Bowburn Colliery, Bowburn, 5/1945.
(7) to Sherburn Hill Colliery, Sherburn Hill, by 11/5/1940; returned to Mainsforth Colliery by 11/1944 (one source gives 1/1944); to NCB No. 4 Area, with colliery, 1/1/1947.

The colliery also had four steam cranes, not necessarily all at the same time. One, built by Smith, is listed here in mid 1929.

NEWLANDSIDE QUARRY, Stanhope V125
Bolckow, Vaughan & Co Ltd until 1/11/1929 (which see) NY 995383 approx.

This limestone quarry latterly consisted of two huge areas on the south side of the Wear Valley, served by a self-acting incline down to the LNER Wearhead branch, ¼ mile east of Stanhope Station. It was closed by Dorman, Long & Co Ltd in May 1945, although dismantling work did not begin until 1950 and was not completed until the summer of 1951. Since then it has been worked by several owners, using road transport.

Gauge : 4ft 8½in

	WITTON	0-4-0ST	OC	BH	367	1877	(a)	(1)	
	NEWLANDSIDE	0-4-0ST	OC	BH	365	1876			
			reb	Wake	2432	?	(a)	(2)	
	HAVERTON	0-4-0ST	OC	AB	656	1890	(a)	(3)	
	NEWPORT	0-4-0ST	OC	K	5115	1914	(b)	(4)	
	HARE	0-4-0ST	OC	GH		1908	(c)	(5)	
	HUGO	0-4-0ST	OC	MW	1517	1900	(d)	(6)	
	ELSA	0-4-0ST	OC	MW	1328	1898			
			reb	Ridley Shaw		1925	(e)	(7)	
137	ARGYLE	0-4-0ST	OC	HL	3140	1915	(f)	(8)	

(a) ex Bolckow, Vaughan & Co Ltd, with quarry, 1/11/1929.
(b) ex Newport Works, Middlesbrough, Yorkshire (NR), by 10/1930.
(c) ex Auckland Park Colliery, Coundon Grange, by 8/1931.
(d) ex Parson Byers Quarry, Stanhope, by 5/1932.
(e) ex Parson Byers Quarry, Stanhope, by 6/1935.
(f) ex Parson Byers Quarry, Stanhope, after c/1935.

(1) to Parson Byers Quarry, Stanhope, c/1937.
(2) to Leasingthorne Colliery, Coundon, at an unknown date; may have been before take-over by Dorman, Long & Co Ltd; returned to Newlandside Quarry, Stanhope; to Cleveland Works, Middlesbrough, Yorkshire (NR); returned by 5/1939; to Parson Byers Quarry, Stanhope, 12/1939.
(3) to Parson Byers Quarry, Stanhope, after 6/1935; returned to Newlandside Quarry; believed dismantled by 10/1941; scrapped.
(4) to Newport Works, Middlesbrough, Yorkshire (NR).
(5) to Eston Mines, Yorkshire (NR), c/1932 (after 5/1932); returned by 6/1935, when it was seen dismantled; to Parson Byers Quarry, Stanhope, after 24/9/1950, by 4/1951.
(6) to Cleveland Works, Middlesbrough, Yorkshire (NR), and returned; to Parson Byers Quarry, Stanhope, c11/1951, by 22/12/1951.
(7) to Cleveland Works, Middlesbrough, Yorkshire (NR), after 6/1935, and returned; to Parson Byers Quarry, Stanhope, c11/1951, by 22/12/1951.
(8) to Parson Byers Quarry, Stanhope, /1945.

No.2 0-4-0ST OC BH 992/1890 from Parson Byers Quarry, Stanhope, may also have worked here and been returned.

47. Bell Bros Ltd purchased five Manning Wardle 'H' class locomotives, with 12in x 18in cylinders and 3ft 0in diameter wheels, around the turn of the 20th century. The first of them, ELSA, MW 1328/1898, pauses at Newlandside Quarry, Stanhope, in June 1935.

PAGE BANK COLLIERY, Page Bank, latterly called **SOUTH BRANCEPETH COLLIERY** N126
Bell Brothers Ltd until 2/5/1923; see also below NZ 230259

This colliery was begun by **J. Robson & Co** (which see), apparently using the title the Page Bank Colliery Co, and one source records that after three years' work, the Harvey seam here was reached on 19th April 1856, which is believed to be the opening date of the colliery. Its two names are both commonly found in the 19th century, though in the twentieth century South Brancepeth predominated. Immediately west of Spennymoor Station on the north side of the Byers Green branch of the West Hartlepool Harbour & Railway Company (WHH&R) lay both Merrington and Whitworth Collieries (NZ 252338), another example of two collieries sharing the same surface site but under different owners and working different seams. To serve the new Page Bank Colliery, which lay on the opposite bank of the River Wear, a 2¼ mile branch was built past this site. Curiously, the railway company only owned the branch as far as the bridge over the river, plus a half share in the bridge; the colliery company owned the other half share in the bridge and the remaining 300 yards or so to the colliery. By 1870 a **brickworks** had also been started at the colliery, and a large number of **beehive coke ovens** was also built. About 1870 M. Pernolet of Paris developed modifications to beehive ovens which saved the gas, sulphate of ammonia and tar produced during the coke making. These modifications were tried on 36 ovens of several different shapes, but they were abandoned after two years because of the inferior quality of the coke produced and the cost of repairing the ovens.

On 31st March 1861 the colliery became one of those leased by Ralph Ward Jackson, the chairman of the WHH&R, which subsequently took over the lease from Jackson. The colliery was offered for sale on 15th March 1864 and was purchased by Bell Brothers on 10th May 1864.

By an agreement dated 28th February 1868 but back-dated to 1st January 1868, Bell Brothers rented the branch from the NER (who had taken over the WHH&R) at £3 per week so that they could convey their workmen between Spennymoor and the colliery. The NER supplied the two vans used but it is not clear who supplied the motive power. By 1891 the number of workmen travelling daily had risen to 250 and the charge was increased to £9 per week from 1st July 1892. By this time the NER was certainly providing the loco, together with three coaches and a brake van, but Bell Bros Ltd supplied the guard, with an NER guard as assistant. The NER loco took the trains down to the Whitworth Park sidings alongside the River Wear, where it came off in order for the colliery loco to be attached (latterly MARY BELL) and work the train across the river to a platform at the colliery. The train was kept at Spennymoor overnight, making its first run to the colliery at 2.30 a.m. and running to suit the shifts before making the last run from Page Bank at 4.30 p.m. The line up from Whitworth Park sidings to the main line included 1¼ miles at 1 in 38, and the NER loco could manage only five wagons at a time, having to make several journeys to make up a train to go to Bell Brothers' works at Port Clarence.

A 'tank' engine was 'sold' to the colliery from Bell Brothers Ltd's Clarence Ironworks at Port Clarence in 1867. The *Mining Journal* Supplement for 15th October 1870 describes the colliery as having "one locomotive engine". Another locomotive was purchased for the colliery in the half year ending September 1873. Curiously, DCOA Return 102, covering 1871 and 1876, does not list any locomotives here. The Return for March 1890 lists two locomotives, and a locomotive was also purchased for here in the half year ending September 1890. Nothing is known about any of these locomotives.

The last of the beehive ovens closed in February 1922. The colliery ceased production in December 1930 and was officially abandoned in June 1931. The brickworks was subsequently taken over by Hartley Main Collieries Ltd (which see).

Gauge : 4ft 8½in

	MARY BELL	0-4-0ST	OC	MW	1422	1899	New	(1)
	COLONEL BELL	0-4-0ST	OC	MW	1697	1906	New	(2)
	HUGO	0-4-0ST	OC	MW	1517	1900	(a)	(3)

(a) ex Parson Byers Quarry, Stanhope, by 10/1921.

(1) to Mainsforth Colliery, Ferryhill Station, /1931.
(2) to Tursdale Colliery, near Cornforth, 1/1927.
(3) to Parson Byers Quarry, Stanhope, by 2/1924.

SHERBURN COLLIERY, Sherburn Colliery, also known as **LADY DURHAM COLLIERY**　　　　**K127**
Sir B. Samuelson & Co Ltd until 2/5/1923　　　　　　　　　　　　　　　　　　　　　NZ 315425
The Lambton & Hetton Collieries Ltd until 1/1/1914 (which see)

Sir Bernhard Samuelson & Co Ltd, who owned the Newport Iron Works in Middlesbrough, Yorks (NR), had previously owned Hedley Hope and East Hedley Hope Collieries (see The Bearpark Coal & Coke Co Ltd). The firm was founded in 1887, probably on 1st July 1887. The firm's share capital was purchased by Dorman, Long & Co Ltd in 1917.

Sherburn Colliery was opened in 1854 and was situated at the end of a long branch of the Earl of Durham's Railway, later the Lambton Railway, from Rainton. The northern section of this was abandoned when Lambton & Hetton Collieries Ltd sold its Sherburn collieries for £180,000 in 1914, and all coal was then dispatched southwards via the remaining section of line to the NER Leamside - Ferryhill line, to which there was a link immediately north of Sherburn Colliery Station. This two mile line also served **SHERBURN HOUSE COLLIERY** (NZ 324417) **(K128)**, sunk in 1844, and **SHERBURN HILL COLLIERY** (NZ 336427) (see below), sunk in 1835. Traffic over the line was worked by locomotives from a shed at Lady Durham Colliery. Under Lambton ownership the workshops serving the Sherburn group of collieries were at Littletown Colliery (NZ 339435), which lay about a mile to the north-east of Sherburn Hill Colliery and had been closed in 1913; the machinery here was subsequently dismantled and re-erected at Sherburn Hill. Under Dorman Long the Sherburn royalty also included the closed collieries of Shincliffe (see Joseph Love & Partners) and Whitwell (see Edward Robson & Co).

Production ceased at Lady Durham Colliery in 1918, though it continued in use as a pumping station. Sherburn House Colliery ceased production in May 1931, though it was not abandoned until September 1935. It was clearly inconvenient to have locomotives at a shed two miles from where most of the work was, though it would appear that a loco shed was not built at Sherburn Hill until the 1930s. However, the shed at Lady Durham was still standing in May 1940 and housed the remains of H&C 72 below.

48. When Sir B. Samuelson & Co Ltd purchased the Sherburn collieries from The Lambton & Hetton Collieries Ltd in 1914, three former Lambton Railway locomotives came with them. The tender engine, 2, H&C 72/1866, stands magnificent after overhaul, probably in 1914.

Gauge : 4ft 8½in

1	later No.1	0-6-0ST	IC	H&C	78	1866	(a)		
		reb	HL	9294	1914				
		reb	LG		c1923		(1)		
2		0-6-0	IC	H&C	72	1866	(b)		
		reb	Ridley Shaw		1928		(2)		
No.3		0-4-0ST	OC	H&C	79	1866	(c)		
		reb	Wake	1013	1914		Scr by mid 1929		
4	VICTORIA	0-6-0ST	IC	HE	484	1889			
		reb	Wake		?	(d)	(1)		
"No.4"		0-6-0T	OC	HC	1335	1918	(e)	(1)	

(a) ex The Lambton & Hetton Collieries Ltd, with collieries, 1/1/1914; Lambton Railway, 14.
(b) ex The Lambton & Hetton Collieries Ltd, with collieries, 1/1/1914; Lambton Railway, 2.
(c) ex The Lambton & Hetton Collieries Ltd, with collieries, 1/1/1914; Lambton Railway, 13.
(d) ex Hutchinson's Trustees Ltd, Widnes, Lancashire, by 14/7/1916, via J.F. Wake, dealer, Darlington.
(e) ex Ministry of Munitions, Inland Waterways & Docks, Richborough, Kent (probably c/1919).

(1) to Sherburn Hill Colliery, Sherburn Hill.
(2) last worked between 1930-1932; was seen derelict in the shed here, without its tender, on 11/5/1940; scrapped c/1940.

SHERBURN HILL COLLIERY, Sherburn Hill
Sir B. Samuelson & Co Ltd until 2/5/1923
The Lambton & Hetton Collieries Ltd until 1/1/1914 (which see)

K129
NZ 336427

This colliery, opened in 1835, was another of those purchased from The Lambton & Hetton Collieries Ltd on 1st January 1914. When Sherburn Hill became the only colliery served by the two mile line to the LNER Leamside – Ferryhill line immediately north of Sherburn Colliery Station, the locomotives formerly kept at the Lady Durham Colliery (see above) were transferred to a new shed here. This was almost certainly not opened until the 1930s, though the locomotives were officially allocated here in the 1920s, certainly by mid 1929. The colliery was vested in NCB Northern Division No. 4 Area on 1st January 1947.

49. No.26, P 629/1896, at Sherburn Hill Colliery in 1946, one of the makers' X class.

Gauge: 4ft 8½in

1		later No.1	0-6-0ST	IC	H&C	78	1866	(a)	
			reb		HL	9294	1914		
			reb		LG		c1923		(1)
4		VICTORIA	0-6-0ST	IC	HE	484	1889		
			reb		Wake		?	(a)	(2)
"No.4"			0-6-0T	OC	HC	1335	1918	(a)	(3)
No.10			0-4-0ST	OC	BH	1095	1896	(b)	(1)
No.26		(JOHN EVANS)	0-6-0ST	IC	P	629	1896	(c)	(1)
No.30		(JOHN EVANS)	0-4-0ST	OC	P	669	1897		
			reb		Ridley Shaw		1930	(d)	(4)
113		PLUTO	0-6-0ST	OC	HL	2655	1906	(e)	(5)
(153)		PATRIOT	0-6-0T	IC	MW	153	1865		
			reb		B Vaughan		1920	(f)	(6)
No.1		"CLARENCE"	0-4-0ST	OC	BH	985	1890		
			reb		AB	2847	1904	(g)	(7)
		CARLTON No.7	0-6-0ST	OC	HL	2607	1905	(h)	(8)

(a) ex Lady Durham Colliery, Sherburn Colliery.
(b) ex Mainsforth Colliery, Ferryhill Station, /1931.
(c) ex Dean & Chapter Colliery, Ferryhill, c/1931.
(d) ex Tursdale Colliery, near Cornforth, /1933.
(e) ex Auckland Park Colliery, Coundon Grange (after 1/11/1929).

Durham Part 1 Page 134

(f)   ex Leasingthorne Colliery, near Coundon, by 24/3/1940.
(g)   ex Mainsforth Colliery, Ferryhill Station, by 11/5/1940.
(h)   ex Dean & Chapter Colliery, Ferryhill, /1943.
(1)   to NCB No. 4 Area, with colliery, 1/1/1947.
(2)   s/s; not listed here, or anywhere else, in Dorman Long's *Schedule of Properties* of 1929; (one source reported on 11/5/1940 loco "sent away for repairs", but this would seem more likely to be a confusion with HC 1335/1918).
(3)   to Chilton Colliery, Chilton Buildings, and returned; to NCB No. 4 Area, with colliery, 1/1/1947.
(4)   to Mainsforth Colliery, Ferryhill Station.
(5)   to Bowburn Colliery, Bowburn, 7/4/1937.
(6)   to Bowburn Colliery, Bowburn, 2/1945.
(7)   to Mainsforth Colliery, Ferryhill Station, by 11/1944 (one source gives 1/1944).
(8)   to Dean & Chapter Colliery, Ferryhill.

TURSDALE COLLIERY, near Cornforth   N130
Bell Brothers Ltd until 2/5/1923   NZ 301360

This colliery was sunk in 1859 and had been acquired by Bell Bros Ltd by October 1870. It was served by sidings east of the NER Leamside-Ferryhill line 2½ miles north of Ferryhill Station. For some time about the turn of the century it was also connected to the NER Coxhoe Branch by a ½ mile NER link. There were also over 200 beehive coke ovens here, the last of which closed in June 1918.

The colliery was situated towards the southern edge of the royalty, so to reduce underground haulage time and cost, Bowburn Colliery was developed from 1906 (which see). Eventually Tursdale was merged with Bowburn, coal winding ceasing at the former in June 1931, though the shaft remained open for pumping and a locomotive was kept here for use as required. The shaft was vested into NCB Northern Division No. 4 Area on 1st January 1947.

A locomotive was ordered new from Henry Hughes & Co, Loughborough, for here in January 1876, and was apparently delivered on 20th June 1876; this may well be the second HH loco recorded later at Parson Byers Quarry, Stanhope (which see). Another loco, probably new and four-coupled, was purchased for the colliery in the half year ending September 1876 (unless this is one and the same with the HH loco). The DCOA Returns do not list any locomotive here until November 1896, when one locomotive driver is listed. No confirmed details about locomotives are known before 1927, nor is any information known about a locomotive here between 1933 and 1940.

Gauge : 4ft 8½in

	COLONEL BELL	0-4-0ST	OC	MW	1697	1906	(a)	(1)
	CHARLIE	0-4-0ST	OC	HC	1402	1922	(b)	(2)
No.30		0-4-0ST	OC	P	669	1897	(c)	
			reb	Ridley Shaw		1930		(3)
	MARY BELL	0-4-0ST	OC	MW	1422	1899	(d)	(4)
"No.2"		0-4-0ST	OC	BH	992	1890	(e)	(5)

One source states that HUGO 0-4-0ST OC MW 1517/1900 worked here, but gives no further information.

(a)   ex Page Bank Colliery, Page Bank, 1/1927.
(b)   ex Frank Edmunds, dealer, Stoke-on-Trent, Staffordshire (probably following an advertisement in the *Colliery Guardian*, 1/1928); previously Bradford Corporation, Esholt Sewage Works, Bradford, Yorkshire (WR), until 12/1927.
(c)   ex Cleveland Works, Middlesbrough, Yorkshire (NR), after 1/11/1929.
(d)   ex Mainsforth Colliery, Ferryhill Station, /1933.
(e)   ex Parson Byers Quarry, Stanhope, c/1939 (by 23/3/1940).
(1)   to Bowburn Colliery, Bowburn, by 8/1932.
(2)   to Mainsforth Colliery, Ferryhill Station, by mid 1929.
(3)   to Bowburn Colliery, Bowburn, /1933, and returned, /1933; to Sherburn Hill Colliery, Sherburn Hill, /1933.
(4)   to Bowburn Colliery, Bowburn, by 5/1933.
(5)   to NCB No. 4 Area, with colliery, 1/1/1947.

WESTERTON COLLIERY, near Westerton   N131
Bolckow, Vaughan & Co Ltd until 1/11/1929 (which see)   NZ 235308

This colliery was served by a branch 2¾ miles long from Binchester Junction, which lay one mile east of Spennymoor Station on the LNER Bishop Auckland-Ferryhill branch. From the same sidings alongside the LNER a ½ mile branch served **DEAN & CHAPTER COLLIERY**, which allowed Dean & Chapter locomotives to work the Westerton traffic. Coal production had ceased in August 1924. Dorman Long re-opened the colliery in 1930, only to close it again in September 1931, although it would appear that a shaft was retained for ventilation and perhaps pumping purposes. According to information from Auckland Park (see above), the loco below left here to go there about 1935, with presumably a Dean & Chapter loco then going down when needed.

Gauge : 4ft 8½in

| | HENRY CORT | 0-4-0ST | OC | BH | 607 | 1881 | (a) | (1) |

(a) ex Bolckow, Vaughan & Co Ltd, with colliery, 1/11/1929 (not confirmed).

(1) to Auckland Park Colliery, Coundon Grange, c/1935.

DORMAN LONG (STEEL) LTD
Dorman, Long & Co Ltd until 2/10/1954

In 1954 Dorman, Long & Co Ltd separated its main manufacturing activities into three new companies, this one, Dorman Long (Bridge & Engineering) Ltd and Dorman Long (Chemicals) Ltd. The Dorman Long group of companies was vested into the British Steel Corporation on 1st July 1968.

NEWFIELD COLLIERY & BRICKWORKS, Newfield M132
Bolckow, Vaughan & Co Ltd until 1/11/1929 (which see) NZ 205332

This complex was served by a one mile branch, with a reverse, from Hunwick Station on the BR Durham-Bishop Auckland line. The colliery shafts were eventually closed and coal worked from a nearby drift, the output servicing the adjacent brickworks, which manufactured firebricks. Because the output was below 150 tons of coal per day the drift became a licensed mine when the coal industry was nationalised in 1947. Coal production ceased in December 1959. The brickworks replaced its rail transport by road transport about 1962, and was eventually closed. The drift mine was much later re-opened by **J. Bartlett** (which see).

50. Another firm to group its designs into classes was Peckett. P 916/1901, seen here at Newfield Colliery & Brickworks in June 1939, was an example of the R1 class.

Gauge : 4ft 8½in

| | No.2 | | | 0-4-0ST | OC | BH | * | 1877 | | |
| | | | | | reb | B Vaughan | | 1893 | (a) | (1) |

(No.1)		0-4-0ST	OC	P	916	1901	(a)	(2)
105	KELVIN	0-4-0ST	OC	CF	1211	1901	(a)	Scr by 7/1939
118	BELMONT	0-6-0ST	OC	HL	2909	1912	(a)	Scr by 7/1939
	BETTY	0-4-0ST	OC	AE	1769	1917	(b)	Scr c11/1960
43		0-4-0ST	OC	DL		1949	(c)	(2)

* in later years the worksplates were very worn. Various visitors between 1939 and 1951 reported that whilst 1877 was visible, their best guess was that there was a 6 in the works number. Given the building date, the most likely possibility is that the loco was BH 368/1877, originally New to Bolckow Vaughan's Witton Park Iron Works (which see).

(a) ex Bolckow, Vaughan & Co Ltd, with colliery and brickworks, 1/11/1929.
(b) ex War Department, Tidworth, Hampshire, 3/1948.
(c) ex Britannia Works, Middlesbrough, Yorkshire (NR), c11/1960.

(1) scrapped, /1953 (after 10/4/1953).
(2) scrapped, /1962 (by 4/7/1962).

A report of 27th July 1939 recorded that No.46, formerly owned by the War Department, an 0-4-0ST with 10in cylinders built by AB, had been broken up here. This may be AB 1415/1917, which fits the description and which is known to have been in the North-East by June 1920. However, it cannot be definitely confirmed as being here.

PARSON BYERS QUARRY, near Stanhope   **V133**
Bell Brothers Ltd until 2/5/1923   NY 998378 approx.

On the southern side of the Wear valley east of Stanhope the limestone bed was 12 feet thick, lying beneath 50 feet of cover (overburden). Bell Brothers acquired the lease of Parson Byers, about half way between Stanhope and Frosterley, on 9th August 1870, and production began in 1872. The quarry was served by a self-acting incline (¾ mile) down to the NER Wear Valley branch, one mile east of Stanhope Station, with the loco shed situated near the top of the incline. The original lease comprised 204 acres and as demand increased, **NEW WOODCROFT QUARRY** (NZ 005374 approx) and **RIDDING HOUSE QUARRY** (NZ 006367 approx.) were developed to the east, which added a further 261 acres, with the railway system extended to serve them. However, the name 'Parson Byers' was retained collectively for the three sites. By the early 1920s Parson Byers itself was largely worked out, and probably in 1920 a new two-road loco shed was built at the entrance to Ridding House Quarry,

51. HAWK – the painted name is just visible on the tank – built by Hawthorns of Leith, at Parson Byers Quarry, Stanhope, allegedly photographed between 1910 and 1920. Industrial locomotives with outside valve gear were rare.

52. *Another Scottish builder – Gibb & Hogg of Airdrie. HARE, built in 1908, awaits scrap at Parson Byers on 25th June 1958, after the closure of the quarry.*

53. *The bank head of the Parson Byers self-acting incline, which connected with the BR Stanhope branch, on 25th June 1958. The drumhouse is situated in front of the trees, with frames and the gantry to bring the ropes down to rail level. Note the 'bulls' on the top of the two humps, or 'kips', which 'caught' each axle to prevent any runaways, and the signal, probably used to tell locomotive drivers when to stop propelling; the line from the quarry approached the bank head on a very sharp bend from the left.*

although the old shed continued in use. Dorman Long closed down production on 10th May 1958, but it was subsequently resumed by other owners, using road transport. Quarrying finally ceased in 1974.

A loco, probably new and four-coupled, was purchased in the half year ending September 1876, although this may be a confusion with the HH loco, which was apparently delivered in June 1876. The "remains" of a locomotive were sold in the half year ending September 1881. No further details are known. A locomotive built by Brush is reported to have been here in September 1918, but again no further details are known.

Reference : *The Parson Byers Limestone Company Railway*, L.E.Berry, Backtrack, Vol.4 No.1 (January/February 1990).

Gauge: 4ft 8½in

	-		0-4-0ST	OC	HH		1876	New	(1)
	-		0-4-0ST	OC	HH		1876	(a)	(1)
No.2			0-4-0ST	OC	BH	992	1890	New	(2)
	HAWK		0-4-0WT	OC	H(L)	?	?	(b)	
		reb	0-4-0ST	OC	?		?		(3)
	HUGO		0-4-0ST	OC	MW	1517	1900	New	(4)
	PAULINE		0-4-0ST	OC	HL	2941	1912	New	Scr c/1945
	ELSA		0-4-0ST	OC	MW	1328	1898	(c)	
		reb			Ridley Shaw		1925		(5)
	OLIVIA		0-4-0ST	OC	HL	3426	1919	New	(6)
	AILEEN		0-4-0ST	OC	HL	3572	1923	New	Scr c9/1958
235	WASP		0-4-0ST	OC	MW	813	1881	(d)	Scr c/1943
	HAVERTON		0-4-0ST	OC	AB	656	1890	(e)	(7)
137	ARGYLE		0-4-0ST	OC	HL	3140	1915	(f)	(8)
	WITTON		0-4-0ST	OC	BH	367	1877	(g)	Scr
	NEWLANDSIDE		0-4-0ST	OC	BH	365	1877		
		reb			Wake	2432	?	(h)	Scr c6/1958
136	SALTBURN		0-4-0ST	OC	HL	3139	1915	(j)	(9)
	HARE		0-4-0ST	OC	GH		1908	(k)	(10)
22			0-4-0ST	OC	MW	756	1880	(m)	(10)

It is also possible that there was a 0-4-0VBT loco here.

(a) may well be the HH loco said to have been built in 1876 for Tursdale Colliery (see above).
(b) origin unknown; one source claimed that it may well have been a 0-4-0WT originally, while another gave a building date of 1890.
(c) ex Browney Colliery, near Meadowfield, by 10/1918.
(d) ex J.F.Wake & Co Ltd, dealer, Darlington, by 1/1928; previously Lucas & Aird Ltd, contrs. This loco was used for removing overburden at the quarry.
(e) ex Newlandside Quarry, Stanhope, c/1934.
(f) ex Cleveland Works, Middlesbrough, Yorkshire (NR), c/1934-1935.
(g) ex Newlandside Quarry, Stanhope, c/1937.
(h) ex Newlandside Quarry, Stanhope, 12/1939.
(j) ex Cleveland Works, Middlesbrough, Yorkshire (NR), /1941.
(k) ex Newlandside Quarry, Stanhope, /1945.
(m) ex Warrenby Works, Middlesbrough, Yorkshire (NR), 12/1954.

(1) a locomotive exploded here in February 1898, believed to have been one of these two; both were subsequently scrapped.
(2) to Tursdale Colliery, near Cornforth, c/1939 (by 23/3/1940). It may be that this locomotive spent periods away from Parson Byers before 1939. Visits to Newlandside Quarry, Browney and Bowburn Collieries and to Clarence Ironworks at Port Clarence are all said to have taken place; but accounts conflict and there is no supporting evidence. Another report says that the locomotive was converted to drive a boring machine in 1919, this apparatus being removed prior to the transfer to Tursdale Colliery.
(3) according to the reference above (L.E. Berry), HAWK was sold on 1/1/1899 "surplus to requirements, having been much unused for several years". However, a photograph exists of a 0-4-0ST here with what appears to be the faded painted name HAWK on the saddletank and which probably dates from about 1910-1920; it is included here. This loco has clear similarities with Hawthorns of Leith designs, so perhaps the sale fell through. Oral tradition said that HAWK was used latterly as a stationary boiler and was scrapped c/1923.

(4) to Page Bank Colliery, Page Bank, by 10/1921; returned by 2/1924; to Browney Colliery, near Meadowfield; ex Bowburn Colliery, Bowburn, by 4/1929; to Newlandside Quarry, Stanhope, by 5/1932; ex Newlandside Quarry, Stanhope, c11/1951; scrapped, 4/1953.
(5) to Newlandside Quarry, Stanhope, by 23/3/1940; returned by 11/1951; scrapped, /1956.
(6) to Browney Colliery, near Meadowfield, after 2/1921, by 11/1921; returned by 9/1922; scrapped 9/1958.
(7) to Newlandside Quarry, Stanhope, by 10/1941.
(8) to Newlandside Quarry, Stanhope, after /1935; returned, /1945; scrapped c6/1958.
(9) to Newlandside Quarry, Stanhope, and returned by 10/1949; scrapped c9/1958.
(10) scrapped, c7/1958 (after 28/6/1958).

DOXFORD & SUNDERLAND LTD
Doxford & Sunderland Shipbuilding & Engineering Co Ltd until 23/3/1970
This former company was set up on 14/6/1961 with the amalgamation of three shipbuilding companies in Sunderland, William Doxford & Sons Ltd, J.L. Laing & Co Ltd and J.L. Thompson & Sons Ltd. However, the subsidiary companies running the individual shipyards were allowed to continue trading until 1/4/1966, when they were abolished; see also below.

PALLION YARD, Sunderland   H134
NZ 377577 approx.

This yard was originally owned by William Doxford, subsequently **William Doxford & Sons** and **William Doxford & Sons Ltd** from 1/1/1891. In 1956 this became a parent holding company, setting up **William Doxford & Sons (Shipbuilders) Ltd** to run the shipyard and **William Doxford & Sons (Engineers) Ltd** to operate the Engine Works (see below). On 14/6/1961 the parent company amalgamated with two others as shown above, though the subsidiary companies continued their functions until 1/4/1966.

William Doxford opened a shipbuilding yard at Cox Green in 1840, one of a large number of small yards on the River Wear then building wooden ships. However, Cox Green was over six miles from the mouth of the river, and in 1857 he moved his business to a site east of Pallion village, also on the south bank of the river but only two miles from the sea. He was probably also influenced by the opening of the Pensher branch in 1852 (NER from 1854), though the river shore premises at Pallion

54. Andrew Barclay also sold a handful of crane tanks in Durham. One came here, but it later became one of three that had their cranes removed. GRINDON, AB 1305/1912, stands with the Queen Alexandra Bridge in the background on 25th August 1960. Note the re-railing jacks fixed at each end of the front buffer beam.

55. A crane tank's typical job – HENDON, RSHN 7007/1940, at work on 3rd April 1969. These cranes had a lifting capacity of four tons. This was one of a pair originally ordered by 'The New Russia Co' from Hawthorn Leslie in 1918; the order being cancelled, the parts were stored until utilised by RSH to form ROKER and HENDON in 1940.

58. Doxford's main fleet in their shed : HENDON, RSHN 7007/1940; ROKER, RSHN 7006/1940; MILLFIELD, RSHN 7070/1942; SOUTHWICK, RSHN 7069/1942, and the second GENERAL, P 2049/1944.

had difficulty in linking to the branch because of the severe fall from the railway to the river. The yard was eventually served by a ½ mile branch from the NER Deptford Branch, which was linked to the Penshaw Branch at Pallion Station and opened on 1st October 1865, although when the branch to 'Doxfords' was built is not known.

In 1870 the firm expanded the yard, developing what later became known as the West Yard, and in 1878 an Engineering Works was added; the firm became a world leader in the development of marine diesel engines. In 1903 the yard was expanded again by the addition of the East Yard (see Walter Scott & Middleton in the Contractors' section in Part 2), so that by 1904 the combined yards covered 38 acres. It would seem extremely unlikely that the yard owned only two locomotives before 1899 (see below).

The Doxford family sold their interest in 1919, but the new owners retained the old title. Regular rail working ceased at the end of January 1971 and was abandoned completely on 19th February 1971. The yard was extensively modernised in later 1970s to build ships completely under cover; but having become the last shipbuilding yard left on the River Wear and one of the most modern in Europe it was closed after pressure from the European Union in 1988. The site, with some buildings still retained, is now used for various different purposes.

As noted above, there may well have been more than two locomotives before 1900. The yard was noted for its extensive use of crane tank locomotives. The main loco shed (NZ 382557) was situated immediately upstream of the Queen Alexandra Bridge, although one or more locomotives were also stabled in the arches of the viaduct leading to the bridge. The names Pallion, Deptford, Hylton, Millfield, Grindon, Roker, Hendon and Southwick are all districts of Sunderland.

Gauge : 4ft 8½in

-		0-4-0ST	OC	BH	424	1877	New	(1)
GENERAL		0-4-0ST	OC	P	703	1899	New	Scr 5/1951
PALLION		0-4-0CT *	OC	HL	2517	1902	New	(2)
DEPTFORD		0-4-0CT	OC	HL	2535	1902	New	
	reb	0-4-0T#	OC			?		Scr 4/1949
WEAR		0-4-0CT	OC	HL	2551	1903	New	
	reb	0-4-0T#	OC			?		Scr 9/1952
HYLTON		0-4-0CT	OC	HL	2594	1905	New	Scr 9/1952
MILLFIELD		0-4-0CT	OC	HL	2632	1906	New	Scr 6/1938
GRINDON		0-4-0CT	OC	AB	1305	1912	New	
	reb	0-4-0T#	OC			?		Scr 6/1969
BROWNIE		0-4-0CT	OC	HL	2550	1903	(a)	Scr 6/1969
ROKER		0-4-0CT	OC	RSHN	7006	1940	New	(3)
HENDON		0-4-0CT	OC	RSHN	7007	1940	New	(4)
SOUTHWICK		0-4-0CT	OC	RSHN	7069	1942	New	(5)
MILLFIELD		0-4-0CT	OC	RSHN	7070	1942	New	(6)
2		0-4-0ST	OC	HL	2989	1913	(b)	(7)
33		0-4-0ST	OC	HL	2827	1910	(c)	(8)
GENERAL		0-4-0ST	OC	P	2049	1944	(d)	(9)

* crane removed for a time   # crane removed

The company is also said to have hired a locomotive from the LNER sometime between 1939 and 1945.

(a) ex R. & W. Hawthorn, Leslie & Co Ltd, Hebburn, /1940.
(b) ex Wallsend & Hebburn Coal Co Ltd, Rising Sun Colliery, Wallsend, Northumberland, on loan, in the period 1939-1945.
(c) ex The Lambton, Hetton & Joicey Collieries Ltd, Lambton Railway, Lambton Staiths shed, Sunderland, on loan, in the period 1939-1945.
(d) ex Thos W. Ward Ltd, Grays, Middlesex, 4/1951; previously Morris Motors Ltd, Cowley, Oxfordshire.

(1) to CF and used by them as a hire loco.
(2) to D.W. Fickes, Dalescroft Railfans Club, Britannia Steelworks, Middlesbrough, Yorkshire (NR), for preservation, 5/1971.
(3) to D.W. Fickes, Dalescroft Railfans Club, Britannia Steelworks, Middlesbrough, Yorkshire (NR), for preservation, 3/1971.
(4) to Blaydon Metal Co Ltd, Blaydon, for scrap, 3/1971; purchased for preservation by Stephenson & Hawthorn Locomotive Trust, Marley Hill, near Sunniside, 1/1974.
(5) to Dinting Railway Centre, Dinting, Derbyshire, for preservation, 4/1971.

(6) to A. Bloom, Bressingham Steam Museum, Bressingham, Norfolk, for preservation, 1/1971.
(7) returned to Wallsend & Hebburn Coal Co Ltd, Rising Sun Colliery, Wallsend, Northumberland.
(8) returned to The Lambton, Hetton & Joicey Collieries Ltd, Lambton Railway, Lambton Staiths shed, Sunderland.
(9) to Blaydon Metal Co Ltd, Blaydon, for scrap, 3/1971.

DUNN & MILLER
NEWTON CAP COLLIERY & BRICKWORKS, Toronto　　　　　　　　　　　　　　Q135
NZ 213308

Herbert Dunn and Alexander W. Miller bought this brickworks and its colliery from Henry Stobart & Co Ltd (which see) on 17th October 1933. Both operations are believed to have been in production at this time. They were served by a ¼ mile branch from the LNER Durham – Bishop Auckland line, 1¼ miles north of Bishop Auckland Station. The men sold the brickworks as a going concern to The Fir Tree Coal Co on 2nd September 1935, but retained control of the colliery until March 1937 when its lease passed to Newton Cap Brickworks Ltd, which by this time had also taken over the brickworks. By April 1937 this firm had itself been taken over, by The North Bitchburn Fireclay Co Ltd (which see).

Gauge : 4ft 8½in

THE COLONEL	0-4-0ST	OC	HG	251	1867	(a)	Scr c/1934
COMET	0-4-0ST	OC	MW	467	1873	(a)	
	reb		LG		1902	(a)	(1)

(a) ex Henry Stobart & Co Ltd, with site, 17/10/1933.
(1) to the successors to Dunn & Miller (which of the successors is not known) and then to The North Bitchburn Fireclay Co Ltd, c4/1937 (this locomotive's presence here is not confirmed under these owners, but it was here in the periods before and after their ownership).

THE DUNSTON GARESFIELD COLLIERIES LTD
Dunston Garesfield Collieries until 4/1905
Dunston, Norwood & Swalwell Garesfield Coal Co until /1903
Dunston & Swalwell Garesfield Coal Co until /1901
Dunston & Swalwell Collieries until /1900

SWALWELL GARESFIELD COLLIERY, Swalwell　　　　　　　　　　　　　　B136
Swalwell Garesfield Coal Co until /1899　　　　　　　　　　　　　　　　NZ 205625
formerly **G.R. Ramsey**

This colliery was opened in 1887 by **G.R. Ramsey**, the son of G.H. Ramsey, the owner of Swalwell Brickworks (see Adams Pict Furnace Brick Co Ltd) and Blaydon Main Colliery (see The Stella Coal Co Ltd), who had died in 1880. The NER Swalwell Branch was extended by about 100 yards to serve it. There may have been a short period of working by the NER before the arrival of the first loco below. At the previous end of the branch lay the firebrick works which continued to operate under the name of G.H.Ramsey; it had running powers over the NER to work to the colliery to collect fireclay mined at the pit. Whether the two businesses effectively shared the same ownership after The Dunston Garesfield Collieries Ltd was set up is not known. The colliery, sometimes also known as **SWALWELL COLLIERY**, finally closed in August 1940, after at least one earlier period of closure.

The site was later developed by the National Coal Board Opencast Executive (see Part 2) as a Disposal Point for opencast coal.

Gauge : 4ft 8½in

-	0-4-0ST	OC	BH	298	1875	(a)	(1)
PROSPECT	0-4-0ST	OC	HL	2479	1901	New	(2)

(a) ex G.H. Ramsey, Swalwell Brickworks, Derwenthaugh.
(1) to Lingford, Gardiner & Co Ltd, Bishop Auckland, by /1925; re-sold to Casebourne & Co Ltd, Haverton Hill (see Cleveland & North Yorkshire Handbook).
(2) to Imperial Chemical Industries Ltd, Dalbeattie, Kirkcudbrightshire, Scotland, via George Cohen, Sons & Co Ltd, /1941.

The company also owned **DUNSTON COLLIERY** (later **DUNSTON & ELSWICK COLLIERY**) (NZ 231626) **(C137)**, which passed to NCB Northern Division No. 6 Area on 1st January 1947, and **NORWOOD COLLIERY**, Dunston (NZ 235618) **(C138)**, which was closed in July 1931 and its workings divided between Dunston and Swalwell Garesfield Collieries. Both had been purchased in July 1899 from John Bowes & Partners Ltd (which see); both were worked by the NER.

DURHAM & YORKSHIRE WHINSTONE CO LTD (registered in 1930)
GREENFOOT QUARRY, near Stanhope   **V139**
originally **Richard Summerson & Co Ltd** (which see)   NY 983392 approx.

This whinstone quarry, begun by 1897, was taken over from its previous owners about 1948. It was served by sidings north of the BR Wearhead Branch 1¼ miles west of Stanhope Station. These went into the quarry, where the narrow gauge tramway brought tubs to the loading point. The quarry was re-purchased by Richard Summerson & Co Ltd at a date unknown.

Gauge : 2ft 3½in

-	4wDM	RH	175420	1936	(a)	(1)

(a)   ex Richard Summerson & Co Ltd, with quarry, c/1948.

(1)   to Richard Summerson & Co Ltd, with quarry, date unknown.

ROGERLEY QUARRY, Frosterley   **U140**
formerly **Bradley's (Weardale) Ltd; Pease & Partners Ltd** until /1947 (which see)   NZ 015377 approx.

This was a huge, elongated limestone quarry on the northern slopes of Weardale, stretching from Frosterley almost all the way to Stanhope, a distance of about two miles. It would seem that by the late 1930s, while the quarry and the locomotives were still owned by Pease & Partners Ltd, the crushing plant was owned and operated by Bradley's (Weardale) Ltd, who then took over the quarry and the locomotive as well when Pease & Partners Ltd gave up the lease in 1947. Probably in the late 1940s the rail link to Frosterley Station on the BR Wearhead Branch was abandoned and replaced by a road, but rail traffic within the quarry continued between the quarry face and a lorry loading platform. The quarry and its plant were taken over by the Durham & Yorkshire Whinstone Co Ltd in 1951. Rail working within the quarry ceased in 1954, and the system was dismantled in 1961. The quarry was subsequently closed, but has since been worked or used by various owners, latterly for high quality fluorspar specimens via a small drift mine into the quarry face.

Gauge : 4ft 8½in

22	FROSTERLEY	0-4-0ST	OC	HL	2563	1903	(a)	s/s c1/1960

(a)   ex Pease & Partners Ltd, with quarry, /1947.

DURHAM COUNTY COUNCIL

From soon after the end of the First World War Durham County Council incorporated the use of locomotives in various schemes, usually land reclamation or road building. Initially locomotives were purchased for specific schemes, but later a fleet of locomotives was kept centrally and then some of them were sent out where they were needed.

The first batch of locomotives was used in the construction of a new East Coast road, approximately eight miles long, between Easington Colliery and West Hartlepool (now A1086) **(L141)**, hauling excavated materials by means of skips. Construction began in the spring of 1921, and a locomotive was recorded here in March 1921, although whether it was on hire and then purchased (i.e., became one of those listed below) or not is not known. The Council's Works Committee minutes of 19th June 1922 record the purchase from Fleming & Co of Glasgow of three locomotives; these were inspected at Ashford in Kent, almost certainly at the premises of Kent Construction Co Ltd, which handled many sales of ex War Department locomotives. The Committee approved the purchase of two more on 19th March 1923 and a sixth on 28th May 1923, again all from Fleming & Co. The road was officially opened on 27th September 1924 except for the Castle Eden section, which was opened on 21st August 1925. Remedial work on culverts at Crimdon Dene and Castle Eden Dene took place in 1926-27. Five locomotives were sold in July 1927, apparently the only ones remaining, but which these were is not recorded. The locomotives known or believed to have worked on this construction are as follows:

Gauge : 2ft 0in

-	4wPM	MR	1216	1918	(a)	(1)
-	4wPM	MR	288	1917	(b)	(1)
-	4wPM	MR	1728	1918	(c)	(1)
-	4wPM	MR	1792	1918	(d)	(1)
-	4wPM	MR	1016	1918	(e)	(1)
LR 2587	4wPM	MR	866	1918	(f)	(1)

(a) here by 30/3/1921; see note above; originally War Department Light Railways, 2937.
(b) here by 23/12/1922; see note above; originally War Department Light Railways, 288.
(c) here by 25/7/1926; see note above; originally War Department Light Railways, 2449.
(d) here by 25/6/1927; see note above; originally War Department Light Railways, 2513.
(e) here by 25/6/1927; see note above; originally War Department Light Railways, 2737.
(f) a photograph at Beamish, The North of England Open Air Museum, Beamish, shows this locomotive with the caption "near Castle Eden", presumably on this contract; formerly War Department Light Railways, 2587.

(1) five locomotives, presumably from those on this list, were sold to G.Cohen & Sons Ltd, Canning Town, London, 7/1927, who advertised five locomotives for sale in *Machinery Market*, 22/7/1927; the disposal of the other is unknown.

SPENNYMOOR SLAG WORKS, Spennymoor   N142
NZ 229337

This scheme reclaimed part of the slag heaps created by the Tudhoe Iron Works (see The Weardale Steel, Coal & Co Ltd). The slag works was served by sidings north of the LNER Bishop Auckland - Ferryhill line east of Spennymoor Station. The council eventually closed down its operation and handed it over to the Golightly Tar & Stone Co (which see).

Gauge : 4ft 8½in

| No.1 | 0-6-0T | IC | JF | 1539 | 1871 | (a) | (1) |
| D C C No.2 | 4wPM | | MR | 2096 | 1922 | (b) | (2) |

(a) ex Lingford, Gardiner & Co Ltd, dealers, Bishop Auckland; previously Pease & Partners Ltd, Broadwood Quarry, Frosterley.
(b) New, ex works, 28/6/1922, initially on a three-month hire, but purchased by Durham County Council per their letter dated 6/10/1922.

(1) to The Witton Park Slag Co Ltd, trading as Broadwood Limestone Co, Broadwood Quarry, Frosterley.
(2) to Golightly Tar & Stone Co, with works.

57. To alleviate the high unemployment of the 1930s, Durham County Council undertook a whole series of road improvement schemes, using its locomotives and tubs to remove earth, with a lot of the digging being done by hand. Here one of the 2ft 0in gauge MR locos works on a diversion of the A695 road at Crawcrook, with a very simple layout.

The final group of locomotives was also used on road construction work. Schemes on which locomotives are known to have been used include what was then the A1 Chester-le-Street bypass in 1930; the B6277 at Ashgill Heads, Teesdale, in 1930; the A6076 Causey New Road, near Stanley, 1930-32; the A695 Crawcrook diversion in 1935-36; the A690 Brancepeth diversion in 1935-36; the A690 Houghton Cut widening; the A1 Stonebridge diversion and the A1 Birtley by-pass in 1939. RH 186322 and 186342 were used initially on road construction and then in the late 1940s at a **slag works** operated on the site of the works formerly owned by the Linthorpe-Dinsdale Smelting Co Ltd (which see) (NZ 348137) **(S143)** near Middleton St.George. When not in use the locomotives were kept at the Council's Central Repair Depot, which was originally at **Elvet Waterside** in Durham City (NZ 276424) **(K144)** but which was moved about 1945-46 to **Framwellgate Moor** (NZ 270452) **(K145)**.

58. One of the Council's 2ft 0in gauge RH locomotives.

Gauge : 2ft 0in

1	4wPM	KC?		1923	New	(1)
2	4wPM	KC?		1923	New	(1)
3	4wPM	FH	1652	1930	New	(2)
4	4wPM	FH	1655	1930	New	(2)
-	4wPM	HU	38384	1930	New	(3)
-	4wPM	MR	5067	1930	(a)	s/s
5	4wDM	RH	186322	1937	(b)	(4)
6	4wDM	RH	186342	1937	(b)	(5)
-	4wDM	MR	7604	1939	New	(6)
-	4wDM	MR	7605	1939	New	(7)
-	4wDM	MR	7606	1939	New	(8)
-	4wDM	RH	195844	1939	(c)	(9)
-	4wDM	RH	195849	1939	(d)	(9)

Note : one of MR 7604-7606 carried No.9

(a) ex Durham County Water Board, Burnhope Reservoir, Upper Weardale, c/1932.
(b) New, per J.C.Oliver Ltd, Leeds, Yorkshire (WR).
(c) ex J.C. Oliver Ltd, Leeds, Yorkshire (WR), hire.
(d) ex J.C. Oliver Ltd, Leeds, Yorkshire (WR), possibly on hire.

(1) to Bell & Co (location unknown), for scrap, c4/1948.
(2) it is believed to be these two which were offered for sale in *Evening Chronicle*, Newcastle upon Tyne, 14/8/1950, and in *Contract Journal* on 23/8/1950; to Northumberland County Council, Northumberland (location not known), 9/1950.
(3) later with G. Cohen & Sons Ltd, contractors, Leeds, Yorkshire (WR).
(4) to A.M. Coke, Sleaford, Lincolnshire, by 1/1949; to J.C. Oliver Ltd, Leeds, dealers; to Dinorwic Slate Quarries Ltd, Caernarvonshire, 21/7/1950.
(5) to J.C. Oliver Ltd, Leeds, dealers; to G.W. Bungey Ltd, dealer, Hayes, Middlesex, 1/1950; to Dinorwic Slate Quarries Ltd, Caernarvonshire, 17/8/1950.
(6) to G.W. Bungey Ltd, dealer, Hayes, Middlesex, 1/1950; to Norcon Ltd, Wombourne, Staffordshire, by 25/9/1951.
(7) to G.W. Bungey Ltd, dealer, Hayes, Middlesex, 1/1950; to Elkington Copper Refiners Ltd, Walsall, Staffordshire, by 9/11/1953.
(8) to G.W. Bungey Ltd, dealer, Hayes, Middlesex, 1/1950; to NCB North East Division No.3 (Rotherham) Area, Denaby Colliery, Rotherham, Yorkshire (WR), by 29/9/1952.
(9) possibly returned to J.C.Oliver Ltd, Leeds, Yorkshire (WR), possibly s/s.

William Douglas auctioned two RH locomotives and two petrol locomotives, the latter described as 'scrap', at the Middleton St.George slag works and the Framwellgate Moor Depot on 24/6/1948. Whilst presumably in the list above, their identity and disposal is unknown.

DURHAM COUNTY WATER BOARD

The Durham County Water Board (DCWB) was incorporated in 1920 to acquire from 31st December 1920 the Weardale & Consett Water Co (see Consett Waterworks Co). The Board began by drawing up a programme to link up its existing reservoirs, following this with plans for its new reservoir at Burnhope. The Board's assets were transferred to the Northumbrian Water Authority on 1st April 1974.

BURNHOPE RESERVOIR, near Wearhead   V146
NY 845387 approx.

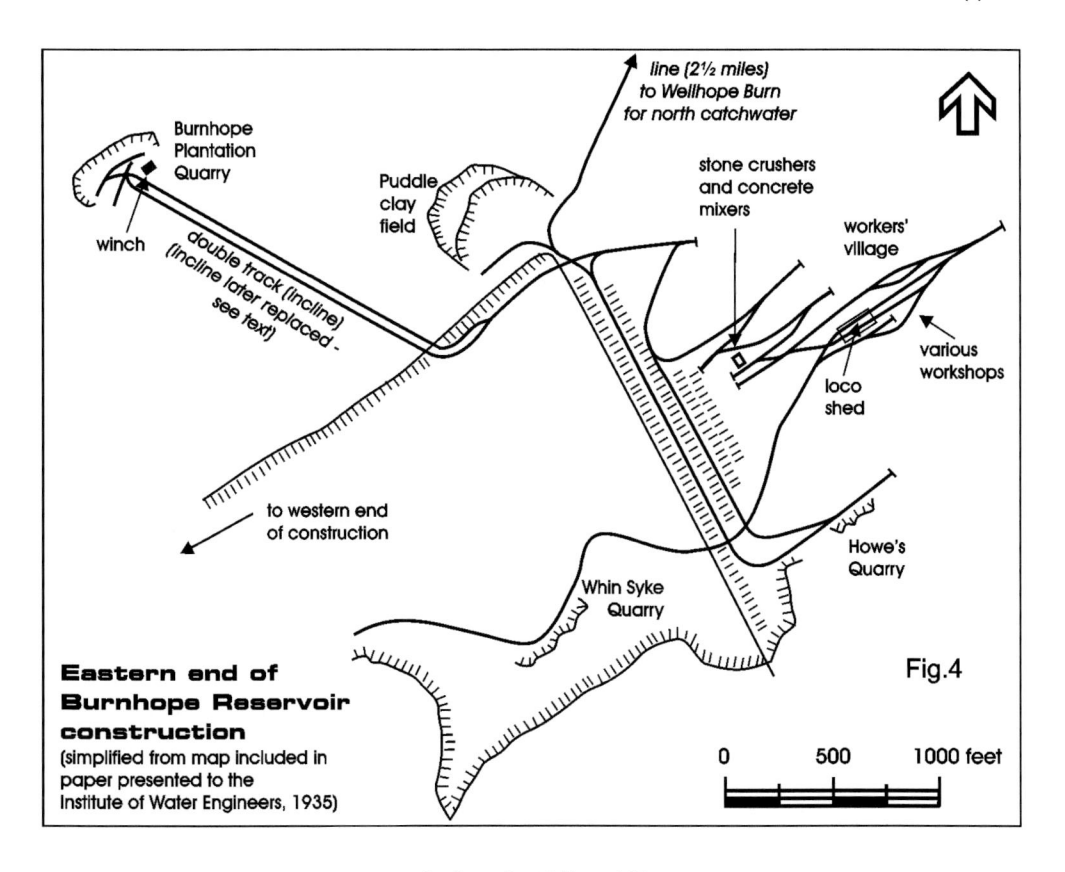

Fig.4

Eastern end of Burnhope Reservoir construction
(simplified from map included in paper presented to the Institute of Water Engineers, 1935)

The minutes for the Board of Directors and the various committees are all deposited in the Durham County Record Office (D/NWA). Plans for this huge reservoir were approved by Parliament in 1922. The reservoir was to be 2200 yards long, with a retaining earth embankment 1770 feet long and hold 1357 million gallons, with 46% going to DCWB and 54% to the Sunderland & South Shields Water Company. The DCWB water would travel by a main and a tunnel (already constructed) to **WASKERLEY RESERVOIR** (see below). This fed the **TUNSTALL RESERVOIR** (NZ 065410 approx.) and also, via the **MUGGLESWICK TUNNEL** (see below) the **SMIDDY SHAW RESERVOIR** (see below), which also drew from the **HISEHOPE RESERVOIR** (NZ 025465 approx.)(see Part 2).

The Act of 1922 provided for a rail link from Wearhead Station on the LNER Wearhead Branch, but when work nominally commenced, on 6th January 1930, the first job was to construct a mile-long road in place of this link, and so the work was never rail-connected. The first sod was officially cut on 18th July 1930. Throughout the scheme employment was restricted to married men from Durham who had been unemployed for at least two years, with preference for those who had served in the armed forces. A hutted village, with a canteen and leisure facilities, was built for the tradesmen, with other men being brought in by bus. At its peak, at the end of 1934, the labour force totalled 700. Electricity had to be brought for 15 miles to service the site, and a small domestic reservoir (40,000 gallons) and a sewage works also had to be constructed.

The main source of stone for the aggregate to be used in making the vast quantity of cement needed was to be **Burnhope Plantation Quarry** (1570 feet), to the north-west of the area (see Fig.4). This was connected to the main site by a double track standard gauge incline about a mile long on a gradient of 1 in 5 and operated by a winch at the top, driven by compressed air, the supply of which was also used to drive machinery in the quarry. At the foot of the incline standard gauge locomotives took the wagons to the crushers and cement plant. The initially small 2ft 0in gauge system around the site itself was operated by petrol locomotives. These clearly proved inadequate, and in May 1932 the Board agreed to dispense with them and use only steam locomotives. Soon afterwards the Board decided to dispense with the standard gauge system as well and convert it to 2ft 0in gauge, so that they had the same gauge throughout, all worked by steam locomotives. Narrow gauge locomotives were subsequently used in Burnhope Plantation Quarry, although mixed gauge track was retained for a time.

On the southern side of the area **Whin Sike Quarry** (see Fig. 4) was developed, both to provide 'fill' for the embankment and stone for its facing side. This quarry was served by the narrow gauge system, although some of the gradients were as severe as 1 in 35 in places. For building the boundary walls around the hillsides a third quarry was also opened on the south side of the valley, known as **Howe's Quarry** from the man in charge of it. This too had narrow gauge rail access. To provide the large amount of puddle clay needed a **Puddle Clay Field** was excavated in the hillside immediately to the north of the embankment, with work beginning in June 1932. Originally the clay was brought down by a winch-operated rope incline, but as the height of the bank rose, this was replaced by locomotive working. Stone for masonry work was brought from Shap in Westmorland.

Besides the general impounding of streams in the local area, water for the new reservoir was also impounded from catchwater areas to both north and south. The north catchwater, 2½ miles long, originated at an intake on the Wellhope Burn. Construction involved excavation, cut and cover and laying pipes, and for this a narrow gauge line ran alongside the route. This work began about May 1930 and lasted until September 1931. The south catchwater was 5½ miles long, but the ground was considered unsuitable for a railway, and instead the pipes were fitted with wooden discs and hauled by diesel tractors. In addition to all the DCWB work, the Sunderland & South Shields Water Company built its own filter plant downstream of the reservoir and laid its own mains.

It appears from various accounts that traffic from Whin Sike Quarry was worked by Avonside or Barclay locomotives or the Kerr Stuart 0-4-2STs. MIDGE, GNAT and WASP ran between the clay field and the pug mill, while the Sentinel, the Fowler and a Barclay were used on clay traffic from the pug mill. It was reported that at busy periods every locomotive would be needed. They were housed in a four-road timber shed near the workshops. The number 71 seems to have never been allocated. There were also five steam cranes here.

The filling of the reservoir was completed in April 1936, although the official opening did not take place until 15th September 1937, by which time most of the buildings had been cleared and the site landscaped. Incredibly, five of the locomotives subsequently went to North America, two to India and one to Africa.

Reference : *Dam Builders' Railways from Durham's Dales to the Border*, H.D. Bowtell, Plateway Press, 1994. (Note : some of Bowtell's work has been superceded by subsequent research into DCWB minutes.)

59. The biggest unemployment scheme in Durham in the 1930s was the construction of Burnhope Reservoir. AB supplied three 2ft 0in gauge well tanks in 1931. Here No.3, AB 1855/1931, passes alongside the short-lived standard gauge system.

60. In 1933, five new AE 0-4-0Ts arrived, an interesting comparison with the AB design. 85, SUNDERLAND, AE 2073/1933, awaits re-sale at Burnhope in 1937 after the work was completed.

Gauge : 4ft 8½in

1		0-6-0ST	OC	AE	2000	1930	New	(1)
2		0-4-0ST	OC	AB	1988	1931	New	(2)

(1) offered for sale in *Contract Journal*, 29/11/1933; to Newbiggin Colliery Co Ltd, Newbiggin, Northumberland, 10/1934.
(2) offered for sale in *Contract Journal*, 29/11/1933; to Cambrian Wagon Works Ltd, Maindy Works, Cardiff, Glamorgan, /1933, via Central Wagon Co Ltd, Wigan, Lancashire.

Gauge : 2ft 0in

1		4wPM		MR	5067	1930	New	(1)
2		4wPM		MR	5232	1930	(a)	(2)
No.3	"GREEN"	0-4-0WT	OC	AB	1855	1931	New	(3)
4	"RED"	0-4-0WT	OC	AB	1991	1931	New	(4)
No.5	"GREY"	0-4-0WT	OC	AB	1994	1931	New	(5)
6	"SENTINEL"	4wVBT	VCG	S	6902	1930	(b)	(6)
No.70	"BLACK"	0-4-0WT	OC	AB	1995	1931	New	(3)
72	MIDGE	0-4-0ST	OC	KS	4290	1923	(c)	(7)
73	GNAT	0-4-0ST	OC	KS	4291	1923	(d)	(7)
74	BURNHOPE	0-4-2ST	OC	KS	1144	1911	(e)	(8)
75	WELLHOPE	0-4-2ST	OC	KS	1145	1912	(f)	(9)
76	IRESHOPE	0-4-2ST	OC	KS	1142	1911	(g)	(8)
77	KILLHOPE	0-4-2ST	OC	KS	1047	1908	(h)	(10)
78	HARTHOPE	0-4-2ST	OC	KS	1291	1915	(j)	(8)
79	"R.A.F."	0-4-0WT	OC	AB	1453	1918	(k)	(11)
80	WASP	0-4-0ST	OC	KS	4001	1918	(k)	(7)
81	DURHAM	0-4-0T	OC	AE	2066	1933	New	(4)
82	WEAR	0-4-0T	OC	AE	2067	1933	New	(4)
83	LANCHESTER	0-4-0T	OC	AE	2071	1933	New	(12)
84	AUCKLAND	0-4-0T	OC	AE	2072	1933	New	(3)
85	SUNDERLAND	0-4-0T	OC	AE	2073	1933	New	(3)
86	STANHOPE	0-4-2ST	OC	KS	2395	1917	(m)	(13)
87	"FOWLER"	0-6-0T	OC	JF	16991	1926	(n)	(14)

(a) ex MR, 27/1/1932; previously Petrol Loco Hirers (subsidiary of MR).
(b) Parts ordered for stock by S, 1/1927. New here, probably in 8/1930; loco was ordered by J.C. Oliver Ltd, Leeds, Yorkshire (WR), though DCWB minutes suggest the Board's direct involvement with S; presumably J.C. Oliver Ltd was acting as a agent to DCWB.
(c) ex Nelson Corporation, Upper Coldwell Reservoir construction, Lancashire, 5/1932.
(d) ex Nelson Corporation, Upper Coldwell Reservoir construction, Lancashire, 7/1932.
(e) ex J. Pugsley & Sons Ltd, Beaufort Plant Depot, Stoke Gifford, Gloucestershire, by 7/1932.
(f) ex J. Pugsley & Sons Ltd, Beaufort Plant Depot, Stoke Gifford, Gloucestershire, 7/1932.
(g) ex J. Pugsley & Sons Ltd, Beaufort Plant Depot, Stoke Gifford, Gloucestershire, 6/1932.
(h) ex J. Pugsley & Sons Ltd, Beaufort Plant Depot, Stoke Gifford, Gloucestershire, c1931 (?).
(j) ex Royal Arsenal, Woolwich, London, c7/1932, possibly via J. Pugsley & Sons Ltd, Stoke Gifford, Gloucestershire.
(k) ex ?, c5/1932; formerly Air Ministry, Works & Buildings Store, West Drayton, Middlesex, and auctioned on 20/3/1931.
(m) ex ?, 12/1933; formerly Holloway Bros (London) Ltd, contractors, whose last known use of it was on roadworks at Swanley, Kent, 1922-24 and who advertised it for sale in 4/1930.
(n) ex W. Dennis & Sons Ltd, Nocton Estates Light Railway, Lincolnshire, via George Cohen, Sons & Co Ltd, 12/1933.

(1) DCWB Minute dated 24/5/1932 states "petrol locomotive plant to be disposed of"; to Durham County Council, Central Repair Depot, Elvet Waterside, Durham City.
(2) see Minute in (1); subsequently sold to "Golightly, contractors, Spennymoor"; this is almost certainly Alfred Golightly, who operated the Golightly Tar & Stone Co (which see) at Spennymoor.
(3) offered for sale in *Contract Journal*, 2/6/1937; to Corby (Northants) & District Water Company, Eyebrook Reservoir construction, Caldecot, Northamptonshire, whose offer was accepted on 20/7/1937.
(4) to Lord Penrhyn's Slate Quarries, Penrhyn, Caernarvonshire, 10/1936.
(5) offered for sale in *Contract Journal*, 2/6/1937; to Lord Penrhyn's Slate Quarries, Penrhyn, Caernarvonshire, 1/1938.

(6) to Cliffe Hill Granite Co Ltd, Markfield, Leicestershire, 1935.
(7) to South Essex Waterworks Company, Abberton Reservoir construction, near Colchester, Essex; offer accepted by DCWB, 21/1/1936 (minute specifically states 'three Wren locomotives'); probably moved, 2/1936.
(8) to South Essex Waterworks Company, Abberton Reservoir construction, near Colchester, Essex; offer accepted by DCWB, 31/12/1935, locomotives moved, 1/1936.
(9) advertised for sale in *Contract Journal*, 2/6/1937; to HE by 2/1938; re-sold to an East African sugar factory, per I. Gundle, 10/1940.
(10) advertised for sale in *Contract Journal*, 2/6/1937; to HE by 2/1938; HE had also acquired KS 1145 (see (9) above) and it is believed that parts from KS 1047 were used to repair this, with the remainder being scrapped.
(11) advertised for sale in *Contract Journal*, 2/6/1937; to Walton's Alston Limestone Co Ltd, Alston, Cumberland, between 9/1937 and 2/1938.
(12) to The Birtley Brick Co Ltd, Union Brickworks, Birtley, c2/1937.
(13) to Lord Penrhyn's Slate Quarries, Penrhyn, Caernarvonshire, 12/1934, per H. Stephenson & Sons, dealers.
(14) advertised for sale in *Contract Journal*, 2/6/1937, but unsold; sold by 15/2/1938, apparently for scrap, and cut up on site.

MUGGLESWICK TUNNEL T147
This was part of the Board's overall scheme (see above), driven to link up the **SMIDDY SHAW RESERVOIR** (NZ 043463) **(T148)** (see Consett Waterworks Co) to the **WASKERLEY RESERVOIR** (NZ 020442) **(T149)**. Work on the 2,265 yard tunnel started at the Smiddy Shaw end in May 1923, using direct labour, and was completed in 1926.

Gauge : narrow

-	4wBE	BEV	459	1923	New	s/s

DURHILLS LTD (registered in 1929)
COLLIER LAW, later PARKHEAD, QUARRY V150
NZ 004429

This was a sand quarry, served originally by sidings from the LNER Stanhope & Tyne Branch at Blanchland Goods Station, but after the closure of the Crawleyside and Weatherhill Inclines on 20th April 1951 it was situated at the end of the line, five miles from Burnhill Junction, where there was a reverse for trains to and from the quarry's loading hopper.

The locomotives below were used within the quarry, but were disused by April 1954, the internal rail system having been replaced by lorries. The Goods Station was closed on 2nd August 1965 and the line itself on 29th April 1968, though the quarry continued in use under different owners.

Gauge : 2ft 0in

-	4wDM	RH	166012	1932	(a)	Scr c/1955
-	4wDM	RH	177640	1936	(b)	Scr c/1955

(a) ex East Midland Gravel Co Ltd, Fengate Pits, Cambridgeshire, following auction on 19/10/1943.
(b) ex County Borough of Derby, Derbyshire.

THE EASINGTON COAL CO LTD
registered 5/1/1899; a subsidiary of **The Weardale Steel, Coal & Coke Co Ltd** from 1908
EASINGTON COLLIERY, Easington L151
NZ 438442

The sinking of three shafts for this colliery began on 11th April 1899, but there were severe difficulties in going through 400 metres of water-bearing limestone. Pumping was replacing by freezing the strata in March 1904, only to be abandoned for more pumps in May 1906. These failed, and following these French and Belgian failures, the bankruptcy of the original company and its take-over by the Weardale company, German engineers introduced freezing techniques in August 1908. Coal was eventually reached in April 1909, although production did not begin until 1910, eleven years after work had begun.

The colliery was originally to be linked to Seaham Harbour Docks by a branch of the Londonderry Railway (see Marquis of Londonderry), but eventually it was served by sidings east of the NER Seaham-West Hartlepool line, four miles south of Seaham Station. It was vested in NCB Northern Division No. 3 Area on 1st January 1947.

Gauge : 4ft 8½in

	VERNON	0-6-0ST	IC	HC	530	1899	(a)	(1)
1	(EASINGTON)	0-6-0ST	OC	AB	912	1901	New	(2)
2		0-6-0ST	OC	?	?	?	(b)	
		reb		Thornley		1920		Scr
14		0-4-0ST	OC	Joicey		c1875		
	reb	0-6-0ST	OC	Tudhoe		c1899	(c)	
		reb		Thornley		1928		(3)
6		0-4-0ST	OC	KS	4027	1919	(d)	(2)
19		0-6-0ST	OC	BH	704	1882		
		reb		Tudhoe I.W.		1909	(e)	(2)
	-	0-6-0ST	IC	VF	5305	1945	(f)	(2)

(a) ex ?; formerly Sir Robert McAlpine & Sons Ltd, Lanarkshire & Ayrshire railway contract for Caledonian Railway (1898-1903).
(b) ex The Weardale Steel, Coal & Coke Co Ltd, Tudhoe Iron Works, Tudhoe.
(c) ex The Weardale Steel, Coal & Coke Co Ltd, Tudhoe Colliery, Tudhoe.
(d) ex The Weardale Steel, Coal & Coke Co Ltd, Thornley Colliery, Thornley, by 5/1925.
(e) ex The Weardale Steel, Coal & Coke Co Ltd, Heights Quarry, Eastgate, /1943.
(f) ex War Department, 75315, Longmoor Military Railway, Hampshire, 4/1946.

(1) to J.F. Wake, dealer, Darlington, c/1910; re-sold to Edmund Nuttall, Sons & Co Ltd for Walker Naval Yard contract, Newcastle upon Tyne.
(2) to NCB No. 3 Area, with colliery,1/1/1947.
(3) to The Weardale Steel, Coal & Coke Co Ltd, Tudhoe Colliery, Tudhoe, by 18/6/1934.

EAST HETTON COLLIERIES LTD
subsidiary of **Pease & Partners Ltd** from 20/8/1944
This company was set up to acquire the collieries below in 12/1935 from the receivers of **Walter Scott Ltd**, which had gone into liquidation in 1933.
Walter Scott until 4/12/1900 (although one source shows Walter Scott Ltd in 1897)

Walter Scott (1826-1910), who was raised to the baronetcy in 1907, was a local entrepreneur, and his purchase of the East Hetton Coal Co in 1880 seems to have been his first major venture. He subsequently acquired Leeds Steel Works and had interests in Smith, Patterson & Co Ltd (which see), the Seaton Burn Coal Co Ltd in Northumberland and the Tyne Brass & Tube Manufacturing Co Ltd at Jarrow (which is not known to have used locomotives). He also developed a successful contractor's business, latterly as Walter Scott & Middleton (see Part 2). This work is not to be confused with the contracting business of his son, (Sir) John Scott (1854-1922).
East Hetton Coal Co Ltd until 19/5/1880
previously **East Hetton Coal Co**

EAST HETTON COLLIERY, Kelloe   N152
sometimes called **KELLOE COLLIERY** in the nineteenth century   NZ 346370

The sinking of this colliery was begun in 1836. It was linked by a two mile branch to Kelloe Bank Foot on The Great North of England, Clarence and Hartlepool Junction Railway, which was opened on 18th March 1839, presumably the date of the first East Hetton traffic. The branch was originally worked by a stationary engine at NZ 342361. The 1st edition O.S. map shows the track on both sides of the engine house to be single line, so presumably the engine hauled waggons up from the colliery and then lowered them down what is named as Kelloe Bank (O.S. map) to the engine house and then lowering them down to the GNofE, C&JHR. This line was linked at Wingate to the Hartlepool Dock & Railway, with coal being shipped at Hartlepool. However, on both railways the coal owners ran their own trains using their own locomotives, and it is almost certain that the East Hetton company did this. If so, given that the route was rope-worked as far as Kelloe Bank Head, then presumably the locomotives worked from this point eastwards to Hesleden Bank Head on the HD&R, and perhaps also at Hartlepool itself. When the York & Newcastle Railway took over the two railways in 1846 and stopped privately-run trains, the East Hetton company received over £7300 in compensation, a figure so large that locomotives must be included (see the entry for HD&R).

However, the details below appear to show that the colliery was also using locomotives in the 1850s. How they were used is unknown, unless perhaps the incline down to the colliery had been abandoned in favour of locomotive haulage. The hauler is still listed in the April 1871 section of DCOA Return 102, but it is not shown in the November 1876 section of the Return, while two locomotives are, presumably the two BH locomotives below. Rather unusually, the loco shed was situated on the branch, near the bottom of the former incline.

For many years the colliery was essentially owned by the Forster family, which was also involved in the ownership of Trimdon Grange Colliery (see below), South Hetton Colliery and Raisby Quarry (both which see). In the mid 1870s the family's affairs fell into the Court of Chancery, which ordered the sale of East Hetton Colliery and Raisby Quarry in September 1878, though Walter Scott did not acquire both East Hetton and Trimdon Grange Collieries until 1880. How the first three locomotives listed below were used is not known. The colliery royalty was extended at various times, and because these additional royalties included collieries found elsewhere in this book, it may be helpful to list them. The Garmondsway royalty to the south was leased in 1871; Garmondsway Colliery had closed in 1845. In 1890 the Coxhoe royalty to the west was added; this included Coxhoe Colliery, which had closed in 1877, while other collieries included West Hetton Colliery, Crow Trees Colliery and Heugh Hall Colliery (see William Hedley & Sons). In 1895 the Cassop royalty to the north-east was added, including the closed collieries of Cassop, Cassop Vale, Cassop Moor and Whitwell Collieries (see entry for Edward Robson & Co).

The colliery, which had eliminated pit ponies and was cutting all coal mechanically from 1937, was vested in NCB Northern Division No. 4 Area on 1st January 1947. There was also one steam crane here.

Gauge : 4ft 8½in

	CARADOC	0-6-0	OC	?	?	?	(a)	
	reb	0-6-0T+t	OC	TR	182	c1851		(1)
	-	?		TR	255	1855	New	s/s by 4/1871
	-	?		TR	256	1855	New	s/s by 4/1871
	EAST HETTON	0-4-0ST	OC	BH	267	1873	New	s/s
No.2	RAISBY	0-4-0ST	OC	BH	318	1874	New	s/s
	EAST HETTON	0-4-0ST	OC	HL	2279	1893	New	(2)
	"WALTER SCOTT"	0-6-0ST	IC	HL	2484	1900	New	(2)
	KELLOE	0-6-0ST	OC	P	525	1892	(b)	(3)
	KILMARNOCK	0-4-0ST	OC	AB?	?	?	(c)	s/s
	-	0-4-0ST	OC	AB	698	1891	(d)	(4)
	"KELLOE"	0-4-0ST	OC	P	560	1893		
	reb			Ridley Shaw		1930	(e)	(2)

(a) said to have been a York, Newcastle & Berwick Railway engine.
(b) possibly ex Leeds Steel Works, Leeds, Yorkshire (WR) (owned by firm), after 3/1905.
(c) origin unknown, and identity not confirmed; here by 26/11/1900, when spares ordered from AB.
(d) ex Trimdon Grange Colliery, Trimdon Grange.
(e) ex Brunner, Mond & Co Ltd, Winnington, Cheshire, via Ridley Shaw & Co Ltd, Middlesbrough, Yorkshire (NR), /1930.

(1) to Marquis of Londonderry, Londonderry Railway, Seaham, 9/1855.
(2) to NCB No. 4 Area, with colliery, 1/1/1947.
(3) to Ridley, Shaw & Co Ltd, Middlesbrough, Yorkshire (NR), /1930.
(4) to Leeds Steel Works, Leeds, Yorkshire (WR) ?

There is evidence to suggest that from time to time a locomotive was sent from here to work at Trimdon Grange Colliery (see next entry), but no evidence of the transfers or the identity of the locomotives survives.

TRIMDON GRANGE COLLIERY & COKING PLANT, Trimdon Grange L153
Trimdon Grange Coal Co until 24/7/1880 NZ 366357
J. Forster until /1871; **M. Forster** until /1870; **Joseph Smith** until /1851

The sinking of this colliery was begun on 21st May 1845. It was served by sidings north of The Great North of England, Clarence & Hartlepool Junction Railway, ¾ mile west of the later Trimdon Station. At Wingate this line joined the Hartlepool Dock & Railway Co, giving coal traffic access for shipment at Hartlepool. The colliery began sending coal to Hartlepool on 30th June 1846, and as coal owners were responsible for their own trains, either Joseph Smith owned a locomotive or he hired the services of another coal owner. When the York & Newcastle Railway took over the two railways in October 1846 the practice of private trains was ended. However, locomotive 131 in the York, Newcastle & Berwick Railway list is shown as built by 'Smith' in 1847, and as Joseph Smith also owned the Trimdon Ironworks in Sunderland, it is possible that he was its builder and that the locomotive was under construction, perhaps at Sunderland, at the date of the take-over (see entry for Hartlepool Dock & Railway).

The colliery remained directly with the Forster family until 1871, when it was taken over by the **Trimdon Grange Coal Co**. It was also acquired by Walter Scott in 1880. It was the site of one of the great explosions in County Durham colliery history, when on 16th February 1882 68 people died, together with another six at East Hetton Colliery after gas had seeped in there, eight of the dead being only 13 years old.

It seems that at first the colliery was shunted by horses. The DCOA Returns record no locomotives here until March 1890, when the colliery was said to have two. Their identity is unknown. In November 1896 the Return describes the colliery as being shunted by a locomotive summoned from East Hetton Colliery by telephone, so that the colliery company presumably had running powers between the two pits. However, in later years shunting by horses was resumed. There were beehive coke ovens here, closed in September 1913 when 55 Otto-Hilgenstock waste heat ovens were opened, retrieving tar and other by-products. Because of the depression in trade the colliery and coke works were closed in September 1930 (per the Inspectorate Annual Report, though one source gives the date of this as 1925). With new owners in charge the colliery was re-opened in 1937. In 1939 the old coke ovens were demolished and replaced by 18 Simon-Carves twin regenerative ovens, started up on 9th November 1939; because of the small number of ovens the carbonisation time had to be only 16 hours to make the plant economic. The coke car was a combined unit built by James Buchanan of Liverpool. The associated by-products plant produced tar, sulphate of ammonia, rectified benzole and gas for local use. The colliery, still using horses for shunting, and the coke works were vested in NCB Northern Division No. 4 Area on 1st January 1947.

Clearly there were more locomotives here than the one below, but their identity and movements between here and East Hetton are unknown.

Gauge : 4ft 8½in

| - | | 0-4-0ST | OC | AB | 698 | 1891 | New | (1) |

(1) to East Hetton Colliery, Kelloe.

ELDON BRICKWORKS LTD

ELDON BRICKWORKS, Eldon, near Shildon Q154
Pease & Partners Ltd until c5/1933 NZ 237279

This brickworks lay adjacent to Eldon Colliery, owned by Pease & Partners Ltd (which see). It was opened in 1920 solely for the manufacture of common bricks. Davison says that the works was closed from 1926, but Pease & Partners' records show this not to be the case. Eldon Colliery closed in September 1932 and was sold for demolition in May 1933. However, the brickworks reverted to the Eldon Estate as landowners, and the Estate decided to re-open it, using this new company, to try to relieve some of the severe unemployment in the area. It was served by a one mile branch from a junction ½ mile north of Shildon Tunnel on the LNER Bishop Auckland-Shildon line. The works continued to produce only common bricks.

Rail traffic was subsequently discontinued, though the works remained in production (owned by J. Crossley & Sons Ltd from 1966).

Gauge : 4ft 8½in

| - | | 4wDM | | HE | 1737 | 1935 | New | (1) |

(1) to North Eastern Trading Estates Ltd, Aycliffe Estate, Aycliffe, c/1949.

JAMES W. ELLIS ENGINEERING LTD

Huwood-Ellis Ltd until 27/4/1970
James W. Ellis & Co Ltd until 8/3/1968 (subsidiary of **Huwood Ltd** from 1/1/1967)
James W. Ellis & Co until /1916
Hannington & Co Ltd until c /1903
HANNINGTON WORKS, Swalwell B155
NZ 204627

This foundry and engineering works was probably opened just before the turn of the twentieth century, almost certainly on the site of a former brickworks owned by Hannington & Co Ltd. It was served by sidings south of the NER Redheugh Branch, 1½ miles east of Blaydon Station. The works was shunted by crane – there was one steam crane here - in 1968 and 1969. It passed to its parent company, Huwood Ltd (which see) on 1st September 1971.

Hannington & Co Ltd also owned **Axwell Park Colliery** nearby; this passed to The Priestman Collieries Ltd (which see).

61. *In the late 1920s the LNER bought a considerable fleet of Sentinel locomotives. This one, S 7852/1929, was purchased direct from British Railways five months before this photo was taken, and was seen on 28th August 1957, still carrying its BR logo and number. Note the long pole used for 'parallel shunting' (see photograph 27).*

Gauge : 4ft 8½in

	-	0-6-0ST	IC	BH	716	1882		
		reb	?			1906	(a)	(1)
	-	0-4-0CT	OC	HL	2606	1905	(b)	(2)
	-	0-4-0ST	OC	AE	1055	1874		
		reb		Sdn		1892	(c)	(3)
	PRESTON No.3	0-6-0ST	IC	HL	2737	1907	(d)	(4)
	MARY *	0-4-0ST	OC	HL	3894	1936	New	(5)
8088		0-4-0T	IC	Dar		1923	(e)	(6)
(68159)	MARY	4wVBT	VCG	S	7852	1929	(f)	Scr /1962
	-	4wDM		RH	305323	1951	(g)	(7)

* nameplates transferred to S 7852 from 1957 to 1962.

(a) ex The Harton Coal Co Ltd, South Shields, Marsden & Whitburn Colliery Railway, /1910.
(b) ex J. Spencer & Sons Ltd, Newburn, Newcastle upon Tyne, c/1923.
(c) ex Synthetic Ammonia & Nitrates Ltd, Billingham, c/1930 (by 25/6/1932) (see Cleveland & North Yorkshire Handbook); at one time GWR 1332; previously GWR 2178; originally South Devon Railway, LARK (broad gauge) (see also Ministry of Munitions).
(d) ex U.A. Ritson & Sons Ltd, Preston Colliery, North Shields, Northumberland, by 25/6/1932.
(e) ex LNER, Dairycoates Motive Power Depot, Hull, Yorkshire (ER); the hire started either in 2/1947 or 6/1947.
(f) purchased from BR, 68159, 3/1957, but previously on hire; originally LNER Y3 class, 8159.
(g) ex South Western Gas Board, Cheltenham Gas Works, Cheltenham, Gloucestershire, c8/1969, via RH.

(1) to Leversons Wallsend Collieries Ltd, Usworth Colliery, Usworth.
(2) offered for sale, 10/1921; scrapped.
(3) derelict by 6/1933; s/s.
(4) to The Horden Collieries Ltd, Horden Colliery, Horden, c/1939.
(5) advertised for sale in *Machinery Market*, 14/11/1957; scrapped, 11/1968.
(6) returned to LNER, Dairycoates Motive Power Depot, Hull, Yorkshire (ER), either in 4/1947 or in 11/1947, depending on date of start of hire.
(7) to Huwood Ltd, with works, 1/9/1971

In *The Engineer* for 29th July 1904 a new 3ft 6in gauge locomotive, AB 969/1903 (though actually built by Dick, Kerr & Co Ltd), was offered for sale by this firm under its Newcastle upon Tyne address. There is no evidence that this locomotive was ever at Swalwell; it may have been ordered for an overseas contract that was subsequently cancelled. It was subsequently sold to APCM Ltd, Gillingham Works, Aylesford, Kent, by May 1905.

ENGLISH ELECTRIC CO LTD
STEPHENSON WORKS, Darlington   S156
Robert Stephenson & Hawthorns Ltd until 1/1/1963 (which see)   NZ 300166

This locomotive works, opened in 1902, was served by sidings east of the BR Durham-Darlington line, 1¾ miles north of Darlington (Bank Top) Station. It was closed in March 1964. The buildings were subsequently taken over by other firms, without rail traffic.

Gauge : 4ft 8½in

D0227 "THE BLACK PIG"	0-6-0DH	EE	2346	1956			
		VF	D227	1956	(a)	(1)	

(a)   ex Robert Stephenson & Hawthorns Ltd, with works, 1/1/1963.

(1)   scrapped c/1964 (after 12/10/1964).

THE FELLING COAL, IRON & CHEMICAL CO LTD
FELLING (later ALBION) IRON WORKS, Felling   D157
    NZ 279623

This company is one of those involved in the complicated and obscure history of the coal, iron and chemical industries at Felling; see also John Bowes & Partners Ltd for Felling Colliery, previously owned by Sir George Elliot, and H.L. Pattinson & Co for Felling Chemical Works, Felling Iron Works and Felling Brick Works.

Felling Iron Works was built by **H.L. Pattinson & Co** and came into production in 1854. The following year it was provided with a new link to the NER, east of Felling Station. It passed, probably in the early 1860s, into the hands of Frazers, Roberts & Co and three members of that firm separately, this ownership being combined into **Frazers, Roberts & Co Ltd**, registered on 2nd February 1866, although *Mineral Statistics* continue to give the owners as H.L. Pattinson. Frazers, Roberts & Co Ltd's second purpose was to build a chemical works at Felling. This was almost certainly built to the north of H.L.Pattinson & Co's works and was probably quite small; it may well have been railway-served via a link from the iron works, but this is not confirmed. On 26th August 1871 **The Felling Coal, Iron & Chemical Co Ltd** was formed to acquire the business of Frazers, Roberts & Co Ltd and also to purchase Felling Colliery from George Elliot. This company re-named the works the **Albion Iron Works**.

The Felling Coal, Iron & Chemical Co Ltd spent several years trying to agree terms with George (later Sir George) Elliot for Felling Colliery, but there is no evidence that this ever changed hands, and it was Sir George who sold it to John Bowes, Esq., & Partners in 1883 (which see). According to *Mineral Statistics* the iron works did not produce iron after 1864, and it seems that instead it concentrated on foundry and general engineering work, similar to the Birtley Iron Works and the Grange Iron Works near Durham.

The Felling Coal, Iron & Chemical Co Ltd went into liquidation on 23rd June 1877. The iron works is said to have re-opened in April 1880, but clearly this was not for long, as shortly afterwards it was reported as derelict. The chemical works was re-opened by **The Tyneside Chemical Co** about 1881. This firm continued to operate the works until about 1894, following which the whole area was cleared.

The two locomotives below were ordered by The Felling Coal, Iron & Chemical Co Ltd. How the two works were shunted before 1871 is not known, nor whether The Tyneside Chemical Co used locomotives.

Gauge : 4ft 8½in

FELLING No.2	0-4-0ST	OC	BH	203	1871	New	(1)
FELLING No.1	0-4-0ST	OC	BH	223	1871	New	s/s

(1)   disposal unknown; the loco eventually passed back into the hands of BH, who are said to have used it as a works shunter before disposing of it to a contractor undertaking a dock contract on the River Tyne, who in turn sold it on to the Marquis of Londonderry, Londonderry Railway, Seaham Harbour.

FENCE HOUSES BRICKWORKS LTD
LUMLEY QUARRY & BRICKWORKS, Lumley Thicks

J158
NZ 307503

This brickworks and quarry were opened in 1937 by **Lumley Brickworks Ltd**. Initially coal for the kilns was supplied by a tramway direct from the adjacent Lumley 6th Colliery (see The Lambton, Hetton & Joicey Collieries Ltd). When these new owners took over the brickworks is not known. The rail system ran from the quarry to a grinding plant at the brickworks. Prior to the introduction of the locomotive below, the tubs ran by gravity to the foot of an incline, whence they were hauled to the plant by an electric winch. In the latter part of 1955 loco working ceased due to the uneven shape of the quarry floor through faulting, and transport from the quarry was taken over by dumper trucks. There was no main line rail connection.

Gauge : 2ft 0in

-		4wDM	RH	223716	1944	(a)	(1)

(a) ex Charles Wall Ltd, Globe Works, Essex, via Grayston Plant & Engineering Co Ltd, Essex, c7/1953.

(1) to Thornton Engineering Co Ltd, Whitley Bay, Northumberland, c/1959, acting as dealers; re-sold to W.T. Bathgate (Limeworks) Ltd, Greenleighton Quarry & Limeworks, near Netherwitton, Northumberland, c/1960.

FERENS & LOVE (1937) LTD
CORNSAY COLLIERY & BRICKWORKS, Cornsay Colliery
Ferens & Love Ltd until /1937
Ferens & Love until /1922
originally **Joseph Love & Partners**

K159
NZ 169433

Joseph Love (1784-1867) was also a partner in Strakers & Love (which see for a longer biography), and owned Shincliffe Colliery and Brickworks (which see). Robinson Ferens (1822-1890) was Love's son-in-law. This partnership began by acquiring **Inkerman Colliery**, near Tow Law, (which see) in 1865, and then began to develop Cornsay.

This colliery was served by a private branch 2¼ miles long from Flass Junction on the NER Deerness Valley branch, the same line also serving Hamsteels Colliery (see Sir S.A. Sadler Ltd.) It was unusual in having no shaft, the coal always being worked from drifts and transported by narrow gauge

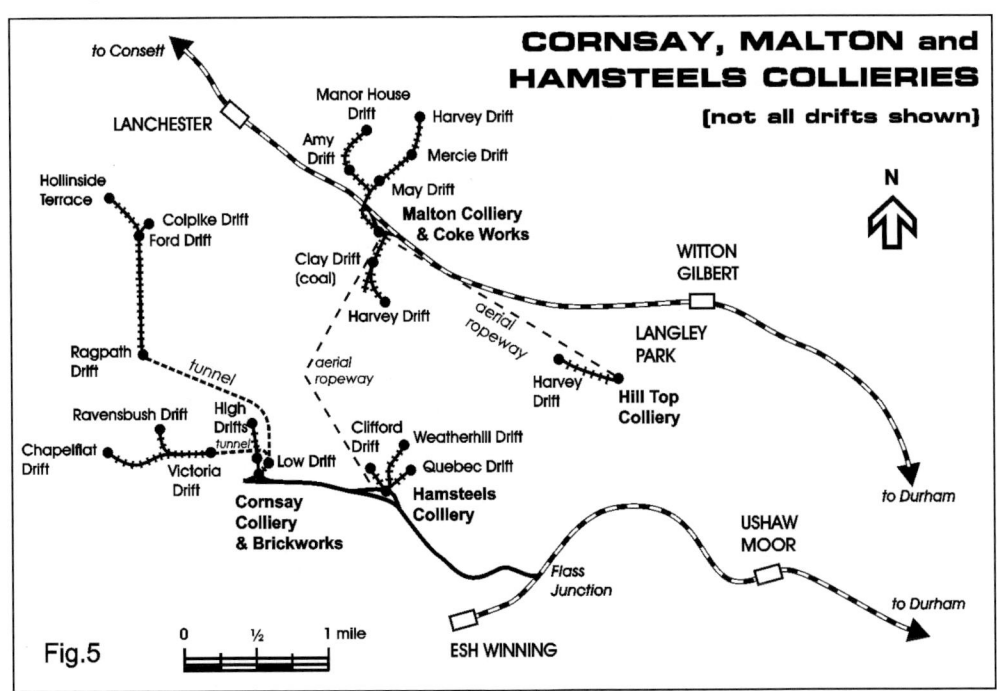

Fig.5 **CORNSAY, MALTON and HAMSTEELS COLLIERIES** (not all drifts shown)

62. No.1, a Manning Wardle 'M' locomotive and presumed to be MW 148/1865. Another of Herbert Coates' pictures, the postcard was franked 1906.

63. Photographed on 11th May 1946 outside her shed, JOHN OWEN, FW 170/1872, already lacks her dome and cylinder covers and has a gaping gash in her bunker. Soon afterwards she broke a crosshead, and then worked on only one cylinder. Locomotives were often not high on the list of priorities for spending money. Note the GWR 'Dean cab' and closed coal rails on the bunker from her rebuild at Swindon in 1894.

tubways. The first two, dating from 1869, served the **Low Drift** (NZ 168433) and the **High Drift** (NZ 168436), about ½ mile from the screens. The latter was subsequently replaced by a new **High Drift** (NZ 168438), apparently after a bad roof fall which happened during the General Strike of 1926. The tubway serving the High Drift ran between the houses of West Street, and perhaps because of this was worked by endless rope haulage, which was slower than main-and-tail haulage. The second tubway ran north westwards for 1½ miles in a tunnel, emerging to reach **Ragpath Drift** (NZ 154446), this line being worked by engine houses at each end of the tunnel, presumably operating main-and-tail rope haulage. From Ragpath Drift it was extended northwards to serve **Colpike Drift** (NZ 154459) and then north-west to reach **Hollingside Terrace** (NZ 150463), a row of cottages built for colliery officials in 1892, a total distance of 2½ miles. In 1900 **Ford Drift** was opened near Colpike Drift. The whole of this system had closed by 1921. The third line, which branched from the Low Drift tunnel a short distance in from the entrance, ran westwards for 1½ miles to serve **Chapelflat Drift** (NZ 150434) near Old Cornsay, with a short branch serving **Ravensbush Drift** (NZ 156436). This route too was worked by main-and-tail rope haulage, initially with sets of 30-40 tubs, but latterly with only 20-25 tubs. In the 1940s the **Victoria Drift** (NZ 165433) was driven, on the north side of this tubway about 150 yards from the western end of the tunnel, to reach the Victoria seam.

Excellent supplies of seggar clay were found with two of the coal seams, and by 1879 brick-making had begun. Eventually three yards were developed, the High Yard, the furthest west of the three, which made common bricks; the Low Yard, the middle of the three, which made firebricks and refractories, and the Pipe Yard, the most easterly of the three, which made salt-glazed sanitary pipes. Latterly the High Yard was little used.

Information gathered in 1949 suggests that at some period, perhaps before the First World War, a "Paddy's train" was operated, presumably a workmen's train consisting of one or two coaches, with a loco supplied by the company, which worked between Cornsay and Waterhouses Station, though this would have involved a run-round or reverse at Flass Junction. Later this train seems to have been replaced by a system allowing anyone connected with the colliery to travel in the brakevans of mineral trains working between Cornsay and Esh Winning; whether this was official is not known. Latterly trains from the colliery were worked down only to Flass Junction, without a brake van at the rear.

By 1932 only the High, Ravensbush and Chapelflat Drifts were being worked. In 1937 the company was acquired by the Holiday family of Durham City, who already had interests in brickmaking. The colliery became a licensed mine on 1st January 1947; but the Victoria seam was badly watered, and all coal production ceased in 1951. The Low Yard closed in 1953, leaving only the Pipe Yard. Rail transport ceased about 1961, and production of pipes withered away, finally ending in 1975; an order to wind the company up was made on 10th June 1975.

The DCOA Returns give one loco here in April 1871 and November 1876, and two in March 1890.

Gauge : 4ft 8½in

-		0-6-0ST	IC	BH	244	1873	New *	s/s
JOHN HARRIS		0-4-0ST	OC	Harris?	?		(a)	(1)
(later No.1?) JOHN BELL		0-6-0ST	IC	MW	148	1865	(b)	(2)
No.3	CORNSAY					?		
(formerly No.1)		0-4-0ST	OC	HL	2478	1901	New	(3)
(No.2)	JOHN OWEN	0-6-0ST	OC	FW	170	1872		
		reb		Sdn		1894	(c)	(4)
PRINCE		0-4-0ST	OC	Harris?	?		(d)	(5)
BRITON		0-4-0ST	OC	?	?		(e)	(6)
KILMARNOCK		0-6-0ST	OC	AB	1497	1916	New	(7)
THORNCLIFFE		0-6-0ST	IC	MW	241	1867		
		reb				1902		
		reb Newton, Chambers & Co				1932	(f)	(8)
PLANET		4wDM		FH	3374	1950	New	
		reb		FH		1953		(9)

* this loco was originally part of an order of five for the NER, but it was diverted here before delivery and replaced in the NER order by BH 285.

(a) JOHN HARRIS and the loco named SHINCLIFFE, which came here from Shincliffe Colliery & Brickworks, Shincliffe, are believed to be the one and the same.
(b) ex Shincliffe Colliery & Brickworks, Shincliffe.
(c) ex Bute Works Supply Co, Cardiff, Glamorgan, /1914; until 8/1912 was GWR 1385; originally Whitland & Cardigan Railway, No.1, JOHN OWEN.
(d) ex Lingford, Gardiner & Co Ltd, Bishop Auckland, dealers, hire, by /1917.

(e)   builders, origin and date of arrival unknown.
(f)   ex Newton, Chambers & Co Ltd, Thorncliffe Ironworks, Chapeltown, Yorkshire (WR), 7/1946.

(1)   sold before 1914 to a "repairing company"; this loco allegedly became the hire loco named PRINCE of Lingford, Gardiner & Co Ltd, Bishop Auckland (see footnote A above).
(2)   sold before 1914 to a "repairing company" and believed went to Lancashire.
(3)   to Hawthorn Limestone Co Ltd, Hawthorn Quarry, near Seaham Harbour, by 24/3/1940.
(4)   in 1946 this loco broke a crosshead, and for about six months it worked on only one cylinder, the men using a pinchbar if it stopped in a position preventing steam being admitted to the remaining cylinder; it ceased work on the arrival of THORNCLIFFE; to a Bishop Auckland firm for scrap, c3/1952.
(5)   returned to Lingford, Gardiner & Co Ltd, Bishop Auckland, ex hire.
(6)   to Lingford, Gardiner & Co Ltd, dealers, Bishop Auckland.
(7)   offered for auction by W.W. Holmes, 7/7/1943; to The South Durham Steel & Iron Co Ltd, Stockton Works, /1943 (see Cleveland and North Yorkshire Handbook).
(8)   to a Bishop Auckland firm for scrap, c3/1952.
(9)   to Smith, Patterson & Co Ltd, Blaydon, 7/12/1961 or 2/1/1962 (FH lists give both dates).

Although the company used seggar from underground in its brickworks, it is believed that there were several small clay pits to the west of the colliery, at least two of which are believed to have later been converted into reservoirs. Whether the locomotives below worked in these pits, or on the line to Ravensbush and Chapelflat Drifts is not known.

Gauge : 2ft 0in

-	4wPM	MR or FH	?	(a)	(1)
-	4wPM	MR or FH	?	(a)	(1)

(a)   origin and date of arrival unknown.

(1)   offered for auction by W.W. Holmes, 7/7/1943, with one listed as 40hp, one as 20hp; they were described as "Simplex" locomotives and so could have been built by either MR or FH; s/s

64. *A five-plank, side door, wagon built by McLachlan & Co of the Haughton Lane Wagon Works, Darlington, about whom very little is known. Note the grease axle boxes and the curious absence of any brakes! A Ferens & Love wagon, with its box extended, and probably used for carrying coke, stands behind.*

FINLEY & WILKINSON
THE HARPERLEY COLLIERY CO

In November 1906 a sale notice, now untraceable, advertised for sale the plant, including a locomotive, of Craig Lea Colliery, near Crook, apparently also including Harperley Colliery nearby.

CRAIG LEA COLLIERY, near Crook	M160
HARPERLEY COLLIERY (Craig Lea Pit) is also found	NZ 137364

The colliery had been worked in the nineteenth century but had been abandoned and the site cleared by 1897, only for it to be re-opened. By 1906 it was owned by **FINLEY & WILKINSON**, according to the Annual Reports of the Mines & Quarries Inspectorate; however, the registration in the National Archives of **The Harperley Collieries Co Ltd**, which took over the business in 1908, gives the owners as Thomas Finley and J.G.Sinclair. The colliery was served by sidings south of the NER Crook – Tow Law line, 2½ miles north of Crook Station.

HARPERLEY COLLIERY, near Crook	M161
	NZ 135356 (main site)

This colliery was opened in 1845, and seems to have had various owners. There were coke ovens here, together with a fire-brick and a sanitary pipe works. By 1897 the main site was owned by **Messrs. Hall & Finley**, and was linked by two narrow gauge tramways to drifts (names unknown) slightly further south at NZ 135355 and NZ 134355. Ownership passed to **THE HARPERLEY COLLIERY CO**, perhaps set up by the same men, in 1906. By this date it was linked to Craig Lea Colliery by a narrow gauge tramway ½ mile long. This was rope-worked, apparently by the main-and-tail system.

There is no evidence in the Inspectorate Reports of either firm going out of business or ceasing production in 1906. If the reference to the locomotive in the opening paragraph is correct, then it would seem almost certain to have been a standard gauge locomotive working at Craig Lea Colliery. If it was indeed disposed of, then presumably the NER took over the shunting.

The Inspectorate Report for 1907 gives the same two owners, but in 1908 the two collieries and the business passed to **The Harperley Collieries Co Ltd** (registered 10/10/1908). By 1919 the southern drifts had been closed, and three new ones opened, two near Craig Lea, at NZ 138363 and 300 yards east of the "main line at NZ 142362, via a tunnel under a minor road, with a third (NZ 135357) near the main site at Harperley. These branches were very probably also worked by rope haulage. The firm appears to have ceased operations in 1923; it went into voluntary liquidation on 19th February 1925, although Craig Lea was not formally abandoned until April 1929.

Curiously, half way along the tramway between Craig Lea and Harperley and presumably linked to it was a third colliery, **Cold Knot Colliery** (NZ 135359) **(M162)**. Its ownership appears to have been entirely separate (see The Woodifield Coal Co Ltd).

FORDAMIN COMPANY (SALES) LTD

this firm was a member of the **English China Clays Group**

CLOSEHOUSE MINE, near Middleton-in-Teesdale	W163
Athole G. Allen (Stockton) Ltd until 21/5/1981	NY 848227

This drift mine was situated about six miles south-west of Middleton-in-Teesdale. A lead mine in the 19th century, it was later worked for barytes, with Athole G. Allen (Stockton) Ltd taking the lease in 1939. Until 1974 it lay in the North Riding of Yorkshire, but under the re-organisation of local government in April 1974 this part of Teesdale was transferred to Co. Durham. The mine was served by a track from the B6276 road between Middleton-in-Teesdale and Brough.

In March 1981 Athole G. Allen Ltd went bankrupt, the business being purchased by Fordamin Company (Sales) Ltd two months later. Mining underground was replaced by opencast working in 1983. Locomotive haulage ceased in 1984. This firm sold out in 1989 and working ceased in 2001.

Gauge : 2ft 0in (never used here)

-	4wDM	HE	4569	1956	(a)	(1)

(a) ex The Owners of Settlingstones Mine Ltd, near Newbrough, Northumberland, 10/1969.

(1) to Ayle Colliery Co Ltd, Alston, Cumbria, via J.A. Lister & Sons Ltd, Consett, c5/1983 (after 19/3/1983, by 19/11/1983).

Gauge : 1ft 7½in

-	0-4-0BE	WR	5655	1956	New	(1)
-	0-4-0BE	WR	D6754	1964	New	(2)
-	0-4-0BE	WR	4149	1949	(a)	(3)
-	0-4-0BE	WR	P7664	1975	New	(4)
-	0-4-0BE	WR	Q7731	1976	New	(4)

(a) ex The Owners of Settlingstones Mine Ltd, near Newbrough, Northumberland, 10/1969.
(1) to F. Shepherd, Flow Edge Colliery, Alston, Cumbria, by 6/1980.
(2) to Ayle Colliery Co Ltd, Alston, Cumbria, via J.A. Lister & Sons Ltd, Consett, c1/1985 (after 10/9/1982, by 13/7/1985).
(3) to F. Shepherd, Flow Edge Colliery, Alston, Cumbria, by 6/1981.
(4) to Ayle Colliery Co Ltd, Alston, Cumbria, via J.A. Lister & Sons Ltd, Consett, c1/1985 (after 19/3/1983, by 13/7/1985).

65. 0-4-0BE WR D6754/1964, having just come up out of the drift entrance (left), stands on the little turntable which turns to give access to her shed, 27th July 1979. She was a 4 h.p. machine and weighed just 1½ tons.

FORESTRY COMMISSION
Dinsdale, near Darlington   S164

The locomotive below is recorded in MR records as delivered new to the Forestry Commission, Dinsdale, near Darlington, but was owned by the Ministry of Supply.

The parish of Low Dinsdale is situated about four miles south-east of Darlington, and includes a number of plantations where the locomotive below might have been used in the war-time drive for felling timber. Equally, the record might mean that the locomotive was delivered to Dinsdale Station (actually situated in the village of Middleton St.George), on the LNER Darlington-Stockton line, three miles east of Darlington (Bank Top) Station, for onward delivery.

Gauge : 2ft 0in

-	4wPM	MR	9103	1941	New	(1)

(1) returned to MR, rebuilt as 4wDM MR 9341 and sold to J. Cochrane & Sons, Cannich-by-Beauly, Invernesshire, Scotland, 13/10/1948.

FOSTER, BLACKETT & WILSON LTD (registered in 1913)
TYNE LEAD WORKS, Hebburn   E165
                                                                                          NZ 322656

This was a long-established paint works served by sidings from the Pontop & Jarrow Railway of John Bowes & Partners Ltd (which see), west of the staithes at Jarrow used between 1883 and 1936. Probably P&JR locos brought the firm's traffic to and from the Pontop & Jarrow Railway's link with the NER at Pontop Junction, Jarrow. The firm continued in business for many years after disposing of its locomotive.

Gauge : 4ft 8½in

| | | 0-4-0ST | OC | P | 1508 | 1918 | New | (1) |

(1) to Priestman Whitehaven Collieries Ltd, Wellington Colliery, Whitehaven, Cumberland, 4/1932, via Robert Frazer & Sons Ltd, Hebburn, dealers.

THE FOWNES FORGE & ENGINEERING CO LTD

This company was formed in 7/1898 to take over the business of George Fownes & Co (formed in 1895) at East Moors, Cardiff, and to open the works below.

ST.BEDE'S WORKS, Tyne Dock D166
NZ 350650

This works was constructed on the site of the former Jarrow Chemical Works, latterly owned by The United Alkali Co Ltd (see entry for **Jarrow Chemical Co**). It was opened about August 1898 and was served by a ½ mile branch north of the NER Gateshead-South Shields line, 1¼ miles west of Tyne Dock Station. The works manufactured marine forgings. It was closed about 1909 when the business was transferred to the Cardiff works, the plant here being auctioned on 27th April 1910 by A.T. & E.A. Crow.

Gauge : 4ft 8½in

| | FOWNES | 0-4-0CT | OC | HL | 2499 | 1901 | New | (1) |

At the auction in April 1910 a 9in cylinder standard gauge loco was offered (though the loco above, which had 12in cylinders, was not mentioned). This was almost certainly a 0-4-0ST, but its identity is unknown.

(1) to Clarke, Chapman & Co Ltd, Gateshead.

THE FRAMWELLGATE COAL & COKE CO LTD

The Framwellgate Coal & Coke Co until 13/1/1885
originally **The Framwellgate Coal Co** (see also below)
FRAMWELLGATE COLLIERY, Framwellgate Moor K167

This colliery consisted of two pits, known as the **CATER HOUSE PIT**, or **Dryburn Grange Pit** or **High Pit** (NZ 255454), west of the village of Pity Me, and **FRAMWELLGATE COLLIERY**, or the **Low Pit** or **Old Pit** (NZ 271455), a mile to the east, and north of Framwellgate Moor village. Coal had been won at the latter on 5th November 1838 by the short-lived **Northern Coal Mining Co**, and production began in March 1840, with the Cater House Pit following about the same time. The colliery's branch was linked to the Lambton Railway's branch from Frankland Colliery (which see), though in fact the traffic was worked by the Londonderry Railway (which see). However, in 1857 the NER opened its line from Leamside to Bishop Auckland, and soon afterwards a ½ mile link was built to join it at Frankland.

After the failure of The Northern Coal Mining Co in 1848, the colliery passed through various hands, until in August 1859 the colliery was acquired by the **Marchioness of Londonderry** for £11,000, and of course it continued to be worked as part of the Londonderry Railway. On 27th September 1873 it was sold by the 5th Marquis of Londonderry to John Stevenson and Richard Jacques, the owners of Acklam Iron Works in Middlesbrough, for £45,500. Stevenson and Jacques then set up **The Framwellgate Coal Company** to operate the colliery. The link to the Lambton Railway was abandoned and all traffic went out via the NER at Frankland. Whether the company remained a subsidiary of subsequent owners of the Acklam Works is not known.

The new owners thus took over a branch 2¼ miles long from the Cater House Pit to the NER. The first mile, between Cater House and the Low Pit, was fairly flat. At the Low Pit there were also beehive coke ovens, which closed in October 1913. Immediately to the east of the pit a **self-acting incline** one mile long took full waggons down to Low Newton, from where the ½ mile link curved round to the NER.

The purchase of 1873 did not include any locomotives or waggons, and no locomotives are known before 1878, but DCOA Return 102 includes one loco for November 1876. Latterly the loco shed was at the Cater House Pit; whether a locomotive went down the incline each day to work at Low Newton, or whether the NER worked this section, is not known.

The Cater House Pit was closed in 1910, but re-opened in 1921. The company went into voluntary liquidation in July 1924, and the plant was auctioned on 11th -12th February 1925 by Wheatley Kirk.

Gauge : 4ft 8½in

| | - | 0-4-0ST | OC | HCR | 203 | 1878 | New | (1) |
| | CETEWAYO | 0-4-0ST | OC | RWH | 1789 | 1879 | New | (2) |

		0-4-0ST	OC	?	?	?	(a)	s/s c /1925

(a) builders, origin and date of arrival unknown.
(1) a 10in x16in saddle tank loco, quite possibly this loco, was the one loco offered in the auction of 2/1925; local accounts said that in later years "the loco" was called CETEWAYO, so perhaps the name from RWH 1789 was transferred to this loco; s/s.
(2) to Cochrane & Co Ltd, Ormesby Ironworks, Middlesbrough, Yorkshire (NR), by /1888.

There was also one steam rail crane here.

THE FRANKLAND COAL CO (1934) LTD
THE ABBEY WOOD COAL CO
FINCHALE COLLIERY and ABBEY WOOD DRIFT, near Durham   G168

Both firms were owned by F.W. Blacklock, who purchased the Frankland Coal Co (1934) Ltd in 1948. This firm owned **FINCHALE COLLIERY** (NZ 293471), a licensed mine (see NCB Section in Part 2 for explanation) opened in 1934 near Finchale Abbey alongside the River Wear, though in fact the shaft was only used for ventilation; the coal was drawn from a drift entrance about 200 yards downstream from the Abbey, and the tubs were hauled up a narrow gauge rope incline to screens near the shaft. In 1952-53 Mr. Blacklock opened **ABBEY WOOD DRIFT** (NZ 283466) under the title of Abbey Wood Coal Co, and linked it to the screens at Finchale Colliery by a line 1¼ miles long. Tubs were hauled up a 200-yard incline, made by a wooden gantry on a gradient of 1 in 3, and then taken to the screens by locomotive.

Both locomotives were constructed from various non-locomotive parts at Mr. Blacklock's scrap metal yard at Sheriff Hill in Gateshead (the site of the King Pit of Sheriff Hill Colliery). The second was unusual in that the engine drove a drum around which a rope was wound once; as the drum revolved the loco pulled itself and its load along. Both locomotives also worked underground as required, though they were not flameproofed. The system had no main line rail connection, coal being taken by road to a siding at Belmont on the BR Durham Gilesgate branch. When coal working on the eastern side of the "take" ceased, almost certainly in July 1953, the Frankland title was given up. Abbey Wood Drift was abandoned in October 1955.

Gauge: 2ft 0in

		4wDM	F. Blacklock	c1952	New	(1)
		4wDM	F. Blacklock	c1953	New	(2)

(1) loaned to Lanchester & Iveston Coal Co Ltd, Greenwell Wood Drift, Lanchester, for a fortnight c/1954; believed stolen c/1956.
(2) believed stolen c/1956.

GATESHEAD COUNTY BOROUGH COUNCIL
UNKNOWN LOCATION(S)   C169

It is not known when or for what reason the council acquired these locomotives, though it would seem likely that they were bought either for a road building or for a site clearance scheme. The council advertised for a 10/12hp locomotive and suitable track in *Contract Journal* on 2nd December /1931.

Gauge : 600mm

		4wPM	MR	914	1918	(a)	s/s
		4wPM	MR	2195	1922	(b)	s/s

(a) originally War Department Light Railways, 2635; owned by Leeds City Tramways, Leeds, Yorkshire (WR) by 1/1922; spares dispatched by MR to Gateshead County Borough on 5/3/1932.
(b) loco was rebuild of MR 933/1918; possibly ex Sir Robert McAlpine & Sons Ltd, contrs, /1930; spares dispatched by MR to Gateshead County Borough on 27/6/1933.

SALTMEADOWS CLEARANCE SITE, Gateshead   C170
NZ 2663 approx.

This was a large area of derelict property near the River Tyne, including the former Allhusen's Chemical Works (see Imperial Chemical Industries Ltd), which the council cleared for re-development. Work began in 1936 and was completed in the summer of 1937. To facilitate the removal of materials, temporary sidings were laid from the LNER Gateshead-South Shields line. Part of the site is now occupied by the Gateshead International Stadium.

Gauge : 4ft 8½in

	JUNO	0-4-0ST	OC	BH	606	1881	(a)	(1)
	TOGO	0-4-0ST	OC	MW	1659	1905	(b)	(2)
54		0-6-0ST	IC	MW	1664	1905	(c)	(3)

(a) ex Darlington Rolling Mills Co Ltd, Darlington, via Rolling Stock Co Ltd, Darlington, 1/1936.
(b) ex Cochrane & Co Ltd, Middlesbrough, Yorks (NR), via Rolling Stock Co Ltd, Darlington, 1/1936.
(c) ex Sir W.G. Armstrong, Whitworth & Co (Engineers) Ltd, Elswick Works, Newcastle upon Tyne, by 8/9/1936.

(1) to Steel & Co Ltd, Crown Works, Pallion, Sunderland, 9/1939.
(2) a circular dated 7/1937 inviting tenders for surplus plant lists only what must be the two MW locomotives; however, they were still on site on 6/12/1937; s/s.
(3) to Sir Lindsay Parkinson Ltd, contractors, for the construction of Royal Ordnance Factory at Euxton, Chorley, Lancashire, c12/1937 (after 6/12/1937).

GOLIGHTLY TAR & STONE CO
SLAG & TARMACADAM WORKS, Spennymoor

N171
NZ 260337

Durham County Council (which see) had begun the reclamation of slag from the former Tudhoe Ironworks, formerly owned by The Weardale Steel, Coal & Coke Co Ltd (which see), but handed the work over to Alfred Golightly, operating under the title of the Golightly Tar & Stone Co, apparently in the early 1930s. The works, which also manufactured tarmacadam, was situated at the western end of the ironworks site, and was served by sidings from the LNER Bishop Auckland – Ferryhill line, ¼ mile east of Spennymoor Station.

Like some others working old slag heaps, Alfred Golightly described himself as a road contractor. However, whether the 2ft 0in gauge locomotives below were used at the works or on road contracting work is not known. At the same time that he owned this works at Spennymoor he was also working **FINEBURN QUARRY** near Frosterley (NZ 033373 approx.) (see Jacob Walton & Co Ltd).

Gauge : 4ft 8½in

	D.C.C. No.2	4wPM	MR	2096	1922	(a)	(1)

(a) ex Durham County Council, reclamation scheme at Tudhoe Ironworks, Spennymoor (the same site at which this firm was operating).

(1) to Witton Park Slag Co Ltd, trading as Broadwood Limestone Co, Broadwood Quarry, Frosterley.

Gauge : 2ft 0in

	-	4wPM	MR	5232	1930	(a)	(1)
	-	4wPM	FH	1886	1934	New	(2)

(a) ex Durham County Water Board, construction of Burnhope Reservoir, near Wearhead, probably c6/1932.

(1) to J McColville, contractor, Abergavenny, Monmouthshire, and possibly used on his Talybont Reservoir contract, Brecknockshire, for Newport Corporation, 1932-39.
(2) returned to FH, c/1935; rebuilt as a 4wDM and re-sold to River Medway Catchment Board, Gillingham Sea Wall works, Gillingham, Kent, 4/1935.

WILLIAM GRAY & CO LTD
William Gray & Co (1918) Ltd until 31/12/1922
EGIS SHIPYARD, Pallion, Sunderland
Egis Shipyard Ltd until 1/1/1919

H172
NZ 373580

This shipyard was opened in 1917 on a site that included the former High Yard at Pallion of Thomas Walter Oswald (which see). The name of the company and its yard was taken from the names of its four directors, Sir John Ellerman, William Gray, Lord Inchcape and Frank Strick, a mixture of British shipbuilders and shipowners. When the yard opened it was the most up-to-date on the River Wear. It was served by a ½ mile branch from Pallion Station on the NER/LNER Penshaw branch, the branch also serving the shipyard of Short Brothers Ltd (which see).

The yard closed in 1930 because of the Great Depression. In 1936 it was taken over by National Shipbuilding Securities Ltd, the company set up to acquire shipyards regarded as surplus to capacity and to clear their sites, and the yard was dismantled during 1938-1939. The site was then taken over by Steel & Co Ltd (see the entry for British Crane & Excavator Corporation Ltd).

Gauge : 4ft 8½in

| | | 0-4-0ST | OC | KS | 3126 | 1918 | New | (1) |

(1) to William Gray & Co Ltd, Old Dockyard, West Hartlepool, 3/1932 (see Cleveland & North Yorkshire Handbook).

GREENSIDE SAND & GRAVEL CO LTD
FOLLY QUARRY, Ryton Woodside   A173
                                                                         NZ 153625

The history of this sand quarry is confused. It would appear to have been opened, or perhaps taken over, about 1958 by **A. Braithwaite & Co**, when it was known as Burnhills Quarry. This firm may well have been a subsidiary of the contractors Sir Robert McAlpine & Sons Ltd, and the subsequent owners above certainly were. The rail system was only used for a short time. The quarry was abandoned in July 1970.

Gauge : 2ft 0in

	4wDM		MR	8717	1941	(a)	(1)
	4wDM		MR	8931	1944		
		reb	MR	8995	1946	(a)	(1)

(a) ex Sir Robert McAlpine & Sons Ltd, Stargate Quarry, c /1958.

(1) locos were subsequently used on contract work by Sir Robert McAlpine & Sons Ltd; they were stored at the quarry from 5/1965; to Sir Robert McAlpine & Sons Ltd, Dunston Depot, c6/1970; to Pleasurerail Ltd, Knebworth House, Hertfordshire, c5/1975; both have since been converted to brake vans.

THE HAMSTERLEY COLLIERY LTD
The Hamsterley Coal Co until /1900
HAMSTERLEY COLLIERY, near Ebchester   F174
                                                                         NZ 117565

This colliery was initially a small landsale colliery, actually a drift, started in 1867 by **F. Watson**. When it was developed on a larger scale, with at least one additional drift and also beehive coke ovens, it was linked to sidings alongside the NER Blackhill branch, ½ mile east of Westwood Station. Tubs were hauled from the drift entrances to the screens by a ¼ mile endless rope system. The coal was then discharged into standard gauge wagons via the screens situated in the sidings. The locomotive was used here, and was kept in a shed ¼ mile west of the screens. Prior to the company acquiring the first loco below in 1902 the screens were shunted by the NER. The colliery was vested into NCB Northern Division No. 6 Area on 1st January 1947.

Gauge : 4ft 8½in

BURNOPFIELD	0-4-0ST	OC	MW	1557	1902	New	(1)
PRINCE	0-4-0ST	OC	Harris?	?	?	(a)	(2)
HAMSTERLEY No.1	0-4-0ST	OC	HL	3467	1920	(b)	(3)

(a) ex Lingford, Gardiner & Co Ltd, Bishop Auckland, dealers, hire, /1917, while MW 1557 was under repair.
(b) ex Thos W. Ward Ltd, /1933, via repairs at HL; originally Bombay Harbour Trust, India.
(1) scrapped on site by James W. Ellis & Co Ltd, Derwenthaugh, /1933.
(2) returned to Lingford, Gardiner & Co Ltd, Bishop Auckland, /1917.
(3) to NCB No. 6 Area, with colliery, 1/1/1947.

S. HANRATTY & SONS
previously **S. Hanratty**; formerly **Hanratty Bros**; originally **Hanratty's**
(it is not known whether any of these were limited companies)
WHESSOE ROAD SCRAPYARD, Darlington   S175
                                                                         NZ 289158

This scrap metal business was begun in 1947 at Union Street in Darlington, with no rail connection. About 1956 it added this yard at Hopetown Sidings in Whessoe Road, situated at Darlington (North

66. *The company's first locomotive, BURNOPFIELD, MW 1557/1902, stands proudly soon after delivery in September 1902.*

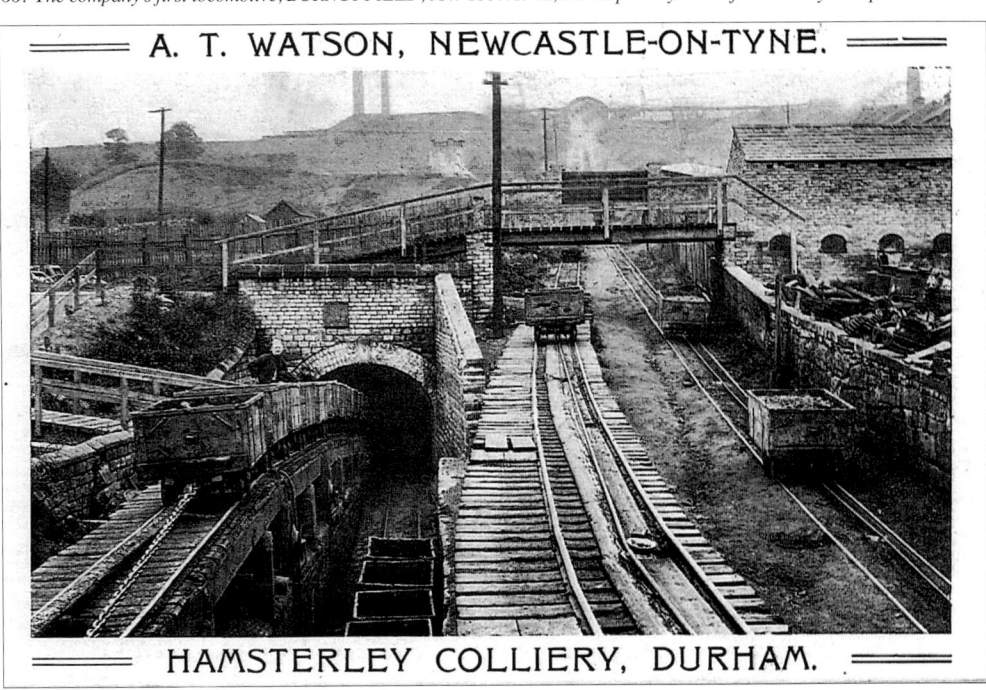

67. *The colliery comprised a number of drifts, which were worked by narrow gauge rope haulage. On the left the rope is attached to the front of the set, a system known as 'direct haulage', with trains of 20 or 30 tubs being run. On the right can be seen 'bottom endless haulage', in which tubs were clipped to the constantly-moving rope individually by a Smallman clip, with the left-hand road being used for out-going empties and the right-hand road for in-coming fulls. Compare this with the 'top endless haulage' illustrated in Photograph 140.*

Road) Station on the BR Darlington - Bishop Auckland line. The firm went into liquidation in April 1982, the site being demolished and cleared during the first half of 1983.

Gauge : 4ft 8½in

-	0-4-0DM	HE	2839	1943	(a)	(1)
1	4wDM	FH	3583	1952	(b)	s/s c4/1983

(a)   ex Henry Williams Ltd, Darlington, c /1966.
(b)   ex Adamson Butterly Ltd, Ripley, Derbyshire, after 13/3/1979, by 16/11/1979.

(1)   last reported here 20/2/1982; gone by 4/6/1982.

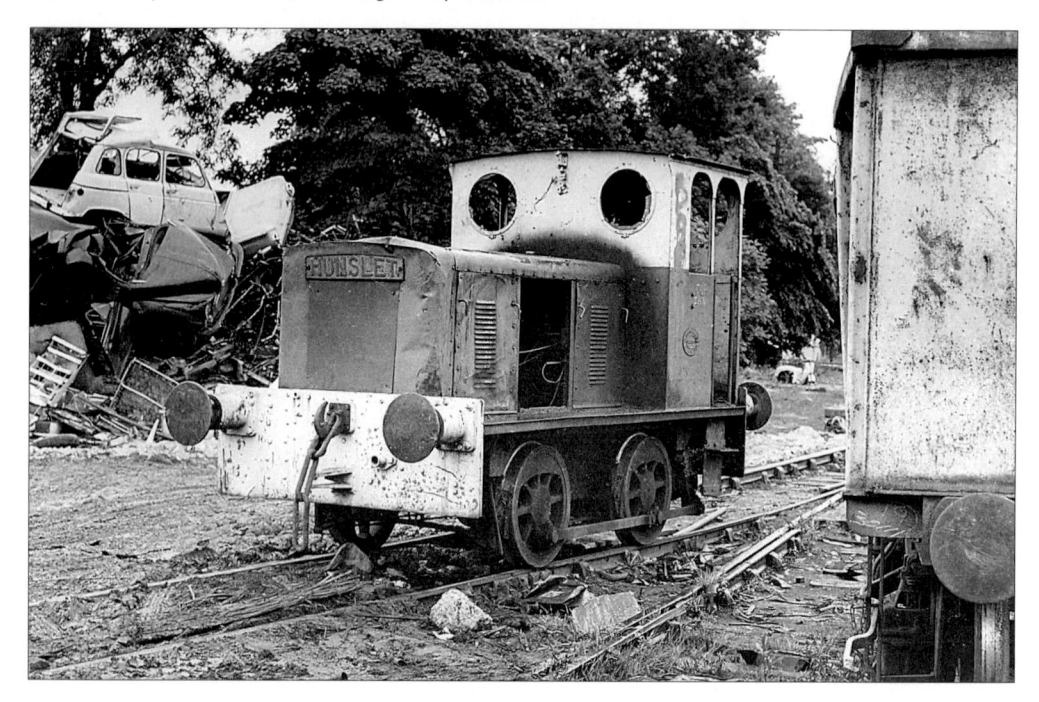

68. HE 2839/1943 was an early Hunslet diesel loco, built with a 40/44 h.p. Fowler engine and weighing only 10 tons. She ended her long life working at Hanratty's scrapyard, pausing on 1st July 1978.

HAREHOPE GILL MINING & QUARRYING CO LTD
HAREHOPE QUARRY, near Frosterley   **U176**
                                                                                NZ 038364

In the nineteenth century there was a mine here, **Harehope Gill Mine**, which was worked for lead and ganister, latterly by Harehope Gill Mining Co Ltd, which was registered in 1875. The quarry suffered various closures, the last abandonment in the nineteenth century being in July 1891. Quarrying here for limestone and ganister was begun by this company in 1901. The quarry was served by a branch (3/4 mile) from the NER Bishopley Branch near Bishopley Junction. The company ceased operations in December 1914, but in 1915 the quarry was taken over by Pease & Partners Ltd (which see).

Gauge : 4ft 8½in

-	0-4-0ST	OC	BH	848	1887	(a)	(1)

(a)   ex CF, c /1901; previously a CF hire loco.

(1)   to Plenmeller Colliery Co, Haltwhistle, Northumberland, between /1909 and /1915.

JAMES H. HARRISON
SPRINGWELL QUARRY, Springwell   **C177**
                                                                                NZ 283586

The quarrying of very hard sandstone to make into grindstones west of the village of Springwell dates well back into the nineteenth century. The first quarry lay between the village and the Pontop & Jarrow Railway of John Bowes, Esq., & Partners, to which it was linked. By the 1890s a new quarry, known as **Springwell Quarry**, had been started to the west of the railway by different owners, and was also served by a link to the P & JR. The original quarry then became known as Old Springwell Quarry (see entry for The Springwell Brick & Tile Co Ltd).

About the time of the First World War the newer quarry was acquired by Richard Kell & Co Ltd, which was still operating it in 1937. It is believed that this owner above took over shortly afterwards. However, his operation was not a success, probably because the exports on which the quarry had long depended were prevented by the Second World War. The plant was auctioned by W.W. Holmes on 17th September 1942 and included the locomotive below. The quarry continues to be operated to the present day, under different owners and now using road transport.

Gauge : 4ft 8½in

    -    ?   MR/FH ? ? (a) (1)

(a)   origin and date of arrival unknown, but presumably about /1938.

(1)   auctioned by W.W. Holmes, 17/9/1942, when described as a "Simplex" locomotive; s/s.

JOHN HARRIS
HOPE TOWN FOUNDRY, Darlington

S178
NZ 264172

This engineering works was founded by William and John Lister in 1840, building a small number of locomotives amongst its general work. In 1853 the lease was taken over by John Harris (1812-1869), a native of Cumberland. He had been amongst the first to recommend, in 1839, that rails be carried on wooden sleepers instead of stone blocks. He became engineer to the Stockton & Darlington Railway for permanent way and new works, and the Foundry, which lay ½ mile to the north of the first Darlington (North Road) Station on the S&D, concentrated mainly on supplying permanent way materials. He was also consulting engineer for a number of railways in the north and for a time operated a contractors business as well. During his ownership the Foundry continued to build locomotives from time to time.

After his death the business was purchased by his former chief foreman, Thomas Summerson (see **Summerson's Foundries Ltd**), although he subsequently moved to premises at Albert Hill and sold the Hope Town Foundry to its neighbours Charles l'Anson & Co Ltd (see **Whessoe Products Ltd**).

The foundry is believed to have had a small locomotive for shunting, presumably standard gauge.

HARTLEPOOL DOCK & RAILWAY COMPANY

This company obtained its first Act of Parliament in June 1832, enabling it to construct the first dock at Hartlepool, known as the Tide Harbour, and a railway to serve it; a second Act amending the first was obtained in June 1834. As finally constructed, its main line ran for about 12½ miles between Haswell and Hartlepool, with branches to Thornley and Wingate. It was built to carry coal from collieries that were either planned or hoped for along the route. **Thornley Colliery** (NZ 365395) sent its first coal along the line on 1st January 1835, but only as far as Castle Eden, and the line did not open fully until 23rd November 1835, with the handling of the first coal from **South Hetton Colliery** (NZ 383453) . Part of the railway was operated by rope haulage, with at least one stationary engine and a self-acting incline from Hesleden towards Hartlepool. But the HD&R did not run the coal traffic, other than on the inclines; this was left to the colliery owners. Some certainly owned their own locomotives; whether they also handled traffic for those who did not is not known. Locomotive-hauled trains certainly ran eastwards to Hesleden bank head, but whether these trains were locomotive-hauled from all of their originating collieries is unclear. It would seem that initially trains east of Hesleden bank foot were not hauled by locomotives, for in January 1840 the Railway began a three-month trial using coal owners' locomotives here to establish the cost of doing this.

To assist Messrs. Hawthorn & Robson, the contractors for the Harbour and also one section of the railway, the company in 1834 purchased a locomotive, which it subsequently sold to the Thornley Coal Co (which see). Apart from this, the Railway had no locomotives of its own until 1840, when it decided to take over the passenger service which it had allowed others, using horses to begin. In October 1843 this service was contracted out to Thomas Richardson of the Castle Eden Ironworks, although the Railway continued to own the locomotives and carriages. Richardson also contracted to handle the coal traffic of The Haswell Coal Co, but whether he or the coal company supplied the locomotives, or used HD&R locomotives, is unknown, as is whether he took over any of the other smaller companies' traffic. By this time it would seem that there was also a small quantity of general traffic.

THE HARTLEPOOL DOCK & RAILWAY and its associated railway systems, 1846

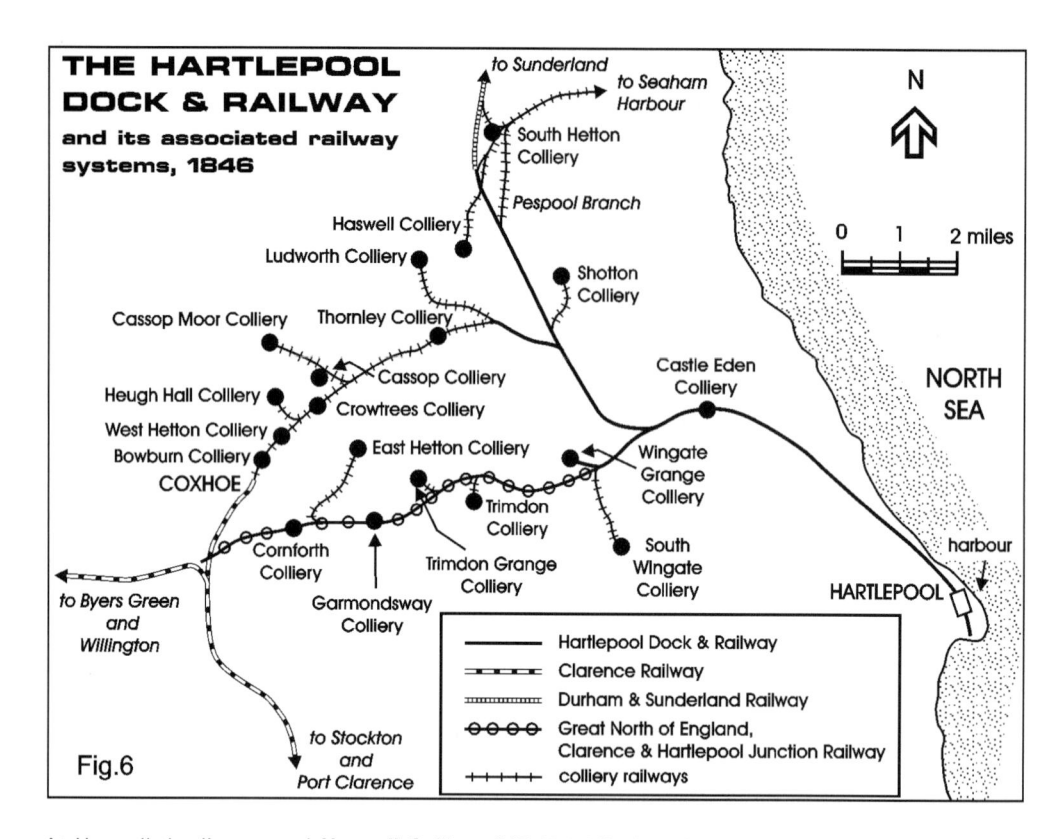

Fig.6

At Haswell the line served **Haswell Colliery** (NZ 374423), though only by a reverse, and most of Haswell's coal did not go to Hartlepool. Initially the Haswell stationary engine also worked South Hetton coal on to the Hartlepool line, but the HD&R subsequently provided the South Hetton company with its own stationary engine, commissioned about May 1837. Six years later South Hetton began the construction of a completely new route, known as the Pespool branch and opened about August 1845, which allowed it to run locomotive-hauled trains directly from South Hetton. **Wingate Colliery**, also known as **Wingate Grange Colliery** (NZ 398373), joined in 1839, **Shotton Colliery** (NZ 398413) and **Castle Eden Colliery** (NZ 437381) in 1842 and **South Wingate Colliery** (NZ 416348) in 1843.

The Hartlepool Railway was subsequently joined by two other railways. The first was the grandly-named **THE GREAT NORTH OF ENGLAND, CLARENCE & HARTLEPOOL JUNCTION RAILWAY**. This seven mile line was intended to tap the West Durham coal traffic by linking the Clarence Railway at West Cornforth to the Hartlepool Dock & Railway at Wingate; but due to an omission in its Act, the Clarence Railway was able to obstruct the construction of the link to its Byers Green branch. The line was opened between Wingate and Kelloe bank foot in March 1839, where it was joined by a branch from **East Hetton Colliery**, also known as **Kelloe Colliery** (NZ 346370), and to West Cornforth by July 1839, when **Cornforth Colliery** (NZ 327347) sent coal down, although only for one month! This line was subsequently joined, from the west, by **Garmondsway Colliery** (NZ 344349), although no traffic from here is recorded, **Trimdon Grange Colliery** (NZ 366357) and **Trimdon Colliery** (NZ 378369). So far as is known, the GNofE,C&HJR never owned any locomotives and the coal owners ran their own trains, presumably using locomotives as far as Kelloe bank foot. From here a stationary engine hauled the fulls up to Kelloe bank head, whence locomotives took over for the remainder of the route and on to the Hartlepool Railway. Thus the traffic from collieries linked to the GnofE,C&HJR is shown in HD&R records. The second railway was the **Stockton & Hartlepool Railway**, opened in February 1841, which joined the Hartlepool Railway near the Tide Harbour, where its traffic was shipped. This line need not concern us.

Coal also came on to the HD&R via the coal owners' own private railways, with The Thornley Coal Co the prime mover. In 1839-40 it built a line to serve its new collieries at **Cassop** (NZ 341382) and **Cassop Moor** (NZ 320392). At the same time a link about ¾ mile long was built from Cassop to **Crowtrees Colliery** (NZ 334379) at Quarrington Hill. Crowtrees and the other collieries around Coxhoe had previously sent coal westwards on to the Clarence Railway, via its Coxhoe branch, for shipment at Haverton Hill, near Stockton, but with the new link, **Bowburn Colliery** (NZ 317366), **Crowtrees Colliery**,

Durham Part 1 Page 170

Heugh Hall Colliery (NZ 323379) and **West Hetton Colliery** (NZ 325370), all served by the same line between Coxhoe and Crowtrees, began sending it eastwards to Hartlepool. The Thornley Coal Co also built a branch to **Ludworth Colliery** (NZ 363415), which was almost certainly opened in 1845. This extremely complex system of lines is shown on Fig.6.

On 12th October 1846 the operation of both the Hartlepool Dock & Railway Co and The Great North of England, Clarence & Hartlepool Junction Railway was taken over by the York & Newcastle Railway, although the HD&R did not finally cease to exist until 1857, while the GNofE,C&HJR lasted until 1923! The new owners not only terminated Thomas Richardson's contract to operate the passenger service, but also ended the operation of their own trains by the coal owners, taking over their locomotives and rolling stock and paying compensation.

It has been known for many years that The South Hetton Coal Co and The Wingate Coal Co owned tender locomotives in the 1830s, though knowledge of their work was very limited. It is now clear that they were used on the Hartlepool Dock & Railway Co, and not just their locomotives, but those of other colliery owners too. The Directors' minutes of the HD&R survive in the National Archives at Kew (RAIL 294/3, 4 and 5), as do the company's Annual Reports (RAIL 1110/212 and 213), day books covering the monthly traffic of the coal owners (RAIL 294/31 and 36) and ledgers (RAIL 294/35 and 37) listing payments. In addition, the Report of the Gauge Commission in 1846, held in the British Library, contains in its appendix the evidence submitted to it by the HD&R, giving, amongst other things, details of some annual reports, of shipments from Hartlepool, and of locomotives and rolling stock which operated over its line.

Coal shipments by collieries

The day books, together with published Local Records, show that the collieries served directly by the Hartlepool Dock & Railway Co shipped coal via its line as follows:

Colliery	Start	Finish	Comments
Thornley	9/7/1835	12/10/1846 (a)	(b)
South Hetton	23/11/1835	6/1846	(c)
Haswell	Late 8/1838	12/10/1846 (a)	
Wingate (Grange)	10/1839	12/10/1846 (a)	
Crowtrees	12/12/1839	12/10/1846 (a)	
West Hetton	Last week of 1/1840	12/1843	(d)
Cassop Moor	6/1840	28/2/1843	(e)
Cassop	2/1841	12/10/1846 (a)	
Bowburn	10/1841	8/1843	(d)
Heugh Hall	12/1841	1/1844	(d)
Castle Eden	4/1842	12/10/1846 (a)	
Shotton	6/1842	12/10/1846 (a)	(f)
South Wingate	30/9/1843	6/1845	(d)

Collieries which shipped coal on to the Hartlepool Dock & Railway Co via The Great North of England, Clarence & Hartlepool Junction Railway:

East Hetton	18/3/1839	12/10/1846 (a)	
Cornforth	7/1839	7/1839	(g)
Trimdon	25/2/1843	12/10/1846 (a)	
Trimdon Grange	30/6/1846	12/10/1846 (a)	

(a) date of take-over of Hartlepool Dock & Railway by York & Newcastle Railway.
(b) Thornley Colliery had sent coal from Thornley to Castle Eden on 1st January 1835, before the construction of the railway was completed. As Ludworth Colliery is not shown in the day books, it is assumed that its coal was included with Thornley.
(c) the end of traffic in June 1846 is probably very significant; see below.
(d) this date does not necessarily mean that the colliery was closed then, though it may do so. Even if it did close then, it may have subsequently re-opened; see Index 2, Locations.

(e) from this date this colliery's traffic was included with Cassop Colliery.
(f) this colliery's main shipments of coal began in February 1843, and it is known that the HD&R carried noteworthy quantities of general traffic to the site, known as the 'New Winning', prior to February 1843.
(g) Tomlinson, in *History of the North Eastern Railway*, dates the first shipment of coal to 11th July 1839. Latimer, in his *Local Records*, gives a week later. This colliery appears to have closed in July 1839 and was never worked again.

Colliery owners
In the Annual Reports for the HD&R, all of the collieries above are listed as separate 'coal companies', for example, Cassop Coal Company, Shotton Coal Company. Unfortunately, this is untrue, at least in part, as it is known that some companies owned more than one colliery. Unfortunately, information about colliery owners in other surviving records covering the 1830s and 1840s is at best uneven. The best available summary would seem to be:

1. Thornley (and Ludworth), Cassop, Cassop Moor and Trimdon Collieries were owned by **The Thornley Coal Co** (which see).
2. South Hetton was owned by **The South Hetton Coal Co** (which see).
3. Haswell, Shotton and Castle Eden were owned by **The Haswell Coal Co** (which see).
4. Wingate Grange was owned by **The Wingate Coal Co** (which see).
5. Crowtrees was owned in 1835 by **William Hedley & Sons** (which see). Hedley died in 1843. At some point the Hedleys sold this colliery (see Index 2, Locations); whether this happened before or during its shipments via the HD&R, and whether the Hedleys were succeeded by The Crowtrees Coal Co, is unknown.
6. West Hetton Colliery's first coal was shipped via the Clarence Railway on 30/1/1834; Latimer's *Local Records* give the first coal shipped at Sunderland via the Durham & Sunderland Railway on 13/10/1836, when the colliery was owned by **Col. Braddyll & Partners**. Braddyll was also one of the partners in the South Hetton Coal Co. Who owned the colliery during its shipments via the HD&R is unknown, so West Hetton Coal Co may be correct.
7. No owners of Bowburn Colliery are known, so Bowburn Coal Co may be correct. (Note: this was a different colliery from that of the same name opened later by Bell Brothers Ltd).
8. No owners of Heugh Hall Colliery at this period are known, so Heugh Hall Coal Co may be correct; see Index 2, Locations.
9. When it was opened, South Wingate Colliery was owned by **Seymour & Partners** (which see).
10. East Hetton Colliery was owned by **The East Hetton Coal Co** (which see).
11. Cornforth Colliery is shown in Latimer's *Local Records* (see (g) above) as owned by **Messrs. Rippon**, about whom nothing else seems to be known.
12. Trimdon Grange, when opened, was owned by **Joseph Smith** (which see).

It may well be that some of the larger coal owners handled the traffic of the smaller coal owners, who would undoubtedly have had their own waggons but not their own locomotives. If such arrangements were made, their details do not survive.

Colliery owners' locomotives
In a return to the Board of Trade dated 6th March 1841 of engine drivers employed on the HD&R and the collieries which had engines on the line, the following details are given, as written:

HD&R	4
South & East Hetton Coal Co	5
Thornley Coal Co	6
Wingate Grange Coal Co	4
Crow Trees & West Hetton Coal Co	3
Total	22

The HD&R Directors' minutes record the following locomotive names, other than those of the HD&R's own engines:

BRADDYLL	1/7/1841	[owned by The South Hetton Coal Co]
NELSON	12/8/1841	[owned by The South Hetton Coal Co]
EXILO	16/12/1841	[owned by The Crowtrees Coal Co]. William Hedley & Sons, the owners of Crowtrees Colliery, are known to have owned a locomotive called EXILE, and the two are believed to be one and the same.
CORNWOOD	28/7/1842 and 11/8/1842; the second half of the handwritten name is definitely WOOD, but the first half may not be CORN; no owners given or known.	

In its evidence to the Gauge Commissioners, the HD&R stated that in 1845 'the coal owners using the line have 24 engines with outside cylinders of various forms and sizes' but that they did not have specific details. The Railway valued them at £24,000. It also stated that the owners had about 2000 coal wagons (£24,000) and about 100 trucks (£1000).

The details of the compensation paid to the coal owners by what was then the York, Newcastle & Berwick Railway survive (RAIL 772/107), but in two lists, for July 1847 (p.106) and December 1847 (p.161). They do not entirely square with each other, and no details of locomotives or waggons are included. The Y&N did not take over any of the colliery lines operated by the companies. The final total in December 1847 was £59,510, considerably more than the value estimated two years earlier. Each of the following companies received over £3,500 (names as shown in the lists, with separate totals to the same company in the list aggregated):

Thornley Coal Co	£12752
Wingate Grange Colliery	£7323
East Hetton Coal Co	£7304
Cassop Coal Co	£6428
Crowtrees Colliery	£5209
Castle Eden Colliery	£4072
Trimdon Grange Colliery	£4067
Haswell Coal Co	£3840

Given an estimate per waggon of about £25 and second-hand locomotives roughly averaging at about £800, it would seem reasonable to suggest that the size of the compensation paid to the companies above included estimates for locomotives. There is also a considerable gap between the last entry above and the next, being the South Wingate Coal Co, which received £2,571, although even this may have included an element for at least one locomotive. Note that the list above includes separate entries for Thornley and Cassop, despite both collieries being owned by the same company.

Details of the actual locomotives

No contemporary records giving details of the locomotives taken over by the York & Newcastle Railway survive. In 1883 the North Eastern Railway drew up a list from records then available, which was revised in a second list dated 12th September 1896, drawn up on the instruction of the NER Locomotive Superintendent, Wilson Worsdell. A copy of this list was made by the noted historian E.L. Ahrons in the 1920s, now in the Stephenson Locomotive Society's Library, and this has been copied and published by others since. Even Worsdell himself is reported to have said that many of the dates were wrong and that the dimensions of much later engines had often been quoted against earlier ones and often he could not give the wheel arrangement (!). The list clearly presents difficulties:

No	Type (a)	Maker (b)	Date (c)	Wheels	Cyls	Orig Cost
112	0-6-0	Hawthorn	6.1841	4'0"	14x18"	975.14.3
113	0-6-0	Hawthorn	6.1840	4'0"	14x18"	1000
114	0-6-0	Richardson	5.1835	4'0"	13½x20"	825
115	0-6-0	Richardson	5.1835	4'0"	13½x20"	730
116	0-6-0	Hawthorn	6.1840	4'0"	14x18"	1000
117	0-6-0	South Hetton	6.1840	4'0"	15x16"	1055
118	0-6-0	South Hetton	6.1841	4'0"	15x16"	1200
119	0-6-0	Fossick & Hackworth	6.1842	4'0"	14x18"	935
120	0-6-0	Fossick & Hackworth	6.1842	4'0"	14½x24"	737
121	0-6-0	Richardson	6.1840	4'0"	14x18"	1005
122	0-6-0	Nesham & Welsh	6.1840	4'0"	14¾x18"	730
123	0-6-0	Fossick & Hackworth	6.1839	4'0"	14½x18"	838
124	0-6-0	Heron Wilkinson	1842	4'0"	14x18"	675
125	0-6-0	Nesham & Welsh	6.1839	4'6	14½x20"	710
126	0-4-0	Nesham & Welsh	6.1840	4'0"	14x20"	765
127	0-4-0	Nesham & Welsh	6.1840	4'0"	14x20"	825
128	0-4-0	Hartlepool Iron Co	6.1840	4'0"	14½x18"	820
129	0-4-0	Hawthorn	1840	4'6"	14x16"	915
130	0-4-0	Deptford Iron Co	6.1841	4'0"	14½x20	820
131	0-6-0	Smith	1847	4'0"	16 x 24	1404

The numbers are those given by the York, Newcastle & Berwick Railway, whose numbers 1-244 were adopted without change by the NER when it was formed in 1854. Despite this, it seems certain that at least some of these locomotives did not survive to pass to the NER. All of them were replaced by the NER with new locomotives between 1856 and 1858.

(a) The numbers appear to have been allocated on the basis of six-coupled engines first, followed by four-coupled engines, a total of 19, five fewer than the HD&R stated were working over the line in 1845. This discrepancy is almost certainly due to the withdrawal in June 1846 of the South Hetton Coal Co traffic and presumably its locomotives. The final locomotive seems to have been included after the original list had been drawn up; possibly it was under construction at the date of the take-over.

(b) There are major difficulties with this section. R. & W. Hawthorn's Engine Book shows no locomotives supplied new in 1840-41 to any of the collieries or owners listed. Thomas Richardson (see Index 3) is not known to have built any locomotives before his move from Castle Eden to Hartlepool in 1847, when he took a lease of the Hartlepool Iron Co's premises. This firm had been set up in 1838, but it is not known to have built or repaired any locomotives. The South Hetton Coal Co is not known to have built any locomotives at this period, when it was spending a huge sum of money on the sinking of Murton Colliery. It has been suggested that these two were sold by South Hetton when they withdrew from working over the HD&R, but no evidence of this is known. The firm of Heron & Wilkinson operated an ironworks at Elswick in Newcastle upon Tyne, but is not known to have built any locomotives. There was a Deptford Iron Co in Sunderland at this date, which may be the firm listed here – the HD&R had dealings with it – but it is not known either to have built or repaired any locomotives. 'Smith' may well be Joseph Smith, the owner of Trimdon Grange Colliery and also of the Trimdon Ironworks in Sunderland, where this locomotive may perhaps have been built; the date of 1847 suggests that it was under construction at the date of the York & Newcastle take-over in October 1846.

(c) It is clearly unusual that so many locomotives should have been built in the month of June. It has been suggested that the dates might be the end of accountancy periods, but in the light of problems with the listed builders, it is probably fairer to place little reliability on this column.

In summary, very little reliance can be placed on the accuracy of this list and it gives no indication of which locomotives came from which owners. For details of the locomotives that the companies did own, see the entries for those noted above under 'colliery owners'.

HARTLEPOOL GAS & WATER CO
HURWORTH BURN RESERVOIR, near Wingate   L179
NZ 407338 approx.

This company was created by an Act of Parliament in 1846, and took over the Hartlepool Gas Works two years later. The Act to permit the construction of this reservoir was passed in 1867, to provide additional supplies for the growing towns of Hartlepool and West Hartlepool. Situated about eight miles north-west of Hartlepool, it was formed by damming the River Skerne. Initial construction began in 1870, and its enlargement, to hold 160 million gallons, was authorised in 1874, after which the company decided to adopt locomotive haulage at the site. The construction works were linked to the NER Stockton & Wellfield branch (Wingate South Junction to Carlton East Junction), at what was to be Hurworth Burn Station (opened in March 1880). Work was completed about the end of 1877. Between 1900 and 1904 a second reservoir, Crookfoot Reservoir, two miles south-east of Hurworth Burn Reservoir, was built by contractors (see Part 2).

Reference: *Dam Builders' Railways from Durham's Dales to the Border*, H.D. Bowtell, Plateway Press, 1994.

Gauge: 4ft 8½in

     HURWORTH   0-4-0ST   OC   BH   306   1875   New   (1)

(1) to Middleton Road Gas Works, Hartlepool (c/1877) (see Cleveland & North Yorkshire Handbook).

THE HARTON COAL CO LTD
The Harton Coal Co (formed in 1842) until 6/8/1885
subsidiary of **Stephenson, Clarke & Associated Companies Ltd** from 1927,
and of **Powell Duffryn Steam Coal Co Ltd** from 1928

Nineteenth century operations
In 1819 the brothers John and Robert Brandling from Gosforth, north of Newcastle upon Tyne, took over the lease of a large area to the south and south-west of South Shields, and in 1825 they opened

8		2-2-2WT		IC	SS	1501	1864	(j)	Scr /1907
8		0-6-0	*	IC	RS	1973	1870	(k)	Scr /1929
8		0-6-0	*	IC	Ghd	3	1889	(m)	(1)
9		0-6-0		IC	Blyth & Tyne		1862	(n)	Scr /1913
10		0-6-0		IC	Blyth & Tyne		1862	(p)	Scr /1914
10		0-6-0		IC	RWH	1564	1873	(q)	Scr /1931
11		0-4-0ST		OC	MW	?	?	(r)	Scr /1920
	LALEHAM	0-6-0ST		OC	AB	1639	1923	(s)	(1)
	BOWES No.9	0-6-0ST		IC	SS	4051	1894		
	reb	0-6-0PT		IC	Caerphilly		1930	(t)	(5)

* fitted with Westinghouse brakes for the passenger service.

(a) ex Robert Frazer & Sons Ltd, Hebburn, dealers, 10/1929; LNER, 396, formerly NER, 398 class, until 8/1925.
(b) ex NER, 786, 398 class, per Robert Frazer & Sons Ltd, Hebburn, dealers, 12/1907; sold by NER, 18/12/1907.
(c) ex NER, 827, 398 class, per Robert Frazer & Sons Ltd, Hebburn, dealers, 3/1912.
(d) ex LNER, 1486, J22 class, formerly NER, 59 class, 21/1/1930.
(e) ex LNER, 1509, J21 class, formerly NER, C class, 2/8/1935; withdrawn by LNER, 8/1935.
(f) ex LNER, 1616, J21 class, 10/10/1929; withdrawn by LNER, 10/1929.
(g) ex LNER, 776, J21 class, 17/5/1935; withdrawn by LNER, 5/1935.
(h) ex LNER, 1953, J24 class, 20/5/1939; previously NER, P class.
(j) ex Robert Frazer & Sons Ltd, dealers, Hebburn, c/1899; previously Furness Railway, 22A, until 8/1899.
(k) ex NER, 718, 708 class, per Robert Frazer & Sons Ltd, dealers, Hebburn, 12/1907; sold by NER, 18/12/1907.
(m) ex LNER, 869, J21 class, 29/1/1931; withdrawn by LNER, 1/1931.
(n) ex NER, 2255, 4/1900; built by Blyth & Tyne Railway at Percy Main, North Shields, Northumberland, as 3; railway acquired by NER, 7/8/1874, and loco renumbered 1303; re-numbered 1923 in 8/1891, 1733 in 1/1894 and 2255 in 3/1899.
(p) ex NER, 1712, 11/1900; built by Blyth & Tyne Railway at Percy Main, North Shields, Northumberland, as 14; re-numbered by NER, 1313, 8/1874, 1729 in 7/1892 and 1712 in 1/1894.
(q) ex NER, 827, 398 class, 31/8/1914.
(r) identity and origin unknown; possibly acquired in 1908; it was fitted up by 10/1914 as part of an armoured train on the Railway to guard against a German invasion (*The Locomotive Magazine*, 15/10/1914).
(s) ex Boldon Colliery, Boldon Colliery, 7/1931, to share with No. 4 the Marsden Quarries duty.
(t) ex Boldon Colliery, Boldon Colliery, c4/1943; on hire from John Bowes & Partners Ltd, Bowes Railway, Springwell Bank Foot loco shed, Wardley.

(1) to NCB No. 1 Area, with the Railway, 1/1/1947.
(2) withdrawn from traffic in 1927; scrapped, 1/1930.
(3) to Robert Frazer & Sons Ltd, dealers, Hebburn, /1912 (after 22/7/1912); to John Bowes & Partners Ltd, Pontop & Jarrow Railway, Marley Hill loco shed, Marley Hill, /1912.
(4) to Boldon Colliery, Boldon Colliery, c/1937, and returned; dismantled at Whitburn loco shed by 3/8/1939; scrapped, /1939.
(5) to Boldon Colliery, Boldon Colliery, c7/1943.

Underground locomotive
Men who worked underground at Westoe Colliery in the 1960s recall a four-wheeled compressed air locomotive being used in the Yard 'G' seam near to the workings of Harton Colliery. This locomotive is said to have carried a maker's plate with the information 'Kilmarnock' and '1923' on it. Westoe Colliery did not begin production until after the National Coal Board took over, and a locomotive of that age is unlikely to have been acquired second-hand, so one is left to assume that the locomotive came new to St. Hilda Colliery. What work it was used for before National Coal Board ownership, especially remembering that St. Hilda Colliery closed in 1940, is likely to remain unknown.

Gauge : 2ft 6in (?)

-		4wCA	?	1923	(a)	(1)

(a) said to have been built at Kilmarnock; of the possible manufacturers, KE seems the most likely; probably New to St. Hilda Colliery, South Shields.

(1) to NCB No. 1 Area, 1/1/1947.

HARTON ELECTRIC SYSTEM

By 1890 the company had four collieries, but was facing severe transport and shipping problems. None of the collieries had any connection with any of the others; the St. Hilda Staiths were unable to handle all the coal to be shipped and large sums were being paid to the NER to handle the traffic, including shipment at Tyne Dock.

Then came a way forward, offered by the demise of the Tyne Plate Glass Co Ltd (which see). This company's works had a quay on the river about 300 yards downstream of St. Hilda Staith and, equally importantly, a single track railway which ran from the quay, then through a narrow tunnel before passing alongside St. Hilda Colliery to a sand quarry on the Bents. This had latterly been operated by two locomotives, very much cut down to pass through the tunnel. The Harton Coal Co Ltd purchased all of this in 1892. The company then opened negotiations with South Shields Corporation, which led in February 1896 to the "Harton Agreement", incorporated later that year in the South Shields Corporation Act. The huge modernisation and railway programme agreed may be summarised as follows, with the stages of completion:

Rebuilding of the four collieries, 1896-1910.
Reconstruction of the Tyne Plate Glass Co's railway, 1902-1903.
Construction of new staiths on the Tyne Plate Glass Co's quay, 1903-1904.
Conversion of the St. Hilda waggonway to railway operation, 1906-1908.
Widening of part of the SSMWCR to accommodate double track, 1907-1908.
Construction of a new railway from Harton Colliery to Whitburn Junction, 1908-1910.
Construction of a new colliery at Bent House Farm, 1909-1913.
To this new St. Hilda Staiths had to be built when the old staiths were destroyed by a fire spreading from a ship in 1894.

The new staiths on the site of the glass works were opened in March 1904 and were called **Harton Low Staiths**, the former St. Hilda Staiths then being re-named **Harton High Staiths**.

Electrification

However, the developing railway was not best suited to steam locomotives, with its tight curves, severe gradients and adjacent domestic housing, and soon after the rebuilding began, the company took the decision to introduce electrification, not only on much of the railway but also in the winders, etc., at the collieries. Siemens-Schuckert Elektricitats A.G. of Munich in Germany received the first contract, to electrify the lines between the Benthouse sidings and the Low Staiths using the continental tramway type overhead system, together with the supply of six locomotives, the specification requiring the haulage of 110 ton trains up the 1 in 38 Erskine Road bank without slipping, regardless of weather. The first electric train ran on 14th December 1908. The section between the Bents and Dean Sidings (see above) and also the branch to Harton Colliery, a distance of just under two miles, followed in 1911, with two more locomotives and a further two in 1913 to cope with the increase in traffic. The former steam shed at St. Hilda Colliery became the main depot for the electric locos, though a small shed was also built at Harton Colliery for the loco used to shunt there. The system was operated as if it was entirely single line, double track sections being regarded as a main line and a loop. The single line working was controlled by dividing the system into five sections, and operating the One Engine in Steam system and train staffs. Erskine Street Tunnel was additionally protected by signals and staffed single line working strictly observed; even maintenance staff walking through the tunnel had to be in possession of the staff. In later years it would seem that staff working over the Harton Colliery branch was abandoned.

In 1909 the company began the sinking of the new fifth colliery at Benthouse, north of the sidings. When this was completed in 1911 it was used only as the upcast shaft for St. Hilda Colliery; but in April 1914 it was decided to develop coal working, and to mark this the colliery was re-named **WESTOE COLLIERY** (NZ 373368). However, because of the First World War no coal was actually drawn. Interestingly, by 1913 a quarter of all electricity generated in the North East was being fed to the company. In 1920 a new loco shed, **WESTOE LOCO SHED** (NZ 374667), was built near Westoe Colliery to re-house some of the locos from St. Hilda. In 1933 this was enlarged to a four-road shed, with an additional section housing the railway's sub-station equipment. All loco maintenance was now carried out here, the St. Hilda shed being retained only for locomotives on duties in its immediate area.

On 6th September 1940 St. Hilda Colliery and the Westoe Shaft were closed, leaving Harton Colliery and the electric system to pass to NCB Northern Division No. 1 Area on 1st January 1947.

References : *The Harton Electric Railway*, William J. Hatcher, Oakwood Press, 1994.
The Private Railways of County Durham, Colin E. Mountford, Industrial Railway Society, 2004.

Gauge : 4ft 8½in

E1	4wWE	Siemens	451	1908	New	(1)
E2	4wWE	Siemens	455	1908	New	(1)
E3	4w-4wWE	Siemens	456	1909	New	(1)
E4	4w-4wWE	Siemens	457	1909	New	(1)
E5	0-4-4-0WE	Siemens	458	1909	New	(1)
E6	0-4-4-0WE	Siemens	459	1909	New	(1)
E7	4w-4wWE	KS	1202	1911	(a)	(1)
E8	4w-4wWE	KS	1203	1911	(a)	(1)
E9	4w-4wWE	AEG	1565	1913	New	(1)
E10	4wWE	Siemens	862	1913	New *	(1)

* At Harton Colliery, 1/1/1947; here from new.

(a) ordered from Siemens Bros Ltd, Stafford, who manufactured the motors and electrical parts, with the remainder sub-contracted by Siemens to Kerr Stuart; both new to Harton Colliery.

(1) to NCB No. 1 Area, with railway, 1/1/1947.

MARSDEN QUARRIES   D182

Not far from the site for Whitburn Colliery lay huge deposits of limestone, which The Whitburn Coal Co Ltd also took over in 1874. An area to the east of the colliery site was already worked out and earmarked for colliery waste, and production was initially concentrated on **Marsden Old Quarry**, sometimes called **Marsden Harbour Quarry** (NZ 397646 approx.). This was at first served by a short branch from the Marsden Railway north of Marsden Station; but later a link was built from the branch to the lime kilns and hoppers built alongside the line to the south of the station to allow stone to be brought there without having to travel on the "main line".

Much larger deposits lay about ¼ mile to the north of Whitburn Colliery, and these were being developed by 1900. Two quarries were developed here (NZ 405642), known as **Marsden Lighthouse Quarry** and **Marsden No. 2 Quarry**. The quarries produced dimension stone for the building and masonry industries (discontinued in 1908), crushed stone for the steel industry and lime for agriculture, seven kilns for producing this being built into the hillside overlooking the railway and the sea. The Old Quarry was eventually closed.

In 1937, in order to develop the crushed stone trade for uses other than the steel industry, the company installed a new stone crushing and screening plant here, and replaced the 33 horses previously used with narrow gauge locomotives, for which a shed and workshop facilities were also provided. These locos hauled tubs between the faces and the plant and on to the standard gauge railheads in the two quarries, from where the quarry locomotive, supplied from the shed at Whitburn Colliery, hauled them to the weighbridge opposite the Lighthouse Signal Box, which in turn controlled the quarry sidings here. SSMWCR locos normally used here were No.4 and LALEHAM.

The quarries passed to NCB Northern Division No. 1 Area on 1st January 1947.

Gauge : 2ft 0in

-	4wDM	RH	187059	1937	(a)	(1)
-	4wDM	RH	189959	1938	(b)	(1)
-	4wDM	RH	189963	1939	(c)	(1)

(a) ex Thomas Mosdale & Son Ltd, Urmston, Lancashire, /1937.
(b) ex P. Caulfield & Son Ltd, Balloch, Dunbartonshire, Scotland, c/1944.
(c) ex Charles Brand & Sons Ltd, Avonmouth contract, Somerset, by 31/12/1946.

(1) to NCB No. 1 Area, with quarries, 1/1/1947.

BOLDON COLLIERY, Boldon   D183

As has been noted above, this colliery was opened on 26th July 1869 and lay alongside the NER Pontop & South Shields branch near to Brockley Whins South Junction. Initially its traffic was worked by the NER, some for shipment at St. Hilda Staith and some to the NER's Tyne Dock; but by 1910 its coal was being worked to the new Dean Sidings for shipment at the Low Staiths.

The history and details of the locomotives here is far from clear. DCOA Return 102 lists one locomotive here in both April 1871 and November 1876, which would appear to rule out a loco from the SSMWCR list above, and its/their identity is therefore unknown. No confirmed locomotive allocation is known before 1922. Locomotives here that are known were allocated their own numbering system.

Gauge : 4ft 8½in

		0-4-0ST	OC	HE	286	1883	(a)	s/s
1		0-6-0	IC	Ghd		1882	(b)	Scr /1944
4		0-6-0	IC	Ghd		1883	(c)	(1)
	LALEHAM	0-6-0ST	OC	AB	1639	1923	(d)	(2)
3		0-4-0T	IC	Ghd	38	1888	(e)	(1)
2		0-6-0	IC	Ghd	43	1889		
		reb		?		1904	(f)	(3)
1		0-6-0T	IC	HC	332	1889	(g)	(1)
	BOWES No.9	0-6-0PT	IC	SS	4051	1894	(h)	(4)

(a) ex Robert Frazer & Sons Ltd, Hebburn, dealers, 10/1922; previous owners unknown; new to Lucas & Aird, contractors, Kingston-upon-Hull, Yorkshire (ER).

(b) ex LNER, 1453, 3/1927; previously NER, 398 class.

(c) ex Robert Frazer & Sons Ltd, Hebburn, dealers, /1928; LNER, 1333, formerly NER 398 class, until 9/1925.

(d) ex George Cohen & Sons Ltd, dealers, /1929, who offered the locomotive for sale in the *Colliery Guardian* on 8/3/1929; previously S. Pearson & Sons Ltd, Queen Mary Reservoir contract, Sunbury, Middlesex.

(e) ex LNER, 24, Y7 class, via Robert Frazer & Sons Ltd, Hebburn, dealers, 2/2/1931.

(f) ex South Shields, Marsden & Whitburn Colliery Railway, Whitburn Colliery, Whitburn, 7, c/1937 (see above).

(g) ex GWR, 782, 11/1939; previously Barry Railway, 34, E class.

(h) ex John Bowes & Partners Ltd, Bowes Railway, Springwell Bank Foot Loco Shed, Wardley, 20/2/1943, on loan.

(1) to NCB No. 1 Area, with colliery, 1/1/1947.

(2) to South Shields, Marsden & Whitburn Colliery Railway, Whitburn Colliery, Whitburn, 7/1931.

(3) returned to South Shields, Marsden & Whitburn Colliery Railway, Whitburn Colliery, Whitburn, by 3/8/1939.

(4) to South Shields, Marsden & Whitburn Colliery Railway, Whitburn Colliery, Whitburn, c4/1943; returned to Boldon Colliery, c7/1943; returned to John Bowes & Partners Ltd, Bowes Railway, Springwell Bank Foot Loco Shed, Wardley, 14/8/1943.

THE HASWELL, SHOTTON & EASINGTON COAL & COKE CO LTD
possibly **The Haswell & Shotton Coal Co Ltd** before the above
The Haswell Coal Co until 22/8/1865
HASWELL COLLIERY, Haswell Plough　　　　　　　　　　　　　　　　　　　　　　L184
　　　　　　　　　　　　　　　　　　　　　　　　　　　　　　　　　　　　　　NZ 374423

The sinking of this colliery was begun on 28th February 1832 and after considerable difficulties coal was reached in 1834. It was then connected by a 1¼ mile line to South Hetton Colliery (see South Hetton Coal Co Ltd; several of the partners who formed this company were also partners in the Haswell Coal Co), in order for Haswell coal to be shipped at Seaham Harbour (see Marquis of Londonderry), where the first shipment was made on 2nd July 1835. To this point the colliery had cost £100,000. To operate the line to South Hetton a **stationary engine** (NZ 373432) (name not known) was built 800 yards north-east of the colliery, which hauled full wagons up from the colliery yard and then, using the same rope, lowered them down for a mile to Low Fallowfield, about ¼ mile short of South Hetton, from where they were collected by South Hetton locomotives. The line was known as the Haswell & Seaham Waggonway, and most, if not all, of it between the engine house and South Hetton was owned by the South Hetton Coal Company, the two companies making an agreement regarding the handling of the traffic.

At the same time another line was being constructed towards Haswell from the south by the **Hartlepool Dock & Railway Company** (which see). This line, opened on 23rd November 1835, crossed the original line at right angles ¼ mile below the engine house on its north-east side, the intention being to continue to Littletown, though this never happened. Instead the two lines were joined by a short curve immediately east of the crossing, with the stationary engine working traffic to and from the interchange sidings. The South Hetton company also decided to send some traffic to Hartlepool, which meant the Haswell stationary engine now had to haul South Hetton fulls from Low Fallowfield up to the junction and then lower them into the interchange sidings. This led to a dispute between the two companies, as a result of which South Hetton first built its own stationary engine and then an entirely new link, known as the Pespool branch, to the Hartlepool Dock & Railway.

The first Haswell coal was not shipped at Hartlepool until late in 1838, and seems to have been confined to small and irregular quantities. The operation of Haswell traffic was latterly contracted to Thomas Richardson of the Castle Eden Ironworks, apparently using Haswell locomotives and rolling stock. This ended in October 1846.

Meanwhile the Durham & Sunderland Railway was also planning to tap the South Hetton and Haswell traffic from the north by constructing a branch from its main line at Murton Junction. This joined the original line just north of the level crossing. This branch, also initially worked by a stationary engine at Murton Junction, was opened on 9th August 1836, and it would seem that most Haswell coal was shipped at Sunderland. It was not until 1877 that the NER finally constructed a short curve between the former D&SR and the former HD&R to allow through running. The railway from the colliery was then linked to this curve, with the link to South Hetton being abandoned. However, despite having three outlets for its coal the colliery continued to send a considerable proportion of its output to Seaham Harbour under its agreement with Lord Londonderry. A new settlement known as Haswell grew up around the railway junction, while the settlement around the colliery became called Haswell Plough, after the name of a nearby public house.

The April 1871 section of DCOA Return 102 lists one stationary engine and no locomotive at Haswell, but in November 1876 the stationary engine is not entered and one locomotive is listed instead. On this evidence it would appear that the stationary engine was abandoned in the early 1870s and a locomotive introduced to shunt the colliery.

It is believed that the NER took over the colliery shunting, together with two MW locomotives and the wagons, in 1883. However, another source claimed that the NER seized the locomotives in lieu of a debt, and R.H. Inness' NER register shows them numbered into NER stock on 1st February 1885. If the NER did indeed handle the traffic in the 1880s, it would seem the colliery company subsequently resumed at least some control. The NER Sidings map of 1894 shows the colliery served by a ¾ mile branch, including a headshunt back to sidings north of Haswell Station. The colliery was closed on 25th September 1896.

Gauge : 4ft 8½in

HASWELL	0-6-0ST	IC	MW	242	1867	New	(1)
SHOTTON	0-6-0ST	IC	MW	479	1874	(a)	(2)
-	0-6-0ST	IC	MW	23	1861	(b)	(3)

(a) ex Shotton Colliery, Shotton Colliery (in 1877?).
(b) ex Owners of Backworth & West Cramlington Collieries, Backworth, Northumberland, probably after 1890.

(1) to NER and became NER 1723; for possible dates see text above.
(2) to NER and became NER 1724; for possible dates see text above.
(3) apparently disposed of by closure in 1896, given that it was not listed in the auction of plant described below.

In the advertisement for the auction of equipment at Haswell Colliery by A.T. & E.A. Crow from 3rd-6th August 1897 in the *Colliery Guardian* of 23/7/1897 three four-wheeled coupled locomotives with 12in cylinders were included, presumably 0-4-0STs. Nothing is known about these.

SHOTTON COLLIERY, Shotton Colliery (originally New Shotton)   L185
NZ 398413

The sinking of this colliery was begun in 1835, initially by a different group of partners from Haswell Colliery; it is not known when the two collieries came under a united management. The colliery, originally called **Shotton Grange Colliery**, was served by a company-owned ¾ mile branch to the Hartlepool Dock & Railway, 1¾ miles from what was later Wellfield Junction. Besides the colliery, 260 beehive coke ovens were built and a **brickworks**, open by 1857, was developed to the south-east of the colliery. Its associated village was at first called New Shotton, to distinguish it from the original Shotton village to the east, but eventually it took the colliery's name.

As it was linked to the HD&R, the colliery's coal went to Hartlepool, the first shipment being in June 1842. As on the HD&R the coal owners ran their own trains, the company must almost certainly have had to buy its own locomotives, and this would seem to be borne out by the fact that the owners received £3840 in compensation from the York, Newcastle & Berwick Railway after the private traffic was ended in October 1846, for such a large total must have included one or more locomotives. Apart from the list of locomotives given in the HD&R entry (which see), nothing is known of them.

The company had always shipped some of its output at Seaham Harbour, and in planning his new line to Sunderland in the early 1850s, the Marquis of Londonderry was anxious to secure a guaranteed coal traffic over the new route. Under a memorandum of agreement between the Marquis and the company dated 5th February 1853, the former agreed to build a new line between Shotton Colliery and the Pespool branch of the South Hetton Coal Company (which see) in order for Shotton coal to be shipped at Sunderland. A second agreement dated 17th March 1860 describes the branch as "recently built". It ran north from the colliery for 1¼ miles (rather than the 600 yards in the 1853 agreement) to join the South Hetton line just north of that line's Pespool Junction with the Hartlepool line. Under the 1860 agreement the company was responsible for repairs to the branch and to find the locomotives and waggons to work the traffic. However, it would seem that the company did not find the locomotives, for in his history of the Londonderry Railway (which see) George Hardy writes that "at this period [about 1860] No.9 Engine was leading coals from Shotton Colliery by the Pespool branch to South Hetton". This would suggest that the Londonderry Railway loco worked over the South Hetton line into South Hetton. Here the Shotton traffic was taken forward by the South Hetton company's locomotives and inclines to the link with the Londonderry Railway at Swine Lodge Bank Head, 1320 yards from Seaham Harbour, from where it was taken on by Londonderry Railway locomotives to Sunderland. Presumably the Londonderry loco was housed and serviced at South Hetton.

This agreement was short-lived, for under an agreement dated 24th July 1868, but backdated to 1st January 1868, this branch was sold to the company (all the above may be found in D/Lo/E/378), though with the traffic agreement continuing. The company now did have to acquire locomotives of its own to work the line. The description of the colliery in the supplement to the *Mining Journal* of 17th September 1870 says that "two locomotive engines are employed", with the same number being given in DCOA Return 102 for both April 1871 and November 1876. So far as is known, at least some of the Shotton traffic continued to travel to Seaham, though whether the interchange point between the Haswell and South Hetton companies was near Pespool Junction or at South Hetton is not known.

The colliery was closed in November 1877 and the site was cleared. After the demise of the company in 1896 the royalty was divided between neighbouring colliery owners. This brought the site into the hands of The Horden Collieries Ltd (which see), which re-opened a new Shotton Colliery in 1901.

The company also owned **CASTLE EDEN COLLIERY** (NZ 437381) **(L186)**, where sinking began on 28th September 1840. It would seem very likely that it was one of the collieries which operated its own trains and locomtives over the Hartlepool Dock & Railway Co (which see), a practice which ceased in October 1846. DCOA Return 102 does not record any locomotives here in either 1871 or 1876. This colliery was sold to The Castle Eden Coal Co (which see) between 1878 and 1881.

From the note above, it is clear that there was more than one locomotive here. No further information is known.

Gauge : 4ft 8½in

    SHOTTON   0-6-0ST   IC   MW   479   1874   New   (1)

(1)   to Haswell Colliery, Haswell (in 1877?).

The partners who formed The Haswell Coal Company also formed, with additional partners, The Ryhope Coal Company (which see)

HAWKS, CRAWSHAY & SONS

GATESHEAD IRON WORKS, Gateshead   **C187**
Hawks, Crawshay & Sons by 7/6/1849   NZ 258638
previously **Hawks, Stanley & Co**
Hawks, Son & Co until 1841
Hawks & Son until 1829
originally **William Hawks**

William Hawks was a blacksmith, setting up his first small ironworks about 1747 at New Deptford, near Sunderland. Eventually he owned various works in Co. Durham which puddled and rolled iron. The Gateshead works became his chief "factory", but when it was first begun is uncertain. The works' most famous contract was the wrought ironwork for Robert Stephenson's High Level Bridge only a short distance away, a combined rail and road bridge opened in 1850, and the firm became well known for its bridge work. A system of horse and carts served the Gateshead works for many years, but latterly it was served by sidings from the NER Allhusen's Branch (see Imperial Chemical Industries Ltd), from the NER Gateshead-South Shields line. The works was closed in September 1889 and part of the site and its plant were eventually acquired by the firm's neighbours, John Abbot & Co Ltd (which see). The site was subsequently used for the Baltic Flour Mill of Joseph Rank Ltd (no rail traffic), itself now converted into the multi-million pound Baltic Centre for Contemporary Art.

There may have been other locomotives besides the one listed.

Gauge : 4ft 8½in

HAWK	0-4-0ST	OC	BH	682	1883	(a)	(1)

(a) New; believed to have been in part exchange for an earlier unidentified loco of the same name.
(1) to John Abbot & Co Ltd, Gateshead, c8/1891.

HAWTHORN AGGREGATES LTD

Hawthorn Limestone Co Ltd until 18/9/1967, which was latterly a subsidiary of **F.W. Dobson & Co Ltd** (which see) until the same date

HAWTHORN QUARRY, near Seaham Harbour   **J188**
NZ 437465

This limestone quarry was opened in 1909. It was served by sidings, 2½ miles south of Seaham Station, to the west of the NER line from Seaham Harbour to Hart Junction, which had been opened on 1st April 1905. Rail traffic ceased in March 1971, and the quarry was subsequently closed.

Gauge : 4ft 8½in

No.	Name	Type	Cyl	Mkr	No.	Date	Notes	Disp
No.1	WEAR	0-4-0ST	OC	BH	368	1877		
		reb		LG		?	(a)	
		reb		HL	7102	?		Scr by /1947
No.2		0-4-0ST	OC	HL	2684	1907		
		reb		Ridley Shaw		1937	(b)	Scr by /1947
	"CORNSAY"	0-4-0ST	OC	HL	2478	1901	(c)	(1)
	BRILL	0-6-0ST	IC	MW	1691	1907	(d)	(2)
	HAWTHORN	0-4-0ST	OC	RSHN	7308	1946	New	(3)
24		0-4-0ST	OC	HL	2453	1900	(e)	
		reb		TIW		Dec 1935		(4)
	HAWTHORN No.5	4wDM		RH	326062	1952	New	Scr 3/1971
	FRED DOBSON	4wDM		RH	326071	1954	(f)	(5)

(a) ex Lingford, Gardiner & Co Ltd, Bishop Auckland, dealers; previously Bolckow, Vaughan & Co Ltd, Middlesbrough, Yorkshire (NR).
(b) ex Imperial Chemical Industries Ltd, Billingham (see Cleveland Handbook), via Ridley, Shaw & Co Ltd, Middlesbrough, Yorkshire (NR), /1937.
(c) ex Ferens & Love Ltd, Cornsay Colliery & Brickworks, Cornsay.
(d) ex Sir Lindsay Parkinson & Co Ltd, contractors, /1941, hire; previously used on contract for Royal Ordnance Factory No. 6, Risley, Lancashire.
(e) ex F.W. Dobson & Co Ltd, Chilton Quarry & Limeworks, Ferryhill Station, by 10/8/1950.
(f) ex Cornforth Limestone Co Ltd, Cornforth Quarry, Cornforth, c/1965.
(1) recorded here on 12/5/1951; scrapped.
(2) returned to Sir Lindsay Parkinson & Co Ltd.
(3) to F.W. Dobson & Co Ltd, Chilton Quarry and Limeworks, Ferryhill Station, by 26/3/1949, and returned, after 12/5/1951; offered for sale in *Machinery Market*, 7/11/1952; to Northern Gas Board, Thompson Street Gas Works, Stockton, via Ridley Shaw & Co Ltd, Middlesbrough, Yorkshire (NR), dealers, /1953.
(4) to F.W. Dobson & Co Ltd, Chilton Quarry & Limeworks, c/1955.
(5) to Cornforth Limestone Co Ltd, Cornforth Quarry, Cornforth, c5/1970.

R. & W. HAWTHORN LESLIE (SHIPBUILDERS) LTD

R. & W. Hawthorn, Leslie & Co Ltd until 1/7/1954
R. & W. Hawthorn, Leslie & Co until 7/4/1886
Andrew Leslie & Co until 1/7/1885
HEBBURN SHIPYARD, Hebburn   **E189**
NZ 305652 approx.

This shipyard was opened by Andrew Leslie in 1854, but had no rail connection until 1872, when it was linked by a ½ mile branch to the NER Gateshead-South Shields line, ½ mile east of Hebburn Station. This line, which for much of its length ran alongside Lyon Street, also served the adjacent shipyard latterly owned by Vickers Ltd (which see). In the autumn of 1957 the two companies joined to set up the **Lyon Street Railway Co Ltd** with a capital of £200 to acquire the branch from its erstwhile owners,

69. Hawthorn Leslie's initial crane tank design, with 12in x 15in diameter cylinders, 2ft 10in wheels and capable of lifting up to four tons. Compare the detail differences with the RSH examples at Doxfords built nearly fifty years later. SANDOW, HL 2273/1893, stands in the firm's shipyard at Hebburn on 15th June 1960, painted middle green. Note that the end of the jib is painted black.

70. APOLLO, RH 319288/1953 on 29th June 1955, travelling from Hawthorn Leslie's yard to British Railways along Lyon Street in Hebburn, the route taken over by The Lyon Street Railway Co Ltd two years later.

the Carr-Ellison Estates. The line was closed and the track lifted by the early 1960s. The yard was among those absorbed into Swan Hunter & Tyne Shipbuilders Ltd (which see) on 1st January 1968.

Over the years the yard had eight rail cranes, seven steam and one diesel.

Gauge : 4ft 8½in

HEBBURN	0-4-0ST	OC	BH	202	1871	New	s/s
HEBBURN No.1	0-4-0ST	OC	BH	418	1877	New	(1)
HEBBURN No.2	0-4-0ST	OC	BH	569	1881	New	s/s
-	0-4-0ST	OC	HL	2039	1885	New	(2)
SANDOW	0-4-0CT	OC	HL	2273	1893	New	Scr 1/1961
(BROWNIE)	0-4-0CT	OC	HL	2550	1903	New	(3)
ATLAS	0-4-0ST	OC	HL	2917	1912	New	(4)
AJAX	0-4-0ST	OC	HL	3319	1918	New	(5)
HERCULES	0-4-0CT	OC	HL	2272	1893	(a)	Scr by /1939
KELLY	0-4-0CT	OC	RSHN	7126	1943	New	Scr 1/1961
-	0-4-0DM		RH	281270	1951	(b)	(6)
TRIUMPH	0-4-0DM		RH	304472	1951	New	(7)
APOLLO	0-4-0DM		RH	319288	1953	New	(7)

(a) ex St. Peter's Engine Works, Walker, Newcastle upon Tyne.
(b) New, for trials, 2/1951.

(1) recorded here, 20/5/1932; s/s.
(2) to Darlington Rolling Mills Co Ltd, Darlington, /1913.
(3) to William Doxford & Sons Ltd, Sunderland, /1940.
(4) to Thos W. Ward Ltd, Templeborough Works, Sheffield, Yorkshire (WR), 10/1951.
(5) to Thos W. Ward Ltd, Templeborough Works, Sheffield, Yorkshire (WR), 3/1953.
(6) to St. Peter's Engine Works, Walker, Newcastle upon Tyne, 2/1951.
(7) to Swan Hunter & Tyne Shipbuilders Ltd, with site, 1/1/1968.

WILLIAM HEDLEY & SONS

William Hedley & Co is also recorded; originally **William Hedley** (see also The Holmside & South Moor Collieries Ltd)

In the 1820s William Hedley (1779-1843), the enginewright from Wylam in Northumberland who played an important part in the development of early locomotives, began to acquire colliery royalties in various parts of Co. Durham. Amongst these was the Coxhoe royalty, which Hedley is believed to have acquired in 1824.

What happened then in railway terms is far from clear, and to help follow the account below the reader should refer to Fig.6. As part of its Act in 1828, revised in 1829, the Clarence Railway was allowed to construct a branch from its main line at Ferryhill via Coxhoe to Sherburn. This was certainly being built northwards during 1833, but it was abandoned soon afterwards after the completion of about ½ mile of earthworks north of Coxhoe. At the same time Hedley was constructing a 1¼ mile branch from the Clarence branch just north of Coxhoe to serve his new colliery at Quarrington Hill called **CROW TREES COLLIERY** (NZ 334377) **(N190)**. This dispatched coal for the first time down the line, with the Clarence's branch open, to be shipped at Stockton, or, more accurately, Haverton Hill, on 16th January 1834. A fortnight later **WEST HETTON COLLIERY** (NZ 325370) sent its first shipment along the line. One source says that Hedley acquired – or opened – this colliery in 1832.

However, away to the north-east the Hartlepool Railway & Dock Co opened its Thornley Branch in January 1835, soon afterwards lengthened by a privately-owned extension to Cassop Colliery and beyond by The Thornley Coal Co (which see). About the same time a link about ¾ mile long was also built westwards from about ¼ mile east of Cassop to Crowtrees; whether this was done by the Thornley company or by Hedley or his successors is not known. One source states that this link was worked by a stationary engine near Cassop Colliery, but the land between the two points is relatively level, well within the capability of locomotives. The first coal from Crowtrees was sent to Hartlepool via this link on 12th December 1839, with coal from West Hetton Colliery following in the last week of January 1840. In 1841 two more collieries were joined to the line between Coxhoe and Crowtrees Colliery – **Bowburn Colliery** (NZ 317366), north of Coxhoe and not to be confused with the later Bowburn Colliery sunk by Bell Bros Ltd (which see), and **Heugh Hall Colliery** (NZ 323379) at Old Quarrington, which was linked to the branch by a ½ mile self-acting incline. Bowburn Colliery's first coal along the line eastwards to Hartlepool was sent in October 1841, with Heugh Hall Colliery following two months later.

It is not known how long Hedley owned Crowtrees and West Hetton Collieries. The records of the Hartlepool Dock & Harbour Railway consistently show 'Crowtrees Coal Co' and 'West Hetton Coal Co', separately, and also Bowburn Coal Co and Heugh Hall Coal Co. Bowburn ceased sending coal to

Hartlepool in August 1843, West Hetton in December 1843 and Heugh Hall in January 1844, but this might be due to their reverting to using the Clarence Railway, rather than the collieries closing (see below).

Sinking at **Coxhoe Colliery** (NZ 329366) was being undertaken in the late 1820s. This resumed on 4th April 1836 and the site linked to the Clarence Railway's branch, though the first shipment of coal from the colliery at Stockton is said not to have taken place until 7th February 1844. Subsequently this short line also served **Coxhoe Iron Works** (NZ 319364), **Clay Hole Colliery** (NZ 320624), opened in 1839, and **Coxhoe Bank Quarry** (NZ 327363), which was worked between 1853 and 1879 and then was re-opened by The Steetley Lime Co (see Steetley Dolomite (Quarries) Ltd). In Steetley days the line from Coxhoe Colliery (on one side of the line) and Coxhoe Bank Quarry (on the other side) down to Coxhoe village was a **rope incline**, so it also may have been rope-worked at this earlier date. This line too was extended, in this case to **South Kelloe Colliery** (NZ 348370), which was opened in 1844. The O.S. map series above describes this extension as the South Kelloe Incline, so presumably it was also **rope-worked**, very probably **self-acting**. The ownership of these various locations is far from clear.

All the information known about Hedley's use of locomotives on the Clarence Railway comes from that railway's Minute Books, held at the National Archives at Kew. In this connection it ought perhaps to be said that the Clarence was financed by a group of businessmen in London, who employed a manager in the North East to run the railway. The directors met weekly and their minutes are very detailed indeed.

On 11th November 1834 the minutes record that 'Mr. Hedley had had an eligible offer of a Steam Engine and was anxious to receive the sanction of the Committee to be allowed to use it on the Road.' The minutes of 10th January 1835 give more details (see footnote (a) below) and describe it as similar to the Clarence's own (first) locomotive, completed in December 1834. However, the directors feared that both locomotives would be too heavy for their track. Their engine was finally approved on 6th May 1835, and it would seem unlikely that Hedley's engine was allowed on the line before this date. The second locomotive below was obtained later in 1835. However, in November 1835 Hedley wrote to the Clarence directors stating that his Small Locomotive Engine was inadequate for the volume of his traffic, and asking if he could hire a loco from them, a rather optimistic request, as the directors only had one locomotive themselves, and indeed soon afterwards handed over coal haulage to contractors.

As stated above, only the title 'Crowtrees Coal Co' appears in HD&R records. However, HD&R Directors' minutes for 16th December 1841 record an instruction to the Crowtrees Coal Co to remove its EXILO engine, which would seem very likely to be the EXILE listed below. When the York & Newcastle Railway took over the HD&R in October 1846 it stopped the coal owners' trains, and in 1847 the Crowtrees Coal Co was paid £5209 in compensation, the size of which strongly suggests that one or more locomotives were included.

By 1854 Bowburn, Coxhoe, Crow Trees, Heugh Hall and South Kelloe Collieries had all passed into the hands of **J.Robson & Co** (which see). They came later into the ownership, illegally, of the **West Hartlepool Harbour & Railway Co** (which see) and then **The Rosedale & Ferry Hill Iron Co Ltd** (which see).

Gauge : 4ft 8½in

WYLAM	0-6-0	OC	Hackworth		?	(a)	(1)
TYNESIDE	0-4-0	OC?	RWH	192	1835	(b)	(2)
EXILE	0-6-0	OC	N & W		1837	(c)	s/s

(a) the minutes of the Clarence Railway dated 10/1/1835 records that 'The locomotive purchased by Mr. Hedley from Mr. Parr had been built by Hackworth & Co.' This locomotive might be the same as that advertised by Mr. Richard Parr to be sold at South Shields on 9th August 1834 (ref: *Early Railways*, ed. Andy Guy & Jim Rees, Newcomen Society, 2001, p.294). It is unlikely to have started work until 5/1835.

(b) New to Hedley, but almost certainly this was the locomotive which had been ordered by the Clarence Railway from RWH in 1834 but then cancelled, for which RWH claimed compensation from the Clarence Railway in April 1835.

(c) apparently built to an order that was cancelled; offered to Clarence Railway, 2/1839, but declined; used by the builders between 3/1839 and 8/1839 on a contract to shunt the staiths at Port Clarence, near Middlesbrough, Yorkshire (NR); first recorded with Hedley in 12/1841.

(1) it was offered to the Clarence Railway in 12/1836, but apparently declined, as it was never taken into Clarence Railway stock. It is believed to have been used by Hedley from 1839 on the Hartlepool Dock & Railway Co; s/s.

(2) almost certainly it was this locomotive which Hedley offered to sell to the Clarence Railway on 26/5/1836, an offer which was declined. In 6/1836 Hedley offered it again, but it was again declined. It is believed to have been used by Hedley from 1839 on the Hartlepool Harbour & Railway Co; s/s.

THE HEDWORTH BARIUM CO LTD (registered in 1891)
COW GREEN MINES, near Langdon Beck, Middleton-in-Teesdale W191

The construction of a new road in Upper Teesdale in the early 1890s stimulated mining in the area, initially by the Teesdale Mineral Co Ltd in 1892, though this firm went bankrupt four years later. In 1898 the royalties were taken over by The Hedworth Barium Co Ltd, seeking mainly the production of barytes.

The firm began operations at Cow Green, about eight miles north-west of Middleton-in-Teesdale and about 1500 feet above sea level, by driving the **LOW LEVEL** (NY 810305), which was horse-worked. Later, a short distance away, the company sank the **WRENTNALL SHAFT**, to begin the working of the lower strata. Probably about the turn of the century the company re-opened the **DUBBYSIKE MINE** (NY 795319), actually a drift, about 1¼ miles north-west of Cow Green. Subsequently the company opened workings in the East Cow Green area, worked for barytes, and the **ISABELLA MINE** (NY 816317), about ¾ mile north-east of Cow Green but at 1900 feet, worked for lead and later barytes.

In such a remote area transport was always a difficult problem. For many years the barytes was carried to Middleton-in-Teesdale Station by horse and cart. According to the article below, the company's first attempt at improving this was to build a railway from the Isabella Mine around the contour of the south end of Herdship Fell to a rise from the Hopkins Mine at East Cow Green; the ore was tipped down the rise to be transported underground and brought out via the Hopkins Level to be washed and dressed at **COW RAKE** (NY 829313). This line, about ¾ mile long, was almost certainly horse-worked.

Then in 1918 the company introduced a much more radical solution, aiming to eliminate the horse and cart work. A narrow gauge railway was laid from the Low Level to Cow Rake, which also became the central collecting point for all the mines. From here an aerial ropeway took the ore down to two storage hoppers alongside what is now the B6277 road about half a mile north of the Langdon Beck Hotel. From here the ore was taken by steam lorry down to Middleton-in-Teesdale Station.

The mines closed in 1920, though the Cow Green and Dubbysike Mines were later re-opened by Anglo-Austral Mines Ltd (which see).

Gauge : 1ft 6in

A locomotive is believed to have been used on the line between the Low Level and Cow Rake, but no details are known.

References : *Out of the Pennines*, ed. B.Chambers, 1997: *Early Barytes Mining at the Cow Green Group of Mines including Dubby Sike and Isabella*, Harold L. Beadle; *The Mines of Upper Teesdale*, R.A. Fairbairn, 2005.

JARROW WORKS, Jarrow Slake D192
NZ 349653

This works was served by a one mile NER branch north of the NER Gateshead-South Shields line, 1¼ miles west of Tyne Dock Station, and lay immediately north of the works of Fownes Forge & Engineering Co Ltd. Partly due to a fall in demand and also because of a charge of polluting the River Tyne with barium sulphide sludge, the company was forced into liquidation in 1922. The plant, including a saddletank locomotive, was auctioned by Wheatley Kirk on 17th and 18th January 1923.

Gauge : 4ft 8½in

| - | 0-4-0ST | OC | CF | 1202 | 1900 | New | (1) |

AB supplied the company with spares for a locomotive named ABERDARE in 1918, but its identity is unknown.

(1) to Ridley & Young, engineers and dealers, Darlington, by 4/1923; re-sold to Cleeve's Western Valleys Anthracite Collieries Ltd, Tir-y-dail Colliery, Glamorgan, /1923.

THE HETTON COAL CO LTD
The Hetton Coal Co until 4/7/1884

The opening of **HETTON COLLIERY** (NZ 360469) on 18th November 1822 was perhaps the most significant development ever in the history of the Durham coalfield, and its associated railway was equally a major landmark in the history of railways.

In the western half of County Durham the coal seams are close to the surface and were relatively easy to work; but in the eastern half of the county there is a thick stratum of magnesian limestone and it was believed that no coal lay underneath. In 1810 Robert Lyon, a landowner in this area near Hetton-le-Hole, began sinking a shaft on his land, only for the cost to bankrupt him. However, in 1819 one of the leading mining engineers of the time, Arthur Mowbray, left his employment as Chief Viewer to the Marquis of Londonderry and with a group of entrepreneurs took over the royalty (the Marquis had hoped to acquire it, but in the end he became a shareholder). The sinking of a new **HETTON COLLIERY**, often called **HETTON LYONS COLLIERY** after Robert Lyon (NZ 360469), began on 19th December 1820, about ¼ mile south of the village of Hetton-le-Hole. The Main Coal seam was reached on 3rd September 1822 and the Hutton seam on 6th January 1823. The Hetton Coal Company, created earlier in 1820, was the first important joint-stock coal company in Durham; its success in finding coal below the magnesian limestone led to the opening up of the whole of the eastern area of the coalfield, while its ever-increasing output, necessary to meet borrowings and provide a return on their investment, challenged the "Limitation of the Vend", the cartel of controlled output and high prices which for so many years the aristocratic Durham coal owners had imposed on the London trade. In 1836 Nicholas Wood (see John Bowes & Partners Ltd and Harton Coal Co Ltd) gained an interest in the company and in 1844 he became Managing Partner, being succeeded by his son Lindsay Wood (1834-1920), later Sir Lindsay Wood, who for many years was also chairman of the Harton Coal Co Ltd (which see) and one of the most powerful men in the Durham coal industry.

A new inland colliery needed a waggonway to transport its coal to the nearest river. But for the Hetton Company constructing staiths upstream on the River Wear and the cost and inconvenience of then taking coal by keel down to the river mouth to load on to sea-going colliers, as both Lambton and Londonderry were doing, was unacceptable; from the outset the Hetton company was determined to ship its coal at Sunderland directly into colliers and thus eliminate the keels.

Mowbray became interested in linking the Vane Tempest waggonway to the Newbottle line as early as 1813. When he became the leading promoter in the Hetton Coal Co he seriously considered linking Hetton Colliery to the Newbottle line (1819); but the most famous colliery engineer of his day, John Buddle, disliked Mowbray intensely and was determined to keep the Newbottle line out of his hands, first by proposing to link the Vane Tempest system to it (1820) and subsequently by proposing a division of the area under which John George Lambton, later the 1st Earl of Durham, got Newbottle and Londonderry got Seaham (DCRO, D/Lo/142/4), which is more or less what eventually happened. In the event the Newbottle Railway was incorporated into the Lambton Railway (which see). Having failed to gain access to the Newbottle line, the company was thus compelled to build its own, which was called the **HETTON RAILWAY (H/J 193)**.

The design, construction and operation of the Railway
To engineer the line the company engaged, probably in the late summer of 1820, one of the area's leading railway engineers, George Stephenson, then employed at Killingworth in Northumberland by Lord Ravensworth & Partners ("The Grand Allies"), who allowed him to design this new line while still working for them. The result was the first railway in the world to be designed to use steam locomotives – a major pioneering step in railway development.

However, between Hetton and Sunderland lay the very real obstacle of Warden Law at 636 feet, whilst at Sunderland there were steep cliffs down to the river. So the design for the 7¾ mile line incorporated two locomotive-worked sections, two rope inclines worked by stationary engines and five self-acting (gravity-worked) inclines. Construction began in March 1821. A month later George Stephenson was appointed Engineer to the new Stockton & Darlington Railway, so his younger brother Robert was appointed resident engineer at Hetton; and as stated earlier, the line was opened, with due ceremony, on 18th November 1822.

In 2004 there came to light a document entitled *Observations on Railways......*, prepared by George Dodds, the Railway's superintendent, in December 1824. In it he lays out a very detailed description of the Railway, its methods of operation and their costs. From this it has been possible to reconstruct Stephenson's original design. There were numerous valuations of the colliery and reports on the railway produced for the company in the next ten years (now in the Northumberland Record Office), as well as the report of a visit in 1827 by two Prussian engineers, Oeynhausen and Dechen, and from these it is possible to deduce how the railway was changed. When the Railway opened the first section, from Hetton Colliery to Copt Hill Bank Foot, just under 1½ miles, was worked by locomotives. To reach the summit at Warden Law the waggons travelled over two powered inclines, the **Copt Hill Incline** (832 yards), worked by the **Byer Engine** (**Byre** is also found) (NZ 355497), and the **Warden Law Incline** (1550 yards), worked by the **Warden Law Engine** (NZ 358505). Both inclines were single line, with fulls and empties being worked alternately; but on the Warden Law Incline Stephenson adopted a very unusual system, whereby a set of fulls at the bank foot was attached by a ½ mile rope to a second set of fulls at the middle, with the engine hauling both sets at the same time. Then followed **four self-acting inclines, Nos.1-4**, which took the line for just under 2½ miles down to North Moor. The next 2½ miles were worked by more locomotives, with the final 325 yards down to the staiths a further self-acting incline.

This system did not last long. By 1827 the Warden Law incline had been divided into two equal sections, that northwards from the Byer Engine being worked by the **Flat** (later **Flatt**) **Engine** (NZ 362500), while the locomotives on the northern section were replaced by rope haulage, almost certainly to enable a higher volume of traffic to be handled. Three stationary engines were built, the **North Moor Engine** (NZ 373540), the **Winter's Lane Engine** (NZ 385555) and the **Staith Engine** (NZ 388571), just south of Hylton Road in Sunderland. The first was merely used to brake the full set and then haul back the empties, but the other two operated main-and-tail haulage; a set of fulls at the Winter's Lane Engine was hauled to Hylton Road with the rope being wound on to the Staith Engine drum while the Winter's Lane rope, attached to the rear end of the set, unwound, the process being reversed for a set of empties to be hauled southwards.

Soon after the opening of Hetton Colliery the company began to plan further developments. On 25th March 1825 the sinking of **ELEMORE COLLIERY** (NZ 346356) was begun at Easington Lane, with coal being reached in February 1827. This was linked to the Railway by a **self-acting incline** a mile long. On 23rd May 1825 the company began the sinking of **EPPLETON COLLIERY** (NZ 364484), but due to difficulties with water coal was not reached until 1833. This was linked to the main line about a mile north of Hetton by another **self-acting incline**, ¾ mile long. At Hetton itself the company established its **ENGINEERING SHOPS** (NZ 360469) and **WAGON SHOPS** (NZ 360468), and also **HETTON BRICKWORKS**, making common bricks. In later years the loco shed was near the shafts, but it is not known whether this site dates from the opening in 1822.

The Londonderry Railway's line between Rainton and Seaham (which see), opened in 1831, crossed under the Hetton line in a tunnel just below the Byer Engine, and in 1843 a link was put in between the two to allow coal from North Hetton Colliery (see The North Hetton Coal Co Ltd) to come on to the Hetton Railway for shipment. In Sunderland itself several works were built alongside the Railway to gain access to its coal, including the **Bishopwearmouth Iron Works** (NZ 388569), the **Wear Glass Works** (NZ 389573) and the **Sunderland Glass Works** (NZ 389574). The last was subsequently replaced by the Trimdon Iron Works, which was itself replaced in 1900 by Sunderland Corporation's **Electricity Generating Station**.

Although after the opening of Eppleton Colliery nothing further was added or closed, the Railway underwent various improvements over the years. South of Warden Law the ropes were taken off the Elemore Incline about 1895 and off the Eppleton incline, latterly known as **Downs Bank**, in 1902, allowing locomotive haulage to be used. In 1836 the engine at Warden Law was replaced, so it may have been then that the incline was rebuilt to allow full and empty sets to be worked simultaneously. In 1876 the powered inclines underwent major rebuilding. The Copt Hill incline was altered to allow fulls and empties to be run simultaneously, and the Flatt Engine was removed. Instead a third drum in the Byer Engine and a return wheel at the Flatt were used to work the wagons to Warden Law bank foot. However, if a gale was blowing the wagons would not run, and for this the Byer Engine was provided with a fourth drum in order to implement a form of main-and-tail rope haulage utilising the two drums and the return wheel.

On the northern section the main-and-tail system would seem to have been replaced by locomotives by the mid 1850s, though the North Moor Engine appears to have survived some years longer. Locomotive haulage was eventually extended right through to the staiths, but again the date of this is not known. To house the locomotives on this section a **two-road shed** (NZ 387564) was built 100 yards to the south of the bridge carrying the Railway over the Chester Road in Sunderland. This still left several level crossings in Sunderland, which became an increasing nuisance over the years. About 1890, to remove the Hylton Road crossing and improve locomotive access to the staiths, a new line was built from north of the loco shed, passing through **Farringdon Row Tunnel** to reach the staiths; the old route, now a branch, was retained to serve the works on it and a large coal depot. A similar solution, a diversion with a cutting and a tunnel, was adopted to remove the Durham Road crossing, apparently undertaken about 1910.

In 1911 Lord Joicey, who had acquired the Earl of Durham's collieries and the Lambton Railway in 1896 and set up Lambton Collieries Ltd, purchased The Hetton Coal Co Ltd and The North Hetton Coal Co Ltd, despite the opposition of Sir Lindsay Wood. The two latter companies were wound up on 29th June and Joicey took possession on 3rd July. He then merged the three businesses in August 1911 to form **The Lambton & Hetton Collieries Ltd** (see The Lambton, Hetton & Joicey Collieries Ltd).

The locomotives

It is now clear that the reference to five locomotives in the account of the opening day in 1822 in the *Newcastle Courant* is incorrect, and that there were only three. Dodds (see above) gives a further two, clearly new, as entering traffic in December 1823, although there is a reference to them earlier than this. Where they were built is uncertain, except that it is unlikely to have been at Hetton; possibilities are the workshops at Killingworth in Northumberland of Lord Ravensworth & Partners (Stephenson's erstwhile employers) or the works at Walker in Newcastle upon Tyne of Losh, Wilson & Bell, which also cast the rails and may have manufactured the original stationary engines. Jim Rees, in *Early Railways: The Strange Story of the Steam Elephant*, argues that the illustration of a six-coupled locomotive on Hetton documents in the 1830s suggests that a six-coupled engine worked on the line, and that the locomotive concerned was almost certainly the six-wheeler built at Wallsend in 1814-15. If she was here, she was additional to the original five. How long these locomotives lasted, and their relationship with those described at Hetton in the 1850s, is also open to question. A newspaper account reported that 'an old engine', the 'oldest on the line, or owned by the company', exploded at Hetton on 20th December 1858. The report that this was the LADY BARRINGTON below may therefore be incorrect, and the locomotive involved may have been one of the original locomotives from the 1820s.

References : *The Private Railways of County Durham*, Colin E. Mountford, The Industrial Railway Society, 2004.
Early Railways 3, ed. Dr. M.R. Bailey, *The Hetton Railway – Stephenson's original design and its evolution*, Colin E. Mountford, Newcomen Society, 2006. Note: the Dodds report, not available when *The Private Railways of County Durham* was published, has changed some of the information given in it.

Gauge : 4ft 8½in

For a discussion of the early locomotives here, see above.

(LYON or LYONS)	0-4-0	VC	Hetton		c1852	New	
	reb		Hetton		1877	(a)	(1)
LADY BARRINGTON	0-4-0	VC	Hetton		c1854	New	(2)
No.1 later 1	2-4-0	OC	RS	1100	1857	New	
reb	2-4-0T	OC	Hetton		?		(3)
WASHINGTON	?	?	?	?	?	(b)	s/s
4	0-6-0	IC	RS	1649	1865	New	
reb	0-6-0ST	IC	Hetton		?		(3)
5	0-6-0ST	IC	RWH	1422	1867	New	
reb		HL	2378	?			(3)
6	0-6-0ST	IC	RWH	1430	1868	New	(3)
7	0-6-0ST	IC	RS	1919	1869	New	
reb		HL	1182	1930			(3)
8	0-6-0ST	IC	RWH	1478	1870	New	(3)
2	0-6-0ST	IC	RWH	1969	1884	New	(3)
LYONS	4wVBT#	VC	Hetton		c1900	(c)	(3)
EPPLETON	4wVBT#	VC	Hetton		c1900	(c)	(3)

\# photographs of LYONS show the wheels on the offside coupled together with a chain, whilst on the nearside the chain connects only one of the wheels with what is presumably a cog (hidden on the photographs), presumably on the end of a drive shaft.

(a) rebuilt with vertical frame plates and new boiler.
(b) arrived in 1860; origin and identity unknown.
(c) assembled from various parts, not all locomotive in origin, with chain drive to one axle outside the frame. An article in *The Locomotive* in 1901 suggests that LYONS had not long been built, but another source claimed that it was built about 1870. They worked as colliery shunters.

(1) one course claimed that the loco was withdrawn from traffic about 1908-09, while another recorded that it worked until 9/7/1911; when it was withdrawn from traffic it was put to work driving machinery in the saw mill at Hetton Workshops; to The Lambton & Hetton Collieries Ltd, officially from 8/1911 but actually from 3/7/1911.
(2) it has been stated that it was this locomotive which exploded at Hetton on 20/12/1858, but the press accounts of the explosion state that the locomotive involved was 'the oldest on the line', so possibly it was one of the original locomotives from the 1820s.
(3) to The Lambton & Hetton Collieries Ltd, officially from 8/1911 but actually from 3/7/1911.

G. HODSMAN & SONS LTD
G. Hodsman & Sons (1910) Ltd until 1916
G. Hodsman & Sons Ltd by 1910
formerly **G. Hodsman & Sons**; originally **G. Hodsman**
GREENGATES QUARRY, near Laithkirk, near Middleton-in-Teesdale W194
NY 935235

George Hodsman was a York man who became one of the chief mining entrepreneurs in Upper Teesdale. He began by taking over **LUNEDALE QUARRY** (NY 954239) **(W195)** at Laithkirk, which had been worked for limestone and whinstone until the bankruptcy of its owners, **The Lunedale Whinstone Co.** (which see), in 1885. This quarry, on the left bank of the River Lune, was linked by a 2ft 6in gauge line about 300 yards long, which passed under what is now the B6276 road by a short tunnel to a discharge point at sidings alongside the NER Middleton-in-Teesdale branch at Lunedale Quarry Signal Box, about half way between Mickleton and Middleton. The loco shed was situated near the discharge point. How the traffic was operated when Hodsman became the quarry's owner is not known.

Soon after re-opening Lunedale Quarry Hodsman opened a new quarry for basalt at 1300 feet on the side of Lune Moor, called **GREENGATES QUARRY**. To reach this he laid a two mile branch from the discharge point sidings, through the original tunnel using the original route and then branching off through a second tunnel under the same road before climbing steeply to reach the quarry, the gradient in parts being as steep as 1 in 20. Using the basalt, road setts, roadstone and eventually tarmacadam were produced. Lunedale Quarry was closed, probably at the same time that Greengates came into production. Photographic evidence (see photo no.71) shows that a passenger service was provided for the quarrymen, travelling in the open tubs.

Eventually a third tunnel had to be made to give access to the lower part of the quarry, and as the quarry became deeper an aerial ropeway was constructed to raise the stone up to the loco level. In 1917 the aerial ropeway was blown down in a gale, and the quarry never re-commenced production, the firm choosing instead to concentrate on its High Force Quarry, alongside the B6277 five miles north-west of Middleton-in-Teesdale. The company also owned tarred slag works at Port Clarence (see Cleveland Handbook) and at East Ardsley, Yorkshire (WR). It was dissolved in 1928.

71. This loco carries an inverted saddletank, and so is presumed to be the 2ft 6in gauge EBENEZER, WB 1002/1888, waiting for quarrymen to climb into the tubs to travel between Middleton-in-Teesdale and Greengates Quarry.

ZURIEL (below) had a boiler inclined at 1 in 20 to keep the firebox crown covered on the steep gradients of the branch, and its height was restricted to 6 feet to allow it to pass through the tunnel into the lower quarry.

Reference : *Out of the Pennines*, ed. B. Chambers: *The Quarry Industry in Teesdale*, H.L. Beadle, 1997.
Gauge : 2ft 6in

LOTTIE	0-4-0T	OC	BH	629	1881	(a)	(1)
(EBENEZER)	0-4-0IST	OC	WB	1002	1888	New	(2)
ZURIEL	0-4-0T	OC	WB	1917	1910	New	(3)

(a) ex The Lunedale Whinstone Co, with the quarry, c/1885.

(1) a 2ft 6in gauge saddletank loco with 5½in cylinders was offered for sale by this firm from its York address in *Contract Journal* dated 7/3/1917; it is believed to have been this loco, but if it was, then it had presumably been rebuilt as a saddletank from a tank loco; s/s.

(2) offered for sale by firm in *Contract Journal* dated 17/9/1917, when it was said to be at Port Clarence, near Billingham; s/s.

(3) offered for sale by firm in *Contract Journal* dated 17/9/1917, when it was said to be at Port Clarence, near Billingham; to T.D. Ridley & Sons, Middlesbrough, Yorkshire (NR), for storage, /1921; to Cradock, Allison & Co Ltd, Eaglescliffe Brickworks, Eaglescliffe, c/1922 (see Cleveland & North Yorkshire Handbook).

THE HOLMSIDE & SOUTH MOOR COLLIERIES LTD
(formed by the amalgamation of **Thomas Hedley & Bros Ltd** and **The South Moor Colliery Co Ltd** on 3/1/1925)

The large royalty of about 4000 acres in North-West Durham that this company worked was very much dominated by the Hedley family. William Hedley, the former enginewright at Wylam Colliery in Northumberland, who played a significant part in the development of the steam locomotive, and had then risen to become the Viewer of Wylam Colliery, left there in 1827. Either in 1828 or the following year he entered a partnership headed by William Bell of Sunderland (who was also at the time the leading owner of Shincliffe Colliery (see Joseph Love & Partners), which had taken over the lease of an area of about 4,000 acres near Stanley, previously held by Lord Ravensworth & Partners (which see). This partnership eventually became The South Moor Colliery Co. In 1838 the partners decided to offload the southern area of their royalty, and this was leased by William Hedley alone (for his other colliery interests see the entry for William Hedley). After Hedley's death in 1843 his interests passed to his four sons, headed by Thomas Hedley, still in partnership with others at South Moor, but alone at Craghead. The two companies remained financially separate until the amalgamation of 1925.

The collieries in both areas are examples of various shafts some distance from each other being given the collective name of the colliery, a practice rare in the North-East coalfield though much more common in other coalfields. However, this is then complicated by changes in the collective name. With the operating history and its effects on the railway systems also going through various changes, the history of these two collieries is exceptionally complicated.

HOLMSIDE COLLIERY/CRAGHEAD COLLIERY, Craghead   G196
Thomas Hedley & Bros Ltd until 3/1/1925
Thomas Hedley & Bros until 24/7/1889
formerly **William Hedley & Sons**; originally **William Hedley**

Hedley began here by sinking the **William Pit** (NZ 216508), work beginning on 14th January 1839, and linked it by a waggonway, 1¾ miles long, to the western end of the Pelton Level, between the Eden and Waldridge Inclines on the Stanhope & Tyne Railway. Horses worked the first fifty yards from the pit to the top of a **self-acting incline**, latterly known as the Craghead Incline. This was about 1¼ miles long on an average gradient of 1 in 26. At the bottom horses were used to shunt the link with the Stanhope & Tyne. In 1854 the Hedleys sank the **George Pit** (NZ 219510), close to the incline bank head, and for quite some time these two pits were named **CRAGHEAD COLLIERY**, the nearby pit village also taking the name Craghead.

Meanwhile to the south west of the village Hedley sank the **Thomas Pit** (NZ 213507) in 1841, which was joined by the **Oswald Pit** (NZ 213508) in 1878. These were given the name **HOLMSIDE COLLIERY**, taking the name from Holmside Hall some miles away where Thomas Hedley lived. They were served by a short ¼ mile branch running south from the waggonway.

In 1845 Hedley's sons opened **BURNHOPE COLLIERY** (NZ 191482), about 1½ miles north-west of the village of Lanchester, extending the waggonway from Craghead by a further 2½ miles to serve it. This extension was operated by the **Burnhope Engine** (NZ 194499), about a mile north of the colliery, which hauled waggons up to it and then lowered them down to Craghead. However, the family only retained this colliery for a fairly short time before leasing it (see the entry for Ritsons (Burnhope Collieries) Ltd).

In 1855 James Joicey (see James Joicey & Co Ltd) opened **West Pelton Colliery** (also known as the **Alma Pit**), which had a ½ mile branch to join the waggonway at Craghead Bank Foot in order to gain access to what was now the North Eastern Railway.

The date when a locomotive was first used at Craghead is not known. Eventually the locomotives kept at Craghead shunted the section between the foot of the incline from Burnhope, together with Holmside Colliery, as far as the bank head of the Craghead Incline, and a locomotive also went down the incline to shunt the sidings at the bank foot which linked with the NER. The locomotive shed was situated near the William Pit, left on its own after the William Pit site was cleared.

The Thomas Pit closed first, followed by the William Pit, soon after the beginning of the twentieth century, and the George Pit in June 1914. However, to reach the lower seams the firm sank two new shafts immediately to the west of Holmside Colliery, the **Edward Pit** (NZ 212508) in 1909 and the **Busty Pit** nearby in 1916, extending the Holmside branch by another ¼ mile to serve them. A **brickworks** was begun between the William and George Pits in 1872 to make common bricks; it closed early in 1908.

Sometime about the late 1930s the title Holmside Colliery gave way, a little confusingly, to **CRAGHEAD COLLIERY (G197)**. About the same time it would seem that the witherite seam was also mined here, as at Morrison North above, albeit briefly. The colliery and its railway were vested in NCB Northern Division No.6 Area on 1st January 1947.

Reference : *The Private Railways of County Durham*, Colin E. Mountford, Industrial Railway Society, 2004.

The first locomotive below was originally thought to have been one of those that William Hedley had helped to construct at Wylam Colliery in Northumberland about 1813. However, the Wylam Waggonway was re-laid with stronger cast-iron edge rails in 1827-1832, and this locomotive is now believed to have been one of a pair built at this time to replace the original engines. Wylam Colliery closed in 1868, and Hedley's sons purchased it at the subsequent auction. Once at Craghead it was renovated and was occasionally steamed on a short length of track.

Gauge : 5ft 0in

"WYLAM DILLY"	4w	VCG	Wylam	1827-1832	(a)	(1)

(a) ex The Owners of Wylam Colliery, Wylam, Northumberland, 1/1869; purchased at auction by the Hedley brothers for £16.10s 0d.

(1) exhibited at the North East Coast Exhibition of Naval Architecture & Marine Engineering, Tynemouth, Northumberland, 9-10/1882; to Edinburgh Museum of Science & Art, Edinburgh, Scotland, (now the Royal Museum of Scotland), 10/1882; the oldest preserved locomotive in the world.

Gauge : 4ft 8½in

HOLMSIDE	0-6-0ST?		HH		?		
		reb	Dunston Engine Works Co			(a)	s/s
BURNHOPESIDE	0-6-0ST	OC	BH	888	1887	(b)	
		reb	HL		1931		(1)
CRAGHEAD	0-6-0ST	OC	BH	971	1890	New	
		reb	HL		1911		(2)
HOLMSIDE No.2	0-6-0ST	OC	CF	1204	1901	New	(3)
HOLMSIDE No.4 (formerly SOUTH MOOR No.4)	0-6-0ST	OC	HL	3528	1922	(c)	(1)

(a) origin and date of arrival unknown; described in the *Colliery Guardian*, 4/9/1896.
(b) ordered by BH for stock, 9/1886, and exhibited at the Newcastle Exhibition of 1887 before delivery here; one source says that the locomotive was named VICTORIA when it was delivered here, but a BH catalogue photograph of the locomotive in fully lined-out livery shows the name BURNHOPESIDE.
(c) ex Morrison Busty Colliery, Annfield Plain, by 1/1/1947.

(1) to NCB No. 6 Area, with colliery, 1/1/1947.
(2) to Morrison Busty Colliery, Annfield Plain, by 30/4/1937.
(3) to Morrison Busty Colliery, Annfield Plain, by 8/1937.

SOUTH MOOR COLLIERY, Stanley and Annfield Plain G198
The South Moor Colliery Co Ltd until 3/1/1925
The South Moor Colliery Co until 24/7/1889

This colliery has the most complicated history of any Durham colliery. Here William Hedley was one of a number of partners.

The "Grand Allies" had worked collieries here and developed waggonways to carry their coal, initially to the River Tyne, but latterly to the River Wear at Fatfield by linking to the Beamish Waggonway, later the Beamish Railway. The new owners here began the sinking of their first pit (NZ 191509) on the western area of the royalty, like Craghead above, in 1839. Since it lay some 1½ miles to the west, it was at first called **WEST CRAGHEAD PIT**, but soon afterwards it was re-named **SOUTH MOOR COLLIERY** (the 1st edition O.S. maps give Southmoor), and then as the collective South Moor Colliery expanded, it became the **William Pit**. This was a new sinking at an eighteenth century pit; in addition the 1st edition O.S. map of 1858 appears to show that the pit was at one time served by a branch of the Shield Row waggonway, though equally this section could also have been part of the Tanfield Way's Causey branch, one of the most important of the eighteenth century waggonways in north-west Durham (see James Joicey & Co Ltd). To serve the new colliery a one mile line was built southwards from Oxhill on what was then the Stanhope & Tyne Railway, later the NER Pontop & South Shields branch. This line was worked by two stationary engines. The first, at the William Pit, lowered the wagons down ¼ mile to a landing near the Stanley Burn, from where they were hauled up to Oxhill by the **Oxhill Engine** (NZ 184523). This pit was sunk to the Hutton seam, and the first coal was dispatched on 25th July 1841. The Oxhill Engine gave the owners considerable trouble, exploding on 25th October 1845 and exploding again in 1853.

In 1845 the partners sunk a new pit nearby to the Brass Thill seam. This was at first called the **Quaking House Pit** (NZ 183517) after a nearby house, but subsequently it became the **Charley Pit**. It was linked by a branch running eastwards to the original line, just under ½ mile south of Oxhill. Given that this junction was part way up the Oxhill incline it would seem likely that the Oxhill Engine ran wagons in and out of the Charley Pit as required.

In 1863 the firm sank the **Louisa Pit** (NZ 194526) to the Low Main seam, taking its name from the wife of William Bell, then the principal partner. This lay alongside Front Street in Stanley on one side and the Pontop & South Shields branch on the other, to which it was connected. The partners then

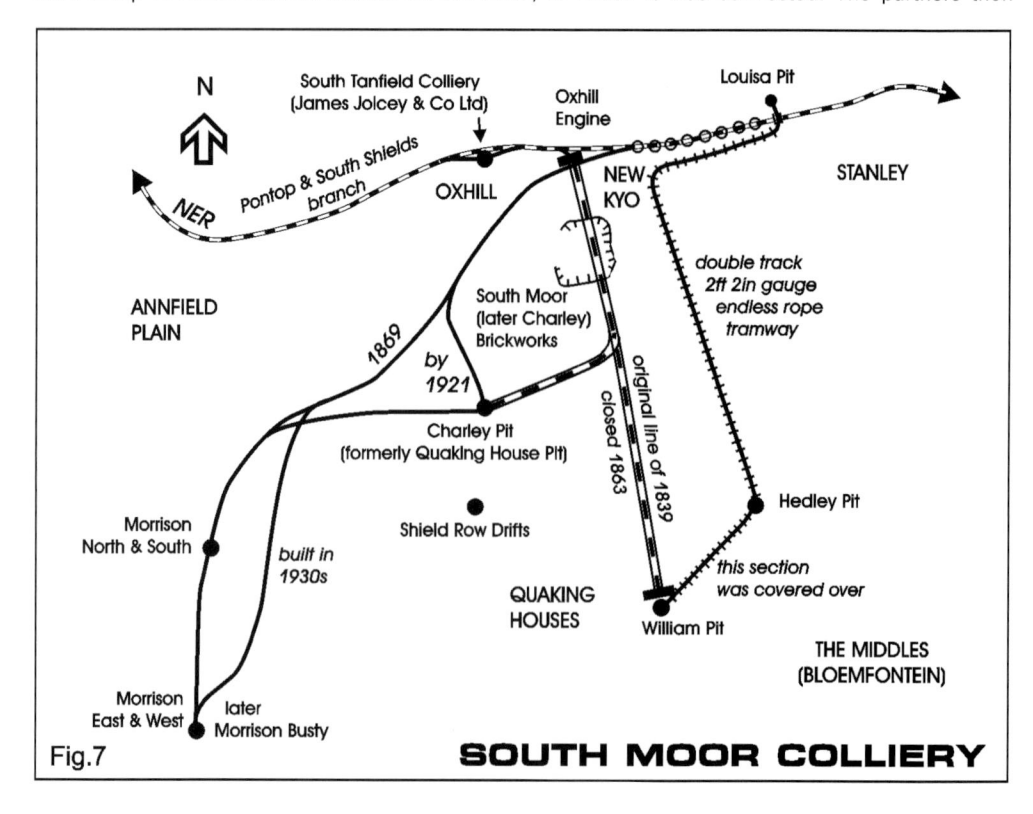

Fig.7 **SOUTH MOOR COLLIERY**

72. *This accident happened on Derby Day, 2nd June 1920, when SOUTH MOOR No.2, BH 1034/1892, pushed her train of wagons off the end of the landsale depot in Stanley. However, the photographer has doctored his picture, by removing the depot and also the locomotive's chimney, which is shown on other pictures! Nevertheless, it is rather a mess!*

73. *Chapman & Furneaux built only a handful of 0-6-0STs. HOLMSIDE No.2, CF 1204/1901, stands, probably at Oxhill, on 14th August 1933.*

connected it underground to the Charley Pit, so that all of that pit's coal could be drawn at the Louisa Pit, and then closed the William Pit, enabling them also to close the whole of the surface rope-worked system. Subsequently a second shaft was sunk here, this becoming known as **Louisa New Pit** and the original shaft the **Louisa Old Pit**.

In 1869 the partners decided to open out the area south of Annfield Plain, and there sank the **Morrison North** and **Morrison South Pits** (NZ 174511) to the Hutton seam, and also built two batteries of beehive coke ovens. To serve this site a railway one mile long was built south-westwards from the original junction at Oxhill, together with a ¼ mile branch to the Charley Pit, serving it now from the west. Presumably this was used to transport materials to the Charley Pit, as no coal was being drawn here. Almost certainly this new system was locomotive-worked, but no details survive. Probably about the same time the firm opened the **North** and **South Shield Row Drifts** (NZ 184509), just north of the Stanley Burn, to work the Shield Row seam (re-named the High Main seam under the NCB). These had no rail connection, the coal being taken underground to the Charley Pit and then on underground to the Louisa Pit. At some date unknown the firm acquired running powers over the NER eastwards from Oxhill to give its locomotives access to the Louisa Pit.

In 1885 the partners returned to the area of their firm's first mining activity and just to the north-east of the former William Pit sank the **Hedley Pit** (NZ 193510). This was linked to the Louisa Pit by a double track endless rope tramway 1700 yards long, almost certainly of 2ft 2in gauge and crossing over the Pontop & South Shields branch on a gantry to reach the Louisa Pit screens. There was further development in the 1890s. In 1893 a new central ventilation shaft was sunk at the Charley Pit to provide better ventilation for all the pits. Early in 1897 a **brickworks** was established here, utilising the boulder clay that overlay the coal seams. Then in 1898 the former William Pit was re-opened, once again called the **William Pit**, though this time after a younger member of the Hedley family (locally, however, it was called the **Billy Pit**). To serve it the endless rope tramway was extended from the Hedley Pit about 200 yards. According to the O.S. map of the period this extension was "covered". With this development of mining in an area over two miles from the centre of Stanley the company decided to build a new colliery village called Quaking Houses, and in 1900 erected an aerial ropeway from the Charley Pit to the village site to carry bricks from the brickworks.

It would seem that the Shield Row drifts had closed by 1914. The Charley Pit was also shut down during the First World War due to shortage of men, and never re-opened. Despite this, a loop had been built to the Charley Pit site by 1921, presumably to provide a by-pass route to a section of the "main line" to Oxhill. The Charley Pit brickworks was subsequently replaced by the much larger **South Moor Brickworks**, south of the village of New Kyo, and as quarrying expanded north-eastwards it spread over the route of the original 1839 line from Oxhill. It manufactured only common bricks. Later, confusingly, the old name of **Charley Pit Brickworks** was re-adopted.

By the end of the First World War the higher seams were nearing exhaustion, so the company decided to work the lower seams in this area of the royalty by developing a major new "colliery" on a greenfield site near Annfield Plain 500 yards south of Morrison North and South Pits At first the two new shafts were known as **Morrison East** and **West** (NZ 176508), and sinking began in June 1923. This was completed in 1925, but because of the General Strike of 1926 no coal was drawn until 1927. At first the new site was served by an extension from Morrison South Pit, but as the volume of traffic increased a ¾ mile loop was built east of the Morrison North & South Pits to avoid East and West Pit traffic passing through them. The loco shed for the system was situated alongside the route to Oxhill, the various duties including shunting the collieries and the sidings at Oxhill.

In 1932 a seam of witherite (barium carbonate) was discovered in the Morrison North Pit, thick enough to mine commercially, the only place in the world where this was done other than at Settlingstones Mine in Northumberland, as well as briefly at Craghead Colliery. A new shaft for man-riding was sunk at Morrison North.

From 1943 all coal from the upper seams was drawn via the Louisa Pit. In 1945 Morrison South Pit closed, to be followed in 1946 by the Hedley and William Pits and their endless rope tramway to the Louisa Pit, while Morrison North Pit was now winding only witherite. However, the Shield Row Drift had been re-opened, at a date not known. By this time the name South Moor Colliery had been abandoned; the Louisa Pit had become the **LOUISA COLLIERY**; Morrison North Pit became **MORRISON NORTH COLLIERY**, while the Morrison East and West Pits had become **MORRISON BUSTY COLLIERY**.

The company was always regarded as progressive, constructing for its employees the first purpose-built medical centre in the Durham coalfield, while underground it was the first to install German coal-cutting ploughs, at Morrison Busty Colliery in 1946, and was considering the installation of underground locomotives when the industry was nationalised.

On 1st January 1947 Louisa Colliery (**G199**), Morrison North Colliery (**G200**), the Shield Row Drift,

Morrison Busty Colliery (**G201**) and the Charley Pit brickworks were vested in NCB Northern Division No. 6 Area.

As described above, the company was almost certainly using locomotives from 1869 between Oxhill and Morrison North/South Pits, but no details survive.

Gauge : 4ft 8½in

SOUTH MOOR No.1	0-4-0ST	OC	I'Anson		?	(a)	
	reb	HL		1903			s/s
SOUTH MOOR No.2	0-6-0ST	OC	BH	1034	1892	New	
	reb	HL	3668	1904			(1)
SOUTH MOOR No.3	0-6-0ST	OC	HL	2956	1912	New	(2)
SOUTH MOOR No.4	0-6-0ST	OC	HL	3528	1922	New	(3)
CRAGHEAD	0-6-0ST	OC	BH	971	1890		
	reb	HL		1911	(b)	(2)
HOLMSIDE No.2	0-6-0ST	OC	CF	1204	1901	(c)	(2)
HOLMSIDE	0-6-0ST	OC	RSHN	6943	1938	New	(2)

(a) ex Dowson (no further information).
(b) ex Craghead Colliery, Craghead, by 30/4/1937.
(c) ex Craghead Colliery, Craghead, by 8/1937.

(1) to Sir S.A.Sadler Ltd, Malton Colliery, Lanchester, c/1923.
(2) to NCB No. 6 Area, with colliery, 1/1/1947.
(3) to Craghead Colliery, Craghead, by 1/1/1947.

The locomotive below seems to have had a very varied working life with this company. Given the known history of the colliery, the only place where the locomotive below is likely to have been used when it arrived in 1887 is on the **beehive coke ovens at Morrison North/South Pits**, Annfield Plain. The company is next known to have employed a narrow gauge locomotive to assist in the construction of the collliery village officially called **The Middles**, but by the end of the twentieth century better known as **Bloemfontein**, between the villages of South Moor and Craghead, in 1903-04. Almost certainly it was used to bring bricks to the site from the **Charley Brickworks**. It was next used in the construction of new houses in the village of **Quaking Houses** nearby, and then during the First World War it was brought in to help with a timber-felling project at **Waldridge Fell**, near Chester-le-Street, which produced pit props for the company. Its fate after this is unknown.

74. BH developed two designs of narrow gauge locomotives, one with outside frames, the other with inside frames. One of the former, 3ft 0in gauge BH 567/1880, poses at the Charley Brickworks near Oxhill. For the other, see Sir S.A. Sadler Ltd.

Gauge : 3ft 0in

-	0-4-0ST	OC	BH	567	1880	(a)	s/s after c/1918

(a) ex BH; rebuilt from 2ft 0in gauge; Birmingham Corporation Gas Department, Windsor Street Gas Works, Birmingham, until c3/1887.

HOPPER, RADCLIFFE & CO
by 1870; previously **George Hopper & Sons**
according to *The Engineer* in 1866, p.332, George Hopper & Sons were about to convert their company into The North Eastern Iron & Wagon Co Ltd. This company was indeed registered in 1866, but there is no evidence that it actually traded, nor is it known how long it lasted.
originally **George Hopper**
BRITANNIA IRON WORKS, Colliery Row, near Fence Houses   J202
NZ 324498

George Hopper (c1799-1875) began in business in Houghton-le-Spring, probably in the 1820s, firstly as a timber merchant and then as a brick manufacturer. Then in the early 1850s he decided to develop an ironworks and general engineering business, choosing a site at Colliery Row, some two miles west of Houghton; the works was certainly in existence by 1854. This was served by an extension of what was by that time the Chilton Moor branch of the Londonderry Railway's Rainton & Seaham section. This had once been the Londonderry Railway's through route to staiths on the River Wear at Pensher, as Penshaw was then called, but by the early 1850s its main purpose was to serve the Railway's Chilton Moor Workshops; Hopper's Britannia Works lay at what now became the end of the line, ½ mile further north. One source says that the works had its own link to the NER, at "Black House Bridge", presumably near Fencehouses, but this is not confirmed. It would appear that the firm used "a tankie" to shunt the works, but no details of any locomotive(s) employed are known.

Besides foundry work, which included the casting of the bell 'Big Ben' for the Houses of Parliament in 1856, the works undertook the repair and construction of steam locomotives, possibly about six and all saddle-tanks, one six-coupled and five four-coupled. At least one appears to have been built without a customer's order: in the *Iron & Coal Trade Review* of 24th September 1873 the firm offered for sale a new four-wheeled tank locomotive with 12in x 20in cylinders. The works also manufactured a wide range of railway equipment, including wagons and wagon wheels (four of the firm's wheels can be seen in the chaldron waggon preserved at the entrance to Beamish, The North of England Open Air Museum). On 1st March 1856 the Marchioness of Londonderry entertained 3000 of her workmen here. By 1870 the firm had expanded to become Hopper, Radcliffe & Co, and between December 1870 and 1874 the firm also owned the Seaham Foundry, founded by Robert

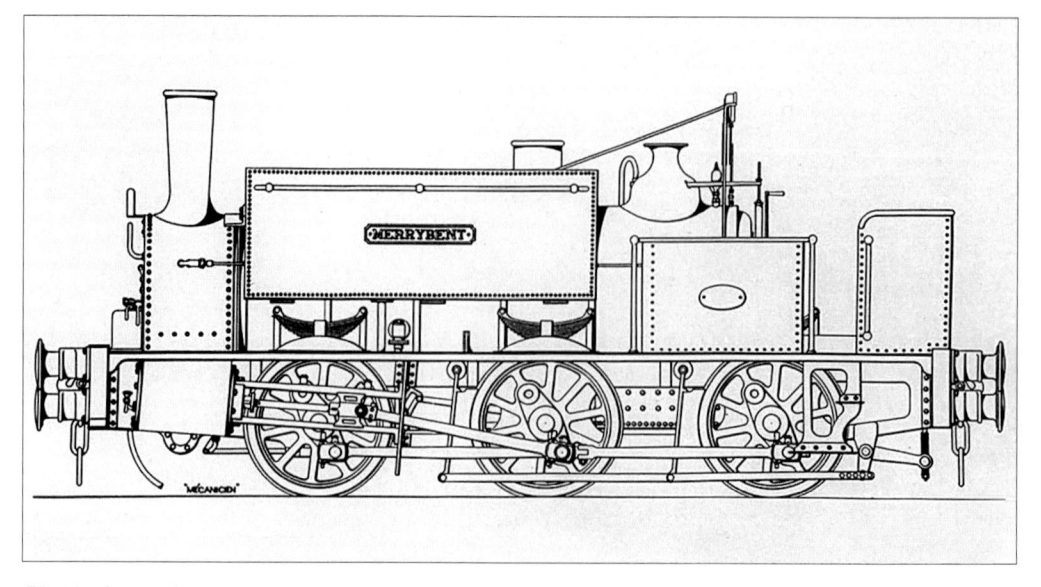

75. The drawing by R.H. Inness ('Mécanicien') of MERRYBENT, the 0-6-0ST built by Hopper, Radcliffe & Co in 1870 for the Merrybent & Darlington Railway – the only known illustration of a Hopper locomotive as built.

Wright in 1843 (the information that it owned the Vane & Seaham Ironworks at Seaham appears to be incorrect).

The firm made heavy losses in 1874, and on 11th November 1874 the Seaham works, "lately occupied by Hopper, Radcliffe & Co", was offered for sale. George Hopper died in September 1875. The Britannia Works was closed, and on 17th July 1876 an advertisement appeared in *Iron & Coal Trades Review* offering for sale plant that included wagons and three tank locomotives, these being "one new loco in course of construction 12in cylinders 4 wheel coupled [whether this was the same locomotive as described in the advertisement of 1873 above is unknown], one secondhand Manning Wardle locomotive 12in cylinders 6 wheel coupled and one secondhand loco 10in cylinders 4 wheel coupled, just extensively overhauled". This does not seem to have produced a sale, for on 14th September 1876 A.T. Crow of Sunderland auctioned the plant on 14th September 1876, the sale including "one tank locomotive by Manning Wardle and two by the owners, one new" (*Colliery Guardian*); the Manning Wardle loco is believed to be that listed below.

Gauge : 4ft 8½ in

MUDLARK	0-6-0ST	IC	MW	171	1865	(a)	(1)
-	0-4-0ST?	OC?	?	?	?	(b)	(2)
-	0-4-0ST	OC?	Hopper Radcliffe		?	(b)	(3)

(a)   ex ?; had been used by Morkhill & Prodham, contractors for the construction of the North Eastern Railway's Blackhill Branch (Blaydon-Consett) between 1865-67.
(b)   included in the auction on 14/9/1876 (see above).
(1)   to ?, presumably on 14/9/1876; the next owner known was Thomas Nelson & Co, contractor for the Bethesda branch for the London & North Western Railway, Bangor, Caernarvonshire, between 1881-84.
(2)   s/s after 14/9/1876.
(3)   unsold on 14/9/1876; offered for sale again on 27/10/1876 (*Iron & Coal Trades Review*); s/s.

Despite this collapse and the auction in 1876, the works was re-opened. However, this was only for a short time, and it has not proved possible to establish who the new owners were. Certainly locomotive repairs were resumed. The business failed again in 1883. This time the works was dismantled by Thomas Bowman of Darlington, who in a sale on 18th and 19th July 1883 offered a "four coupled tank locomotive, 12½ in cylinders, recently repaired". Its identity is unknown.

THE HORDEN COLLIERIES LTD

This company, registered on 30th January 1900, was the last major new colliery company to be formed in the Durham coalfield, other than by amalgamation. It was set up by a very powerful alliance of local iron and coal trade businessmen: Sir Hugh Bell of Bell Brothers Ltd, which owned collieries in Durham and ironworks in Middlesbrough; David Dale, managing director of The Consett Iron Co Ltd; Arthur Pease, of Pease & Partners Ltd and members of the Stobart family, which controlled Henry Stobart & Co Ltd and had other similar interests. The company took over a huge royalty of 28 square miles, largely unworked apart from an area in the west. This area included **Shotton Colliery** (see The Haswell, Shotton & Easington Coal & Coke Co Ltd), **Castle Eden Colliery** (see Castle Eden Coal Co Ltd) and **Hutton Henry** and **South Wingate Collieries** (see Hulam Coal Co Ltd), all at the time closed. The coastal area in the east was untouched, though the sinking of Dawdon Colliery at Seaham (see The Londonderry Collieries Ltd) and Easington Colliery (see The Easington Coal Co Ltd) immediately to the north had both been started in 1899.

The company began at Horden, with
HORDEN COLLIERY & COKE WORKS, Horden   **L203**
                                                                                                                     NZ 442418

The sinking of the North Pit at Horden was begun on 6th November 1900. It was to be served by sidings west of the NER's new line from Seaham to Hart Junction, ¾ mile north of Horden Station. This line was opened for traffic on 1st April 1905, so although coal was reached at Horden on 22nd July 1904 it is unlikely to have begun production until the following year. A new colliery village also had to be built. Subsequently a large coke works and by-products plant was added. The first 60 Koppers regenerative ovens, together with a by-products plant manufacturing tar, gas for industrial use, sulphate of ammonia and crude benzole, were opened about February 1920, a further 60 being added in 1924. The colliery was developed on a huge scale, at one time employing over 6,000 men and being acclaimed as the largest in the world. It, together with the coke works, was vested into NCB Northern Division No. 3 Area on 1st January 1947.

At the same time as sinking Horden Colliery the company set about re-opening :
SHOTTON COLLIERY & COKE WORKS, Shotton Colliery **L204**
NZ 398413

This colliery had been closed since 1877 (see The Haswell, Shotton & Easington Coal & Coke Co Ltd). It was served by a ¾ mile branch to the NER Sunderland & Hartlepool branch, 1¾ miles north of Wellfield Junction. Work to re-open the colliery began in July 1902 and coal was first drawn in 1904, the profits helping to met the major cost of developments at Horden. The oral tradition at the colliery was that it was shunted by shire horses until after the First World War; the stables were still standing in 1970. South of the colliery a large brickworks was developed (see below). At the colliery coke was first made in beehive ovens; but these were closed in the quarter ending March 1915 and replaced by 60 Koppers regenerative ovens, with a by-products plant manufacturing tar, sulphate of ammonia, crude benzole and gas for town and industrial use, which began production in February 1915. The plant suffered periods of closure due to lack of trade. The colliery and coke works were vested in NCB Northern Division No. 3 Area on 1st January 1947.

With these two collieries in production the company turned to the development of what was the southernmost colliery on the Durham coast,
BLACKHALL COLLIERY, Blackhall **L205**
NZ 461396

Sinking began here in October 1909, still being done by men using hammers and hand-held drills to bore the rings of holes used for explosives to blast out the rock. This colliery was also served by sidings adjacent to the NER Seaham-Hart Junction line, east of the line and 1½ miles south of Horden Station. The first coal was wound in 1913. Like the other collieries, it was vested in NCB Northern Division No. 3 Area on 1st January 1947.

In the 1930s the company set about re-opening **CASTLE EDEN COLLIERY** (NZ 437381) **(L206)** to mine coal in the southern area of the royalty. This colliery had been closed since 1894, and was served by sidings immediately west of Hesleden Station on the LNER Sunderland & Hartlepool branch. A great deal of money had been spent, and the colliery was about to begin production when the Second World War broke out. No coal was ever drawn, although the colliery became an important site for the pumping of the underground water that bedevilled all the coastal collieries.

The company moved its locomotives about as required. However, there was normally only one locomotive at Shotton and one at Blackhall.

Gauge : 4ft 8½in

		-	0-4-0ST	OC	BH?	?	?	(a)	s/s
		(HORDEN)	0-6-0ST	OC	AB	1015	1904	New	(1)
	No.1	(HORDEN)	0-6-0ST	OC	P	1310	1914	New	(1)
	No.2	HORDEN	0-6-0ST	OC	HL	3440	1920	New	(1)
	No.3	HORDEN	0-6-0ST	OC	HL	3568	1923	New	(1)
	No.4	HORDEN	0-6-0ST	OC	HL	2737	1907	(b)	(1)
	71524	5	0-6-0ST	IC	RSHN	7178	1944	(c)	(2)
	494		0-6-0T	IC	Dar		1887	(d)	(3)
	No.5	HORDEN	0-6-0T	IC	RSHN	7305	1946	New	(1)

On 1st January 1947 AB 1015/1904 was at Shotton Colliery, P 1310/1914 was at Blackhall Colliery and the remainder were at Horden Colliery. In addition to the locomotives above, a steam crane is known to have worked at Castle Eden Colliery.

(a) origin unknown and identity unconfirmed; it is shown on an early photo of Horden Colliery, and may be a contractor's locomotive.
(b) ex James W. Ellis & Co Ltd, Swalwell, c/1939.
(c) New, 1/1945; owned by Ministry of Fuel & Power.
(d) ex LNER, Darlington Motive Power Depot, Darlington, prior to 3/11/1946; LNER J71 class; used at Horden Colliery.

(1) to NCB No. 3 Area, 1/1/1947.
(2) to Bentinck West Hartley Colliery Ltd, Pegswood Colliery, Pegswood, Northumberland, /1946 (still on loan from Ministry of Fuel & Power).
(3) returned to LNER, Darlington Motive Power Depot, c/1946.

Underground locomotives
In 1944 the company began the development of locomotive haulage underground, firstly at Blackhall and then at Horden. As the workings of the coastal collieries moved further and further out under the

sea the travelling time for the men between the shaft and the faces, included as part of the shift time, lengthened. To tackle this the company drove a new underground roadway in both collieries specifically for man-riding trains, not using the 2ft 0in gauge in use elsewhere in the collieries but the larger gauge of 2ft 8½in. The first two locomotives below were the first diesels to be used underground in Durham.

Gauge : 2ft 8½in

-	0-4-0DMF	HE	2979	1944	New	(1)
-	0-4-0DMF	HE	2980	1944	New	(1)
-	0-4-0DMF	HE	3330	1946	New	(2)

(1) to NCB No. 3 Area, 1/1/1947; at Blackhall Colliery.
(2) to NCB No. 3 Area, 1/1/1947; at Horden Colliery.

SHOTTON BRICKWORKS, Shotton Colliery   **L207**
NZ 397404

This brickworks, which had worked in the 19th century, was re-opened in 1904 and was served by a ¼ mile branch from the line between the colliery and the NER. By the mid 1920s a narrow gauge steam locomotive was being used to pull full tubs of clay from the pit, which lay to the east of the works, up to the kilns for processing, and this seems to have continued until just before the Second World War. It is possible that there may have been more than one locomotive, though not necessarily at the same time, but no details are known.

Further note
In the *Contract Journal* dated 14th March 1906 a Mr. Firth of Shotton Colliery advertised for sale a 2ft 0in gauge locomotive "good as new". Neither this man nor such a locomotive are believed to have had any connection with The Horden Collieries Ltd, and the origin, identity and fate of this locomotive are unknown.

THE HUTTON HENRY COAL CO LTD
The Hutton Henry Coal Co (?) until 1873
HUTTON HENRY COLLIERY, Station Town   **L208**

The sinking of this colliery was begun in 1871, with two shafts, the **Marley** (NZ 414368) and the **Perseverance** (NZ 415366). They were served by a ½ mile branch from the link between Wingate Junction on what was now the NER Ferryhill & Castle Eden branch (formerly the Clarence Railway) and Wingate South Junction on the Stockton & Wellfield branch, the colliery line dividing to serve the two pits. The first coal was drawn on 7th April 1876. A coke works (beehive ovens) was established, and soon 1,000 people were employed. The first locomotive known to have worked at the colliery was JF 3155/1880, below; DCOA Return 102 records no locomotive here in either April 1871 or November 1876.

SOUTH WINGATE COLLIERY (also known as RODRIDGE COLLIERY)
near Hutton Henry   **L209**
formerly **The Hartlepool & Hutton Henry Coke & Firebrick Co** (formed in 1855)   NZ 416348
originally **Seymour & Partners**

The sinking of this colliery, also known as **Hartbushes Colliery** from the nearby Hartbushes Hall, began in 1840. During the sinking it was discovered that the Harvey seam had been washed out and replaced by a thick seam of fireclay, so that a large **brickworks** was immediately established. The colliery was served by a 1½ mile branch from a junction ¼ mile east of Wingate on The Great North of England, Clarence & Hartlepool Junction Railway, giving its access to the Hartlepool Dock & Railway and shipment at Hartlepool. Coal traffic began on 30th September 1843. It was the practice on both railways for the coal owners to run their own trains, but whether the company owned any locomotives for this is unknown. Traffic ceased in June 1845, very probably because of the severe water problems that were to plague the colliery. Latterly the line to the colliery had a short branch near the pit curving away north-eastwards to serve the brickworks, which lay beside what is now the B1280 road. Who worked the traffic, and with what locomotives, is not known.

The new company formed in 1855 re-opened the colliery, but it was closed again in 1857. The northern end of its branch line was soon buried under the expanding village of Station Town.

Despite its previous history, the Hutton Henry company spent a major part of its profits in trying to re-open South Wingate Colliery. The 3rd edition O.S. map of 1897 does not show a railway to the site, so presumably one was not built. In the same year Hutton Henry Colliery flooded and the company was forced into liquidation. The two collieries and their plant, which included three tank locomotives and a 3ft 0in gauge locomotive, almost certainly those below, were auctioned by A.T. and E.A. Crow of Sunderland between 19th and 23rd September 1898.

On 7th October 1898 **The Hulam Coal Co Ltd** was formed to take over the two collieries. Its efforts to re-commence production were unsuccessful, and it went into liquidation on 5th May 1900. Later in 1900 the royalty passed to The Horden Collieries Ltd (which see), but the firm never re-opened this colliery.

It is believed that, subject to the proviso above, all the locomotives below worked at Hutton Henry Colliery.

Gauge : 4ft 8½in

-	0-4-0ST	OC	JF	3155	1880	New	(1)
DUKE	0-4-0T	OC	HG	276	1870	(a)	(2)
-	0-4-0ST	OC	BH	613	1881	(b)	(1)

(a) ex Lloyd & Co, Linthorpe Ironworks, Middlesbrough, Yorkshire (NR).
(b) ex Castle Eden Coal Co Ltd, Castle Eden Colliery (probably acquired in the auction of their plant in April 1894).
(1) almost certainly one of the locomotives offered for sale in the auction of 19-23/9/1898; s/s.
(2) almost certainly one of the locomotives offered for sale in the auction of 19-23/9/1898; to Lingford, Gardiner & Co Ltd, dealers, Bishop Auckland.

HUTTON HENRY COKE OVENS L210

The locomotive below was used on the beehive coke ovens at Hutton Henry Colliery, hauling the tubs used to fill them.

Gauge : 3ft 0in

| - | 0-4-0ST | OC | BH | 559 | 1880 | (a) | (1) |

(a) ex Bolckow, Vaughan & Co Ltd; former location and date of arrival unknown.
(1) almost certainly the 3ft 0in gauge loco offered for sale in the auction of 19-23/9/1898; s/s.

HUWOOD LTD
HANNINGTON WORKS, Swalwell

B211
NZ 204627

This engineering works, begun about 1900, was taken over from James W. Ellis Engineering Ltd (which see), a subsidiary company of Huwood Ltd, on 1st September 1971. It was closed in the following month for re-organisation, and rail traffic was not resumed when the works re-opened in 1972. The works was subsequently closed, and in 2000 the site was re-developed as an extension to the MetroCentre shopping complex.

Gauge : 4ft 8½in

| - | 4wDM | RH | 305323 | 1951 | (a) | (1) |

(a) ex James W. Ellis Engineering Ltd, with site, 1/9/1971.
(1) scrapped on site by J.J. Stanley Ltd, Swalwell, c6/1979.

IBBOTSON & CO
GATESHEAD STEELWORKS, Gateshead

C212
NZ 257636

This steelworks, which actually comprised only rolling mills, with presumably the means of re-heating steel, was served by sidings north of the NER branch to Allhusen's Chemical Works (see Imperial Chemical Industries Ltd). The premises were quite small and appear to have commenced production in 1894. Nothing is known of Mr. Ibbotson; he does not seem to have had any connection with any other local steelworks. However, the works was short-lived, for an advertisement in the *Glasgow Herald* of 5th September 1896 announced an auction of the plant to be held by Fuller Horsey on 24th September 1896, the list including two locomotives and a loco shed.

Curiously, it is clearly the same works which features in a second auction ten years later. An advertisement in *Machinery Market* dated 21st September 1906 announced an auction by Hillman & Douglas on behalf of "Carr Bros, Ibbotson's Old Steel Works, Coulthard's Lane, Gateshead" on 27th September 1906, the plant including a 0-6-0ST Black Hawthorn loco.

The Gateshead directories for the period 1896-1906 consistently make no mention of the steelworks, although in 1905 the firm of R.H. Longbotham & Co Ltd is recorded as operating an engineering business in Coulthard's Lane; so it might have been this firm's plant that was being auctioned. However, there is a dealer's advertisement of 1902 on behalf of "Carr Bros" (see Part 2), which describes this firm as machinery merchants, Darlington; so this firm could have acquired the contract for dismantling the plant; but the Darlington directories for this period do not mention it.

The premises did not survive; by the O.S. map of 1915 they had been replaced by lines of sidings.

IMPERIAL CHEMICAL INDUSTRIES LTD, General Chemicals Group

Imperial Chemical Industries Ltd was registered on 7th December 1926 and began trading on 1st January 1927. It was formed by an amalgamation of Brunner, Mond & Co Ltd, Nobel Industries Ltd, The United Alkali Co Ltd and The British Dyestuffs Corporation, the new firm being the largest manufacturing company in Britain at that time. The company history strongly indicates that the new company took immediate steps to eradicate the identities of its constituent companies. In 1929 it divided its industrial activities into operating units called "Groups", which survived until replaced by "Divisions" in 1944.

ALLHUSEN'S WORKS, Gateshead   **C213**
The United Alkali Co Ltd until 7/12/1926   NZ 2662

This company was formed on 1st November 1890 by the amalgamation of ten different companies, including **The Newcastle Chemical Co Ltd** (see here) and **Charles Tennant & Partners Ltd** (see below) and **The Jarrow Chemical Co Ltd** (which see).
The Newcastle Chemical Co Ltd until 1/11/1890 (set up by C. Allhusen)
C. Allhusen & Son Ltd until 1872
previously **C. Allhusen & Son**
formerly **C. Allhusen & Co**; see also below

This works, on the south bank of the River Tyne east of the river crossing, was begun by **Charles Attwood & Partners** (see The Weardale Steel, Coal & Coke Co Ltd) in 1834 to manufacture soap and alkali. It was not successful, and from 6th January 1840 it was taken over by **Christian Allhusen, Turner & Co**. Allhusen (1806-1890) had come to England from Germany at the age of 25 to trade as a corn merchant. He had no scientific training, but a great flair for developing the inventions of others into practical businesses. On 29th December 1845 Wilton Turner retired from the partnership, and the works was then owned by **Christian Allhusen & Co**. The subsequent changes of title are shown above.

76. SODIUM, RS 2554/1885, with the ogee-shaped saddletank sometimes adopted by the builders in the last twenty years of the nineteenth century. Note the unusual buffers.

The works had no rail connection until January 1850, when the York, Newcastle & Berwick Railway completed the construction of a ½ mile branch to the works from a junction ½ mile east of Gateshead Station, as allowed under its Act of 1847. Since the branch was owned by the YN&BR, and the NER from 1854, it may well be that the NER shunted the works until 1872, when the first known locomotive arrived. However, there were also **two self-acting inclines** on the east side of the works serving quays on the River Tyne.

The works expanded to occupy 137 acres and ¾ mile of river frontage and become one of the major chemical works in the North of England. It was a leading manufacturer of soda ash and crystals, bleaching powder, and later sulphuric and hydrochloric acid. An overhead railway, known as "Penman's gears", was used to transport materials between different sections of the works. After 1890 it expanded into the production of caustic soda and the recovery of pure sulphur. It was the last works in the North-East to manufacture alkali, brine being brought for processing by sea from the company's works at Port Clarence, near Middlesbrough.

After the First World War the works was steadily run down, some of its premises being sub-let for other uses. It would seem that what was left closed either in 1924 or 1926, and although the works passed to ICI, it was never re-started and became derelict. In 1936 Gateshead Borough Council (which see) began a major reclamation scheme. The Gateshead International Stadium now stands on part of the site.

Gauge : 4ft 8½in

In *The Engineer* of 8/12/1871 the firm advertised to hire a four-wheeled tank locomotive for four months, possibly to handle the shunting until the arrival of the first locomotive below. No details are known of whether a locomotive was actually acquired.

		CHEMIST	0-4-0ST	OC	RS	2017	1872	New	s/s
	2	CALCIUM	0-4-0ST	OC	RS	2124	1873	New	(1)
		*	0-6-0ST	OC	RS	2381	1880	New	s/s
		*	0-6-0ST	OC	RS	2554	1885	New	(2)
		CARBON	0-4-0ST	OC	RS	2637	1888	New	(3)
		SULPHUR	0-4-0ST	OC	RS	2668	1889	New	(3)
		SODIUM	0-4-0ST	OC	BH	935	1889	(a)	(4)
		JED	0-4-0ST	OC	P	1468	1917	New	(5)

*   some dispute exists concerning the names carried by these two locomotives. RS 2381 is believed to have been originally ALFRED; it may later have been SODIUM, and latterly it is believed to have run without any name. RS 2554 appears to have been nameless for a time and then been named ALFRED, but by 1902 it was named SODIUM.

(a)   ex South Stockton Iron Co Ltd, South Stockton, Yorkshire (NR), by 3/1909.

(1)   to Billingham Works, Billingham (see Cleveland & North Yorkshire Handbook).
(2)   to Billingham Works, Billingham, 5/1927 (see Cleveland & North Yorkshire Handbook).
(3)   to Castner, Kellner Alkali Co Ltd, Wallsend, Northumberland.
(4)   to The Owners of the Middlesbrough Estate Ltd, Middlesbrough, Yorkshire (NR).
(5)   to Edward Collins & Sons Ltd, Kelvindale Paper Mills, Glasgow.

TENNANT'S WORKS, Hebburn   E214
Charles Tennant & Partners Ltd until 1/11/1890   NZ 301645
Charles Tennant & Partners until 1884(?)
originally **Charles Tennant & Co**

Charles Tennant entered business in Scotland in 1786, and began making soda in 1810. This works was opened in 1865, initially manufacturing alkali by the Leblanc process. It was served by sidings west of the NER Gateshead-South Shields line, ¼ mile west of Hebburn Station. It also had a **self-acting incline** to a wharf on the River Tyne. The works was closed in 1932, but the bauxite section, which lay at the eastern end of the site, was subsequently taken over by International Aluminium Co Ltd (which see).

Gauge : 4ft 8½in

No.1		0-4-0ST	OC	BH	204	1871	New	s/s
No.1		0-4-0ST	OC	MW	881	1884	(a)	s/s
No.2		0-4-0ST	OC	BH	305	1874	New	(1)
	TENNANT'S No.3	0-4-0ST	OC	BH	881	1886	New	
		reb	MW			1911		(1)
No.2		0-4-0ST	OC	P	452	1889	(b)	s/s

(a) ex Isaac Miller, contractor, Carlisle, Cumberland, following Gretna contract for London, Midland & Scottish Railway, /1927.
(b) ex Fleetwood Works, Fleetwood, Lancashire.
(1) to International Aluminium Co Ltd, with the bauxite section of the site, /1932.

77. TENNANT'S No.3, BH 881/1889, a standard small Black Hawthorn design with 10in by 17in cylinders and 2ft 10in wheels and the distinctive Black Hawthorn chimney, tank and cab.

IMPERIAL CHEMICAL INDUSTRIES LTD, Nobel Division
Imperial Chemical Industries, Nobel Group, until 31/12/1943; however, a map of the site dated 4/6/1943 gives the title as I.C.I. (Explosives) Ltd
HASWELL WORKS (formerly TUTHILL FACTORY), Haswell L215
The Northern Sabulite Explosives Co Ltd until 31/12/1942; this firm was owned by **Imperial Chemical Industries Ltd** from the early 1930s NZ 388426

The Northern Sabulite Explosives Co Ltd was formed in 1923, took over the Tuthill Limestone Quarry previously owned by Pease & Partners Ltd (which see) and established a works on the quarry floor to manufacture the explosive sabulite, for use in mines and quarries. It was served by a ½ mile branch from the South Hetton Coal Co Ltd's Pespool branch, later operated by the National Coal Board, which joined the LNER Sunderland-Stockton line one mile south of Haswell Station.

Only the production areas and magazines were in the quarry; the remainder of the premises, including road distribution facilities, were situated at normal ground level. To move 2ft 0in gauge wagons between the two sites **two rope-worked inclines**, known as 'lifts' were installed, one rising 70 feet, the other 30 feet. Photographs of the site also suggest that mixed gauge track ran from the standard gauge reception and departure sheds down on to the quarry floor, although it is difficult to judge from the photographs whether one or both gauges were in use when they were taken. In the Second World War the works was used to manufacture smoke bombs and practice grenades. The production of explosives continued until 1961, when the narrow gauge rail system was closed. The works was used as a storage depot and distribution centre (and re-named **Haswell Depot**), with the standard gauge rail traffic being worked by British Railways until its closure in October 1966. The site was purchased in the same month by Sludge Disposals Ltd, with the disused rail connection being taken out in May 1967. The quarry was taken over by the National Coal Board about 1985 for the disposal of washery waste from the Hawthorn Combined Mine at Murton (see Part 2).

78. No.1, 0-4-0DM Bg 3048/1941, as repainted shortly after the Second World War.

79. Bg 3051/1942, incorporating the petrol engine used latterly in the Broadbent locomotive, and as built except for the addition of handrails. Note the camouflage applied because of the Second World War.

80. *The 30ft lift used between the edge of the quarry and its floor.*

Gauge : 4ft 8½in

			2-2-0PM		Broadbent		1924	New	
		reb	0-4-0PM *		Broadbent		1928		(1)
No.1			0-4-0DM		Bg	3048	1941	New	(2)
-	later No.2		4wPM		Bg	3051	1942	(a)	
		reb	4wDM		Bg		1947		(2)

*   Fitted with a second-hand Austin 20hp car engine, acquired from Pease & Partners Ltd.

(a)   New, but incorporating the Austin engine from the Broadbent loco above.

(1)   its Austin engine was re-used in Bg 3051/1942; remains s/s, /1942.

(2)   to Thos W. Ward Ltd, Haswell Station, for scrap, /1961.

The 2ft 0in gauge system was used between various buildings on the quarry floor, and was operated by hand or by horse until the arrival of the locomotives below.

Gauge : 2ft 0in

1	4wDM	RH	280865	1949	New	(1)
2	4wDM	RH	280866	1949	New	(1)

(1)   to Shevington Works, near Wigan, Lancashire, c/1962.

INKERMAN COLLIERY CO LTD (registered in 1932)
INKERMAN COLLIERY, Tow Law

M216

NZ 114397 (main site)

This was quite a long-established small company, working various drifts in the Tow Law area. Inkerman was an old colliery, once owned by Ferens & Love Ltd (which see). The original colliery was served by sidings north of the NER line from Consett to Tow Law, one mile north of Tow Law Station, and its traffic always appears to have been worked by the main line company. Subsequently a drift, latterly known as **INKERMAN No.1**, was opened near the main colliery, and a second drift, known as **INKERMAN No.2** (NZ 123407) was opened about a mile to the north-east and a narrow gauge tramway, presumably rope-worked, laid to serve it. The colliery became a licensed mine under the National Coal Board. Subsequently opencast workings were developed to the north of the colliery, and the locomotive below was used to haul the skips from here to the main site.

Gauge : 2ft 0in

| | | 4wDM | RH | 226268 | 1944 | (a) | Scr c7/1967 |

(a)   ex Ministry of Supply, Ruddington Depot, Nottinghamshire, c/1962.

THE INTERNATIONAL ALUMINIUM CO LTD
originally **The Bauxite Refining Co Ltd** (registered in 1908); name of company changed in 12/1926
TENNANT'S WORKS, Hebburn   **E217**
                                                                                                                                              NZ 306647

This bauxite works was formerly the eastern section of the premises owned latterly by Imperial Chemical Industries (General Chemicals) Ltd (which see), which had been closed in 1932 apart from this section. It was served by sidings north of the LNER Gateshead-South Shields line, ¼ mile west of Hebburn Station. It was closed in 1942 and the firm itself went into voluntary liquidation on 15th January 1948. The works was acquired by G. Cohen, Sons & Co Ltd in 1946 and was dismantled in 1948.

Gauge : 4ft 8½in

BAUXITE No.1	0-4-0ST	OC	BH	881	1886		
		reb	MW		1911	(a)	(1)
BAUXITE No.2	0-4-0ST	OC	BH	305	1874	(a)	(2)

(a)   ex Imperial Chemical Industries (General Chemicals) Ltd, with site, /1932.
(1)   to G. Cohen, Sons & Co Ltd, /1946 and sold to a firm at Crook who utilised parts for their own purposes; the firm may well be Walton Bros, who owned a sawmill at Crook and who are said to have used a locomotive without its wheels as a stationary boiler until a sale in the early 1960s.
(2)   to G. Cohen, Sons & Co Ltd, /1946; presented to the North East Historical Engineering & Industrial Locomotive Society for preservation, /1946; stored at the National Coal Board's Killingworth loco shed, Northumberland, from 11/1947; presented to the Science Museum, South Kensington, London, 10/1953.

JARROW CHEMICAL CO LTD
Jarrow Chemical Co until 1885

This company was formed by Messrs J. and W. Stevenson, J. Tennant and J.C. Williamson in 1845 to take over the **JARROW CHEMICAL WORKS** (NZ 354655 approx.) **(D218)**. This had been started by the Cookson family in 1822 (also owners of the Tyne Plate Glass Co (which see) in South Shields and the antimony factory in North Shields in Northumberland) to make soda from salt, and then to manufacture alkali using the Leblanc process, but had been closed in 1844 following claims for damages to crops. It lay on the eastern side of Jarrow Slake, just west of Templetown Colliery, and then on the eastern side of Tyne Dock after this was opened in 1859. By 1885 it was served by a high level link from Hilda Junction, ½ mile north-east of Tyne Dock Station on the NER Gateshead-South Shields line and by low level links to the sidings serving the eastern side of Tyne Dock.

In 1858 the firm took over the **FRIAR'S GOOSE CHEMICAL WORKS** (NZ 274633) **(C219)**, on the south bank of the River Tyne between Gateshead and Felling Shore. This works had been built by Anthony Clapham of Newcastle upon Tyne. It began making soap in 1829 and soda two years later. For many years it was served by a link to the waggonway system serving Tyne Main Colliery, one of whose pits lay immediately to the south of the works. This system had no link to the NER, so presumably its traffic was handled on the river. After expanding into the production of alkali, bicarbonate of soda and Epsom salts, and passing through various hands, it was offered for auction on 27th May 1858 and was subsequently bought by the Jarrow company. By 1885 the works was served by a ¼ mile extension of the branch serving Allhusen's Chemical Works (see Imperial Chemical Industries Ltd, General Chemicals Group). For a time the two works employed 1,400 men, making the firm the largest alkali undertaking on the Tyne.

The 1st edition O.S. map shows what appears to be quite an extensive internal tramway system at the Jarrow Works, which may well be where the locomotive below worked; no similar system is shown at the Friar's Goose Works.

The firm was one of those which merged on 1st November 1890 to form The United Alkali Co Ltd. The Jarrow Works was shut down and dismantled almost immediately, though the Friar's Goose Works survived into the mid-1890s.

Reference : *The Old Tyneside Chemical Trade*, University of Newcastle, 1961.

At Jarrow Chemical Works ?

Gauge : 3ft 6in gauge

-	?	?	Lill	1863	(a)	(1)

(a) supplied new; described in the makers' records as a "tram tank" locomotive (probably meaning a side tank locomotive) with 5in x 12in cylinders; almost certainly it had four-coupled wheels.

(1) it is not known whether the locomotive survived to be taken over by The United Alkali Co Ltd on 1/11/1890; s/s.

JAMES JOICEY & CO LTD

James Joicey & Co until 8/1/1886; see also below

James Joicey (1806-1863), the eldest son of a colliery deputy at Backworth in Northumberland, began his adult career by opening a school in his home village. A biography in the *Newcastle Weekly Chronicle* of 19th June 1880 states that he then moved to Durham, where with an engineer called Joseph Smith he was involved in the winning of South Hetton Colliery (1833), the construction of its railway to Seaham, and 'undertook the formation of a new railway to Hartlepool', presumably the Hartlepool Dock & Railway. He then purchased a collier brig to enter the London coal trade, and prospered so well that in either 1837 or 1838 he was able to lease his first colliery, **Oxhill Colliery**, near Annfield Plain, later called **South Tanfield Colliery**. At first this was worked jointly with Smith, but the latter left after nine months. On 1st January 1850 he went into partnership with his brother John (1816-1881), having by then also acquired **Tanfield Lea Colliery**, opened in 1831, **East Tanfield Colliery**, opened in 1844 and **Twizell Colliery**, near West Pelton, also opened in 1844. These collieries were operated under the titles of **The Owners of South Tanfield Colliery**, etc. His youngest brother Edward (1824-1879) joined the partnership on 1st January 1853, and it may be from this date that the business took the title **James Joicey & Co**; it was certainly the title used when the partnership was revised four years later. Meanwhile, in 1849 James and his second brother George (1813-1856) set up the company of **J. & G. Joicey**, initially to operate a foundry in Pottery Bank, Newcastle upon Tyne; but from 1854 the firm entered the manufacture of winding engines and colliery and general engineering equipment, together with an unknown number of locomotives. For many years this firm was controlled by Jacob Joicey (1843-1899), George's eldest son; it was closed down in 1925.

In 1862 James, John and Edward Joicey were joined by William James (1836-1912), James' illegitimate son. In the following year James died, having just seen George's second son James (1846-1936) begin working for the firm as a clerk. In 1867 he became a partner, eventually coming to control the firm and see it converted to a limited company on 8th January 1886. On 3rd July 1893 James Joicey was created a baronet, and was raised to baron on 13th January 1906. On 16th May 1896 he manoeuvred the Earl of Durham into selling to him all the Lambton collieries, railways and other industrial activities (see Lambton Railway), taking possession on 1st July 1896 and setting up a new company, The Lambton Collieries Ltd, to run the business. In 1911 he similarly left Sir Lindsay Wood with no alternative but to sell him The Hetton Coal Co Ltd and The North Hetton Coal Co Ltd (see Hetton Railway), absorbing them to form The Lambton & Hetton Collieries Ltd. In November 1924 James Joicey & Co Ltd was put into voluntary liquidation and its business amalgamated with The Lambton & Hetton Collieries Ltd to form a new company, **The Lambton, Hetton & Joicey Collieries Ltd**, still with Lord Joicey as chairman. This new firm was registered on 26th November 1924, although it did not purchase the assets of James Joicey & Co Ltd, for £293,246, until 18th August 1925. At this date Lambton & Hetton controlled 20,000 acres raising 2,860,000 tons and Joicey 6,000 acres raising 2,060,000 tons, making the new company acclaimed as the largest coal company in the world, with Lord Joicey as the world's most powerful coal owner.

ALMA COLLIERY, Grange Villa   G220
                                                                                               NZ 232515

Although latterly managed as a separate colliery, this was originally the **Alma Pit** of **WEST PELTON COLLIERY**, the other being the **Handenhold Pit** (NZ 233526). This colliery, which had been abandoned in 1802, was leased by the Joiceys on 8th November 1853. It was re-opened in 1855, with the Handenhold Pit following two years later. It was served by a ½ mile branch south from the Pelton Level of the NER Pontop & South Shields Branch, the line running alongside the private railway from Craghead Colliery (see The Holmside & South Moor Collieries Ltd).

Coal production here ceased on 19th March 1921. However, about the end of 1922 the firm opened **TWIZELL BURN DRIFT** (NZ 226515) **(G221)**, which was served by a tramway about ¾ mile long and almost certainly rope-worked, laid from the Alma Pit, whose buildings were kept open to handle the drift's coal. A document in the Durham Record Office (NCB 7/5/118) dated 16th August 1923 shows that the company considered introducing a 2¼ ton battery locomotive underground to work 100 tons per shift over a 1000 yard section, presumably near to the drift entrance. This would have been the first-ever use of a battery locomotive underground in a Durham colliery, but it would appear that after consideration the proposal was dropped. Both the original pit and the Drift passed to The Lambton, Hetton & Joicey Collieries Ltd on 26th November 1924.

The locomotive below is said to have worked here, and the 3rd edition O.S. map shows what appears to be a locomotive shed here. It is also said that shunting was at some point undertaken by the NER. How the coal was being worked when The Lambton, Hetton & Joicey Collieries Ltd took over is not known.

Gauge : 4ft 8½in

ALMA	0-4-0ST	OC	Joicey	?	?	(a)	(1)

(a) origin and date of arrival unknown .

(1) scrapped; one source states that it was cut up at Handen Hold Colliery, Pelton Fell.

An important group of the collieries were served by the
BEAMISH RAILWAY G222

This system can be traced back to a waggonway built in 1763, starting west of the hamlet of Beamish and, unusually for Durham, running west to east to staiths on the River Wear at Fatfield. After various changes it passed into the hands of the Joiceys, probably about 1853. Further changes followed, including the abandonment of the section to the staiths in favour of a link to the NER Pontop & South Shields branch at Durham Turnpike Junction, north of Chester-le-Street.

In its final form it ran from **BEAMISH MARY COLLIERY** (NZ 211536), where the sinking was completed in 1886 (with **BEAMISH AIR COLLIERY**, closed in 1911 and merged with Beamish Mary) for 4¾ miles to Durham Turnpike Junction, serving via short branches at Beamish **EAST STANLEY COLLIERY** (NZ 231529) (sunk by 1833), **BEAMISH SECOND COLLIERY** (NZ 221537) (also known as the **CHOP HILL PIT**), where sinking had begun in 1824, and **BEAMISH ENGINE WORKS** (NZ 222537), which included both repair shops and the locomotive shed built around a courtyard. It would seem that the entire system was operated by rope haulage apart from the ½ mile or so between the end of rope haulage at Pelton bank foot and the junction, which was apparently worked by horses. The arrival of the first locomotives below appears to have replaced the stationary engines at the Air Pit and East Stanley. The stationary engine working the steeply-graded bank down to Pelton may not have been replaced until the 1890s.

It was probably in the 1880s that a link was put in at Pelton with the Pelaw Main Railway, owned by The Birtley Iron Company (which see). This allowed the Iron Company to bring in traffic from the NER for destinations as far north as Birtley, including the independent firms that the Pelaw Main Railway served, although very little coal traffic was dispatched this way.

The Railway and its collieries passed to The Lambton, Hetton & Joicey Collieries Ltd on 26th November 1924.

Reference : *The Private Railways of County Durham*, Colin E. Mountford, Industrial Railway Society, 2004.

Gauge : 4ft 8½in

1	BEAMISH	0-6-0ST	IC	RS	2013	1872	New	(1)
2	STANLEY	0-6-0ST	IC	RS	2014	1872	New	(1)
3	TWIZELL	0-6-0T	IC	RS	2730	1891	New	(1)
	HARPERLEY	0-6-0ST	?	Joicey?	?	?	(a)	(2)
4	LINHOPE	0-6-0T	IC	RS	2822	1895	New	(1)
No.5	MAJOR	0-6-0T	IC	K	4294	1905	New	(1)

(a) ex Tanfield Lea Colliery, 11/1892.

(1) to The Lambton, Hetton & Joicey Collieries Ltd, with the Railway, 26/11/1924.
(2) s/s, although a new 0-4-0ST locomotive called HARPERLEY was acquired in 1894; see below.

EAST TANFIELD COLLIERY, near Tanfield G223
NZ 194552

This colliery was opened in 1844 and was acquired by Joicey between 1850 and 1852. It was served by sidings west of what became the NER Tanfield Branch. It was closed down at the end of 1913.

However, it was re-opened in 1915 by the East Tanfield Colliery Co Ltd (see South Derwent Coal Co Ltd).

The DCOA Return 102 records one locomotive here in April 1871 and again in November 1876, but no details are known. It would appear from the NER Sidings maps of 1894 that the company had running powers between East Tanfield and Tanfield Lea Collieries, although how these were used is not known.

Gauge : 4ft 8½in

| TANFIELD | 0-4-0ST | OC | Joicey | 377 | 1885 | New | (1) |
| HARPERLEY | 0-4-0ST | OC | Joicey | 429 | 1894 | New | Scr |

(1)   to Twizell Colliery, near West Pelton.

HANDEN HOLD COLLIERY, West Pelton
G224
NZ 233526

This was the other pit which formed part of **WEST PELTON COLLIERY** (see Alma Colliery above). It was re-opened in 1857, and was served by ½ mile branch from the Pelton Level on the NER Pontop & South Shields Branch. The last coke ovens to be owned by the company worked here, to the south of the colliery area, and were closed on 6th June 1923. It passed to The Lambton, Hetton & Joicey Collieries Ltd on 26th November 1924.

Once again, it would seem likely that locomotives were used here before the first listed below.

Gauge : 4ft 8½in

| WHITEHALL | 0-4-0ST | OC | AE * | 1387 | 1898 | New | (1) |
| KYO | 0-6-0ST | IC | RS | 2993 | 1901 | New | (1) |

*   over the AE works plate was fixed a second plate, reading Joicey 457/1898.

(1)   to The Lambton, Hetton & Joicey Collieries Ltd, with colliery, 26/11/1924.

SOUTH TANFIELD COLLIERY, near Annfield Plain
G225
The Owners of South Tanfield Colliery, until 1/1/1853?; see also above and below

NZ 179521

This colliery was opened in July 1837, when it was known as **OXHILL COLLIERY**; its name had been changed by 1850. Initially it was leased by Joicey and his partner Joseph Smith (see biographical note above), but whether from the opening of the colliery or from a date in 1838 is uncertain. Smith left the partnership after nine months here. From 1st January 1850, when the partnership of James and John Joicey began, the colliery was operated under the title of **The Owners of South Tanfield Colliery**. This continued until 1853, or possibly 1857. On 27th November 1856 James Joicey purchased the colliery from Thomas Swinburne, the partnership continuing to work it although James was its sole owner.

It was served by sidings south of the NER Pontop & South Shields Branch ¾ mile east of Annfield Plain Station. There were also **beehive coke ovens** and a **brickworks** here, both begun in the mid-1850s. The colliery was closed in November 1914.

There was a locomotive here by November 1896, but its identity is unknown.

TANFIELD LEA COLLIERY, Tanfield Lea
G226
The Owners of Tanfield Lea Colliery, until 1/1/1853?; see above and below

The sinking of the original shaft, the **Bute Pit** (NZ 188544) at this colliery was begun on 3rd March 1830 and was completed on 25th November 1831. This was served by sidings south of the NER Tanfield Branch. The sinking of another shaft, the **New Pit**, 600 yards to the west, was begun on 30th September 1839, and subsequently a third, the **Margaret**, was sunk on the opposite side of the railway. The colliery was eventually concentrated on the Bute and Margaret Pits. It is not clear from whom and when Joicey acquired the colliery, but it was probably in the late 1840s. Under the partnership of James and John between 1850 and 1853 (possibly 1857) it was operated under the title of **The Owners of Tanfield Lea Colliery**. Beehive coke ovens were built here, the last being put out in May 1914.

It would appear from the NER Sidings maps of 1894 that the company had running powers between Tanfield Lea and East Tanfield Collieries, but how these were used is not known. The colliery passed to The Lambton, Hetton & Joicey Collieries Ltd on 26th November 1924.

Gauge : 4ft 8½in

| HARPERLEY | 0-6-0ST | ? | Joicey | ? | ? | (a) | (1) |
| LEIGH | 0-4-0ST | OC | Joicey | ? | ? | (a) | s/s |

EDEN	0-4-0ST	OC	HL	2481	1900	New	(2)

(a) origin and date of arrival unknown.
(1) to Beamish Engine Works, Beamish, 11/1892.
(2) to The Lambton, Hetton & Joicey Collieries Ltd, with colliery, 26/11/1924.

81. LEIGH, an unidentified locomotive built by J & G Joicey & Co, at Tanfield Lea Colliery.

TWIZELL COLLIERY, near West Pelton
The Owners of Twizell Colliery, until 1/1/1853?; see above and below

G227
NZ 223524

The sinking of this colliery began in March 1843 and it was opened in April 1844. Although owned solely by James Joicey it was worked by the partnership, between 1850 and 1853 (possibly 1857) under the title of **The Owners of Twizell Colliery**. The colliery was served by sidings south of the NER Pontop & South Shields Branch at the top of the Eden Incline. It passed to The Lambton, Hetton & Joicey Collieries Ltd on 26th November 1924.

Again, there may well have been other locomotives here over the years besides the one listed below.

Gauge : 4ft 8½in

TANFIELD	0-4-0ST	OC	Joicey	377	1885	(a)	(1)

(a) to East Tanfield Colliery, near Tanfield.
(1) to The Lambton, Hetton & Joicey Collieries Ltd, with the colliery, 26/11/1924 (not confirmed).

James Joicey also owned **TANFIELD MOOR COLLIERY** (NZ 171553) **(G228)** at Whiteley Head (later White-le-Head), which he purchased from John Bowes, Esq., & Partners in September 1850. This colliery, sunk in 1769, was served by sidings south of the NER Tanfield Branch. It retained a Newcomen type pumping engine, installed in 1756, until 1878. It too passed to The Lambton, Hetton & Joicey Collieries Ltd on 26th November 1924.

LAFARGE CEMENT UK
Blue Circle Industries plc; latterly this firm was owned by **Lafarge Aggregates Ltd**, trading under the title of Lafarge Cement UK, which commenced trading here under this name on 28/1/2002; Lafarge itself is a French company.
Blue Circle Industries Ltd until 8/10/1981
Associated Portland Cement Manufacturers Ltd (formed 10/7/1900) until 31/5/1978

WEARDALE WORKS, near Eastgate   **V229**
NY 947385

This cement works, situated alongside the A689 road between Eastgate and Westgate, was opened in 1965, being linked by conveyor to quarries on the moors to the south. It was served by sidings north of the BR Wearhead Branch, ½ mile west of the former Eastgate Station, but latterly forming the end of the branch. Dispatch of cement by rail ceased on 19th March 1993, but the locomotives listed below continued to be used on the internal rail system to shunt wagons kept for storing cement awaiting road transport.

Despite many years of mineral reserves, its owners decided not to make the major investment needed in the light of over-capacity in the industry generally, preferring instead to supply the North-East by bringing cement from elsewhere to the CargoDurham Distribution Centre of The Seaham Harbour Dock Co. The works was closed on 9th August 2002 and demolished during 2004-05. Part of the site is planned to be the western end of the Weardale Railway preservation scheme and part is to be re-developed using sources of alternative energy.

82. 4wDH TH 148V/1965 brings the cement wagons through the loading hopper on 27th October 1971. Like much of Durham's industry, no trace of this works now survives.

Gauge : 4ft 8½in

-	4wDH	TH	148V	1965	(a)	(1)
2 (by 27/8/1991) ELIZABETH	4wDH	RR	10232	1965	(b)	(2)
-	4wDH	RR	10197	1965	(c)	(2)

(a) New, via International Engines Exhibition, Olympia, London, 4/1965.
(b) ex Swanscombe Works, Kent, 25/7/1981.
(c) ex TH, 3/9/1984; previously Resco (Railways) Ltd, dealers, Erith, Kent; Esso Petroleum Co Ltd, Fawley, Hampshire, until c6/1983.

(1) sold for scrap to Booth Roe Metals Ltd, Rotherham, South Yorkshire, 11/5/1990; cut up at Rotherham, 21-22/6/1990.
(2) to Weardale Railway Trust, /2006, for preservation.

THE LAMBTON, HETTON & JOICEY COLLIERIES LTD

This was the largest of the County Durham colliery companies. The core of its activities was formed by the colliery interests of the Earls of Durham, the Lambton family. On 1st July 1896 these were purchased by Sir James Joicey, later Lord Joicey, himself an important colliery owner (see James Joicey & Co Ltd), and converted into a limited company with the title **The Lambton Collieries Ltd**. In

1911 Joicey purchased The Hetton Coal Co Ltd and its subsidiary, The North Hetton Coal Co Ltd (both which see), taking possession on 3rd July 1911 and combining them with The Lambton Collieries Ltd in August 1911 to form **The Lambton & Hetton Collieries Ltd**. On 26th November 1924 he merged his original company with Lambton & Hetton to form **The Lambton, Hetton & Joicey Collieries Ltd**, the name being changed in December 1924. The family retained control until nationalisation, firstly under James' elder son, Arthur (1880-1940), and then his younger son, Hugh (1881-1966), respectively 2nd and 3rd Baron Joicey.

The largest area owned by the company was served by the

LAMBTON RAILWAY H/J/K230

The Lambton family had been prominent in County Durham for centuries. The Railway that came to bear their name traced its origins from three waggonways. The first, the Lambton Waggonway, ran from various pits about three miles north-east of Chester-le-Street on the eastern side of the River Wear northwards to staiths on the river at Pensher (later Penshaw). The Lumley Waggonway served pits around Lumley, about 2½ miles south-east of Chester-le-Street, and joined the Lambton line at Bourn Moor (later Burnmoor). The greatest development came under John George Lambton (1792-1840), created the 1st Earl of Durham in 1833. In 1822 he purchased at auction the adjacent estate of John Nesham, which not only gave him the Newbottle pits, only just over a mile to the east of the original Lambton pits, but much more importantly, included the Newbottle Waggonway, a direct route 5¾ miles long to deep water at Sunderland.

The beginnings of the system and its operation
At this date the working pits served by Lambton's original system were **MORTON WEST PIT**, near Bourn Moor (NZ 318497), **LUMLEY 3RD PIT** (NZ 296503) at the end of the Lumley line and the **7TH PIT** (NZ 307507), while from near the latter a two mile branch served the **8TH PIT** (NZ 308492) and the **GEORGE PIT** (NZ 302486). Newbottle Colliery comprised the **SUCCESS PIT** (NZ 331520), the **MARGARET PIT** (NZ 332519) and the **DOROTHEA PIT** (NZ 335524). On the western side of the River Wear Lambton owned **HARRATON COLLIERY**, whose **5TH PIT** (NZ 291539) was linked to the Beamish Waggonway and its staith at Fatfield, on the opposite bank of the river from Pensher.

Having acquired the Newbottle line, Lambton wasted no time in joining his original system to it by a one mile link between Bourn Moor and Philadelphia, always known as Junction Bank. He also sank new collieries, built new branches, began to develop mining in the Littletown area, east of Durham City and rebuilt and re-aligned the Newbottle route to Sunderland. A survey of the Earl's collieries and their railway in 1835 by the leading viewer, John Buddle (1773-1843) (Northumberland Record Office, 3410/Bud/28), shows that horses were used on a few level sections, all of the rest was worked by rope haulage, including no fewer than sixteen stationary engines. From **LITTLETOWN COLLIERY** (NZ 339435), opened in 1833 or 1834, the **Littletown Engine** worked traffic for ½ mile down to a junction with the Londonderry Railway's line to Pittington Colliery. His traffic then travelled over the Londonderry line for about 6¾ miles to Newbottle, joining the Newbottle line at the top of Junction Bank. At the end of the Lumley branch was now the re-opened **LUMLEY 2ND PIT** (NZ 295504), from which the **Black Row Engine** hauled waggons up to the engine house and then lowered them down to the junction with the Cocken branch. This began at **COCKEN COLLIERY** (NZ 298474), opened in 1823. This was worked by the **Cocken Engine**, which also hauled up fulls from the **CHARLES PIT** (NZ 296481), another part of Cocken Colliery, and also probably from a third link, from **LUMLEY WEST PIT** (NZ 293485). From the Cocken Engine the waggons were hauled up to the **Pea(l) Flatts Engine**, from where the **New Lambton self-acting incline** took them down to the junction with the Lumley branch. The original line to the staiths at Pensher appears to have been worked by horses for most of its length, the **Pensher Engine** taking over for the final ¼ mile and another stationary engine assisting the horses at the staiths themselves.

From the junction of the Lumley and Cocken branches, first the **Bourn Moor Engine** and then the **Junction Bank Engine** hauled the fulls up to the top of Junction Bank, where the line was joined by a one mile long branch from **HOUGHTON COLLIERY** (NZ 338504) opened in 1827 and worked by the **Houghton Engine**, near the pit. Also joining here was traffic from the Success and Margaret Pits and also the Londonderry Railway. The next section, to the foot of Herrington West Bank at Philadelphia, was worked by horses. This served the Dorothea Pit and on the opposite side of the line what was called in 1835 **Philadelphia Yard** but later **Lambton Engine Works**, developed into an extensive central workshops, not just for the railway but the collieries and the later shipping fleet, with ferrous and non-ferrous foundries, rolling mills and boiler shop. The **Herrington Engine** hauled fulls up to the engine house and then lowered them down to the bottom of Herrington East Bank, where the **Hayston Hill Engine**, later the Foxcover Engine, took them to the top of Hasting Hill. Then in succession the **Grindon self-acting incline**, the **Grindon Engine**, the **Arch Engine**, **Ettrick's self-acting incline** (eliminated soon after 1835), the **Glebe Engine** and the short **Low self-acting incline** took fulls down to the Lambton

Staiths at Sunderland. Near the foot of the Glebe Incline the railway served **Bishopwearmouth Ironworks** (NZ 388569 (see The Derwent Iron Co).

The branch from Rainton to Sherburn
The development of further mines to the south soon brought a major extension to the system. The opening of Littletown Colliery was followed by **SHERBURN**, later **SHERBURN HILL COLLIERY** (NZ 336427) in 1835, **SHERBURN HOUSE COLLIERY** (NZ 322415) in 1844 and **SHERBURN COLLIERY** (NZ 315425) in 1854. To avoid this new coal going on to Lord Londonderry's system, a new line was built southwards from Bourn Moor. The extension to Sherburn Colliery was worked by locomotives stationed there, but **a stationary engine at Sherburn Hill** hauled the full waggons up from Sherburn House, and then a **self-acting incline** took them down to a junction, called Bird-in-the-Bush Junction, with what was now the branch from Littletown Colliery. From here it would appear that the **Buddle Engine** hauled them up to the engine house before lowering them down to the bottom of the Belmont Bank. Next the **Belmont Engine** hauled them up to the summit of the line, before the **Belmont self-acting incline** took them down to Rainton, where again locomotives were soon introduced to work them through Fencehouses up to Bourn Moor. There were links with the NER at both Sherburn and Fencehouses. Near the latter **LAMBTON D COLLIERY** (NZ 318508) was re-opened in 1854, and about the same time this was joined by **LADY ANN COLLIERY** (NZ 318507), an extensive **brick and sanitaryware works** being developed here too.

From this new line another branch two miles long was built from Rainton to serve **FRANKLAND COLLIERY** (NZ 296452), opened in 1842. A report in 1871 states that a beam engine that worked the pit also worked the first ¾ mile of the railway. Here, at the top of a **self-acting incline** the line was joined by a 1½ mile branch from **Framwellgate Colliery**, latterly owned, with its branch, by the Marchioness of Londonderry (which see), while at the bottom of the incline a second **self-acting incline** ½ mile long brought fulls down from **BRASSIDE COLLIERY** (NZ 305458), opened in 1846. From the junction the **Beesbanks Engine** hauled them over the River Wear for about ¼ mile, before locomotives, probably replacing horses at first, took them up to Rainton.

Other nineteenth century branches
During the second half of the nineteenth century various minor branches or links were added. On the Pensher line there was a branch serving **Cross Rigg Quarry** (NZ 325539 approx), **Robertson's Slate Yard** and a store for **Tadcaster Brewery**; a ¾ mile branch serving Pensher Quarry (NZ 321517) and a branch which first served the hamlet of Wapping but was subsequently diverted to serve **Bowes House Farm** (NZ 310520), almost certainly to carry 'choppy' from the farm to the collieries for the pit ponies. The line to Frankland Colliery was extended by ¼ mile to serve the colliery village there, and a similar line was put in to serve Sherburn Hill village. Nearby a link served **SHERBURN HILL QUARRY**, adjacent to the colliery. Workshops for these southern collieries were developed at Littletown, and it would seem that similar facilities were developed on the Lumley branch – **"LUMLEY SHOPS"** (NZ 296482) are shown on the 2nd edition O.S. map. This branch also served **TOAD HOLE QUARRY** (NZ 295482) and then ran northwards for a mile to serve **Great Lumley village**. Between 1841 and 1855 a one mile branch served **LUMLEY 9TH COLLIERY** (NZ 292512), and in 1864 a rebuilt **LUMLEY 6TH COLLIERY** was re-connected to the line. At Newbottle a short branch from the top of Junction Bank served the Pensher Foundry, after the Londonderry Railway ceased operating in this area.

The original gauge of the Lambton and Lumley waggonways was 4ft 2in/4ft 3in, and the Newbottle line was converted to this from 4ft 0in after it was acquired. In the early 1840s, when the system totalled 28 miles and used 1100 waggons, it was converted to 'Parliamentary' gauge, and it was almost certainly soon after this that locomotives were first used. Initially they probably just replaced horse working, on the sections indicated above, but gradually they replaced some of the rope haulage too. Then in 1852-53 the York, Newcastle & Berwick Railway opened its Pensher Branch, from Washington to Sunderland. A link from this line to the Lambton Railway was put in just east of Millfield Station and the Earl applied for permission to run his locomotives over the Y&NBR. Twelve years later the North Eastern Railway built its Deptford branch from Pallion Station on the Pensher branch. This transformed the Earl's northern rail operations. He drove a short tunnel from Lambton Staiths to link to the Deptford branch and then obtained running powers over the NER between Pensher and Deptford, and also latterly between Pallion and Hendon to gain access to the River Wear Commissioners' Hudson Dock, which in turn allowed him to close in 1867 the old rope-hauled route between Philadelphia and Sunderland. He subsequently extended these running powers to Harraton Colliery and then to **NORTH BIDDICK COLLIERY** (NZ 307542), which was also served by the NER Pontop & South Shields branch, ¼ mile south of Washington South Junction; this colliery was acquired from Sir George Elliot in 1891. For all of these workings over the NER, tender locomotives were used. The closure of the line between Herrington and Sunderland allowed the opening on the route of **NEW HERRINGTON COLLIERY**, later shortened to **HERRINGTON COLLIERY** (NZ 341533), in 1874.

The final change under the Earl's ownership occurred on what was now called the Littletown & Sherburn branch. Developments and closures, including the lines to Frankland, Brasside and Cocken between 1878 and 1881, had already left no rope haulage north of the bank foot of the Belmont self-acting incline. Then probably in the early 1890s a new link was built between Sherburn Hill and Littletown Collieries. This eliminated rope haulage on this section as far as the bottom of the incline up to the Belmont Engine, and led to a tender locomotive being kept at Sherburn Colliery for the heavy work which the line now presented.

The early years of Joicey's ownership – coke ovens, signalling, locomotives, electricity
Shipments at Pensher ceased soon after 1897. About the same time the Lumley branch was converted to locomotive haulage throughout, and in 1908 a modernised **LUMLEY 3RD PIT** (NZ 295503), also officially known as **LUMLEY NEW WINNINGS**, was opened. The company was never a major contributor to the Durham coking industry, but in 1902 it opened six of Kay's Patent ovens near Lambton D Pit. These were replaced by 30 Semet-Solvay waste heat ovens in 1907, and supplemented by 35 Simplex regenerative ovens in 1918. Although some signalling was in use in the nineteenth century, the first modern signal box was installed at Burnmoor level crossing in 1904, and this was followed by a further six, possibly seven, boxes over the next twenty years. The same year saw the purchase of the first 0-6-2T for the work over the NER, and more were purchased over the next thirty years. In 1905 the company began the construction of **PHILADELPHIA POWER STATION** (NZ 334520), served by a ¼ mile branch near Philadelphia level crossing. Two years later the station was leased to the **Newcastle upon Tyne Electric Supply Co Ltd**, although in 1918 the colliery company bought the first-ever battery electric locomotive to work in the county to shunt the works.

The acquisition of The Hetton Coal Co Ltd and The North Hetton Coal Co Ltd
The 7-mile long **HETTON RAILWAY** (see below) used locomotives at its southern and northern ends and seven rope inclines for the 3¾ miles between them. The North Hetton Colliery branch (see below) ran from a junction with the Hetton Railway southwards to serve the two North Hetton Colliery pits and then on to join the NER Durham & Sunderland branch ¾ mile west of Hetton Station. It was clearly in Joicey's interest to link the two systems in order for them to be operated as one unit. He had already acquired an area once occupied by Lord Londonderry's **NICHOLSONS PIT** (NZ 327484) to develop it as a central waste disposal area, linking it to Fencehouses by a ½ mile branch. In 1916 a new line 1¼ miles long was built between a junction with this branch and a junction with the North Hetton branch, and running powers were obtained over the NER Durham & Sunderland branch between North Hetton and Hetton Colliery Junctions to enable traffic to pass between the two systems at their southern ends. At their northern ends the two railways were joined by simply driving a tunnel through the rock dividing the two sets of staiths. In 1919 a large new locomotive shed was built at Lambton Staiths, allowing the eventual closure of the Hetton Railway's loco shed in Sunderland near Chester Road. New locomotives were soon ordered for the Hetton Colliery shed, and between 1929-31 the company bought five ex-GWR 0-6-2Ts to help with the traffic at the northern end and allow the withdrawal of the remaining Hetton Railway chaldron waggons.

Further changes, 1914-20
With the Hetton company absorbed, Joicey next decided to dispose of the southern collieries, whose distance from Sunderland had always caused problems. Littletown Colliery closed in 1913, leaving the three Sherburn collieries to be sold to the Middlesbrough steel firm, Sir B. Samuelson & Co Ltd, on 1st January 1914. This firm kept only the line between Sherburn Hill and the NER at Sherburn Colliery, although the redundant line from Rainton to Littletown remained until after the First World War. Then in 1920, probably on 1st February, Joicey purchased **SILKSWORTH COLLIERY** (NZ 377541) from Londonderry Collieries Ltd, consolidating it with the Hetton royalties and eventually linking it to the Hetton Railway (see below).

The final period
Lumley New Winnings closed in 1924, to be followed by the Margaret Pit at Newbottle in 1927, North Biddick Colliery in 1931, North Hetton Colliery in 1935, the Philadelphia Generating Station in 1936 and Lumley 6th Brickworks in 1937. In 1936 the former Londonderry **MEADOWS PIT** (NZ 324481) was acquired, and ten years later it was converted for coal-drawing. Although its rolling mills had long gone, Lambton Engine Works was developed into an extensive workshops area, with a large new **WAGON SHOPS** (NZ 337524) built between the Wars. Most of the facilities at **HETTON ENGINE WORKS**, including its **WAGON SHOPS**, were also retained and rebuilt, as were both the Lambton and Hetton Staiths at Sunderland. At Fencehouses all the early by-product coke ovens were replaced between 1935 and 1938 by 60 Collin regenerative ovens.

On 1st January 1947, when the system and its collieries were vested in NCB Northern Division No. 2 Area, the 'main line' ran from **HERRINGTON COLLIERY** to the LNER at Penshaw, with branches serving **LAMBTON ENGINE WORKS** and the adjacent **WAGON WORKS** at Philadelphia, where was also situated the **DOROTHEA COLLIERY**; a branch at Newbottle serving **HOUGHTON COLLIERY**; a long branch from Burnmoor Junction serving **LAMBTON D COLLIERY, LAMBTON COKE WORKS**, the **MEADOWS PIT** and **NICHOLSONS DISPOSAL POINT**, together with a second branch from Burnmoor serving **LUMLEY SIXTH COLLIERY**. The running powers over LNER lines to serve **HARRATON COLLIERY, LAMBTON STAITHS** and **South Dock** (River Wear Commissioners) in Sunderland were also transferred to the NCB.

Reference : *The Private Railways of County Durham*, Colin E. Mountford, Industrial Railway Society, 2004.

Early locomotives

Gauge : 4ft 8½in

PRINCE ALBERT	0-6-0	OC	Hackworth		c1842	New	s/s
EARL GREY	?	?	?	?	?	(a)	s/s
LAMBTON CASTLE	?	?	?	?	?	(a)	s/s
ALBERT EDWARD	0-6-0	?	RWH	308	1839	(b)	s/s
-	0-6-0	IC	?	?	?	(a)	
	reb		TR	213	1852		(1)
-	0-6-0	IC	TR	236	1853	New	(1)
-	0-6-0	IC	TR	251	1854	New	(1)
-	0-6-0ST	IC	H&C	21	1865	New	(2)
DURHAM	0-6-0ST	IC	MW	152	1865	New	(3)
-	0-6-0ST	IC	H&C	76	1866	New	(4)
-	0-6-0	IC	?	?	?	(c)	
	reb 0-6-0ST	IC	H&C	98	1870		(3)

(a) origin and identity unknown.
(b) ex The Wingate Coal Co, Wingate Grange Colliery, Wingate.
(c) origin and identity unknown; H&C records for this order give "converting 17 x 24 into a Tank Engine for Earl of Durham", presumably from a tender engine.
(1) presumably included in numbering scheme below, which see.
(2) presumably included in numbering scheme below; to HCR; rebuilt as HCR 170/1875 and re-sold to John Crossley, Commondale Brickworks, Commondale, Yorkshire (NR), 7/1875.
(3) presumably included in numbering scheme below; s/s.
(4) presumably included in numbering scheme below; to HCR; rebuilt as HCR 171/1875 and re-sold to Stanton Iron Works Co Ltd, Teversal Colliery, Nottinghamshire, 11/1875.

1	0-6-0	IC	H&C	71	1866	New	(1)
2	0-6-0	IC	H&C	72	1866	New	(2)
2	0-6-0ST	IC	RWH	1969	1884	(a)	Scr c/1939
3	0-6-0	IC	BP	550	1865	New	(1)
4	0-6-0	IC	BH	17	1866	(b)	(1)
5	0-6-0	IC	H&C	30	1864	New	Scr c/1909
5	0-6-2T	IC	RS	3377	1909	New	(1)
6	0-6-0	IC	?	?	?	(c)	(1)
7	0-6-0	IC	?	?	?	(c)	(1)
8	0-6-0	IC	?	?	?	(c)	(1)
9	0-6-0	IC	?	?	?	(d)	s/s
9	0-6-0	IC	ED		1877	New	(1)
10	0-6-0	IC	TR?	?	?	(e)	Scr c/1909
10	0-6-2T	IC	RS	3378	1909	New	(1)
11	0-6-0	IC	?	?	?	(f)	(3)
11	0-4-0ST	OC	HC	1412	1920	New	(1)
12	0-6-0	IC	Blair		1865?	(g)	Scr c/1912
12	0-4-0ST	OC	HL	2789	1912	New	(1)
13	0-4-0ST	OC	H&C	79	1866	New	(2)
13	0-4-0ST	OC	HL	3055	1914	New	(1)
14	0-6-0ST	IC	H&C	78	1866	New	(2)
14	0-4-0ST	OC	HL	3056	1914	New	(1)
15	?	?	?	?	?	(h)	s/s
15	0-4-0ST	OC	HCR	169	1875	New	(1)
16	?	?	?	?	?	(h)	s/s
16	0-4-0ST	OC	H&C	96	1870	New	(1)

17		?	?	?	?	?	(h)	s/s
17		0-4-0ST	OC	HCR	130	1873	New	(4)
17		0-4-0ST	OC	MW	2023	1923	New	(1)
18		0-6-0	OC	BH	32	1867	(j)	
		reb 0-6-0ST	OC	?		?		
			reb	HL	1491	1935		(1)
19		0-4-0ST	OC	MW	344	1871	New	(1)
20		0-6-0	IC	RS	2260	1876	New	(1)
21		0-4-0ST	OC	RS	2308	1876	New	(1)
22		0-4-0ST	OC	HC	230	1881	New	(1)
23		0-4-0ST	OC	BH	688	1882	New	(1)
24		0-4-0ST	OC	BH	832	1885	New	(1)
25		0-6-0	IC	ED		1890	New	(1)
26		0-6-0	IC	ED		1894	New	(1)
27		2-4-0	IC	RS	491	1846		
		reb 0-6-0	IC	Ghd		1864		
		reb 0-6-0ST	IC	Ghd		1873	(k)	
		reb 0-6-0T	IC	LEW		1904		(1)
28		0-4-0ST	OC	HL	2530	1902	New	(1)
29		0-6-2T	IC	K	4263	1904	New	(1)
30		0-6-2T	IC	K	4532	1907	New	(1)
31		0-6-2T	IC	K	4533	1907	New	(1)
32		0-4-0ST	OC	HL	2826	1910	New	(1)
33		0-4-0ST	OC	HL	2827	1910	New	(5)
34		0-4-0ST	OC	HL	2954	1912	New	(1)
35		0-4-0ST	OC	HL	3024	1913	New	(1)
36	(HAZARD)	0-4-0ST	OC	P	615	1896	(m)	(1)
37		0-6-0ST	IC	RWH	1430	1868	(n)	(6)
38		0-6-0ST	IC	RWH	1478	1870	(p)	(7)
39		0-6-0ST	IC	RWH	1422	1867		
		reb	HL	2378			(q)	(8)
40	(later No.40)	0-6-0ST	IC	RS	1919	1869	(r)	
		reb	HL	1182		1930		(9)
41		0-6-0T	OC	KS	3074	1917	(s)	(1)
42		0-6-2T	IC	RS	3801	1920	New	(1)
43		0-4-0ST	OC	GR	769	1920	New	(1)
44		0-6-0ST	OC	MW	1934	1917	(t)	(1)
45		0-6-0ST	IC	HL	2932	1912	(u)	(1)
46		0-6-0T	IC	MW	1813	1913	(v)	(1)
47		0-4-0ST	OC	HL	3543	1923	New	(1)
48		0-4-0ST	OC	HL	3544	1923	New	(1)
49		0-4-0ST	OC	MW	2035	1924	New	(1)
50		0-4-0ST	OC	MW	2036	1924	New	(1)
51		4wBE		DK		1918	(w)	Scr c/1937
52		0-6-2T	IC	NR	5408	1899	(x)	(1)
53		0-6-2T	IC	Cdf	302	1894	(y)	(1)
54		0-6-2T	IC	Cdf	311	1897	(z)	(1)
55		0-6-2T	IC	K	3069	1887	(aa)	(1)
56		0-6-2T	IC	K	3580	1894	(ab)	(1)
57		0-6-2T	IC	HL	3834	1934	New	(1)
58		0-6-0ST	IC	VF	5299	1945	(ac)	(1)
59		0-6-0ST	IC	VF	5300	1945	(ad)	(1)

At Vesting Day on 1st January 1947 it is believed that Nos. 11, 12, 13, 14, 32, 34, 48 and 58 were at Hetton loco shed, and Nos. 17, 23, 35, 49, 50, 53, 55 and 56 were at Lambton Staiths loco shed. All of the remainder were allocated to the Lambton loco sheds at Philadelphia except for those working at Harraton and Silksworth Collieries; which these were is not known.

(a) ex Hetton Railway, 2; included in Lambton Railway list, 1919.
(b) BH records give "constructed from material supplied by the Earl of Durham"; R.H. Inness suggested that it might be a rebuild of a RS locomotive.
(c) almost certainly second-hand, origin unknown, in /1864; R.H. Inness suggested that the locomotives were built between 1853 and 1855.

(d) existence assumed, but identity and origin unknown; could have been one of the TR locomotives above.
(e) precise identity and origin unknown; Samuel Tulip, the company's Chief Engineer from 1897 to 1935, interviewed in 1950, said that "some Richardson engines were at work in his day"; photographic evidence suggests that this loco could well have been one of them; one source suggests that it was TR 236/1853.
(f) identity and origin unknown; Samuel Tulip, interviewed in 1950, said that "some Richardson engines were at work in his day", and one source suggests that it was TR 251/1854; however, photographs of the locomotive would appear not to show any known Richardson features.
(g) one source stated that 12 was a Blair loco, while another stated that the loco was built for a Spanish railway, but that it was not delivered, and was acquired by the Earl of Durham c/1868; equally, given Samuel Tulip's remark quoted in (f) and (g), this could be another Richardson engine, perhaps the one rebuilt as TR 213/1852.
(h) the existence of these locomotives is assumed, based on assumptions about the numbering scheme.
(j) BH records give "rebuild of loco for Earl of Durham"; the origin and identity of the original loco are unknown.
(k) ex NER, 8/1898. This loco was built new as a 2-4-0 for the Newcastle & Darlington Railway, No.22. It became NER 30 when the NER was formed in 1854, subsequently being re-numbered 1899 and then 1761. It was rebuilt as an 0-6-0 in 1864 and as an 0-6-0ST in 1873. Built new with 14in x 22in cylinders and 5ft 7¼in wheels, it had 15in x 20in cylinders and 4ft 0½in wheels when purchased.
(m) ex The North Hetton Coal Co Ltd, with North Hetton Colliery, 3/7/1911; it may have run as NORTH HETTON No.1 for a time; re-numbered in 1919.
(n) ex Hetton Railway, 6; re-numbered in 1919.
(p) ex Hetton Railway, 8; re-numbered in 1919.
(q) ex Hetton Railway, 5; re-numbered in 1919.
(r) ex Hetton Railway, 7; re-numbered in 1919.
(s) ex War Office, Railway Operating Department, 606, c/1920.
(t) ex War Office, Inland Waterways & Docks Dept, Sandwich, Kent, 23, c/1920.
(u) ex The Londonderry Collieries Ltd, with Silksworth Colliery, No.1 SILKSWORTH, c2/1920.
(v) ex Ebbw Vale Steel, Iron & Coal Co Ltd, Monmouthshire, (exact location uncertain), 1/1923.
(w) almost certainly New to The Lambton & Hetton Collieries Ltd; purchased to shunt the Philadelphia Power Station; carried No.1; re-numbered in 1925. Note: the "works number" 9537 quoted in Lambton records is erroneous.
(x) ex GWR, 426, withdrawn in 1/1927 and put on Sales List; acquired per R.H. Longbotham & Co Ltd, 4/1929; originally Taff Vale Railway O2 class, 85.
(y) ex GWR, 448, sold from running stock to R.H. Longbotham & Co Ltd, 2/1930, for LHJC; originally Taff Vale Railway, O class, 26.
(z) ex GWR, 475, sold from running stock to R.H. Longbotham & Co Ltd, 2/1930, for LHJC; originally Taff Vale Railway, O1 class, 64.
(aa) ex GWR, 159, sold from running stock to R.H. Longbotham & Co Ltd, 2/1931, for LHJC; originally Cardiff Railway, 28.
(ab) ex GWR, 156, sold from running stock to R.H. Longbotham & Co Ltd, 2/1931, for LHJC; originally Cardiff Railway, 1.
(ac) ex War Department, Longmoor Military Railway, Hampshire, 75309, 4/1946; arrived by 11/5/1946.
(ad) ex War Department, Longmoor Military Railway, Hampshire, 75310, 4/1946; arrived by 11/5/1946.

(1) to NCB No. 2 Area, with Railway, 1/1/1947.
(2) to Sir B. Samuelson & Co Ltd, with the Sherburn collieries, 1/1/1914.
(3) last recorded repair at Lambton Engine Works was 6/1914; s/s.
(4) last recorded repair at Lambton Engine Works was 7/1921; s/s.
(5) loaned to William Doxford & Sons Ltd, Pallion, Sunderland, during 1939-1945 War, and returned; to No. 2 Area, with Railway, 1/1/1947.
(6) to Tanfield Lea Colliery, Tanfield Lea (see below).
(7) to Tanfield Lea Colliery, Tanfield Lea, 5/1938 (see below).
(8) hired to John Bowes & Partners Ltd, Springwell Bank Foot loco shed, Bowes Railway, 18/12/1937; returned, 28/3/1938; loaned to War Department, Central Ordnance Depot, Derby, Derbyshire, 10/1940; overhauled at LMS Derby Works, 1941-1942 (here on 12/5/1941); returned to Lambton Railway, Philadelphia; to NCB No. 2 Area, with Railway, 1/1/1947.
(9) to Handen Hold Colliery, West Pelton, by 4/6/1937 (see below).

HETTON RAILWAY H/J231

When this railway was taken over in 1911, it ran for seven miles between Hetton and Sunderland, serving at its southern end **HETTON COLLIERY, COKE OVENS** (beehive) and **BRICKWORKS, ELEMORE COLLIERY, EPPLETON COLLIERY** and **EPPLETON QUARRY** (limestone). Locomotives worked the first 1½ miles, after which two stationary engines handled traffic to the summit of the line at Warden Law From here four self-acting inclines took fulls down to North Moor, Sunderland, from where locomotives took over to Hetton Staiths on the River Wear. For precise details see the entry for The Hetton Coal Co Ltd.

In 1920 the company took over **SILKSWORTH COLLIERY** (NZ 377541) (see above), which was served by a 2½ mile extension, owned by the company, of the NER branch to Ryhope Colliery. This lay close to the Hetton Railway, but it was not until 1938-39 that the company constructed a ½ mile link between the two. The colliery was then worked by locomotives from Lambton Staiths shed (see above), which also supplied one or more four-coupled locomotives to shunt the colliery itself.

Eppleton Quarry closed in the late 1930s, but all the collieries passed to NCB Northern Division No.2 Area on 1st January 1947.

Reference : *The Private Railways of County Durham*, Colin E. Mountford, Industrial Railway Society, 2004.

The list below shows the locomotives taken over on 3rd July 1911. Those that survived to 1919 were then re-numbered into the Lambton Railway list (see above).

Gauge : 4ft 8½in

1		2-4-0T	OC	RS	1100	1857	(a)	s/s by /1919
2		0-6-0ST	IC	RWH	1969	1884	(a)	(1)
-		0-4-0	VC	Hetton		c1852		
		reb		Hetton		1877	(b)	(2)
4		0-6-0ST	IC	RS	1649	1865	(a)	(3)
5		0-6-0ST	IC	RWH	1422	1867	(a)	(1)
6		0-6-0ST	IC	RWH	1430	1868	(a)	(1)
7		0-6-0ST	IC	RS	1919	1869	(a)	(1)
8		0-6-0ST	IC	RWH	1478	1870	(a)	(1)
	EPPLETON	4wVBT	VC	Hetton		c1900	(c)	Scr c/1914
	LYONS	4wVBT	VC	Hetton		c1900	(c)	Scr c/1914

(a) ex The Hetton Coal Co Ltd, 3/7/1911.
(b) ex The Hetton Coal Co Ltd, 3/7/1911; being used as a stationary boiler in the sawmill at Hetton Shops.
(c) ex The Hetton Coal Co Ltd, 3/7/1911; see the notes on these locomotives under the entry for The Hetton Coal Co Ltd.

(1) re-numbered into the Lambton Railway list, 1919 (see above).
(2) headed the Stockton & Darlington Railway centenary procession in steam, 27/9/1925; presented to York Railway Museum for preservation, 7/1926.
(3) last recorded repair was 6/1912; s/s.

The North Hetton branch

This ran from a junction with the Hetton Railway at Copt Hill southwards to serve the **Hazard Pit** (NZ 340477), from which a short branch served a coal depot, and then the **Moorsley Pit** (NZ 344466), which together comprised **NORTH HETTON COLLIERY**. From the Moorsley Pit the line continued to a junction with the NER Durham & Sunderland branch ¾ mile west of Hetton Station, a total length of 1½ miles. Shunting was done by the one locomotive below, housed at the Hazard Pit. All of this was taken over on 3rd July 1911. Oral tradition stated that this locomotive left in 1914, never to return, after which a Lambton Railway locomotive was allocated there (see *The Private Railways of County Durham*, p.142). The Moorsley Pit was closed in 1915. The branch was used to link the southern ends of the Lambton and Hetton Railways in 1916.

Gauge : 4ft 8½in

HAZARD	0-4-0ST	OC	P	615	1896	(a)	(1)

(a) ex The North Hetton Coal Co Ltd, 3/7/1911.

(1) re-numbered into the Lambton Railway list, 1919 (see above); may have run as NORTH HETTON No.1 for a time before this.

COMPRESSED AIR LOCOMOTIVES FOR USE UNDERGROUND

Very little has been written on the design, construction and use of compressed air locomotives for work underground in Britain, perhaps because their adoption was so limited. What seems clear is that there were probably more of them at the Earl of Durham's collieries than the whole of the rest of Britain combined. There is a report of a compressed air locomotive built in 1850, and another that William Horsley of the Hartley Engine Works at Seaton Sluice in Northumberland built one in 1862, but these were very much isolated examples.

The development of the Earl's fleet was very much a local initiative. They were designed by two men called William Lishman, a mining engineer, and James Young, at the time almost certainly the Manager of the Earl's Lambton Engine Works, Young's enthusiasm no doubt being the factor which persuaded the management of the Earl's collieries to make the investment needed. There is no evidence that they knew of anyone else's work. The locomotives themselves were built nearby at Belmont by The Grange Iron Co Ltd, a general engineering and foundry business established in 1866. This firm occasionally manufactured stationary steam engines and a few steam locomotives, but it was certainly not a specialist locomotive builder.

Fortunately, James Young described some of his work in a paper entitled "Compressed Air Engines" which he presented to the Cleveland Institution of Engineers at Middlesbrough on 25th April 1881. He returned there on 23rd May 1881 for a discussion on it, and both reports were subsequently printed in the Institution's *Transactions* for 1880-81. Even more fortunately, the illustrations which Young showed his audience were also included in the *Transactions*.

Young's first aim was to replace the pit ponies which were used in their thousands to bring the coal from the coal faces to the main roadways. For this he designed a little locomotive unbelievably basic in design. Built to 2ft 0in gauge, it is shown in illustrations Nos. 83 and 84. A rectangular wooden frame with dumb buffers supported an air receiver with a capacity of about 25 cubic feet. The two 3in x 6in cylinders were mounted between the frames, and drove coupled 12in cast iron wheels with a wheelbase of only 17 inches, to enable the locomotive to go round curves of six or seven feet radius. There was no reversing gear, only slip eccentrics worked by hand, no springs and only a simple brake, worked by hand at one end and foot at the other, and the total weight was just under one ton. It could be driven from both ends, but when hauling empty tubs the driver sat in the first tub to work the controls. On gradients of up to 1 in 48 it could haul up to four tons for up to 500 yards on one charge of air, a longer distance if the ground was more level. The larger design (also illustration No. 83) was more sophisticated. This had 4½in x 10in cylinders and 20in steel wheels and a receiver of 56 cubic feet, and could pull 30 tons 500 yards on one charge of air. A strong cab was provided to protect the driver from falls of rock. The engines had to be re-charged 4-5 times an hour.

In his paper Young said that the steam engine on the surface installed to provide the compressed air had 30in x 84in horizontal cylinders and delivered air at 210 lbs p.s.i, and was designed to "serve two collieries close together". However, at the time of speaking only one was using his locomotives, 12-14 of the small ones and one large one. However, earlier in his paper he states that his locomotives had "been in use in some of the Earl of Durham's collieries for the last three years". In the following month's discussion he twice says that his locomotives had been in use for "between three and four years". This clearly indicates that the first were in use in 1878 and may have been introduced in 1877. This would seem to be borne out by a reference to them in the American trade publication, the *Engineering & Mining Journal*, which in January 1878 stated that air-powered locomotives were "being used at the Earl of Durham's colliery at Philadelphia, England, and that such units had been used in collieries on Tyne and Wear rivers." (If this last sentence is correct the authors have found no evidence of them).

The American reference is clearly to Newbottle Colliery, though whether the locomotives were used at the Margaret Pit or the Dorothea Pit or both (see Map J) is unknown. An unidentified English source in 1880 claimed that seven were in use here by that date. Young's reference to 'two collieries close together' is clearly to Lambton D Pit and the Lady Ann Pit at Bournmoor (see Map J), and he infers that the reason for the very larger compressor serving them was that locomotives were soon to be installed in the pit then not currently using them. Taking all this together it would appear to show that there were at least seven locomotives at Newbottle, a further 12-14 small locomotives and one large one at one of the Bournmoor pits and presumably a similar number required for the other, giving between 35 and 40 in all. Young says that he also made a compound engine, but that he had found great difficulty in getting it to run steadily because of its small size. He believed the problem was caused by the "difference of area in the cylinders, for when it got away at speed it rocked tremendously and I was obliged to discontinue its use on that account".

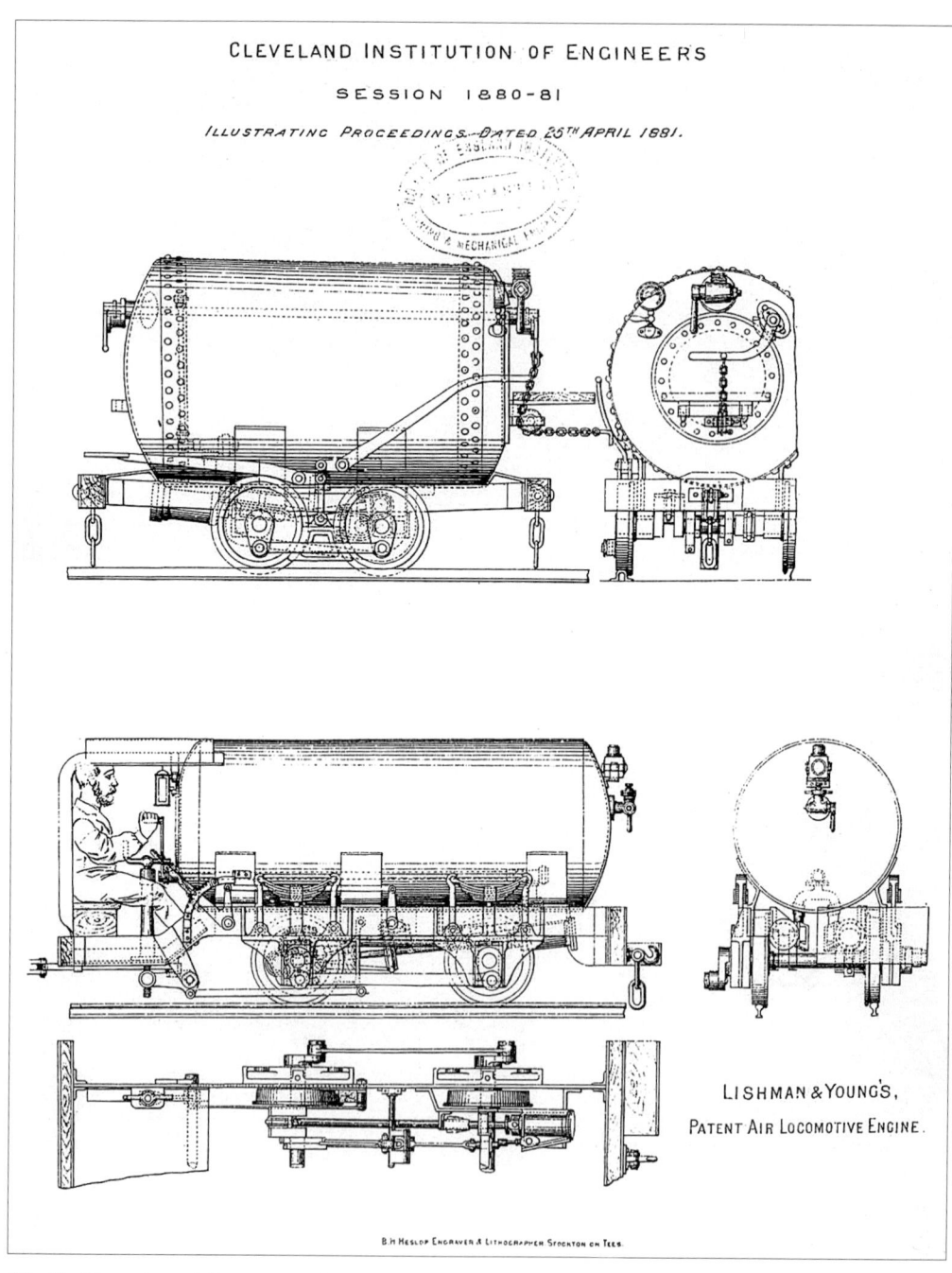

83. The illustrations of his compressed air locomotives which James Young presented to the Cleveland Institution of Engineers on 25th April 1881.

They were still in use at Lambton D Pit in 1887, for on 4th August 1887 a visit was made to see them working by distinguished visitors to the International Exhibition held in Newcastle upon Tyne that summer. In addition another locomotive was exhibited at the Exhibition itself. This was said to have been built to 2ft 9½in gauge, with 4in x 7in cylinders, to have been named JUBILEE and to have been destined for the Earl of Durham's collieries afterwards, though a rather distant photograph of it shows a locomotive more similar to surface outlines than to the earlier illustrations.

GRANGE IRON COMPANY, LIMITED,
FOUNDRY, ENGINEERING & BOILER WORKS, DURHAM,
MANUFACTURERS OF
COLLIERY HEAPSTEADS, SCREENS, AND PULLEYS, PUMPING MACHINERY,

Air-Compressing, Winding, Hauling, and Steam Engines of all kinds
The Grange Patent Automatic Expansion Gear;
Steam Cranes
Gas Apparatus;

Ventilating Fans,
Endless Chain Plant,
Boring Tools
Chemical Works Plant,
Tank Locomotives for Collieries and Iron Works.

Sole Makers of Lishman & Young's Patent Air Locomotive Engines
For Underground Haulage, Fitted with Patent Sparkless Wheels for use in Firey Mines.
EDWARD'S NEW PATENT COAL-WASHING MACHINE,
Continuous in its action and uses little water.
BARTRAM & POWELL'S Patent "COMET" PUMP. No Valves. Slow Speed. Draws 28 feet.

84. *The Grange Iron Co Ltd's advertisement in* Engineering, *7th April 1882. Besides the compressed air locomotive, note the very wide range of other items which the company offered.*

The remains of two of them also happen to be included on the photograph of the interior of the Fitting Shop at Lambton Engine Works taken about 1890-1891, and included in *The Private Railways of County Durham*, photo 121. They aroused some national and indeed international interest, and it is believed that six were sent to California in 1879, the first mines locomotives in the United States of America. Their demise would seem to be linked to the introduction of electrically-driven equipment in mines: the Earl himself had installed electrically-driven endless rope haulage underground at the Margaret Pit by 1892.

THE JOICEY GROUP OF COLLIERIES

This title was used by the company. These collieries were physically separate from the Lambton & Hetton group, as were their railway operations. Engineering work and locomotive repairs were either carried out at Beamish Engine Works or at the collieries. No Joicey locomotives ever came to Lambton Engine Works for repair, though some of the former Hetton Railway locomotives were sent here in their declining years.

The largest component of this group was:

BEAMISH RAILWAY G232
This ran from **BEAMISH MARY COLLIERY** (NZ 211536) to Durham Turnpike Junction on the LNER Pontop & South Shields Branch, a distance of 4¾ miles, serving, via short branches from Beamish, **BEAMISH 2ND**, later **SECOND**, **COLLIERY** (NZ 221537), sometimes called the Chop Hill Pit, **EAST STANLEY COLLIERY** (NZ 218529) and **BEAMISH ENGINE WORKS** (NZ 222537). The loco shed for the Railway formed one side of the Engine Works courtyard. Latterly the Railway also controlled locomotives at Handen Hold Colliery at West Pelton (see below). So far as is known, no locomotives from here ever went to Lambton Engine Works (see above) for repairs.

In 1932 the company began to mine coal in the grounds of Beamish Hall via two drifts at NZ 215552 and NZ 218543, collectively called **BEAMISH PARK DRIFT**. Both they and Beamish Second Colliery were "discontinued" in March 1934, and when they re-opened in 1935 it was as **BEAMISH SECOND & PARK COLLIERY**. The two drifts were linked by a narrow gauge tramway about ¾ mile long, very probably rope-worked, but the system was not linked to the standard gauge system, and how the coal was transported is unclear. Both the colliery and the drifts were closed again in 1939, as was East Stanley Colliery. Beamish Second & Park Colliery was re-opened in 1944, and it, with Beamish Mary Colliery and Beamish Engine Works, were vested in NCB Northern Division No. 5 Area on 1st January 1947.

Reference : *The Private Railways of County Durham*, Colin E. Mountford, Industrial Railway Society, 2004.

Gauge : 4ft 8½in

1	BEAMISH	0-6-0ST	IC	RS	2013	1872	(a)	(1)

	2	STANLEY	0-6-0ST	IC	RS	2014	1872	(a)	(2)
	3	TWIZELL	0-6-0T	IC	RS	2730	1891	(a)	(3)
	4	LINHOPE	0-6-0T	IC	RS	2822	1894	(a)	
				reb	RS		1934		(3)
	No.5	MAJOR	0-6-0T	IC	K	4294	1905	(a)	
				reb	HL	2812	1931		(3)
		TANFIELD	0-4-0ST	OC	Joicey	377	1885	(b)	Scr c/1935

(a) ex James Joicey & Co Ltd, at the amalgamation, 26/11/1924.
(b) ex Twizell Colliery, near West Pelton, by 5/6/1935.

(1) said to have been scrapped c/1939, but the frames were reported to have been seen outside the shed in 7/1946; frames to NCB No. 5 Area, 1/1/1947.
(2) to Handen Hold Colliery, West Pelton, by 29/7/1939.
(3) to NCB No. 5 Area, with the Railway, 1/1/1947.

ALMA COLLIERY, Grange Villa G233
NZ 232515

This colliery, sometimes found on maps as West Pelton Colliery (Alma Pit), was sunk in 1855 but had ceased coal production in March 1921. It was served by a ½ mile branch from the Pelton Level on the LNER Pontop & South Shields branch. However, at the end of 1922 James Joicey & Co Ltd had opened **TWIZELL BURN DRIFT** (NZ 226515) **(G234)**, whose coal was brought by a narrow gauge tramway ¾ mile long to the Alma Pit buildings, which were retained to service the drift. It is believed that the tramway was rope worked. In 1928 The Lambton, Hetton & Joicey Collieries Ltd opened a second drift about 100 yards south of the original one (NZ 226514). In December 1929 the company obtained running powers over the LNER to allow locomotives and wagons to be worked between the Handen Hold Colliery (see below) and the Alma Colliery sidings. After at least two periods of closure Twizell Burn Drift passed to NCB Northern Division No.5 Area on 1st January 1947.

HANDEN HOLD COLLIERY, West Pelton G235
NZ 233526

This colliery, sometimes shown on maps as West Pelton Colliery (Handenhold Pit), was opened in 1857 and served by sidings at the foot of the Eden Incline on the LNER Pontop & South Shields branch. In December 1929 the company obtained running powers over the LNER Pontop & South Shields

85. KYO, RS 2993/1901 and STANLEY, RS 2014/1872, standing out of use at Handen Hold Colliery in the late 1930s. Both carried box-shaped saddletanks and neither ever worked again, although STANLEY survived until about December 1961.

branch to allow locomotives and wagons to be worked between the Handen Hold and Alma Colliery sidings (see above). Photographs show that KYO and STANLEY lay derelict for many years. The colliery passed to NCB Northern Division No.5 Area on 1st January 1947.

Gauge : 4ft 8½in

	WHITEHALL	0-4-0ST	OC	AE *	1387	1898	(a)	Scr /1939
	(KYO)	0-6-0ST	IC	RS	2993	1901	(a)	Scr /1943
40		0-6-0ST	IC	RS	1919	1869		
			reb	HL	1182	1930	(b)	(1)
(2)	(STANLEY)	0-6-0ST	IC	RS	2014	1872	(c)	(2)
38		0-6-0ST	IC	RWH	1478	1870	(d)	(1)

    * over the AE plate was fixed a second plate, reading Joicey 457/1898.

(a)   ex James Joicey & Co Ltd, with amalgamation, 26/11/1924.
(b)   ex Lambton Railway loco sheds, Philadelphia, by 4/6/1937.
(c)   ex Beamish Engine Works loco shed, Beamish Railway, by 29/7/1939.
(d)   ex Tanfield Lea Colliery, Tanfield Lea.

(1)   to NCB No. 5 Area, with colliery, 1/1/1947.
(2)   disused by 29/7/1939; to NCB No.5 Area, with colliery 1/1/1947.

TANFIELD LEA COLLIERY, Tanfield Lea   G236
NZ 188544

This colliery was opened in 1831 and served by sidings south of the LNER Tanfield Branch. It was vested in NCB Northern Division No. 6 Area on 1st January 1947.

Gauge : 4ft 8½in

	EDEN	0-4-0ST	OC	HL	2481	1900	(a)	(1)
37		0-6-0ST	IC	RWH	1430	1868	(b)	(1)
38		0-6-0ST	IC	RWH	1478	1870	(c)	(2)

(a)   ex James Joicey & Co Ltd, with amalgamation, 26/11/1924.
(b)   ex Lambton Railway loco sheds, Philadelphia, after 7/1935.
(c)   ex Lambton Railway loco sheds, Philadelphia, 5/1938.

(1)   to NCB No. 6 Area, with colliery, 1/1/1947.
(2)   to Handen Hold Colliery, West Pelton.

TWIZELL COLLIERY, near West Pelton   G237
NZ 223524

This colliery was opened in 1844 and served by sidings south of the LNER Pontop & South Shields branch at the top of the Eden Incline. It may have had other locomotives besides the one below, or been worked by the NER; the O.S. maps do not show a loco shed. It suffered at least one lengthy closure before it was finally closed in 1938.

Gauge : 4ft 8½in

TANFIELD	0-4-0ST	OC	Joicey	377	1885	(a)	(1)

(a)   ex James Joicey & Co Ltd, with amalgamation, 26/11/1924.

(1)   to Beamish Engine Works loco shed, Beamish Railway, by 5/6/1935.

The company also owned **TANFIELD MOOR COLLIERY** at White-le-Head (NZ 169545) **(G238)**, which was taken over from James Joicey & Co Ltd on 26th November 1924 and was situated at the end of the LNER Tanfield Branch. So far as is known no locomotives ever worked here. It was vested in NCB Northern Division No. 6 Area.

The Earl of Durham and his successors also operated a sizeable fleet of both colliers and tugs, distinguished by their black funnel tops with three red rings. Their mechanical parts, including their engines, were also maintained by Lambton Engine Works.

LANCHESTER & IVESTON COAL CO LTD

This company was formed in 1929 by the amalgamation of two small companies employing only 14 men between them, the Lanchester Coal Co, which was then developing Lanchester Drift, and the Iveston Coal Co, which owned Iveston New Drift. They worked an area of the upper seams within the huge royalty owned by The Consett Iron Co Ltd. Full details of the company's activities would be inappropriate here, except a list of its mines: Lanchester Drift - never went into production; Lanchester No.2 Drift - closed 1945; Lanchester No.3 Drift - closed 1949; Iveston New Drift, near Leadgate - closed 1940; Iveston No.2 Drift - closed 1953, and two mines taken over by the company, Kyo Colliery, near Leadgate and Rumby Hill Colliery, Crook, both closed in 1944. The company's final new mine was

GREENWELL WOOD DRIFT, near Lanchester G239
NZ 168465

Like its other mines that survived into the days of the National Coal Board, this was a licensed mine. It had no main line rail connection. The locomotives below were used both on the surface and underground. The drift closed with the voluntary liquidation of the company in April 1966.

Gauge : 2ft 0in

-	4wDM	F. Blacklock	c1952	(a)	(1)
-	0-4-0DMF	HE	4991	1955	New	(2)
-	0-4-0DMF	HE	4979	1955	(b)	(2)

(a) ex Abbey Wood Coal Co Ltd, Finchale Colliery, near Durham, loan, c/1954.
(b) ex Weardale Lead Co Ltd, Stotsfield Burn Mine, Rookhope, /1961.

(1) returned to Abbey Wood Coal Co Ltd, Finchale Colliery, near Durham, after about a fortnight, c/1954.
(2) to Ayle Colliery Co Ltd, Alston, Cumberland, 5/1966.

86. *Either HE 4979/1955 or HE 4991/1955 with the superstructure hinged back to show the engine, controls and cab well. These small 2ft 0in gauge locomotives had 15hp engines and 1ft 5in wheels.*

LINTHORPE-DINSDALE SMELTING CO LTD
MIDDLETON IRONWORKS, near Middleton St. George S240
NZ 348137

This works had an extremely complicated history. It was opened in 1865 by the **Middleton Iron Co Ltd** and was called the Middleton Iron Works. Two blast furnaces were built. The choice of site was unusual, not being near to supplies of either coal or iron ore, or at a port, or even within a major centre of population. It was served by sidings ¼ mile west of Oak Tree Junction on the NER Fighting Cocks branch (formerly the Stockton & Darlington Railway), four miles east of Darlington (North Road) Station.

It closed down in 1866 and remained idle until 1870, when it was acquired by **George Wythes & Co**. This firm advertised in *The Engineer* of 24th November 1871 for a 12in four-coupled tank locomotive, but nothing is known of any locomotive(s) acquired in this period. This firm had built two more furnaces by 1874, but was then devastated by the economic depression after 1876. It ceased production during 1876 and did not resume until 1881.

At this point the records conflict. *Blast Furnace Statistics* gives The Executors of the late George Wythes as the owners between 1882 and 1891, when they were replaced by Joseph Torbuck, who is listed until 1900. The Statistics also list the furnaces as out of blast between 1884 and 1900. However, local directories at Darlington say that by 1887 the works was open again, now called the Dinsdale Moor Iron Works and owned by William Richards. By 1894 the owners are listed as Richards & Tutt, but soon afterwards ownership reverted to Richards alone. It may be Richards or new owners who set up the **Dinsdale Smelting Co**, which became a limited company in 1900. On 3rd April 1903 this company amalgamated with Edward Williams, the owner of the Linthorpe Iron Works in Middlesbrough, to create the **Linthorpe-Dinsdale Smelting Co Ltd**, which in turn was re-registered on 25th May 1920. By now the works had resumed the use of its original name of Middleton Iron Works. The company is believed to have become a subsidiary of Dorman, Long & Co Ltd in 1927. The Linthorpe Works was closed in May 1930, leaving just the Middleton works in operation.

The firm went into voluntary liquidation in August 1946 and about March 1947 W.H. Arnott, Young & Co Ltd began the dismantling of the works: the blast furnaces were demolished in April 1948. They may well have used the site as a plant yard (see Contractors' Section). The slag heaps, some of which lay south of the Darlington-Saltburn line, were later worked by Durham County Council (which see), following which the main site was used by the British Railways Board as a rail welding yard (which see).

87. *Cochrane & Co Ltd, which owned the Ormesby Ironworks in Middlesbrough, built nine vertical-boilered locomotives for themselves. One of them was sold to the Linthorpe-Dinsdale company, and is seen here on 8th July 1933. Note the unusual wagon on the right.*

Gauge : 4ft 8½in

VICTORIA	0-4-0ST	OC	P	634	1897	(a)	(1)
-	0-4-0ST	OC	MW	756	1880	(b)	(2)
DINSDALE No.1	0-4-0ST	OC	P	845	1900	New	(3)
DINSDALE No.2	0-4-0ST	OC	P	880	1901	New	(4)
DINSDALE No.3	0-4-0ST	OC	P	1058	1906	New	(3)
-	0-4-0VBT	VC	Cochrane		?	(c)	(5)

One source states that the works had two GW/HG locomotives, presumably second-hand, since none came here new. There may also have been locomotives on loan from the Linthorpe Works in Middlesbrough.

(a)   ex H. Lovatt Ltd, contractor.
(b)   ex E. Williams, Linthorpe Iron Works, Middlesbrough, Yorkshire (NR).
(c)   ex Cochrane & Co Ltd, Ormesby Iron Works, Middlesbrough, Yorkshire (NR), by 8/7/1933.

(1)   to Egglescliffe Chemical Co Ltd, Urlay Nook, near Stockton, by 5/8/1936 (see Cleveland & North Yorkshire Handbook).
(2)   to Dorman, Long & Co Ltd, Warrenby Works, Redcar, Yorkshire (NR), /1937.
(3)   to Gjers, Mills & Co Ltd, Ayresome Iron Works, Middlesbrough, Yorkshire (NR), after 25/8/1946, by 14/4/1949.
(4)   to Dorman, Long & Co Ltd, Middlesbrough, Yorkshire (NR), after 10/8/1949; reported as sent to Lackenby Works, but construction of this works did not begin until c/1951.
(5)   sold for scrap, /1947; scrapped c/1950.

THE LINTZ COLLIERY CO
LINTZ COLLIERY, near Burnopfield   **G241**
                                                                              NZ 164561

This colliery and the lines that served it had a complicated history. The colliery was opened in 1855 by two men named McLean and Prior, about whom little is known. A firebrick works was also developed. The two sites were linked by a line 1½ miles long running south-east to join the NER Tanfield Branch at Whiteley Head (later White-le-Head), but with a south-facing connection, so that the coal must have gone down to join the NER Pontop & South Shields branch. The first ¾ mile from the colliery was downhill, and this may have been a self-acting incline. The remainder, steeply uphill, was worked by a **stationary engine** at Whiteley Head. Near the foot of this incline, at Pickering Nook, the line passed through a short tunnel under the Burnopfield to Dipton section of the Pontop & Jarrow Railway of John Bowes, Esq., & Partners, itself also opened in 1855.

In 1867 a link between the two lines was installed at Pickering Nook, the Lintz traffic coming on to the Pontop & Jarrow Railway from March 1867. This made the incline to White-le-Head redundant, and it is possible that P & JR locomotives worked up to the colliery, though there is no way of proving this. In February 1870 the owners failed to sell the colliery to Palmers Shipbuilding & Iron Co Ltd, (which had a link with the P &JR at Jarrow), and within six weeks the Lintz traffic was taken off the P & JR. Instead a new line, comprising a self-acting incline about one mile long, was built down to the NER Consett branch, which had been opened in December 1867, joining the NER about ¼ mile north-east of Lintz Green Station.

By 1871 McLean was the sole owner, though the actual working of the colliery was in the hands of George Gooch, who traded under the name of The Lintz Colliery Company. Encouraged perhaps by the "coal famine" of the early 1870s, the **HIGH** or **SOUTH PIT** (NZ 163557) **(G242)** was sunk between 1871 and 1873 and linked to screens at Lintz Colliery by a 2ft gauge tramway. This was followed in 1874 by a drift known as the **WEST LINTZ** or (from the name of the seam worked) the **BRASS THILL DRIFT** (NZ 159550) **(G243)**. The tramway was extended to serve this, and unusually for the period in Durham, a locomotive was purchased to work it. The two-mile line was quite heavily graded, about half being at 1 in 50 and the remainder between 1 in 64 and 1 in 82, all against the load. Interestingly, DCOA Return 102 lists a stationary engine at Lintz Colliery in both April 1871 and November 1876; what its function was is unknown.

Meanwhile down alongside the NER **South Garesfield Colliery** (NZ 158573) **(G244)** had been opened in 1875 by Thomas Richardson & Co about one mile north-east of Lintz Green Station. This was at first only a small landsale drift, but in the previous year the NER built a "branch" alongside its line to link this and the foot of the self-acting incline to Lintz Green Station, thus preventing the shunting of the coal traffic here from interrupting other traffic on the "main line".

The West Lintz Drift appears to have been closed about 1878, and the remainder of the colliery was closed in 1885, while South Garesfield Colliery closed in 1879/1880. The latter was re-opened in 1887 by **The South Garesfield Colliery Co Ltd** (which see) and by 1891 this company had also taken over the Lintz Colliery royalty.

References: *Industrial Railway Record*, No.15 (1967), pp. 95-99. For a detailed description of BH 258, see *Engineering*, 18th September 1874.

Gauge : 4ft 8½in (worked at Lintz Colliery)

-	0-4-0ST	OC	JF	2849	1876	New	(1)

(1)   offered for sale in *Colliery Guardian*, 8/7/1887; photographs at Beamish, The North of England Open Air Museum, suggest that this loco must have passed with the colliery to The South Garesfield Colliery Co Ltd.

Gauge : 2ft 0in (worked between Lintz Colliery and West Lintz Drift)

| | LINTZ | 0-4-2ST | OC | BH | 258 | 1873 | New | (1) |

(1) offered for sale in *Colliery Guardian*, 8/7/1887; s/s.

LLOYD'S (DARLINGTON) LTD
ALBERT HILL FOUNDRY, DARLINGTON   S245
NZ 267337

This long-established foundry had been owned latterly by **Summerson's Foundries Ltd** (which see). This company, together with the other Summerson companies, went into receivership on 25th October 1967. Sometime late in 1968 or early in 1969 the works was purchased from the receivers by F.H. Lloyd & Co Ltd of Wednesbury in Staffordshire, which set up Lloyd's (Darlington) Ltd to re-open the works. The locomotive below passed to the new owners with the site, though it had not been used since about 1960. The new owners did not re-introduce rail traffic. The works was closed in 1976 and the site was subsequently re-developed.

Gauge : 4ft 8½in

| | - | 4wVBT | VCG | S | 6076 | 1925 | (a) | Scr c2/1970 |

(a) formerly Summerson's Foundries Ltd; to new owners, with site, when purchased from the receivers, c/1968.

LONDON & NORTH EASTERN RAILWAY
Although the vast majority of the company's locomotives were included within its capital stock numbering scheme, the company also inherited from the North Eastern Railway a number of narrow gauge petrol locomotives under a programme of up-grading the handling of materials at its various permanent way works in the North-East. These were not included in the general numbering schemes.

CROFT JUNCTION PERMANENT WAY WORKS, Darlington   S246
North Eastern Railway until 1/1/1923   NZ 294129

This works was situated within the triangle of lines at Croft Junction, south of Darlington (Bank Top) Station, where the Darlington - Saltburn line left the Darlington - London main line.

Gauge : 2ft 0in

| | - | 4wPM | MR | 2103 | 1921 | New | (1) |

(1) to Shanks & McEwan Ltd, contractors, Corby, Northamptonshire.

PARK LANE PERMANENT WAY WORKS, Gateshead   C247
North Eastern Railway until 1/1/1923   NZ 262633

This works was situated at the northern end of the large Park Lane Yard ¼ mile east of Gateshead East Station, bordered by Park Lane and Albany Street.
Gauge : 2ft 0in

| | - | 4wPM | MR | 2104 | 1921 | New | s/s |

WHESSOE LANE PERMANENT WAY SHOPS, Darlington   S248
North Eastern Railway until 1/1/1923   NZ 286164

These shops were situated at the northern end of the North Road Works area, adjacent to the boundary with the Rise Carr South Works of Darlington & Simpson Rolling Mills Ltd (which see).

Gauge : 2ft 0in

| | - | 4wPM | MR | 2077 | 1923 | New | s/s |

LONDON LEAD CO
CORNISH HUSH MINE, Whitfield Brow, near Frosterley   U249
NZ 001335

For most of the nineteenth century lead mining in the Northern Pennines was dominated by W.B. Lead (see The Weardale Lead Co Ltd) and the London Lead Company. This company, established in October 1692 and backed by London entrepreneurs, entered the area about 1750 and established its headquarters at Middleton-in-Teesdale, with much of its mining being undertaken in Teesdale and in

Allendale in Northumberland, centred at Nenthead. About 1861 the company began to develop a new mine it called the Cornish Hush Mine in a remote area of Upper Weardale near the Howden Burn, a tributary of the Bollihope Burn, which joined the River Wear at Frosterley. The ore from the mine was taken down the valley of the burn about a mile to the **WHITFIELD BROW CRUSHING MILL** (NZ 006348) east of the confluence of Howden Burn with Bollihope Burn. At first horses were used between the mine and Whitfield Brow, probably on a tramway; but in 1874 the company decided to replace them with a small locomotive and placed an order with Stephen Lewin of Poole in Dorset. How Lewin, who had not previously built a locomotive, came to be given this order is not known. There has also been confusion because the address for delivery was given as Middleton-in-Teesdale, where, as noted above, its headquarters were situated, together with the firm's main workshops and stores.

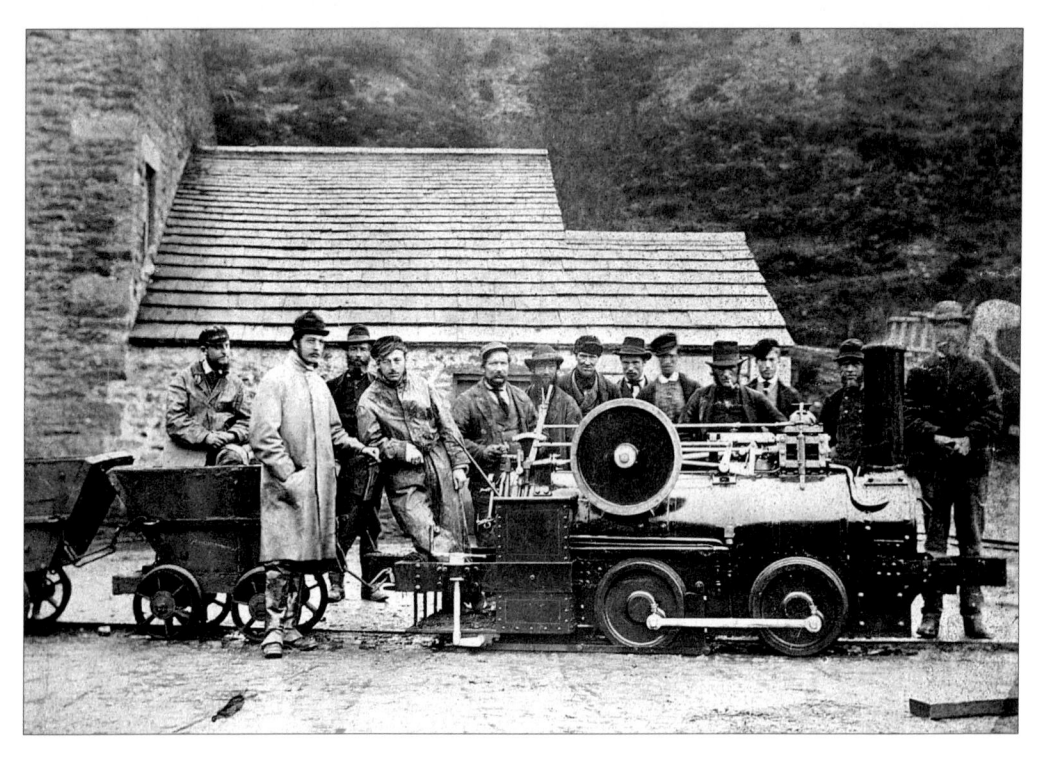

88. *This little 1ft10in gauge loco was the first to be built by Stephen Lewin of Poole in Dorset, and was purchased to haul lead ore from Cornish Hush Mine to the crushing mill at Whitfield Brow, a distance of about a mile. Clearly taken soon after it arrived in 1874, note the valve gear mounted on top of the boiler, the position of the hand brake, the coupling of the train and not least the men.*

The locomotive attracted attention in the technical press of the day, and is discussed in *Stephen Lewin and the Poole Foundry* by Russell Wear and Eric Lees, published jointly by the Industrial Railway Society and the Industrial Locomotive Society in 1978. The tramway also served other mining levels en route, including the Whitfield Brow Lead Mine immediately south of the crushing mill.

The London Lead Company surrendered its leases in 1883, though it is possible that the mine continued to be worked in a small way until about 1900. However, the 2nd ed. O.S. map dated 1898 shows the crushing mill as closed and the tramway partially lifted. The London Lead Company itself was wound up in 1905.

In 1970 Cornish Hush Mine was re-opened to prospect for fluorspar (see Part 2).

Gauge : 1ft 10in

        "SAMSON"   0-4-0WTG OC   Lewin   1874   New   (1)

(1)   a Lewin locomotive, almost certainly this one, was dispatched from the NER's Middleton-in-Teesdale Station in 1904; s/s.

THE MARQUIS OF LONDONDERRY
The 6th Marquis of Londonderry from 5/11/1884
The 5th Marquis of Londonderry until 5/11/1884 (entitled **Earl Vane** until 25/11/1872)
The Marchioness of Londonderry until 20/1/1865
The 3rd Marquis of Londonderry until 6/3/1854

In 1819 Charles Stewart (1788-1854) married Francis Anne Vane Tempest, and so gained control of the pits which formed the Pensher and Rainton Collieries south east of Chester-le-Street, together with the Vane Tempest Waggonway, which took their coal for four miles to Pensher on the River Wear for shipment. He immediately appointed John Buddle (1773-1843), the leading Viewer of his day, to take charge of them. In 1821 he bought the Milbanke estate at Seaham and in 1822 he became the 3rd Marquis of Londonderry. To serve his growing and powerful industrial empire he developed the

LONDONDERRY RAILWAY   H/J/K250
This divided into two sections, known latterly as Rainton & Seaham and Seaham & Sunderland.

The Rainton & Seaham section
This was developed from the Vane Tempest waggonway above. The southernmost extent of this line was **PITTINGTON COLLIERY** (NZ 333443), and a document in the extensive Londonderry Collection in the Durham Record Office (D/Lo/B/306/22) describes the operation of most of it as it was in 1839. The first section, the 1007 yards from the colliery to the bank head of the first incline, was worked by horses; latterly they were replaced by the **Pittington Engine**, built at the colliery using parts of an old locomotive and also working the colliery as well as the railway. The bank head was also the junction of the Earl of Durham's line from Littletown Colliery, whose traffic travelled over the Londonderry line to Newbottle until probably about 1844. The incline itself was a self-acting bank 600 yards long on a gradient of 1 in 26, which took the line down to the road through Pittington village. Here there was a stationary engine called the **Flatts Engine** (NZ 326448), which worked the 1377 yard branch which trailed in from **BROOMSIDE COLLIERY** (NZ 317437), opened about 1835, and also the next section of the main line, 726 yards on a rising gradient of 1 in 576. Both of these gradients were worked by main-and-tail haulage. The full waggons were then hauled up for 858 yards on a gradient of 1 in 32 by the **Pittington Bank**, or **Hindmarch's, Engine** (NZ 328463).

How the next 1½ miles was worked is uncertain. From Hindmarch's Engine a short bank of 484 yards at 1 in 96 took the waggons down to Hetton Lane and the trailing branch from the **ALEXANDRINA**, or **LETCH, PIT** (NZ 334464), opened in 1824, where the **Robney Engine** (NZ 331467) was situated. The 1839 document states that waggons were worked onwards from the junction by a stationary engine, presumably the Robney Engine, which also worked the Letch branch, but other accounts appear to give a different account, not least that horses worked on the branch. Next came **Benrish**, or **Benridge Bank**, a self-acting incline 995 yards long, which dropped the fulls down to Rainton alongside the **MEADOWS PIT** (NZ 324479), also opened in 1824. Here the line was joined by a branch 1333 yards long from the **ADVENTURE PIT** (NZ 315471), opened in 1817, which was worked by a **stationary engine**, almost certainly at the pit end of the line. Next came the **Meadows Engine** (NZ 321486), which worked the next 1920 yards northwards on a descent of 1 in 192, following which the **Dubmire self-acting incline** of 990 yards dropped the fulls down to Sedgeletch. From here the **Jane Pit Engine** (NZ 327516) hauled them up a 1 in 34 incline for 680 yards to Newbottle, where there was a junction with the Lambton Railway (see The Lambton, Hetton & Joicey Collieries Ltd). From here the Londonderry fulls had once gone on to the staiths at Pensher, but these had closed by 1831, and by 1839 the only traffic using the Lambton Railway link besides the Earl's own traffic was that from North Hetton Colliery.

The North Hetton branch
This began as a 1¾ mile branch from Rainton serving the North Pit, which was short-lived, the **DUNWELL PIT** (NZ 338480) and the **HAZARD PIT** (NZ 340477), opened in 1818. These were originally part of Rainton Colliery, but in 1825 Lord Londonderry sold them to a partnership which called itself **The North Hetton Coal Co** and re-named the pits **NORTH HETTON COLLIERY**. The line was extended southwards for ½ mile to serve the new **Moorsley Pit** (NZ 344466), opened in 1828, and by 1831 the Dunwell Pit had been closed. The line was initially worked by horses, but in 1831 a report recommended that a **stationary engine** should be installed at the Hazard Pit, to haul waggons up from the Moorsley Pit and lower them down to the junction with the Londonderry Railway. From here the waggons were transferred on to the Lambton Railway at Newbottle for the coal to be shipped at Sunderland. In 1836 the colliery was sold to a new partnership of Lord Londonderry, the Earl of Durham and The Hetton Coal Co, still trading as The North Hetton Coal Co. In this year the **Rainton Bridge Engine** (NZ 340485) was built in order for North Hetton coal to be shipped at Seaham (see below). In 1843 the branch was re-routed to join the Hetton Railway at Copt Hill, presumably for the

coal to be shipped through the Hetton company's staiths at Sunderland. By the late 1850s the Hetton company was the sole owner (see the entry for The Lambton, Hetton & Joicey Collieries Ltd).

The construction of Seaham Harbour and the railway to it
With the development of new collieries, the output was far beyond the capacity of the staiths at Pensher to handle it. So Londonderry decided to build himself a completely new harbour on the coast at Seaham, 5½ miles south of Sunderland and work began in September 1828. But he could not finance the construction of a railway to it as well, and so in November 1828 he came to an agreement with Mr. Shakespear Reed of Sunderland under which the latter would build and operate the railway, including the provision of waggons, for a period of nine years, subsequently altered to begin on 1st July 1831.

The new line began officially at Rainton Bridge. The link between the old system and the new was built by the Marquis and was presumably worked by what is shown on the 1st edition O.S. map as '**Rainton Old Engine**' (NZ 336486). Three inclines, all worked by stationary engines, then brought the line up to its first summit. The first, worked by the **Rainton Engine** (NZ 351493), was 1271 yards long at a gradient of 1 in 33; the second, worked by the **Copt Hill Engine**, was 715 yards at 1 in 15½; the third was 957 yards at 1 in 120 and was worked by the **Warden Law Engine** (NZ 366498), which also worked the bank on its eastern side, known latterly as the **Long Run**, which was 2970 yards on a falling gradient, again 1 in 120. The first two banks were double line inclines, but between Copt Hill and Warden Law a form of main-and-tail haulage was used, while on the Long Run two sets of 8 waggons were formed into 16 and run alternately.

At the end of the Long Run the line was crossed on the level by the Durham & Sunderland Railway. From here the self-acting **Seaton Incline** took the fulls down for 1177 yards at 1 in 22½, whence the **Londonderry Engine** (NZ 408495) hauled them up a short bank of 330 yards on a gradient of 1 in 36, before lowering them down the Carrhouse Incline, 902 yards at 1 in 40. Next came the self-acting **Seaham Incline**, 1276 yards long, starting at 1 in 30 but flattening to 1 in 52, which brought them down to about ½ mile from the dock. A final **self-acting incline** took the fulls down into the dock, where they were shunted by another **stationary engine**. From Rainton Bridge to Seaham was just under 7¼ miles. The Marquis purchased the line at the end of 1840, and the gauge was converted from 4ft 2in/4ft 3in to 4ft 8½in during May 1843. Two years later the North Hetton Coal Co began the sinking of **Seaton Colliery** (NZ 409496), to the north of the Carrhouse Incline, although coal was not won until 1854. In the meantime the Marquis began the sinking of **SEAHAM COLLIERY**, a few yards to the east, begun in 1849 and won in 1852. There were obvious advantages in combining the two collieries under one ownership, which the Marchioness achieved in 1860, although the Seaton title was not dropped until 1864.

Extensions to this system
Understandably, the Marquis wanted to attract as much coal to his port as possible, and to this end he sought agreements with other coal owners. The biggest of these was **The South Hetton Coal Co** (which see), which opened a line four miles long between **South Hetton Colliery** (NZ 383453) and the dock at Seaham in August 1833. An extension of 1¼ miles served **Haswell Colliery** (NZ 374423), opened in July 1835 and owned by **The Haswell Coal Co**. Some eight years later the South Hetton company opened **Murton Colliery** (NZ 399473), which was served by two branches to its original line. It also built branches from South Hetton to join the Hartlepool Dock & Railway Co (the Pespool branch) and to join the Durham & Sunderland Railway. The Marquis subsequently utilised the first of these from which to build, about 1860, a Londonderry branch 1½ miles long to serve **Shotton Colliery** (NZ 398412), also owned by The Haswell Coal Co. Another complicated arrangement was agreed to serve **Framwellgate Colliery**, which consisted of two pits, the Framwellgate Pit (NZ 271455) and the Cater House Pit (NZ 255454), a mile to the west. Sinking began here in 1838, and The Northern Coal Co built a branch 2½ miles from the Cater House Pit to join the Lambton Railway's Frankland branch. The waggons them travelled for nearly two miles over the Lambton Railway to Rainton, where they passed on to a specially-built Londonderry line, known as the Framwellgate branch, ½ mile long which joined the Seaham line at Rainton Bridge. There were also other developments here. A new **ADVENTURE PIT** was sunk some yards to the south of the old one, and a completely new branch was built to serve it. By 1827 the **CHILTON MOOR WORKSHOPS** (NZ 323493) had been set up north of Rainton alongside the line to Pensher, but latterly at the end of the line. Probably in the early 1850s this line was extended by about ½ mile to serve the **Britannia Iron Works** (NZ 324498) near Fence Houses, established by George Hopper (which see).

The introduction of locomotives
Locomotive working in the Rainton area had superceded horse and rope haulage by 1860. The area stretched to Chilton Moor Shops, where there was almost certainly a **loco shed** and the Britannia Iron

Works to the north, as far east as Rainton Bridge on the Seaham line, to Benridge Bank Foot on the Pittington line and certainly as far as the junction with the Lambton Railway in the west, perhaps right down to that line's Beesbanks Engine.

The Seaham & Sunderland section
With the increasing development of the east Durham coalfield, the Sunderland Dock Company had opened the Hudson Dock at Sunderland in 1850. The Marquis first shipped coal there in 1852, but soon decided that he needed his own rail access to it and that he would operate a public passenger service over his line. The first turf was cut in February 1853, it was formally opened for goods traffic in August 1854 and for passengers in July 1855. It was not constructed under an Act of Parliament, though this was eventually obtained in 1863. In addition to the main line, a second self-acting incline, known as the **Polka Bank**, was constructed from Seaham Colliery to Seaham Harbour. Five passenger stations were eventually provided, **Seaham Harbour** (NZ 424494), later called **Seaham**, and **Hendon Burn** (NZ 409567) initially; **Ryhope** (NZ 414527) in 1858 and **Seaham Colliery** (NZ 421496) the following year, and **Seaham Hall Station** (NZ 414505), opened for the Marquis' private use in 1875. In 1859 a one mile branch was opened to serve **Ryhope Colliery** (NZ 399575), owned by The Ryhope Coal Co Ltd, this branch being extended by 2½ miles (and by-passing Ryhope Colliery) to serve the Marquis' **SILKSWORTH COLLIERY** (NZ 377541), opened in 1873. In 1868 Hendon Burn Station was abandoned in favour of running into the NER's Hendon Station, which was in turn abandoned for Sunderland Central Station when the NER opened this in 1879.

Although the new line was locomotive-worked from its beginning, it opened with very primitive locomotive repair facilities at Polka Bank Foot, Seaham, while all waggon repairs were still handled at Chilton Moor Shops. This situation was not rectified until the mid-1860s, by which time the Londonderrys were also developing their own shipping fleet. Opposite Seaham Station was built **LONDONDERRY ENGINE WORKS** (NZ 423494), which was developed into a fine facility capable of building its own locomotives, although foundry work was handled by **Robert Wright's** works nearby. A **Grease Works**, **Central Stores** and **Granary** were also built here, while the **LONDONDERRY BRICKWORKS** was opened at Seaham Colliery in 1866. Near Polka Bank Foot was constructed the **LONDONDERRY WAGON WORKS** (NZ 428490), which handled a wide range of work besides waggon repairs, including chain and spring making. All these developments brought about the closure of Chilton Moor Shops, probably in 1868.

The mid-nineteenth century saw other major industries developed at Seaham. The first was the **Londonderry Bottle Works**, later the **Seaham Bottle Works**, which opened in 1855 by **John Candlish**, later **Robert Candlish & Son Ltd**. Next came the **VANE & SEAHAM IRON WORKS** (NZ 436478), a mile to the south of the harbour and owned by the Marchioness. This was initially served by a branch from the harbour, but in 1862 a new branch was built from near the junction with the South Hetton line, known as the **Blastfurnace branch**. The works opened in 1863, but was not a success and closed in 1865. It had a faltering history (see Hopper, Ratcliffe & Co) and finally closed in 1883. The next development was the **Seaham Chemical Works** (NZ 434486), latterly owned by **Watson, Kipling & Co Ltd**. This is believed to have closed in 1881. Immediately to the west of the Bottle Works a new **Seaham Gas Works** (NZ 431489) was opened by the **Seaham Gas Company** in 1866. Finally, in the mid-1890s an **electrozone works** (NZ 436483) was opened by **The British Electrozone Corporation** near the end of the Blastfurnace branch. 'Electrozone' comprised sea-water through which an electric current had been passed, for which antiseptic, germicidal and disinfectant properties were claimed. All of these premises were served and shunted by the Londonderry Railway.

Changes to and the closure of the Rainton & Seaham section
Framwellgate Colliery was sold to Stevenson & Jacques of Middlesbrough in 1873, while Shotton Colliery was closed in 1877, resulting in the closure of those branches. Probably in the same year the Warden Law Engine and its two inclines were replaced by locomotive haulage, part of the engine house being converted into a **loco shed**. Broomside Colliery was abandoned in September 1890 and Pittington Colliery closed about the same time, both having previously been sold by the Marquis. This meant the closure of the system south of the Robney Engine. By 1892 the **NICHOLSON'S PIT** (NZ 328483, sunk in 1817) at Rainton had closed, to be followed by the **MEADOWS PIT** in March 1893 and the end of coke making there a month later. In 1892 the gradient on the west side of the Londonderry Engine was eased, allowing locomotive working to be adopted here, probably with a loco housed at Seaham Colliery, and shortly afterwards the Rainton Engine was also abandoned, with the Copt Hill bank being extended half way down the Rainton bank and the remainder handed over to locomotive working. With the closure of the **ADVENTURE** and **ALEXANDRINA PITS** in 1896, the Railway was closed and soon lifted.

It was abundantly clear to the 6th Marquis by the mid-1890s that a major re-organisation of his activities was crucial, both for him personally and for his town of Seaham. To construct the new dock that the port needed meant obtaining an Act of Parliament, and this set up **The Seaham Dock Company** (which see) on 1st January 1899. The two remaining collieries, Seaham and Silksworth, were transferred on 31st May 1899 to **The Londonderry Collieries Ltd**, which immediately began the sinking of Dawdon Colliery on the site of the Vane & Seaham Iron Works. This left the Railway, with its main line to Sunderland and the various branches from it, and after some complicated manoeuvring, this was sold for £400,000 to the North Eastern Railway, which took possession on 6th October 1900. The NER did not take over the lines to Seaham Colliery, to the Dawdon Colliery sinking nor between Ryhope and Silksworth; neither the Engine Works or the Wagon Works were included in the sale, the former soon closing and the latter passing to The Londonderry Collieries Ltd; only thirteen of the locomotives were included and none of the goods and coal wagons.

References
The Private Railways of County Durham, Colin E. Mountford, Industrial Railway Society, 2004.
The Londonderry Railway, George Hardy, ed Charles Lee, Goose & Son, 1973. George Hardy was in charge of locomotive maintenance from 1855 and Railway Manager from 1883, not finally retiring until 1902; a copy of his original, and much longer, draft can be seen in the library of the Regional Resource Centre, Beamish, The North of England Open Air Museum.

Gauge : 4ft 8½in

1	(STOCKTON)		0-4-0	OC	RS	753	1849	(a)	(1)
1			0-6-0	IC	BH	34	1868	New	(2)
2	(CHEAPSIDE)		0-4-2	IC	HF	46	1841	(b)	(3)
		reb			Seaham		1860	(c)	(4)
2			2-4-0T	IC	Seaham		1889	New	(5)
3			0-6-0	IC	TR	254	1855	New	
		reb	0-6-0ST	IC	Seaham		1876		(6)
4	"NORTH BRITISH"		0-6-0	IC	RWH	479	1846	(d)	
		reb	0-6-0ST	IC	Seaham		1875		(7)
5	SEFTON		0-4-2	OC	VF	320	1848	(e)	(8)
5			0-6-0	OC	F&H		1861	New	(9)
5			0-6-0	IC	Seaham		1885	New	(10)
6	(CARADOC)		0-6-0T+t	OC	reb TR	182	c1851	(f)	(11)
6			0-4-0ST	OC	Harris		1867	New	(12)
6			0-6-0ST	IC	Seaham		1883	New	(13)
7			0-6-0	IC	RS	1073	1856	New (g)	
		reb	0-6-0ST	IC	Seaham		?	(h)	(14)
8			0-6-0	IC	RS	1075	1856	New (g)	
		reb	2-4-0T	IC	Seaham		1879		(15)
9			0-6-0	IC	RS	1096	1857	New	
		reb	2-4-0T	IC	Seaham		1880		(16)
-			4-4-0	OC	RS	1206	1859	(j)	(17)
10			0-6-0	IC	RS	1217	1859	New	(18)
11			0-6-0	IC	RS	1326	1860	New	(19)
12			0-6-0	IC	RS	1327	1860	New	
		reb	0-6-0ST	IC	Seaham		1877		(20)
13			0-6-0	IC	RS	1416	1862	New	(21)
14			0-6-0	IC	RS	1417	1862	New	(22)
15			0-6-0	IC	Blair		1868	New	(23)
16			0-4-0VBT	VC	HW	21	1870	New	(24)
17			0-4-0VBT	OC	HW	33	1873	New	(24)
18			0-4-0WT	OC	Lewin	683	1877	New	(24)
19			0-4-0ST	OC	BH	203	1871	(k)	(24)
20			0-6-0	IC	Seaham		1892	New	(25)
21			0-4-4T	IC	Seaham		1895	New	(26)
-			0-6-0T	IC	Seaham		1902	(m)	(27)

Additional notes
According to NER records, a locomotive was hired from the NER in August 1856.

A locomotive is believed to have been hired by the River Wear Commissioners between 1895 and 1899, presumably for working at the South Dock in Sunderland. Whether the same locomotive was used for the whole period, and which locomotive(s) were involved, is not known.

The rebuilds of Nos. 2, 8 and 9, together with new loco 21, were all carried out for the passenger service.

Sales of some of the locomotives were held in December 1896 and March 1897, but some of the sales appear to have fallen through; see details in the disposal footnotes.

(a) ex Forster & Lawton, contractors for the Seaham-Sunderland section, 2/1854; described as a "firetube type with diagonal cylinders".

(b) ex Lancashire & Yorkshire Railway, 139, 5/1854; originally Manchester & Leeds Railway, 35.

(c) "renewal" of 2 Haigh 1841, but actually a new locomotive, including new boiler, motion and tender, and (mentioned by Hardy only in his summary) "new framing".

(d) ex North British Railway, 32, 8/1855.

(e) ex J. Blundell, dealer, Warrington, Lancashire, 10/1855; originally Liverpool, Crosby & Southport Railway, who sold the loco in 1/1850.

(f) ex East Hetton Coal Co, East Hetton Colliery, Kelloe, 9/1855; makers unknown, but believed to have been a York, Newcastle & Berwick Railway engine; described as a tank engine with a tender at the back, a firetube type with diagonal cylinders.

(g) RS 1073 was originally ordered by C.M. Palmer of John Bowes, Esq., & Partners for the Pontop & Jarrow Railway, an order subsequently cancelled in favour of a larger engine, RS 1074. John Bowes, Esq., & Partners also owned as a subsidiary company The Northumberland & Durham Coal Company in London, which operated what when sold in 1859 became the North London Railway. Hardy claimed that RS 1075 was originally ordered for the "North London Railway", a confusion with RS 1073 and also an error.

(h) there is no mention in any of Hardy's writing that this loco was rebuilt as a saddletank; however, when C.H.A. Townley was collecting information about the locomotives at Seaham in the late 1940s he was twice told that 7 had been rebuilt as a saddletank – which would also appear to be confirmed by the description in the sales of 1897.

(j) hired from RS, /1859, whilst waiting for delivery of RS 1217; this loco was ordered by the Ottoman Railway Company of Turkey; Hardy gives Cairo in Egypt as its destination, which could be true as Turkey ruled Egypt at this time.

(k) one version says ex a contractor (unknown) building a dock on the River Tyne, c/1890; another states that it was acquired from a scrapyard at Jarrow; the loco was originally owned by The Felling Coal, Iron & Chemical Co Ltd, Felling.

(m) new locomotive, probably begun during 1900.

(1) frame, cylinders and motion used in the construction of a stationary engine on the Railway at Pittington about 1863; remainder scrapped.

(2) sold to 'The Millfield Grange Co' in 3/1897, but the sale obviously fell through; to NER, 7/10/1900; allocated NER 2267; to Robert Frazer & Sons (Newcastle) Ltd, Hebburn, dealers, 29/3/1901; apparently not re-sold and presumably scrapped.

(3) withdrawn from traffic in 1856 and used in 1857-1858 to drive the machinery in the workshops being developed at Polka Bank Foot, Seaham; this use was subsequently terminated and the locomotive "renewed" (see footnote (c) above).

(4) withdrawn from traffic in 1884; cylinders and motion incorporated into a horizontal engine operating a crane to draw ballast from ships at Seaham Harbour; later this was modified and installed in the Engine Works to drive machinery; the tender was used as a water tank at the Harbour, and later as a tank to store sand; all eventually scrapped.

(5) to NER, 7/10/1900; became NER 1113; withdrawn from traffic, 28/12/1906 and cut up at Percy Main workshops, Percy Main, North Shields, Northumberland, 13/1/1910.

(6) sold to Godfrey & Liddels, London, 3/1897, but the sale obviously fell through; to The Londonderry Collieries Ltd, date unknown.

(7) believed to be sold to The North Walbottle Coal Co Ltd, Northumberland, 3/1897.

(8) dismantled by 1858; scrapped after 27/7/1860.

(9) described as "a firetube type with diagonal cylinders"; withdrawn from traffic by 1873 and parts used to drive a sawmill at Seaham Harbour, with the remainder being scrapped.

(10) to NER, 7/10/1900; became NER 2268; sold, /1902.

(11) almost certainly it was this locomotive ("a six wheeled loco & tender with 14½in x 18in cylinders") which was offered for sale in the *Colliery Guardian* on 25th May 1867; to John Harris, Hopetown Foundry, Darlington, in part payment for new 0-4-0ST built by him and delivered c9/1867.

(12) scrapped by /1883.

(13) to Sir B. Samuelson & Co Ltd, Newport Ironworks, Middlesbrough, Yorkshire (NR), either in 12/1896 or 3/1897.

(14) sold to J.H. Denton, Middlesbrough, Yorkshire (NR), 3/1897, but the sale obviously fell through; to Sir B. Samuelson & Co Ltd, Newport Ironworks, Middlesbrough, Yorkshire (NR), c/1901.

(15) to NER, 7/10/1900; became NER 2269, and worked at Scarborough, Yorkshire (ER); replaced, /1920.
(16) to NER, 7/10/1900; became NER 2270; withdrawn from traffic, /1902.
(17) returned to RS; RS 1217 was delivered new in 11/1859.
(18) to NER, 7/10/1900; allocated NER 2271; to Robert Frazer & Sons (Newcastle) Ltd, Hebburn, dealers, 29/3/1901; apparently not re-sold and presumably scrapped.
(19) to NER, 7/10/1900; allocated NER 2272; withdrawn from traffic, /1902.
(20) to NER, 7/10/1900; allocated NER 2273; to Robert Frazer & Sons (Newcastle) Ltd, Hebburn, dealers, 29/3/1901; to The Seaton Burn Coal Co Ltd, Seaton Burn, Northumberland, almost immediately; scrapped, /1923.
(21) to NER, 7/10/1900; allocated NER 2274; replaced, /1920.
(22) to NER, 7/10/1900; allocated NER 2275; withdrawn from traffic, /1902.
(23) to NER, 7/10/1900; allocated NER 2276; withdrawn from traffic, /1902.
(24) to The Seaham Harbour Dock Co, 1/1/1899.
(25) to NER, 7/10/1900; became NER 1335; withdrawn from traffic, 30/6/1906; to The Seaton Delaval Coal Co Ltd, Seaton Delaval, Northumberland, 4/1907.
(26) to NER, 7/10/1900; became NER 1712; withdrawn from traffic, 21/12/1906; to Isle of Wight Central Railway, Isle of Wight, 8/6/1909, where it became WCR No.2.
(27) remained unused and the personal property of the Marquis of Londonderry until sold to The Seaham Harbour Dock Co, Seaham, 1/1906.

The Marquis also owned **BELMONT COLLIERY** (NZ 319452) **(K251)**, opened in 1835; **GRANGE COLLIERY** near Belmont (NZ 301447) **(K252)**, where sinking began in 1844; **OLD DURHAM COLLIERY** (NZ 292415) **(K253)**, east of Durham City, opened in December 1849, and **KEPIER GRANGE COLLIERY** (NZ 299442) **(K254)**, north-west of Carrville. All of these were officially abandoned in December 1885.

THE LONDONDERRY COLLIERIES LTD

This company was registered on 31st May 1899 by the 6th Marquis of Londonderry (which see), who was re-organising the management and financing of his industrial interests at this time. The Marquis was the principal shareholder and he and his successors held the chairmanship until nationalisation in 1947. The company took over the Marquis' two collieries at that time, Seaham and Silksworth. The company began the development of the new Dawdon Colliery immediately, where sinking started on 26th August 1899. Although The Seaham Harbour Dock Company owned the branch between the harbour and the colliery, the Marquis continued to own personally all the railway repair facilities and rolling stock, the company paying him for their services. This continued until 30th December 1911, when the **LONDONDERRY ENGINE WORKS** (NZ 423494) **(J255)** (valued at £7,500), the **LONDONDERRY WAGON WORKS** (NZ 428490) (see below) (£11,442) and 2,426 wagons, 1,965 of them 4-ton black chaldron waggons, (£41,057) were acquired by the company, being paid for by a new share issue to the Marquis. According to the Seaham directories, the Engine Works was owned by The Seaham Harbour Engine Works Co Ltd; but the Annual Reports of The Londonderry Collieries Ltd make it clear that it was owned by the colliery company; perhaps The Seaham Harbour Engine Works Co Ltd was a subsidiary company. The works was closed in the autumn of 1925.

In 1920 the company sold Silksworth Colliery to The Lambton & Hetton Collieries Ltd, but in 1926 it began the sinking of a new colliery to the north of Seaham, Vane Tempest Colliery. The three collieries, all of which developed workings out under the North Sea, the Londonderry Brickworks and the Wagon Works were all vested into NCB Northern Division No. 2 Area on 1st January 1947.

SEAHAM COLLIERY, New Seaham J256
NZ 412496

This colliery was a combination of **Seaton Colliery**, the sinking of which began in 1845 and later known as the **High Pit**, and **Seaham Colliery** a few yards to the east, begun in 1849 and later known as the **Low Pit**. Both the High and Low Pits had their own links to parallel bank heads, the westernmost being that of the Seaham Incline, a self-acting incline about ¾ mile long down to Seaham Docks, owned by the NER, which joined the Seaham-Sunderland line just north of Seaham Harbour Station, and the Polka Bank, owned by The Seaham Harbour Dock Company, a second self-acting incline alongside, which ran down to Seaham Docks. Empty waggons were worked into the two pits by a stationary engine at the Low Pit, with a separate drum and rope for each pit. The rope run between the Low Pit and Polka Bank Head was straightforward, but to reach the High Pit the rope had to be carried over the houses and the post office (according to one authoritative account) and then round a return wheel. Both ropes used a box van, known to the men as a "dilly", which ran by gravity down to the bank head, dragging the rope behind it, in order for the empties to be coupled on. Once at the pit the empties were run through the screens by gravity, and then run similarly down to the bank head. Between the two pits lay the **LONDONDERRY BRICKWORKS**, established in 1866 to manufacture common bricks.

At some date before the First World War this system was altered. Instead of running High Pit coal down to the bank head, its tubs were worked via a gantry to the Low Pit and screened there (so all coal was now being dispatched from the Low Pit site). This allowed the rope and dilly at the High Pit to be replaced by horse working. Soon after the War the Polka Bank was modified to permit locomotive working over it, and this in turn allowed a locomotive to replace the horses at the High Pit, with a shed being built on the extreme west of the pit yard. The rope and dilly at the Low Pit continued into NCB days, although an electric hauler replaced the steam engine in 1932, while a second locomotive was brought in at the High Pit from 1943.

DAWDON COLLIERY, Dawdon   J257
NZ 436478

This colliery was developed on the site of the former Vane & Seaham Iron Works (see Marquis of Londonderry), the remains of which were demolished. After the sod-cutting ceremony on 26th August 1899, sinking began on 19th April 1900. Like all sinkings in this area, it proved very difficult because of the water-bearing strata, and after being suspended in December 1902 it was contracted to a German firm using freezing techniques in March 1903. The sinking was finally completed on 5th October 1907 and the first coal was shipped, from Seaham, on 23rd October 1907. The colliery was served by three lines. The first ran from Seaham Harbour southwards along the coast. This was the former "ballast railway" of the Londonderry Railway, which, like the Seaham branch, was taken over from the Marquis in January 1901 by The Seaham Harbour Dock Co, whose locomotives worked traffic between the colliery and the harbour. The second was a short branch direct to the colliery from the Londonderry Railway's former Blastfurnace branch, which from October 1900 was part of the NER. In addition, the branch itself was extended round the east side of the colliery and a connection put in, the line then taken southwards to join the Hartlepool line near Hawthorn, about 2¼ miles south of Seaham (Colliery) Station. The Hartlepool line was opened on 1st April 1905. Both of these lines were of course fully signalled, and brought NER locomotives into the colliery. However, it would seem that from the beginning the colliery had at least one locomotive for its own shunting work, the first locomotive being DAWDON. The locomotive shed was situated on the north-west side of the colliery yard.

The colliery disposed of its waste by tipping it down the cliff at Nose's Point on to the beach, almost certainly the first colliery on the coast to use this method. Later waste from other collieries was also brought here for disposal.

VANE TEMPEST COLLIERY, Seaham Harbour   J258
NZ 425503

The development of this colliery, to work the coal between Seaham and Ryhope north of a major fault line, became possible once the Marquis had vacated the nearby Seaham Hall as one of his residences. Following the freezing of the sand feeder below the limestone, the sinking of the Tempest shaft began on 16th February 1926, and after delays caused by the General Strike, the Vane shaft was begun on 18th May 1927. Work was completed on 18th June 1928, but production did not begin until 20th June 1929. The colliery was originally to be named Londonderry Colliery, but it had taken the names of the two shafts by the time it was opened. It was connected by a ½ branch to the LNER Seaham - Sunderland line, ½ mile north of what was now called Seaham Station.

Early in the 1930s a company called **Coal & Allied Industries Ltd** built a low temperature coal carbonisation plant at the extreme southern end of the Vane Tempest site (NZ 425498), which was presumably linked to the colliery rail system and utilised the colliery's coal. This venture soon failed, and in 1938 a new firm, **Modern Fuels Ltd**, built 51 Gibbons Cellan-Jones low temperature coke ovens on part of the site to manufacture a solid fuel marketed as "Burnbrite", whilst also utilising the former company's by-product recovery plant to manufacture chemicals. This works seems to have closed during the Second World War. Neither company is known to have had its own locomotives, leaving the assumption that their shunting was undertaken by locomotives from the colliery.

Gauge : 4ft 8½in

	DAWDON		0-6-0	IC	TR	254	1855			
		reb	0-6-0ST	IC	Seaham		1876	(a)	(1)	
	SEAHAM		0-4-0ST	OC	HL	2701	1907	New	(2)	$
No.1			?	?	?	?	?	(b)	s/s	
	DAWDON No.2		0-4-0ST	OC	HL	3492	1921	New	(2)	$

LONDONDERRY	0-4-0ST	OC	AB	1724	1922	New	(2)	*
CASTLEREAGH	0-4-0ST	OC	AB	1885	1926	New	(2)	#
VANE TEMPEST	0-6-0ST	OC	RS	4112	1936	New	(2)	#
STEWART	0-4-0ST	OC	AB	2160	1943	New	(2)	*
WYNYARD	0-4-0ST	OC	AB	2165	1944	New	(2)	$

$ At Dawdon Colliery, 1/1/1947   *   At Seaham Colliery, 1/1/1947
#   At Vane Tempest Colliery, 1/1/1947

These allocations seem to have been their normal locations, having been delivered new to them, with little or no transfer movement.

(a)   ex The Marquis of Londonderry (unconfirmed) at an unknown date; previously Londonderry Railway, 3.
(b)   identity and origin unknown, about c/1920; described as "a little tin thing".
(1)   scrapped early in the 1940s.
(2)   to NCB No. 2 Area, 1/1/1947.

A photograph taken during the sinking of Vane Tempest Colliery in 1927 shows an 0-4-0ST, unfortunately unidentifiable. This would seem unlikely to be a locomotive belonging to the company, and is more likely to have been owned by a contractor, presumably the firm (unknown) which carried out the sinking.

Underground locomotive
The company purchased the locomotive below for Vane Tempest Colliery:

Gauge : 2ft 0in

-	4wBEF	Atlas	2456	1946	New	(1)

(1)   to NCB No. 2 Area, 1/1/1947.

SILKSWORTH COLLIERY, Silksworth   H259
     NZ 377541

The sinking of this colliery was begun in 1869 and completed in 1873. It was served by a steeply-graded 1½ mile extension of the one mile branch of the Londonderry Railway to Ryhope Colliery. When the NER took over the Londonderry Railway in October 1900 the Ryhope branch was included, but the extension to Silksworth remained in the Marquis of Londonderry's ownership and was then transferred to The Londonderry Collieries Ltd. However, the NER worked all the traffic.

In 1911-1912 the colliery underwent major rebuilding, as a result of which it was given its own locomotive for shunting at the pit, though with the NER continuing to work the traffic over the branch.

In 1920, almost certainly from 1st February, the colliery was sold to The Lambton & Hetton Collieries Ltd, which see for its subsequent history.

Gauge : 4ft 8½in

SILKSWORTH No.1	0-6-0ST	IC	HL	2932	1912	New	(1)

(1)   to The Lambton & Hetton Collieries Ltd, with the colliery, /1920 (probably 1/2/1920).

LONDONDERRY WAGON WORKS, Dawdon   J260
later **SEAHAM WAGON WORKS**
Marquis of Londonderry until 30/12/1911   NZ 428490

This was originally called the Londonderry Wagon Works, the name changing in the first half of the twentieth century. It was built in 1865 by the then Chief Agent to the Londonderry Estate, John Daglish, to provide a central repair and build facility for all the waggons on the Londonderry Railway (which see). It was situated in Seaham on the Blastfurnace branch north of the junction with the South Hetton Railway. It remained the personal property of the Marquis until 30th December 1911.

Also included in the transfer were 1,965 4-ton chaldron waggons, whose ownership was shown by a white L on the central top plank. Gradually these began to be replaced by wooden hopper wagons carrying up to 16 tons. These were too large to go into the original repair bays of the 1865 building and new facilities had to be built. However, a considerable number of the 4-ton waggons survived to be taken over by the NCB in 1947. Although it has not been confirmed, it would seem very likely that the works also handled repairs for the waggons owned by The Seaham Harbour Dock Co.

The Blastfurnace branch was transferred to the NER, and indeed formed the beginning of the new line that the NER built to Hartlepool and opened in 1905. Although both Londonderry Collieries and the Dock Company possessed running powers over the NER/LNER at Seaham, workings in and out of the Wagon Works had to include a guard's brake van, the former LR brake vans doing this duty for many years.

The works was shunted by horses until 1939, when the unusual machine below was obtained. The works passed to NCB Northern Division No.2 Area on 1st January 1947.

Gauge : 4ft 8½in

| - | 4wPM | MH | L116 | 1939 | New | (1) |

(1) to NCB No. 2 Area, with the works, 1/1/1947.

JOSEPH LOVE & PARTNERS

SHINCLIFFE COLLIERY & BRICKWORKS, Shincliffe   K261
previously **Spark & Love**   NZ 299399
Bell, Davidson, Morrison & Spark until c/1865
William Bell & Co until 1841

The sinking of this colliery began on 11th September 1837 and coal was won on 8th November 1839. The colliery was linked by a one mile branch running north to join the Durham & Sunderland Railway at a junction ¾ mile east of Shincliffe Town Station. This line was worked by a stationary engine (NZ 298407) near Manor Farm, about ½ mile from the colliery, almost certainly hauling waggons up from the colliery and then lowering them down to the junction. The brickworks, making common bricks, lay north east of the colliery.

Subsequently a second line, about ½ mile long, was built southwards from the colliery to join the NER Leamside branch at Shincliffe Station. When this link was built is not known, though it was in place by 1855. It was presumably loco worked. The NER Sidings map drawn in 1885 shows both links in place, though whether the rope haulage on the original line had been replaced by locomotive working is unknown; it may well have been, as DCOA Return 102 does not list a stationary engine here in April 1871.

Joseph Love (1796-1875) was also a partner in Ferens & Love (see Ferens & Love (1937) Ltd), in Strakers & Love (which see for his biographical note) and in Love & Son (see Sir S.A. Sadler Ltd)

The owners of Shincliffe Colliery for many years also owned **HOUGHALL COLLIERY** (NZ 282406) **(K262)**, which was served by a privately-owned branch from Shincliffe Town Station which went on to serve **(Old) Croxdale Colliery** (NZ 267373) **(K263)**, not owned by this company. From 1867 Houghall and Shincliffe Collieries were worked together. Joseph Love & Partners also owned Cornsay Colliery and Brickworks (see Ferens & Love (1937) Ltd).

Both Shincliffe and Houghall Collieries were officially abandoned in May 1886, the closure date presumably being some months earlier.

Gauge : 4ft 8½in

| JOHN BELL | 0-6-0ST | IC | MW | 148 | 1865 | New | (1) |
| SHINCLIFFE | 0-4-0ST | OC | ? | ? | ? | (a) | (1) |

(a) its origin and date of arrival are unknown and its identity is disputed. It may possibly have been the John Harris loco listed under Cornsay Colliery & Brickworks. The name may also have been carried at some point by BH 244 from Cornsay, which may therefore have also come here.

(1) to Cornsay Colliery & Brickworks, Cornsay Colliery.

In the *Colliery Guardian* of 11/2/1887 it was announced that Thomas Bowman, dismantling Shincliffe Colliery, would auction this colliery's plant on 23-24/2/1887, amongst which was a "tank locomotive, 4-wheeled coupled with 12in cylinders". If this was not SHINCLIFFE above, then its origin, identity and disposal are unknown.

LOW BEECHBURN COAL CO LTD (registered 16/10/1889)

LOW BEECHBURN COLLIERY, near Crook   M264
    NZ 162347

This colliery was originally called **Thistleflat Colliery**. The borings for it were undertaken in 1848, and the colliery was working by 1851, together with a brickworks and 74 beehive coke ovens. It was served by sidings east of the NER Crook – Bishop Auckland line, ¾ mile south of Crook Station. It is believed to have been shunted by the NER before the arrival of the locomotive below. The colliery was closed in January 1925 and sold to F. Hindley of Manchester in October 1930, but it was abandoned on 9th December 1930 and the buildings dismantled.

Gauge : 4ft 8½in

-	0-4-0ST	OC	P	644	1896	(a)	s/s

(a) ex The Earl of Bradford's Collieries, Great Lever, Lancashire, via Cudworth & Johnson, dealers, c/1922.

THE LUNEDALE WHINSTONE CO
LUNEDALE QUARRY, Middleton-in-Teesdale W265
NZ 954239

Soon after the NER opened its line to Middleton-in-Teesdale in 1868 this company, headed by a man called Unwin, began developing this quarry. A ½ mile branch from a siding ½ mile east of Middleton Station was driven via a tunnel to deposits of both limestone and whinstone on the northern bank of the River Lune. Apart from the production of roadstone, setts were produced from the whinstone and there was also an attempt to produce kerbs and paving stones by sawing the limestone, though this was not successful.

The line was worked by horses for many years, but in 1881 a steam locomotive was introduced. The company ran into financial difficulties in 1884, and the whole of the plant and machinery was offered for sale by auction under distress for rent on 18th May 1885. Shortly afterwards the quarry was taken over by George Hodsman (see George Hodsman & Sons (1910) Ltd).

References: *Dam Builders Railways from Durham's Dales to the Border* by H.D.Bowtell, Plateway Press, 1994.
The Quarry Industry in Teesdale by H.L. Beadle, from *Out of the Pennines*, ed. B. Chambers, 1997.

Gauge : 2ft 6in

-	0-4-0T	OC	BH	629	1881	(a)	(1)

(a) New; the BH works list gives the gauge as 2ft 1in, but this is believed to be an error; see note below.

(1) offered for sale, 18/5/1885; to G. Hodsman, with quarry, c/1885.

BH 768 was ordered by this company on 13/9/1883, to be another 0-4-0T and to be fitted with a snowplough. The gauge recorded is 2ft 6in. The order appears to have been cancelled before delivery, and in 1885 the loco was sold to Kerr, Stuart & Co Ltd for export.

THE MAINSFORTH COAL CO
MAINSFORTH COLLIERY, Ferryhill Station N266
NZ 307316

The sinking of this colliery was begun in 1873 and coal was reached in June 1877. It was served by a ½ mile branch from the NER Ferryhill – Stockton line, one mile south of Ferryhill Station. Because of the severe economic depression at the time production ceased soon after coal was reached, and after standing for several years the colliery equipment was auctioned by C. Willman on 25th-26th August 1881.

In 1900 the site was taken over by The Carlton Iron Co Ltd (which see).

Gauge : 4ft 8½in

THRISLINGTON	0-6-0ST	OC	BH	354	1875	(a)	(1)

(a) ex The Rosedale & Ferry Hill Iron Co Ltd, West Cornforth (according to one source, but not confirmed).

(1) to South Durham Iron Co, Albert Hill Ironworks, Darlington.

Three locomotives were included in the auction of 25th-26th August 1881. It seems unlikely that BH 354/1875 could have been one of them, as its purchasers, the South Durham Iron Co, went into liquidation in June 1877, so their identity, origin and disposal are unknown.

MAWSON, CLARK & CO LTD
Mawson, Clark & Co until /1926
DUNSTON WORKS, Dunston

C267
NZ 222628

This company manufactured grease, candles and oil. It was established in Newcastle upon Tyne in 1869 and later moved to Heaton in the eastern part of the city. It moved to Dunston some time after 1889, where it was served by sidings south of the NER Redheugh branch. From 1953 until the works closed down in 1955 it was shunted on an "as required" basis by a locomotive from the nearby Dunston Power Station of the British Electricity Authority.

Gauge : 4ft 8½in

-		0-4-0ST	OC	AB	730	1893	(a)	(1)
FRED		0-6-0ST	OC	HE	580	1893	(b)	(2)

(a) ex Robert McAlpine & Sons Ltd, contractors, by 4/1910; believed to have come from this company's Dunston Power Station contract, Gateshead.
(b) ex A.J. Keeble, Wissington, Norfolk, by 12/1916.
(1) to J.F. Wake, dealer, Darlington, /1916; re-sold to Gas, Light & Coke Co Ltd, Beckton By-Products Works, Beckton, Essex, /1921.
(2) scrapped by Clayton & Davie Ltd, Dunston, /1953.

SIR ROBERT McALPINE & SONS LTD
STARGATE QUARRY, Stargate

A268
NZ 166631

This sand quarry, which was also known as Summerhill Quarry, was worked by this firm of contractors in the 1950s and 1960s. The locomotives below were only used within the quarry, and were out of use by May 1958. There was no main line rail connection, the sand being removed by road.

Gauge : 2ft 0in

-	4wDM		MR	8717	1941	(a)	(1)
-	4wDM		MR	8931	1944		
		reb	MR	8995	1946	(a)	(1)

(a) New to Sir Robert McAlpine & Sons Ltd, Hayes, Middlesex; to this location at an unknown date.
(1) to A. Braithwaite & Co Ltd, Burnhills Quarry, Ryton Woodside, c/1958 (see Greenside Sand & Gravel Co Ltd).

Note: Burnhills Quarry, also known as Folly Quarry, was operated by Sir Robert McAlpine & Sons Ltd through its subsidiary companies, A. Braithwaite & Co Ltd and Greenside Sand & Gravel Co Ltd (which see).

A photograph exists of a narrow gauge steam outline 4-4-0 tender locomotive at a quarry in the Ryton area. Its origin and disposal are not known, and it is also not known whether the photograph was taken in this quarry or one nearby.

THE MID-DURHAM CARBONISATION CO LTD
Stella Gill Coke & Bye-Products Co Ltd until 3/10/1928
STELLA GILL COKE WORKS & BY-PRODUCTS PLANT, Pelton Fell

G269
NZ 260522

This works was opened in 1920, another of the rare examples, like the Norwood Coke Works at Dunston, of a modern coking plant not directly attached to a colliery or iron & steel works. The initial works incorporated 50 Simon-Carves regenerative ovens and was served by sidings on the north side of the large Stella Gill Yard on the NER Pontop & South Shields branch, ½ mile south of South Pelaw Junction. The Mid-Durham Carbonisation Co Ltd was another of the companies owned by the Kellett family (see also the Pelton Brick Co Ltd and The Washington Coal Co Ltd). The original ovens were closed down in 1932-1933 and were replaced by 50 Still regenerative ovens, which were started up in 1935. Various tar products, sulphate of ammonia and crude benzole were manufactured as by-products, together with gas, which was supplied to The Newcastle upon Tyne & Gateshead Gas Co Ltd (which see) for use at Chester-le-Street and Durham City. The works was vested in NCB Northern Division No. 5 Area on 1st January 1947.

It is believed that the company hired a locomotive from the Bank Foot, Waldridge, loco shed of The Charlaw & Sacriston Collieries Co Ltd (which see) when its own locomotive was out of commission.

Gauge : 4ft 8½in

-	0-4-0ST	OC	AB?	?	1879	(a)	s/s c/1923
STELLA No.1	0-4-0ST	OC	HL	3504	1923	New	(1)
-	0-4-0ST	OC	P	1460	1916	(b)	(2)

(a) ex a contract in Wales, /1921. If the building date of 1879 is correct, then it would seem to be one of AB 196, 197 and 199; of these, only 199 is known to have worked in Wales, but it is believed not to have survived a boiler explosion. It may have been a locomotive built by Barclays & Co, or the building date may be incorrect.
(b) ex Washington Coal Co Ltd, Usworth Colliery, Usworth, loan.
(1) to NCB No. 5 Area, with works, 1/1/1947.
(2) returned to Washington Coal Co Ltd, Usworth Colliery, Usworth.

MINISTRY OF DEFENCE, ARMY DEPARTMENT
War Department until 1/4/1964
ROYAL ORDNANCE FACTORY, Birtley　　　　　　　　　　　　　　　　　　　　　　　　**G270**
originally **Ministry of Supply**　　　　　　　　　　　　　　　　　　　　　　　　　　NZ 265564

By coincidence the site of this factory, ROF 3, had once been a National Projectile Factory (see Ministry of Munitions). Construction was begun in 1936 and it was opened in May 1937. It was served by sidings east of the LNER Newcastle – London main line, ½ mile north of Birtley Station.

In 1984 all internal rail lines were removed except for one siding and its accompanying facilities, which were retained in order for rail traffic to take place if required. The factory became part of Royal Ordnance plc on 2nd January 1985. It has since had various owners, but continues (2006) in production.

Gauge : 4ft 8½in

-	0-4-0DM	JF	22061	1937	(a)	(1)
-	0-4-0DM	JF	22137	1937	New	(2)
853	0-4-0DM	JF	22976	1942	(b)	(3)
	0-4-0DM	AB	352	1941	(c)	(4)
123 A7	0-4-0DM	EE	1188	1941		
	0-4-0DM	DC	2157	1941	(d)	(5)

(a) New; ex JF on loan prior to delivery of JF 22137, 9/1937.
(b) ex Bicester Workshops, Bicester, Oxfordshire, 17/10/1962.
(c) ex Puriton Ordnance Depot, Puriton, Somerset, by /1966.
(d) ex Bicester Workshops, Bicester, Oxfordshire, 17/12/1969.

(1) returned to JF, 12/1937.
(2) to Bicester Workshops, Bicester, Oxfordshire, 6/10/1962.
(3) to J.A. Lister & Sons Ltd, Consett, for scrap, 12/1970.
(4) to North Eastern Iron Refining Co Ltd, Stillington, 2/1962 (see Cleveland & North Yorkshire Handbook).
(5) to Royal Ordnance Factory, Radway Green, Crewe, Cheshire, 6/1984.

BURNHILL STORAGE DEPOT, near Salter's Gate　　　　　　　　　　　　　　　　　　**T271**
formerly **Ministry of Public Building & Works**　　　　　　　　　　　　　　　　　　NZ 073433
previously **Ministry of Works**
originally **Ministry of Supply**

This depot, in a remote area of the northern Pennines, was built in 1931 by Sir Robert McAlpine & Sons Ltd for storing ammunition. It was served by sidings east of the LNER line from Consett to Tow Law, 1¼ miles south of Burnhill Junction, but with the closure of this line between the depot and two miles north of Tow Law on 18th May 1939, the depot became the terminus of the remaining northern section.

The locomotives below were used to move ammunition between the two transfer buildings alongside the line to six large stores, all eight buildings being protected by high earth banks. No locomotives are known before 1942.

The depot was closed on 12th March 1969, the railway to it following on 1st May 1969. The eight buildings above survive, now used to store classic road vehicles.

Gauge : 2ft 0in

		4wDM	HE	2842	1942	New	(1)
		4wDM	HE	2843	1942	New	(2)
		4wDM	HE	2844	1942	New	(3)
	LOD 758188	4wDM	HE	1835	1937	(a)	(4)
(758079)	ARMY No.20	4wDM	RH	211641	1942	(b)	(5)
(758115)	ARMY No.23	4wDM	RH	226278	1944	(b)	(5)

(a) ex Bicester Workshops, Bicester, Oxfordshire, after 9/1961, by 5/1963.
(b) ex East Riggs Depot, Dumfriesshire, Scotland, 25/2/1966.

(1) to Bicester Workshops, Bicester, Oxfordshire, 9/1961.
(2) to Harrison's Limeworks Ltd, Flusco Quarry, near Penrith, Cumberland, by 6/1954.
(3) to Bicester Workshops, Bicester, Oxfordshire, 9/3/1966.
(4) to East Riggs Depot, Dumfriesshire, Scotland, 15/1/1969.
(5) to Bicester Workshops, Bicester, Oxfordshire, 18/10/1965.

MINISTRY OF MUNITIONS (existed from 5/1915 to 3/1921)
CARTRIDGE CASE FACTORY, Birtley
G272
NZ 265565

The construction of this works was undertaken by Sir W.G. Armstrong, Whitworth & Co Ltd of Newcastle upon Tyne, at the same time as the construction of the National Projectile Factory (see next entry) adjacent to it, with construction beginning in August 1915. However, unlike that factory, this works was operated throughout its time by Armstrong Whitworth. It was served by sidings east of the NER Newcastle – London main line, ½ mile north of Birtley Station.

It is not known when production ceased, but it was probably early in 1919. The premises were auctioned by Anderson & Garland of Newcastle upon Tyne on 23rd/24th May 1922, the plant including the locomotive below.

Gauge : 4ft 8½in

	-	0-4-0ST	OC	BH	?	?	(a)	(1)

(a) origin and date of arrival unknown; one source gives loco as WD [War Department] No.140; a notice of the sale in *Surplus* dated 1/3/1922, which gives its reference number as 75719, describes it as having 12in x 20in cylinders; this was rare for BH, and it may well be that the loco was BH 5/1866, which did have cylinders this size and was built for Sir W.G. Armstrong & Co's Elswick Works in Newcastle upon Tyne.

(1) offered for auction, 23-24/5/1922; s/s.

NATIONAL PROJECTILE FACTORY, Birtley
G273
NZ 265564

The construction of this works, National Projectile Factory No.9, was begun for the Ministry by Sir W.G. Armstrong, Whitworth & Co Ltd of Newcastle upon Tyne in August 1915. It was intended from the beginning to employ Belgian workers, at least 4,000 of them, for whom an enclosed village named Elisabethville was constructed nearby. The factory was operated by Armstrong Whitworth until 11th February 1916, when control passed to the Belgian government. The factory was served by the same sidings east of the NER Newcastle – London main line, ½ mile north of Birtley Station, which served the adjacent Cartridge Case Factory (see above), though the two works were shunted by different locomotives. Production began in July 1916 and almost certainly ceased in December 1918, with the Belgians immediately returning home.

The Factory was sold at the end of 1919 to Sir William Angus, Sanderson & Co Ltd (which see) for the manufacture of the firm's Angus-Sanderson car. This firm closed the factory down in 1921, after which the premises became a Government Training Centre. Then in 1936, by a curious co-incidence, the site was taken over for a new Royal Ordnance Factory (see Ministry of Defence).

Gauge : 4ft 8½in

No.8	0-4-0ST	OC	AE	1054	1874	(a)	(1)
No.9	0-4-0ST	OC	AE	1055	1874	(b)	(2)
No.10	0-4-0ST	OC	AE	1056	1875	(c)	(1)

(a) ex Powlesland & Mason, contractors, Swansea, Glamorgan, c/1915; formerly GWR 1331, and previously GWR 2177; originally South Devon Railway, broad gauge (7ft 0in), CROW; taken over by GWR in 2/1876 and rebuilt to standard gauge at Swindon in 8/1892.
(b) ex Powlesland & Mason, contrs, Swansea, Glamorgan, c/1915; formerly GWR 1332, and previously GWR 2178; originally South Devon Railway, broad gauge (7ft 0in), LARK; taken over by GWR in 2/1876 and rebuilt to standard gauge at Swindon in 7/1892.
(c) ex Ministry of Munitions, Heath Town Phosphorous Factory, Wolverhampton, Staffordshire; formerly GWR 1333, and previously GWR 2179; originally South Devon Railway, broad gauge (7ft 0in), JAY; taken over by GWR in 2/1876 and rebuilt to standard gauge at Swindon in 5/1892.

(1) to Sir William Angus, Sanderson & Co Ltd, with site, c12/1919.
(2) to Ministry of Munitions, Hilltop Farm, Billlingham, after 6/1919, before 12/1919, as it is believed not to have been taken over by Sir William Angus, Sanderson & Co Ltd (see Cleveland & North Yorkshire Handbook).

89. *This locomotive had an incredible life. Built as a broad gauge (7ft 0in) engine called LARK for the South Devon Railway, she passed to the Great Western Railway in 1876, saw out the broad gauge and was rebuilt to standard gauge (4ft 8½in) in 1892, and eventually was sold into industry. Here she still carries the GWR brass safety valve bonnet.*

MINISTRY OF SUPPLY
HARRINGTON SHORE WORKS, COXHOE BRANCH, Coxhoe　　　　　　　　　　　　N274
NZ 330367 approx

In 1943, during the Second World War, the Ministry entered into a contract with The British Periclase Co Ltd, a subsidiary of The Steetley Lime & Basic Co Ltd, to erect a plant known as the Harrington Shore Works near Workington in Cumberland for the manufacture of calcined magnesium. Dolomite limestone was required for this process, and the Steetley company's Coxhoe Quarry in Co.Durham (see entry under Steetley Dolomite (Quarries) Ltd) was chosen to supply this. It would seem almost certain that a separately-defined part of the quarry was used for this, very probably operated by the Steetley company for the Ministry, but with the Ministry supplying the plant, which included a locomotive to shunt the traffic. This operation went under the curious official title of "Harrington Shore Works, Coxhoe Branch"!

Perhaps understandably, details about the Ministry locomotives here are very sketchy. Work ceased soon after the end of the War, and a plant disposal schedule dated 13th November 1945 lists items for disposal at "Harrington Shore Works", but correspondence attached to it says the items were at Coxhoe. This gives details of the last two locomotives below, together with 20 Hudson tipping wagons and other plant. Unfortunately, these papers in the National Archives do not give details of the final disposals.

Gauge : 4ft 8½in

PEGGY	?	?	?	?	?	(a)	(1)
H.S.W. No.1	0-4-0ST	OC	HC	1734	1943	(b)	(2)
-	0-4-0ST	OC	P	2032	1942	(c)	(3)

(a) said to have come to Coxhoe during the Second World War under the auspices of the Ministry of Supply, and so may well have been the first locomotive here, though this is unconfirmed.
(b) although owned by the Ministry, this loco is believed to have been ex The Steetley Lime & Basic Co Ltd's Shireoaks Works, Worksop, Nottinghamshire, after 3/1943.
(c) ex Harrington Shore Works, near Workington, Cumberland (?); a Ministry plant inventory gives the date of acquisition as 3/1944, which may be the date when it arrived at Coxhoe.

(1) local oral tradition said this loco was "scrapped".
(2) to The British Periclase Works Co Ltd, Palliser Works, Hartlepool, still owned by the Ministry of Supply.
(3) to Birmingham Corporation Gas Department, Swan Village Gas Works, West Bromwich, Staffordshire, 1/1947.

ROYAL ORDNANCE FACTORY, Aycliffe   R275
NZ 278235 approx.

The construction of this huge factory, Royal Ordnance Factory No.9 and covering four square miles, was begun on 18th May 1940. Opened in April 1941, it was served on its north side by a ½ mile link to the LNER Shildon – Newport (Middlesbrough) line, and there was also a 1¼ mile branch on its western side to Heighington Station on the LNER Shildon – Darlington line. It was a shell Filling Factory, making and filling various bombs and shells, as well as manufacturing fuses and detonators. The Factory eventually comprised over 1,000 buildings and employed over 16,000 people. To accommodate passenger trains for the workers, two passenger stations were built inside the factory. The one at the northern end was named Simpasture East and had two platforms and electric light signalling. The other, called Demon's Bridge, lay at the north-eastern corner of the site for people travelling from Teesside, while travellers from the Darlington area used Heighington Station.

The loco shed lay on the north-eastern side of the factory, though the Hudswell Clarke loco below was kept at the Heighington side of the complex to shunt the Small Arms Factory. The complex, later ROF No.59, was closed in September 1945 and the premises were transferred to the Board of Trade for subsequent development as a Trading Estate by **North Eastern Trading Estates Ltd** (which see).

Gauge : 4ft 8½in

ROF 9 No.1	0-4-0DM		JF	22934	1941	New	(1)
ROF 9 No.2	0-4-0DM		JF	22943	1941	New	(2)
ROF 9 No.3	0-4-0ST	OC	HC	1722	1941	New *	(3)
ROF 9 No.4	0-4-0DM		JF	22947	1941	New	(4)
ROF 9 No.5	0-4-0ST	OC	RSHN	7046	1941	New *	(5)
ROF 9 No.6	0-4-0DM		JF	22948	1941	New	(6)
ROF 9 No.7	0-4-0ST	OC	P	2016	1941	New *	(7)
ROF (No.8)	0-4-0ST	OC	P	2042	1943	New *	(7)

* originally oil-burning, but later converted to burn coal

In addition to the above, an AB loco, identity unknown, came here to be fitted with oil-burning equipment.

(1) to Board of Trade, Thorp Arch Depot, near Wetherby, Yorkshire (NR), 18/11/1945; to Southern Railway, Southampton Docks, Hampshire, 400S, /1946.
(2) to Board of Trade, Thorp Arch Depot, near Wetherby, Yorkshire (NR), 18/11/1945; to Vickers-Armstrongs Ltd, Squires Gate, Lancashire, 2/1946.
(3) to Board of Trade, Thorp Arch Depot, near Wetherby, Yorkshire (NR), 18/11/1945; to Pilkington Brothers Ltd, Ravenhead Works, St. Helen's, Lancashire, 3/1946.
(4) to Ministry of Supply, East Riggs Depot, Dumfriesshire, Scotland, 14/9/1945.
(5) to Board of Trade, Thorp Arch Depot, near Wetherby, Yorkshire (NR), 18/11/1945; to Pilkington Brothers Ltd, Ravenhead Works, St. Helen's, Lancashire, 6/1946.

(6) to Board of Trade, Thorp Arch Depot, near Wetherby, Yorkshire (NR), 18/11/1945; to R.S.Hayes Ltd, dealers, Bridgend, Glamorgan; re-sold to Fred Watkins (Engineers) Ltd, Milkwall, Gloucestershire.

(7) to Board of Trade, with the site, for use on the planned Trading Estate, 27/10/1945; to North Eastern Trading Estates, Aycliffe, with site.

ROYAL ORDNANCE FACTORY, Spennymoor N276
NZ 267337 approx

The construction of this Engineering factory, ROF 21, was begun in December 1940. It was served by sidings from the LNER Spennymoor – Ferryhill line, one mile east of Spennymoor Station. It was closed in March 1946 and handed over to the Board of Trade, which then developed the site as a Trading Estate (see Thomas Summerson & Sons Ltd).

Gauge : 4ft 8½in

-	0-4-0ST	OC	HC	1507	1923	(a)	(1)

(a) said to have been ex Shaw's Glazed Brick Co Ltd, Darwen, Lancashire, via Thos W Ward Ltd, 2/1938, although the factory was not opened until 1941.

(1) to Ministry of Supply, King's Newton Depot, Derbyshire.

WILLIAM MURPHY
Fairground operator, South Shields D277

William Murphy was a nineteenth century showman, who with other showmen families, travelled round various sites in the north of England operating fairground attractions. One of these was a simple railway system which offered rides to visitors, on which the locomotive below was used. The precise location of his base in South Shields is not known, nor for how long he operated the locomotive.

Gauge : 2ft 6in

ST. LINA FEAR NO FOE	0-4-2T	OC	TG	24*	1895	New	s/s

* as given in one source, but this number is suspect; it could be an error for 224.

NEWALLS INSULATION CO LTD
Newall's Insulation & Chemical Co Ltd until 1/6/1971
The Washington Chemical Co Ltd until 1/10/1964, which was a subsidiary of **Turner & Newall Ltd** from 12/2/1920
formerly **The Washington Chemical Co**
see also below

WASHINGTON CHEMICAL WORKS, Washington Station H278
NZ 324557 approx.

This works was begun in 1837 by **Hugh Lee Pattinson**, who also founded the Felling Chemical Works (which see). In 1842 he set up **The Washington Chemical Co**. The works was served by sidings alongside what became the NER Pontop & South Shields branch, just north of Washington Station.

At first the works manufactured lead carbonate for paint, and later added the production of sulphuric acid. Pattinson died in 1858 and the works passed to his four son-in-laws, one of whom, Robert Stirling Newall (1812-1889) bought it outright in 1872, but continued to trade under the original title. Under his son Frederick Newall (1855-1930) the works began production of magnesia from dolomite (magnesian limestone) and magnesia chemicals, becoming the largest manufacturer in the world by 1900. Newall subsequently developed the process of combining magnesium carbonate with short-fibre chrysotile asbestos (white asbestos). In 1903 he set up Magnesia Coverings Ltd to act as contractors for the application of white asbestos, and in 1908 this firm's name was changed to Newall's Insulation Co Ltd. A further innovation during the First World War was the production of granulated cork for cold store insulation, a new plant being built to the south of the main works.

When Newall joined with Sir Samuel Turner to set up Turner & Newall Ltd in 1920 both The Washington Chemical Co Ltd and Newall's Insulation Ltd became subsidiary companies, together with Turner Brothers Asbestos Co Ltd of Rochdale and J.W. Roberts Ltd of Leeds.

In 1939 a plant was built to extract magnesium from seawater, the latter being pumped from Sunderland in a large pipe laid along the LNER Penshaw branch. In 1953 the company began the production of calcium silicate insulation, and the production of magnesia insulation was phased out.

By the early 1960s the production of insulation materials had outstripped the production of chemicals, and to recognise this the two companies were amalgamated in 1964 under the new title of Newall's Insulation & Chemical Co Ltd (see above). The works, by now covering 139 acres, now concentrated on the manufacture of calcium silicate and glass fibre. The production of magnesia from seawater and dolomite was ended in June 1970, and with the end of chemical production the company's name was changed again in 1971, rail traffic coming to an end at the same time. On 8th October 1980 the works was sold to Cape Insulation Ltd, who operated on the site, albeit on a reduced scale, for over twenty years, before the site was closed and reclaimed.

No locomotives are known before 1872.

Gauge : 4ft 8½in

-		0-4-0ST	IC	JF	1572	1872	New	(1)
-		0-4-0ST	OC	FW	375	1878	New	s/s
-		0-4-0ST	OC	HL	2247	1892	New	(2)
	SYLVIA	0-4-0ST	OC	HL	2645	1906	New	(3)
No.2		0-4-0ST	OC	HL	2780	1909	New	(4)
No.1	(formerly No.3)	0-4-0ST	OC	HL	3349	1918	New	(5)
No.2		0-4-0ST	OC	RSHN	7068	1943	New	(5)
No.3		0-4-0ST	OC	RSHN	7117	1943	New	(5)
-		4wDH		S	10097	1962	(a)	(6)
	MURIEL	0-4-0DH		EEV	D1123	1966	New	(7)
	MARGARET	0-4-0DH		EEV	D1126	1966	New	(8)

(a) owned by S and used for trials; ex NCB Durham No. 2 Area, Lambton Railway, Philadelphia, for trial, 19/6/1962.
(1) offered for sale, 1/1880; it was later at Raine & Co Ltd, Delta Works, Derwenthaugh (which see), but almost certainly not until the 1890s.
(2) to Pease & Partners Ltd, Bishop Middleham Quarry, Bishop Middleham (c/1909?).
(3) to Turner Brothers Asbestos Co Ltd, Trafford Park Works, Manchester, /1924.
(4) to Turners Asbestos Cement Ltd, Trafford Park Works, Manchester, /1933.
(5) scrapped by R.R. Dunn, Bishop Auckland, 3/1967, two on site at Washington and the third at Railway Sidings, Bishop Auckland.
(6) to NCB Durham No. 6 Area, Derwenthaugh loco shed, Swalwell, for trial at Derwenthaugh Coke Works, Winlaton Mill, 13/8/1962.
(7) to British Sugar Corporation, Wissington Works, Norfolk, 1/1970.
(8) to Andrew Barclay, Sons & Co Ltd, Kilmarnock, Ayrshire, Scotland, 4/1972; re-sold to Wiggins, Teape & Co Ltd, Fort William, Invernessshire, Scotland, /1973.

Gauge : 2ft 0in

This was an internal system used within the works complex.

-	4wDM	MR	8747	1942	(a)	Scr c8/1967
-	4wDM	MR	40S273	1966	New	(1)

(a) ex ?, by 11/11/1964; new to Sir Robert McAlpine & Sons Ltd, contractors.
(1) to R.R. Dunn, Whorlton Lido, Whorlton, near Barnard Castle, /1972; to Sir Robert McAlpine & Sons Ltd, Dunston Plant Depot, Gateshead, /1972.

BARMSTON BRICKWORKS, Washington Station   H279
NZ 324556 approx.

The company began the production of insulation bricks at what became known as the Old Yard in 1911. This lay close to the works on its south side. It was closed in 1937 and replaced by this new yard, further to the south-east. The works was served by the site's overall system, but the locomotive below normally worked here, though it was occasionally found elsewhere. Production ceased in 1965.

Gauge : 4ft 8½in

No.2	0-4-0BE	HL	3584	1924	(a)	Scr 1/1966

(a) ex Robert Stephenson & Hawthorns Ltd, Newcastle upon Tyne, 3/1939; loco was built by HL for the Wembley Exhibition in 1924, where it was named WEMBLEY, and it was also exhibited at the North East Coast Exhibition in Newcastle upon Tyne in 1925, before becoming the HL works shunter; at Washington it carried HL 3584/1939 works plates.

90. *This experimental 0-4-0BE, HL 3584/1924, of which this is HL's official photograph, was built for the Wembley Exhibition of 1924, before becoming HL's works shunter. She was sold to Newalls in 1939. The plate below the works plate read 'Built to Durnall's Patent'.*

FORD QUARRY, South Hylton, Sunderland

H280
NZ 362572

Quarrying for limestone here had begun on a small scale by 1894, but large-scale quarrying, by The Washington Chemical Co Ltd, did not begin until the early years of the Second World War. The quarry was served by sidings south of the LNER Penshaw branch, ½ mile north-east of Hylton Station, though following the complete closure of the section between Penshaw and the quarry in 1972 it lay at the end of the line from Sunderland.

Much of the quarrying took place below normal ground level, and by the 1930s two rope-worked inclines were used to haul the tubs out from the quarry floor. The locomotives below were introduced with the major increase in production after 1939, probably for use on the floor, rather than to replace the inclines. Their use had been discontinued by 1966 and stone was brought out via a covered conveyor belt to hoppers at surface level, from which the standard gauge wagons were loaded. This traffic was always shunted by the main line company. Production ceased in 1980 with the sale of the Washington site to new owners.

Gauge : 2ft 0in

| No.2 | 4wDM | HE | 2982 | 1943 | New | Scr 2/1967 |
| No.1 | 4wDM | HE | 3098 | 1944 | (a) | Scr 2/1967 |

(a) ex Ministry of Supply, probably during the Second World War.

THE NEW BRANCEPETH COLLIERY CO LTD

subsidiary of **The Weardale Steel, Coal & Coke Co Ltd** from 12/1933; latterly a subsidiary jointly owned by **The South Durham Steel & Iron Co Ltd** and **The Cargo Fleet Iron Co Ltd**

NEW BRANCEPETH COLLIERY & COKE WORKS, New Brancepeth

K281

Cochrane & Co Ltd until 12/1933
Cochrane & Co until 27/12/1889

NZ 222421

The sinking of this colliery was begun in 1856 by Alex Brodie Cochrane, trading as Cochrane & Co, the owners of Ormesby Ironworks in Middlesbrough. The company had been established two years earlier. The colliery, also known at first as **Sleetburn Colliery**, was served by sidings south of the NER Deerness Valley branch, about two miles west of Deerness Valley Junction. The Busty seam was reached in 1858, but the colliery closed in 1860. A new shaft, to the Brockwell seam, was begun in 1865 and production began again in 1867. Beehive coke ovens, which lay to the north-west of the colliery, were built in the 1860s. A battery of 40 Otto-Hilgenstock waste heat by-product ovens was opened about April 1905, with 40 more added in the second half of 1906. More ovens continued to be built, but it is difficult to determine specific dates; the total had reached 95 by the end of 1907 and 135 three years later. Tar, sulphate of ammonia and rectified benzole were manufactured as by-products. 11 Otto waste heat ovens were added in 1934. However, various rebuilds and shutdowns meant that the number of ovens in use varied considerably. In 1935-1936 nine low-temperature coke ovens – three each of Kemp, Lecocq and Cellan-Jones designs – were built at the works by Gas Chambers & Coke Ovens Ltd, Coking & By-Products Ltd and Gibbons Bros Ltd respectively, for the production of domestic smokeless fuel. Two more were added subsequently, but to which design is not known.

The colliery locomotives shunted the coke ovens, and also the **brickworks** adjacent to the colliery, started in 1908, which manufactured common bricks. Barytes was also mined at the colliery around the turn of the nineteenth and twentieth centuries.

In the twentieth century Cochrane & Co Ltd became associated with both The South Durham Steel & Iron Co Ltd and Dorman, Long & Co Ltd, but in 1933 it became a wholly-owned subsidiary of The Stanton Ironworks Co Ltd in Derbyshire. This resulted in the colliery, coke works and brickworks being sold to The Weardale Steel, Coal & Coke Co Ltd in December 1933, which then set up this subsidiary company to operate them, though it was subsequently sold on. The colliery, coke works – 70 of the old Otto-Hilgenstock ovens, the 11 built in 1934 and the 11 low temperature ovens - and brickworks were vested in NCB Northern Division No. 5 Area on 1st January 1947.

There may well have been other locomotives besides those listed below.

Gauge : 4ft 8½in

	DESPATCH	0-4-0ST	OC	RWH	1019	1857	(a)	(1)
	COOMASSIE	0-4-0ST	OC	RWH	1635	1873	(a)	(2)
	-	0-4-0ST	OC	HL	2358	1896	(b)	(3)
	POWERFUL	0-6-0ST	OC	MW	1602	1903	(c)	(4)
	THE ALLY	0-6-0ST	OC	HL	3185	1916	New	(4)
4		0-4-0ST	OC	Joicey	210	1869		
		reb 0-6-0ST	OC	?		1876		
			reb	HL	7409	1897		
			reb	Bolt's Law		1924		
			reb	Thornley		1931	(d)	(4)
18		0-6-0ST	OC	RWH	1622	1874	(e)	(5)

(a) ex Ormesby Works, Middlesbrough, Yorkshire (NR).
(b) ex Ormesby Works, Middlesbrough, Yorkshire (NR), 6/1905.
(c) ex Ormesby Works, Middlesbrough, Yorkshire (NR), /1923 (one source gives c/1928).
(d) ex The Weardale Steel, Coal & Coke Co Ltd, Thornley Colliery, Thornley, after 12/1933.
(e) ex The Weardale Steel, Coal & Coke Co Ltd, Tudhoe Coke Ovens, by 24/3/1940, loan.

(1) s/s; however, the nameplate survived over the entrance to the shed for many years.
(2) to Ammonia-Soda Co Ltd, Plumley, Cheshire.
(3) to HL, /1916 and re-sold to Brunner, Mond & Co Ltd, Middlewich, Cheshire, 12/1916.
(4) to NCB No. 5 Area, with colliery and coke works, 1/1/1947.
(5) returned to The Weardale Steel, Coal & Coke Co Ltd, Tudhoe (said to have worked only one day here).

NEW BRANCEPETH COKE OVENS K282

These lay on the north-west side of the colliery. As stated above, the first beehive ovens here were built in the 1860s, and the locomotives below would have done their usual task on such ovens by pushing the tubs along the top of the ovens to the correct charging hole. These ovens were superceded by the Otto ovens built in 1904 (see above).

In the JF works list the gauge of JF 5661 is given as 3ft 0in? and the gauge of JF 5883 as 2ft 6½in. The latter is uncommon, and given that both locomotives were almost certainly built to the same gauge, both are shown as 2ft 6½in.

Gauge : 2ft 6½in

	-	0-4-0WTG OC	JF	5661	1888	New	(1)
	-	0-4-0WTG OC	JF	5883	1889	New	(1)

(1) two drivers were working here daily (= two locomotives), according to Durham Coal Owners Association Return 427, 4/1901; s/s.

THE NEW BUTTERKNOWLE COLLIERY CO LTD
BUTTERKNOWLE COLLIERY, Butterknowle P283
The Butterknowle Colliery Co Ltd until 1/6/1907
The Butterknowle Colliery Co until 23/10/1885; see also below

Coal mining in the far south-west of the county around the villages of Butterknowle, Evenwood, Cockfield and Woodland is the most complicated in Co.Durham. The upper seams are all missing and the lower seams outcrop; but the closeness of coal to the surface led to a myriad of shallow mines and drifts. Tracing the history of these is made more difficult because official maps, by the Ordnance Survey, the North Eastern Railway and the National Coal Board frequently disagree over the nomenclature of these workings ("Butterknowle Colliery" is attributed to six different places on an NCB map, some several miles apart), and also because few of the companies, many of them very small, were members of the Durham Coal Owners Association and so do not feature in the DCOA Returns. The description below is derived primarily from the Annual Reports of the Mines & Quarries Inspectorate.

By the end of the eighteenth century the Butterknowle royalty, including pits at Butterknowle and Copley, was leased to Robert Lodge, a gentleman at Barnard Castle. In 1824 Lodge's daughter married Rev. William Prattman, a minister in the Congregational Church at Barnard Castle. By 1828 Prattman had taken over control of the two collieries, and in 1830 he persuaded the Stockton & Darlington Railway to build its Haggerleazes branch (taking its name from a nearby farm). This was opened on 1st October 1830, and ran for 4¾ miles from West Auckland to a station about ½ mile east of the village of Butterknowle. Prattman had persuaded the Stockton & Darlington to extend their line to Haggerleazes in order for his collieries to be linked to it by ½ mile branch. The whole line was horse-worked, and it seems possible that Prattman's horses worked his traffic over the S & D branch, at least initially. Faced with geological problems by the mid-1830s, Prattman sank the Black Diamond Pit, later just the **DIAMOND PIT** (NZ 109256), and the beginning of the "modern" **BUTTERKNOWLE COLLIERY**; Grewburn Colliery, the name taken from the Grewburn Beck alongside the pit, is also found. This failed to save Prattman, and in 1841 he went bankrupt.

With Prattman's bankruptcy the two collieries passed into the hands of his creditors, or Trustees, who then worked Butterknowle for the next thirty years, though they disposed of **Copley Colliery** (NZ 111255) **(P284)** in 1853. The title "Butterknowle Colliery" then became a collective name for new workings developed on the same royalty. Almost certainly the next development was a branch from just east of the Diamond Pit north-westwards for ¾ mile up the narrow valley of the Salter's Burn to **MARSFIELD DRIFT** (NZ 107264) **(P285)**. However, the NCB map gives this name to another working (NZ 106273) **(P286)** at the end of a ¾ mile extension of this branch, marked on the 2nd ed. O.S. map as "old railway", so it may well be that this line was extended to serve this pit. Almost certainly this Marsfield branch was built about 1870. As there was rather more space alongside this burn than the Grewburn Beck, **beehive coke ovens** were built along the line's eastern side, while at its southern end a **brickworks** was developed (NZ 111257). Probably about the same time the line to the Diamond Pit was extended for ¾ mile to serve the **QUARRY PIT** (NZ 102254) **(P287)** and locomotive haulage replaced horses, the loco shed being situated near the brickworks on the opposite side of the line (NZ 111258).

In March 1880, probably because of the economic depression, the Diamond Pit was abandoned. However, four years later the company sank a new pit halfway between the Diamond Pit and Haggerleazes Station. This was called at first the **SLACKS PIT** (NZ 112254), but in 1888 it was re-named the **GORDON PIT (P288)**. Also in 1888 the company replaced the Marsfield Drift by what the Inspectorate Reports term the **MOOR HILL DRIFTS**; the 1898 O.S. map names these as **SALTER'S BURN DRIFT** (NZ 105263) **(P289)** to the west and linked to the screens at Marsfield by a tramway ¼ mile long, and **WHAM DRIFT** (NZ 111265) **(P290)**, linked by a slightly longer line from the east.

In August 1894 the Gordon Pit was abandoned, and by 1897 the Salter's Burn Drift had also closed, leaving only the Quarry Pit and the Wham Drift. However, the Diamond Pit was re-opened by 1904, almost certainly to work coal in the upper two seams, left when it had been abandoned in 1880.

By 1873 the ownership was vested in the hands of **The Owners of Butterknowle Colliery**. In 1879 **The Butterknowle Colliery Company** was set up, becoming a limited company in 1885. The company got into difficulties about 1905 and the colliery was closed; but due to company records in the National Archives we have very detailed information about the next six years. On 26th May 1906 the company went into liquidation and the colliery and its plant were offered for auction by A.T. & E.A. Crow on 13th November 1906 (*Colliery Guardian*, 9th November 1906). **The New Butterknowle Colliery Co Ltd**, with some of the same directors involved, was registered on 6th January 1907, acquired the assets and production resumed in May 1907, though curiously the colliery was not transferred between the two companies until 6th February 1908. On 1st February 1910 The New Butterknowle Colliery Co Ltd also went into liquidation. However, from 1908 the Inspectorate's reports show J.D. Chipchase and W. Walker, both Durham men, as the owners, and it was they who on 10th June 1910 set up yet another company, **The Butterknowle Marsfield Collieries Ltd**. However, the liquidator sold some equipment for scrap and held an auction of most of the rest via John Glover & Co Ltd on 26th-27th July 1910. The liquidator was still active in November 1910, but eventually the new company took over, though apparently now without any locomotives. It too ran into difficulties, and its operations had closed down by 1913. The company survived dormant for some time, but by the 1930s was operational again, working Grewburn Colliery, the old Diamond Pit. The colliery became a licensed mine following the creation of the NCB on 1st January 1947, and was eventually abandoned in March 1950.

Note : only BUTTERKNOWLE and FIREFLY were offered in the auction of 26th-27th July 1910.

Gauge : 4ft 8½in

BUTTERKNOWLE	0-4-0ST	OC	HCR	107	1871	New	(1)
SHOTLEY	0-4-0ST	OC	BH	264	1873	New	(2)
COPLEY	0-4-0ST	OC	?	?	?	(a)	(2)

(a) identity, origin and date of arrival unknown.

(1) offered for sale in *Colliery Guardian*, 30/11/1871 (when only two months old) by the trustees of the late Rev William Luke Prattman; not sold; offered for auction, 26-27/7/1910; to Lingford, Gardiner & Co Ltd, dealers, Bishop Auckland The liquidator reserved the right to use BUTTERKNOWLE to clear materials after the auction. He paid for locomotive coal in October 1910, and paid for gauge cocks and oil from Lingford Gardiner in November 1911, so it may well have been several months after the auction that this locomotive actually left for Lingford, Gardiner & Co Ltd.

(2) according to the liquidator's accounts (PRO BT34 2159/93572) a locomotive, presumably one of these two, was sold on 16/3/1910 to William Wetherell, who was clearly a local scrap merchant; the disposal of the other one is unknown.

BUTTERKNOWLE or MARSFIELD COKE OVENS   P291
NZ 107263 approx.

These were the beehive coke ovens mentioned above. Local tradition maintained that coal from the various pits was taken to the Marsfield Pit (see above) for washing. The washed coal was then discharged from two hoppers into small tubs, which the locomotive took down along the tops of the ovens to whichever oven was to be charged. Following the company failure in 1910 the ovens were never re-started and the standard gauge track alongside them was lifted.

Gauge : 3ft 0in

FIREFLY	0-4-0ST	OC	BH	252	1873	New	(1)

(1) offered for auction, 26-27/7/1910; to J. Best & Co, contractors, Angram Reservoir contract, Nidderdale, Yorkshire (NR).

THE NEWCASTLE SHIPBUILDING CO LTD
HEBBURN SHIPYARD, Hebburn   E292
NZ 301643

This company was registered on 30th August 1919 to take over the ship-repairing firm of Huntley Shipbuilding Co Ltd of Hebburn. Its yard was apparently served by an extension of the railway system within the Tennant's Works of The United Alkali Co Ltd (which see), which lay immediately to the south and which in turn was served by sidings west of the NER Gateshead – South Shields line west of Hebburn Station. How the yard was shunted under the previous ownership is not known.

In 1920 the company acquired a further 1500 yards of river frontage, but it went into liquidation in 1921.

Gauge : 4ft 8½in

-	0-4-0ST	OC	AB	866	1900	(a)	s/s

(a) ex Easton Gibb & Co Ltd, contractors; spares for this loco were ordered in 6/1920 by Gustavus Bailey Ltd of Newcastle upon Tyne to be delivered here.

THE NORTH BITCHBURN COAL CO LTD
The North Bitchburn Coal Co until 2/7/1890

The North Bitchburn Coal Co was formed in 1839 to develop a royalty 1½ miles south of Crook. Fortunately, a wide range of company documents from the whole of its existence was donated to the Durham County Record Office by one of its leading shareholders (ref : D/NBCC), and these have proved invaluable in piecing together the company's history. The chief partners were members of various Darlington families, and from 1840-1857 the Managing Partner was Henry Stobart (see Henry Stobart & Co Ltd). However, an expected branch of the West Durham Railway was not built, and the company had to wait until 8th November 1843 for the Bishop Auckland & Weardale Railway, a subsidiary of the Stockton & Darlington Railway, to open its line between Bishop Auckland and Crook. Thereafter the company gradually expanded, initially in south-west Durham, but in 1910 taking over The Thrislington Coal Co Ltd in south-east Durham.

The company was re-registered with the same title on 14th February 1903. In 1920 it became a subsidiary of Pease & Partners Ltd; North Bitchburn papers give the date as 26th February 1920, Pease & Partners' papers give 1st April. Henry Stobart & Co Ltd also became a subsidiary of Pease & Partners Ltd in 1920. At this point the North Bitchburn company was profitable, but by 1930 its aggregated losses totalled over £100,000 and with Pease & Partners Ltd also indebted, both Pease & Partners and its Associated Companies had to be put under the control of the High Court's Chancery Division. Of the North Bitchburn assets, Randolph and Gordon House Collieries were closed in July 1932. Only Thrislington Colliery was profitable, and in August 1932, without the knowledge of the company's debenture holders, the North Bitchburn Board of Directors agreed to sell it to Henry Stobart & Co Ltd, who in turn agreed to pay £119,751 to Pease & Partners Ltd to pay off North Bitchburn's debt. This transfer had been completed by 27th September 1932, and on 5th October 1932 the North Bitchburn company was put into the hands of receivers. In January 1933 Randolph Colliery was taken over by The Randolph Coal Co Ltd (which see), although Randolph Coke Works waited until about 1938 to be taken over by Sadler & Co Ltd. The receiver ran North Bitchburn and Rough Lea Brickworks at a profit, but failed to find a purchaser, and in February 1934 he advised the debenture holders that the only way that they could hope to recoup their money was to set up a new company to run the two works. From this came the decision to set up **The North Bitchburn Fireclay Co Ltd** (which see). The North Bitchburn Coal Co Ltd was finally wound up on 27th July 1934.

NORTH BITCHBURN COLLIERY & BRICKWORKS, near Howden-le-Wear   M293
NZ 167325

Although the company was set up in 1839, no work was done until 1841 and the sinking of North Bitchburn Colliery, on the eastern side of Beechburn Beck, did not begin until 27th August 1845. It was linked by a ½ mile branch to the line between Bishop Auckland and Crook (see above), ½ mile south of Beechburn Station. Note : the alternative spellings of Beechburn and Beachburn are quite commonly found, for the colliery, the village and the beck (stream) which ran through them; the company always used Bitchburn. In the 1850s **beehive coke ovens** were built on the north, south and west sides of the colliery and a **brickworks** was started, using fireclay mined from the colliery. In 1867 a **sanitaryware works** (NZ 166324) was also started, on the western side of Beechburn Beck, and eventually the making of bricks was discontinued.

In 1889 the company opened **HOWDEN COLLIERY** (NZ 164331) **(M294)**, about ½ mile to the north of North Bitchburn Colliery, linking it to the latter by a short branch. When this colliery closed is not known. The coke ovens at both collieries seem to have closed about the beginning of the twentieth century. In later years coal production at North Bitchburn was continued by means of drift mines. The colliery was closed (by the receiver) about December 1932, and did not re-open. On 5th May 1934 the drift mine and sanitaryware works passed to The North Bitchburn Fireclay Co Ltd (which see).

DCOA Return 102 shows one locomotive here in April 1871 and again in November 1876, but details are not known.

Gauge : 4ft 8½in

CROSFIELD	0-4-0ST	OC	FJ	112	1873		
reb	0-4-0T	IC	LE		1886	(a)	(1)
JESMOND	0-6-0T	IC	JF	1540	1871	(b)	(2)
-	0-4-0ST	OC	P	677	1897	(c)	(3)
HUSTLER	0-4-0ST	OC	P	1337	1913	(d)	(4)

(a) ex Carnforth Hematite Iron Co Ltd, Carnforth, Lancashire, /1894.
(b) ex Lingford, Gardiner & Co Ltd, dealers, Bishop Auckland; previously North Cleveland Ironstone Co, Lofthouse, Yorkshire (NR).
(c) ex Lingford, Gardiner & Co Ltd, dealers, Bishop Auckland, /1911; previously North Lincolnshire Iron Co Ltd, Scunthorpe, Lincolnshire.
(d) ex Randolph Colliery, Evenwood.

(1) to Rough Lea Colliery by /1926.
(2) to The North Bitchburn Fireclay Co Ltd, with site, 5/5/1934.
(3) to Thrislington Colliery, West Cornforth, /1932.
(4) to Randolph Colliery, Evenwood, /1932.

On 31st August 1900 the company purchased **VICTORIA COLLIERY** (NZ 165320) **(M295)**, about ½ mile south-east of North Bitchburn Colliery, on the opposite side of the railway. Whether a company locomotive was used here is not known. The colliery was closed by 1914.

ROUGH LEA COLLIERY & BRICKWORKS, New Hunwick   M296

NZ 195330

The NER opened its Bishop Auckland branch between the Leamside line, Durham and Bishop Auckland on 1st April 1857. On 23rd September 1857 the company leased the New Hunwick royalty, and the sinking of Rough Lea Colliery was completed in June 1858. It was linked to the NER line ½ mile north of Hunwick Station. As at North Bitchburn, **beehive coke ovens** were built and a **sanitaryware works** was developed. In later years this concentrated on making glazed sanitary pipes and sinks.

The colliery ceased production about the end of 1923 and was abandoned in June 1924. The sanitaryware works passed to The North Bitchburn Fireclay Co Ltd on 5th May 1934.

Rough Lea is not listed in the DCOA Returns, and when it first began using its own locomotives for shunting is not known.

Gauge : 4ft 8½in

ARROW	0-4-0ST	OC	MW	498	1874	(a)	s/s after 2/1895
LILY	0-4-0ST	OC	MW	540	1875	(b)	(1)
ARROW	0-4-0ST	OC	BH	991	1890	(c)	s/s
THE COLONEL	0-4-0ST	OC	HG	251	1867	(d)	(2)
CROSFIELD	0-4-0ST	OC	FJ	112	1873		
reb	0-4-0T	OC	LE		1886	(e)	(3)
COMET	0-4-0ST	OC	MW	467	1873	(f)	(4)

(a) ex Casebourne & Co Ltd, Haverton Hill (see Cleveland & North Yorkshire Handbook).
(b) ex Rotherham, Masboro' & Holmes Coal Co, Holmes Colliery, Rotherham, Yorkshire (WR).
(c) ex North Eastern Steel Co Ltd, Middlesbrough, Yorkshire (NR).
(d) ex Henry Stobart & Co Ltd, Newton Cap Colliery & Brickworks, Toronto, loan, after 5/1/1924.
(e) ex North Bitchburn Colliery, near Howden-le-Wear, by /1926.
(f) ex Henry Stobart & Co Ltd, Newton Cap Colliery & Brickworks, Toronto, by 15/8/1931, presumably a loan.

(1) to Lingford, Gardiner & Co Ltd, dealers, Bishop Auckland.
(2) returned to Henry Stobart & Co Ltd, Newton Cap Colliery & Brickworks, Toronto.
(3) to The North Bitchburn Fireclay Co Ltd, with the sanitaryware works, 5/5/1934.
(4) returned to Henry Stobart & Co Ltd, Newton Cap Colliery & Brickworks, Toronto, by 17/10/1933.

On 2nd December 1882 the company acquired **STOREY LODGE COLLIERY** (NZ 137253) **(P297)** about a mile north-west of the village of Evenwood. This was served by sidings alongside the NER Haggerleazes branch about two miles from Spring Gardens Junction, and so far as is known it was shunted by the NER. In 1891 the company greatly added to its holdings in this area by purchasing the Randolph royalty, around the villages of Cockfield and Evenwood, for £55,000. At this time there were four collieries on the royalty, **GORDON HOUSE COLLIERY** (NZ 138260) **(P298)**, which lay to the north of the Haggerleazes branch (and was another example in this area of the same name being used for different shafts – see The New Butterknowle Colliery Co Ltd), to which it was connected by a ¾ mile branch, possibly a rope incline; **EVENWOOD COLLIERY**, more commonly known as **THRUSHWOOD COLLIERY** (NZ 146254) **(P299)**, opened in 1835 and served by sidings south of the Haggerleazes branch about a mile west of Spring Gardens Junction; **NORWOOD COLLIERY** (NZ 148259) **(P300)**, opened by 1833 and served by sidings north of the Haggerleazes branch about ½ mile from Spring Gardens Junction; and **TEES HETTON COLLIERY** (NZ 160255) **(P301)**, another old colliery and served by a short branch ¼ mile west of Spring Gardens Junction. Of these, only Evenwood and Tees Hetton were working when the royalty was purchased.

To work the western side of the royalty more efficiently the company decided to sink a completely new **GORDON HOUSE COLLIERY** on a new location (see below), which was opened in 1893. This was followed in the same year by the sinking of a second new colliery on the south-eastern area of the royalty to gain access to the deeper seams, to be called **RANDOLPH COLLIERY** (see below). New coke ovens were to be built here also, together with a largely new line to the NER. Randolph Colliery began production in 1895, resulting in the closure of both Tees Hetton and Evenwood that year. Storey Lodge was also closed. Norwood Colliery (not to be confused with the colliery of the same name in Gateshead) was re-opened briefly, but closed finally in 1904.

GORDON HOUSE COLLIERY, Cockfield   NZ 131244   P302
GORDON HOUSE COLLIERY, near Esperley Lane Ends   NZ 133240   P303

The new Gordon House Colliery (NZ 131244) was situated just outside Cockfield village, quite a distance from the original Gordon House Colliery. It was opened in 1893. From it a tramway ran north-westwards and crossed over Cockfield Quarry (see Richard Summerson & Co Ltd). It then went immediately into a tunnel about 200 yards long and emerged near screens on sidings alongside the NER Haggerleazes branch at Low Lands, about 2¼ miles west of Spring Gardens Junction. The tramway was almost certainly worked by main-and-tail rope haulage, and was about ½ mile long.

Although opened only in the 1890s, it had been abandoned by 1921 and replaced by yet another Gordon House Colliery (NZ 133240), south-west of Esperley Lane Ends. To serve it the tramway was extended by about ¼ mile, and converted into double track for endless rope haulage, the tunnel too having to be widened. Although working in 1923, it was closed on 11th June 1927, only to be re-opened in May 1929. To reduce its operating costs, it was linked underground to Randolph Colliery in 1931, with its coal being wound there; but there was no market for its coal and the two collieries were closed about May 1932. They were taken over by a new company, **The Randolph Coal Co Ltd**, in January 1933.

According to local tradition the locomotive below "worked at Gordon House". A photograph exists (included here) showing a 0-4-0ST, very possibly the loco below, working at the **Gordon House screens** alongside the NER, in 1909. Possibly it had not been there long, as what is believed to be a loco shed is under construction nearby.

Gauge : 4ft 8½in

    CARBON   0-4-0ST   OC   Grange   1866   (a)   (1)

(a)   origin unknown; arrived c /1909?
(1)   to Randolph Colliery, Evenwood.

RANDOLPH COLLIERY & COKE WORKS, Evenwood   P304
                                                                                             NZ 157248

The sinking of this colliery began in 1893 and the first coal was drawn in 1895. To serve it the branch to Tees Hetton Colliery (see above) was extended, by constructing at the new colliery a self-acting incline 1260 yards long with, unusually, a drum-house straddling the track at the bank head. Also late in 1895 or early in 1896 60 Coppee by-product ovens were started up, with a further 26 added by 1899. Despite this, they were closed in June 1909 and replaced by a new coke ovens and by-product plant, based on 50 Koppers waste heat ovens; a further 10 were added in 1917. Locomotives were used to shunt the whole site as far as the incline bank head.

The colliery was linked underground to Gordon House Colliery (see above) in 1931 to reduce operating costs by winding Gordon House coal here, but despite this there was no market for the coal and the two collieries were closed about May 1932. There are conflicting reports about what happened next. One says that on the orders of the receiver pumping here ceased on 4th November 1933 and the site was handed back to the royalty owners. However, local sources state that Randolph Colliery, still combined with Gordon House, was taken over by **The Randolph Coal Co Ltd** (another venture of the Summerson family – see their entries) in January 1933, and this would appear to be supported by the Annual Reports of the Mines & Quarries Inspectorate. The new firm did not take over the coke ovens, which appear to have passed to the ownership of Pease & Partners Ltd and lain idle until about 1938, when they were re-started by Sadler & Co Ltd, a firm believed to be linked to the very similar business of Sir S.A. Sadler Ltd (which see).

There may well have been other locomotives in addition to those below.

91. These two photographs are the only known illustrations of a locomotive built by the Grange Iron Co Ltd near Durham. Both show her at Lands sidings, near Cockfield, which served the screens of Gordon House Colliery. Note the large pole and the long chain at the front end, used in difficult shunting operations.

92. Tubs from the shaft came in from the right to the main building to be tippled and screened. This view, said to date from about 1910, seems to show new buildings, with what might be a loco shed under construction.

Gauge : 4ft 8½in

MOSTYN	0-4-0ST	OC	MW?	?	?		
		reb	LG		1906	(a)	(1)
HUSTLER	0-4-0ST	OC	P	1337	1913	New	(2)
CARBON	0-4-0ST	OC	Grange		1866	(b)	(1)

(a) ex Lingford, Gardiner & Co Ltd, Bishop Auckland, /1906.
(b) ex Gordon House Colliery screens, Low Lands, near Cockfield.
(1) to The Randolph Coal Co Ltd, with site, 1/1933.
(2) to North Bitchburn Colliery, near Howden-le-Wear; returned, /1932; to The Randolph Coal Co Ltd, with site, 1/1933.

THRISLINGTON COLLIERY & COKE WORKS, West Cornforth N305
The Thrislington Coal Co Ltd until 1/1/1910; see also below NZ 309338

This colliery had been opened in 1867 by **The Rosedale & Ferryhill Iron Co Ltd** (which see). This company failed in 1879. The Thrislington Coal Co Ltd re-opened the colliery in 1889. It was originally served by a ½ mile branch from Ferryhill Iron Works Junction on the NER Spennymoor – Hartlepool line (formerly the Clarence Railway). The complex of sidings here served not only the colliery and its adjacent coke ovens, opened by 1894, but also the derelict iron works (see The Carlton Iron Co Ltd) and the slag breaking works of William Barker & Co (which see). By 1894 a second outlet had been provided, a ¾ mile link to the NER Newcastle – Darlington line, ¾ mile north of Ferryhill Station. This was a rope incline, worked by a stationary engine despite the full wagons travelling downhill. Which of the companies had this installed is not known, but certainly later it belonged to the colliery company. In 1899 the company opened a battery of 25 Semet-Solvay waste heat coke ovens. These were closed down in June 1902 for repairs and extensions, a battery of 30 being re-started in 1903.

The Thrislington Coal Co Ltd was taken over by The North Bitchburn Coal Co Ltd in April 1910, though the official change was back-dated to 1st January 1910. It would clearly appear to be a move by the bigger company to move into the newer part of the coalfield. 55 Koppers regenerative ovens were opened about November 1915, the Semet-Solvay ovens closing at the same time. The site of the old ovens was cleared and replaced by a tar distillation works, owned at first by a Mr. Dent, latterly by **Dent, Sons & Co Ltd**. At first this was called the **Albion Chemical Works**; but in 1920 it caught fire, and when it was rebuilt in 1921 it was renamed the **West Cornforth Chemical Works**. The coke ovens were closed in July 1930, though they were to be re-opened after the sale below.

As described above, with the company in serious financial difficulties the Directors agreed on 16th August 1932 to sell Thrislington Colliery to Henry Stobart & Co Ltd, who would then pay £119,751 to Pease & Partners Ltd in lieu of The North Bitchburn Coal Co Ltd's debt. This transfer was reported as completed at the Directors' next meeting on 27th September 1932. Although this was done without the knowledge of the company's debenture holders, the action was subsequently approved by the Chancery Division of the High Court. For the colliery's subsequent history see Henry Stobart & Co Ltd.

The locomotives also shunted the coke ovens and the tar distillation plant of Dent, Sons & Co Ltd.

Gauge : 4ft 8½in

ISABELLA	0-4-0ST	OC	BH	976	1889	New	(1)
THE SIRDAR	0-4-0ST	OC	CF	1187	1899	New	(1)
-	0-4-0ST	OC	HL	2247	1892	(a)	(1)
-	0-4-0ST	OC	AB	1085	1907	(b)	(1)
-	0-4-0ST	OC	P	677	1897	(c)	(1)

(a) ex Pease & Partners Ltd, Bishop Middleham Quarry, Bishop Middleham, in the 1920s.
(b) ex Pease & Partners Ltd, Chilton Colliery, Chilton Buildings, /1929.
(c) ex North Bitchburn Colliery & Brickworks, near Howden-le-Wear, /1932.
(1) to Henry Stobart & Co Ltd, with the site, c8/1932.

THE NORTH BITCHBURN FIRECLAY CO LTD
Following the collapse of The North Bitchburn Coal Co Ltd (see above), the receiver ran the North Bitchburn and Rough Lea Works at a profit, but was unable to find a buyer for them. He therefore advised the former company's debenture holders to set up a new company themselves in the hope that by this means they could recoup some of their losses. This new company was registered on 5th May 1934. In December 1962 it became a subsidiary of the Hepworth Iron Co Ltd of Sheffield, Yorkshire (WR). It closed down about 1975.

NORTH BITCHBURN COLLIERY & BRICKWORKS, near Howden-le-Wear   M306

By the 1930s the colliery was actually a **drift mine** (NZ 167325) serving the **sanitaryware works** (NZ 166324), the two being divided by Beechburn Beck and served by a ½ mile branch from the LNER Crook – Bishop Auckland line, ½ mile south of Beechburn Station. The sanitaryware works had been started in 1867, and also manufactured firebricks.

On 1st January 1947, with the nationalisation of the coal industry, the mine became a Licensed Mine. Rail traffic ceased in 1965, with British Railways closing the Crook – Bishop Auckland line to all traffic on 5th July 1965. The drift mine was closed in 1966, and the making of firebricks ended in 1967, though the production of sanitary pipes continued for a few years. The works was demolished about 1975.

Gauge : 4ft 8½in

JESMOND		0-6-0T	IC	JF	1540	1871	(a)	(1)
(CROSFIELD)		0-4-0ST	OC	FJ	112	1873		
	reb	0-4-0T	OC	LE		1886	(b)	(2)
DAVID		0-4-0ST	OC	RSHN	6940	1938	New	(3)
(GRETA)		0-4-0ST	OC	HL	2139	1889	(c)	Scr /1950
SAMSON		0-4-0DM		JF	4000013	1947	New	(4)
CONSTANTINE		0-4-0DM		JF	4110001	1949	(d)	(5)

93. The former CROSFIELD, FJ 112/1873, awaiting scrap at North Bitchburn Brickworks about 1950. Note the curious circular hole in the smokebox below the chimney.

(a) formerly owned by The North Bitchburn Coal Co Ltd; taken over from receiver, with site, 5/5/1934.
(b) ex Rough Lea Brick & Pipe Works, Hew Hunwick.
(c) ex Rough Lea Colliery & Brickworks, New Hunwick, by 27/2/1949.
(d) ex Newton Cap Colliery & Brickworks, Toronto.
(1) to Rough Lea Brickworks, New Hunwick.
(2) dismantled by 27/2/1949; scrapped c1950.
(3) to Rough Lea Brickworks, New Hunwick, by 7/9/1947; returned by 3/1948; to Newton Cap Colliery & Brickworks, Toronto, 3/1948.
(4) to Newton Cap Colliery & Brickworks, Toronto, 6/5/1948; ex Rough Lea Brickworks, New Hunwick, by 27/2/1949; to Rough Lea Brickworks, New Hunwick, before 16/5/1951; returned by 16/5/1951; scrapped, /1968.
(5) to John Brown (Land Boilers) Ltd, Dumbarton, Ayrshire, Scotland, c/1965.

ROUGH LEA COLLIERY & BRICKWORKS, New Hunwick   M307
NZ 195330

When the new company took possession the former colliery was not working, and the sanitaryware works was producing glazed pipes and sinks. The works was served by sidings west of the LNER Bishop Auckland – Durham line, ½ mile north of Hunwick Station, but the LNER handled the traffic and there was no company locomotive for shunting.

It would seem that the company began to use a locomotive to shunt within the works in the mid-1930s. Latterly coal and fireclay was brought from Newton Cap Colliery & Brickworks, 1½ miles to the south (see next), though the mine itself was not officially abandoned until 1976. A narrow gauge endless rope system was installed between the clay heap and the grinding mill.

Rail traffic ended about 1954. This works is believed to have closed before the company's other brickworks.

Gauge : 4ft 8½in

(CROSFIELD)		0-4-0ST	OC	FJ	112	1873		
	reb	0-4-0T	OC	LE		1886	(a)	(1)
(COMET)		0-4-0ST	OC	MW	467	1873		
	reb			LG		1902	(b)	(2)
JESMOND		0-6-0T	IC	JF	1540	1871	(c)	Scr c/1940
GRETA		0-4-0ST	OC	HL	2139	1889	(d)	(3)
DAVID		0-4-0ST	OC	RSHN	6940	1938	(e)	(4)
SAMSON		0-4-0DM		JF	4000013	1947	(f)	(5)
BALLARAT		0-4-0DM		JF	4110002	1950	(g)	(6)

(a) formerly owned by The North Bitchburn Coal Co Ltd; taken over from the receiver, with the site, 5/5/1934.
(b) ex Newton Cap Colliery & Brickworks, Toronto, by 15/10/1937.
(c) ex North Bitchburn Colliery & Brickworks, near Howden-le-Wear.
(d) ex Cochrane & Co Ltd, Ormesby Ironworks, Middlesbrough, Yorkshire (NR), via Geo. Cohen, Sons & Co Ltd, dealers; here by 27/6/1938.
(e) ex North Bitchburn Colliery & Brickworks, Howden-le-Wear, by 7/9/1947.
(f) ex North Bitchburn Colliery & Brickworks, Howden-le-Wear, before 16/5/1951.
(g) ex Newton Cap Colliery & Brickworks, Toronto, by 16/5/1951.

(1) to North Bitchburn Colliery and Brickworks, Howden-le-Wear, by 27/2/1949.
(2) to Newton Cap Colliery & Brickworks, Toronto, by 25/1/1949.
(3) to North Bitchburn Colliery & Brickworks, Howden-le-Wear, by 2/1949.
(4) returned to North Bitchburn Colliery & Brickworks, Howden-le-Wear, by 3/1948; ex Newton Cap Colliery & Brickworks, Toronto, by 27/2/1949; to NCB Durham No. 5 Area, Stella Gill Coke Works, Pelton Fell, 4/1954.
(5) to North Bitchburn Colliery & Brickworks, Howden-le-Wear, by 16/5/1951.
(6) returned to Newton Cap Colliery & Brickworks, Toronto.

NEWTON CAP COLLIERY & BRICKWORKS, Toronto   Q308
for ownership see below
NZ 213308

This colliery and its adjacent brickworks had been bought from Henry Stobart & Co Ltd (which see) by Herbert Dunn and Alexander Miller in 1933 (see **Dunn & Miller**). On 2nd September 1935 they sold the brickworks as a going concern to **Fir Tree Coal Co**, but they retained control of the colliery until March 1937, when its lease passed to the **Newton Cap Brickworks Ltd**, which by this time had also taken over the brickworks. In April 1937 this firm in turn was itself taken over by The North Bitchburn Fireclay Co Ltd.

The site was served by a ¼ mile branch from the LNER Durham – Bishop Auckland line, 1¼ miles north of Bishop Auckland Station. It lay about 1½ miles south of Rough Lea Brickworks above. The colliery became a Licensed Mine on 1st January 1947. The works was substantially rebuilt between 1947 and 1949, and also sent coal and clay to Rough Lea Brickworks. The colliery was closed in 1966, with rail traffic ending the same year. The brickworks continued into the 1970s, being demolished in 1975.

The first locomotive below is known to have been here under the ownership of both Henry Stobart and North Bitchburn Fireclay. It is therefore assumed to have also been here throughout the ownerships above, though which of the companies owned it after Dunn & Miller is not known.

Gauge : 4ft 8½in

(COMET)	0-4-0ST	OC	MW	467	1873	(a)	
	reb		LG		1902		(1)
DAVID	0-4-0ST	OC	RSHN	6940	1938	(b)	(2)
SAMSON	0-4-0DM		JF	4000013	1947	(c)	(3)
CONSTANTINE	0-4-0DM		JF	4110001	1949	New	(4)
BALLARAT	0-4-0DM		JF	4110002	1950	New	(5)

It is possible that the names CONSTANTINE and BALLARAT were reversed for a time.

(a) from previous owners (see above), c4/1937.
(b) ex North Bitchburn Colliery & Brickworks, Howden-le-Wear, 3/1948.
(c) ex North Bitchburn Colliery & Brickworks, Howden-le-Wear, 6/5/1948.

(1) to Rough Lea Brickworks, New Hunwick, by 15/10/1937; returned by 25/1/1949; scrapped, /1950.
(2) to Rough Lea Brickworks, New Hunwick, by 2/1949.
(3 to Rough Lea Brickworks, New Hunwick, by 5/1951.
(4) to North Bitchburn Colliery & Brickworks, Howden-le-Wear.
(5) to Rough Lea Brickworks, New Hunwick, by 16/5/1951, and returned; scrapped, /1964.

THE NORTH BRANCEPETH COAL CO LTD
The North Brancepeth Coal Co until 2/7/1890
LITTLEBURN COLLIERY & COKE WORKS, near Meadowfield K309
NZ 255395

The sinking of this colliery, two miles to the south-west of Durham City, was begun on 20th April 1870 and completed on 31st March 1871. It was served by sidings east of the NER Durham – Darlington line, 2¾ miles south of Durham Station. Beehive coke ovens were also built here. In September 1913 the company opened a modern coke works and by-product plant, incorporating 64 Huessener waste heat ovens, the beehive ovens closing at the same time.

The colliery and coke works closed down in August 1931 and the company went into liquidation, but about two months later working was re-commenced by The Bearpark Coal & Coke Co Ltd (which see), which owned the adjacent royalty. However, the coke works was not re-opened.

Gauge : 4ft 8½in

-	0-6-0T	IC	JF	1541	1871	New	(1)
-	0-4-0ST	OC	JF	2076	1873	New	(2)
-	0-4-0ST	OC	JF	2079	1874	New	(2)
WHITWELL	0-4-0ST	OC	BH	1096	1895	New	(3)
-	0-4-0ST	OC	KS	4143	1919	New	(3)

(1) DCOA Return 102 lists only two locomotives here in November 1876; if this is correct then one of the JF locomotives had presumably been disposed of by then; this loco s/s; however, A.R. Bennett's *Chronicles of Boulton's Siding* records that this locomotive was driving electric light plant at Brancepeth Colliery [owned by Strakers & Love] in 1924, but this has not been confirmed, and it may be that Bennett became confused between Brancepeth and North Brancepeth.
(2) DCOA Return 102 lists only two locomotives here in November 1876; if this is correct then one of the JF locomotives had presumably been disposed of by then; s/s.
(3) to The Bearpark Coal & Coke Co Ltd, with colliery, c10/1931.

The firm also owned **NORTH BRANCEPETH COLLIERY**, more commonly known as **BOYNE COLLIERY** (NZ 248407) **(K310)**, which had closed by 1898, and **BROOMPARK COLLIERY** (NZ 253415) **(K311)**, both on the same royalty as Littleburn Colliery. It is not known whether any of the locomotives above worked at either of these collieries.

NORTH EASTERN PAPER MILLS CO LTD (registered in 1892)
NORTH EASTERN PAPER MILLS, Whitburn D312
NZ 411632

A photograph taken on 1st April 1934 at Marsden, on the South Shields, Marsden & Whitburn Colliery Railway of The Harton Coal Co Ltd, shows a small 0-4-0ST locomotive, probably built by Andrew Barclay, out of use. Six weeks later it was no longer there. No such locomotive is known to have been owned by the Railway, and it has been speculated that the locomotive may have worked at the North Eastern Paper Mills, owned by this firm, about a mile to the south.

94. The unidentified 'Barclay?' locomotive photographed on the South Shields, Marsden & Whitburn Colliery Railway at Marsden on 1st April 1934, which may have worked at The North Eastern Paper Mills Co Ltd's factory at Whitburn, about 1/2 mile to the south. The Harton Coal Co Ltd's van No.4 is an interesting vehicle, while the tender almost certainly came from SSM&WCR 8, an 0-6-0 formerly NER 718 of the 708 class, which was scrapped in 1929.

In 1889 a small chemical works was set up on the site, about 1/2 mile south of Whitburn Colliery. In 1892 (one source gave 1895) the site was converted to a paper works owned by this company, which was a wholly-owned subsidiary of The Harton Coal Co Ltd, a very unusual activity for a colliery company. The works produced newsprint, which was distributed all over the British Isles. It closed in 1934, which may well be significant in view of the date of the photograph. The site was later re-used for the Whitburn Central Workshops of NCB Durham No.1 Area (see Part 2).

Other than the loco being built to 4ft 8½in gauge, no further information about its origin, date of arrival or details of its disposal are known.

NORTH EASTERN TRADING ESTATES LTD (registered in 1936)
AYCLIFFE TRADING ESTATE, Aycliffe R313
NZ 2723

This estate was developed on the site of the Royal Ordnance Factory at Aycliffe after it was closed by the Ministry of Supply and handed over to the Board of Trade in October 1945. It was served by sidings from the BR Bishop Auckland – Darlington line at Heighington Station. The shunting was contracted to British Railways during 1961. The estate continues (2006), under different owners and without rail traffic.

Gauge : 4ft 8½in

ROF 9 No.7	0-4-0ST	OC	P	2016	1941	(a)	(1)
No.8	0-4-0ST	OC	P	2042	1943	(a)	(2)
-	4wDM		HE	1737	1935	(b)	(3)

(a) ex Board of Trade; ex Ministry of Supply, from the Royal Ordnance Factory for use on this site, 27/10/1945.
(b) ex Eldon Brickworks Ltd, Coundon Grange, Bishop Auckland, c/1949.

(1) to Team Valley Trading Estate, Gateshead, 8/1963.
(2) to Team Valley Trading Estate, Gateshead, 7/1949 (by 30/7/1949).
(3) to J.A. Lister & Sons Ltd, Consett, for scrap, c/1963.

TEAM VALLEY TRADING ESTATE, Gateshead C314
NZ 2459 and 2460

This huge estate of some 700 acres was the first trading estate in the country and was developed in response to the high unemployment on Tyneside in the 1930s and the Jarrow March in 1936. It was opened in September 1937, and was served by sidings west of the LNER Newcastle – Durham line, three miles south of Newcastle upon Tyne. It incorporated a re-routed section of the Pelaw Main Railway (see The Pelaw Main Collieries Ltd) which connected with the LNER Redheugh Branch at Dunston. National Coal Board working over this route ceased in the 1960s. Such shunting as was left was contracted to BR in June 1966. The estate continues (2006), under different owners and without rail traffic.

95. No.1, HL 3934/1937, purchased new for the opening of the Team Valley Trading Estate, here on 2nd September 1959. She illustrates the later standard HL 0-6-0ST with 14in x 22in cylinders and 3ft 6in wheels, and an enlarged, side-window cab. Note the combined TV monogram on the rear of the bunker, which was also carried on the smokebox door.

Gauge : 4ft 8½in

No.1	TEAM VALLEY ESTATE	0-6-0ST	OC	HL	3934	1937	(a)	Scr c/1965
	No.8	0-4-0ST	OC	P	2042	1943	(b)	Scr 1/1964
	ROF 9 No.7	0-4-0ST	OC	P	2016	1941	(c)	(1)

(a) New; built to an order to HL, but delivered by RSH in 1/1938 after the amalgamation of RS with the locomotive-building business of HL.
(b) ex Aycliffe Trading Estate, Aycliffe, 7/1949 (by 30/7/49).
(c) ex Aycliffe Trading Estate, Aycliffe, 8/1963.

(1) hired by NCB Northumberland & Durham No.5 Area, to shunt the Shop Pit section of the Pelaw Main branch of the Bowes Railway, but working from its Team Valley loco shed, 8/1964 – 5/1966 (except between 6/9/1965 and 21/9/1965); scrapped, 6/1966.

On 3rd May 1949 Robert Stephenson & Hawthorns Ltd brought 0-6-0DM RSHN 7410/1949 here for several days of demonstration runs and then trials in the presence of representatives of the North West Division of the British Electricity Authority, which had ordered it.

NORTHERN GAS BOARD
incorporated into the **British Gas Corporation**, 1/1/1973 (which see). This Board covered a very large area, stretching from Northumberland down to Harrogate and Huddersfield in Yorkshire.

DARLINGTON GAS WORKS, Darlington S315
Darlington Corporation, Gas Department, until 1/5/1949 NZ 292154

Gas production in Darlington was started in 1830 by a private company, whose assets were taken over by the Darlington Gas & Water Company, set up by Act of Parliament in 1849. This in turn was taken over in 1854 by the Darlington Local Board of Health, later absorbed into Darlington Corporation.

The gas works was served by sidings south of the NER Darlington – Bishop Auckland line, ¼ mile south east of Darlington (North Road) Station. How the works was shunted before the arrival of the first locomotive below is not known. Rail traffic ceased in 1965, and the works was subsequently closed and demolished.

Gauge : 4ft 8½in

TOM BARRON	0-4-0ST	OC	AB	1287	1914	New	(1)
KING GEORGE VI	0-4-0ST	OC	RSHN	7013	1941	(a)	(2)
-	4wDM		RH	305320	1951	New	(3)
NORTHERN GAS BOARD No.1	0-4-0ST	OC	P	2142	1953	(b)	(4)

(a) ex The Darlington Forge Ltd, Darlington, loan.
(b) ex Thompson Street Gas Works, Stockton, c9/1964.

(1) hired to The Darlington Forge Ltd, Darlington, sometime during 1939 and 1945, and returned; hired to Thomas Summerson & Sons Ltd, Darlington, /1940, and returned; how the works was shunted during these hire periods is not known; under repair at LNER Darlington Works, 10/1945; scrapped, /1951 (by 1/9/1951).
(2) returned to The Darlington Forge Co Ltd, Darlington.
(3) to Thompson Street Gas Works, Stockton, /1964.
(4) to St. Anthony's Tar Works, Walker, Newcastle upon Tyne, 9/1967.

HENDON GAS WORKS, Hendon, Sunderland H316
Sunderland Corporation until 1/5/1949 NZ 408555

This works was established about 1860 and was served by sidings west of the NER Durham – Sunderland line 1½ miles south of Sunderland South Dock. It was shunted by the NER and its successors, until the arrival of the locomotive below. It was closed in March 1969 and demolished by Golightly (Developments) Ltd, Ferryhill, in 1970-1971.

Gauge : 4ft 8½in

| A.M.No.187 | 4wDM | RH | 198325 | 1940 | (a) | (1) |

(a) ex Charles Jones (Aldridge) Ltd, Staffordshire, dealer, /1958; previously Air Ministry, Broadheath, Cheshire.

(1) to Howdon Gas Works, Willington Quay, Northumberland, for repairs, 7/1959; returned, 9/1959; to Golightly (Developments) Ltd, Ferryhill, with works, 12/1970; resold to Doxford & Sunderland Ltd, Wolsingham Works, Wolsingham, 5/1972.

REDHEUGH GAS WORKS, Gateshead C317
The Newcastle upon Tyne & Gateshead Gas Co until 1/5/1949 NZ 237625

The Newcastle upon Tyne & Gateshead Gas Company was established by Act of Parliament in 1864. This works was opened in October 1876 and was served by sidings south of the NER Redheugh branch. It grew to cover a considerable area. The coal carbonisation plant was closed in March 1967, but locomotive working was retained until 1972 to handle fuel oil and purification materials. The works passed to the British Gas Corporation (which see) on 1st January 1973.

Gauge : 4ft 8½in

WILLIAM BROWN	4wVBT?		?	?	?	(a)	s/s
-	0-4-0WT	OC	?	?	?	(b)	s/s
BLACK PRINCE	0-4-0ST	OC	BH	354	1875	(c)	(1)
W B WILKINSON	0-4-0ST	OC	BH	1025	1891	New	
	reb		HL		1902		(2)
L W ADAMSON	0-4-0ST	OC	HL	2387	1897	New	Scr /1938
SIR W H STEPHENSON	0-4-0ST	OC	HL	2514	1901	New	(3)
JAMES W ELLIS	0-4-0ST	OC	HL	3573	1923	New	Scr 9/1963
LT COLONEL W H RITSON	0-4-0ST	OC	HL	3576	1923	(d)	(4)
SIR CECIL A COCHRANE	0-4-0ST	OC	RSHN	7409	1948	New	(5)
-	0-4-0DM		RSHN	7869	1956	(e)	(6)
-	0-4-0DM		RSHN	7899	1958	New	(7)
-	4wDM		RH	476140	1963	New	(8)

96. Another Hawthorn Leslie solution to restricted clearance (compare photo 31) - a dropped cab and cut down mountings. SIR W.H.STEPHENSON, HL 2514/1901, at Redheugh Gas Works on 29th July 1959.

97. Robert Stephenson & Hawthorns never really established themselves in the diesel locomotive market, and this design, with only a 107 h.p. Gardner engine, 2ft 7½in wheels and weighing a mere 14 tons, was too small for a market now handling much larger wagons; RSHN 7869/1956, on hire to Redheugh Gas Works, 7th November 1956.

(a) oral tradition remembered this locomotive; it was nicknamed the "Coffee Pot", which would suggest that it was a vertical-boilered design; no other details are known.
(b) origin, builders and date of arrival unknown; a photograph of this loco suggests that the maker could well be FJ.
(c) loaned from The Darlington Forge Co Ltd, Darlington, in the late 1880s-1890.
(d) ex Elswick Works, Newcastle upon Tyne, c/1947.
(e) ex Robert Stephenson & Hawthorns Ltd, Newcastle upon Tyne, on trial, 8/1956.

(1) returned to The Darlington Forge Co Ltd, Darlington.
(2) to Elswick Works, Newcastle upon Tyne, and returned, by 19/8/1950; scrapped, 1/1952.
(3) to Elswick Works, Newcastle upon Tyne, and returned; scrapping begun in 4/1957 and completed in 6/1957.
(4) to Elswick Works, Newcastle upon Tyne, c/1948.
(5) to Stephenson Hawthorn Locomotive Preservation Group, NCB Northumberland Area, Backworth Loco Sheds, Backworth, Northumberland, for preservation, 5/1971.
(6) returned to Robert Stephenson & Hawthorns Ltd, Newcastle upon Tyne, /1958.
(7) to Elswick Works, Newcastle upon Tyne, 6/1964.
(8) to British Gas Corporation, with site, 1/1/1973.

NORTHERN INDUSTRIAL IMPROVEMENT TRUST LTD (registered in 1929)
WASHINGTON BRICKWORKS, Washington   **H318**
NZ 303575

This was an old brickworks, adjacent to Washington F Colliery and re-opened in the late 1920s by the Kellett family under this title after they had taken control of The Washington Coal Co Ltd, which owned the colliery. Although the colliery passed to NCB No. 1 Area, the brickworks remained with M.H. Kellett, who used the company above to run it. The presence of this locomotive here has not been confirmed and it would seem that locomotive haulage was only used for a few months. The works was closed in 1970.

Gauge : 2ft 0in

       -        4wDM   RH   213836   1942   (a)   (1)

(a) loaned from Pelton Brick Co Ltd, Pelton, /1952 (Note : this was another firm owned by the Kellett family).

(1) returned to Pelton Brick Co Ltd, Pelton, /1952.

NORTHERN SAND & GRAVEL LTD (registered in 1933)
BARLOW LANE QUARRY, near Stargate   **A319**
NZ 167632

This company is believed to have been another of the subsidiary quarry companies set up or acquired by Sir Robert McAlpine & Sons Ltd to provide their quarrying requirements. This quarry had no main line rail connection, locomotive haulage being used only within the quarry, between the face and the loading point for the road haulage. Its dates of operation are not known.

Gauge : 2ft 0in

   PN 5037   4wDM   MR   8725   1941   (a)   (1)

There may have been a second MR loco here.

(a) ex Sir Robert McAlpine & Sons Ltd, contractors, whose last known use of it was at its Ossington Airfield contract, Nottinghamshire, /1941.

(1) to Sir Robert McAlpine & Sons Ltd, contractors, Hayes Plant Depot, Hayes, Middlesex, by 8/5/1955, when it was moved to National Physics Laboratory contract, Feltham, Middlesex.

THE NORTH HETTON COAL CO LTD
formerly **The North Hetton Coal Co**, probably until 4/7/1884
The history of this company and its collieries is complicated. The first colliery that it owned was
NORTH HETTON COLLIERY, near Hetton-le-Hole   **J320**

This colliery comprised a group of three pits. The first two, the **Dunwell Pit** (NZ 338480) and the **Hazard Pit** (NZ 340477), completed in 1818, were originally part of Rainton Colliery, owned by the Marquis of

Londonderry (which see). They were served by what up to 1825 was the Londonderry Waggonway's main line, the branch for these two pits being worked by horses.

On 24th March 1825 they were sold to a partnership headed by William Russell (later Viscount Boyne) of Brancepeth Castle, which re-named them North Hetton Colliery. Their traffic continued to be handled by the Londonderry system. The new owners began the sinking of the **Moorsley Pit** (NZ 344466), ½ mile to the south of the Hazard Pit, on 19th April 1826, completing it on 28th May 1828. A report dated 12th December 1831 (NRO 3410/Bud/56) states that the Dunwell Pit was then not in production, and that 18 horses were employed in conveying the coal to the junction with the Londonderry Railway, whence the waggons were taken to Pensher and then transferred on to Lord Durham's Railway to be shipped at Lord Durham's Low Lambton Staiths on the River Wear. The report recommended that a stationary engine should be installed near the Hazard Pit, the summit of the branch, to work the waggons up from the Moorsley Pit and then down to the junction with the Londonderry line. It would seem that this was done soon afterwards. It would also seem that the partnership adopted the title The North Hetton Coal Company, though when this was done is not known.

In July 1831 the Marquis of Londonderry opened his new railway between Rainton Bridge and Seaham, but the North Hetton branch was not connected to it and continued to ship its coal at Low Lambton. However, on 17th July 1836 the colliery was sold on again, this time to a partnership of the Earl of Durham, the Marquis of Londonderry and The Hetton Coal Company (which see), and they continued the title of The North Hetton Coal Company. It would seem that the Earl was the leading shareholder. Sometime soon after this a ½ mile link was built north-eastwards from the site of the former Dunwell Pit to join the Londonderry line to Seaham at Rainton Bridge to allow its coal to be shipped at Seaham, though about ½ mile of the old route from the Dunwell Pit was retained to serve a coal depot. The length of the branch was then 1½ miles. However, in 1843 a short link was built between Copt Hill bank head on the Londonderry Railway and just north of the Byer Engine on the Hetton Railway to allow North Hetton coal to be shipped at the Hetton Staiths in Sunderland. There was also some form of "common user" arrangement for the waggons, Londonderry, Hetton and North Hetton waggons being used indiscriminately. The construction of this link perhaps presaged a change in the ownership, for by the late 1850s, perhaps earlier, the Earl and the Marquis had withdrawn from the partnership, leaving the company solely owned by The Hetton Coal Company.

North Hetton Colliery continued thus until 1896, when the Londonderry line to Seaham was closed. It would seem that to allow North Hetton coal to continue to be shipped at Sunderland a new link to the Hetton Railway was built, this time between Rainton Bridge and the meetings of the Copt Hill Incline (worked by the Byer Engine) on the Hetton Railway, though the O.S. maps of the period show no evidence of this. It would seem that the stationary engine at the Hazard Pit worked until 1902, then to be replaced by a locomotive, its shed too being at the Hazard Pit.

On 3rd July 1911 the company, together with The Hetton Coal Co Ltd, was taken over by Lord Joicey, and in August 1911 they were merged with The Lambton Collieries Ltd to form The Lambton & Hetton Collieries Ltd (which see). In 1916 this firm built a new line from Rainton Meadows to link to the North Hetton line, and thenceforth all North Hetton coal was shipped at Lambton Staiths in Sunderland (see Lambton Railway).

The locomotive of about 1833
The Engine Book of R. & W. Hawthorn of Newcastle shows that their Works number 171 was a "Locomotive" ordered by The North Hetton Coal Company. The book does not record order dates at this period, but it is likely to date from about 1833. It was one of the first locomotives ordered from the firm, but sadly no other details are given. Given the notes above, one can only speculate on the purpose for which it was ordered. Perhaps the intention was to see if it could replace the horses on the line more efficiently than the stationary engine recommended in 1831. If so, presumably it proved unsatisfactory. There has also been considerable speculation over what happened to it, one suggestion being that it was transferred to the nearby Hetton Railway, but in fact there is no evidence or oral tradition about it at all.

Gauge : 4ft 8½in

       HAZARD   0-4-0ST   OC   P   615   1896   (a)   (1)

(a)   ex J.F. Wake, dealer, Darlington, /1902; previously Aberpergwm Collieries Ltd, Aberpergwm Colliery, Glyn Neath, Glamorgan.

(1)   to The Lambton & Hetton Collieries Ltd, with the colliery and its branch, 3/7/1911.

The company was also involved with four other collieries. On 12th August 1845 it began the sinking of **SEATON COLLIERY** (NZ 409496) **(J321)** near Seaham. This was situated north of the Carrhouse

Incline on the Londonderry Railway line between Rainton Bridge and Seaham, and linked to Carrhouse bank foot. There seem likely to be hidden reasons why the North Hetton company undertook this sinking, rather than the Marquis himself, especially when a few yards to the east the Marquis himself began the sinking of Seaham Colliery in 1849. Seaton Colliery was subsequently taken over by The Hetton Coal Company, who sold it to the Marchioness of Londonderry under an agreement dated 20th June 1860, backdated to 31st March 1860.

By 1859 the company had acquired two more collieries from either the Marquis of Londonderry, who had died in 1854, or his widow, the Marchioness. These were **GRANGE COLLIERY** (NZ 301447) **(K322)** and **KEPIER GRANGE COLLIERY** (NZ 299442) **(K323)**, both served by the NER Gilesgate branch. Grange Colliery had closed by 1866 and the The Grange Iron Co Ltd developed an ironworks on the site from 1867; Kepier Grange subsequently passed into new hands.

Finally, on 8th October 1864 the company acquired **PITTINGTON COLLIERY** (NZ 333443) **(K324)** from the Marchioness. This lay to the south of the North Hetton royalty, and was situated at the end of the Londonderry Railway's line from Rainton Meadows. The Railway continued to carry Pittington traffic. The colliery was almost certainly closed about 1889-1890; it was officially abandoned in June 1891.

NORTHUMBRIAN WATER AUTHORITY

This body was created on 1st April 1974 to take over from the local district councils (abolished on 31st March 1974) the sewage works that they had previously operated. It also took over the reservoirs previously operated by the various water boards (also abolished on 31st March 1974).

At the works listed below monorail systems were for the handling of sewage sludge. The units used were powered trucks, not a locomotive with skips. All were subsequently removed, some possibly before the formation of this authority in 1974.

BARKER'S HAUGH SEWAGE WORKS, Frankland Lane, Durham K325
Durham Urban District Council until 31/3/1974 NZ 278432

The use of the monorail system ceased about 1972-1973, though the equipment remained on site.

Monorail

-		1w1PH	RM	12438	1964	(a)	(1)
-		1w1PM	RM	4904	1956	(b)	(1)

(a) ex Seghill Plant Sales, Seghill, Northumberland.
(b) ex Chester-le-Street Urban District Council, Sacriston Works, Sacriston.
(1) to R.P. Morris for preservation, and moved to Bowburn Works, Bowburn, 5/7/1980.

BEARPARK SEWAGE WORKS, Bearpark N326
Durham Rural District Council until 31/3/1974 NZ 248432

At least one monorail powered wagon was used here, but no further details are known. The system had been removed by 1980.

BELMONT SEWAGE WORKS, Carrville, near Durham J327
Durham Rural District Council until 31/3/1974 NZ 304452

At least one monorail powered wagon was used here, but no further details are known.

BOWBURN SEWAGE WORKS, Bowburn N328
Durham Rural District Council until 31/3/1974 NZ 302370

The use of the monorail system ceased about 1972-1973, though the equipment remained on site.

Monorail

-		1w1PH	RM	11451	1963	New	(1)
-		1w1PH	RM	11836	1963	(a)	(2)
-		2wPH	RM	14753	1966	(b)	(1)
-		1w1PM	RM	4904	1956	(c)	(1)
-		1w1PH	RM	12438	1964	(c)	(1)

(a) ex Seghill Plant Hire, Seghill, Northumberland.
(b) ex Witton Gilbert Works, Witton Gilbert, the property of R.P. Morris, 4/7/1980.
(c) ex Barker's Haugh Works, Durham, the property of R.P. Morris, 5/7/1980.

(1)   to The Monorail Collection, Blaenau Ffestiniog, Gwynedd, for preservation, 18/4/1981.
(2)   to The Monorail Collection, Blaenau Ffestiniog, Gwynedd, for preservation, 21/7/1981.

LANCHESTER SEWAGE WORKS, Durham Road, Lanchester   G329
Lanchester Rural District Council until 31/3/1974   NZ 164466

Monorail

| | - | 2wPH | RM | 13629 | 1965 | (a) | (1) |

(a)   ex Tarmac Civil Engineering Ltd, contractors.
(1)   to The Monorail Collection, Blaenau Ffestiniog, Gwynedd, for preservation, 1/9/1985.

SACRISTON SEWAGE WORKS, Cross Lane, Sacriston   G330
Chester-le-Street Urban District Council until 31/3/1974   NZ 247477

The use of the monorail system ceased about 1970.

Monorail

| | - | 1w1PM | RM | 4904 | 1956 | New | (1) |

(1)   to Durham Urban District Council, Barker's Haugh Works, Durham.

SHERBURN SEWAGE WORKS, Sherburn   K331
Durham Rural District Council until 31/3/1974   NZ 317419

At least one monorail powered wagon was used here, but no further details are known. The system had been removed by 1980.

WITTON GILBERT SEWAGE WORKS, Witton Gilbert   K332
Durham Rural District Council until 31/3/1974   NZ 231452

Monorail

| | - | 2wPH | RM | 14753 | 1966 | (a) | (1) |

(a)   ex Tyneside Plant & Spares Ltd (no information known about this firm).
(1)   to R.P. Morris for preservation, and moved to Bowburn Works, Bowburn, 4/7/1980.

Unknown location

Monorail

| | - | 1w1PH | RM | 12330 | 1964 | (a) | (1) |

(a)   ex Seghill Plant Sales Ltd, Seghill, Northumberland.
(1)   the engine of this machine was still at the Northumbrian Water Authority's maintenance depot at Pity Me, north of Durham City, in the early 1980s; s/s.

There was also a Hatz diesel engine from another RM diesel-hydraulic machine at the Pity Me depot at this time; its origin is unknown.

ORD & MADDISON LTD

Ord & Maddison until /1897

This company was one of the leading quarry owners in Weardale and Teesdale in the second half of the nineteenth century, though information about its activities and locomotives is unfortunately limited. It was begun by J.R. Ord and H. Maddison, two Darlington men, probably in the 1850s when they look over Lanehead Quarry near Stanhope in Weardale. The firm also owned a number of small collieries, both in Durham and Northumberland. The firm remained in the control of the Maddison family until the mid 1960s and about this time became a subsidiary of Tarmac Roadstone Ltd (which see). The company ceased operating in 1971, though it was not finally dissolved until 31st December 1981.

The company's quarries fell into groups, and it seems sensible to deal with them thus:

LANEHEAD QUARRY, Stanhope   V333
for previous owners see below   NY 990403

When the promoters of the Stanhope & Tyne Railway were planning their line from 1831 onwards they saw the manufacture and transport of lime along their line as one of their major sources of revenue.

The first section of the railway, between Stanhope and Annfield, opened on 15th May 1834, beginning with the Crawleyside Incline at Stanhope with its fearsome gradient of 1 in 7¼. Its bank foot was situated at the entrance to the quarry workings. From the same point a line ran to the west and then turned north up Stanhope Burn for about a mile to reach the Stanhope Lead Smelter, latterly serving iron and lead workings en route. At the southern end of the quarry large lime kilns were built, which were soon dispatching 50-60 waggons a day. The quarry passed through several hands, and by the 1850s it was owned by the Wear Valley Railway, which also owned the quarries at Bishopley (see below). Most, if not all, of the output went to the Consett Iron Works under a 99-year agreement, taken over by the NER. The NER thus owned the quarry, but sub-let its operation. It is believed that Ord & Maddison took on this work in the 1850s, and certainly by 1859, and that Lanehead was the first quarry which the firm worked.

The quarry was exhausted by 1890, and to replace it the company took over Ashes Quarry (see below) to the north of the village of Stanhope. The two quarries were linked by a ½ mile line built south-westwards from Lanehead, with what would appear to be a locomotive shed alongside it. The connection with Lanehead was retained, as the lime kilns were still in use; but the Crawleyside Incline was re-modelled and lengthened, in order to provide a direct link to Ashes Quarry with its own bank foot. It may be that this work was undertaken in 1886-1887 (see below).

It is virtually certain that in the latter part of the nineteenth century one or more locomotives were used within the quarry, but no details survive.

ASHES QUARRY, Stanhope V334
NY 995400 approx.

This quarry, which took its name from a nearby house, lay about ½ mile to the south-east of Lanehead Quarry (see above). Working seems to have begun in the 1870s and to serve it the Crawleyside Incline was re-modelled, with an extension and a separate bank foot provided to serve Ashes Quarry. After Ord & Maddison took over its working, probably in the late 1880s-early 1890s, a link was built between the two quarries. Alongside this link was what is believed to be a locomotive shed. It would seem that untreated limestone was dispatched via the bank foot serving Ashes Quarry, while limestone to be converted to lime was propelled to the kilns in the latterly-closed Lanehead Quarry and the lime dispatched from there.

The quarry itself was owned by the NER, and it seems clear that by 1900 the NER wanted to divest itself of the ownership of quarries and its agreement to supply limestone to The Consett Iron Co Ltd; so on 1st October 1900 the quarry was sold to **The Consett Iron Co Ltd**.

When the Consett company took possession it found a locomotive which it declared 'unsafe', and it refused to allow it to be included in the valuation of the quarry equipment. This was clearly a standard gauge locomotive, because the Consett company immediately sent a locomotive from Consett to replace it. For the quarry's subsequent history see the entry for The Consett Iron Co Ltd.

Gauge : 4ft 8½in

- ? ? ? ? ? (a) (1)

(a)   identity, origin and date of arrival all unknown; very probably a 0-4-0ST. It is not known whether this locomotive was owned by Ord & Maddison Ltd or by the NER.

(1)   not included in the sale to The Consett Iron Co Ltd, 1/10/1900; s/s.

BISHOPLEY QUARRY, near Frosterley U335
NZ 025361 approx.

NORTH BISHOPLEY QUARRY, near Frosterley U336
NZ 026364 approx.

BROWN'S HOUSE(S) QUARRY, also known as SOUTH BISHOPLEY QUARRY, near Frosterley U337
NZ 021364 approx.

South of Frosterley lay very large deposits of carboniferous limestone, and to develop them the Wear Valley Railway, a subsidiary of the Stockton & Darlington Railway, opened its Bishopley branch, 1¼ miles long from Bishopley Junction, 1¼ miles east of Frosterley Station, in 1847. The first quarry to be opened was Bishopley Quarry, which was served by a ½ mile line near the end of the branch. Lime kilns were also built here. The next quarry to be developed was North Bishopley Quarry, and to link these new workings to Bishopley Quarry, a **stationary engine** was purchased second-hand from the NER (from the former Stockton & Darlington Railway's Adelaide Colliery incline). This raised up full waggons to the engine house and then lowered them down into Bishopley Quarry. Later this system was replaced by locomotive working. Although owned by the S&D and then the NER, the actual working of the quarry was let out on contract.

Ord & Maddison took over the contract about 1875. In the early 1880s the firm began a new quarry to the west called Brown's House or South Bishopley Quarry, which was served by a link from North Bishopley Quarry, via a tunnel. The records of Pease & Partners Ltd (which see) show that by 1898 they owned this quarry, but that it was sub-let, in fact to Ord & Maddison Ltd. Both the records of Pease & Partners Ltd and the reports of the Mines & Quarries Inspectorate give the quarry's name as Brown's Houses; the O.S. maps give Brown's House.

In 1886-1887 all traffic from Stanhope via the NER Stanhope & Tyne branch ceased for several months whilst the NER replaced the Crawleyside Engine. This compelled Ord & Maddison to concentrate the production of lime at its Bishopley kilns while the Stanhope kilns (see Lanehead Quarry below) could not be operated, and to assist this the NER contributed towards the cost of building a new ½ mile link between North Bishopley Quarry and Frosterley Station on the NER Wear Valley branch. This link seems to have been abandoned and lifted by 1914.

It is believed that Ord & Maddison closed down their operations here in March 1920, leaving the quarry to revert to Pease & Partners Ltd (which see).

At least two or three locomotives are believed to have been used on the system at any one time, but no details are known. It would appear that two "tank engines" were hired from the NER about May 1881, possibly on a long-term basis, but again no details are known.

In *Contract Journal* for 15/1/1920 the company advertised for sale a four coupled locomotive with 9in x 16in cylinders, rebuilt in 1920. In the edition of *Machinery Market* dated 19/8/1921 the company advertised two four coupled engines for sale. The first had 9in cylinders and 2ft 9½in wheels, and may well be the same locomotive as advertised the year before. The second had 10in cylinders and 3ft 1in wheels. This advertisement came from the firm's Darlington office, but the locomotives may well be connected to Bishopley. Nothing is known of the disposal of the locomotives.

AYCLIFFE QUARRY & LIMEWORKS, Aycliffe   R338
NZ 286223 approx.

There were two quarries and limeworks at Aycliffe, on opposite sides of the NER Durham – Darlington line at Aycliffe Station. Ord & Maddison owned the quarry and works on the eastern side, which they were certainly working by 1894. The quarry and works on the western side had been started in the 1880s by a local farmer, George Chapman (see The Aycliffe Lime & Limestone Co Ltd). Ord & Maddison Ltd were still working their quarry in 1937, though when it was closed is not known.

It is possible that one or more locomotives were used here, but no details are known.

The firm also owned **FERRY HILL LIME WORKS** (NZ 304323) **(N339)**; this had closed by 1894.

MIDDLETON QUARRY, Middleton-in-Teesdale   W340
NY 947245 approx.

PARK END QUARRY, near Middleton-in-Teesdale   W341
NY 925256

CROSSTHWAITE QUARRY, near Middleton-in-Teesdale   W342
NY 926255 approx.

Middleton Quarry was begun by the company soon after the NER opened its branch to Middleton in 1868. It lay to the south-east of Middleton Station, to which it was linked by a branch, with a reverse, about ½ mile long. The locomotive shed was situated nearby, just over the Mickleton road. It was a whinstone quarry, and in addition to large quantities of road materials, thousands of tons of setts were manufactured and sold all over the North of England. Probably soon after the turn of the twentieth century the company began Park End Quarry on the eastern side of Crossthwaite Common as an eventual replacement for Middleton Quarry. The new quarry lay about 1¼ miles further up Teesdale, and the line to Middleton Quarry was extended to serve it. However, this was soon abandoned in favour of a new quarry ¼ mile back down the line, Crossthwaite Quarry, where the stone was of much better quality. Middleton Quarry, latterly called the Low Quarry, was worked until about 1930, after which production was concentrated at Crossthwaite, though the Low Quarry housed a whinstone washing plant and also for a time a plant manufacturing tarmacadam. Crossthwaite Quarry closed down about April 1971 and the plant was subsequently dismantled.

Rail traffic was replaced by road transport in 1952, the track being lifted by W.H. Arnott, Young & Co Ltd.

References : *Dam Builders' Railways, from Durham's Dales to the Borders*, H.D. Bowtell, Plateway Press, 1994. *Out of the Pennines*, ed. B. Chambers: *The Quarry Industry in Teesdale*, H.L.Beadle, 1997.

Gauge : 4ft 8½in

	LIONEL	0-4-0ST	OC	HH	?	?	(a)	s/s
No.2		0-4-0ST	?	?	?	?	(b)	(1)
No.3	"GREENSIDES"	0-4-0ST?	?	?	?	?	(c)	Scr
	-	0-6-0ST	IC	MW	126	1864	(d)	s/s
	-	0-4-0ST	OC	BH	?	?	(e)	(2)
	DERWENT	0-4-0ST	OC	?	?	?	(f)	(3)
898		0-4-0T	IC	Ghd	34	1888	(g)	(4)
1302		0-4-0T	IC	Ghd	32	1891	(h)	Scr 9/1952
	(MABEL)	0-4-0ST	OC	T.D.Ridley	2	?	(j)	s/s by /1950

(a) origin unknown; arrived in /1887 and was first loco here.
(b) identity, origin and date of arrival unknown.
(c) said to have once worked at Frosterley Quarry, Frosterley, so possibly ex Pease & Partners Ltd, who owned this quarry; no other information known.
(d) ex E. Radcliffe. dealer, Hawarden, Cheshire; previously Manston Coal Co Ltd, West Yorkshire Colliery, Manston, Yorkshire (WR), until c10/1882.
(e) ex T.D. Ridley, dealer, Middlesbrough, Yorkshire (NR), by 1910.
(f) ex " a colliery in the Newcastle upon Tyne area", c/1920; only possible local locomotive named DERWENT would be MW 112/1864, last recorded at Skerne Ironworks, Darlington (which see).
(g) ex LNER, 898, Y7 class, 19/4/1929; withdrawn from traffic by LNER, 4/1929.
(h) ex LNER, 1302, Y7 class, 19/3/1930; withdrawn from traffic by LNER, 3/1930.
(j) origin unknown; here by 4/1936.

(1) said to have been scrapped in the company's yard at Darlington.
(2) to Ridley, Shaw & Co Ltd, dealers, Middlesbrough, Yorkshire (NR), during 1939-1945 War, and said to have "gone south".
(3) to Ridley, Shaw & Co Ltd, dealers, Middlesbrough, Yorkshire (NR), c/1927-1929.
(4) used by W.H. Arnott, Young & Co Ltd in the dismantling work in 1952; scrapped after 10/1952.

98. LIONEL, an 0-4-0ST built by Henry Hughes at Loughborough, is believed to be the first locomotive to arrive at Crossthwaite Quarry, in 1887. Note the top-hinged smokebox door, favoured by a few builders around the middle of the nineteenth century.

99. *MABEL, one of the handful of locomotives said to have been built by T.D.Ridley & Sons of Middlesbrough, at Crossthwaite Quarry, probably in the 1930s.*

The firm also operated **GREENFOOT QUARRY** at Stanhope (NY 983392 approx.) **(V343)**, but apparently without its own locomotives (see Richard Summerson & Co Ltd).

OSMONDCROFT LTD
member of **Europa Minerals Group PLC**
OSMONDCROFT MINE, near Winston   Q344
NZ 127161

This drift mine was started in the nineteenth century and worked a small isolated area of coal known as the Yordale seam in the far south-west of the county. It was situated south of the A67 Darlington – Barnard Castle road, one mile south west of Winston and took its name from a nearby farm. It suffered periods of closure, for example being abandoned in October 1931, when it was owned by The Barcus Close Coal Co Ltd. It was re-opened by these new owners in 1988. Rope haulage was used to haul tubs from underground to the surface, and the locomotive below was never put to use. The mine ceased production in August 1991 and was put on to a 'care & maintenance' basis. On 16th April 1992 it was acquired by the Burslem Engineering Association, but it was not re-opened, and the buildings were demolished in 1998.

Reference : *British Small Mines (North)*, A.J. Booth, Industrial Railway Society, 2000.

Gauge : 2ft 0in

26201/10206	4wBE	WR	D6686	1964	(a)	(1)

(a)   ex Thyssen (GB) Ltd, contrs, Pontefract, North Yorkshire, 5/1988.
(1)   last reported here, 12/1/1989; sold for scrap early in /1989.

OSWALD & CO
Thomas Walter Oswald, born in 1836, had a chequered career. At the age of 23 he took over from his father the shipbuilding yard called Pallion High Yard, on the south bank of the River Wear, but went bankrupt in 1861. He started again soon afterwards, re-opening the Pallion Yard. He is credited with establishing iron shipbuilding on the Wear, and as part of this he added an engine works at the yard. He also had a shipbuilding yard at the North Dock. In 1870 he opened the Wear Rolling Mills at Castletown on the north bank of the river, by which time he also had a shipbuilding yard at the South

Dock. He went bankrupt for the second time with the collapse of the rolling mills in 1872, and for the third time with the collapse of the Pallion Yard in 1875. He then moved to Southampton to open a yard there, after which he went to Milford Haven.

PALLION HIGH YARD & ENGINE WORKS, Pallion, Sunderland   **H345**
NZ 377577 approx.

This yard is believed to have been served by sidings from the NER Deptford branch. The business, including three four-wheeled locomotives with 9in, 10in and 12in cylinders, was advertised for sale in the edition of *Iron* dated 21st April 1877, and the plant was also advertised in *The Engineer* of 8th May 1877, where "Pallion Ironworks" is used; but so far as is known there was no iron works here. The plant was auctioned by Wheatley Kirk on 17th May 1877. The site was later incorporated in William Doxford & Sons' yard (which see).

Gauge : 4ft 8½in

-	0-4-0VBT	VCG	Chaplin	708	1866	(a)	(1)
WEAR	0-4-0ST	OC	HCR	101	1870	New	(2)
HYLTON	0-4-0ST	OC	BH	120	1871	New	(3)
PALLION	0-4-0ST	OC	BH	174	1871	New	(4)

(a)   origin unknown, by 9/7/1870; new to the construction of Leibig's Extract of Meat Co Ltd, Fray Bentos, Uruguay.
(1)   to Kilsyth Coal Co, Solsgirth Colliery, Dollar, Clackmannanshire, Scotland.
(2)   had 9in cylinders, so was very probably the 9in loco mentioned in 1877; offered for sale, 17/5/1877; s/s.
(3)   had 12in cylinders, so was possibly the 12in loco mentioned in 1877; s/s.
(4)   had 12½in cylinders; whether it was included in the 1877 sale is not known; to C.D.Phillips, dealer, Newport, Monmouthshire; to Griff Colliery Co Ltd, Nuneaton, Warwickshire, /1882.

WEAR ROLLING MILLS, Castletown, Sunderland   **H346**
NZ 366583

This works was built before the railway through Castletown had been constructed, so at face value it would appear not to have had a rail connection, an unusual situation, the more so since in the edition of *The Engineer* for 1st December 1871 Oswald advertised for a locomotive with 13in x 20in cylinders for it.

THE OWNERS OF PELTON COLLIERY LTD

This company owned two collieries about half a mile apart. The sinking of the original **PELTON COLLIERY** (NZ 253517) **(G347)**, at Pelton Fell, west of Chester-le-Street, was begun on 12th August 1835 and it was opened in the following year. By 1854 it was owned by **Fairless & Co**. It was linked by sidings linked to the southern side of Stella Gill Yard on the NER Pontop & South Shields branch, its line running alongside the railway from Waldridge Colliery (see The Priestman Collieries Ltd). After passing through the hands of **W.C. Curteis & Co**, in 1866 it was acquired by **Lord Dunsany & Partners**. In 1873 they opened a new pit, presumably to reach deeper seams, about ½ mile to the north-east, on the opposite side of the NER line (NZ 251522) **(G348)**. As this was close to the village of Newfield it was called **NEWFIELD COLLIERY** (not to be confused with the Newfield Colliery owned by Bolckow Vaughan and later by Dorman Long, near Willington). It was linked to the north side of Stella Gill Yard by a self-acting incline about ½ mile long.

Given the potential for confusing the two Newfield Collieries it is perhaps not surprising that by 1897 the original colliery had been renamed **PELTON FELL COLLIERY** and Newfield Colliery had been renamed **PELTON COLLIERY**. In later years Pelton Fell Colliery had a tramway 1¼ miles long, almost certainly rope-worked, running westward to the **TRIBLEY PIT** (NZ 237513) **(G349)**.

Lord Dunsany & Partners converted themselves into a limited company in 1876, and owned the two collieries until June 1901, when they were taken over by **The Owners of Pelton Colliery Ltd**. This company went into liquidation in March 1928. The two collieries and their plant were auctioned by order of the Chancery Division of the High Court by Wheatley Kirk between 2nd and 4th October 1928. Pelton Colliery, combined with a drift at Newfield, was subsequently re-opened by **The Mid-Durham Coal Co Ltd**, one of the companies owned by M.H. Kellett, in 1929, with the LNER shunting the traffic.

It is not known at which of the two collieries the locomotives below worked. Two locomotives were included with the plant that was auctioned between 2nd and 4th October 1928.

Gauge : 4ft 8½in

No.1		0-4-0WT	OC	FJ	125	1874	New		(1)
	PELTON	0-4-0ST	OC	HL	2073	1887	New		(2)
	-	0-4-0ST	OC	?	?	?			
			reb	LG		?	(a)		(2)

(a) origin, original builders and date of arrival all unknown.
(1) to John Bowes, Esq., & Partners, Pontop & Jarrow Railway, Marley Hill, c/1880.
(2) possibly the two locomotives offered for sale, 2-4/10/1928; s/s.

THE OWNERS OF REDHEUGH COLLIERY LTD
The Owners of Redheugh Colliery until 8/11/1907
The Executors of the late John Fleming until /1894
John Fleming & John Milling until /1893
Redheugh Coal Co until /1876
REDHEUGH COLLIERY, Gateshead　　　　　　　　　　　　　　　　　　　　　　　**C350**
　　　　　　　　　　　　　　　　　　　　　　　　　　　　　　　　　　　　　　NZ 247630

This colliery was sunk in 1872 and was served by sidings west of the NER Newcastle – Durham line, ½ mile north of Bensham Station. It would seem that normally the colliery was shunted by the NER, but the locomotives listed below are believed to have worked here. The colliery was closed in 1924 and abandoned in May 1927, though the company was not actually wound up until 27th July 1937.

Gauge : 4ft 8½in

-		0-4-0ST	OC	BH?	?	?	(a)	(1)
WYE		0-4-0ST	OC	MW	1037	1887	(b)	s/s
LOCKE		0-4-0ST	OC	HC	338	1889		
			reb	HC		1903	(c)	(2)

(a) origin and date of arrival unknown; possibly built by another maker and sold by BH after it had been rebuilt by the firm.
(b) ex Charles Chambers, contractor for the extension of the Golden Valley Railway in Herefordshire between Dorstone and Hay, who hired this loco to the Railway about 1889; believed to have arrived here c/1891.
(c) ex Locke & Co (Newland) Ltd, St.John's Colliery, Normanton, Yorkshire (WR).
(1) to Wallsend Slipway & Engineering Co Ltd, Wallsend, Northumberland, c/1895?
(2) to C.A. Parsons & Co Ltd, Heaton, Newcastle upon Tyne, c/1925.

The firm advertised for a second-hand locomotive with 14in cylinders in the *Colliery Guardian* of 6th October 1916. No further information is known.

PAGE BANK BRICK CO LTD (registered in 1933)
PAGE BANK BRICKWORKS, Page Bank　　　　　　　　　　　　　　　　　　　　**N351**
　　　　　　　　　　　　　　　　　　　　　　　　　　　　　　　　　　　　　　NZ 228354

This brickworks lay adjacent to Page Bank Colliery, which was closed by Dorman, Long & Co Ltd (which see) in December 1930 (one source gives July 1931). It would seem that in 1933 the yard was taken over by The Hartley Main Collieries Ltd – the only example of a Northumberland colliery company having an operation in County Durham – which then set up this company to run it. It would seem that the parent company relinquished its interest about 1935.

It is believed that the LNER Page Bank Branch was lifted soon after the closure of the colliery. If so, then traffic to and from the brickworks must have been handled by road. The yard had to be closed in 1940, apparently because of the wartime blackout restrictions. The locomotives lay derelict for many years before being scrapped.

Gauge : 2ft 8½in

-	4wPM	FH	1782	1931	(a)	Scr c/1963
-	4wPM	FH	1892	1934	New	Scr c/1963

(a) ex The Hartley Main Collieries Ltd, Cramlington Colliery, Cramlington, Northumberland (very probably the company's Stores Department there); previously 2ft 4in gauge.

There is a possibility that 4wPM Hu 38788/1, built in 1930, may also have come here from Hartley Main Collieries Ltd (see above), but evidence is lacking.

PALMERS SHIPBUILDING & IRON CO LTD
Palmer Bros & Co until 21/7/1865
JARROW SHIPYARD & STEELWORKS, Jarrow

E352
NZ 318657 to NZ 330657

This shipyard ranks amongst the most famous in the world. Developed by George and Charles Mark Palmer from 1851, its second vessel was the first iron, screw-driven collier, built for the London trade and using water as ballast, immediately rendering obsolete the wooden sailing ships with the chalk ballast used hitherto. Launched on 30th June 1852, she was named the John Bowes and was built for the colliery owners John Bowes, Esq., & Partners, for whom C.M. Palmer was managing partner. The first premises were situated immediately to the west of the Springwell Staiths and were linked to the Springwell Colliery Railway (the Pontop & Jarrow Railway from 1853), both owned by John Bowes, Esq., & Partners. In 1872 the NER opened its line between Pelaw and South Shields, and the yard was linked to it, ¼ mile west of Jarrow Station. In 1859 the brothers took over another shipbuilding yard at Willington Quay, Howdon, on the north bank of the Tyne. George Palmer retired from the business in 1862.

From its initial success, the Jarrow yard expanded rapidly. Blast furnaces were added in 1857, with six by 1863, and a steel works and rolling mills were added in 1865. The riverside location enabled the firm to import its iron ore from its Grinkle Park mines near Staithes in the North Riding of Yorkshire, as well as to export iron and later steel all over Britain and beyond.

100. H&C 32/1864, almost certainly brand new. Note the copper-capped chimney – and the complete absence of any protection for the loco crew. Hudswell & Clarke records give the customer as the 'Jarrow Iron Co', which is believed to be Palmer Bros & Co; there was no other iron works at Jarrow.

Initially the majority of the ships' engines were built by R. & W. Hawthorn of Newcastle upon Tyne, but as early as 1853 the yard built its first engine, and subsequently a large engine works was developed. By 1880 the yard had eight berths, a 440ft long graving dock and a 600ft-long slipway. It became the firm's boast that it took raw materials in and turned ships out without anything leaving the site. By the beginning of the twentieth century the yard, its furnaces and works had spread over 140 acres, occupied a mile of river frontage and was served by a rail system totalling eight miles. By this period a slag heap had been established about a mile to the south of the yard (NZ 323645), the yard and the heap being connected by a branch which ran alongside the Pontop & Jarrow Railway, crossing it near to the heap. For a time the company also owned a small yard at Amble in Northumberland and a dry dock at Swansea in South Wales. In 1911 it leased and a year later purchased the Hebburn Yard of Robert Stephenson & Co Ltd (see below). Besides a large quantity of merchant ships, the firm built a wide range of warships, notably torpedo-boat destroyers for the British Admiralty, but also undertook work for foreign governments, as well as passenger liners and oil tankers.

101. This magnificent study of the Foundry area at Palmers was taken in 1912 by the South Shields photographer William Parry. It shows no fewer than five, perhaps six, locomotives and a steam rail crane, and a wide miscellany of items lying on the ground.

102. Locomotive photographs taken inside Palmers are rare, perhaps because it closed in the 1930s. HL 2135/1889 was a standard Hawthorn Leslie design of her period, with 12in x 18in cylinders and 3ft 0in wheels – and a very long overhang at the rear end. Note the extension to the cab roof, supported by a central pillar. Note also the background behind her, not least the unusual steel 'chaldron' wagon T19.

The firm became perhaps the most famous victim of the Depression after 1929. The steelworks, which latterly had five furnaces, had been closed in 1921, Palmers' steel plate requirements henceforth coming from The Consett Iron Co Ltd. Few orders were received after 1926, and Palmers' last ship, the 982nd built by the firm, was launched at Jarrow on 19th July 1932. On 20th June 1933 a receiver was appointed and the following year he sold the firm to National Shipbuilders' Security Ltd, a company set up by other shipbuilders to reduce over-capacity in the industry by acquiring and closing down yards. The yard was demolished by Thos W. Ward Ltd in 1934-35, with a covenant on the land that it could not be used again for shipbuilding for forty years. Jarrow became 'the town that was murdered', with thousands left unemployed, 200 of whom formed the 'Jarrow March' in October 1936, walking to London to petition the government for work. This led to the introduction of Trading Estates, beginning with the Team Valley Trading Estate in Gateshead in 1937 (which see). New construction on the Palmers site began in the same year (see the entries for Armstrong Whitworth Rolls Ltd and The Consett Iron Co Ltd).

Information about the locomotives here is quite sketchy, and undoubtedly there were more here than those listed below. Nos. 1, 2, 4-14 were all said to have been "Hawthorns", presumably BH, RWH and HL. Assuming that there was initially a No.3, it is curious that this number was never used again after it became vacant. A "crane engine, a very old type Hawthorn, worked in the melting shop" (possibly HL 2113/1888). The yard also had at least five steam cranes and one electric rail crane.

Reference : *Palmers of Jarrow*, Jim Cuthbert and Ken Smith, Newcastle City Libraries & Information Service, 2004.

Gauge : 4ft 8½in

	-	0-4-0ST	OC	H&C	32	1864	New	s/s
	-	0-4-0T	OC	RS	1619	1865	New	s/s
	-	0-4-0T	OC	RS	1801	1866	New	s/s
No.5	JARROW	0-4-0ST	OC	BH	62	1868	New	s/s
No.6		0-4-0ST	OC	BH	128	1869	New	s/s

No.7		0-4-0ST	OC	BH	216	1871	New	s/s	
No.8		0-4-0ST	OC	BH	176	1872	New	s/s	
No.9		0-4-0ST	OC	BH	315	1874	New	s/s	
No.10		0-4-0ST	OC	BH	476	1879	New	s/s	
	MIDGE	0-4-0ST	OC	BH	231	1874	(a)	s/s	
	SLAVE	0-4-0CT/WT	OC	RWH	1877	1880	(b)	(1)	
No.4		0-4-0ST	OC	RWH	2026	1885	New	(2)	
No.11?		2-2-2CT	IC	HL	2113	1888	New	s/s	
No.12		0-4-0ST	OC	HL	2135	1889	New	s/s	
No.1		0-4-0ST	OC	HL	2169	1889	New	s/s	
	TEAM VALLEY	0-4-0ST	OC	HL	2489	1901	(c)	s/s	
No.13		0-4-0ST	OC	HL	2666	1906	New	(3)	
No.14		0-4-0ST	OC	HL	2667	1906	New	(3)	
No.15		0-4-0ST	OC	P	1392	1915	New	(2)	
No.16		0-4-0ST	OC	P	1413	1915	New	(2)	
No.17		0-4-0ST	OC	HL	3237	1917	New	(4)	
No.18?	FLOSSIE	0-4-0ST	OC	?	?	?	(d)	(5)	
No.19		0-4-0ST	OC	KS	3097	1918	New	(6)	
No.20		0-4-0ST	OC	KS	4029	1919	New	(2)	
	HECTOR	0-4-0CT	OC	HL	2447	1900	(e)	s/s	

(a) ex Thos. Nelson, contractor.
(b) ex William Denny & Brothers, Dumbarton, Ayrshire, Scotland, /1885.
(c) ex T.D. Ridley & Sons, contract for NER widening, Ouston to Low Fell, c/1901.
(d) said to have been purchased from a contractor and to have been built by either VF or WB.
(e) ex Hebburn Shipyard, Hebburn (after 1912).

(1) in *Machinery Market* dated 9/6/1916 an HL 0-4-0CT with 10in x 15in cylinders was advertised by the Grange Iron Co Ltd, Durham City. The only loco which fits the description given would appear to be this one; s/s.
(2) to Thos W. Ward Ltd, contractors for dismantling the site, /1934.
(3) to Thos W. Ward Ltd, contractors for dismantling the site, /1934; to Palmers Hebburn Co Ltd, Hebburn.
(4) to Thos W. Ward Ltd, contractors for dismantling the site, /1934; to Darlington Rolling Mills Co Ltd, Darlington, after repair at HL, /1934.
(5) said to have been scrapped after the jib of a crane crushed it.
(6) s/s after 12/1928.

In an advertisement in *Machinery Market* dated 6th July 1934 Fuller Horsley announced that they would be auctioning six four-coupled locomotives from this yard "soon". Their identities and disposal are not known.

How the loco below was used within the works is not known.

Gauge : 3ft 0in

CHARGER	0-4-0IST	OC	WB	1381	1891	New	(1)	

(1) to Sir John Jackson Ltd, Blackwater Reservoir contract, Argyll, Scotland (contract completed in 1910).

One oral tradition recorded that Palmers built two locomotives themselves with cast iron boilers for working the slag on to the heaps. These were followed by two others, attributed to Hawthorn Leslie, although no information has come to light.

HEBBURN SHIPYARD, Hebburn E353
Robert Stephenson & Co Ltd until 5/6/1912 NZ 306653

This yard was begun in the early 1880s by **McIntyre & Co Ltd**, who were forced to close down in 1884. It was taken over by **Robert Stephenson & Co Ltd** in November 1885, in an attempt by the locomotive-building firm to break into the shipbuilding market. The yard had five berths, and also had a foundry. It was served by a ½ mile branch from the NER Gateshead – South Shields line, ½ mile east of Hebburn Station. The yard survived the collapse of Palmers Shipbuilding & Iron Co Ltd, and was taken over by a new company, **Palmers Hebburn Co Ltd** (see Vickers Ltd) in 1934.

There may well have been more locomotives used here than the three below, and also more transfers between the Hebburn and Jarrow Yards after 1912. The yard also had 11 steam cranes.

Gauge : 4ft 8½in

COMET	0-4-0T	OC	RS	2326	1888	New	(1)
EGBERT	0-4-0CT	OC	HL	2173	1890	New	(2)
HECTOR	0-4-0CT	OC	HL	2447	1900	New	(3)

(1)　disposed of after being taken over from Robert Stephenson & Co Ltd, 6/1912.
(2)　to Robert Stephenson & Co Ltd, Darlington Works, Darlington (between 1900 and 1912).
(3)　to Jarrow Shipyard, Jarrow.

H.L.PATTINSON & CO
John Lee & Co until c/1851
FELLING CHEMICAL WORKS, Felling
FELLING IRON WORKS, Felling
FELLING BRICKWORKS, Felling　　　　　　　　　　　　　　　　　　　　　　　　all D354

Felling Chemical Works (NZ 279626 approx.) was begun in 1833 by John Lee of Alston in Cumberland and his nephew Hugh Lee Pattinson (1796-1858), who first discovered how to extract silver from lead and then developed processes for including white and red lead in paint and how to produce alkali. The works was linked by ¼ mile branch to the Felling waggonway, which ran from Felling Colliery (NZ 275623) to the River Tyne (for the operation of this line see the entry for John Bowes & Partners Ltd); Lee & Company's quay lay immediately downstream of the Felling Staiths. With the opening of the Brandling Junction Railway's line from Gateshead to South Shields in August 1839 the colliery was linked to it immediately west of Felling Station, which presumably gave the chemical works access to it too. Meanwhile in 1837 Pattinson opened another chemical works at Washington (see Newalls Insulation Co Ltd).

In the early 1850s the firm built its own line to its quay, but to overcome the higher land north of the works and then the steep fall to the river virtually all of this 300-yard line was in a tunnel. It was worked by a **stationary engine** at the works. Then in 1854 the firm opened **Felling Iron Works** (NZ 279623), to the east of the chemical works on the opposite side of Brewery Lane. This had two blast furnaces. About the same time it also opened **Felling Brickworks** (NZ 278622), to the south of the iron works. Both were commercially linked to the needs of the chemical works, which by now had expanded to over 20 acres, and both were initially served by extensions from the chemical works' system. However, about 1855 the iron works was also connected to the NER via a short link east of Felling Station.

Like the Allhusens' works at Gateshead, the chemical works had an overhead railway, here at a height of 25 feet, for conveying materials within the sections of the works. Thomas Bell describes the system: "A railway for half chaldron wagons [built at the iron works] has been formed from the quay to the works, partly underground or in a tunnel arched with brick [from the brickworks] and having at the works a stationary engine with (an) engine house, by which the wagons of materials are brought to or taken from the works. The railway on reaching the manufactories is carried in various directions on strong gears over the several buildings constituting the manufactory, depositing the materials where they are required." (p.22 in the reference below). How this overhead system, presumably operated by some form of rope haulage, was co-ordinated with the surface system, is not known.

Probably in the early 1860s the iron works was sold to Frazers, Roberts & Co and three members of the firm separately, though Mineral Statistics continues to list the owners as H.L. Pattinson. It produced no more iron after 1864, and the iron works' continuing complicated history can be found under the entry for The Felling Coal, Iron & Chemical Co Ltd.

By 1880 the chemical works had grown to cover more than 30 acres, but in the early 1880s a cheaper method of alkali production using soda and ammonia was developed in Belgium, and these developments and the severe trade depression forced the closure of the works, putting 1400 people out of work. Books about the chemical industry in Britain give the closure date as October 1886, but the local press shows this to be an error. It probably closed about December 1885, as in the edition of *The Engineer* for 5th March 1886, the auctioneers Snowball & Co of Newcastle upon Tyne advertised the works' plant for sale.

This sale advertisement included three tank locomotives, but nothing is known of their origin, identity or disposal.

About 1889 at least part of the works was re-opened by **Pattinson's Pearl Hardening Co**. This was a short-lived venture and had closed by 1893. By 1895 part seems to have come into the ownership, again short-lived, of the **Northumbrian Chemical Co**. It is not known whether either of these firms used locomotives here.

Reference: *The Old Tyneside Chemical Industry*, published by University of Newcastle, 1961.

PEASE & PARTNERS LTD

This company was formed on 19th August 1882 by an amalgamation of Joseph Pease & Partners, which had previously operated the colliery interests of the Pease family, and J.W. Pease & Co, which had operated ironstone mines and limestone quarries. A major company then, it grew further to dominate the industrial activity of the southern half of Durham, together with major interests in Teesside, North Yorkshire and beyond. More than any other company of its time in North East England, it also bought out other companies or purchased the share capital to operate them then as subsidiary companies. The Pease family also had important or controlling share interests in other companies, some involving members of the family, some in firms involving relatives or other Quaker families in Darlington, and some with none of these, for example, as directors of the Stockton & Darlington Railway and then the North Eastern Railway, and also the locomotive builders Robert Stephenson & Co. They had huge industrial interests in North Yorkshire too, as well as commercial and social interests: J. & J. W. Pease was an important local bank until it stopped payments in 1902, while one of the family built the seaside resort of Saltburn-by-the-Sea.

Fortunately, various records of the company, including its Annual Reports, are now deposited at the Edward Pease Free Library in Darlington, while other important company documents can be found in the North Bitchburn Coal Company collection held by the Durham County Record Office.

The Pease family was only one of a number of prominent Quaker families living in Darlington in the nineteenth century. Others included the Backhouse, Fry, I'Anson and Kitching families. The first important member of the family at the beginning of the nineteenth century was Edward Pease (1767-1858), a leading promoter of the pioneering Stockton & Darlington Railway, and the family were directors and shareholders both in this and the NER. The family's industrial and commercial interests were greatly developed by his second son, Joseph Pease (1799-1872). He first became a colliery owner in 1830 when he began St. Helen's Colliery near West Auckland in partnership with his brother-in-law, Henry Burbeck. His interests were developed further by his eldest son, Sir Joseph Whitwell Pease (1828-1903); but of the later members of the family, only Sir J.W. Pease's eldest son, Sir Arthur Pease (1857-1939), played a significant part in the businesses. From the 1930s this control passed to others.

Joseph Pease & Partners was formed in the 1850s, while J.W. Pease & Co was formed in 1852. When the new company was formed in 1882 its prospectus lists the two companies as owning the following:

Joseph Pease & Partners	J.W. Pease & Co
Adelaide Colliery	Crag's Hall Mine
Brandon Colliery	Lingdale Mine
Esh Colliery	Lofthouse Mine
Pease's West	Upleatham Mines
St.Helen's Colliery	Whitcliffe Mine
Tindale Colliery	Broadwood Quarry
Windlestone Colliery	Frosterley Quarry

The names of the collieries above, all in County Durham, need further explanation, which is given below. Both of the quarries were in Weardale. The mines were all ironstone mines in the North Riding of Yorkshire (and so outside the scope of this book); in 1882 the company was the largest producer of ironstone in Cleveland.

The company was re-registered as a public company on 11th October 1898. At this date it was producing annually 1,300,000 tons of coal, 715,000 tons of coke from 1,866 coke ovens, 1,196,000 tons of ironstone, 260,000 tons of limestone, owned a chemical works and firebrick works and employed 6,000 people. It already controlled The Skinningrove Iron Co Ltd in Yorkshire, and was to acquire other ironworks on Teesside in the years ahead.

The troubles that were to come to a head in 1930 can be traced back to two important policy decisions. The company saw its position in the coal trade in the longer term threatened by the gradual exhaustion of its west Durham pits. Its first reaction was to acquire in 1909 an untouched royalty at Thorne, north-east of Doncaster in Yorkshire, the only North-East coal company besides Dorman, Long & Co Ltd later to invest south of the River Tees. The sinking of Thorne Colliery proved difficult and protracted; coal was not reached until July 1924 and was an enormous drain on the company's finances. Secondly, it also began acquiring what became subsidiary companies both in Durham and Yorkshire, and these proved increasingly shaky financially in the depressed financial cycle that set in after 1921. Some collieries and quarries had to be closed in 1925 and more closures followed early in 1930. Despite these, the company and its subsidiary companies were still employing over 20,000 people at this time. With losses approaching £1½ million, an Advisory Committee was appointed in September 1930. Eventually a Scheme of Arrangement regarding its debts was agreed with the

Chancery Division of the High Court on 24th November 1932. This brought more closures and the liquidation of one of the subsidiaries, The North Bitchburn Coal Co Ltd, but was successful in ensuring the survival of the parent company and the other subsidiaries. By the mid 1930s the slimmed-down company and its subsidiaries were profitable again.

In 1947 the company underwent a major reorganisation. It lost its collieries and withdrew from quarrying, converted its remaining activities in ironworks and wharves into subsidiary companies, and became a holding company. By 1958 the only activity left was the Tees Ironworks at Cargo Fleet, near Middlesbrough, Yorkshire (NR), which then reverted to the direct ownership of Pease & Partners Ltd. Its main contract was to cast rail chairs for British Railways, and when BR indicated that this contract was not going to be renewed, the company decided early in 1959 to close the works and to cease trading. In the event the Franco-British & General Trust Ltd made an offer for the shares, and Pease & Partners Ltd ceased to exist about August 1959.

It may be helpful to list the various subsidiary companies that Pease & Partners Ltd acquired or created, with dates given where appropriate. Some are involved with locomotive transfers listed below:

The Skinningrove Iron Co Ltd, Carlin How, Yorkshire (NR).
T. & R.W. Bower Ltd, Allerton Main Collieries, Yorkshire (WR), in 1917-1918.
Henry Stobart & Co Ltd (Durham) (which see), from 1st January 1920; to National Coal Board.
The North Bitchburn Coal Co Ltd (Durham) (which see), from 1st April 1920 (North Bitchburn papers give 26th February 1920); this company was liquidated on 27th July 1934.
East Hetton Collieries Ltd (Durham) (which see), from 20th August 1944; to National Coal Board.
Pease & Partners Normanby Iron Works Co Ltd, Cargo Fleet, Yorkshire (NR) and its predecessors.
Pease & Partners Tees Ironworks Co Ltd, Cargo Fleet, Yorkshire (NR), and its predecessors.
Deepwater Wharf Ltd, Cargo Fleet, Yorkshire (NR).

The company's locomotives

The company was easily the biggest colliery and quarry company not owning its own railway system in Co. Durham. Equally, there does not seem to have been what might be called a central locomotive repair facility, though some major overhauls were undertaken at the Tees Ironworks at Cargo Fleet in Yorkshire (NR), and there were small engineering shops at Pease's West at Crook. Much of its colliery and quarry business did not survive 1930, many of the quarries being in remote parts of Weardale. For these reasons information about the locomotives that the company owned is very sketchy. There were undoubtedly more locomotives than those listed below, while there are a few locomotives, listed at the end, which the firm is known to have owned but where they worked is not known.

A number of valuations of the company survive, unfortunately without giving any details of locomotives. The valuation for 1891 lists the number of coke ovens, beehive type and by-product, at each location, and as an example for readers of the vast scale of beehive ovens in County Durham, the numbers are included below (the reader should note that the term 'beehive' included both various sizes of 'beehive oven' and also other shapes, their common factors being that they were all built of brick and all the by-products, except sometimes the heat generated, were lost to atmosphere). The centre of the company's colliery business was at Crook, especially in the area to the north-west of the town which became called "Pease's West", a name which became so common that not only is it shown on Ordnance Survey maps but it was for a time used as a prefix for all of the company's collieries. For this reason Pease's West is dealt with first in the list below, followed by the other locations alphabetically. The company's head office was at Northgate in Darlington.

The company's first locomotive numbering scheme was probably introduced towards the end of the nineteenth century. A new numbering scheme replaced this about 1944.

PEASE'S WEST, Crook M355
Joseph Pease & Partners until 19/8/1882

The first railway to reach Crook was the Stockton & Darlington Railway's line from South Church near Bishop Auckland and was opened on 8th November 1843. Next to be constructed was the Weardale Extension Railway, from Crook to a junction with the Pontop & South Shields Railway at Waskerley, the Stockton & Darlington Railway, in which of course the Peases were heavily involved, being behind this line too. It was opened on 14th May 1845, and included the huge Sunniside Incline out of Crook, 1¾ miles long with the severest gradient being 1 in 16 in places. The bank foot of the incline was situated to the north west of Crook, and it was here that Joseph Pease developed the huge industrial area that became known as Pease's West.

The sinking of the first colliery here began in April 1844, before the Extension Railway was even open. This was the **(NEW) RODDYMOOR PIT**, whose site is uncertain, which Pease acquired in

PEASE & PARTNERS LTD near Crook

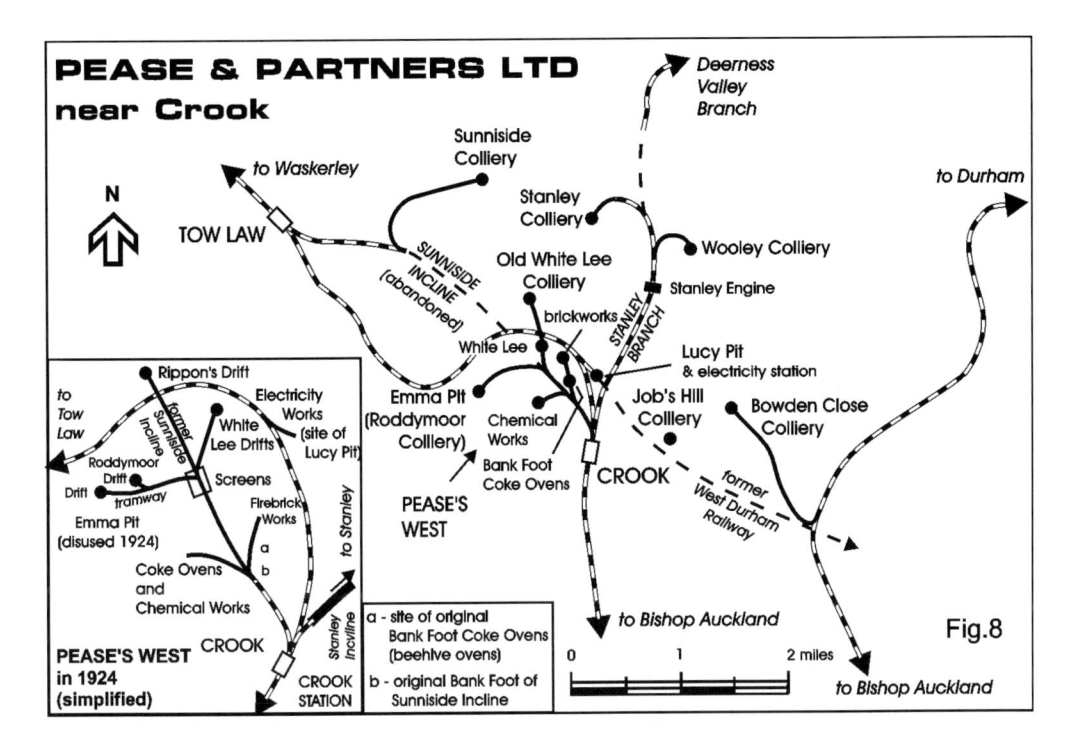

Fig.8

January 1846. This may be the same location as the **EMMA PIT** (NZ 155365), whose sinking began in 1846. This was followed in 1849 by the **LUCY PIT** (NZ 163368) (see below) and according to Fordyce (1860), the **EDWARD PIT**. The Lucy Pit appears to have been linked to the main area by the ½ mile long **Roddymoor Incline**, probably a self-acting incline. The Edward Pit is not shown in the area on the 1st edition O.S. map, though a pit with this name is shown some two miles away near the top of the Sunniside Incline at **HEDLEY HOPE COLLIERY** (NZ 135394), which Joseph Pease & Partners owned briefly before selling it on (see The Bearpark Coal & Coke Co Ltd). Various other drifts are known to have operated in this area, including the **EMMA FIVE QUARTER DRIFT** (abandoned in December 1891), the **RIPPON DRIFT** and the **RODDYMOOR DRIFT**. Some of these became possible after the Sunniside Incline was closed on 2nd March 1868. Some are known to have been served by "tramways", presumably narrow gauge and rope-hauled. In 1889 Pease & Partners Ltd acquired two collieries in this area from Bolckow, Vaughan & Co Ltd, **STANLEY COLLIERY** (see below) and **WHITE LEE COLLIERY** (NZ 155375), to the north-east of Pease's West. The latter's former link to the foot of the Roddymoor Incline on the former West Durham Railway was abandoned and instead it was linked to the Pease's West complex. It was subsequently called Old White Lee Colliery. It was closed by the mid-1890s and the site re-developed for workshops. In its place the **WHITE LEE DRIFTS** (NZ 158369), or (New) White Lee Colliery, were opened.

Latterly production was concentrated at the Emma Pit, renamed **RODDYMOOR COLLIERY**. This survived to pass to NCB Northern Division No.5 Area on 1st January 1947.

The coal at Crook and nearby proved to be the best in the world for making into coke for use in the iron furnaces of the day - pure, strong and with a high calorific value. This led from 1850 onwards to the construction of large numbers of "beehive" coke ovens (there were varying examples of size and shape). These were built to the east of Sunniside Incline bank foot, and became known as the **BANK FOOT COKE OVENS** (NZ 161364). By 1891 there were 633 such ovens here, with a further 76 at the Emma Pit. However, in 1882 the company built a battery of recovery retort ovens here, the first by-product coke ovens ever built in Britain. They were designed by the Frenchmen Henry Simon and Francois Carves. This was a battery of 25 ovens, lit in October 1882, with a further 25 added in 1883. More were built in 1885-1886 and in 1891-1892, making 108 in all. These ovens saw the beginning of the technical developments to save the waste heat, the gas and the various chemicals that had previously all been discharged to atmosphere. To process the latter the company next built a **CHEMICAL WORKS** (NZ 158364) to the west of the former bank foot, initially to manufacture ammonium sulphate and benzole, but latterly for tar and other chemicals too. In the 1880s and 1890s about half of the coke produced was sent to the ironworks in Cumberland, notably at Barrow, for which the company had 636 coke wagons in 1891. The waste heat from the coke ovens was also used here to support an **ELECTRICITY WORKS** (NZ 163368), the heat being used to produce steam which in turn drove the turbines. This works was built on the site of the Lucy Pit and was opened about 1905. It was soon supplying much of Crook, to which were subsequently added the collieries further afield. In 1908 the company formed the **Bankfoot Power Co Ltd** jointly with the Newcastle upon Tyne Electric Supply Co Ltd to handle and distribute the output surplus to Pease & Partners' requirements. How long this arrangement lasted, and when the works closed, is not known, although it probably did not survive the mid 1930s. Latterly the site was a sub-station.

In 1908 the company began the construction of Otto waste heat ovens, planned to produce 4,000 tons of coke per week. 60 of these began production in October 1908, and a further 30 were ordered in 1909, with 26 of the Simon Carves ovens being rebuilt in 1910. The completion of all 120 Otto ovens made the site the largest coke works in the county. In 1915 the company opened a sulphuric acid plant here, which continued until 1932, and then a plant for manufacturing colour paint. A coke blending plant was built in 1939.

Fifty-six of the Simon Carves ovens were closed during the Second World War and a further 26 had ceased production by 1947. Despite this, the ovens were second in Durham only to the Norwood Coke Works of The Priestman Collieries Ltd in the quantity of coke produced in 1946 (200,248 tons, compared with 226,387 tons). The Bank Foot Coke Ovens passed to NCB Northern Division No.5 Area on 1st January 1947.

Meanwhile, to the north east of the coke ovens a **FIREBRICK WORKS** (NZ 161365) had been opened in 1860 (firebricks were used to line the interior of a beehive coke oven). The collieries also produced large quantities of seggar clay, which was made into distinctive yellow common bricks. Latterly the clay was brought to the brickworks by a "telfer", a low level ropeway carrying skips. **A SANITARY PIPE WORKS** was later added. Latterly clay was brought to the works by an aerial flight from Roddymoor Colliery. The brickworks closed in 1935.

To bypass the Sunniside Incline the NER eventually built a new line between Crook and Tow Law, which opened for freight traffic on 10th April 1867, although, as noted above, the Sunniside Incline remained open until the beginning of passenger traffic on 2nd March 1868. The opening of this new line cut the Lucy Pit off from the rest of the area, though when it was closed is not known. After the opening of the deviation line, only the first ¼ mile of the former line, as far as Pease's West sidings, remained owned by the NER, all of the remainder of the system being owned by the company. It would seem likely that the company obtained running powers over the NER to provide access to the Lucy Pit/Electricity Works. Latterly there was a loco shed near the Bank Foot Coke Ovens, though whether it was sited here back in the nineteenth century is not known. It would seem absolutely certain that more locomotives worked here than those shown below.

Gauge : 4ft 8½in

-		0-4-0ST	OC	JF	2078	1874	New	s/s
35		0-4-0T	OC	HL	2774	1909	New	Scr c/1920
14	(formerly 45)	0-4-0ST	OC	P	1467	1917	New	(1)
11	(formerly 32)	0-4-0T	OC	HL	2685	1906	(a)	(1)
13	(formerly 42)	0-4-0ST	OC	HL	2823	1910	(b)	(2)
10	CAROLINE	0-4-0ST	OC	BH	998	1891	(c)	Scr c/1939
12	PATRICIA	0-4-0ST	OC	HL	2993	1913	(d)	(1)

(a) ex Bowden Close Colliery & Coke Ovens, Helmington Row.
(b) ex Bowden Close Colliery & Coke Ovens, Helmington Row (at a different date from HL 2685 and HL 2913).
(c) ex Ushaw Moor Colliery, Ushaw Moor.
(d) ex Bowden Close Colliery & Coke Ovens, Helmington Row (at a different date from HL 2685 and HL 2823).

(1) to NCB No. 5 Area, with site, 1/1/1947.
(2) to Henry Stobart & Co Ltd, Thrislington Colliery, West Cornforth, c3/1945.

The next two locomotives were also acquired for use at Pease's West, and the credit that both were subsequently preserved is due to the company. LOCOMOTION, which hauled the first train on the Stockton & Darlington Railway on 27th September 1825, is said to have been used at the Lucy Pit; DERWENT, according to the Darlington Railway Centre & Museum, where she is now displayed, worked between Job's Hill Colliery (see below), Billy Row at Crook and the Sunniside Incline.

Gauge : 4ft 8½in

1	LOCOMOTION	0-4-0	VC	RS	1825	(a)	(1)
25	DERWENT	0-6-0	OC	Kitching	1845	(b)	(2)

(a) ex Stockton & Darlington Railway, /1850; used as a pumping engine.
(b) ex NER, c/1869; originally Stockton & Darlington Railway, 25, DERWENT (with two tenders); this loco was a replacement for former Stockton & Darlington Rly, 12, TRADER, 0-6-0 OC Hackworth 1842 (with two tenders), which itself had been a rebuild of No.12, BRITON, 0-6-0 VC Shildon Works, 1838, which had been sold to Joseph Pease & Partners about 1867 but which exploded in 1868.

(1) returned to Stockton & Darlington Railway, restored and placed on exhibition near North Road Station, Darlington, 6/1857; transferred to Bank Top Station, Darlington, /1890.
(2) hired to Consett Waterworks Co for construction of Smiddy Shaw reservoir, near Waskerley, 1870; returned in 1872; displayed at 50th anniversary celebrations of Stockton & Darlington Railway, 9/1875; returned to NER, 3/1898, and placed on Bank Top Station, Darlington, with LOCOMOTION above.

All the narrow gauge steam locomotives below are believed to have worked on the batteries of beehive coke ovens, many of which continued in production for many years after the first waste recovery ovens were built.

Gauge : 3ft 6in

-		0-4-0ST	OC	JF	5822	1888	New	(1)

(1) s/s by 4/1901; one source says 'to Al...t [Albert?] Dixon'.

103. The maker's photograph of 3ft 0in gauge JF 5822/1888, supplied to Pease & Partners Ltd for working on the system serving their beehive coke ovens at Pease's West, Crook.

The next group of locomotives worked on a system that carried coal from coal bunkers along a low gantry to the Simon-Carves coke ovens for charging. The system was subsequently re-organised and shunting of the tubs was undertaken by the Booth battery loco, with its shed on a higher level, and with No.15 being retained as a stand-by loco. This loco-worked system was replaced by conveyors for the remaining ovens in 1946, but neither loco had been scrapped when the ovens passed to NCB Northern Division No. 5 Area on 1st January 1947.

Gauge : 3ft 0in

15	(MIRIAM)	0-4-0ST	OC	?	?	?	(a)	(1)
-		0-4-0ST	OC	?	?	?	(a)	Scr /1935
-		0-4-0ST	OC	?	?	?	(a)	Scr
-		0-4-0ST	OC	?	?	?	(a)	Scr
-		0-4-0BE		J.Booth	164	?	New	(1)

(a) It has been held in the past that these four locomotives were acquired c/1900 from Lingford, Gardiner & Co Ltd, dealers, Bishop Auckland, and that they were MW 97/1863, MW 98/1863, MW 113/1864 and BH 447/1878, all previously standard gauge locomotives, which LG had bought from Bolckow, Vaughan & Co Ltd at Middlesbrough, Yorkshire (NR) and had rebuilt to 3ft 0in gauge. The authors have been unable to verify this information. DCOA Return 427 records that four drivers (= four locomotives) were working here daily in 4/1901.

(1) to NCB No. 5 Area, 1/1/1947, with ovens, but out of use.

ADELAIDE COLLIERY, South Church, near Bishop Auckland
Q356
Joseph Pease & Partners until 19/8/1882; see also below
NZ 225276

This is another example of a site where two collieries shared the same surface. The first colliery here was **DEANERY COLLIERY** (NZ 233275), which was sunk about 1810 by a Mr. Brown. **ADELAIDE COLLIERY** (NZ 232275), a few yards to the west, was sunk in 1825 by Jonathan Backhouse, trading as the **Black Boy Coal Co** (see Black Boy Colliery) and was acquired by Joseph Pease in 1830. Known collectively as Shildon Bank Collieries, they were initially served by a ½ mile branch from the Stockton & Darlington Railway's Black Boy branch, but after the opening of Shildon Tunnel in 1842, Adelaide was served by a ¼ mile link from just north of the tunnel. The two collieries became one with the closure of Deanery Colliery in 1840. By 1891 the colliery had 139 beehive coke ovens, built to the north of the colliery's pit heap. These were served by a different link, from the NER Haggerleazes branch, which skirted the north of the colliery. Soon after this the colliery was closed.

However, in 1901 the colliery was re-opened by the **West Durham Wallsend Coal Co Ltd** (registered 5/12/1892). By an unusual quirk of history, Pease & Partners Ltd bought their former colliery back, taking over the assets of the West Durham Wallsend Coal Co Ltd with effect from 31st January 1923; the latter firm was liquidated on 10th March 1924. Pease & Partners Ltd's second tenure was short-lived, for the colliery was closed in June 1924, though it was not finally abandoned until October 1932.

It would seem very likely that locomotives worked here, but nothing is known.

BISHOP MIDDLEHAM QUARRY, Bishop Middleham
N357
NZ 333320 approx.

This limestone quarry was almost certainly started by **The Rosedale & Ferryhill Iron Co Ltd** (which see) in the 1870s. This firm failed in 1879, and almost certainly the quarry was not worked again until Pease & Partners leased it in 1902-1903 (Pease & Partners' Annual Reports). There had once been a colliery here, served by a 1½ mile branch from the NER Ferryhill – Stockton line, 2¼ miles south of Ferryhill Station. **Bishop Middleham Colliery** was never owned by Pease & Partners Ltd, its last owners being Dorman, Long & Co Ltd, who never worked coal from it, but later developed it as a man-riding shaft for Mainsforth Colliery. To serve the new quarry the branch was extended by another mile. According to reports in *The Quarry*, the quarry was being opened out in April 1903, and by October 1903 there was a "tank locomotive" here, together with a new engine house; it is possible that this may refer to the locomotive shed, rather than a stationary engine. The quarry was one of the casualties of the crisis period, closing about 1932.

The first two locomotives were said to have been delivered new here, but this is not possible, given the information included above.

Gauge : 4ft 8½in

6		0-4-0ST	OC	RS	2325	1877		
			reb	RS		1894	(a)	(1)
16	(form. 24)	0-4-0ST	OC	HL	2453	1900	(a)	(2)
25	(form. 22)	0-4-0ST	OC	HL	2456	1900	(a)	(3)
20	(form. 33)	0-4-0ST	OC	HL	2713	1907	New	(4)
9		0-4-0ST	OC	HL	2247	1892	(b)	(5)
10	(form. 25)	0-4-0ST	OC	HL	2559	1903	(c)	(6)

(a) New to Pease & Partners Ltd; location prior to here and date of transfer unknown; see note above; one source claimed that HL 2456/1900 was the first loco here.
(b) ex Washington Chemical Co Ltd, Washington Works, Washington (c/1909?).
(c) ex Bowden Close Colliery & Coke Ovens, Helmington Row.

(1) to Bowden Close Colliery & Coke Ovens, Helmington Row.
(2) to Chilton Quarry & Limeworks, near Ferryhill Station.
(3) to Chilton Quarry & Limeworks, near Ferryhill Station (at a different date from HL 2453/1900).
(4) to Henry Stobart & Co Ltd, Fishburn Colliery & Coke Works, Fishburn.
(5) to The North Bitchburn Coal Co Ltd, Thrislington Colliery, West Cornforth, in 1920s, and returned; from Henry Stobart & Co Ltd, Thrislington Colliery, West Cornforth, and returned there.
(6) to Henry Stobart & Co Ltd, Thrislington Colliery, West Cornforth, /1938.

BOWDEN CLOSE COLLIERY & COKE OVENS, Helmington Row   M358
Joseph Pease & Partners until 19/8/1882   NZ 181362

This colliery was sunk in 1844-45 by **The Northern Coal Mining Co** and was acquired after that company's bankruptcy in 1848. It was linked to the West Durham Railway by a branch one mile long from Sunny Brow. This was worked by a **stationary engine** at the colliery, which hauled the empty waggons up from the junction and presumably lowered the full waggons down. In 1857 the NER opened its Bishop Auckland branch, from Leamside to Bishop Auckland, and a link to the Bowden Close branch was put in where the line crossed the former West Durham line. Despite this, the stationary engine was still working in 1870, and is recorded on both sections of DCOA Return 102, the second in November 1876. In 1891 the NER decided to close a major section of the former West Durham line, from which time all the Bowden Close traffic was handled by the Bishop Auckland branch. Whether the stationary engine was also abandoned at this time is not known. There may well have been locomotives in use here before 1903, the date of the first shown below.

In 1878 the nearby **JOB'S HILL COLLIERY** (see below) was closed and its working absorbed into Bowden Close Colliery. By 1891 the colliery had 137 beehive coke ovens, which ceased work in July 1911. 120 Otto-Hilgenstock waste heat ovens began production here in June 1912. The colliery was closed in May 1930, but the coke ovens continued briefly; 80 were closed during the first half of 1932 and the final 40 closed in April 1933. The plant and the colliery were dismantled almost immediately.

Gauge : 4ft 8½in

25		0-4-0ST	OC	HL	2559	1903	New	(1)
No.32		0-4-0T	OC	HL	2685	1906	New	(2)
40		0-4-0ST	OC	HL	2799	1909	New	(3)
16	(formerly 6)	0-4-0ST	OC	RS	2325	1877		
		reb		RS		1894	(a)	(4)
43		0-6-0T	OC	HL	3104	1915	New	(5)
	PATRICIA	0-4-0ST	OC	HL	2993	1913	(b)	(6)
42		0-4-0ST	OC	HL	2823	1910	(c)	(7)

(a) ex Bishop Middleham Quarry, Bishop Middleham.
(b) ex Tees Iron Works, Cargo Fleet, Yorkshire (NR).
(c) ex Thorne Colliery, Moorends, near Doncaster, Yorkshire (ER).

(1) to Bishop Middleham Quarry, Bishop Middleham.
(2) to Pease's West, Crook.
(3) to Harehope Quarry, near Frosterley.
(4) to Ushaw Moor Colliery, Ushaw Moor.
(5) to Henry Stobart & Co Ltd, Fishburn Colliery & Coke Ovens, Fishburn.
(6) to Pease's West, Crook (at a different date from HL 2685 and HL 2823).
(7) to Pease's West, Crook (at a different date from HL 2685 and HL 2913).

BROADWOOD QUARRY, near Frosterley   U359
J. W. Pease & Co until 19/8/1882   NZ 033366 approx.

This quarry was being worked by 1871, as the locomotive below is recorded as being delivered here. The Wear Valley Railway, nominally a subsidiary of the Stockton & Darlington Railway, was opened from west of Witton Park to Frosterley on 3rd August 1847, and it may be that the quarry was begun soon afterwards, almost certainly by the Stockton & Darlington Railway to meet its obligations to supply limestone to various iron companies. When it was taken over by J. W. Pease & Co is not known. It lay to the south of the River Wear, near its confluence with the Bollihope Burn, and was served by ½ mile branch one mile east of Frosterley Station.

The quarry was closed about 1921, but was taken over about 1926 by the **Broadwood Limestone Co** (which see), a trading name for **The Witton Park Slag Co Ltd** (which see).

It would seem almost certain that there were more locomotives here than just the one listed below.

Gauge : 4ft 8½in

EGYPT	0-6-0T	IC	JF	1539	1871	New	(1)

(1) believed loaned to Auckland Rural District Council, Witton Park Slag Works, Witton Park, c/1910 and returned c/1914; to Lingford, Gardiner & Co Ltd, dealers, Bishop Auckland, possibly at the closure of the quarry about 1921.

BROWN'S HOUSES QUARRY, also known as SOUTH BISHOPLEY QUARRY, near Frosterley   U360
Ord & Maddison Ltd until 3/1920   NZ 021364 approx.
NORTH BISHOPLEY QUARRY, near Frosterley   U361
Ord & Maddison Ltd until 3/1920   NZ 026364 approx.

Brown's Houses Quarry, listed in Pease & Partners Ltd's prospectus of 1898, was at that date sub-let, almost certainly to Ord & Maddison Ltd (which see), who were also working other limestone quarries in this area south of Frosterley. This company ceased operations here in March 1920, and it is believed that Pease & Partners Ltd began production themselves later that year.

North Bishopley Quarry was served by a ½ mile branch from near the end of the NER Bishopley Branch, with Brown's Houses Quarry served by a ½ mile extension from North Bishopley, via a tunnel; the former NER ½ mile branch direct to Frosterley Station had been abandoned about 1914. There was a loco shed at Brown's Houses Quarry; see also the entry for Harehope Quarry below. An article about North Bishopley Quarry in *The Quarry* of November 1903 records a locomotive here at that time.

All of these quarries ceased production in 1925, though North Bishopley was re-opened in 1938 by different owners, without rail traffic.

CHILTON COLLIERY & COKE OVENS, near Chilton Buildings   N362
Henry Stobart & Co Ltd until 1924; see also below   NZ 278308

This colliery had been started in 1835; for its earlier history see the entry for The South Durham Coal Co Ltd. It was served by a ¾ mile branch from the LNER Chilton Branch. It was transferred from Henry Stobart & Co Ltd in 1924 in order to join its coal to Pease & Partners Ltd's **WINDLESTONE COLLIERY** (NZ 287296) (see below), which was served by sidings south of the LNER Chilton Branch at Chilton Buildings, a mile to the south-east of Chilton Colliery. The merger completed, Windlestone Colliery was closed in October 1924. Henry Stobart & Co Ltd had established engineering workshops here of a reasonable capacity, and in 1928 53 people were employed here.

The original 50 Simplex waste heat coke ovens here had been closed in 1921, but 52 new Simplex waste heat ovens, together with a by-products plant for manufacturing tar, sulphate of ammonia and crude benzole, had been started in the quarter ending 31st December 1923. They and the colliery became an early victim of the firm's financial crisis, the ovens being closed in April 1930 and the colliery following on 24th May 1930. However, in 1934 both were re-opened by Dorman, Long & Co Ltd (which see).

Gauge : 4ft 8½in

CHILTON No.1	0-6-0ST	OC	AB	961	1903	(a)	(1)
WINDLESTONE	0-4-0ST	OC	AB	1085	1907	(a)	(2)
CHILTON No.2	0-6-0ST	OC	AB	1097	1907	(a)	(1)
CHILTON No.3	0-6-0ST	IC	P	1219	1910	(a)	(3)

(a) ex Henry Stobart & Co Ltd, with colliery, /1924.
(1) to Henry Stobart & Co Ltd, Fishburn Colliery & Coke Works, Fishburn, c/1930.
(2) to North Bitchburn Coal Co Ltd, Thrislington Colliery, West Cornforth, /1929.
(3) to Thorne Colliery, Moorends, near Doncaster, Yorkshire (ER), /1931.

CHILTON QUARRY & LIMEWORKS, Chilton, near Ferryhill Station   N363
    NZ 300315 approx.

This quarry and limeworks was established on the site of the former Little Chilton Colliery, probably by Frederick Mildred, a local farmer, who was operating the site by 1897. It was served by a link to the NER Durham – Darlington line, ¼ mile south of Ferryhill Station. The Pease & Partners Ltd Annual Reports do not give specific details of when the firm took it over, though "additional limestone resources" were acquired in 1913-1914, which may well refer to here; the List of Mines & Quarries for

1914 gives the quarry as owned by the Earl of Eldon, the land-owner, and not working.

The quarry and limeworks are listed in Pease & Partners Ltd's list of assets of 1929 as 'working'. The List of Quarries for 1937 shows the quarry producing limestone, dolomite, clay, gravel and sand. They are believed to have been taken over by **F.W. Dobson & Co Ltd** (which see) about 1947, and this may well be correct, although there is no mention of this in the Pease & Partners' Annual Reports.

Gauge : 4ft 8½in

25	(formerly 22)	0-4-0ST	OC	HL	2456	1900	(a)	(1)
17	GEORGE	0-4-0ST	OC	K	1705	1871		
			reb	J.Tait	90	1920	(b)	(2)
24	(formerly 16)	0-4-0ST	OC	HL	2453	1900	(c)	(3)

(a) ex Bishop Middleham Quarry, Bishop Middleham.
(b) ex Tees Ironworks, Cargo Fleet, Yorkshire (NR), /1932, loan.
(c) ex Bishop Middleham Quarry, Bishop Middleham (at a different date from HL 2456/1900).

(1) to Henry Stobart & Co Ltd, Fishburn Colliery & Coke Ovens, Fishburn, /1925; returned by 19/3/1934; to Henry Stobart & Co Ltd, Thrislington Colliery & Coking Plant, West Cornforth, after 12/5/1946, by 1/1/1947.
(2) returned to Tees Ironworks, Cargo Fleet, Yorkshire (NR).
(3) to F.W. Dobson & Co Ltd, with quarry and limeworks, c/1947.

ELDON COLLIERY, Eldon, near Shildon

There were two different collieries with this name, both of them passing through Pease hands, though at different times.

ELDON COLLIERY (John Henry Pit)   Q364
NZ 244286

This colliery was certainly working by 1821, and after the opening of the Stockton & Darlington Railway in 1825 it was linked to the Railway's Black Boy branch, opened in July 1827. Joseph Pease then began a major development of the colliery, with the sinking of the John Henry Pit started on 20th August 1829 and the sinking of a second shaft begun on 29th December 1831. Later the colliery had its own branch 1½ miles long to the Stockton & Darlington Railway, 1½ miles north of Shildon Station. How this branch was worked is not known. In 1858 the colliery was being shunted by contractors (identity unknown) using a hired NER locomotive (identity also unknown). In May 1859 the colliery was sold to Samuel Smithson & Partners.

ELDON COLLIERY & COKE OVENS; also known as SOUTH DURHAM COLLIERY   Q365
The South Durham Coal Co Ltd until 1903
The South Durham Coal Co until /1888   NZ 238279

On 22nd December 1862 the sinking of a new colliery was begun about ½ mile to the south-west of the original Eldon Colliery and upon the branch line to it, ¾ mile from the junction with the NER. A second shaft was begun on 6th May 1865. For many years this colliery was known as South Durham Colliery. Latterly the colliery was owned by **The South Durham Coal Co Ltd** (see the entry for this firm). The original colliery became called **OLD ELDON COLLIERY**.

Pease & Partners Ltd bought the colliery in 1903, and the title **ELDON COLLIERY** was soon used in preference to South Durham Colliery, so that one is left to assume that the old Eldon Colliery had closed, perhaps some considerable time earlier. Two years later the colliery was linked underground to **WINDLESTONE COLLIERY** (see below), according to the company's Annual Report, and eventually the John Henry Pit was also returned to coal-winding. A battery of 50 Otto-Hilgenstock waste heat by-product ovens was opened about February 1905, and a further battery of 50 ovens was started up in the half year ending 30th June 1913. A **brickworks** had been started here in 1877, but it seems that it might have been closed when Pease & Partners acquired the colliery, for it would appear that in 1920 the company opened (re-opened?) the brickworks (NZ 237279) here for the manufacture of common bricks. This was closed at the beginning of the General Strike in April 1926. Davison, in his history of North East brickworks, claims that it was mothballed until 1933, but the Pease & Partners' List of Assets in 1929 records it as working. Due to the financial crisis, coke production was stopped in May 1931 and coal production in August 1931. With no hope of recovery, the site was abandoned in October 1932. However, pumping continued, and this was taken over by Dorman, Long & Co Ltd in January 1933 to protect that company's Auckland Park Colliery nearby. The colliery and coke ovens were sold to Thos W. Ward Ltd for dismantling in May 1933 (see Part 2). The brickworks reverted to the Eldon Estate as landowner, and was re-opened later in 1933 by **Eldon Brickworks Ltd** (which see).

104. ELDON No.2, MW 1566/1902, which Pease & Partners Ltd acquired with Eldon Colliery from The South Durham Coal Co Ltd in 1903.

Gauge : 4ft 8½in

	ELDON	0-6-0ST	IC	MW	926	1884	New	s/s after 2/1916
18	(formerly 3)	0-4-0ST	OC	HL	2185	1890	New	s/s
	ELDON No.2	0-6-0ST	IC	MW	1566	1902	New	(1)
30	(form. ELDON No.3)	0-6-0ST	OC	P	1092	1906	New	(2)
34		0-4-0ST	OC	AB	1085	1907	New	(3)
	ELDON No.1	0-4-0ST	OC	T.D.Ridley 74	1920	New	(2)
	RAKIE	0-4-0ST	OC	P	583	1894	(a)	(4)

(a) ex Skinningrove Iron Co Ltd, Carlin How, Yorkshire (NR), loan, /1924.

(1) to Thos W. Ward Ltd, contractors, with site, 5/1933.
(2) to St.Helen's Colliery, St.Helen Auckland, /1932.
(3) to Henry Stobart & Co Ltd, Chilton Colliery, Chilton Buildings.
(4) returned to Skinningrove Iron Co Ltd, Carlin How, Yorkshire (NR), /1925.

The closure of Eldon Colliery did not end the company's operations in the area. During the 1930s the company opened **ELDON DRIFT** (NZ 253281) **(Q366)**, about ½ mile north-east of the hamlet of Old Eldon. This used a narrow gauge tramway, almost certainly rope-worked. The drift passed to NCB Northern Division No. 4 Area on 1st January 1947.

ESH COLLIERY, Esh Winning (formerly Esh)   K367
Joseph Pease & Partners until 19/8/1882   NZ 195423

This colliery was the second of the firm's developments in the valley of the River Deerness. The sinking was begun on 12th January 1857, four months after the start of sinking Waterhouses Colliery (see below), though it was not until 1866 that the colliery began production. It was served by ¼ mile branch from Flass Junction on the NER Deerness Valley branch, opened in 1858. There were 201 beehive coke ovens here in 1891. One source was informed that narrow gauge locomotives were used on them, but there is no evidence in the DCOA Returns, or from any other source, to support this.

The colliery was closed in November 1930, but after standing idle for twelve years, the company re-opened it in 1942. It passed to NCB Northern Division No. 5 Area on 1st January 1947.

The LNER worked the traffic from 1942; but whether this was also the case up to the closure in 1930 or whether a locomotive was used is not known.

FINE BURN QUARRY, near Frosterley
for previous owners see below

U368
NZ 010349 approx.

This limestone quarry had been begun by **Bolckow, Vaughan & Co Ltd** in the 1860s. It was served by a ¾ mile private extension of the NER Bishopley branch. In 1871 it was taken over by **Jacob Walton & Co Ltd** (which see), who closed it about 1907. It re-commenced production under Pease & Partners' ownership in June 1910. It may well have been combined with the adjacent Bollyhope Quarry, worked by William Spencer & Co Ltd (which see) until 1909.

When it was re-opened it had its own locomotive, but later it was worked by a locomotive running down from Harehope Quarry (see below). The quarry was closed in 1925, together with the others served by the Branch.

Gauge : 4ft 8½in

NORMANBY No.4	0-4-0T	OC	K	1790	1872	(a)		s/s

(a)   ex Normanby Iron Works Co Ltd, Normanby, Yorkshire (NR).

FROSTERLEY QUARRY, Frosterley
J. W. Pease & Co until 19/8/1882

U369
NZ 033373 approx.

This limestone quarry on the northern side of the valley was begun in the 1850s by the local landowners, the Rippon family. It was served by a ½ mile branch from the Stockton & Darlington Railway's Wear Valley Branch, ½ mile east of Frosterley Station. It was taken over by J.W. Pease & Co about 1875. The locomotive shed (NZ 033369) was sited just south of the tunnel under what is now the A689 road. The quarry was closed in 1921.

However, it would appear that production was re-started towards the end of the 1930s, with the last two locomotives below there by 1940. Production appears to have ceased soon after the end of the Second World War.

Gauge : 4ft 8½in

26	FROSTERLEY	0-4-0ST	OC	HL	2563	1903	New	(1)
26	HAREHOPE	0-4-0ST	OC	HL	2799	1909	(a)	
		reb		TIW	April	1938		(2)
No.1	MERRYBENT	0-4-0ST	OC	HL	3053	1914	(b)	(2)

(a)   ex Rogerley Quarry, Frosterley, by 15/8/1931.
(b)   ex Rogerley Quarry, Frosterley, by 23/3/1940.
(1)   to Rogerley Quarry, Frosterley (c/1921?).
(2)   to Tees Ironworks, Middlesbrough, Yorkshire (NR), by 5/1947.

HAREHOPE QUARRY, near Frosterley
The Harehope Gill Mining & Quarrying Co Ltd until 1915

U370
NZ 038364

The Harehope Gill Mining & Quarrying Co Ltd (which see for the site's earlier history) re-opened this quarry for limestone and ganister in 1901. It was served by a ½ mile branch from sidings at Bishopley Junction, 1¼ miles east of Frosterley Station on the NER Wear Valley branch. The firm ceased operations in December 1914, and Pease & Partners Ltd took over in 1915. The NER Bishopley Branch ran for 1¼ miles from Bishopley Junction, serving various quarries (see Map U). In later years a locomotive from here is said to have also travelled along the branch to shunt **FINE BURN QUARRY** (see above), presumably using running powers over the NER. The quarry is listed as "idle" in the 1929 list of assets, and appears to have been abandoned in 1931.

The quarry is known to have had several different locomotives, but only one can be positively identified.

Gauge : 4ft 8½in

26 (formerly 40)	HAREHOPE	0-4-0ST	OC	HL	2799	1909	(a)	(1)

(a)   ex Bowden Close Colliery & Coke Ovens, Helmington Row.
(1)   to Rogerley Quarry, Frosterley, /1931.

JOB'S HILL COLLIERY, near Helmington Row
Joseph Pease & Partners until 19/8/1882

M371
NZ 174354

This colliery was acquired with New Roddymoor Colliery (see Pease' West above) in January 1846 and subsequently re-opened, although perhaps not until 1855. It was served by sidings north of the former West Durham Railway about ¾ mile south-east of Crook. It was closed in 1878. Its workings may well have been absorbed into Bowden Close Colliery (see above) nearby. It is not known whether locomotives were used here, although see the entry for DERWENT under Pease's West.

ROGERLEY QUARRY, Frosterley

U372
NZ 015377 approx.

This became a very large limestone quarry, eventually two miles long, on the northern slopes of Weardale between Frosterley and Stanhope, eventually reaching the workings of The Consett Iron Co Ltd's Ashes Quarry. The deposits were leased from 1st January 1882 and production began in 1889. It was originally linked to Frosterley Station on the NER Wear Valley Branch by a **self-acting rope incline**.

It may be that originally there were two loco sheds here, one at the interchange sidings with the NER and one in the quarry. Latterly there was just one shed, about half way up the bank between the LNER and the quarry. There may well have been one locomotive or more before the first listed below.

The quarry survived the crisis of the 1930s, but by the late 1930s the operation of the crushing plant had been taken over by Bradley's (Weardale) Ltd. It would appear that production had ceased by March 1940. In 1947 **Bradley's (Weardale) Ltd** took over full ownership and resumed production (see Durham & Yorkshire Whinstone Co Ltd).

Gauge : 4ft 8½in

No.1	MERRYBENT	0-4-0ST	OC	HL	3053	1914	(a)	(1)
22	(formerly 26)							
	(FROSTERLEY)	0-4-0ST	OC	HL	2563	1903	(b)	(2)
39		0-4-0ST	OC	HL	2798	1909	(c)	(3)
26	HAREHOPE	0-4-0ST	OC	HL	2799	1909	(d)	(4)

(a) ex Barton Limestone Co Ltd, Barton Quarry, Barton, Yorkshire (NR), c/1915; said to have been allocated 46 in the first numbering scheme and 21 in the second scheme, but there is no evidence that these numbers were ever carried.
(b) ex Frosterley Quarry, Frosterley (c/1921?).
(c) ex St. Helen's Colliery and Coke Ovens, St. Helen Auckland, /1926.
(d) ex Harehope Quarry, near Frosterley, /1931.

(1) to Frosterley Quarry, Frosterley, by 23/3/1940.
(2) to Bradley's (Weardale) Ltd, with quarry, /1947.
(3) to St. Helen's Colliery and Coke Ovens, St. Helen Auckland, /1927.
(4) to Frosterley Quarry, Frosterley, by 15/8/1931.

ST. HELEN'S COLLIERY & COKE OVENS, St. Helen Auckland
Joseph Pease & Partners until 19/8/1882

Q373
NZ 198272

The sinking of this colliery, originally called **ST. HELEN'S AUCKLAND COLLIERY**, was begun on 24th March 1830 and was completed in January 1831. The colliery was originally served by a one mile branch from the bank foot of the Brusselton East Incline, worked by the Brusselton Engine (NZ 205258), on the Stockton & Darlington Railway east of Shildon. On 2nd June 1835 Joseph Pease began the sinking of **WOODHOUSE CLOSE COLLIERY** (NZ 202283) **(Q374)**, and a extension of the branch one mile long was built from just west of St.Helen's Auckland Colliery to serve it. One can only assume that horses worked this extension and shunted the waggons at the two collieries.

On 8th July 1856 the NER opened its line from Bishop Auckland to Barnard Castle. This passed immediately to the south of St. Helen's Colliery, which was then linked to it, ½ mile east of St. Helen's Auckland Station. However, it would seem that traffic was still dispatched via the original link until the Brusselton Inclines were closed on 13th October 1858.

It would seem that Woodhouse Close Colliery had closed by 1882, as it is not listed in the new company's prospectus. However, it seems not to have been finally abandoned until 1898. How the branch to it was worked in the later part of the nineteenth century is not known.

By 1882 St. Helen's Colliery had become another example of two collieries sharing the same surface, with the opening of **TINDALE COLLIERY**. However, the latter is not included in the 1898 list, so presumably its workings had been combined with St. Helen's by then, though its area was not finally abandoned until 1928.

In 1891 the colliery is listed as having 212 beehive coke ovens, though 82 are described as 'useless'. About May 1905 the company opened a battery of 25 Otto-Hilgenstock waste heat by-product ovens here, to the east of the colliery, with 10 more joining them in July 1906. A battery of 25 Simon Carves regenerative ovens was added in January 1919. By 1924 a **brickworks** had been started, but on the opposite side of the LNER line (NZ 198269).

10 of the Otto ovens were closed in 1920. The 25 Simon Carves ovens closed on 12th July 1924, and the remaining 25 Otto ovens followed about November 1924. The colliery was closed at the beginning of the miners' strike in March 1926 and did not re-open, despite the locomotive transfers below (perhaps these involved the brickworks, which did resume production).

It would seem almost certain that there must have been more locomotives here than those listed below, but no other details are known.

Gauge : 4ft 8½in

No.5		0-4-0ST	OC	BH	37	1867	New	Scr
8	(formerly 39)	0-4-0ST	OC	HL	2798	1909	(a)	(1)
30	(form. ELDON No.3)	0-6-0ST	OC	P	1092	1906	(b)	s/s
	ELDON No.1	0-4-0ST	OC	T.D.Ridley	74	1920	(b)	s/s
18		0-4-0ST	OC	T.D.Ridley	76	1920	(c)	s/s

(a) ex Wooley Colliery, near Crook, by /1925.
(b) ex Eldon Colliery, Eldon, near Shildon, /1932.
(c) origin and date of arrival unknown; here by 25/6/1935.
(1) to Rogerley Quarry, Frosterley, /1926; returned, /1927; to Henry Stobart & Co Ltd, Thrislington Colliery, West Cornforth, /1933.

STANLEY COLLIERY, Mount Pleasant, near Crook K/M375
Joseph Pease & Partners until 19/8/1882 NZ 169389

This colliery was opened in 1858 and was served by the Stanley Branch of the Stockton & Darlington Railway, opened at the same time. The line was 3½ miles long from Crook and included three inclines, two of them (the southern one having a gradient of 1 in 16) worked by the Stanley Engine (NZ 170374). The branch was subsequently extended by the NER to join the Deerness Valley branch at Waterhouses, and the Supplement to the *Mining Journal* of 19th November 1870 gives this description of the operation: "the Stanley stationary engine hauls on the incline from Waterhouses, and the branches from Stanley and Wooley Collieries [for the latter see below] up to the engine....", in other words, the engine worked three inclines on its northern side, as well as the incline down into Crook. However, the first locomotive below is shown as new to the colliery in 1865, and the DCOA Return for March 1890 also lists one locomotive here; presumably it was used to shunt at the colliery itself.

For part of the nineteenth century the colliery was included in "Pease's West", which is the reason why it appears not to be included in the 1882 list above. There were 154 beehive coke ovens here in 1891. It was previously believed that the colliery was closed in November 1911; but the company's Annual Reports list it as being laid in during 1900-1901; almost certainly its workings were combined with Wooley Colliery.

Gauge : 4ft 8½in

No.8	(STANLEY)	0-4-0ST	OC	MW	144	1865	New	s/s after 4/1903
7 (formerly 44)	CLEVELAND	0-4-0ST	OC	MW	744	1880	(a)	(1)

(a) ex W. & J. Lant, contractors; originally Forcett Limestone Co Ltd, Forcett Quarry, Forcett, Yorkshire (NR).
(1) to another Pease & Partners Ltd location; at Thorne Colliery, Moorends, near Doncaster, Yorkshire (ER), by 1/1928.

SUNNISIDE COLLIERY, near Crook M376
Joseph Pease & Partners until 19/8/1882 NZ 147395

The sinking of this colliery began on 19th October 1846, though one source says that it was not opened until 1867. It was served by a branch 1¼ miles long from near the top of the Sunniside Incline on the Stockton & Darlington Railway between Crook and Tow Law. How the branch was worked is not known, nor is whether the firm used its own locomotive here. The colliery is not included on the 1882 list above, and so had presumably been abandoned by this time.

TINDALE COLLIERY, St. Helen Auckland – see the entry for St. Helen's Colliery

TUTHILL QUARRY, near Haswell L377
Tuthill Limestone Co Ltd until 1899 (registered in 1880) NZ 388426

This quarry had opened by 1882 (one source quotes the owners as the Tuthill Limestone Quarry Co, presumably inaccurately) and was acquired by Pease & Partners Ltd in 1899. It was served by a ½ mile branch from the southern end of the South Hetton Coal Co Ltd's Pespool Branch, the junction being at the same point where the Londonderry Railway's Shotton Branch had once diverged. The quarry was closed after the First World War, and was subsequently taken over by **The Northern Sabulite Explosives Co Ltd** (see Imperial Chemical Industries Ltd, Nobel Division).

A notebook begun on 5th January 1895 by the boilersmith at the South Hetton locomotive sheds of The South Hetton Coal Co Ltd (which see), which came to light in 2004, shows that he regularly undertook work on the locomotives at Tuthill Quarry, presumably under contract. He consistently calls them 'No. 1' and 'No. 2', the work including the fitting of a new firebox to No. 2 in 1896; but their identity, whether one or both are included below, and how long this arrangement lasted, are all unknown.

Gauge : 4ft 8½in

	LADY CORNELIA	0-6-0ST	IC	MW	49	1862	(a)	s/s after /1895
	-	0-4-0ST	OC	MW	498	1874	(b)	(1)
	-	0-4-0ST	OC	RS	2875	1897	New	s/s
27		0-6-0ST	IC	P	1040	1905	New	(2)

(a) ex Brecon & Merthyr Railway, 16, 2/1882.
(b) ex J.Whitham & Sons, Perseverance Iron Works, Leeds, Yorkshire (WR).

(1) to Casebourne & Co Ltd, Haverton Hill (see Cleveland & North Yorkshire Handbook).
(2) to another Pease & Partners Ltd location; to T. & R.W. Bower Ltd, Allerton Main Collieries, Swillington, Yorkshire (WR), by 6/1939 (this firm was a subsidiary of Pease & Partners Ltd).

USHAW MOOR COLLIERY & COKE OVENS, Ushaw Moor K378
Henry Chaytor until 12/1893; see also below NZ 220428

This colliery was originally a drift worked only as a landsale colliery in the 1850s, before the railway was built. This, the Deerness Valley branch, was opened by the NER in 1858, and the colliery was served by sidings north of the line, one mile west of where Ushaw Moor Station was subsequently built in 1884. By 1872 the owner was **John Sharp**, who sank a shaft that year, but sold out to **Henry Chaytor** of Witton Castle, Witton Park, in 1874. Its acquisition by Pease & Partners Ltd gave the firm three collieries (Waterhouses, Esh and Ushaw Moor) in a line from the head of the valley.

As elsewhere, beehive coke ovens were built here, but six Bauer coke ovens were opened here in the quarter ending 30th June 1902 and six more were added in January 1903. All were closed in February 1914, to be re-started early in 1918 and then finally closed in April 1918.

The colliery suffered several periods of closure, including from the late 1920s through the crisis period; but it was re-opened in the 1930s and survived to pass to NCB Northern Division No. 5 Area on 1st January 1947.

Once again, there may well have been other locomotives besides those listed below.

Gauge : 4ft 8½in

10	CAROLINE	0-4-0ST	OC	BH	998	1890	New	(1)
16	originally 6 (?)	0-4-0ST	OC	RS	2325	1877		
			reb	RS		1894	(a)	(2)
17	GEORGE	0-4-0ST	OC	K	1705	1871		
			reb	J.Tait *	90	1920	(b)	(3)

* carried a plate 'James Tait Junr & Partners Ltd, Engineers, Middlesbrough, No.90, 1920'

(a) ex Bowden Close Colliery & Coke Ovens, Helmington Row.
(b) ex Tees Ironworks, Cargo Fleet, Yorkshire (NR), /1922.

(1) to Pease's West, Crook.
(2) to NCB No. 5 Area, with colliery, 1/1/1947.
(3) to Tees Ironworks, Cargo Fleet, Yorkshire (NR), after 16/6/1946.

WATERHOUSES COLLIERY, Waterhouses
Joseph Pease & Partners until 19/8/1882

K379
NZ 185411

The sinking of this colliery began in October 1856, the first of the firm's developments in the valley of the River Deerness. Coal was reached in the following year, but production did not begin until about 1861. Initially it was situated at the end of the NER Deerness Valley Branch, opened in 1858, about ½ mile west of Waterhouses Station. For many years the colliery was known as **PEASE'S WEST BRANDON COLLIERY**, often shortened to **BRANDON COLLIERY**, which must certainly have led to confusion with the Brandon Colliery owned by Strakers & Love (which see) a few miles away. This name was still in use in the early 1890s, but had been changed to Waterhouses Colliery by 1898.

There were 229 beehive coke ovens here in 1891. The colliery survived to be vested in NCB Northern Division No. 5 Area on 1st January 1947. Latterly the colliery was worked by the LNER; but whether this was always the case, or whether the company used locomotives here, is not known.

WINDLESTONE COLLIERY

N380

There is some confusion regarding collieries with this name. The name first appears as an alternative for West Auckland Colliery (NZ 184267) (see Bolckow, Vaughan & Co Ltd), which is also described as the Windleston Wallsend Pit; this title is taken from the parish of this name. This colliery was sunk in 1837 and lay about ½ mile west of St. Helen's Colliery, listed above; but it is not known to have been owned by Joseph Pease & Partners.

The better known Windlestone Colliery was situated at Chilton Buildings (NZ 287296). Joseph Pease & Partners began sinking this colliery in 1872, but the results were clearly disappointing. The valuation of the company in 1881 describes the colliery as "re-opened in May 1880", but this clearly refers to sinking work on the two shafts; this work was unfinished and no coal had been drawn. No doubt because the colliery was right on the southern edge of the coalfield, the report describes the coal as "unmerchantable", except in the west, which had been sub-let to the owners of Eldon Colliery and it was still sub-let in 1898. However, in 1903 Pease & Partners Ltd acquired Eldon Colliery and two years later Windlestone at last began production. Finally, after Pease & Partners Ltd had acquired Henry Stobart & Co Ltd it transferred the latter's Chilton Colliery, about ¾ mile to the north-west, between the two, to its own ownership in order to merge the Windlestone workings with Chilton, which was much closer to Windlestone than Eldon. This then made it possible to close Windlestone, which occurred in October 1924, though it was not finally abandoned until November 1931.

It was served by sidings south of the NER Chilton Branch, but whether a locomotive was used here is not known.

WOOLEY COLLIERY, near Mount Pleasant, Crook
Joseph Pease & Partners until 19/8/1882

K/M381
NZ 178385

This colliery lay fairly close to Stanley Colliery (see above). Its sinking was begun in August 1864 and it began production in 1866. It was served by sidings alongside the NER Stanley Branch about 2¾ miles from Stanley Branch Junction at Crook, and it was worked by the NER's Stanley Engine (see Stanley Colliery above), which hauled the wagons up from the colliery before lowering them towards Crook. There were 88 beehive coke ovens here in 1891. Almost certainly it absorbed the workings of Stanley Colliery when it closed in 1900-1901. In 1909 the colliery's winding engine became the first in the North of England to be powered by electricity (supplied from the company's works at Pease's West, described above). The colliery suffered periods of closure from the 1920s onwards; it was re-started on 1st July 1929, but closed again in September 1931. It was re-opened again when trade improved in the mid 1930s, and was vested in NCB Northern Division No. 5 Area on 1st January 1947, still worked by the Stanley Engine. It is reported that both the Wooley drum and the Crook drum could be put into gear together (which seems unlikely) and with the incline to Crook being approximately twice the length of the incline from Wooley, two sets of fulls and empties could be worked on the Wooley bank while one of fulls and one of empties was worked on the Crook side. The usual load to Crook was four 10½-ton wagons or two 20-ton wagons.

The locomotive below was delivered new here; whether any others worked here is not known.

Gauge : 4ft 8½in

39		0-4-0ST	OC	HL	2798	1909	New	(1)

(1) to St. Helen's Colliery, St. Helen Auckland, by /1925.

Other locomotives
The company or its constituent predecessors also acquired the locomotives below, but no record survives of where they were used:

Gauge : 4ft 8½in

	VICTORIA	0-4-0	IC	Bury		1838	(a)	s/s
No.24		0-4-0ST	OC	MW	573	1875		
		reb		LG		c1900	(b)	s/s
41		0-6-0ST	IC	HC	673	1906	(c)	s/s

(a) ex Manchester & Bolton Railway, 120, 3/1854.
(b) ex Lingford, Gardiner & Co Ltd, dealers, Bishop Auckland, c/1901; sold by NER in /1897 to LG; originally NER 959 class, 997; re-numbered 1893 in 1890, 1798 in 1894 and 2054 in 1897.
(c) ex Whitaker Bros, contractors.

In addition to the above, Joseph Pease & Partners also hired a locomotive from the NER in October 1863, but its identity and where it was used are unknown.

Besides the locations above, the company also worked **THROSTLE GILL DRIFT** (NZ 206257) **(Q382)**, west of Shildon, which was opened in the 1930s. A **stationary engine** worked a rope-hauled narrow gauge system, hauling tubs out of the drift and then running them, almost certainly by main-and-tail haulage, to the screens ½ mile north at NZ 206264. The drift was closed in 1943; but it would seem almost certain to be this drift that was re-opened by NCB Durham No. 4 Area as Haggs Lane Drift (see NCB Section in Part 2). The company also for a period in the 1920s-1930s worked **MARSHALL GREEN COLLIERY** at Witton-le-Wear (NZ 148313) **(Q383)**, whose output served the important refractory brickworks (**Slotburn Brickworks**) there using the clay also mined at the pit. So far as is known no locomotives were used here, either by Pease & Partners Ltd or by subsequent owners.

PELAW MAIN COLLIERIES LTD

The Birtley Iron Co until 26/5/1926, probably the date when Pelaw Main Collieries Ltd was registered; see also below.

The Birtley Iron Company was founded in 1827 by Benjamin Thompson (1779-1867), a Sheffield man already involved with the iron trade elsewhere in the country, who in 1811 had come to Co. Durham to manage Urpeth Colliery, near Chester-le-Street, for the assignees of the bankrupt company of Harrison, Cooke & Co (see below). Thompson became a prominent advocate of rope haulage, and a minor figure on the national stage of railway development in the late 1820s and 1830s. He was the engineer for the Brunton & Shields colliery railway in Northumberland, and in 1828 he became the contractor to Shakespeare Reed for the construction of the new section of the Londonderry Railway between Rainton Bridge and Seaham (see Marquis of Londonderry). He was also a director of both the Brandling Junction and Newcastle & Carlisle Railways. At Birtley two blast furnaces were built in 1828-29 and a third was added in 1847, by which time iron ore was being brought from a mine which the company had developed on the Yorkshire coast near Whitby. In 1838 Thompson sold his share of the Birtley Iron Company to a new group of partners headed by a London businessman, Frederick Perkins. Perkins died in 1871 and was succeeded by his son, Charles Perkins, who died in 1905. The blast furnaces ceased production in 1866 and the works then concentrated on foundry and general engineering work. However, it continued to be known colloquially as "the Iron Works", and this use has been continued in these notes. Throughout this period the iron works and colliery businesses, though legally the same, were kept separate for trading purposes, the latter operating under various titles : **Pelaw Main Collieries**, the **Owners of Pelaw Main Collieries**, **Charles Perkins & Partners** and then **Pelaw Main Collieries** for a second time, are all found. The origin of the title "Pelaw Main" is unclear; Pelaw was a hamlet near Bill Quay on the River Tyne, but there was no colliery there. In later years the company controlled a royalty of about 6000 acres.

The Birtley Iron Co fell on bad times in the depression of the early 1920s and the collieries were closed. In 1926 they were acquired by new owners under the title of **Pelaw Main Collieries Ltd**. In 1930 the iron works site was re-opened by another new company, The Birtley Co Ltd; for its subsequent history see the entry for Caterpillar Tractor Co Ltd. About the same time Pelaw Main Collieries Ltd passed into the ownership of the Paris, Lyon & Mediterranean Railway, the only example of foreign ownership in the North-East coalfield. During the Second World War French control could not be exercised, and the collieries were managed by the company's Agent, Henry Hornsby; it resumed with the end of the War.

The iron works and all of the company's collieries, together with quite a number of other businesses, were served by what became called the Pelaw Main Railway.

PELAW MAIN RAILWAY   C/D/G384

The Railway was basically a combination of two old waggonways, the **Ouston Waggonway** and the **Team Colliery Waggonway**. The title 'Pelaw Main Railway' is found as early as 1843, but the old names continued in use till about the end of the nineteenth century.

The first waggonway to be built in the Birtley area, opened about 1805 to carry coal to the River Wear at Fatfield, was the first-ever to incorporate a stationary engine to haul waggons. About 1807-08 the line passed to Harrison, Cooke & Co, who sank a new pit and although keeping the line to Fatfield, built a new line to Bill Quay on the River Tyne, which opened on 17th May 1809. Having re-built part of the line again in 1810, the company went bankrupt and Thompson was brought in. He sank a new colliery at Ouston and completely rebuilt the southern end of the waggonway, which was now re-named the **Ouston Waggonway** and was opened on 17th November 1815. The remnant of the old line to the River Wear was also abandoned about this time. In the early 1830s Thompson developed two new pits at Urpeth, and extended the waggonway to serve them.

The Ouston Waggonway

The line thus began at (giving the names used latterly) at **URPETH C COLLIERY** (NZ 248535), from where the **Urpeth Engine** at **URPETH BUSTY COLLIERY** (NZ 249536) hauled the fulls up to itself and then lowered them using the same rope, down to the junction with the branch from **OUSTON A COLLIERY** (NZ 264536). By 1857 this branch had been extended for ¼ mile to serve the **NEW WINNINGS** (NZ 264532), while ¾ mile to the north of Ouston A Colliery, Thompson sank **OUSTON B COLLIERY** (NZ 266547), completed in 1824. This had closed by 1875 and the site was re-developed as **The Birtley Brick & Tile Works**, whose owners are not known; it was a short-lived venture and by 1914 the site had become the central disposal point for all the local colliery waste and manure from all the pit ponies, as well as a major coal depot. Traffic in this area was worked by the **Ouston Engine** at Ouston A Colliery, but how it did this is not clear.

About ½ mile north-west of Ouston B Colliery lay **BIRTLEY IRON WORKS** (NZ 271550 approx.). From here the waggons had to be hauled up the eastern side of the Team Valley for about a mile. Originally this length was divided into two sections. The first was worked by the **Birtley Church Engine** (NZ 274555), from which the waggons ran by gravity for about forty yards to the bottom of the upper section, worked by the **Blackhouse Engine** (NZ 280560), known for a brief period around the turn of the twentieth century as the **Blackfell Engine**. The next 2000 yards, straight and flat along the side of the valley, was initially worked by horses and was followed by the short Ayton (later Eighton) Banks incline up to the **Eighton Engine** (NZ 281582). However, in 1821 Thompson converted the whole of this section to what he called 'reciprocating haulage', or main-and-tail haulage, with the two engines hauling as required. This did not work as well as he had hoped, and subsequently the section was re-divided as before, with the main-and-tail haulage being retained only on the flat section, co-worked by a new **stationary engine** at Eighton Banks bank foot.

The next mainly flat section, the 1¼ miles from the Eighton Banks Engine to Whitehill, was also probably converted from horse haulage to main-and-tail rope working, but no evidence survives. Then came the self-acting **Whitehill Incline**, 1337 yards long, which took the waggons down to Heworth bank foot. The final 1¼ miles to the River Tyne was a gradual decline, and the **Heworth Engine** at the bank foot ran fulls and empties alternately. Once at the river horses handled the waggons between the sidings and two short, parallel **self-acting inclines**, which took the full waggons down to the loading points.

According to a document in the Northumberland County Record Office (NRO 3410/JOHN/9/145), five collieries not owned by the Birtley company were linked to the line in the late 1820s-early 1830s. The first was **Heworth Colliery** (NZ 284605) **(D385)**, an old colliery where a new shaft was sunk in 1821. This was linked by a **self-acting incline** to the bottom of the Whitehill Incline. This gave two bank foots side by side, and the Heworth Engine was adapted to handle traffic to and from both routes, together with some quite complicated railway working. The Heworth Staith, with a self-acting incline down to it, was situated immediately downstream from the Pelaw Main Staiths. After the opening of the NER line between Pelaw and South Shields in 1872 a link was put in between the two lines, although it handled only Heworth traffic. The colliery was latterly owned by **The Heworth Coal Co**, which became a limited company in 1902.

The second colliery to be linked to the Ouston line was **Eighton Moor Colliery** at Low Fell, near Gateshead (NZ 271575), then at the southern end of the Team Waggonway (see below). Sometime during the 1820s its owners built a new line northwards to join the Ouston line near Whitehill bank head, which eventually led to the two lines being merged into one ownership. The other collieries were the **Blackfell**, or **Mount Moor**, **Colliery** near Eighton Banks (NZ 279577) (see Lord Ravensworth & Partners); **Waldridge Colliery**, near Chester-le-Street (NZ 253502) (see The Priestman Collieries Ltd) and **Stormont (Main) Colliery**, near Gateshead (NZ 276592), which had various owners. None of the links with the last three collieries survived very long.

At Whitehill various companies exploited the extensive sandstone deposits, linking quarries on both sides of the line to the Railway. Eventually these came under the sole ownership of **Richard Kell & Co Ltd** and were known as **Windy Nook Quarries** (NZ 278605 approx and NZ 280602 approx.). They are believed to have been closed about the beginning of the Second World War.

The introduction of locomotives
The first locomotive known to have been owned by The Birtley Iron Company arrived in 1859, probably to work at the iron works. Possibly the rope haulage between Ouston A Colliery and the works was abandoned about this time. In 1869 the stationary engine at the foot of the Eighton Banks Incline exploded, and instead of repairing it, the company introduced locomotive working between Blackhouse and Eighton Banks bank foot, with a wooden **LOCO SHED** at the former. Perhaps about the same time, locomotive working was introduced between Eighton Banks bank head and Whitehill, with another wooden **LOCO SHED** at the former, and also at Pelaw Main Staiths, where a large stone **LOCO SHED** was provided.

Developments around Birtley
This area soon began to see major developments. In August 1862 a new colliery, **BEWICKE MAIN D COLLIERY** (NZ 254556), was opened to work the coal in the north-western area of the royalty, and a one mile branch from Birtley was built to serve it. About half-way along the **Station Brickworks** (NZ 265556) was opened, owned latterly by **Blythe & Sons (Birtley) Ltd**, one of various firms whose traffic was handled by the railway. In 1902, to the west of the colliery, the company opened the **RIDING DRIFT** (NZ 245553) and built a narrow gauge line to serve it, latterly worked by a locomotive. Five years later the **MILL DRIFT** (NZ 246555) was added and the line extended to it. This closed in April 1915, followed by the Riding Drift in April 1926.

Another branch built in the area ran from Ewe Hill near Ouston to join the NER Pontop & South Shields branch. This may have served to bring coal on to the system from **SOUTH PELAW COLLIERY** (NZ 264523), probably opened in the 1840s, which was linked to the Pontop & South Shields line about ½ mile further west. The company closed this colliery in 1886, but it 1892 it was opened by new owners, using a different rail link. By 1890 this link to the NER at Durham Turnpike Junction was also serving two other businesses – the **Birtley Grange Brick & Tile Works**, latterly the **Pelaw Grange Brickworks** (NZ 268541) and owned by **J.O. Scott & Co Ltd**, and the **Pelaw Grange Sawmills** (NZ 266540), latterly owned by **Joseph Smith**. The former closed in 1938, the latter in the late 1920s.

In 1867-68 the NER opened its new line between Gateshead and Durham, which ran down the Team Valley, and this soon led the company to put in a short branch to Birtley Station. This in turn led to the development of yet another brickworks, the **Ravensworth Brick & Tile Works** (NZ 258556), which was situated on the eastern side of this branch. It was latterly owned by the **Ravensworth Brick & Tile Co**, and was closed either in 1938 or 1940.

Because the Iron Works and its area had become a major collecting and distribution point for traffic from different lines, the rail complex here became known, rather oddly, as '**Birtley Tail**'. For many years the loco shed for the work here was situated in the iron works, though there may have been another loco shed at Ouston A Pit.

In 1893 the company added to this area's importance by opening **OUSTON E COLLIERY** (NZ 256548), in the fork between the lines to Ouston and to Bewicke Main Colliery. It also built an **ELECTRICITY STATION** here (later replaced by another), while a **GAS WORKS** (NZ 271552) was built at the iron works, the latter supplying not only the works but the whole of Birtley village. By 1919 the iron works and Ouston loco sheds had been closed, and a new loco shed had been built near the beginning of the Bewicke Main branch, although, despite its location, it was always known as the **E Pit Shed** (NZ 270537). Normally four locomotives were kept here. However, because Birtley Tail sidings lay about 80 yards west of the foot of the Blackhouse Incline, the waggons had to be run down into them by gravity. When a siding was empty the set was prone not to run the full distance, and so a horse, housed in a stable nearby, was kept here to haul them down to the required point. This was the last place on one of Durham's private railways where a horse was used for shunting.

The Team (or Teams) Colliery Waggonway
This waggonway was built in 1669. Its early history, together with others in the area, some of which were later developed into railways, can be found in *A Fighting Trade – Rail Transport in Tyne Coal, 1600-1800, Vol.1 – History, Vol.2 – Data*, by G. Bennett, E. Clavering and A. Rounding, published by the Portcullis Press in 1990. By the first quarter of the nineteenth century the line ran from the **Team Staith** (NZ 237624) for 3½ miles southwards to **Team Colliery** (NZ 265578). En route it also served **Farnacres Colliery** (NZ 234617) and **Derwent Crook Colliery** (NZ 251600), and was subsequently extended a short

distance to serve **Eighton Moor Colliery** (NZ 271575). As mentioned above, in the 1820s the owners of this last colliery subsequently built a link northwards to join the Ouston Waggonway at Whitehill. Twenty years later **Norwood Colliery** (NZ 234618) was sunk alongside the line at Dunston, but not long afterwards both it and Farnacres Colliery were sold and disconnected from the line. The Team Staith had been closed by 1850, but at almost the same place the **Teams Fire Brick Works** (NZ 235621) was established. Further south Derwent Crook Colliery was closed by 1857, but **Allerdean Colliery** (NZ 257586) had been opened. So by the mid 1850s there were only two collieries linked to the line, Allerdean and Team, and both they and the waggonway were owned by **William Wharton Burdon**, whose ancestors had worked Team Colliery since 1796.

The first 2½ miles from the Team Staith lay in the floor of the valley, and was almost certainly worked by horses. But sending coal to Pelaw Main Staiths reversed the coal flow, which now had to be hauled out of the valley, so that stationary engines were unavoidable. The first incline, up as far as Team Colliery, was latterly worked by the **Allerdene Engine** (NZ 264581), which also had to shunt Allerdean Colliery. At Team Colliery itself the link to Whitehill had been altered to join from the north, creating a severe 110 degree curve, although once round this, it was a straight climb, up a second single line incline, to the **Boundary Engine** (NZ 273594) at Wrekenton. The final section to Whitehill was also worked by the Boundary Engine, using main-and-tail haulage with a return wheel. The first locomotive to be ordered by 'The Owners of Teams Colliery' arrived in 1868, presumably to replace horses north of Allerdene bank foot. By this period the northern end of the line had been linked to the NER's Redheugh branch, allowing coal to be dispatched via this link if required. Like the Ouston Waggonway, this line too had another owner's colliery joined to it, **Sheriff Hill Colliery** at Gateshead (NZ 273596), though again how its traffic was worked and for how long it was linked are not known.

In 1872 W.W. Burdon died and after first passing to trustees, the line came into the ownership of A.E. Burdon; but he sold out to the Birtley Iron Co in 1882, allegedly as the result of losing a horse-racing wager. This gave the company a system with two long arms which joined at Whitehill, with links to the NER at Durham Turnpike Junction, Birtley Station and Dunston. It would seem that the title **PELAW MAIN RAILWAY** came into more common use about this time, while **TEAM COLLIERY** became applied collectively to both collieries, with the original colliery being called the **Betty Pit** and Allerdene being called the **Shop Pit**, with the loco shed at Allerdene bank foot being known as the **SHOP PIT LOCO SHED** (NZ 253587).

With the extension of the Gateshead Tramways system towards Wrekenton, the company was pressed to dispense with the main-and-tail haulage between the Boundary Engine and Whitehill, and this was replaced by locomotive working in 1909, with the **STARRS LOCO SHED** (NZ 273594) near the engine house. In 1913 the company opened **BLACKHOUSE H COLLIERY**, known locally as the **Wash Houses Pit** (NZ 280556), building a ½ mile branch behind the Blackhouse Engine to serve it. It was closed in 1914, but re-opened in 1920 and linked to a drift further down the side of the valley by a narrow gauge line a mile long.

The owners who took over in the 1920s made major investments in both its mines and the Railway. To reach the deeper seams the company sank the Ann Pit at Team Colliery, which thereafter took the name **RAVENSWORTH ANN COLLIERY**, while the Shop Pit became **RAVENSWORTH SHOP COLLIERY**. With the coal under the western side of the Team Valley now available, in 1936 the company opened what was officially called **RAVENSWORTH PARK DRIFT** (NZ 242588), though better known as **Lady Park Drift**. This was served by a ¼ mile branch from the Team Valley line. A **DRY-CLEANING PLANT** was built at Ouston E Colliery. On the debit side both Bewicke Main and Blackhouse Collieries closed in 1932. For the Railway, in a scheme far ahead of all of the other local colliery railways, all of the steam haulers in the stationary engines were replaced by electric haulers. The last of these was the Allerdene Engine in 1937, where the bank head and the Ravensworth Ann Colliery yard were re-modelled to allow the Railway to pass under the A1 road. This work meant the shunting of the colliery was henceforth done by a locomotive, with the former engine house being converted into a loco shed. The company also developed both **ENGINEERING WORKSHOPS** and **WAGON SHOPS** at Ouston E Colliery, and replaced all of the Railway's chaldron waggons with 10-ton wooden hoppers, purchased from the NER.

The development of the Team Valley Trading Estate in 1937 resulted in the Pelaw Main line down the Team valley being re-routed, part of it on a long viaduct. Beyond the northern end of the Trading Estate the line was now linked to **Norwood Coke Works** (NZ 238613 approx) (see The Priestman Collieries Ltd), though the Team Firebrick Works, latterly owned by **Lucas Bros Ltd**, closed in 1938.

Because of the fall of France in the Second World War, the French owners were unable to exercise control of the company, and production at both Urpeth C and Ouston E Collieries was suspended in 1941. The section of the Railway between Ouston E Colliery and Whitehill remained in occasional use to allow locomotives and wagons to be repaired at the workshops. Ouston E Colliery re-opened in 1946. However, the absence of any working Pelaw Main locomotives meant that The Birtley Co Ltd had to acquire its own locomotive to work its traffic to and from Birtley Station.

On 1st January 1947 the Railway, together with Urpeth C Colliery (closed), Ouston E Colliery, Ravensworth Park Drift, and Ravensworth Shop and Ann Collieries, was vested in NCB Northern Division No.6 Area (although like the Bowes Railway nearby, its northern end, including Heworth Colliery, was within No.1 Area.

Reference : *The Private Railways of County Durham*, by Colin E. Mountford, Industrial Railway Society, 2004.

Gauge : 4ft 8½in

Name	Type		Maker	No.	Date	Notes	Scr
PELAW	0-4-0WT	OC	H(L)	220	1859	(a)	(1)
BEWICKE	0-6-0ST	OC	BH	52	1868	New	Scr 12/1929
VICTORY	0-4-0ST	OC	Harris		1863	(b)	Scr c/1900
DERWENT	0-4-0ST	OC	Harris		1865	(b)	Scr c/1900
BYRON	0-4-0ST	OC	Harris		1868	(b)	Scr c/1900
BIRTLEY	2-4-0WT	OC	H(L)		?	(c)	
	reb 0-6-0ST	OC	Birtley		1871		(2)
PELAW	0-6-0T	OC	BH	60	1868	(d)	(3)
OUSTON	0-6-0ST	OC	BH	602	1881	New	Scr 12/1929
BURDON	0-6-0ST	OC	BH	48	1868	(e)	Scr c/1920
URPETH	0-4-0ST	OC	AB	277	1884	(f)	
	reb		AB	9410	1915		(4)
TYNE	0-4-0ST	OC	AB	786	1896	New	
	reb		AB		1940		(5)
DERWENT	0-6-0ST*	OC	AB	970	1903	New	
	reb		RSHN		1945		(5)
PELAW II	0-6-0	IC	Todd		1847		
	reb 0-6-0ST	IC	Ghd		1879	(g)	(6)
ROSEBERY	0-6-0ST	IC	RS	2139	1873	(h)	Scr 7/1932
BALFOUR	0-6-0ST	IC	RS	2239	1875	(j)	Scr /1925
GLADSTONE	0-6-0ST	IC	RS	2240	1875	(k)	Scr /1929
SALISBURY	0-6-0ST	IC	RS	2244	1875	(m)	Scr /1925
EAST CLIFF	0-6-0ST	IC	P	774	1899	(n)	Scr /1928
OUSTON	0-6-0ST	IC	RWH	1657	1875	(p)	Scr /1928
BYRON	0-6-0ST	IC	RWH	1662	1875	(q)	(7)
LEAFIELD	0-6-0ST	IC	RWH	1666	1875	(r)	Scr c/1930
VICTORY	0-6-0ST	IC	RWH	1669	1875	(s)	(8)
MOSELEY	0-6-0ST	IC	Don	213	1876		
	reb		Wake	2091	1920	(t)	Scr 8/1932
CHARLES PERKINS	0-4-0T	OC	HL	2986	1913	New	(5)
-	4wVBT	VCG	S	6936	1927	New	Scr /1932
24	4-4-0T	OC	BP	770	1867		
	reb		?		1880		
	reb		?		1900	(u)	(9)
26	4-4-0T	OC	BP	772	1867		
	reb		?		1901		
	reb		Neasden		1920	(v)	(10)
44	4-4-0T	OC	BP	868	1869		
	reb		?		1888		
	reb		?		1902		
	reb		Neasden		1920	(w)	(10)
BUSTY	0-4-0ST	OC	BH		?		
	reb		Ridley Shaw		1928	(x)	Scr /1936
CHARLES NELSON	0-4-0ST	OC	P	1748	1928	New	(5)
HENRY C. EMBLETON	0-6-0T	OC	HL	3766	1930	New	(5)
900	0-4-0T	IC	Ghd	35	1888	(y)	(11)
1308	0-4-0T	IC	Ghd	37	1891	(z)	(5)
1310	0-4-0T	IC	Ghd	38	1891	(aa)	(5)
-	0-4-0ST	OC	RS	3057	1904	(ab)	(12)

* in later years is said to have regularly run as an 0-4-2ST, with the rear coupling rods removed.

In 1946 the company ordered a new 0-4-0ST from Peckett & Sons Ltd, Bristol (Makers No.2093), but this had not been delivered by Vesting Day, 1st January 1947.

On 1st January 1947 locomotives were allocated to the following loco sheds:

Ouston E Pit	AB 970/1903
Blackhouse	BP 868/1869, HL 3766/1930
Eighton Banks	BP 772/1867, AB 786/1896
Shop Pit	Ghd 37/1891
Ann Pit	Ghd 35/1888, Ghd 38/1891
Starrs	P 1748/1928
Pelaw Main Staiths	HL 2986/1913

The horse used for shunting at Birtley Tail was also transferred to NCB No. 6 Area.

(a) New; see also note above in text.
(b) ex B.C. Lawton, contractor for the construction of the NER Team valley line, /1869.
(c) origin unknown. Reputed to have been purchased from the Caledonian Railway, Scotland, /1871, but this appears to be unsupported in fact, although it may have arrived on the Railway in 1871.
(d) a loco very likely to be this one was offered for sale in *The Engineer* of 13/12/1878 by The Darlington Forge Co Ltd, Darlington (see note under this entry).
(e) New to "Owners of Teams Colliery", i.e. W.W. Burdon; ex A.E. Burdon, with Team Colliery and railway, /1882.
(f) ex AB, 1884, as a new loco, but alleged to have been a rebuild by AB of a loco as yet unidentified.
(g) ex Robert Frazer & Sons Ltd, dealers, Hebburn, 1/1904; NER 2289, 287 class, until 2/7/1903; this loco was built new by Charles Todd of Leeds in 1/1847 as a 0-6-0 tender engine with 15in x 24in cylinders and 4ft 9in wheels for the York & Newcastle Railway, No.107; it kept the same number under the York, Newcastle & Berwick Railway and when this railway was merged into the NER in 1854. It was rebuilt at the NER's Gateshead Works about 3/1862 and again there in 12/1879, when it was converted into a 0-6-0ST using miscellaneous parts from other locomotives. It was re-numbered into the NER Duplicate List in 12/1890 as 1903 and to 1716 in 1/1894; it was withdrawn from traffic by 4/1903, but was nevertheless allocated 2289 in another re-numbering in 7/1903; presumably it never carried this last number. It was sold to Robert Frazer & Sons Ltd in 12/1903.
(h) ex Robert Frazer & Sons Ltd, dealers, Hebburn, /1905; previously NER, 1670, 964 class, until 3/1905.
(j) ex Robert Frazer & Sons Ltd, dealers, Hebburn, /1905; previously NER, 1673, 964 class, until 3/1905.
(k) ex Robert Frazer & Sons Ltd, dealers, Hebburn, /1905; previously NER, 1674, 964 class, until 30/6/1904.
(m) ex Robert Frazer & Sons Ltd, dealers, Hebburn, /1905; previously NER, 1676, 964 class, until 30/6/1904.
(n) ex J.F. Wake, dealer, Darlington, 9/1908; previously S. Pearson & Son Ltd, Admiralty Harbour contract, Dover, Kent, until 9/1908.
(p) ex Robert Frazer & Sons Ltd, dealers, Hebburn, 1/1910; previously NER 1350, 1350 class, until 1/1910.
(q) ex Robert Frazer & Sons Ltd, dealers, Hebburn, 2/1910; previously NER 1355, 1350 class, until 22/2/1910.
(r) ex NER, 1359, 1350 class, 9/9/1910.
(s) ex NER, 1362, 1350 class, 9/9/1910.
(t) ex J.F. Wake, dealer, Darlington, /1920; previously Great Northern Railway, 606, J6 class, until 3/1920.
(u) ex Robert Frazer & Sons Ltd, dealers, Hebburn, /1927 (by 7/6/1927); previously Metropolitan Railway, A class, 24; Metropolitan records apparently give the date of sale to Robert Frazer as /1913, which would cast doubt on the date of /1927 above.
(v) ex Robert Frazer & Sons Ltd, dealers, Hebburn, /1927; previously Metropolitan Railway, A class, 26.
(w) ex Robert Frazer & Sons Ltd, dealers, Hebburn, /1927; previously Metropolitan Railway, A class, 44.
Note: whilst on the Metropolitan Railway these locomotives carried a small A to identify the class to which the locomotives belonged; these letters were not a suffix to the number, and so have been omitted here.
(x) ex Ridley, Shaw & Co Ltd, dealers, Middlesbrough, Yorkshire (NR), 17/4/1928; identity and previous owners unknown.
(y) ex Robert Frazer & Sons Ltd, dealers, Hebburn, 5/1932; previously LNER 900, Y7 class, until 2/1931.

(z) ex Robert Frazer & Sons Ltd, dealers, Hebburn, 5/1932; previously NER 1308, Y7 class, sold to Frazer, 2/1931, moved 9/1931.
(aa) ex Robert Frazer & Sons Ltd, dealers, Hebburn, /1933; previously NER 1310, Y7 class, sold to Frazer, 2/1931, moved 9/1931.
(ab) ex U.A. Ritson & Sons Ltd, Burnhope Colliery, Burnhope, loan, in 1930s.
(1) s/s, but a drawing of this loco by R.H. Inness was dated 22/1/1906, so the loco may have survived until this period.
(2) sources disagree on the Scr date, giving /1925, /1928, /1929 and withdrawn /1930.
(3) to United National Collieries Ltd, Risca Colliery, Monmouthshire, /1881.
(4) to Steetley Lime & Basic Co Ltd, Coxhoe, c /1930.
(5) to NCB No.6 Area, with the Railway, 1/1/1947.
(6) Scr /1925, after 5/1925.
(7) sources disagree, giving both /1928 and /1930 as Scr dates.
(8) probably Scr /1932, though /1929 is also found.
(9) Scr /1932; spares from this loco were kept to repair BP 772 and BP 868.
(10) these locomotives did not work after coal traffic ceased on this section in 1940, and were left in their sheds; to NCB No. 6 Area, with the Railway, 1st January 1947.
(11) latterly dismantled at Ravensworth Ann Colliery; to NCB No. 6 Area, with the Railway, 1st January 1947.
(12) returned to U.A. Ritson & Sons Ltd, Burnhope Colliery, Burnhope, in 1930s.

Additional notes

In the edition of the *Colliery Guardian* for 24/7/1903 Philip Kirkup, Agent for The Birtley Iron Co Ltd, advertised for sale a 12in 0-4-0ST "done for work" at "Ravensworth Colliery", possibly meaning the Shop Pit shed. Its identity and disposal are unknown; possibly it was one of the John Harris locomotives.

In the *Colliery Guardian* for 16/2/1906 Kirkup offered for sale a 12in six-wheeled coupled locomotive, lying at Ouston. This loco would appear not to be any of the six-wheeled locomotives listed above, and its identity and disposal are unknown.

About 1908 The South Hetton Coal Co Ltd acquired a Metropolitan Railway A class 4-4-0T, No.6 (which see). There was an oral tradition at South Hetton that its loco had previously worked on the Pelaw Main Railway. However, there was no parallel tradition on the Pelaw Main Railway supporting this.

Gauge : 1ft 11½in   G386

As noted above, a narrow gauge line was built from Bewicke Main Colliery to serve the new Riding Drift when it opened in 1902. How this line was worked before the arrival of the locomotive below is unknown, though it was very likely to have been by some form of rope haulage. This line was extended to the Mill Drift in 1907. This drift was closed in 1915, co-incidentally with the departure of the locomotive below. It is assumed that rope haulage was then resumed, until the Riding Drift closed in April 1926; but see also the footnote below.

        -        0-4-0ST   OC   AB   703   1893   (a)   (1)

(a) ex William Jones, dealer, Greenwich, London, by 5/1905; previously James Nuttall, contractor for the Lynton & Barnstaple Railway, Devon.
(1) to J.F. Wake, dealer, Darlington, c/1915.

Further notes

In the edition of *Contract Journal* for 20/1/1909 Philip Kirkup advertised for what was presumably intended to be a second locomotive for this line, seeking a locomotive as near as possible with 5in x 10in cylinders and a 3ft 0in wheelbase. Nothing further is known.

In June 1924 the Coast Road Joint Committee (which see) acquired from N.E. Potts, described as a dealer at Boldon, a 2ft 0in gauge 4wPM locomotive, MR 429/1917, which was described as being "free on rail at Ravensworth Colliery". Whether this locomotive has any connection with The Birtley Iron Co is not known.

PELTON BRICK CO LTD
PELTON BRICKWORKS, Pelton Fell   G387
NZ 254517

This clay quarry and brickworks was opened in 1938 (according to Davison; another source gives 1940). It was owned by the Kellett family, who also owned Pelton Fell Colliery (The Mid-Durham Coal Co Ltd) and the nearby Stella Gill Coke Works (The Mid-Durham Carbonisation Co Ltd). It had no main line rail connection. About 1950 an internal rail system was installed. This was removed and replaced by a dumper truck about 1967. The works closed in 1975.

Gauge : 2ft 0in

-	4wDM	RH	213836	1942	(a)	(1)
-	4wDM	RH	375696	1954	New	(2)

(a) ex RH, c/1950; previously Ministry of Supply, Home Grown Timber Production Department, Inverness, Scotland.

(1) believed to have been loaned to Northern Industrial Improvement Trust Ltd, Washington Colliery Brickworks, Washington, during 1952 (this firm was another owned by the Kellett family), and returned; to R.P. Morris, Longfield, Kent, 9/1968, but it is very doubtful whether the loco was ever transferred there; s/s.

(2) to R.P. Morris, Longfield, Kent, 9/1968; re-sold, still at Pelton Fell, to M.E. Engineering Ltd, Cricklewood, Middlesex, /1968; to Bord na Mona, Kilberry Works, County Kildare, Eire, 1/1970.

PICKFORD, HOLLAND & CO LTD
Pickford, Holland & Co until 17/4/1913
ECLIPSE BRICKWORKS, Crook   **M388**
NZ 159351

This brickworks and quarry was established in 1899 on the site of the coke ovens of the former Crook Colliery by Randolf Pickford, a metallurgical chemist from Middlesbrough and Job Holland, a Sheffield businessman. The works manufactured silica bricks, used in places where high heat resistance was required, such as the lining of coke ovens. It was served by sidings west of the NER Crook to Bishop Auckland line, ½ mile south of Crook Station. Rail traffic ceased in 1965. The brickworks, latterly owned by Dyson Refractories Ltd, closed early in 1984. The firm also owned a brickworks at Blaenavon in Monmouthshire and Moss Ganister Mine at Dore, near Sheffield, Yorkshire (WR). The company itself continues (2006) to exist.

There may well have been other locomotives besides those listed below, but details are unknown.

Gauge : 4ft 8½in

-	0-4-0ST	OC	BH	?	?	(a)	Scr c/1944
-	4wPM		FH	1829	1933	(b)	(1)
-	4wDM		RH	207102	1941	New	(2)
-	4wDM		RH	443644	1961	New	(3)

(a) origin and date of arrival unknown.
(b) ex FH, 8/1933, as a new locomotive, but said to have been a rebuild of an earlier locomotive.

(1) offered for sale in *Machinery Market*, 6/6/1941; to Dunlop Cotton Mills Ltd, Thornliebank, near Glasgow, Renfrewshire, Scotland.
(2) to RH, 6/1961 for overhaul and then sent to Basic Brick Works, Blaenavon, Monmouthshire, 9/1961 (also owned by the firm).
(3) to Thos W. Ward Ltd, Templeborough Works, Sheffield, Yorkshire (WR), 7/1965; re-sold to C.F. Booth Ltd, Rotherham, Yorkshire (WR), c/1966.

PORT OF SUNDERLAND AUTHORITY
River Wear Commission until 1/10/1972
HUDSON and HENDON DOCKS, (later known as the SOUTH DOCKS), Sunderland   **H389**

The River Wear Commission was created in 1717. The growth of coal shipments led to a demand for docks to be built at the mouth of the River Wear, but at first this was done by private companies. Isambard Kingdom Brunel (1806-1859), to become one of the great railway engineers, was involved with the first, a tidal basin and locked harbour covering 9 acres, planned for the north bank. This became known as the North Dock and was opened on 1st November 1837 by the **Wearmouth Dock Company**, whose chief promoter was Sir Hedworth Williamson (see Sir Hedworth Williamson's Limeworks Ltd). From 1839 it was served by a branch from the Brandling Junction Railway from Gateshead to Monkwearmouth, also on the river's north bank. The dock was purchased on 1st January 1847 by the **York & Newcastle Railway** and subsequently passed into the ownership of the NER. For the record, the final stage of the Dock branch, down into the dock itself, consisted of two parallel rope inclines, served by a **stationary beam engine**. Goods coming out of the dock were hauled up by the engine, but the other, carrying coal down into the dock, was usually self-acting, except in bad weather, when the engine was used for assistance, an unusual arrangement. The engine was also used to handle shunting on the dock side, the rope running round a series of pulleys to reach the various sidings.

The North Dock, also called the Wearmouth Dock but known irreverently as "Sir Hedworth Williamson's bathtub", soon proved inadequate, and this led to the construction of the **HUDSON DOCK** on the opposite bank. This was heavily backed by the famous railway entrepreneur, George Hudson (1800-1871) and the York & Newcastle Railway, though once again it was owned by a private company, the **Sunderland Dock Company** (which see), which obtained the Sunderland Dock Act in 1846. The dock (NZ 412575 approx.) consisted of two sections. There was a Tidal Harbour at the mouth of the River Wear followed by a Half Tide Basin giving access to the dock, and a second Half Tide Basin giving access from the sea near the dock's southern section. The construction employed 13,000 workers. The dock was opened on 20th June 1850, and was served by the Durham & Sunderland branch of what was now the York, Newcastle & Berwick Railway, to be followed in 1852 by that railway's Pensher branch and by the Marquis of Londonderry's Londonderry Railway from Seaham in 1854. Extensions were opened in 1855 and 1856, increasing the enclosed area to 127 acres.

However, although coal shipments increased by over 50% between 1852 and 1858, the Commissioners were able to levy a toll on every vessel leaving the Dock. This made it financially unviable, and the Commissioners took it over under a new Act of Parliament on 1st August 1859. Almost certainly this transfer included a locomotive. It was in poor repair (see below) and every time it broke down the Traffic Manager had to find between four and six horses to replace it. On 2nd May 1860 the Commissioners accepted an offer from Nicholson & Co, contractors from Gateshead, to "find the locomotive power for six months, they taking the old engine for £170" (T&WA 202/1006/93), whilst they ordered MW 17 below. This was delivered in November 1860, co-incidentally at the end of the six months' contract with Nicholson & Co, with the Commissioners resuming direct control of the shunting. An Inventory of Working Plant owned by the Commissioners and dated 31st December 1861 lists the **Moor Incline** and the **Hendon Incline**, but no further information has yet come to light about these.

The **HENDON DOCK** (NZ 412566) (11 acres) was added to the south in 1868 to provide suitable capacity for larger ships. Gradually the Hudson and Hendon Docks became collectively known as the **SOUTH DOCK**. It was claimed in Sunderland for many years that because the NER had no financial interest in the Hudson and Hendon Docks it tried to divert traffic to the docks that it did own at Tyne Dock and Hartlepool. Within the docks there were 32 coal staiths, though not all were operational at the same time.

From the 1860s onwards space or premises around the docks were leased to and developed by a wide variety of companies. So far as is known, none of these had their own locomotives, but sub-contracted their shunting to the Commissioners.

Towards the end of the nineteenth century the Hudson Dock was divided into the Hudson North Dock and the Hudson South Dock by the construction of the pier comprising Nos. 18 and 19 Coal Staiths, and in 1900 the dock was almost doubled in size to 40 acres by an expansion on the seaward side (see Sir John Jackson Ltd in the Contractors section of Part 2). On the west side of the Hudson Dock was the shipbuilding yard of Bartram & Sons Ltd, the last yard in Britain to launch directly into the sea. This closed in 1978. In 1934 the deep-water Corporation Quay was opened on the River Wear. This was financed by Sunderland Corporation but operated by the Commissioners. Meanwhile the North Dock was taken over by the Commissioners in 1922, together with the two inclines. Eleven years later the dock was closed to commercial traffic; the coal incline was abandoned and the steam hauler replaced by a 150 bhp electric hauler, still used for shunting on the dock side. By the mid-1950s the gates had been removed and the dock was extensively re-modelled to provide moorings and fitting-out facilities for the shipbuilding and ship-repairing firm of T.W. Greenwell & Co Ltd, who also had graving docks near the South Dock (see Sir William Arrol & Co Ltd in the Contractors section of Part 2). By the 1990s the North Dock had become a marina. The Hendon Dock was closed during the Second World War and used by the Royal Navy to re-fit minesweepers. By 1964 there was only one coal staith, together with two coal conveyors, in the South Dock, and coal shipment finally ceased in 1986.

From very early days the Commission developed quite extensive **workshops** (NZ 412578), on the seaward side of the North Half Tide Basin. The first loco shed was agreed in 1860, and was constructed to the east of the workshops. By the end of the First World War this had become a cement store and a new shed had been built near the workshops to the south. This in turn was replaced by a third shed near the No.18 coal staith (NZ 410573). By the 1950s, if not for some time previously, the locomotive fleet was divided between the Traffic Department and the Works Department, the latter using its allocation for maintenance work based on the Pier Yard and for repairs to the piers and sea walls. No.9 was the regular locomotive on this work for many years, later joined by No.10. No.13 became a Works

Department locomotive from October 1956. Records show that the "rebuilds" listed below normally involved the fitting of a replacement boiler, undertaken at the RWC workshops; only HL 2589 is known to have been sent away for repairs.

105. No.2, MW 57/1862. Note the personalised lamp on the smokebox, the device, apparently a sextant, on the sandbox (this appeared on other River Wear Commission locomotives), the fluted safety valve cover and the angled front cab plate – to say nothing of the very elaborate decoration on the coupling rods.

106. A very early Andrew Barclay locomotive, No. 4, AB 52/1866, with a box tank, close-fitting splashers over the wheels and another angled front to the cab. She was subsequently rebuilt quite extensively.

The Commissioners' operations were taken over by Sunderland Corporation under the title above on 1st October 1972. By the end of the twentieth century most general cargo was handled at the Corporation Quay. A locomotive was used only if cargo arrived by rail for shipment. This traffic ceased

107. No.7, HC 266/1883, showing the same features described for No.2, and even more elaborate lining out on everywhere that it could be put, and immaculately clean – Victorian pride in all its glory.

108. No.9, RWH 2029/1885, was specifically acquired to give a very light engine for pier repairs, and this shows her at work on 15th August 1919.

109. This shows the opening of Nos. 1 and 2 coal conveyors in the South Dock on 15th May 1924, filling the steam collier BIRTLEY. The wealth of detail is fascinating; besides the locomotive, note the three cars, the vertical-boilered rail crane, the variety of wagon owners and the huge amount of timber.

110. The Port's last two locomotives, 21, RH 395294/1956, and 22, RH 416210/1957, hauling a train of imported coal, on 12th April 1995. Unused since 2001, they and the port await a decision on the future.

in December 2001, after which the locomotives were stored. Oil traffic handled by rail, using English, Welsh & Scottish Railway locomotives, ceased shortly afterwards, since when there has been no rail traffic.

Most of the River Wear Commissioners' records are deposited with the Tyne & Wear Archive Service. Deposit 202 includes various documents and references to locomotives. The Port of Sunderland Authority also retains historic material.

Gauge : 4ft 8½in

-		?	?	?	?	?	(a)	(1)
1		0-6-0ST	IC	MW	17	1860	New	Scr c/1900
No.1		0-4-0ST	OC	HL	2589	1904		
		reb		HL		1933	(b)	(2)
	R.W.C. No.2	0-6-0ST	IC	MW	57	1862	New	(3)
3		0-4-0ST	OC	HE	14	1866	New	(4)
	R.W.C. No.4	0-4-0ST	OC	AB	52	1866	New	(5)
	R.W.C. No.5	0-4-0ST	OC	BH	173	1871	(c)	(6)
No.6		0-6-0ST	OC	FW	140	1872	New	(3)
	R.W.C. No.7	0-6-0ST	OC	HC	266	1883	New	
		reb		?		1908		(7)
No.9	(GNAT)	0-4-0ST	OC	RWH	2029	1885	(d)	
		reb		?		1936		(8)
No.10		0-4-0ST	OC	HC	221	1881	(e)	
		reb		?		1908		
		reb		?		1937		(9)
No.12		0-4-0ST	OC	AB	805	1897	New	
		reb		?		1910		
		reb		K		1938		(10)
No.13		0-4-0ST	OC	AB	1127	1907	New	
		reb		?		1930		(11)
	R.W.C. No.14	0-6-0T	OC	HC	1039	1913	New	
		reb		?		1938		(12)
	R.W.C. No.15	0-4-0ST	OC	P	1761	1929	New	(9)
No.16		0-4-0ST	OC	P	1589	1928	(f)	(2)
No.17		0-6-0T	OC	AB	2029	1937	New	(13)
No.18		0-4-0DM		RH	243081	1948	New	(14)
No.19		0-4-0DM		RH	304471	1951	New	(15)
No.20	later P.S.A. No.20	0-4-0DE		RH	327969	1954	New	(16)
-		4wVBT	VCG	S	9561	1953	(g)	(17)
No.21	later P.S.A. No.21	0-4-0DE		RH	395294	1956	New	
No.22	later P.S.A. No.22	0-4-0DE		RH	416210	1957	New	
		reb		YE	L167	1999		
-		0-4-0DH		AB	548	1967	(h)	(18)
No.2		0-6-0DH		AB	423	1958	(j)	(19)

There have been 10 steam rail cranes, including a 35-ton crane, CoS 3750/1916, originally built for the North Eastern Railway, 4 electric rail cranes and 2 diesel rail cranes here at various times.

A locomotive from South Dock was sent to the new pier works at Roker in 1883 (see below), but which locomotive it was, how long it stayed there, and whether any other locomotives worked there are all unknown.

(a) almost certainly ex Sunderland Dock Company, with the dock, in 1859. In January 1860 it had broken down and the Commissioners' Committee of Management authorised the purchase of a set of four new wheels from Fossick & Hackworth, who may therefore have been the locomotive's builder.
(b) ex Lever Bros (Port Sunlight) Ltd, Port Sunlight, Cheshire, by 8/1932.
(c) hired from BH when new, 22/8/1871; purchased 3/1872.
(d) ordered from R. & W. Hawthorn in 1885, but built after the amalgamation which formed R. & W. Hawthorn, Leslie & Co Ltd and delivered in 7/1886, but with RWH plate; whilst awaiting delivery RWH/HL hired a locomotive to the RWC; its identity is unknown.
(e) ex Wear Rolling Mills Co Ltd, Sunderland, /1888; (the plant of this firm was auctioned on 12/1/1888).

(f) ex Synthetic Ammonia & Nitrates Ltd, Billingham Works, Billingham, per C.J.M. Lowe, dealer, Newcastle upon Tyne; purchased, 18/8/1931, received, 1/9/1931.
(g) S demonstration loco; ex NCB South Wales Division No. 9 Area, Pantyffynnon Colliery, Pantyffynnon, Glamorgan, for trial; arrived about second week of 9/1955 (by 26/9/1955).
(h) ex NCB North East Area, Dawdon Colliery, Seaham, on loan, 17/12/1976.
(j) ex NCB North East Area, Wearmouth Colliery, Sunderland, on loan, 5/1977.
(1) sold to Nicholson & Co, contractors to the Commissioners for the locomotive power for shunting the dock, 2/5/1860 (see above); s/s.
(2) sold to Thos W. Ward Ltd, Middlesbrough, Yorkshire (NR), for scrap, 1/9/1953 and cut up on site.
(3) on 6/7/1904 A.T. & E.A. Crow offered for auction on behalf of the Commissioners an 11in [cylinder] Manning Wardle loco and a 13in Fox Walker loco, which are believed to have been these two; they were presumably unsold, for they survived to be scrapped c/1910.
(4) scrapped in 1909 or 1910, according to RWC records.
(5) scrapped in 1922 or 1923, according to RWC records.
(6) sold for scrap to Thomas Young & Sons (Shipbreakers) Ltd, Sunderland, 3/1930.
(7) condemned in /1931 and sold for scrap to Thos W. Ward Ltd, 6/1933.
(8) sold for scrap to Ellis (Metals) Ltd, Swalwell, 10/1960.
(9) to Thomas Young & Sons (Shipbreakers) Ltd, Sunderland, for scrap, 9/1956; scrapped 5/1957.
(10) sold to Thomas Young & Sons (Shipbreakers) Ltd, Sunderland, for scrap, c12/1955; scrapped /1956.
(11) to Thomas Young & Sons (Shipbreakers) Ltd, Sunderland, for scrap, 10/1959.
(12) sold to Thomas Young & Sons (Shipbreakers) Ltd, Sunderland, for scrap, 5/11/1957, and moved later in 11/1957.
(13) to Thos W. Ward Ltd, Templeborough Works, Sheffield, Yorkshire (WR), 4/5/1950, and dispatched, dismantled, on 2/6/1950; resold to NCB North East Division No.1 Area, Maltby Main Colliery, Maltby, Yorkshire (WR), 7/1952.
(14) to NCB Coal Products Division, Northern Region, Monkton Coking Plant, Wardley, 12/1970.
(15) to Wiggins, Teape & Co Ltd, Scottish Pulp & Paper Mills, Corpach, Fort William, Invernessshire, Scotland, 11/1969, per R.R. Brittain Ltd, Cranmore, Somerset.
(16) to Clarke Chapman-John Thompson Ltd, Gateshead, 6/1973, per Northern Supply Co, Sunderland.
(17) to The Seaham Harbour Dock Co, Seaham, for trial, by 17/10/1955.
(18) returned to NCB North East Area, Dawdon Colliery, Seaham, 5/1977.
(19) returned to NCB North East Area, Wearmouth Colliery, Sunderland, c7/1977.

111. A rare photograph of a large vertical-boilered rail crane in industrial use: the 35-ton Cowans Sheldon crane, CoS 3750/1916, on 21st May 1957.

Additional notes

When quoting for the four new wheels for the existing locomotive in January 1860, Fossick & Hackworth also discussed the supply of a second-hand locomotive; there is no record that this was proceeded with.

When placing the order for MW 57 on 30th April 1862 the Commissioners authorised the hire of a locomotive until it should be delivered. There is no record that this was done.

On 29/8/1885 A.T. & E.A. Crow of Sunderland offered for auction an 11in [cylinder] locomotive, "following pier and dock extensions". Its identity and disposal are unknown. MW 17/1860 and 57/1862 above both had 11in x 17in cylinders, so the loco offered may not have been one of them.

On 22/9/1886 A.T. & E.A. Crow of Sunderland again offered for auction a 11in x 18in Manning Wardle locomotive; see previous note.

A locomotive is believed to have been hired from the Marquis of Londonderry's Londonderry Railway, Seaham, between 1895 and 1899. Whether the locomotive was the same for the whole period, and which locomotive(s) were involved, is not known.

In *Contract Journal* of 16/12/1903 the Commissioners offered for sale a 10in x 18in loco (no other details given). None of the locomotives in the list above built before 1903 had 10in x 18in cylinders. Its identity and disposal are unknown.

ROKER PIER, Roker, Sunderland   H390
NZ 409589

In the early 1880s the Commission began to plan the construction of the Roker Pier on the northern side of the mouth of the river in order to protect a larger area of water for shipping entering or leaving the river. The blocks were manufactured on site, and the minutes of the Works Committee of 30th October 1883 record that to assist in the construction work a "portable railway" was to be provided, together with a narrow gauge locomotive ordered from W.B. Dick & Co of London. The same minutes indicate that an order for a standard gauge locomotive was to be proceeded with to replace the locomotive transferred to the works at Roker. A blockyard to manufacture concrete blocks was established along the sea shore between the North Pier and the new pier, and a short line built to link this to the timber yard of Armstrong, Addison & Co Ltd, which lay to the east of the North Dock and was served by the North Dock branch of the NER. The construction of the pier, which was to be 960 yards long, began in 1885; it was finally opened on 23rd September 1903. It is assumed that the locomotive purchased in 1891 also came here.

However, besides the above very little is known. Although the two locomotives below were actually ordered from Dick, Kerr & Co (as W.B. Dick & Co became in November 1883) this firm then sub-contracted the orders. What work, if any, the locomotives did in the last twenty years of their life is unknown; they do not appear on any of the Commissioners' extensive photographic collection of repair and maintenance work. The pier was provided with a standard gauge track for maintenance work, and a small loco shed was built near to the line's junction with the timber yard, though whether a locomotive was regularly kept here is unknown and probably unlikely. This system was abandoned in the early 1960s.

Gauge : 4ft 8½in

As noted above, a locomotive was transferred here from the South Dock about the end of 1883; but which locomotive(s) worked here, and for how long, is unknown.

Gauge : 1ft 8in

8	(MIDGET)	0-4-0ST	OC	BarrM		1884	(a)	(1)
11	(WASP)	0-4-0ST	OC	HAF	54	1891	(b)	(1)

(a) ordered from Dick, Kerr & Co, who sub-contracted the work to Barr, Morrison & Co; the Works Committee minutes of 30th November 1883 record that the locomotive was to be provided with trailing wheels, i.e., built as a 0-4-2ST, but there is no record of whether this was actually carried out; Barr, Morrison & Co went bankrupt in January 1884 and the order was completed by the firm's Trustees, being delivered on 25th February 1884; the builders' records are lost and no record of any works number survives.

(b) ordered from Dick, Kerr & Co, but built by Hartley, Arnoux & Fanning of Stoke-on-Trent, Staffordshire; this firm was absorbed into KS, whose Register gives the gauge as 2ft 5½, presumably in error.

Note : the supposed Dick, Kerr & Co works numbers of 262 and 266 sometimes quoted for these locomotives are almost certainly spurious.

(1)   sold for scrap to Thomas Young & Sons (Shipbreakers) Ltd, Sunderland, /1929.

In *Contract Journal* for 16/12/1903 the Commissioners advertised for sale a 20in gauge locomotive. Whether it was one of the above is not known.

PORT OF TYNE AUTHORITY
Tyne Improvement Commission until 1/8/1968

The Tyne Improvement Commission was created by Act of Parliament on 15th July 1850, to control the 17-mile tidal section of the river up to Wylam. Until 1848 the river had been controlled by Newcastle Corporation, though in that year North and South Shields were established as separate ports. When the Port of Tyne Authority was created, it took over the quays formerly operated by the councils at Newcastle upon Tyne, Gateshead and North Shields, while the T.I.C. river police were transferred to the South Shields force.

SOUTH PIER BLOCKYARD & TROW ROCKS QUARRY, South Shields   D391

The construction of piers and dredging of the river, together with the removal of the various islands in it, were the major tasks facing the newly-established Commission, essential if the commercial expansion of the region was not to be severely inhibited. Work on the north pier began in October 1855 and on the south pier in March 1856, initially under a contract awarded to B.C. Lawton, although after 1863 the Commission undertook most of the work itself using direct labour. Not until 1895 were both piers complete with their lighthouses, the north pier being 966 yards and the south pier 1800 yards long. To service these works the Commission opened up **TROW ROCKS QUARRY** (NZ 383666) near Trow Point, about a mile to the south-east, in 1855, and established the **SOUTH PIER BLOCKYARD** (NZ 374676) near the south pier's landward end, together with its own staiths (NZ 366682) to handle stone needed for the north pier. To link these locations a railway was laid along the shoreline, and of course laid along the pier as work progressed. The railway was also extended north-westwards for ½ mile to serve the **SOUTH GROYNE** (NZ 369683) when this was constructed. For many years the system

112. A view of the South Shields blockyard on 12th October 1868, with a wealth of fascinating detail. The locomotive, apparently a four-wheeler, looks like a design from the 1840s or 1850s; the long, sloping viaduct has on it what one would assume is a train of waggons carrying limestone aggregate from Trow Rocks Quarry for making concrete; the gantry itself, with the vertical-boilered crane on top, was apparently moved by hand; and the vehicle to the left of and higher than the large concrete block – what was its purpose?

had no link to the NER or LNER, but as part of the development of the river bank in the 1920s and early 1930s a rail link was installed between the LNER's South Shields Station and the Blockyard, a distance of about ¾ mile.

The system was initially worked by horses until steam locomotives were introduced, with locomotive sheds at both the blockyard and the quarry, although horses were used to haul waggons to and from the quarry face. By 1878 the system had over 400 end-tipping waggons and numerous block and other waggons. As South Shields Corporation worked to attract tourists to its beaches, the railway to Trow Rocks Quarry was incorporated into the new promenade. In 1894 The Harton Coal Co Ltd tried to take over the Trow Rocks Quarry and its railway, hoping to link the latter to its own system, but the proposal was rejected by South Shields Corporation because of an allegedly dangerous level crossing.

Trow Rocks Quarry was closed in 1954 and the use of the railway system dwindled, until by the late 1960s it was rarely used.

Reference : *The Tyne Pier Works Railways*, D.G. Charlton, Tanfield Railway News, No.56, December 2000.

113. The same gantry can be seen here, eighty three years later, on 11th July 1951. A small 48 h.p. diesel, No.48, RH 294263/1950, has replaced the steam locomotives. Note the method of coupling the locomotive to the wagon.

Gauge : 4ft 8½in

	-	0-4-0ST	OC?	?	?	?	(a)	s/s
	GERALDINE	0-4-0ST	OC	BH	369	1876	New	Scr /1933
	GROYNE (form. No.1 ETHEL)	0-4-0ST	OC	BH	288	1873	(b)	(1)
	NEWBURN No.1	0-4-0ST	OC	HCR	121	1872	(c)	(2)
	NEWBURN No.2	0-4-0ST	OC	HCR	122	1872	(c)	(2)
6 (46)	PEREYRA	0-4-0ST	OC	MW	1042	1888		
		reb		TJR		1914	(d)	Scr 6/1949
7 (47)	BELLE VUE	0-4-0ST	OC	P	1099	1907		
		reb		Adams		1937	(e)	(3)
	No.48	4wDM		RH	294263	1950	(f)	(4)

Note : the BH works list shows MIRIAM 0-4-0ST OC BH 623/1883 as New to the South Pier Blockyard, but it has been suggested that it was New to the North Pier Yard at Tynemouth in Northumberland.

The system also had at least five steam rail cranes. One, built by Stothert & Pitt of Bath, is shown on a huge wooden mobile crane some 30ft from the ground, on the 1868 photograph; see Photograph 112.

(a) a Commission photograph taken at the South Pier blockyard and dated 12/10/1868 shows a locomotive, unfortunately almost head-on, that appears to be an 0-4-0ST with a very tall chimney, probably with outside cylinders; its appearance suggests that it may well have been built as an 0-4-0 tender engine in the 1840s; its identity and origin are unknown.
(b) ex Albert Edward Dock, North Shields, Northumberland.
(c) ex river works at Heddon, Northumberland, some time after 1910.
(d) ex Topham, Jones & Railton Ltd, contractors, by 8/1928.
(e) ex A.R. Adams & Son, dealers, Newport, Monmouthshire, /1937; previously New Westbury Iron Co Ltd, Westbury Iron Works, Westbury, Wiltshire.
(f) New to South Pier, but delivered via Tyne Dock, Tyne Dock.

(1) to North Pier Blockyard, Tynemouth, Northumberland.
(2) sold c/1923? to a scrap merchant in North Shields, Northumberland, called Blackburn?
(3) Scr c/1949 (after 26/3/1949).
(4) to Yorkshire Dales Railway, Embsay, Yorkshire (NR), by 7/1980 (last reported here 16/3/1979, first reported at Embsay 24/7/1980).

TYNE DOCK, Tyne Dock D392
London & North Eastern Railway until 1/5/1937; **North Eastern Railway** until 1/1/1923   NZ 3465/3565

The suggestion that a dock should be developed from the large area of water known as Jarrow Slake was first made in 1810, and in 1849 the York, Newcastle & Berwick Railway began some excavation, only for it to be suspended. The NER resumed work in 1855, and Jarrow Dock, later re-named Tyne Dock, was opened on 3rd March 1859. It was said to have the largest waterspace of any dock in the world, occupying 49 acres and with space for 500 vessels, with the remaining 45 acres of Jarrow Slake used for storing and seasoning timber. Grain was also imported, as was Spanish iron ore for the iron works at Consett from 1880. By 1900 there were four large coal staiths with 32 loading points, and over 7½ million tons of coal was shipped in 1908. The Dock was served by over 20 miles of track.

Originally the dock was served by a branch from the Gateshead-South Shields line, 1¼ miles west of Tyne Dock Station. In 1885 this was supplemented by a second branch, from the Pontop & South Shields branch ½ mile north of Brockley Whins North Junction.

In 1937 the cash-strapped LNER sold the dock to the Tyne Improvement Commission for £808,000. This meant the TIC had to find its own locomotives, and it hired in LNER locomotives, from the 0-6-0T J71 and J72 classes and the 0-4-0T Y7 class, until its own fleet was up to strength. Two locomotive

114. The interior of Tyne Dock loco shed on 26th March 1949, with (left to right) 22, P 1953/1938, one of the four large locomotives purchased after the Dock was taken over from the LNER, with 26, P 1616/1923, and 29, BH 373/1875, behind her.

sheds were built, one at the northern end of the dock's western side at NZ 350658 and one near the staiths at the south-eastern corner at NZ 355652.

In 1953 the TIC constructed new cranes on the deepwater quay on the river bank to handle the Consett iron ore traffic. Coal shipments from the dock ceased in 1967 and the iron ore traffic ended in 1973, this work having been transferred to the River Tees. After this the PTA locomotives were little used. In 1985 the new Tyne Coal Terminal near the entrance to Tyne Dock to replace the Authority's Jarrow Staiths (see John Bowes & Partners Ltd) was opened, though this was operated by British Rail. However, coal exports ceased in 1998 and coal began to be imported in 2004. The area of the southern bank immediately upstream of Tyne Dock, now called the Riverside Quay, comprises the Authority's deep water berths. Much of Tyne Dock itself has now been filled in and developed for other port purposes as the port has diversified into new cargoes and logistics. The Quay includes a rail terminal, with traffic being handled by the English, Welsh & Scottish Railway.

There may well have been other short-term locomotive transfers between Tyne Dock and Albert Edward Dock, North Shields, besides those listed below.

Gauge : 4ft 8½in

0-6-0T J71 and J72 class and 0-4-0 Y7 class locomotives were hired from the LNER until the TIC fleet was large enough to handle all of the shunting. Details of which locomotives were involved are not known.

11		0-6-0ST	OC	RS	3072	1901	(a)	(1)
13		0-6-0ST	OC	AE	1933	1924	(b)	(2)
21	(1 until 1949)	0-6-0ST	OC	P	1952	1938	New	(3)
22	(2 until c3/1949)	0-6-0ST	OC	P	1953	1938	New	(4)
23	(3 until c31949)	0-6-0ST	OC	P	1954	1938	New	(4)
24	(4 until c3/1949)	0-6-0ST	OC	P	1955	1938	New	(5)
25	(5 until c3/1949)	0-6-0ST	OC	AE	1618	1912	(c)	(6)
8	(originally No.5)	0-6-0ST	OC	BH	645	1881	(d)	Scr /1950
6		0-6-0ST	OC	BH	666	1882	(d)	(7)
26	(6 until c3/1949)	0-6-0ST	OC	P	1616	1923	(d)	(8)
29	(9 until 1949)	0-6-0ST	OC	BH	373	1875	(e)	Scr 10/1950
27	(7 until c3/1949)	0-6-0ST	OC	RSHN	7212	1945	New	(9)
No.1		0-6-0ST	IC	RSHN	7138	1944	(f)	(10)
-		4wDM		RH	294263	1950	New	(11)
50		0-6-0DM		AB	380	1950	New	(12)
51		0-6-0DM		AB	381	1950	New	(13)
52		0-6-0DM		AB	382	1950	New	(14)
53		0-4-0DE		RH	323600	1953	New	(15)
54		0-4-0DE		RH	349087	1954	New	(16)
55		0-4-0DE		RH	381751	1955	New	(17)
56		0-4-0DE		RH	381752	1955	New	(18)
57		0-4-0DE		RH	381753	1955	New	(19)
(TIC No.58)		0-4-0DE		RH	381755	1955	New	(20)
No.35		0-4-0DE		RH	418600	1958	(g)	(21)

There was also one steam rail crane here.

(a) ex Albert Edward Dock, North Shields, Northumberland, 5/5 /1937 (by rail).
(b) ex Albert Edward Dock, North Shields, Northumberland, 24/9/1937 (by barge across River Tyne).
(c) ex Albert Edward Dock, North Shields, Northumberland, 2/3/1939 (by barge across River Tyne); George Cohen & Sons Ltd, Stanningley, Leeds, Yorkshire (WR) until 6/2/1939, and purchased for Tyne Dock; originally named COURTNEY
(d) ex Albert Edward Dock, North Shields, Northumberland, /1940.
(e) ex Albert Edward Dock, North Shields, Northumberland, 20/12/1943 (by barge across River Tyne).
(f) ex Albert Edward Dock, North Shields, Northumberland, by 26/3/1949.
(g) ex Albert Edward Dock, North Shields, Northumberland, 14/3/1984.

(1) returned to Albert Edward Dock, North Shields, Northumberland, /1938.
(2) to Mersey Docks & Harbour Board, Liverpool, /1941.
(3) to Albert Edward Dock, North Shields, Northumberland, by 7/1954; returned by 4/1955; to Albert Edward Dock, /1955.
(4) to Albert Edward Dock, North Shields, Northumberland, by 5/1955.

(5) to Albert Edward Dock, North Shields, Northumberland, by 7/1954.
(6) to Albert Edward Dock, North Shields, Northumberland, by 4/1953.
(7) to Albert Edward Dock, North Shields, Northumberland, /1945.
(8) to Albert Edward Dock, North Shields, Northumberland, c6/1955.
(9) to Albert Edward Dock, North Shields, Northumberland, 7/1955.
(10) to Albert Edward Dock, North Shields, Northumberland, /1949 (after 26/3/1949).
(11) to South Pier Blockyard, South Shields, /1950.
(12) to Thos W. Ward Ltd, Templeborough Works, Sheffield, Yorkshire (WR), 5/1963.
(13) to Albert Edward Dock, North Shields, Northumberland, 10/1955; returned, 10/1955; to Thos W. Ward Ltd, Templeborough Works, Sheffield, Yorkshire (WR), 5/1963.
(14) to Leslie Sanderson Ltd, dealer, Birtley, /1963; re-sold to NCB Kent Area, Chislet Colliery, Hersden, Kent, c5/1967.
(15) to Albert Edward Dock, North Shields, Northumberland, 1/1959; returned, 1/1960; to Albert Edward Dock; returned, /1962; to Thos W. Ward Ltd, Templeborough Works, Sheffield, Yorkshire (WR), 10/1968.
(16) to Albert Edward Dock, North Shields, Northumberland, 1/1959; returned, c/1962; cannibalised as source of spares for other locomotives; remains sold for scrap, c5/1979 (last reported here 16/3/1979, not here 11/7/1979).
(17) to Central Electricity Generating Board, Dunston Generating Station, Dunston, 12/1971.
(18) to Thos W. Ward Ltd, Templeborough Works, Sheffield, Yorkshire (WR), 10/1968.
(19) to Thos W. Ward Ltd, Templeborough Works, Sheffield, Yorkshire (WR), 2/1970; re-sold to Albright & Wilson Ltd, Portishead, Somerset, 6/1971.
(19) to Albert Edward Dock, North Shields, Northumberland, for repairs, c8/1983; returned after repairs, 3/1984; to Robinson & Hannon Ltd, Blaydon, subsequently Robinson Group Ltd, scrap merchants, for scrap, c/1997, but kept on site; eventually cut up c7/1999.
(20) to Tanfield Railway Preservation Society, Tanfield Railway, Marley Hill, 25/1/1992.

115. A scene so typical of ports in the 1960s and 1970s; the huge cranes on the Riverside Quay at Tyne Dock, with 0-4-0DE 35, RH 418600/1958, on 17th May 1985.

RIVER BANK WORKS AT RYTON ISLAND, COUNTY DURHAM and HEDDON-ON-THE-WALL, NORTHUMBERLAND
A393

NZ 144653 approx.

In 1861 the Commissioners approved a comprehensive plan for the improvement of the river itself, including extensive dredging, the removal of numerous islands and the strengthening of the river banks. By 1895 over 61 million cubic yards of silt, gravel and rock had been removed. The major work remaining was the work from Lemington to the Tidal Stone (NZ 142655) on the northern bank and from Stella to west of Ryton on the southern bank, including extensive dredging work and the removal of part of Ryton Island. Work began in 1898 and continued until 1912. To service the work, a short branch was laid to the Tidal Stone from the NER Newcastle-Carlisle line, ¼ mile west of Heddon-on-the-Wall Station in Northumberland. All rail traffic arising from these works used this route, as the southern bank had no rail connection to the NER.

116. During the protracted work to strengthen the river banks of the Tyne in the Ryton (Durham) and Wylam (Northumberland) areas, there was no rail access to the works on the Ryton side; so, incredibly, a locomotive was used to haul wagons between the two banks directly across the river! Four wagons make the crossing, obviously at low tide, about 1912.

An album of photographs maintained by the Commission and now deposited with the Tyne & Wear Archive Service shows that temporary track was laid on both banks, which was moved as required, and that three locomotives were used, working as required on whichever bank they were wanted. The two banks were connected by track laid in the river bed, and contractors' wagons were hauled through the water using a locomotive on each bank connected to the wagons by a chain. Whether the locomotives crossed this way too is not known.

Gauge : 4ft 8½in

NEWBURN No.1	0-4-0ST	OC	HCR	121	1872	(a)	(1)
NEWBURN No.2	0-4-0ST	OC	HCR	122	1872	(b)	(1)
-	0-4-0ST	OC	BH?	?	?	(c)	s/s

(a) per ?, via repairs at CF, /1898; previously The Plymouth Iron Co, Merthyr Tydfil, Glamorgan.
(b) per ?, via repairs at CF, /1899; previously The Plymouth Iron Co, Merthyr Tydfil, Glamorgan.
(c) origin unknown and identity unconfirmed; here by 12/1899.

(1) to South Pier Blockyard, South Shields, some time after 1910; however, another source claims that they were sold for scrap to a scrap merchant named Blackburn in North Shields, Northumberland, about 1913-1914.

THE PRIESTMAN COLLIERIES LTD

The main people behind this company were the Priestman family of Darlington, Quakers like so many of the prominent Darlington industrial families. The first to rise to power was Jonathan Priestman (1826-1888), who was Managing Director of The Consett Iron Co Ltd between 1864 and 1869, and subsequently became Managing Partner of the increasingly-powerful Ashington Coal Co Ltd in Northumberland. He was also the owner of Victoria Garesfield Colliery below, in partnership with George Peile of Shotley Bridge.

The company was formed on 1st January 1899 when the owners of Blaydon Burn Colliery and the nearby High Brickyard and the Lilley and Victoria Garesfield Collieries near Rowlands Gill amalgamated with the Owners of Chester South Moor and Waldridge Collieries near Chester-le-Street to form The Owners of the Priestman Collieries. The limited company with the shorter title was registered on 14th December 1903 upon the absorption of Hannington & Co Ltd. This company also owned a foundry and engineering works at Swalwell known as the Hannington Works. This did not pass to The Priestman Collieries Ltd, but was acquired by James W.Ellis & Co Ltd (see James W.Ellis Engineering Ltd). Joseph Cowen & Co's brickworks at Blaydon was not included in the new company in 1899, but was taken over subsequently.

AXWELL PARK COLLIERY, Swalwell — B394
Hannington & Co Ltd until 14/12/1903 — NZ 200619
Hannington & Co until 11/10/1887; see also below

This colliery, also known as **Axwell Garesfield Colliery**, was for many years a small landsale colliery, started possibly about 1841, or perhaps in the mid 1850s. From the 1870s until about 1882 it was owned by Mrs. Mary Hannington. Half a mile away, near Whickham, was the seasale colliery, **Axwell Colliery** (NZ 195608) **(B395)**, owned latterly by J.S. Bagnall & Sons. By the second half of 1887 'Hannington & Co' was owned by William Rutherford, while Axwell Colliery had passed into the hands of G.H. Snowball. Having registered as a limited company, Hannington & Co Ltd took over Snowball's business on 31st December 1887, immediately closing Axwell Colliery and merging its royalty with Axwell Park. The firm also owned a **brickworks** at Swalwell (NZ 204627), which became the site for the Hannington Works of James W. Ellis & Co Ltd (which see). Following the sale of its assets, the company went into voluntary liquidation on 28th April 1904.

Only one loco is known to have worked at Axwell Park Colliery, so presumably it was shunted by horses initially; what happened when the loco below was under repair is not known.

The colliery, together with a **brickworks** making common bricks adjacent to it begun in 1901, passed to NCB Northern Division No. 6 Area on 1st January 1947.

Gauge : 4ft 8½in

AXWELL	0-4-0ST	OC	HL	2330	1896	New	(1)

(1)   to NCB No. 6 Area, with the colliery, 1/1/1947.

AXWELL PARK COKE OVENS, Swalwell — B396

These beehive ovens lay adjacent to the colliery. An article in *Iron & Coal Trades Review* for January 1906 describes "Messrs Rutherford & McLellan's patent electrically-driven coke oven charging car", built by M. Coulson & Co Ltd of Spennymoor. An accompanying photograph shows a four-wheeled standard gauge vehicle comprising a coal-carrying box and an enclosed cab, surmounted by a trolley pole. The ovens ceased production on 26th April 1930.

BLAYDON BURN WAGGONWAY, Blaydon — B397
Joseph Cowen & Co until 1/1/1899

Joseph Cowen (1831-1899) was born at Blaydon Burn and besides his industrial interests, he was a journalist, newspaper owner, politician and theatre owner. This waggonway was developed to serve **COWEN'S FIREBRICK WORKS** (NZ178635) **(B398)**, which was begun in the 1820s, west of Blaydon. It would seem that at first bricks were dispatched via a ¼ mile link northwards to a quay on the River Tyne; but after the opening of the Newcastle & Carlisle Railway between Blaydon and Hexham on 9th March 1835 it was also linked to this, ¼ mile west of Blaydon Station. The quay continued in use until at least the 1850s, but was subsequently closed.

By 1855 the line to the firebrick works had been extended south-westwards along the Blaydon Burn for 1¼ miles and served the **LOW MILL** (NZ 175633) (a new clay works, opened in 1855), the **FREEHOLD PIT**, later called **COWEN'S PIT** (NZ 174633), **PATH HEAD QUARRY** (NZ 173632) (sandstone), the **EDWARD PIT** (NZ 172631), then what was originally called the **BETSEY PIT** but was later known as

Fig.9

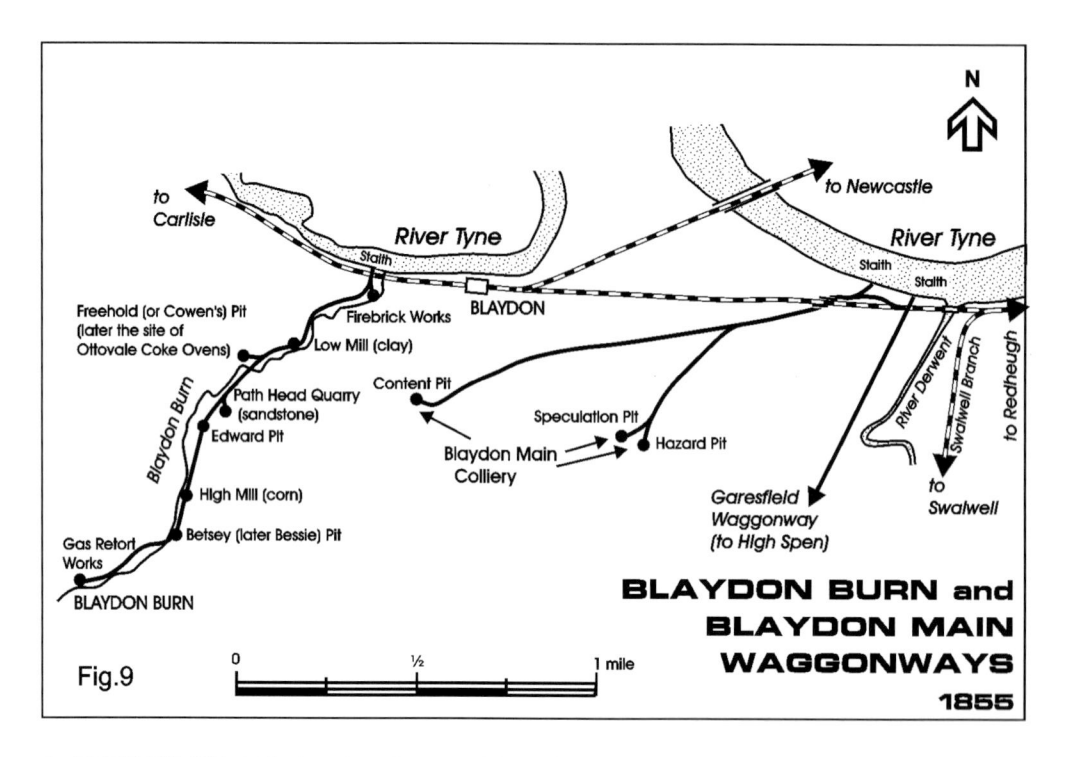

BLAYDON BURN and BLAYDON MAIN WAGGONWAYS 1855

the **BESSIE PIT** (NZ 170624) and ending at a **GAS RETORT WORKS** (NZ 165624). A diagram showing the system at this date can be found at Fig. 9. There were also other businesses near to the line not connected to it. In 1858 Cowen opened the **HIGH YARD** (NZ 169624) and about the same time a new drift, called the **MARY PIT** (NZ 168623). The fireclay for the brickyards was mined in the various coal pits. Horses were used before the introduction of the first locomotive (see note below). The High Yard was closed in 1914; the closure date for the Low Mill is not known. For some time after 1896 the Bessie Pit was linked by an aerial ropeway 2¼ miles long to the firm's Lilley Drift & Brickworks at Rowlands Gill (see below).

By the mid-1890s coal production had been concentrated at the Bessie and Mary Pits, together known as **BLAYDON BURN COLLIERY (B399)**. About June 1904 the new firm opened the **OTTOVALE COKE WORKS** (NZ 174633) **(B400)** on the site of Cowen's Pit, with 40 Otto-Hilgenstock waste heat ovens. 40 more were started up in January 1906 and a further 10 in 1910. The by-products manufactured were sulphate of ammonia, naphthalene, benzole and gas for industrial use.

Blaydon Burn Colliery, the Ottovale Coke Works and Cowen's Firebrick Works were all vested in NCB Northern Division No. 6 Area on 1st January 1947.

In later years the loco shed was situated in a cutting near the Bessie Pit, while the workshops were situated at the extreme end of the line at the Mary Pit.

Gauge : 4ft 8½in

	BEDE METAL						(a)	s/s
	REINDEER	?	?	?	?	?	(b)	s/s
	BLAYDON BURN							
later	BLAYDON BURN No.1	0-4-0ST	OC	RS	2840	1896	New	Scr /1926
	ENTERPRISE	0-4-0ST	OC	CF	1190	1900	New	Scr /1935
	VENTURE	0-4-0ST	OC	CF	1198	1901	New	Scr /1929
	ACTIVE	0-4-0ST	OC	RS	3075	1901	New	(1)
	INDUSTRY	0-4-0ST	OC	HC	749	1906	New	(2)
	ENERGY	0-4-0ST	OC	HC	764	1906	New	
		reb	HL		8860	1934		(2)
	BETTY	0-4-0ST	OC	RS	3376	1909	New	(2)
	-	0-4-0ST	OC	HC	724	1905	(c)	(3)
	CLAUDE	0-4-0ST	OC	HL	2349	1896	(d)	(4)
	GEORGE	0-4-0ST	OC	HC	1190	1916	(e)	(5)

Durham Part 1 Page 316

NELL	0-4-0ST	OC	HC	1191	1916	New	(6)
GERALD	0-4-0ST	OC	HL	2426	1899	(f)	(2)
BLAYDON BURN No.2	0-4-0ST	OC	HC	1514	1923	New	(2)
WALDRIDGE No.2	0-4-0ST	OC	HC	674	1903		
		reb	HL	4161	1925	(g)	(7)

(a) ex ?; the name would strongly suggest that it came from The Bede Metal & Chemical Co Ltd, Hebburn, but there is no obvious fit with any of that firm's locomotives.
(b) origin, identity and date of arrival all unknown.
(c) ex HC, Leeds, Yorkshire (WR), 11/1915, on hire pending delivery of HC 1190 and 1191.
(d) ex Blaydon Main Colliery, Blaydon, by /1916.
(e) ordered by Priestman Collieries Ltd, but diverted by the Ministry of Munitions to Sir W.G. Armstrong, Whitworth & Co Ltd, Lemington, Newcastle upon Tyne, 1/1916; to Blaydon Burn Colliery later in 1916; to Blaydon Main Colliery, Blaydon, and returned from there by /1918.
(f) ex Blaydon Main Colliery, Blaydon.
(g) ex Watergate Colliery, near Sunniside.

(1) to Bolckow, Vaughan & Co Ltd, Auckland Park Colliery, Coundon Grange, via James W. Ellis & Co Ltd, Swalwell, /1927.
(2) to NCB No. 6 Area, with the system, 1/1/1947.
(3) to Waldridge Colliery, Waldridge, /1916; returned, /1917; to Waldridge Colliery, and returned; to Watergate Colliery, near Sunniside, after /1926, and returned; to NCB No. 6 Area, 1/1/1947.
(4) HC hire loco; to Sir W.G. Armstrong, Whitworth & Co Ltd, Scotswood Works, Newcastle upon Tyne, 2/1916.
(5) loaned to The Ashington Coal Co Ltd (another company in which the Priestmans had an interest), Ashington Colliery, Northumberland, /1916; returned to Blaydon Burn Colliery, /1918; loaned to Newcastle Alloy Co Ltd, Newcastle upon Tyne, /1919, and returned; loaned to Brims & Co Ltd, contractors, apparently for a contract at Ashington Colliery, Ashington, Northumberland, and returned; loaned to The Ashington Coal Co Ltd, Ashington Colliery, Northumberland, /1925; returned, /1926; to Watergate Colliery, near Sunniside, and returned; to NCB No. 6 Area, with the system, 1/1/1947.
(6) to Watergate Colliery, near Sunniside, and returned; to Norwood Coke Works, Dunston, c/1930.
(7) to Watergate Colliery, near Sunniside.

117. BLAYDON BURN, RS 2840/1896. Note the use of only one crosshead slide, and the unusually-shaped side sheet on the cab.

BLAYDON MAIN COLLIERY, Blaydon
The Stella Coal Co Ltd until /1908

B401
NZ 188632

This colliery was purchased from The Stella Coal Co Ltd in 1908, the entry for which should be seen for its previous history. This pit, the Hazard Pit, was served by a ½ mile branch to the NER Blackhill branch just south of its junction with the Newcastle – Carlisle line. The colliery closed at the beginning of the national miners' strike in March 1921 and did not re-open.

Gauge : 4ft 8½in

CLAUDE	0-4-0ST	OC	HL	2349	1896	(a)	(1)
GERALD	0-4-0ST	OC	HL	2426	1899	(a)	(2)
GEORGE	0-4-0ST	OC	HC	1190	1916	(b)	(3)

(a) ex The Stella Coal Co Ltd, with the colliery, /1908.
(b) according to the HC Engine Book this locomotive was delivered new here; but according to oral tradition locally, its delivery was diverted by the Ministry of Munitions to Sir W.G. Armstrong, Whitworth & Co Ltd, Lemington, Newcastle-upon-Tyne, 1/1916; it was returned to Blaydon Burn Colliery by Sir W.G. Armstrong, Whitworth & Co Ltd and arrived here later in 1916.

(1) to Blaydon Burn Colliery, Blaydon Burn, by /1916.
(2) to Blaydon Burn Colliery, Blaydon Burn.
(3) to Blaydon Burn Colliery, Blaydon Burn, by /1918.

LILLEY DRIFT & BRICKWORKS, near Rowlands Gill
Joseph Cowen & Co until 1/1/1899

A402

It would appear that Cowen began operations here in the 1880s. What was originally called the **LILY DRIFT** (NZ 167593) was linked by a short narrow gauge tramway, almost certainly rope-worked, to screens at NZ 168592, which in turn were linked to the NER Blackhill (Consett) branch ½ mile north of Rowlands Gill Station. The **brickyard** was opened about 1901 and lay to the north of the screens, alongside what became the A694 road. Originally there were beehive coke ovens here too, but these were closed in March 1921 and subsequently removed to allow an expansion of the brickyard.

Nomenclature here is both confused and complicated. Garesfield Lily is found, and then Lilly and then the final spelling Lilley, while in the first half of the twentieth century the Ordnance Survey uses Lilley Drift Colliery. By 1919 two further drifts were working, the **Barlowfield Drift** (NZ 166594) and the **Alice Drift** (NZ 165595), the tramway from the latter joining the former before reaching the screens; however, on the O.S. 4th edition map of 1939 these names are reversed, with the drift at NZ 165595 shown as closed. The railway yard was also linked via a 2¼ mile aerial ropeway to Blaydon Burn Colliery's Bessie Pit (see above).

The drift and the brickworks, which made common bricks, passed to NCB Northern Division No. 6 Area on 1st January 1947.

How the screens and the brickworks were shunted before the arrival of the first locomotive below is not known, nor how the shunting was done when the loco below was under repair.

Gauge : 4ft 8½in

INDUSTRY	0-4-0ST	OC	?	?	?	(a)	s/s c/1905
ASHINGTON	0-4-0ST	OC	B	303	1883		
		reb	RS	2987	1900	(b)	(1)

(a) origin, identity and date of arrival all unknown.
(b) ex The Ashington Coal Co Ltd, Ashington, Northumberland, via RS, /1900 (?).
(1) to NCB No. 6 Area, with the site, 1/1/1947.

NORWOOD COKE OVENS & BY-PRODUCT PLANT, Dunston
begun by **The Team By-Product Coke Co Ltd** (registered 7/10/1912)
see also below

C403
NZ 238613

This was the first coke ovens and by-product plant in Britain to be independent of a colliery or iron & steel works. The aim was to provide a modern works for the many small collieries in north-west Durham which produced good coking coal but could not afford to build their own by-product ovens to replace the now-obsolete beehive ovens. It was built on a green field site next to good rail facilities and alongside what became the A692 Gateshead-Consett road. The works incorporated 120 Otto regenerative ovens, started up in 1915. Initially it was linked to the Pelaw Main Railway (see Pelaw Main Collieries Ltd) at its southern end, but links at its northern end were soon put in to both the NER

Tanfield branch and to the Dunston-Gateshead line. 10 Simon-Carves Otto combination ovens, the first of their type in Britain to incorporate the underjet firing system with individual regenerators, were added in 1921. The works produced coke nuts and recovered tar, sulphate of ammonia and benzole, while the surplus gas was piped to the Newcastle upon Tyne Electric Supply Co Ltd's Dunston A Power Station to serve boilers to produce steam to generate electricity.

The firm appointed a receiver on 26th February 1930 and went into voluntary liquidation on 24th May 1930. It would seem certain that Priestman Collieries Ltd purchased the works from the receiver, probably in May 1930. The works was vested in NCB Northern Division No. 6 Area on 1st January 1947.

In 1946 The Priestman Collieries Ltd was Durham's largest manufacturer of coke, producing 426,385 tons, of which Norwood produced 226,387 tons, the highest total for an individual plant outside the steel industry.

Gauge : 4ft 8½in

TEAMBY (form. ASHBOURNE)	0-6-0ST	IC	HC	439	1896	(a)	s/s
ERNEST BURY	0-6-0ST	OC	HL	3282	1917	New	(1)
NELL	0-4-0ST	OC	HC	1191	1916	(b)	(1)
VENTURE	0-4-0ST	OC	HL	2152	1889		
	reb		HL	6970	1914	(c)	(2)
VENTURE *	0-4-0ST	OC	HL	2837	1910	(d)	(1)
HASWELL	0-6-0T	IC	HC	1251	1917	(e)	(1)

* it may well be that HL 2837 did not carry the name VENTURE until after HL 2152 ceased working.

(a) ex Naylor Bros, contrs, Harrow, Middlesex, /1913 (by 6/1913).
(b) ex Blaydon Burn Colliery, Blaydon Burn, c/1930.
(c) ex Bede Metal & Chemical Co Ltd, Hebburn, c/1930 (the precise date is not known, and it may be that the loco arrived here before Priestman Collieries Ltd took over the plant).
(d) ex HL, 25/10/1932; previously Marston, Thompson & Evershed Ltd, Burton-on-Trent, Staffordshire.
(e) ex Victoria Garesfield Colliery, near Rowlands Gill.

(1) to NCB No. 6 Area, with the plant, 1/1/1947.
(2) to Ridley, Shaw & Co Ltd, Middlesbrough, Yorkshire (NR), for scrap by 8/1939.

One steam crane was also used here.

VICTORIA GARESFIELD COLLIERY, near Rowlands Gill A404
The Owners of Victoria Garesfield Colliery until 1/1/1899 NZ 146581
Priestman & Peile until c1/1897
previously **Victoria Garesfield Colliery Co**
originally **Thomas Ramsey**

This was initially a small landsale colliery, opened by Thomas Ramsey in 1870. When it was developed on a larger scale it was linked by a 1½ mile branch to the NER Blackhill (Consett) branch, just north of Rowlands Gill Station. Unusually, this branch had a reverse to bring it up to the height of the interchange sidings with the NER.

By 1896 there had been considerable expansion. In addition to the shafts, three drifts had been opened. A few yards to the northwest of the colliery lay what was later re-named the **Coronation Drift** (NZ 143583) and the **Speculation Drift** (NZ 143582), both served by a tramway which ran through a tunnel and then divided to serve the two drifts, and the **Hookergate Drift** (NZ 146582). The tramway serving the Speculation Drift, which worked the Tilley seam, eventually stretched for over three miles westwards into Chopwell Wood, occasionally coming out on to the surface. The track serving the first two drifts is not shown on the 3rd edition O.S. map of 1920, but had been re-instated by the 4th edition in 1940, although working in the Tilley seam was abandoned in 1934.

About half way along the branch to the NER, at Highfield, the **VICTORIA COKE OVENS** (NZ 153583) were built in 1861. These later passed to the owners of the colliery, and their name was changed to the **WHINFIELD COKE OVENS (A405)**. Although these were beehive ovens they were altered in 1915 so that the waste heat was used to make steam, which in turn generated the electricity to power a **CUPROUS OXIDE PLANT**, built immediately to the south of the ovens. Early in the First World War this was rebuilt into an **ALLOY PLANT**, manufacturing very hard steel for the sides of battleships. After this War it was rebuilt again as a **CHEMICAL WORKS**. By 1947 Whinfield's 193 ovens here were the last working beehive coke ovens in the country.

The loco shed was served by a short extension to the west of the colliery, unusually via a short tunnel.

On 1st January 1947 the colliery, the Whinfield Coke Ovens and the works were all vested in NCB Northern Division No.6 Area.

Gauge : 4ft 8½in

VICTORIA No.1	0-4-0T	OC	MW	?	?	(a)	s/s
VICTORIA No.2	0-6-0ST	IC	FJ	167	1879	New	Scr
VICTORIA No.3	0-6-0ST	IC	RS	2620	1887	New	s/s
VICTORIA No.4	0-6-0ST	IC	RS	2847	1896	New	
	reb		?		1904		(1)
VICTORIA No.5	0-6-0ST	IC	RS	2879	1900	New	
	reb		HL	4861	1925		(2)
HASWELL	0-6-0T	IC	HC	1251	1917	New	(3)

(a) origin and date of arrival are unknown.

(1) loaned to The Ashington Coal Co Ltd, Ashington, Northumberland, /1916, and returned, /1918; to NCB No. 6 Area, with the colliery and other plant, 1/1/1947
(2) to NCB No. 6 Area, with the colliery and other plant, 1/1/1947.
(3) to Norwood Coke Ovens, Dunston.

WALDRIDGE COLLIERY, Waldridge   G406
Waldridge Coal Co Ltd until 1/1/1899; this company was registered in 1891, but seems to have traded (curiously) as **The Owners of Victoria Garesfield Colliery** until c/1893 and **The Owners of Waldridge Collieries** until c/1895, according to returns made to the Durham Coal Owners Association.
formerly **Thiedmann & Wallis**: see also below

This colliery (NZ 251502), latterly known as **WALDRIDGE A COLLIERY**, was opened in August 1831. Subsequently beehive coke ovens were built here. The site was originally linked by a 2¾ mile branch running northwards to the Ouston Railway (see Pelaw Main Collieries Ltd), enabling its coal to be shipped on the River Tyne at Bill Quay. However, in September 1834 the Stanhope & Tyne Railway opened its line to South Shields. This crossed the colliery line at Pelton Fell, near the bottom of the incline, so the route to Ouston was abandoned and the remaining section linked to the S&TR.

Oral tradition averred that originally the line between Waldridge Colliery and Waldridge Bank Foot on the Pontop & South Shields branch was worked by horses and that the stables that stood for many years north of the stream called the Cong Burn were built for them. However, by the 1850s the route had been converted to rope haulage. The first ½ mile, crossing the Cong Burn on an embankment, was a **self-acting incline**, ending at the Stables. From here the **Waldridge Engine** (NZ 252505) worked a one mile long incline down to Pelton Fell. This incline was only single line, so presumably the empty and full sets were worked alternately.

The construction of the railway to Waldridge Colliery was followed by the sinking to the south of Sacriston and Charlaw Collieries, and from the bank foot of the incline from Waldridge Colliery their owners constructed a 3¼ mile line to serve them. This line was opened on 29th August 1839. Most of it was worked by two inclines, the first worked by a stationary engine, the Sacriston Engine, and the second a self-acting incline down to join the Waldridge line at the Waldridge Engine. The two bank foots were side by side, and this point became known as Bank Foot, Waldridge, presumably to distinguish it from the nearby Waldridge Bank Foot on the Stanhope & Tyne Railway. Thus the Waldridge colliery owners worked the Sacriston traffic between Bank Foot, Waldridge, and Pelton Fell, better known in later years as Stella Gill, latterly on behalf of The Charlaw & Sacriston Collieries Co Ltd (which see).

It would appear that horses continued to be used at the colliery until 1881, when a locomotive was purchased to work here. The Waldridge Incline continued to operate until 1897, when it was replaced by locomotive working, with the engine house being demolished and replaced with the **Bank Foot, Waldridge, loco shed** on the same site (NZ 252505). By this date the **WALDRIDGE D PIT** (NZ 253500) had been opened a short distance to the south east of the A Pit. Later the **WALDRIDGE SHIELD ROW DRIFT** (NZ 244497) **(G407)** was opened about one mile to the south west, a standard gauge branch being built to screens about half way, with narrow gauge rope haulage between the screens and the drift. Meanwhile, to work the eastern side of the royalty the Waldridge owners opened **CHESTER SOUTH MOOR COLLIERY** (NZ 268494) **(G408)** at Chester Moor. This lay alongside the NER Newcastle – Durham line, 1¼ miles south of Chester-le-Street Station, and was worked by the NER. When The Priestman Collieries Ltd was created in 1899 this royalty was of course separate from the large royalty in north-west Durham.

The beehive coke ovens closed in March 1922, Waldridge D Colliery followed in 1925 and Waldridge Shield Row Drift closed at the beginning of the Miners' strike in 1926 and did not re-open, the company preferring to work the remaining coal to Chester South Moor Colliery. With no Priestman Collieries coal now being worked here, about November or December 1926 the company handed over control of the railway between Bank Foot, Waldridge, and Stella Gill to The Charlaw & Sacriston Collieries Ltd (which see).

In the list below the four-coupled locomotives worked from the shed at the A Pit, while the six-coupled locomotives worked from the shed at Bank Foot, Waldridge.

Gauge : 4ft 8½in

WALDRIDGE	0-4-0ST	OC	BH	546	1881	New	(1)
VICTORIA	0-6-0ST	?	?	?	?	(a)	s/s
CECIL	0-6-0ST	IC	AB	803	1897	New	Scr /1924
WALDRIDGE No.2	0-4-0ST	OC	HC	674	1903	New	(2)
MARGARET	0-6-0ST	IC	AB	1005	1904	New	
	reb		AB	8833	1924		(3)
CLAUDE	0-4-0ST	OC	HL	2349	1896	(b)	(4)
FAITH	0-4-0ST	OC	HC	1201	1916	New	Scr c/1926
CECIL	0-6-0T	IC	HC	1524	1924	New	(3)

(a) ex The Ashington Coal Co Ltd, Ashington, Northumberland; said to have been the first locomotive to work at Bank Foot, Waldridge (i.e., c/1897); it may have been one of the locomotives listed at Victoria Garesfield Colliery above.
(b) ex Blaydon Burn Colliery, Blaydon Burn, /1916.
(1) to HC, Leeds, /1903, in part payment for HC 674.
(2) to HL for repairs, /1925, and then to Watergate Colliery, near Sunniside.
(3) to The Charlaw & Sacriston Collieries Co Ltd, with railway and Bank Foot, Waldridge, loco shed, c11-12/1926.
(4) returned to Blaydon Burn Colliery, /1917; ex Blaydon Burn Colliery, and again returned.

WATERGATE COLLIERY, near Sunniside　　　　　　　　　　　　　　　　　　　　　　　　　**A409**
　　NZ 222599

To work the lower seams in the north-eastern area of the main royalty the company decided to sink new shafts in 1923 and 1924, choosing a site near Sunniside where it could be connected with the LNER Tanfield branch on its western side and the A692 Gateshead – Consett road on the eastern side. The new colliery, named Watergate Colliery, began production in 1926. Only one locomotive worked here at a time, the usual locomotive being CLAUDE. The colliery was vested in NCB Northern Division No. 6 Area on 1st January 1947.

Gauge : 4ft 8½in

NELL	0-4-0ST	OC	HC	1191	1916	(a)	(1)
WALDRIDGE No.2	0-4-0ST	OC	HC	674	1903		
	reb		HL	4161	1925	(b)	(2)
CLAUDE	0-4-0ST	OC	HL	2349	1896	(c)	(1)
GEORGE	0-4-0ST	OC	HC	1190	1916	(c)	(1)

(a) ex Blaydon Burn Colliery, Blaydon Burn.
(b) ex HL, after repairs, /1925; previously Waldridge Colliery.
(c) ex Blaydon Burn Colliery, Blaydon Burn, after /1926 (not necessarily at same date as HC 1191).
(1) returned to Blaydon Burn Colliery, Blaydon Burn.
(2) to Blaydon Burn Colliery, and returned; to NCB No. 6 Area, with the colliery, 1/1/1947.

RAINE INDUSTRIES plc
Raine Engineering Industries Ltd until 7/1/1982
Raine & Co Ltd until 18/11/1969; the firm became a subsidiary of **The Empire Rib Co Ltd** of Sheffield in 12/1963
Raine & Co until 3/10/1891
formerly **B.W. & G. Raine**, originally **N. Raine**

Nicholas Raine had been works manager at the Witton Park Iron Works of Bolckow, Vaughan & Co Ltd, leaving there to set up his own business. It would seem that the firm began by taking over long-established rolling mills at South Hylton, near Sunderland, in 1871, which they gave up about 1881.

WINLATON MILL WORKS, Winlaton Mill   A410
NZ 186604

This had once been the most famous iron and steel works on Tyneside, the steel being used to make swords. It was begun by **Ambrose Crowley** in 1691, and in the eighteenth century it almost certainly obtained the coal it required via the Garesfield Waggonway (see the entry for the Marquis of Bute), which was built past the works. It was later owned by **Crowley, Millington & Co** (this title was in use by 10/3/1862), which sold it in 1863. The Raines acquired it in 1885 and developed it as a rolling mill. Between 1893 and 1899 the waggonway past it was reconstructed by The Consett Iron Co Ltd as their Chopwell & Garesfield Railway (which see). It would seem very likely that the firm transferred materials between the Winlaton Mill Works and their Delta Works at Derwenthaugh, but whether this was done using their own locomotives or was handled by the NER and The Consett Iron Co Ltd, the owners of the Chopwell & Garesfield Railway, is not known. The works was closed in 1918 (one source gives 1915) and lay derelict until demolished in 1936. In the 1950s the site was used by the National Coal Board to construct workshops for its Clockburn Drift (see Part 2).

Gauge : 4ft 8½in

-		0-4-0VBT	?	?	?	?	(a)	(1)
-		0-4-0ST	?	?	?	?	(a)	s/s
-		0-4-0ST	?	?	?	?	(a)	s/s
NORTH STAR		0-4-0ST	IC	JF	1572	1872	(b)	(2)

One source stated that there was also a HH locomotive here.

(a) origin, identity and date of arrival unknown.
(b) ex Lingford, Gardiner & Co Ltd, dealers, Bishop Auckland; it was new to The Washington Chemical Co Ltd, who had offered it for sale in 1/1880, so there would seem to be a gap in its known history.

(1) in *Machinery Market* dated 1/8/1890 the firm advertised for sale a four-wheeled "coffee pot" locomotive, built by Head, Wrightson & Co of Stockton with 6in x 12in cylinders and 2ft 4in wheels; it is not known whether the locomotive at Winlaton Mill and the locomotive advertised are one and the same; if so, it would appear to have remained unsold, for the locomotive at Winlaton Mill was said to have been scrapped about 1900, latterly having been used at the Delta Works below to pump water from the River Tyne.
(2) to Delta Works, Derwenthaugh.

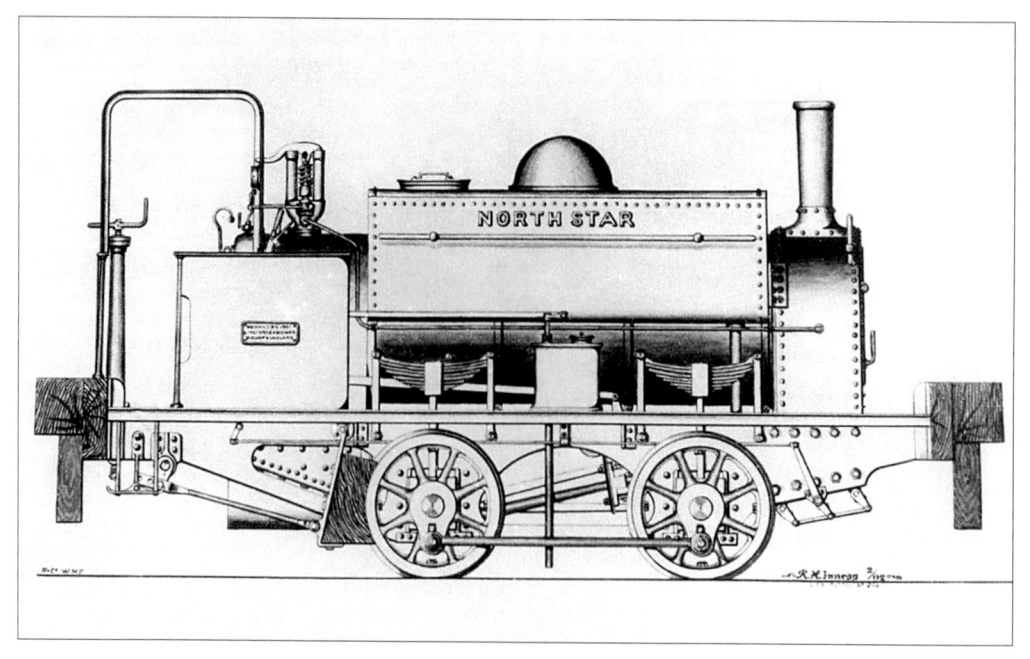

118. R.H. Inness' drawing of NORTH STAR, JF 1572/1872, dated February 1902. Very few 0-4-0STs had inside cylinders. Note the large wooden brake block.

DELTA WORKS, Derwenthaugh

B411
NZ 208633

One source says that the company moved here in 1915, but the company's advertisement in the brochure for the North East Coast Exhibition of 1929 says that the company opened its works at Derwenthaugh in 1890. It was served by sidings north of the NER Redheugh Branch, 1½ miles east of Blaydon Station. About ¼ mile west of the works' junction with the NER was the junction with the Chopwell & Garesfield Railway, and there may have been exchange traffic with the Winlaton Mill Works (see that entry).

Not a great deal is known about the early locomotives here. Rail traffic with BR ceased about 1981, but internal rail traffic continued until May 1984. The works was closed in August 1990.

Gauge : 4ft 8½in

	NORTH STAR	0-4-0ST	IC	JF	1572	1872	(a)	s/s
	BENTON	0-4-0ST	OC	HCR	103	1871	(b)	s/s
	-	0-4-0ST	OC	MW	455	1874	(c)	s/s
	-	0-4-0ST	OC	BH	852	1886	(d)	s/s
	BEATTY	0-6-0ST	IC	MW	1669	1905	(e)	(1)
	DELTA 1924	0-4-0ST	OC	AE	1932	1924	New	Scr 10/1957
	No.2	0-4-0ST	OC	WB	2664	1942	(f)	Scr c/1959
	B No 20	0-4-0ST	OC	HL	3745	1929	(g)	s/s c/1965
	B No 38	0-4-0ST	OC	HL	3496	1921	(h)	(2)
No.7		0-4-0ST	OC	AB	1337	1913	(j)	(3)
13	(MOS No.1L)	0-4-0ST	OC	AB	2118	1941	(k)	s/s c/1965
	-	0-6-0DM		HC	D624	1942	(m)	(4)
27		0-4-0DM		HE	5387	1959	(n)	(4)
	-	0-4-0DM		AB	384	1951	(p)	(4)
41		0-4-0ST	OC	RSHN	7674	1951	(q)	(5)
1	(formerly 21)	0-4-0DE		BP	7946	1961		
				BT	339	1961	(r)	(6)
2	(formerly 22)	0-4-0DE		BP	7947	1961		
				BT	340	1961	(r)	(7)
	SIR WILLIAM	0-4-0DE		BP	7873	1962		
				BT	443	1962	(r)	(8)

Four steam cranes were also used here.

(a) ex Winlaton Mill Works, Winlaton Mill.
(b) ex CF; New to D. Burn, West Stanley Colliery, Stanley (see South Derwent Coal Co Ltd).
(c) ex Benton & Woodwiss, contractors; previously The Charlaw & Sacriston Collieries Ltd, Sacriston Colliery, Sacriston.
(d) ex J. Spencer & Sons Ltd, Newburn, Newcastle upon Tyne.
(e) ex Bradford Corporation, Waterworks Dept, Nidd Valley Light Railway, Yorkshire (NR).
(f) ex C.A. Parsons & Co Ltd, Newcastle upon Tyne, per A.W. Wanless & Co, dealer, Newcastle upon Tyne, 9/1953.
(g) ex The Consett Iron Co Ltd, Consett, 8/1957.
(h) ex The Consett Iron Co Ltd, Consett, loan, /1958.
(j) ex Millom Hematite Ore & Iron Co Ltd, Millom, Cumberland, 6/1959.
(k) ex H. Dunn Plant & Machinery Co Ltd, dealers, Bishop Auckland, c3/1962; Royal Ordnance Factory, Linwood, Renfrewshire, Scotland, until /1957.
(m) ex Thos W. Ward Ltd, Templeborough Works, Sheffield, Yorkshire (WR), c5/1964; previously Shell-Mex & B.P. Ltd, Hull, Yorkshire (ER).
(n) ex The Consett Iron Co Ltd, Consett, loan, 1/1965; returned to Consett, 2/1965; purchased, 2/1965.
(p) ex Leslie Sanderson Ltd, dealer, Birtley, after 25/7/1968; previously Lever Bros Ltd, Port Sunlight, Cheshire.
(q) ex NCB Northumberland Area, Rising Sun Colliery, Wallsend, Northumberland, loan, 4/1969.
(r) ex Parkgate Iron & Steel Co Ltd, Sheffield, Yorkshire (WR), 9/1970.

(1) withdrawn from traffic in 1957 and used as a stationary boiler; Scr /1959.
(2) returned to Consett Iron Co Ltd, Consett.
(3) to Clayton & Davie Ltd, Dunston, for scrap, 5/1962.
(4) to T.J. Thomson & Son Ltd, Stockton-on-Tees, for scrap, 4/1971 (see Cleveland & North Yorkshire Handbook)
(5) returned to NCB Northumberland Area, Burradon Colliery, Burradon, Northumberland, c6/1969

(6) dismantled and used for spares by 10/1977; last reported here, 2/3/1985; s/s.
(7) dismantled and used for spares from 1973; remains scrapped, 3/1981.
(8) to T.J. Thomson & Son Ltd, Dunston, for scrap, 3/1985, and cut up, 4/1985.

119. 1, 0-4-0DE BP/BT 7946/1961, on 1st April 1974. This too was an unusual design, with its outside frames and cranks. Raines had three of these, all coming north from a steelworks in Sheffield.

RAINTON COLLIERY CO LTD
ADVENTURE COLLIERY, Rainton J412
NZ315471

This company re-opened this colliery, which had been closed in 1896 by the Marquis of Londonderry (which see). According to tradition it acquired 0-4-0ST BH 203/1871 from the Marquis to shunt it, only to sell it to The Seaham Harbour Dock Co in 1903. However, research shows that the company did not acquire the colliery until 1912, and a photograph dated 7th March 1914 shows the surface buildings, including the winding house, still under construction, which calls into question the tradition above.

It was served by sidings east of the NER Leamside branch, one mile north of Leamside Station, and for many years it was worked by the NER and then the LNER. It passed to NCB Northern Division No. 2 Area on 1st January 1947.

RAISBY QUARRIES LTD (registered in 1947 and re-registered 29/12/1967)
Raisby Hill Limestone Co Ltd and **Raisby Basic Co Ltd** until 1947; for the relationship of these companies see below.
RAISBY HILL QUARRY, later known as GARMONDSWAY QUARRY, near Coxhoe N413
Matthew Chaytor and Richard Matthews until 10/9/1881 NZ 343354 approx.

Quarrying of the magnesian limestone at Raisby Hill began in 1845, initially just a small quarry and some lime kilns, with an east-facing link to the Great North of England, Clarence & Hartlepool Junction Railway, 1½ miles east of Coxhoe Bridge Station. The owners were also involved with East Hetton and South Hetton Collieries. After the Raisby Hill Limestone Co took over the business in 1881 production was rapidly expanded. This quarry later became known as the **High Quarry** (NZ 347351). It was served by a loco shed not far from the junction with what was now the NER.

In 1919 a new quarry, known as the **Low Quarry** (NZ 338352), was opened to the west, with a separate west-facing link to what was now the NER Ferryhill-Hartlepool line. Production increased rapidly, so that by 1930 the quarry face was 1¼ miles long. It was unique as a quarry, in that four different types of stone were worked, all with different uses. Furnace dolomite, "transition" limestone and "white" limestone were worked and processed by the Raisby Hill Limestone Co Ltd, while Basic dolomite was handled by a sister company, the Raisby Basic Co Ltd (registered 30/9/1882), which converted this into doloma for the steel industry. A visitor to this quarry in 1945 noted it as being owned by F.W. Dobson & Co Ltd (see Index 1), but this has not been confirmed. This quarry had its own loco shed situated amongst the main buildings, although the two quarries were linked internally. From the 1950s much of the output from this quarry went to the Consett iron and steel works.

120. Raisby purchased three of these Peckett R2 class locomotives between 1919 and 1923. Note the recessed cab. Each of them gave more than forty years' service. P 1637/1923 was recorded on 1st May 1960.

The rail link into the High Quarry was closed in May 1965, together with the closure of the railway east of the quarry as far as Trimdon Grange Colliery. This left the Low Quarry at the end of the line from Ferryhill, four miles from Ferryhill Station. In latter years the loco shed in the Low Quarry was abandoned and the second Sentinel loco was moved up to the new quarry buildings. The frame of a new shed was erected but the building was never completed. Rail traffic ceased in the 1970s. On 1st January 2001 the quarry was acquired by **Tarmac (Northern) Ltd** (which see). The company itself continues (2006) to exist.

Details of any locomotives before 1897 are not known.

Gauge : 4ft 8½in

RAISBY (BESSIE till 10/1902)	0-4-0ST	OC	HE	240	1880	(a)	s/s by 6/1928
-	0-4-0ST	OC	AB	1289	1912	New	(1)
10 (originally No.1 VICTORY)	0-4-0ST	OC	P	1544	1919	New	(2)
11 (orig. No.2; later No.1)	0-4-0ST	OC	P	1637	1923	New	Scr c/1965
No.3	0-4-0ST	OC	P	1648	1923	New	Scr /1969
MOS No.2L	0-4-0ST	OC	AB	2119	1941	(b)	Scr c6/1970
-	4wDH		S	10031	1960	(c)	(3)
M14	4wDH		S	10077	1961	New	(4)

(a) ex South Stockton Iron Co Ltd, South Stockton, Yorkshire (NR), by 7/1897.
(b) ex Royal Ordnance Factory, Linwood, Renfrewshire, Scotland, per H. Dunn Plant & Machinery Co Ltd, dealers, Bishop Auckland, c/1957 (by 4/1959).

(c) S demonstration locomotive; ex Central Electricity Generating Board, Blyth Power Station, Blyth, Northumberland, c22/4/1961.
(1) to Wilsons & Clyde Coal Co Ltd, Glencraig Colliery, Glencraig, Fifeshire, Scotland, c/1936.
(2) derelict by 1/8/1958, with parts being used to repair other Peckett locos; remainder scrapped between 1960 and 1964.
(3) sold by S to Dorman Long (Steel) Ltd, Middlesbrough, Yorkshire (NR), and delivered there, 28/4/1961.
(4) to Tarmac Northern Ltd, with the quarry, 1/1/2001.

Gauge : 2ft 0in

A visitor to the Low Quarry in 1945 noted that 'the two petrol locos are kept as spares for the narrow gauge Barclay at the other quarry'. This would suggest that at that date the narrow gauge system was in use only in the High Quarry, where the AB had its own locomotive shed, although track ran through the link into the Low Quarry.

70	0-4-0WT	OC	AB	1995	1931	(a)	(1)
6	4wPM		MR	4045	1926	(b)	s/s
-	4wPM		MR	?	?	(c)	s/s

(a) origin unknown; previously Corby & District Water Company, Eyebrook Reservoir, Northamptonshire; originally Durham County Water Board, Burnhope Reservoir construction, Burnhope.
(b) ex Francois Cementation Co Ltd, Kendal, Westmorland, by 15/4/1940.
(c) origin unknown; here by early in 1945.
(1) to Dinorwic Slate Quarries Co Ltd, Llanberis, Caernarvonshire, /1948 (by 7/1948), per H. Dunn Plant & Machinery Co, dealer, Bishop Auckland.

CORNFORTH LIMEWORKS & QUARRY, West Cornforth N414
Tarmac Roadstone (Northern) Ltd until 1/6/1983 NZ 317347 approx.

This quarry was served by the same line that served Garmondsway Quarry above, 2½ miles north of Ferryhill Station. Rail traffic had ceased on 19th July 1978, but the locomotives below were left on site. The quarry was closed on 18th June 1983.

Gauge : 4ft 8½in

| 6/1698 | R.F. SPALDING | 4wDM | RH | 262996 | 1949 | (a) | Scr c12/1984 |
| 6/1701 | (FRED DOBSON) | 4wDM | RH | 326071 | 1954 | (a) | (1) |

(a) ex Tarmac Roadstone (Northern) Ltd, with site, 1/6/1983; out of use.
(1) scrapped on site by A.C. Metals Ltd, Darlington, 3/1984.

RAMSHAW COAL CO LTD
WEST TEES COLLIERY, Ramshaw, near Evenwood P415
NZ 158263

This colliery, at first known as **Railey Fell Colliery**, was owned for many years by Henry Stobart & Co Ltd, though without its own locomotives, so far as is known. It was served by ¼ mile branch from the NER Haggerleazes/Butterknowle branch, ¼ mile to the west of the branch to Randolph Colliery and passing under the NER Darlington – Barnard Castle line. It was the only colliery owned by the Ramshaw Coal Co Ltd in the 1920s, so it is assumed that the locomotive below worked here, and that shunting was subsequently resumed by the LNER. The company subsequently acquired other collieries, and also re-named this colliery **RAMSHAW No.1 COLLIERY**. It passed to NCB Northern Division No. 4 Area on 1st January 1947.

Gauge : 4ft 8½in

| - | 0-4-0ST | OC | P | 916 | 1901 | (a) | (1) |

(a) origin and date of arrival unknown; New to W.S. Laycock Ltd, Sheffield, Yorkshire (WR).
(1) to Bolckow, Vaughan & Co Ltd, Newfield Colliery, Newfield, /1927.

The company also worked **WEST AUCKLAND COLLIERY**, West Auckland (NZ 184267) **(Q416)**, formerly owned by Bolckow, Vaughan & Co Ltd, which it re-opened in 1944; this too passed to NCB Northern Division No. 4 Area on 1st January 1947.

THE RANDOLPH COAL CO LTD

This company was created following the financial collapse of Pease & Partners Ltd and its Associated Companies, specifically The North Bitchburn Coal Co Ltd, the previous owners of Randolph Colliery and Coke Works.

RANDOLPH COLLIERY, Evenwood   P417
NZ 157248

This colliery had been opened in 1895 and its adjacent coking and by-products plant in 1909. They had been closed in May 1932, in the collapse of The North Bitchburn Coal Co Ltd. The Randolph Coal Co Ltd was another venture of the Summerson family (see their entries) and it took over both Randolph Colliery and the workings of **GORDON HOUSE COLLIERY** (NZ 133240) **(P418)**, near Esperley Lane Ends, with which it was combined, in January 1933. The colliery **brickworks** also passed to a Summerson company, this time W. Summerson Ltd. The colliery company did not take over the adjacent coke ovens, lying to the north-west of the colliery, and comprising 60 Koppers waste heat ovens dating from 1909-1917. These seem to have lain idle until about 1938, when they were taken over by Sadler & Co Ltd, a firm which would appear to be linked with the very similar company of Sir S.A. Sadler Ltd. The coke works traffic was then shunted by the colliery locomotives.

The two premises were linked by a one mile branch to the LNER Butterknowle Branch (formerly the Haggerleazes Branch) ¼ mile west of Spring Gardens Junction, including an self-acting incline 1260 yards long incorporating a drum house at the bank head. By 1939 the majority of the surface at Gordon House, together with the tramway that it once had used, had been cleared.

Randolph Colliery, with Gordon House, was vested in NCB Northern Division No. 4 Area on 1st January 1947. Neither the brickworks nor the coke ovens were nationalised.

Gauge : 4ft 8½in

CARBON	0-4-0ST	OC	Grange		1866	(a)	s/s after 24/5/1935
MOSTYN	0-4-0ST	OC	MW?	?	?	(a)	(1)
		reb	LG		1906		
HUSTLER	0-4-0ST	OC	P	1337	1913	(a)	(2)
RANDOLPH	0-4-0ST	OC	RSHN	7043	1942	New	(2)
WINSTON	0-4-0ST	OC	RSHN	7159	1945	New	(2)

(a)   taken over with the colliery, 1/1933; previously The North Bitchburn Coal Co Ltd.
(1)   to The Bearpark Coal & Coke Co Ltd, East Hedley Hope Colliery, near Tow Law, 4/1945.
(2)   to NCB No. 4 Area, with the colliery, 1/1/1947.

RANDOLPH COKE & CHEMICAL CO LTD

This company was formed in 1957 as a subsidiary of North Eastern Tar Distillers (Sadlers) Ltd to take over the Randolph Coke Works & By-Product Plant from Sadler & Co Ltd, its previous operators (see the entry for The Randolph Coal Co Ltd above). At the same time the locomotives at Randolph Colliery, operated by NCB Durham No. 4 Area, were sold to the new company, which then took on the shunting of the colliery. The locomotives continued to be housed in the colliery's loco shed.

RANDOLPH COKE WORKS & BY-PRODUCTS PLANT, Evenwood   P419
NCB Durham No.4 (South-West Durham) Area until 20/5/1957
NCB Northern Division No.4 (South-West Durham) Area until 1/1/1950   NZ 157249

The coke works and its by-products plant were served by a one mile branch to the BR Butterknowle Branch ½ mile west of Spring Gardens Junction, including a **self-acting incline** 1260 yards long incorporating an overhead drum house at the bank head. Latterly this method of working was replaced by a 70 h.p. electric hauler installed in the drum house. This probably became necessary once coal had to be brought in from elsewhere to be used in the coke ovens. The normal working of the incline was four 21-ton wagons of coke down and either two 21-ton wagons of coal or seven empty wagons up. The works comprised 15 Woodall-Duckham Becker regenerative ovens, started up in 1948, with a further 11 added in 1957. The coke car did not utilise a locomotive, but was a combined car made by James Buchanan of Liverpool; a second one (40hp) was provided in 1962. The shunting of Randolph Colliery ended with its closure on 17th February 1962. The coke works closed in September 1968, but was re-opened in the following month with the old company as a subsidiary of Millom Hematite Ore & Iron Co Ltd, without rail transport. The works, after further changes of main ownership, though still operating under the 1957 title, finally closed on 25th May 1984, the last coke being pushed ten days earlier.

121. WINSTON, RSHN 7159/1944, stands near the top of the Randolph Incline on 7th August 1966. The rather ramshackle building behind her was once the drumhouse for the self-acting incline; but after the closure of the colliery, coal had to be brought in for the coke ovens, and the incline was converted to powered operation by replacing the drums with a 70 h.p. electric hauler.

Gauge : 4ft 8½in

RANDOLPH	0-4-0ST	OC	RSHN	7043	1942	(a)	(1)
(WINSTON)	0-4-0ST	OC	RSHN	7159	1945	(a)	(1)

(a) ex NCB Durham No. 4 Area, Randolph Colliery, Evenwood, 20/5/1957.

(1) to C. Herring & Son Ltd, Hartlepool, for scrap, 3/1969, but not removed to be scrapped at Hartlepool until the autumn of 1969.

LORD RAVENSWORTH & PARTNERS

This partnership of colliery owners, known locally as "The Grand Allies", was at the beginning of the nineteenth century the most powerful in the country. It had been made between the families of Wortley, Ord, Liddell and Bowes, to run for ninety nine years from November 1726. It was renewed when that term expired. Sir Henry Liddell had been created Baron Ravensworth on 26th June 1747. The partnership owned collieries in both Durham and Northumberland.

SPRINGWELL COLLIERY RAILWAY   C/D/E420

On 8th May 1821 the "Allies" began the sinking of **SPRINGWELL COLLIERY** (NZ 285589), about four miles south-east of Gateshead. Not far away lay another "Allies" pit – **MOUNT MOOR COLLIERY** (NZ 279577), which had been won nearly a century earlier and whose coal was carried by a waggonway to staiths on the River Wear near Washington. But this involved trans-shipping the coal in keels down to colliers at Sunderland, and the Allies were determined to end this system. So they engaged John Buddle (see the Lambton and Londonderry entries) to develop proposals for a new waggonway between their Stanley Colliery at Stanley, Mount Moor, Springwell and the River Tyne at Jarrow. Buddle proposed a line 11½ miles long, with six rope inclines and two locomotive-worked sections. However, the Allies decided not to proceed with the section west of Mount Moor, and then commissioned George Stephenson, their former enginewright at Killingworth Colliery in Northumberland, to re-design the remainder. Coal from Mount Moor Colliery was hauled up a 750-yard incline at about 1 in 15 to the **Blackham's Hill Engine** (later spelling) (NZ 283581), which then let the waggons down to Springwell Colliery, 1170 yards at 1 in 70. Both inclines were single line. From here a self-acting incline 1¼ miles long at 1 in 24, the **Springwell Incline**, worked coal down to the Leam Lane, whence locomotives would take over for the final 4¾ miles to Jarrow. A second, short "inclined plane", almost certainly self-acting, took the waggons down to the staiths (NZ 330658). George Stephenson was at this time very busy with the construction of the Stockton & Darlington Railway, and it is likely that Joseph Locke, one of Stephenson's assistants, designed the line, with

George's younger brother Robert, who had recently left his employment as Engineer to the Hetton Railway, was the resident engineer. Two locomotives were ordered from the recently-established firm of Robert Stephenson & Co in Newcastle upon Tyne.

The sinking of Springwell Colliery was finished on 24th February 1824, but construction of the railway seems to have been protracted. Although the line was only completed from Jarrow to Springwell and the locomotives had not been delivered, the "Allies" decided to open it, using horses, on 17th January 1826. The locomotives were finally delivered in April 1826, the loco shed being at **SPRINGWELL BANK FOOT** (NZ 296605). The remaining section, between Springwell and Mount Moor, is reported to have been opened two months later.

In 1842 the line was extended for a further 2½ miles to **Kibblesworth Colliery** (NZ 243562), which was owned by **George Southern** and won in March 1842. This section was opened on 30th May 1842, though one source gives a week later. This required two more inclines, again single line. The **Kibblesworth Engine** lowered the wagons down the 1¼ miles to Long Acre at Lamesley, from where the **Black Fell Engine** hauled them for 1¼ miles up to Mount Moor. Whether Southern provided his own waggons and shipped his coal at Jarrow is not known.

Soon after this a passenger service was begun over part of the line. In August 1839 the Brandling Junction Railway's line between Gateshead and Monkwearmouth (Sunderland) was opened, crossing under the colliery line near Wardley. A link between the two lines was put in on the north-west side of the crossing and a building known as **Springwell Station** (NZ 312624) erected in the triangle of land thus created. There does not seem to have been a similar building at Jarrow at this time; possibly the train stopped in the staiths' sidings. It would seem that the service was begun by the BJR, very probably using horses, though local tradition averred that the coach was allowed to go down to Jarrow by gravity. The earliest-known timetable is dated 1st July 1845, though the service may have begun before that. By 1849 there were five trains on weekdays and four on Sundays.

By 1849 Lord Ravensworth & Partners was in financial difficulties. One of the partners was John Bowes, whose own firm, John Bowes, Esq., & Partners, was developing a powerful colliery empire in north-west Durham. Another partner in this firm was Nicholas Wood, who was also Agent to Lord Ravensworth & Partners. John Bowes' managing partner was the dynamic Charles Mark Palmer. He wanted to acquire the Springwell Colliery railway and the two collieries in order to prevent the line falling into the hands of the "Railway King", George Hudson, and also because he hoped to link Marley Hill to Kibblesworth and so have his own line to ship Bowes & Partners' coal at Jarrow. This duly happened in 1851 under an agreement back-dated to 1st January 1850. Southern surrendered his colliery to Palmer in November 1851, enabling the latter to go ahead with the link and create the Pontop & Jarrow Railway (see John Bowes & Partners Ltd). The older partnership continued until the 1880s, though no longer owning any collieries.

References : *The Private Railways of County Durham*, Colin E. Mountford, Industrial Railway Society, 2004, and *The Bowes Railway*, Colin E. Mountford, Industrial Railway Society, 1976 (2nd edition).

Gauge : 4ft 8½in

No.1		0-4-0	VC	RS	1826	New	(1)
No.2		0-4-0	VC	RS	1826	New	(1)
	STRATHMORE	?	?	?	?	(a)	(1)

The two RS locomotives were almost certainly ordered before the order for the first two locomotives for the Stockton & Darlington Railway, placed on 16th September 1824. However, they were not delivered until after the Stockton & Darlington locomotives, presumably because the latter were required urgently. What RS works numbers the Springwell locomotives were allocated has been the subject of considerable dispute, now impossible to resolve.

(a)   origin and identity unknown; believed to have been here before 1850.

(1)   to John Bowes, Esq., & Partners, with the railway, 1/1/1850.

REDHEUGH IRON & STEEL CO (1936) LTD
This company may well have been a subsidiary of **Thos W. Ward Ltd** of Sheffield, Yorkshire (WR).

REDHEUGH WORKS, Teams, Gateshead   **C421**
NZ 235623

This foundry and engineering works began production in 1872 and was served by sidings north of the NER Tanfield branch about ¾ mile south of its junction with the Redheugh branch. So far as is known the works never had any standard gauge locomotives, the shunting being carried out by the main line company; the locomotive below is believed to have been used for transferring ladles of molten metal between different parts of the works.

The works, latterly not rail-connected and owned by Spartan Redheugh Ltd, was closed on 6th October 1999 after the company was put into voluntary liquidation.

Gauge : 2ft 6in

		4wDM		MR				
-			reb	FH	2064	1937	(a)	s/s

(a)   delivered from FH, 7/1937 (ordered by Thos W. Ward Ltd); its MR identity is unknown.

G. RENNOLDSON
Wapping Street, South Shields   D422
                                                                                           NZ 360679

George Rennoldson was originally a millwright. In 1826 he set up a shipyard at Wapping Street, subsequently adding an engineering works, where allegedly he built a number of locomotives, though the works had no railway connection. In the explosion of the locomotive below his daughter was killed. Subsequently the family concentrated on ship-building, notably of tugs. The company of J.P. Rennoldson & Sons Ltd lasted until 1929.

Gauge : 4ft 8½in

| - | ? | ? | Rennoldson | 1837 | New | (1) |

(1)   on trial here on 20/11/1837; having worked for 15 minutes and being stopped for oiling, the boiler exploded, killing two people.

A. REYROLLE & CO LTD   (registered 16/5/1901)

The firm merged with **C.A. Parsons & Co Ltd** of Newcastle upon Tyne on 5/8/1968 to form **Reyrolle Parsons Ltd**, but continued to trade under its old name.

HEBBURN & NEW TOWN WORKS, Hebburn   E423/E424

Alphonse Reyrolle (1864-1919) was born in France, but began his electrical design work in London in

122. As perhaps befitted a company manufacturing electrical switchgear, Reyrolles had no steam locomotives. Their first locomotive was a 4wBE, EE 512/1920, seen here soon after delivery. She was to work for fifty years.

123. The second was perhaps the most successful design by Armstrong Whitworth's Scotswood works in Newcastle upon Tyne in trying to gain a foothold in the initial years of diesel locomotive development. No.2 was AW D22, built in 1933. She too worked until the end of rail traffic in 1970.

1886 and then chose Hebburn as the site of his new factory, opened in 1901. The **HEBBURN WORKS** (NZ 304644) **(E423)** lay west of the NER Gateshead – South Shields line, ½ mile west of Hebburn Station.

This was the beginning of a decade that was to see enormous development in the production and applications of electricity on Tyneside. Reyrolle was subsequently joined by Norbert Merz, co-founder of the famous electrical engineers, Merz & McClellan. The firm was to become world-famous for its electrical switchgear. In 1922 it opened a second works at Hebburn, known as the **NEW TOWN WORKS** (NZ 308647) **(E424)**, just over ½ mile away, just east of Hebburn Station. Latterly one locomotive was used at each works, and also for shunting between the two works, using running powers over the LNER and British Railways. By the 1960s the two works covered 100 acres and employed 9,000 people. Rail traffic ceased in 1970.

The works continues in production, in 2006 trading as Reyrolle Ltd (registered 27/7/1999).

Gauge : 4ft 8½in

1	4wBE	EEDK	512	1920	New	Scr 1/1971
No.2	0-4-0DE	AW	D22	1933	(a)	(1)

(a) AW demonstration loco; purchased by A. Reyrolle & Co Ltd in /1937.
(1) to W.F. & J.R. Shepherd Ltd, Byker, Newcastle upon Tyne, 12/1970.

R.H.P. BEARINGS LTD
Ransome Hoffman Pollard Ltd until 30/10/1977, the company above being a subsidiary of this firm.
Ransome & Marles Bearing Co Ltd until 1/1/1970

AUTOMOTIVE BEARINGS DIVISION, Greencroft Industrial Estate, Annfield Plain   **G425**
NZ 160508

This works was officially opened on 1st October 1953 (although it began production before this) and was served by sidings south of the BR mineral line from Consett to Annfield Plain (the Pontop & South Shields branch), one mile west of the former Annfield Plain Station. It was closed on 27th March 1981, with the work being transferred to the firm's works at Newark.

Gauge : 4ft 8½in

-	4wDM	RH	305302	1951	New		(1)
-	4wDM	RH	275881	1949	(a)		(2)

(a)   ex Stanley Works, Newark, Nottinghamshire, 3/1971.

(1)   to Stanley Works, Newark, Nottinghamshire, 3/1971.
(2)   to G.G. Papworth Ltd, Rail Distribution Centre, Ely, Cambridgeshire, 28/10/1981.

RICHARDSON BROS
Darlington   S426

This family were fairground operators, and the locomotive below is described in the JF works list as a "fairground locomotive". How it was used is unknown.

Gauge : 2ft 0in

PRIDE OF THE TYNE	0-4-2WT	OC	JF	7529	1895	New	s/s

RITSONS (BURNHOPE COLLIERIES) LTD
U.A. Ritson & Sons Ltd until 1936; **U.A. Ritson & Sons** until 12/1899; originally **U.A. Ritson**
BURNHOPE COLLIERY, Burnhope   G427
                                                                                       NZ 193481

This colliery was sunk in 1845 by the sons of William Hedley (see The Holmside & South Moor Collieries Ltd), and was served by a 2½ mile extension of the railway serving their Craghead Colliery. It was entirely rope-worked by the **Burnhope Engine** (NZ 194499), which hauled waggons up from Burnhope and then lowered them down to Craghead. About 1860 the Hedleys surrendered the lease to Fletcher & Sowerby. It was acquired by Utrick Ritson of Sunderland in 1881. This family also acquired **SOUTH PONTOP COLLIERY** at Annfield Plain (NZ163513) **(G428)**, which closed on 22nd April 1927, and **PRESTON COLLIERY** (NZ 342690) in North Shields, Northumberland.

The section of the Durham Coal Owners Association's Return 102 for April 1871 lists only the stationary engine, but that for November 1876 also includes a locomotive. Whether a shunting locomotive was then maintained permanently is not known. Latterly, besides coal drawn via the shafts, the colliery also developed a number of drifts. The largest of these was the **RABBIT WARREN DRIFT**, normally abbreviated to the **WARREN DRIFT** (NZ 209481). This was begun in 1910 and was served by a narrow gauge tramway from the Annie Pit, 1¼ miles long and presumably rope-worked. The drift lasted until about the Second World War. The colliery was sold to **Halmshaw & Partners** on 1st December 1936, who in turn sold it to **The Bearpark Coal & Coke Co Ltd** on 25th February 1939. The locomotive could not handle many 20-ton wagons on the gradients and so it was disposed of, with the empties being detached from the rope and run to the back of the pit, then run by gravity to the rope end once filled at the screens. An aerial flight was also built to take coal to Bearpark Colliery. It was vested in NCB Northern Division No. 6 Area on 1st January 1947.

Gauge : 4ft 8½in

-	0-4-0ST	OC	?	?	?	(a)	(1)
(BURNHOPE)	0-4-0ST	OC	RS	3057	1904	New	(2)

(a)   origin, identity and date of arrival unknown; photographs in the colliery engineer's office suggested it might have been built by a Scottish manufacturer.

(1)   to RS, /1904, in part exchange for the new locomotive.
(2)   loaned to Pelaw Main Collieries Ltd in the 1930s, and returned; to South Durham Steel & Iron Co Ltd, West Hartlepool (see the Cleveland & North Yorkshire Handbook), /1939.

ROBINSON & HANNON LTD
TYNEDALE WORKS, Blaydon Haugh   B428
                                                                                       NZ 190637

This scrapyard occupies part of the works formerly owned by Smith, Patterson & Co Ltd, (which see), which had gone into voluntary liquidation in September 1964. The plant, including the locomotive, was auctioned in November 1964, but the locomotive remained with the premises, which were taken over by this company from 1st January 1965. It was served by sidings north of the BR Newcastle – Carlisle line, ½ mile east of Blaydon Station. The works continues in use (2006), still run by The Robinson Group Ltd.

Gauge : 4ft 8½in

	4wDM	FH	3374	1950		
	reb	FH		1953	(a)	(1)

(a) ex the liquidators of Smith, Patterson & Co Ltd, with the premises, 1/1/1965.

(1) hired to Cox & Danks Ltd for dismantling the BR Blackhill branch, and returned, 7/1965; to T. Turnbull Ltd, Thornaby, Yorkshire (NR), for scrap, c2/1966.

EDWARD ROBSON & CO
Cassop

The history of the collieries in the Cassop area is extremely difficult and complex. **CASSOP COLLIERY** (NZ 341382) **(K430)**, situated near New Cassop, was sunk by the Thornley Coal Co (which see), with work beginning on 16th September 1836. It was served by a privately-owned, 1¾ mile branch from Thornley Colliery, itself served by what later became the NER Thornley Branch. The colliery first shipped coal on 2nd June 1840. The line to Cassop Colliery was then extended for another 1½ miles to **CASSOP MOOR COLLIERY** (NZ 320392) **(K431)**, where sinking began on 14th June 1840, with **CASSOP VALE COLLIERY** (NZ 334390) **(K432)**, near Old Cassop, between the two, these collieries also being owned by The Thornley Coal Company. The line between Thornley and Cassop was worked by a **stationary engine** just west of Cassop Colliery, but how the extension to Cassop Moor was worked is not known. There was also a link from Cassop Colliery to **Crow Trees Colliery** (NZ 334379) **(N433)** (see William Hedley & Sons). It has been thought that this line was worked by a stationary engine near Cassop Colliery, but as the land between the two is relatively level, this may not have been the case. This link enabled coal from the Coxhoe area to be shipped at Hartlepool instead of Stockton. It was opened in December 1839, but it seems to have been abandoned by the late 1850s.

From 1850 onwards information about both the collieries and the method of working the line becomes very confused. It would appear that Cassop Moor Colliery closed in the mid 1850s, probably followed a few years later by Cassop Vale Colliery, leaving just Cassop Colliery. However, the colliery is referred to later as "Cassop Colliery (Vale Pit)", which might mean that Cassop closed and Cassop Vale stayed open. By 1854 the collieries had been sold by The Thornley Coal Company to R.P. Philipson, a Newcastle upon Tyne solicitor, who was also a partner in The Harton Coal Company. Cassop Colliery is believed to have closed in 1867. It does not appear again in *Mineral Statistics* until 1872, when the owners are given as Robson & Co. Edward Robson also owned Victoria Colliery at Witton-le-Wear and Tyne Main Colliery at Gateshead, both only small collieries. In 1874 he sold Tyne Main to a **Henry Cochrane**, who in the following year purchased Cassop as well. However in 1876 and 1877 **George Wythes** is listed as the owner, presumably the same as George Wythes & Co, the firm that at this time owned the Middleton Iron Works near Darlington (see the Linthorpe & Dinsdale Smelting Co Ltd). Besides all this, the title "Cassop Colliery Co" is found in the Manning Wardle Engine Book for MW 465 below, but is not found in any North-East records.

Equally, at an unknown date, perhaps around the middle of the century, the line above was extended for about another mile from Cassop Moor Colliery to **Whitwell Colliery** (NZ 306404) **(K434)**. This sinking of this colliery began on 2nd May 1836 and was completed on 21st June 1837. The A Pit (NZ 308407) was originally served by a branch from the Durham & Sunderland Railway, while the B Pit was served by a privately-owned ½ mile extension from the A Pit. The colliery, later with a **brickworks** and a **foundry**, latterly had a short link to the NER Leamside branch, ¾ mile north of Shincliffe (YNB) Station. In the 1860s it was owned by **J.M. Ogden**, but from 1872 the owners were the **Whitwell Coal Co**. It would seem that the colliery ceased production in 1877, though it was not abandoned until 1884.

Cassop Colliery, owned by George Wythes, closed in 1877, presumably also closing whatever remained of the branch line above; the colliery was officially abandoned on 9th October 1878. But even that is not the end of the difficulties. In 1921 *Contract Journal* advertised an auction to be held by A.T. & E.A. Crow on 9th September 1921 for the "Owners of Cassop Colliery", in which were included a "2ft 0in gauge saddletank and a coffee pot loco, same gauge".

In 1890 the Cassop royalty had been acquired by Walter Scott for its remaining coal to be worked to the adjacent East Hetton Colliery (see East Hetton Collieries Ltd). There is no evidence in the records of the Durham Coal Owners Association that Walter Scott Ltd ever re-opened Cassop Colliery, and the authors can think of no reason why the title "Owners of Cassop Colliery" should be used 44 years after the colliery was last worked. Nor is there any evidence that Walter Scott Ltd ever owned a 2ft 0in gauge saddletank and a "coffee pot" loco at East Hetton Colliery. We are thus completely unable to explain the advertisement.

Gauge: 4ft 8½in

SWIFT	?	?	F&H		?	(a)	(1)
CASSOP No.1	0-4-0ST	OC	MW	465	1873	New	(2)

A locomotive, identity unknown, was hired from the NER in August 1855. In 1872 "G. Wythes, per J. Marley" purchased 0-4-0ST OC JF 1570/1872 New, but whether this man is the same as George Wythes above, and whether this loco came here or went to the Middleton Iron Works (see above) are all unknown.

(a)   origin and date of arrival unknown.

(1)   this loco, described as four wheeled plus tender, was offered for sale in *Colliery Guardian*, 23/4/1868; s/s.

(2)   disposal uncertain; probably sold to Stanghow Ironstone Co Ltd, Saltburn, Yorkshire (NR), but may have been sold to Clay Lane Iron Co Ltd, South Bank, Middlesbrough, Yorkshire (NR).

J. ROBSON & CO

Between 1840 and the mid 1850s this firm became one of the major colliery owners in the Durham coalfield, yet very little is known about it. It was certainly in existence by 1840, when Newfield was opened, and by 1855 it owned eleven collieries, all in the southern and south-western areas of the coalfield. Only brief notes have been included here, as most of them were later acquired by better-known owners, where a more detailed history is given. The eleven were **BINCHESTER COLLIERY**, near Westerton (NZ 242316) **(N435)**; **BOWBURN COLLIERY**, near Coxhoe (NZ 317366) **(N436)**; **BYERS GREEN COLLIERY** (NZ 223335) **(N437)**; **COXHOE COLLIERY** (NZ 329366) **(N438)**, at Coxhoe; **CROW TREES COLLIERY**, near Coxhoe (NZ 334377) **(N439)**; **HEUGH HALL COLLIERY**, near Quarrington Hill (NZ 323379) (see Fig.6); **HUNWICK COLLIERY** (NZ 210328) **(M440)**; **LITTLE CHILTON COLLIERY**, Ferryhill Station (NZ 302314) **(N441)**; **NEWFIELD COLLIERY** (NZ 209332) **(M442)**; **PAGE BANK COLLIERY** (NZ 230259) **(N443)** and **SOUTH KELLOE COLLIERY**, near Coxhoe (NZ 348370) (see Fig.6). Bowburn, Coxhoe, Crow Trees, Heugh Hall and South Kelloe are included under the entry for William Hedley & Sons. In the 1856 edition of Hunt's *Mineral Statistics* **MERRINGTON COLLIERY**, Spennymoor (NZ 348370) **(N444)** has been added to the list; but John Robson had clearly died, and the owners of all twelve are listed as 'The Executors of the late John Robson & Ralph Ward Jackson'. The latter was chairman of the West Hartlepool Harbour & Dock Railway, which was heavily dependent on the traffic from many of these collieries, and their subsequent history should be followed under that entry. Curiously, in the 1920s many of them were re-united under the ownership of Dorman, Long & Co Ltd.

Several of the collieries were served by the West Durham Railway, which made an end-on connection with the Byers Green Branch of the Clarence Railway, which in turn gave the collieries access to Hartlepool and Port Clarence. On 2nd September 1844 the Clarence Railway was leased to the Stockton & Hartlepool Railway, whose Secretary on 8th October 1844 wrote to T.C. Gibson & Co, 'Agents for Hunwick Colliery', accepting the agents' offer of "two engines to haul the Hunwick coals till such time as the company can obtain additional locomotive power of their own". Nothing is known about the identity of these locomotives, or how long this arrangement lasted. What the traffic arrangements were for the other collieries above is unknown.

JOHN ROGERSON & CO LTD
John Rogerson & Co until 12/5/1887

WOLSINGHAM STEELWORKS, Wolsingham   **T445**
NZ 082372

This works was another industrial venture of Charles Attwood (see The Weardale Steel, Coal & Coke Co Ltd). It was begun in 1864 under the title of the **Stanners Close Steel Co**, with Attwood installing Siemens regenerative furnaces. It was unusual in not being sited near deposits of either iron ore or coal. Initially it produced railway rails, but before long it was concentrating on the manufacture of marine items, such as propellers and anchors. It was served by sidings north of the NER Wear Valley branch, ½ mile east of Wolsingham Station. John Rogerson (1828-1894) was Attwood's nephew. A loco shed to house "three small locomotives" was being built here in March 1893, but the identity of these locomotives is unknown. In later years shunting was done by a steam crane.

John Rogerson & Co Ltd sold out in 1930, and for the later history of the site see the entry for Weardale Steel (Wolsingham) Ltd.

Gauge : 4ft 8½in

No.1	?	?	P	?	?	(a)	s/s
No.2	0-4-0ST	OC	P	1460	1916	New	(1)

(a) origin and date of arrival unknown.
(1) to Leversons Wallsend Collieries Ltd, Usworth Colliery, Usworth, via C.W. Dorking & Co Ltd.

THE ROLLING STOCK & ENGINEERING CO LTD
formerly **The Rolling Stock Co Ltd**

NESTFIELD WORKS, Darlington **S446**
NZ 298156

This company began operating in Darlington as agents for and the owners of private railway wagons; but about 1906 it decided to set up a facility to repair wagons, acquired a site and in December 1906 purchased the Perseverance Engine Works at Stalybridge in Cheshire in order to transfer its plant to the Darlington premises. The works was served by a reverse from the short branch serving the Blake Boiler, Wagon & Engineering Co Ltd (which see), 1½ miles north of Darlington (Bank Top) Station on the NER Durham – Darlington line. The company may have owned more locomotives than those listed below.

Latterly the company gave up its wagon work and concentrated on light engineering. Rail traffic ceased in 1968 and the works was closed in the mid 1970s.

Gauge : 4ft 8½in

	NESTFIELD	0-4-0ST	OC	?	?	?	(a)	s/s c/1931
		reb	HL	9492	1908			
No.4		0-4-0ST	OC	HL	2533	1902	(b)	Scr /1952
	-	4wDM		RH	305303	1951	New	(1)

(a) origin, identity and date of arrival unknown.
(b) ex Samuel Tyzack & Co Ltd, Sunderland, by 14/3/1931.
(1) to Llanelli Steel Co Ltd, Llanelli, Carmarthenshire, 6/1968.

There were also two steam and two diesel rail cranes here.

THE ROSEDALE & FERRY HILL IRON CO LTD
registered 23/6/1864 to take over the business of **James Morrison & Co**

FERRY HILL IRON WORKS, West Cornforth **N447**
NZ 305343 approx.

THRISLINGTON COLLIERY, West Cornforth **N448**
NZ 309338

HIGHLAND QUARRY, near West Cornforth **N449**
NZ 326330

FERRY HILL IRON WORKS was opened in 1859 by **James Morrison**; in fact, the works was not at Ferry Hill, but West Cornforth. It was served by two links to the NER, a private branch (½ mile) from Ferryhill Iron Works Junction on the NER Ferryhill – Wingate line (formerly the Clarence Railway) at West Cornforth Station and a one mile NER branch to the Durham – Darlington line north of Ferry Hill Station. The business expanded rapidly; from three furnaces in 1859 the works had six by 1866 and ten by 1875. Meanwhile in 1867 and adjacent to the works, the firm had re-opened **THRISLINGTON COLLIERY** (NZ 309338) (see the entry for Stephen Walton). This was served by another link to West Cornforth Station, and was presumably shunted by the iron works locomotives indiscriminately. About 1870 the company began to open up the **HIGHLAND QUARRY** (its name taken from a nearby house), and laid a 1¼ mile line to bring its limestone up to the works. However, this was only worked for a few years and was then replaced by **BISHOP MIDDLEHAM QUARRY** (NZ 333320 approx.) **(N450)**, about ¼ mile further to the south-east. What rail connection this quarry had under these owners is not known. The company obtained its ironstone from mines at Rosedale in North Yorkshire.

The works was producing 4000 tons of iron a week in 1877, but faced with the severe economic depression of the period the company failed on 18th January 1879, and an order of the Court of Chancery on 10th February 1879 ordered the winding up of the company under the supervision of the court. The plant, including locomotives, was auctioned by C. Willman on 16th December 1879, though it would appear that the locomotives were not disposed of until a year later. The ironworks was purchased by John Rogerson & Co (see above) in December 1882 and in February 1883 he took possession. *Blast Furnace Statistics* lists Rogerson & Co as the owners until 1895 (Rogerson died in

1894), and show no production. However, local sources suggest that that the works was taken over by The Carlton Iron Co Ltd (which see), which owned East Howle Colliery nearby; that production was re-started, however briefly, and that the works was closed in 1890. It was demolished about 1895, by which time the slag heaps were being worked by William Barker & Co (which see). Thrislington Colliery was re-opened in 1889 by The Thrislington Coal Co Ltd (see The North Bitchburn Coal Co Ltd). Bishop Middleham Quarry was eventually re-opened by Pease & Partners Ltd (which see).

Gauge : 4ft 8½in

ROSEDALE	0-4-0ST	OC	MW	199	1866	(a)	(1)
FERRYHILL	0-4-0ST	OC	BH	22	1867	New	(2)
JAMES MORRISON	0-4-0ST	OC	BH	95	1869	New	(3)
THRISLINGTON	0-6-0ST	OC	BH	354	1875	New	(4)
GEORGE LEEMAN	0-4-0ST	OC	GW	?	?	(b)	(3)
EUGENE	?	?	?	?	?	(b)	(5)
MARTIN MORRISON	?	?	?	?	?	(b)	(6)

(a) ex W. Ritson, contractor for the NER Allendale branch, Northumberland, /1867.
(b) origin and date of arrival unknown.
(c) origin, builders and date of arrival unknown.

(1) to Bolckow, Vaughan & Co Ltd, Cleveland Works, Middlesbrough, Yorkshire (NR).
(2) to John Watson, 12/1880.
(3) to J. Appleyard, 12/1880.
(4) to Mainsforth Coal Co, Mainsforth Colliery, Ferryhill Station.
(5) to George Chapman, 12/1880.
(6) to Storbuck, 12/1880.

Watson, Appleyard, Chapman and Storbuck are all believed to have been dealers.

COXHOE COLLIERY, near Coxhoe **N451**
NZ 328366

The company's increasing need for its own supplies of coal appears to have been met by taking over **BOWBURN** (NZ 317366) **(N452)**, **HEUGH HALL** (NZ 323379) (see Fig.6), **CROW TREES** (NZ 334377) **(N453)**, **COXHOE** (also known as **WEST HETTON**) (NZ 328366) and **SOUTH KELLOE** (NZ 348370) **COLLIERIES** (for South Kelloe see Fig.6). All of these lay near to the villages of Coxhoe and Quarrington Hill and all were sunk by 1844, most under the auspices of William Hedley & Sons (which see). They all eventually came into the ownership, illegally, of the West Hartlepool Harbour & Railway Co, whose Coxhoe branch provided the outlet for their coal. The first, third and last were served by a 1¾ mile private branch from Coxhoe Station, with Heugh Hall linked to this by a ½ mile self-acting incline. Coxhoe and South Kelloe were served by a second private line from Coxhoe 1¼ miles long. The ¾ mile from South Kelloe to Coxhoe Colliery was almost certainly a self-acting incline, but the ½ mile down from Coxhoe Colliery to the outskirts of Coxhoe village was operated by a stationary engine. The WHH&R having been absorbed by the NER, these collieries were advertised for auction on 27th March 1866. According to *Mineral Statistics*, they were acquired by J. Morrison & Co, despite the Rosedale & Ferry Hill Iron Co Ltd being formed two years before to take over this firm; perhaps it was decided to operate the collieries as a business separate from the iron works. On the opposite side of the line from Coxhoe Colliery was **COXHOE BANK QUARRY** (NZ 327363 approx.) **(N454)**, and it may well be that the Rosedale & Ferry Hill Iron Co Ltd acquired this too.

It would appear from *Mineral Statistics* that Bowburn Colliery was closed in 1870 and that Crowtrees, Heugh Hall and South Kelloe followed in 1871, leaving only Coxhoe still working, and owned from 1871 by **The Rosedale & Ferry Hill Iron Co Ltd**. The company offered all five collieries for sale in the *Colliery Guardian* of 6th June 1872, but apparently failed to sell them. It would seem that the company continued to work Coxhoe Colliery, and very probably Coxhoe Bank Quarry too, until its failure in 1879. None of the collieries worked again, though the shaft of Heugh Hall Colliery was re-opened about 1907 as an up-cast and man-riding shaft for the new Bowburn Colliery sunk by Bell Brothers Ltd (see Dorman, Long & Co Ltd) until the new Bowburn up-cast shaft was completed about 1911, while South Kelloe shaft is believed to have been used for ventilation and pumping by East Hetton Collieries Ltd (which see), which also worked the lower seams in this area to its East Hetton Colliery. Coxhoe Bank Quarry was eventually re-opened by The Steetley Lime Co Ltd (see Steetley Dolomite (Quarries) Ltd.

In the advertisement in the *Colliery Guardian* of 6th June 1872, referred to above, the firm also offered for sale two 7in tank locomotives. No further details are known.

THE RYHOPE COAL CO LTD
The Ryhope Coal Co until 1875
RYHOPE COLLIERY, Ryhope
H455
NZ 399575

This area of coal was originally leased in 1831 to the Haswell Coal Co, but the various attempts that it made to find coal all ended in failure, and it gave up the lease. It was re-let to this new company, which included some of the Haswell partners, in 1857. Sinking began in 1858 and was completed in December 1859, with the first coal being drawn on 7th February 1860. The colliery was linked to the Londonderry Railway by a one mile branch from Ryhope Grange. One of the earliest colliery photographs in the North-East shows Ryhope Colliery from the south in 1859, and it includes two locomotives always captioned as "probably two Londonderry Railway locomotives". One is a 0-6-0 tender loco with inside cylinders, the other a 0-6-0ST with inside cylinders, outside frames and a tall, flared chimney. At this date the Londonderry Railway appears not to have had an 0-6-0ST, so perhaps this was actually a Ryhope Coal Co loco.

In 1870 the colliery acquired its first known locomotive for shunting, but as this was numbered No.2 there may well have been an earlier loco. By 1894 the branch had been extended by ¼ mile to serve a **brickworks** and **TUNSTALL QUARRY** (NZ 397536) **(H456)**, producing limestone; the latter closed officially in 1910, though there was little or no production for some years before this. Unusually for a coastal colliery in Durham, beehive coke ovens were also built here; the last of these closed in July 1913. The branch was also extended to join the branch to Silksworth Colliery (see The Marquis of Londonderry), which originally had by-passed Ryhope Colliery on its northern side. The Ryhope branch passed to the NER with the take-over of the Londonderry Railway on 6th October 1900. The colliery was vested in NCB Northern Division No. 1 Area on 1st January 1947.

Gauge : 4ft 8½in

No.2		0-6-0ST	IC	MW	274	1870	New	s/s after 2/1889
-		0-4-0ST	OC	AB	1082	1907	New	(1)
1		0-6-0ST	OC	P	1403	1916	New	(2)
2		0-6-0ST	OC	P	1455	1918	New	(2)
-		0-6-0ST	IC	BH	716	1882		
		reb		?		1906	(a)	s/s
No.3		0-6-0T	IC	Dar	480	1892	(b)	(2)

(a) ex Leversons Wallsend Collieries Ltd, Usworth Colliery, Usworth.
(b) ex LNER, 1144, 7/1938; LNER J71 class, previously NER, withdrawn, 3/1937.

(1) to J.F. Wake, dealer, Darlington, by 3/1916; re-sold to Workington Iron & Steel Co Ltd, Moss Bay Works, Workington, Cumberland, /1916.
(2) to NCB No. 1 Area, with colliery, 1/1/1947.

SIR S.A.SADLER LTD
Samuel A. Sadler until 1901; **Sadler, Forbes, Abbott & Co Ltd** until /1885
Sadler & Co until /1884; **Samuel A. Sadler** until 29/3/1883
This company was one of the major North East manufacturers of chemicals from tar, with a large works in Middlesbrough, Yorkshire (NR).

HAMSTEELS COLLIERY, near Quebec
K457
Joseph Johnson (Durham) Ltd until 1/4/1923
NZ 185431 (Taylor Pit)
The Owners of Hamsteels Collieries until /1917
The Trustees of the late Joseph Johnson until /1908, possibly trading as The Hamsteels Colliery Co
Johnson, Reay & Johnson until /1885

This colliery was begun by Johnson, Reay & Johnson in 1867. It was served by sidings north of the private branch from Cornsay Colliery (see Ferens & Love (1937) Ltd) to Flass Junction on the NER Waterhouses branch, 1½ miles from the junction. Although there were two shafts, most of the coal came from various drifts to the west, north and east of the screens, to which it was worked by narrow gauge rope haulage (see Fig.5, page 157). The most important of these were the **Clifford** or **Busty Drift** (NZ 178434) and the **Main Coal** or **Quebec Drift** (NZ 181434). The standard gauge locomotive, from a shed at the eastern end of the yard near the Taylor Pit, shunted the sidings and the beehive coke ovens, built adjacent to the branch line. DCOA Return 102 records one locomotive here in April 1871 and, curiously, the same in November 1876, indicating that there was at least one locomotive here in addition to those below.

Soon after Sir S.A. Sadler Ltd took over, Hamsteels Colliery and its coke ovens was closed, in July 1924 and most of the site was cleared, including the sidings. However, in May 1932 the firm re-opened the Clifford Drift and the **Weatherhill Drift** (NZ 179434). Rope haulage brought the full tubs to a large hopper, from which an aerial ropeway two miles long took the coal to screens at Malton Colliery (see below).

The colliery was vested in NCB Northern Division No.5 Area on 1st January 1947.

Gauge : 4ft 8½in

HAMSTEELS	0-6-0ST	IC	MW	480	1874	New	(1)
(WHITWORTH)	0-6-0ST	IC	MW	569	1875	New	
HAMSTEELS by 8/1922		reb	MW		1922		(2)

(1) scrapped here, after 10/1910, perhaps in 1924.
(2) to Malton Colliery, near Lanchester, c/1924.

HAMSTEELS COKE OVENS

K458
NZ 184430

These beehive ovens lay alongside the railway line, and the locomotives below shunted conical metal tubs out on top of the ovens to charge them, as noted elsewhere. The company wrote of the operation in DCOA Return 427 (April 1901) "we tip all the coals into two hoppers which the two locomotives load the ovens from." It is believed that the ovens had closed by 1914.

Gauge : 2ft 10in

HAMSTEELS No.3	0-4-0ST	OC	BH	439	1877	New	(1)
-	0-4-0ST	OC	BH	856	1885	(a)	(1)

(a) ex The Bowling Iron Co Ltd, Bradford, Yorkshire (WR), /1897.
(1) two locomotives were reported in daily use in DCOA Return 427, 4/1901; s/s.

124. This was the inside-framed design that Black Hawthorn developed for narrow gauge locomotives (for the other design, see The Holmside & South Moor Collieries Ltd). 2ft 10in gauge BH 439/1877 hauls a train of full conical tubs across the wooden gantry to the beehive coke ovens at Hamsteels Colliery.

MALTON COLLIERY & COKING PLANT, near Lanchester K459
George Love & Son until /1888 NZ 180461 (screens)

This was another colliery without a shaft, comprising entirely of drifts working to central screens served by sidings alongside the NER, in this case south of the Lanchester Valley branch one mile south east of Lanchester Station. It was begun in 1880 by George Love & Son (one source gives begun in 1870 by G.Love alone), apparently another branch of the Love family of Joseph Love mentioned elsewhere (see Ferens & Love (1937) Ltd). At the same time a **brickworks** was established for the manufacture of firebricks. S.A. Sadler Ltd bought the owners out in 1888. 70 Simon-Carves by-product coke ovens were built in 1893 and started up in January 1894. A further 35 were added three years later and 3 more in the quarter ending 30th September 1909. This 'regenerative' design of coke oven produced between 40% and 50% more gas than waste heat ovens, and in October 1906 the Malton ovens became the first in the country to supply gas to the local population, in this case the village of Lanchester, a trade widely developed in Britain over the next forty years. Engines using gas for winding coal were also installed here, but this use was not widely adopted. The by-products plant produced tar, crude benzole and concentrated ammonia liquor. The initial owners of this plant were the **Durham Coke & Bye-Products Co Ltd**, a company set up by Henry Simon, the contractor for its construction, and it is uncertain how long this arrangement continued before Sadlers took over full ownership.

The first "area" of drifts lay to the south east a short distance from the screens, the first being the **Brockwell Drift** (NZ 182456), followed by the **Harvey Drift** (NZ 183453) and then the **Clay Drift** (for coal) (NZ 182457). But certainly by the mid-1880s a second tramway had been laid alongside and then under the NER to reach coal on the north side of the railway. This served the **Harvey Drift** (NZ 179471) and the **Manor House Drift** (NZ 177469), a distance from Malton of about a mile. Later the first was replaced by the **New Harvey Drift** (NZ 178471), while the **Amy Drift** (NZ 178466) was driven just below the A691 road. The third area was reached by a branch from this line running north from near to the tunnel under the NER. This line, also about one mile long, served six drifts in all, the most important being the **May Drift** (NZ 182465), the **Mercie Drift** (NZ 185468) and the **Hutton Drift** (NZ 186473). Almost certainly all of these tramways were narrow gauge lines worked by rope haulage. Latterly these two areas became known as **North Malton**. Working here appears to have ended in 1942.

By 1900 Sadlers had also taken over **HILL TOP COLLIERY** (NZ 208444) **(K460)**, which lay some two miles to the south-east of the village of Langley Park. This was a "small island royalty" – a pocket of coal, in this case the upper seams, surrounded by a much larger royalty, in this case the large Pease & Partners royalty that stretched down to Crook. This too was a drift, initially a landsale operation begun by Roger Halliday, probably a local man, in 1867. Production was greatly increased after Sadlers took over and a two mile long aerial ropeway to carry the coal to Malton was opened in November 1900; it was one of the first in the country and was German-designed. Curiously, the firm operated just two beehive coke ovens here for many years. In 1937 the firm began developing the **Harvey Drift** (NZ 203446) here, with another rope-worked tramway ½ mile long to bring its coal to Hill Top.

Finally, as noted above, in 1932 a second aerial ropeway two miles long was put in to carry coal from Hamsteels Colliery to Malton.

Malton Colliery, comprising the May Drift, the Clay Drift, Hill Top and the Harvey Drift, together with Malton Brickworks and the coking plant, by then reduced to 105 working ovens (another source gives 87), were all vested in NCB Northern Division No. 5 Area on 1st January 1947.

Gauge : 4ft 8½in

BOBS		0-6-0ST	OC	HCR	176	1876	(a)
		reb Worth & Mackenzie, Stockton				1900		Scr /1914
BASIL *							
(formerly WILBERFORCE)		0-4-0ST	OC	I'Anson		1875	(b)	(1)
CECIL		0-4-0ST	OC	RWH	1847	1881	(c)	(2)
(BASIL)		0-6-0ST	OC	BH	1034	1891		
		reb		HL	3668	1904	(d)	(3)
(HAMSTEELS)		0-6-0ST	IC	MW	569	1875		
		reb		MW		1922	(e)	(3)
No.4		0-4-0	?	Butterley		?		
	reb	0-4-0ST	OC	Butterley		1895		
		reb		Ridley Shaw		1942	(f)	(4)

SILKSTONE	0-6-0ST	IC	MW	341	1871			
	reb		?		1903			
	reb Newton, Chambers & Co				1925			
	reb		?		1935	(g)	(3)	

* This was a painted name, which was carried first by the I'Anson loco and was then transferred to BH 1034.
(a) ex Cliffe Coal & Fireclay Co Ltd, Hollingthorpe Colliery, Crigglestone, Yorkshire (WR), c/1887.
(b) this loco was offered for auction in the liquidation of the Skerne Ironworks Co Ltd, Darlington on 9/9/1879; whether it was purchased by Love & Son for use here, or went somewhere else before coming here, is not known.
(c) ex Sir B. Samuelson & Co Ltd, Newport Works, Middlesbrough, Yorkshire (NR), c/1916.
(d) ex The South Moor Colliery Co Ltd, Morrison Colliery, Annfield Plain, c/1923.
(e) ex Hamsteels Colliery, near Quebec, c/1924.
(f) ex Middlesbrough Works, Middlesbrough, Yorkshire (NR): may have come here via Ridley, Shaw & Co Ltd, Middlesbrough, Yorkshire (NR), /1942; previously at Steel & Co Ltd, Sunderland.
(g) ex Newton, Chambers & Co Ltd, Thorncliffe Ironworks, Chapeltown, Yorkshire (WR), 6/1946.

(1) loaned to The East Holywell Coal Co Ltd, East Holywell Colliery, near Backworth, Northumberland (as WILBERFORCE), and returned; to Ridley, Shaw & Co Ltd, dealers and repairers, Middlesbrough, Yorkshire (NR), /1943; re-sold to The South Medomsley Colliery Co Ltd, near Dipton, c/1945.
(2) to Cerebos Ltd, Greatham, near Billingham, c/1933 (see Cleveland & North Yorkshire Handbook).
(3) to NCB No. 5 Area, with colliery and coking plant, 1/1/1947.
(4) to Middlesbrough Works, Middlesbrough, Yorkshire (NR).

SADLER & CO LTD (registered in 1883)
It is believed that there was some link between this company and Sir S.A. Sadler Ltd above, though what this was is not known.

RANDOLPH COKE WORKS, Evenwood   P461
previously **The North Bitchburn Coal Co Ltd** (which see)   NZ 157249

About 1938 the company took over this works from the receiver of The North Bitchburn Coal Co Ltd (which see). The works, opened in 1909, incorporated 60 Koppers waste heat coke ovens, and also manufactured tar, sulphate of ammonia and crude benzole. It was shunted by locomotives from the adjacent Randolph Colliery, owned by the Randolph Coal Co Ltd. Coal from other collieries in this area of south-west Durham was brought here for carbonisation. Unusually, it was not taken over by the NCB Northern Division in 1947. The NCB continued to shunt the coke works. In 1948 the Koppers ovens were replaced by 15 Woodall-Duckham Becker under-jet ovens. In May 1957 the plant and the colliery's locomotives, sold by the NCB, were incorporated into a new company, the **Randolph Coke & Chemical Co Ltd** (which see).

THE SEAHAM HARBOUR DOCK COMPANY
PORT OF SEAHAM, Seaham Harbour   J462
subsidiary of **Victoria Group** from 4/9/2003
subsidiary of **Durham Port Holdings Ltd** from 1/1996; see also below   NZ 431494
(the later workshops & diesel loco shed)

By the end of the 1880s the docks at Seaham were seriously inadequate to handle the larger steam colliers then being built, and the Marquis of Londonderry came under increasing pressure to expand the facilities. By the mid-1890s the 6th Marquis (which see) had decided to re-organise all of his industrial interests in County Durham. The collieries were put under The Londonderry Collieries Ltd (which see) and what was left of the Londonderry Railway was sold to the NER; but the first step was to get through Parliament the Seaham Harbour Dock Act. This set up The Seaham Harbour Dock Company, incorporated on 12th August 1898, and provided for a major expansion of the South Dock (NZ 434492 approx.).

The Seaham Harbour Dock Company, with the Marquis as chairman and largest shareholder, took official possession of the docks and plant, including locomotives and a small number of ballast and flat wagons, from the Marquis on 1st January 1899 for a little over £88,000. Almost certainly the locomotives acquired were Nos.16-19 in the Londonderry Railway list (see Notes on Locomotives in the entry for the Marquis of Londonderry), and probably No.3, though this has not been confirmed. The 4-ton coal wagons used by the Londonderry collieries continued to be owned by

125. One of the jobs for the very small locomotives Nos.16-18 was to collect coal spillage from the bottom of the staiths. Here on 30th May 1952 16, 0-4-0VBT HW 21/1870, is at work at the East Berth, a staith constructed not of wood but of concrete.

126. A famous locomotive, now preserved – but here in the full livery applied in the 1930s; 18, Lewin 683/1877, on 30th June 1936.

Durham Part 1 Page 341

127. For the opening of the South Dock in 1905, the company needed to expand its locomotive fleet. One of two new locomotives purchased from Peckett was SEAHAM, P 1052/1905, seen here on 1st July 1935.

128. One of two small locomotives purchased from the contractors for the construction of the new dock, S. Pearson & Son Ltd, was DICK, HE 628/1895, also seen on 1st July 1935.

129. SEATON was acquired in 1906. When the Londonderry Railway was taken over by the North Eastern Railway in 1900, this locomotive was under construction. She stood unused in the Wagon Shops until the Dock Company bought her. Engraved around the number plate was SEAHAM HARBOUR ENGINE WORKS.

130. MILO, RS 2241/1875, one of two NER 964 class locomotives acquired in 1907-08, on 30th June 1936. They were 88 years old when finally scrapped.

the Marquis (see below). The letting of the contract for the rebuilding of the South Dock to S. Pearson & Son Ltd (see Part 2) soon followed. Construction work began in April 1899 and the new dock, with three coal berths, was officially opened on 11th November 1905. The arrangement under which The South Hetton Coal Co Ltd operated staiths with its own men on the north side of the old South Dock for its coal shipments, with its traffic being worked for it between Swine Lodge Bank Foot and the staiths, continued under the Dock Company. In the extended South Dock the South Hetton company had exclusive rights at No.2 (the "Pole") berth, a southern curve being installed from Swine Lodge bank foot to provide for this traffic, and eventually all South Hetton traffic was dispatched here. However, these rights were officially surrendered to the Dock Company on 9th April 1923, though in practice South Hetton coal continued to be shipped through the Pole Berth.

When the company was formed, the northern side of the three acre North Dock was served by a ½ mile long **self-acting incline**, running from the foot of the Seaham Incline, with the links from the bank foot to the south side of the North Dock and the north side of the South Dock worked by the small locomotives, for which a special **loco shed** was built in the vee between the lines (NZ 431496). The ballast unloaded from ships also had to be removed, and the North Dock had a special incline for this traffic, worked by a stationary engine; it was superceded by "an incline at the south end of the docks", but lay unused for many years before finally being removed in 1927, though the engine house survived until 1929. A second **stationary engine**, on the site of the later locomotive shed, was provided to haul waggons of ballast up from the South Dock and then lower it to the south shore for disposal; this incline was almost certainly removed when the construction of the South Dock commenced in 1899. To help with this a new line was built underneath both approach lines, via two tunnels, to give access to the beach north of the docks, where gravel was collected. The "branch" to the beach was maintained for many years to obtain gravel for the company's **blockyard** (NZ 434496), and during the 1920s a line was built out along the North Pier in order to take out materials needed to repair storm damage; these trains could only be handled by No.18 loco or a crane, hence the company's need to retain No.18 even after the new diesel fleet was purchased in 1967. The Directors' minutes show that a new incline to bring traffic to the South Dock was constructed in 1908, but no further details are known. In the 1950s the floor of the tunnels was lowered by two feet to allow a rail crane through without needing to be dismantled. How long the self-acting incline to the north side of the North Dock survived is not known.

Excluded from the NER take-over of the Londonderry Railway were the branches to **Seaham Colliery** (¾ mile, a **self-acting incline**), the line to the new sinking at **Dawdon** (just under one mile) and the ½ mile branch to **Fox Cover Quarry**, all of which continued to be owned by the Marquis. The anomaly that this situation created was obvious, and in January 1901 the Dock Company leased these lines, almost certainly purchasing them later. The branch to Fox Cover Quarry is believed to have closed, with the quarry, about 1919. In December 1911 the Marquis sold some 2,500 4-ton coal waggons to The Londonderry Collieries Ltd (which see), their ownership being denoted by a white L painted centrally on the waggon. At an unknown date the Dock Company acquired between 800 and 1,000 of these wagons (these carried SH centrally), probably from The Londonderry Collieries Ltd, their main use being to work the traffic between Seaham Colliery and the docks. When Dawdon Colliery began production in 1907 the Dock Company also worked its traffic down to the docks, normally by locomotive but sometimes by gravity if a locomotive was not available and waggons were needed urgently for shipment. As these waggons had no axleboxes they needed constant greasing.

Also excluded from the NER take-over were the **Londonderry Engine Works** (NZ 423494) and the **Londonderry Wagon Works** (NZ 428490), which the Marquis continued to operate. The former was soon closed and some of the equipment was moved to new premises next to the foundry of **Robert White & Co Ltd** (NZ 432494), near to the docks. The new works began by retaining the original name, but soon became known officially as the **Seaham Harbour Engine Works** (NZ 432493). Both engine works and foundry were linked to the Dock Company's system, and repairs to the Dock Company's locomotives continued here. The Wagon Works was reached using running powers over the NER. In addition to the wagons above, the Marquis also sold the Engine Works and the Wagon Works and 2,500 wagons to The Londonderry Collieries Ltd in December 1911. The Engine Works was closed in the autumn of 1925, and thereafter the Dock Company was compelled to develop its own repair facilities, alongside its **main loco shed** (NZ 431494), which was sited not far from Swine Lodge bank foot on the site of the former stationary engine for the ballast line, the water tank for the shed being supported by the walls of the former engine house. The main interchange sidings between the NER and the Dock Company were developed at Seabanks Junction, the point where the NER line to Hartlepool left the former Blastfurnace Branch, though those for Seaham Colliery traffic were situated at Polka Bank Foot. The running powers for traffic

131. A six-coupled tender engine would seem rather unsuitable for dock work, but CLIO, built by the NER at Gateshead in 1875, was purchased in 1911. Seen here on 30th June 1936, she worked until about 1942, while her tender survived for concrete mixing until the early 1960s.

132. In the 1950s the company became enamoured of the new Sentinel designs to replace their ageing fleet, acquiring a 100 h.p. machine and then ordering three 200 h.p. machines, only to find the very heavy work they had to do caused major boiler problems. TEMPEST, S 9618/1956, is seen here on 27th April 1957, not long after delivery, but the order for S 9619 was cancelled.

133. Instead of continuing with buying new traction, the company purchased no fewer than 17 steam locomotives second hand between 1960 and 1963, mostly from steelworks as their fleets were converted to diesel. 173, HL 3919/1937, was a former Dorman Long engine from the Acklam Works at Middlesbrough.

134. Eventually the company 'bit the bullet' and in 1967 purchased five new diesels from English Electric, which joined the Thomas Hill diesel conversion of S 9618 and the Lewin loco, which was kept for repair works to the piers and for bringing shingle from the north beach. D1, EEV D1191/1967, with D5, EEV D1195/1967 behind (facing the opposite way round), and TH 104C/1960, stand near the newly-provided loco shed on 25th August 1970.

to the Wagon Works may have subsequently been transferred to The Londonderry Collieries Ltd.

Besides working traffic for The Londonderry Collieries Ltd (some of it in Dock Company waggons), the South Hetton Coal Co Ltd and the Seaham Foundry Co Ltd's works, the Dock Company also worked traffic for the **Seaham Bottle Works** (NZ 433489) owned by **Robert Candlish & Sons Ltd** until 1913 and, situated slightly to the south of the port, until its closure on 15th April 1921, and for the **Seaham Gas Works** (NZ 431489). Some coal from Easington Colliery (see The Easington Coal Co Ltd) was also shipped, while coal from Londonderry Collieries' Vane Tempest Colliery, to the north of the town, was brought in via the LNER.

To handle yet more traffic the Castlereagh Extension was added to the South Dock in 1925, increasing its size to 13¾ acres, though by this time the North Dock had ceased handling coal traffic. The opening on 26th April 1939 of the East Berth on the western side of this extension raised the number of shipping berths to five, though after the war the number of berths declined, both from falling coal traffic and the company's need to curb expenditure. Until April 1969 the company also operated the last two steam paddle tug boats in Britain, the RELIANT (built 1907) and the EPPLETON HALL (built 1914), both subsequently preserved.

The links to Seaham and Dawdon Collieries, traffic from Easington and Vane Tempest Collieries and coal from the railway to South Hetton, Murton and latterly the Hawthorn Combined Mine gave the port a healthy coal traffic for many years. In 1959 the Seaham Incline was rebuilt, by the NCB, to allow 21-ton wagons to replace the 4-ton black chaldron wagons still then being used. In 1961 the Dock Company still had over 800 of these waggons, and although many were subsequently scrapped, quite a number were retained for dockside work until the late 1970s, a number subsequently being preserved at the North of England Open Air Museum at Beamish. In 1965 rope working on the Seaham incline ceased, with locomotives being allowed over it thereafter, though in practice it became largely disused, with most of the traffic using the adjacent BR link to the colliery. It was leased to the NCB from 25th March 1967, and finally lifted in January 1988.

By the late 1960s the original company was unable to fund the demolition and improvements needed to give the port a viable future. It was taken over by The Dundee, Perth & London Shipping Co Ltd on 1st October 1971, and has been a subsidiary of various companies since. It was taken

135. An aerial view of Seaham Docks in 1975, looking north-east. The North Dock, by then not used for shipping, can be seen in the top left corner, while the NCB line from Seaham Colliery enters near the North Dock and the NCB line from South Hetton and Hawthorn curves in from the west. The staiths still remaining can be seen in the bottom right area.

over in a management buyout in January 1996 by Durham Port Holdings Ltd, which in turn sold out to the Victoria Group in September 2003. The company continues to be limited by private Parliamentary Acts, the most recent in 1972.

The company first handled colliery waste in the 1940s, and in 1965 the volume of coal refuse exceeded the volume of coal shipped for the first time. Latterly this was the only traffic coming down from South Hetton and this ceased on 5th April 1984. Meanwhile both Seaham and Dawdon Collieries went over to British Rail "merry-go-round" trains, thus ending coal shipments and leaving only the colliery waste traffic, now being worked up to Dawdon Colliery for tipping on the beach rather than being dumped at sea. The branch to Seaham Colliery was lifted in January 1988 after this traffic went to BR, while Dawdon Colliery closed on 26th July 1991. Rail traffic at Seaham finally ceased in mid-1992 when steel exports from Teesside went elsewhere, leaving the port dependent on retaining the new traffic flows which it had secured, attracting new traffic flows and providing appropriate facilities.

However, as part of a major economic development plan for Seaham following the closure of the collieries, it was decided to sell the whole of the existing on-shore facilities of the port for retail development and to construct a major new distribution centre, of which the docks would be a part, on the 23 acres of land to the south of the town once occupied by the Seaham Bottle Works, the Seaham Gas Works and the Seaham Chemical Works. These premises, some occupied by tenants, would be served by an up-graded rail link from Seabanks Junction. Work on the construction of the new facilities began at the end of 2000. Under the title of CargoDurham Distribution Centre they were opened on 3rd July 2002.

Reference: The Directors' Minute Books and Annual Reports of The Seaham Harbour Dock Company, examined by kind permission of the company.

Reference: *Industrial Railways of Seaham*, A.J. Booth, Industrial Railway Society, 1994.

Gauge : 4ft 8½in

(3)	DAWDON		0-6-0	IC	TR	254	1855		
		reb	0-6-0ST	IC	Seaham		1876	(a)	(1)
(16)			0-4-0VBT	VC	HW	21	1870	(b)	(2)
(17)			0-4-0VBT	OC	HW	33	1873	(b)	
			reb		Seaham		?		(3)
18			0-4-0WT	OC	Lewin	683	1877	(b)	
		reb	0-4-0T+WT	OC	Seaham?		?		
		reb	0-4-0ST	OC	Seaham?		?		(4)
19			0-4-0ST	OC	BH	203	1871	(a)	(5)
	SEAHAM		0-6-0ST	OC	P	1052	1905	New	Scr /1961
	REX		0-4-0ST	OC	MW	838	1885	(c)	Scr /1939
	DICK		0-4-0ST	OC	HE	628	1895	(c)	Scr 12/1963
1	SEATON		0-6-0T	IC	Seaham		1902	(d)	Scr 5/1962
	SILKSWORTH		0-6-0ST	OC	P	1083	1906	New	(6)
	MILO		0-6-0ST	IC	RS	2241	1875	(e)	(6)
	AJAX		0-6-0	IC	Blyth & Tyne		1867	(f)	Scr /1926
	MARS		0-6-0ST	IC	RS	2238	1875	(g)	(6)
	CLIO		0-6-0	IC	Ghd		1875	(h)	(7)
	JUNO		0-6-0ST	OC	HL	3527	1922	New	(8)
	NEPTUNE		0-6-0ST	OC	HL	3898	1936	New	(8)
	-		4wVBT	VCG	S	9558	1953	(j)	(9)
	-		4wVBT	VCG	S	9561	1953	(k)	(10)
	"SENTINEL"		4wVBT	VCG	S	9575	1954	(m)	(11)
	TEMPEST		4wVBT	VCG	S	9618	1956	New (n)	(12)
	-		0-6-0DE		YE	2668	1956	(p)	(13)
	-		4wDH		S	10003	1959	(q)	(14)
	-		0-6-0DM		WB	3160	1959	(r)	(15)
41			0-6-0T	OC	KS	3074	1917	(s)	(16)
18			0-6-0ST	OC	BH	32	1867		
		reb		HL	1491	1935	(t)	(17)
44			0-6-0ST	OC	MW	1934	1917		
		reb	0-6-0T	OC	LEW		1951	(t)	(6)
	B No 23		0-4-0ST	OC	HL	3744	1929	(u)	(18)
	B No 41		0-4-0ST	OC	RSHD	7016	1940	(v)	(19)

	B No 10	0-4-0ST	OC	HL	3476	1920	(w) *	(20)
	B No 38	0-4-0ST	OC	HL	3496	1921	(x)	(21)
	B No 15	0-4-0ST	OC	HL	3873	1936	(x)	Scr 12/1963
	-	4wDH		TH	104C	1960	(y)	Scr 1/1975
No.1		0-4-0ST	OC	HL	3354	1918	(z)	(22)
No.3		0-4-0ST	OC	HL	3355	1918	(z)	Scr 5/1965
10		0-4-0ST	OC	HL	3352	1918		
		reb		DL		1948	(aa)	(23)
173		0-4-0ST	OC	HL	3919	1937	(aa)	(24)
52		0-4-0ST	OC	RSHN	7340	1946	(aa)	(25)
183		0-4-0ST	OC	RSHN	7347	1947	(aa)	(26)
177		0-4-0ST	OC	RSHD	7036	1940	(ab) *	(27)
24		0-4-0ST	OC	RSHN	7342	1947	(ab)	(20)
25		0-4-0ST	OC	RSHN	7345	1947	(ab)	(25)
54		0-4-0ST	OC	RSHN	7346	1947	(ab)	(20)
	-	0-6-0DH		EEV	D1121	1966	(ac)	(28)
	-	0-6-0DH		RR	10256	1966	(ad)	(29)
D1		0-6-0DH		EEV	D1191	1967	New	(30)
D2		0-6-0DH		EEV	D1192	1967	New	(30)
D3		0-6-0DH		EEV	D1193	1967	New	(31)
D4		0-6-0DH		EEV	D1194	1967	New	(32)
D5		0-6-0DH		EEV	D1195	1967	New	(32)

* converted to oil burning by 7/1965.

6 steam cranes were also used here, together with one diesel rail crane.

- (a) ex Marquis of Londonderry, Londonderry Railway, probably in 1/1899 (these locomotives are not confirmed as being transferred, but evidence would suggest that they were).
- (b) ex Marquis of Londonderry, Londonderry Railway, probably in 1/1899.
- (c) ex S. Pearson & Son Ltd, constractors for the new Seaham South Dock, /1905.
- (d) had allegedly been started as a 0-4-4T for the Londonderry Railway at its Londonderry Engine Works, but it seems more likely that the loco was an 0-6-0T from the beginning; work transferred to the new Seaham Harbour Engine Works, probably during 1901; the loco carried a date of 1902 on its worksplate, but it was not purchased by the Dock Company until 11/1/1906, and then as a new loco never used (Directors' Minutes).
- (e) ex NER, 1662, 8/7/1907; 964 class, previously 972.
- (f) ex NER, 1719, 6/9/1907; previously 1308; originally Blyth & Tyne Railway, 8.
- (g) ex NER, 1661, 11/12/1908; 964 class, previously 969.
- (h) ex NER, 125, 14/12/1911; 120 class.
- (j) S demonstration locomotive; arrived from S, c10/1953.
- (k) S demonstration locomotive; ex River Wear Commissioners, South Docks, Sunderland, by 17/10/1955.
- (m) S demonstration locomotive; possibly came from trials at Scottish Electricity Board, Clydesmill Power Station, Cambuslang, Glasgow, after 11/9/1956; purchased, 2/11/1956; was to be named STEWART, but name never carried.
- (n) two were ordered, the other, S 9619, to be named VANE; but the order for S 9619 was cancelled when serious boiler problems arose with S 9575 and S 9618; S 9619 was completed and used by the makers as a demonstration locomotive.
- (p) YE demonstration loco; ex NCB Northern (Northumberland & Cumberland) No.3 Area, Ashington Colliery, Ashington, Northumberland, 9/5/1958.
- (q) S demonstration loco, almost certainly S 10003; it appears to have arrived at Seaham early in 10/1959, but whether from the S works or from another trial is not known.
- (r) ex ?, 10/1959; demonstration loco.
- (s) ex NCB Durham No.2 Area, Hetton Loco Shed, Hetton (Hetton Railway), 11/1959, hire.
- (t) ex NCB Durham No.2 Area, Lambton Railway loco sheds, Philadelphia, 2/1960.
- (u) ex The Consett Iron Co Ltd, Consett Works, Consett, 11/2/1960.
- (v) ex The Consett Iron Co Ltd, Consett Works, Consett, 13/2/1960.
- (w) ex The Consett Iron Co Ltd, Consett Works, Consett, 17/3/1960.
- (x) ex The Consett Iron Co Ltd, Consett Works, Consett, 28/4/1960.
- (y) rebuild of TEMPEST, S 9618/1956, ex TH, via trials at Dorman Long (Steel) Ltd, Lackenby Works, Middlesbrough, Yorkshire (NR), 10/1960.
- (z) ex South Durham Steel & Iron Co Ltd, West Hartlepool Works, West Hartlepool (see Cleveland & & North Yorkshire Handbook), 11/1961.

(aa)   ex Dorman Long (Steel) Ltd, Acklam Works, Middlesbrough, Yorkshire (NR), 6/1963.
(ab)   ex Dorman Long (Steel) Ltd, Britannia Works, Middlesbrough, Yorkshire (NR), 11/1963.
(ac)   ex EEV for demonstration trial, 6/1966; identity not confirmed, but almost certainly as shown (38 tons, 305 h.p. with Cummins engine); on trial for only 2-3 days.
(ad)   ex RR for trials, 27/6/1966; trials began on 8/7/1966, with a dynamometer test on 13/7/1966.
(1)   to The Londonderry Collieries Ltd, Dawdon Colliery, possibly c/1907.
(2)   to Head, Wrightson & Co Ltd, Teesdale Ironworks, Thornaby, Yorkshire (NR) (builders), for preservation, 6/1959.
(3)   to Head, Wrightson & Co Ltd, Teesdale Ironworks, Thornaby, Yorkshire (NR) (builders), for preservation, 6/1962.
(4)   to North of England Open Air Museum, Beamish, for preservation, 1/1975.
(5)   one source gives scrapped c/1925, another gives scrapped c3/1939.
(6)   scrapped by C.W. Dorkin & Co Ltd, Sunderland, 7/1963.
(7)   subsequently fitted with the cab from AJAX; out of use by /1942 and dismantled by 6/1953; frames cut up, 6/1957, with the tender retained for concrete mixing until at least 19/4/1961.
(8)   scrapped by W. & F. Smith, Ecclesfield, Sheffield, Yorkshire (WR), 1/1967.
(9)   to Armstrong Whitworth (Metal Industries) Ltd, Jarrow, by 13/11/1953.
(10)   to unknown location in Middlesbrough, Yorkshire (NR); returned to S works from there by 24/1/1956.
(11)   scrapped by W. & F. Smith Ltd, Ecclesfield, Sheffield, Yorkshire (WR), 6/1965.
(12)   to TH for conversion to diesel hydraulic loco, /1959.
(13)   for trial to Appleby-Frodingham Steel Co Ltd, Scunthorpe, Lincolnshire, 15/5/1958.
(14)   to Swan Hunter Shipbuilders Ltd, Wallsend Yard, Wallsend, Northumberland, 8/10/1969, for trials.
(15)   almost certainly returned to W.G. Bagnall Ltd, Stafford, Staffordshire, who sold it, from Stafford, on 28/10/1959.
(16)   returned to NCB Durham No. 2 Area, Lambton Railway, Philadelphia, 11/1959 (by 21/11/1959).
(17)   scrapped by C.W. Dorkin & Co Ltd, Sunderland, 11/1963.
(18)   to W. & F. Smith Ltd, Ecclesfield, Sheffield, Yorkshire (WR), for scrap, 4/1967; scrapped, 6/1967.
(19)   to W. & F. Smith Ltd, Ecclesfield, Sheffield, Yorkshire (WR) for scrap, 1/1967; scrapped, 2/1967.
(20)   to W. & F. Smith Ltd, Ecclesfield. Sheffield, Yorkshire (WR) for scrap, 12/1966.
(21)   to W. & F. Smith Ltd, Ecclesfield, Sheffield, Yorkshire (WR) for scrap, 5/1967.
(22)   believed only used as a source of spares for HL 3355/1918; remains scrapped, 2/1964.
(23)   to W. & F. Smith Ltd, Ecclesfield, Sheffield, Yorkshire (WR), for scrap, 3/1967.
(24)   to W. & F. Smith Ltd, Ecclesfield, Sheffield, Yorkshire (WR), for scrap, 9/1967.
(25)   to W. & F. Smith Ltd, Ecclesfield, Sheffield, Yorkshire (WR), for scrap, 11/1966.
(26)   to W. & F. Smith Ltd, Ecclesfield, Sheffield, Yorkshire (WR), for scrap, 1/1968.
(27)   to W. & F. Smith Ltd, Ecclesfield, Sheffield, Yorkshire (WR), for scrap, 4/1967.
(28)   if loco was EEV D1121, then (already sold) to NCB South Northumberland Area, Burradon Colliery, Burradon, Northumberland, 6/1966.
(29)   while at Seaham loco sold to NCB Yorkshire Division No. 7 (Wakefield) Area, Walton Colliery, Walton, Yorks (WR); dispatched to NCB, 26/7/1966.
(30)   scrapped on site by Clive Hornsby Ltd, Wingate, 12/1988.
(31)   to Scottish Grain Distillers Ltd, Cameron Bridge, Fife, Scotland, by 8/1989.
(32)   to Booth Roe Metals Ltd, Rotherham, South Yorkshire, between 10/7/1994 and 7/10/1994, for use as yard shunters.

The Engineer's Report to the Directors dated 10/5/1911 states "the loco we bought from the NER turned out well. I may say we have got a grand bargain at the price paid." Which loco this refers to is unclear; it can hardly refer to the locomotives bought in 1908, yet the date is earlier than the purchase of CLIO on 14/12/1911.

SHELL DIRECT (U.K.) LTD (registered 20/7/1961)
JARROW DEPOT, Jarrow   **E463**
Shell-Mex & B.P. Ltd until 20/7/1961   NZ 336654

This is a large oil storage depot, opened in November 1929, to which a new ½ mile long branch line was built ½ mile east of Jarrow Station on the LNER Gateshead – South Shields line. Fuel oil is brought in by rail and dispatched by road. Except for the period when the locomotive below was used, the rail traffic has always been handled by a main line company, initially the LNER and latterly by the English, Welsh & Scottish Railway. The depot continues (2006) in operation.

Gauge : 4ft 8½in

No.16 (THE EARNEST BRIERLEY)	0-4-0DM	DC	2164	1941		
		EE/VF	1195	1941	(a)	(1)

(a)   ex Trafford Park Works, Manchester, /1950.

(1)   to Salt End Works, Kingston-upon-Hull, Yorkshire (ER), via Bg for repairs, 1/1965.

SHORT BROTHERS LTD (believed registered in 1917)
PALLION SHIPYARD, Pallion, Sunderland   H464
                                              NZ 375580

This shipbuilding business was established in 1850 by George Short (1814-1863) in a yard at South Hylton in 1850, although he is recorded here as early as 1837. The business moved downstream to Pallion in 1871. The business passed down through four generations of the family, finally being controlled by four nephews of the original founder. The yard had no rail connection until about 1898, when a ½ mile branch was built down to it from Pallion Station on the NER Penshaw branch. Because of the difference in height the branch was a large semi-circle, and Pallion Station had to be altered to accommodate it.

It would appear from the list below that the firm shunted its yard with its own locomotive from 1903. However, in 1939 the Sunderland firm of Steel & Co Ltd took over the adjacent premises of Coles Engineering Co Ltd and developed the Crown Works to manufacture cranes (see British Crane & Excavator Corporation Ltd). This firm had its own shunting locomotives, and there is evidence to suggest that this firm's locomotives shunted Short Bros' premises, though whether on a permanent basis or only when P 971 was unavailable is not known. When this loco was under repair in 1949 rail traffic was handled by road lorries.

The last ship was launched in January 1964, after which the firm closed itself down, as its berths were not large enough to accommodate the size of ships then being required; it went into voluntary liquidation on 3rd April 1964.

The firm had various industrial interests elsewhere in Britain, including the Harland & Wolff shipyard in Belfast, the aircraft industry and bodies for buses.

Gauge : 4ft 8½in

-		0-4-0ST	OC	P	971	1903	New	(1)
-		4wVBT	VCG	S	9561	1953	(a)	(2)
YARD No.249		4wVBT	VCG	S	9563	1954	New	s/s c /1964

(a)   S demonstration locomotive; ex Metropolitan-Cammell Carriage & Wagon Co Ltd, Saltley Works, Saltley, Birmingham, after 22/1/1954.

(1)   advertised for sale in the *Evening Chronicle*, Newcastle upon Tyne, 12/7/1954; to J.C. Wight Ltd, Pallion, Sunderland, for scrap, /1954; still in their yard, 29/6/1956; scrapped.

(2)   to Tees Side Bridge & Engineering Works Ltd, Cargo Fleet, Middlesbrough, Yorkshire (NR), by 28/4/1954.

The firm also had one steam rail crane.

SHOTTON BRICK CO LTD
SHOTTON BRICKWORKS, Shotton Colliery   L465
NCB Durham No. 3 (South-East Durham) Area until c/1953
NCB Northern Division No. 3 (South East Durham) Area until 1/1/1950   NZ 397404

This brickworks, which dated from 1903, and its quarry were taken over from the NCB about 1953. They were served by a short link to the NCB branch to Shotton Colliery, which joined the BR Sunderland – Stockton line 1¾ miles north of Wellfield Junction. The works' entire output was common bricks. The ownership of the works changed hands several times after 1953, the company latterly being operated by Hargreaves Ltd of Leeds. The works closed in 1966.

The locomotives below were used to haul tubs between the quarry and the brick yard.

Gauge : 2ft 6in

-	4wDM	HE	4400	1954	New	s/s
-	4wDM	HE	5282	1957	New	s/s

SKERNE IRON WORKS CO LTD

This company was registered on 17/6/1872 to take over the business of **Pease, Hutchinson & Co** with effect from 31/3/1872
Pease, Hutchinson & Ledward until 31/12/1865

SKERNE IRON WORKS, Albert Hill, Darlington

S466
NZ 298158

This works, another venture of the Pease family, was opened in 1864 and was served by a short link from the NER Durham – Darlington line, 1½ miles north of Darlington (Bank Top) Station. It expanded rapidly, soon having 90 puddling furnaces and employing 1000 men, and became famous for rolling plates and constructing bridges. In 1875 the company purchased the Britannia Ironworks at Middlesbrough (see Cleveland Handbook), but this extended the company financially just at a time when industry was about to enter a severe economic depression. The works closed in 1876 or 1877, some of its workforce then forming The Cleveland Bridge & Engineering Co Ltd to continue the bridge-manufacturing business. An order of the Court of Chancery to wind up the company was made on 24th May 1879, and its plant was advertised to be auctioned by Thomas Bowman on 9th September 1879.

Despite this, the works seems to have been back in production by January 1880, under owners of the same name (see below).

Gauge : 4ft 8½in

| - | 0-4-0ST | OC | MW | 112 | 1864 | (a) | s/s |
| - | 0-4-0ST | OC | I'Anson | | 1875 | (b) | (1) |

(a) ex MW, c/1871; previously The Consett Iron Co Ltd, Consett Works, Consett.
(b) ex Fry, I'Anson & Co, Rise Carr Rolling Mills, Darlington.
(1) offered for auction, 9/9/1879; it subsequently arrived at Malton Colliery, near Lanchester, which was owned by Love & Son until 1888, then by S.A. Sadler Ltd.

As noted above, the works resumed production in 1880; the company's records at the National Archives suggest that the Court of Chancery's order was stayed in January 1880. However, by 1889 The Skerne Iron Works Co Ltd was back again in Chancery, which ordered the works to be auctioned by Richard Benson, starting on 21st May 1889. The order, published in the *Colliery Guardian* on 3rd May 1889, lists two locomotives, but no information about them is known.

For the sake of completeness the subsequent history of this works and its site is also included. Following the sale of 1889 the works passed to the **Skerne Steel & Wire Co Ltd** (registered 1/7/1893), only for this firm to go into liquidation on 3/10/1894. It is not known whether this company owned any locomotives. By 1897 the works was owned by John Frederick Wake, and it would seem that in partnership with William Carr he operated his locomotive repair and dealing business here (see the entry for **Wake & Carr** in Part 2). They closed the works in 1899, with an auction in November 1899. After Wake moved to the Geneva Yard, south of Darlington, the site seems to have lain derelict for some time before it was cleared. Then in 1911 part of the area was occupied by the new works of **Henry Williams Ltd** (which see); other parts were developed by other companies, either not using locomotives or without any rail link.

SLATER & CO (LIMESTONE) LTD

MARSDEN QUARRIES, near Whitburn
D467
NCB Northumberland & Durham No. 1 Area until 9/1965
NCB Durham No. 1 (North-East Durham) Area until 1/1/1964
NCB Northern Division No. 1 (North-East Durham) Area until 1/1/1950

NZ 405643 approx.

This group of five quarries, not all of them in production, was purchased from NCB No. 1 Area in September 1965. The internal rail system was immediately replaced by road haulage and the quarries were subsequently closed.

Gauge : 2ft 0in

-	4wDM	RH	177535	1936	(a)	(1)
-	4wDM	RH	187059	1937	(a)	(1)
-	4wDM	RH	189959	1938	(a)	(2)
-	4wDM	RH	189963	1939	(a)	(3)
-	4wDM	RH	287662	1950	(a)	(1)

(a) ex NCB No. 1 Area, with quarries, 9/1965.
(1) sold for scrap, c/1967.
(2) parts sent to Newlandside Quarry, Stanhope (also owned by this firm at the time); remainder sold for scrap.
(3) to North of England Open Air Museum, Brancepeth Store, Brancepeth, for preservation, c/1967; subsequently scrapped.

HOWARD SMITH (HENDON) LTD
HENDON PAPER MILLS, Grangetown, Sunderland H468
Hendon Paper Works Co Ltd until 1/1/1963 NZ 410550

This works was opened in 1872 and was served by sidings west of the NER Durham – Sunderland branch to the South Docks at Sunderland, 1½ miles south of the docks. It may be that horses shunted the traffic before the arrival of the first locomotive below. Rail traffic ceased in 1965 and the works was subsequently closed.

136. The firm's 'Planet' locomotive, HENDON, FH 3865/1958, stands while J27 0-6-0 65885 shunts wagons into the works, in the mid 1960s.

Gauge : 4ft 8½in

HENDON	0-4-0ST	OC	AB	1256	1912	New	Scr by 1/1952
HENDON	4wVBT	VCG	S	7062	1927	New	(1)
HENDON	4wDM		FH	3865	1958	New	(2)

(1) to TH, c/1958; rebuilt as 4wDH TH 101C/1960 and re-sold to Alexander Stephen & Sons Ltd, Linthouse, Glasgow, 25/1/1960.
(2) to FH, and re-sold to NCB Yorkshire No. 7 Area, Old Roundwood Colliery, near Wakefield, Yorkshire (WR), c10/1965.

SMITH, PATTERSON & CO LTD (registered in 2/1873; re-formed, 12/1897)
PIONEER FOUNDRY, Blaydon Haugh

B469
NZ 190637

This firm's first foundry was set up in a watermill at Blaydon Burn, from which it moved to a new foundry here, opened on 17th February 1873. It was served by a short link from the Blaydon Sidings, ½ mile east of Blaydon Station on the NER Newcastle & Carlisle line. The works became famous for the manufacture of rail chairs and the cast iron segments used to line many tunnels on the London Underground system. The firm also had its own shipping fleet, working from the quay alongside the works. How the foundry was shunted before the arrival of the first locomotive below is not known. The firm's chairman for some time was Walter Scott (which see). The company went into voluntary liquidation in September 1964, and the plant, including the FH locomotive, was auctioned by Edward Rushton from 25th-27th November 1964. Part of the premises was taken over on 1st January 1965 by the scrap merchants **Robinson & Hannon Ltd** (which see), who also took over the locomotive.

Gauge : 4ft 8½in

BUTTERFLY	?	?	?	?	?	(a)	(1)
PIONEER	0-4-0ST	OC	HE	18	1867		
	reb		HC		1906	(b)	(2)
JUBILEE	0-4-0ST	OC	HL	3577	1923	New	Scr /1964
TEES-SIDE No.2	0-4-0ST	OC	MW	1327	1897	(c)	(3)
-	4wDM		FH	3374	1950		
	reb		FH		1953	(d)	(4)

(a) identity, origin and date of arrival unknown.
(b) ex Schoen Steel Wheel Co Ltd, Newlay, Leeds, Yorkshire (WR), after 3/1918; as this works was under the control of the Ministry of Munitions during the First World War, the loco may have actually been owned by the Ministry.
(c) ex Tees Side Bridge & Engineering Works Ltd, Cargo Fleet, Middlesbrough, Yorkshire (NR), loan, after 5/1947.
(d) ex Ferens & Love (1937) Ltd, Cornsay Brickworks, Cornsay Colliery, by 12/1961.

(1) according to oral tradition this loco was sold to a quarry in the 1920s.
(2) to Edward Lloyd Ltd, Sittingbourne, Kent, /1943.
(3) returned to Tees Side Bridge & Engineering Works Ltd, Cargo Fleet, Middlesbrough, Yorkshire (NR), /1947 (after 2/9/1947).
(4) to Robinson & Hannon Ltd, with part of the premises, 1/1/1965.

There were also two steam cranes here, the last of which was scrapped in 1964.

THE SOUTH DERWENT COAL CO LTD
The Owners of South Derwent Colliery until c/1900

This firm arose following the decline of **Robert Dickinson & Partners** (which see). Amongst the collieries which the latter owned were **South Derwent Colliery** at Annfield Plain and **West Shield Row Colliery** between Annfield Plain and Stanley, which in 1891 were taken over by **The Owners of South Derwent Colliery**.

EAST TANFIELD COLLIERY, near Tanfield
East Tanfield Colliery Co Ltd until /1929; see also below

G470
NZ 194552

This colliery was sunk in 1844 and had been closed by **James Joicey & Co Ltd** (which see) at the end of 1913. It was re-opened by the **East Tanfield Colliery Co Ltd** in 1915, which closed the colliery on 28th January 1928 and went into liquidation. The colliery was acquired by this company in 1929 and re-opened in January 1930. It was served by sidings west of the NER Tanfield branch just over three miles south of Bowes Bridge Junction (with the Pontop & Jarrow Railway of John Bowes & Partners Ltd). It was vested in NCB Northern Division No. 6 Area on 1st January 1947, by this time the only colliery still owned by the company.

Gauge : 4ft 8½in

-	0-4-0ST	OC	AB	973	1903	(a)	(1)
STANLEY No.2	0-4-0ST	OC	?	?	?		
(formerly SHIELD ROW)	reb		LG		1914	(b)	(2)
STANLEY No.1	0-4-0ST	OC	AB	1659	1920	New	(2)

(a) ex West Stanley Colliery, Stanley.
(b) ex West Shield Row Colliery, near Stanley (which see for possible identity).

(1) to North Grimston Quarries Ltd, North Grimston, near Norton, Yorkshire (NR), by 11/1923.
(2) to NCB No. 6 Area, with the colliery, 1/1/1947.

WEST SHIELD ROW COLLIERY, near Stanley   **G471**
Robert Dickinson & Partners until c/1890   NZ 193530 (see below)

The first **WEST SHIELD ROW COLLIERY** (NZ 193530) was sunk by Robert Dickinson & Partners in 1878 alongside the NER Pontop & South Shields branch, immediately to the north of the Louisa Pit of South Moor Colliery (see Holmside & South Moor Collieries Ltd). Coke ovens were also built here. By 1883 the colliery was regarded as part of South Derwent Colliery. Coal production at this shaft was abandoned by the end of the nineteenth century and instead was concentrated on the **Margaret Pit** (NZ 192536) and the **Fan Pit** (NZ 193537), which were connected to the screens at the original site by a line ¾ mile long which the 2nd edition O.S. map calls a "waggonway"; but in fact this may well have been a narrow gauge rope-worked line.

137. SHIELD ROW, almost certainly a Lingford Gardiner works photograph. Unfortunately, Lingford Gardiner commonly substituted their own plates for the original maker's plates, as here. She may well be HCR 204/1879.

The O.S. maps show a loco shed at the original shaft, and the locomotives below are believed to have worked here, but confirmation of this is lacking. In later years, after the company had acquired West Stanley Colliery nearby, West Shield Row is said to have been shunted by a locomotive from there. The colliery was abandoned in December 1934, having been closed earlier in the year.

Gauge : 4ft 8½in

SHIELD ROW No.1	0-4-0ST	OC	l'Anson	?			
		reb	LG	?	(a)	s/s	
SHIELD ROW No.2	0-4-0ST	OC	?	?			
		reb	LG	1914	(b)	(1)	

(a) ex Lingford, Gardiner & Co Ltd, dealers, Bishop Auckland.
(b) ex Lingford, Gardiner & Co Ltd, dealers, Bishop Auckland; this locomotive is said to have been built by HCR c/1878 and to have had 10in x 16in cylinders and 2ft 9in wheels; if so, then it could well have been HCR 204/1879. supplied new to George Gilbert & Sons, contractors, Salford, Manchester.

(1) to East Tanfield Colliery, near Tanfield.

WEST STANLEY COLLIERY, Stanley   G472
The Owners of West Stanley Colliery until /1909 (J.H. Burn & F.H. Burn)   NZ 193526
J.H. Burn until c/1891; **David Burn** until /1874
Burn, Clark, Swainston & Co until /1858
initially **Clark, Rayne, Burn, Hawthorn & Anderson**

This colliery was sunk, almost certainly in 1843, alongside the Pontop & South Shields Railway, 1¼ miles west of Stanley Bank Head. It was known as **STANLEY (TANFIELD) COLLIERY** until 1865. It was presumably shunted by horses until the arrival of the first locomotive below. On 16th February 1909 it saw the worst disaster in Durham mining history, when an explosion killed 169 men and boys. It is said that after the South Derwent Coal Co Ltd took over West Stanley Colliery the colliery supplied a locomotive to work at West Shield Row Colliery (see above). West Stanley Colliery closed in March 1936.

138. *The Lingford Gardiner hire loco, PRINCE, said to have been rebuilt by Lingford Gardiner from a John Harris loco.*

Gauge : 4ft 8½in

BENTON	0-4-0ST	OC	HCR	103	1871	New	(1)
BENTON	0-4-0ST	OC	BH	1032	1891	New	(2)
ROTHERSYKE	0-4-0T	OC	FJ	187	1882	(a)	(3)
BENTON	0-4-0ST	OC	AB	973	1903	New	(4)
PRINCE	0-4-0ST	OC	Harris?	?		(b)	(5)
-	0-4-0ST	OC	CF	1193	1900	(c)	(6)

(a)   ex Cleator & Workington Junction Railway, No.1, ROTHERSYKE, 3/1897.
(b)   ex Lingford, Gardiner & Co Ltd, dealers, Bishop Auckland, on hire.
(c)   ex South Garesfield Colliery Co Ltd, Lintz Colliery, near Burnopfield, /1929.
(1)   to BH, /1891, in part exchange for BH 1032.
(2)   to Sir W.G. Armstrong, Whitworth & Co Ltd, Puzzulio Shipyard, Italy, possibly c/1897.
(3)   to AB, in part exchange for AB 973, c1/1904; re-sold to C.D. Phillips, dealer, Newport, Monmouthshire, 3/1904.
(4)   to East Tanfield Colliery, near Tanfield.
(5)   returned to Lingford, Gardiner & Co Ltd, dealers, Bishop Auckland.
(6)   to Cowpen Coal Co Ltd, Northumberland, /1937.

As noted above, the company also owned **SOUTH DERWENT COLLIERY (G473)** at Annfield Plain. This comprised two pits, the **Willie Pit** (NZ 169516), which was sunk by Robert Dickinson & Co and began production in 1872, added to subsequently by the **Cresswell Pit** (NZ 164514). They were linked by a narrow gauge tramway, with the Willie Pit served by sidings alongside the Pontop & South Shields branch, though so far as is known the colliery never used locomotives. It was closed on 28th March 1933.

THE SOUTH DURHAM COAL CO LTD
The South Durham Coal Co until /1888
CHILTON COLLIERY, Chilton Buildings   **N474**
The Chilton Coal Co until /1875   NZ 278307

The sinking of this colliery, then called **GREAT CHILTON COLLIERY** to distinguish it from Little Chilton Colliery some two miles away, was begun by **Christopher Mason** on part of the Earl of Eldon's estate about 1834. It was located almost on the southern edge of the Durham coalfield. To serve it the Clarence Railway began the construction of a three mile branch, the Chilton branch, from Chilton Junction on its main line, 1¼ miles south of Ferryhill Station. However, the colliery was abandoned in 1835 before commencing production and its equipment was sold in October 1835. The Clarence Railway had not laid any track when this happened, but subsequently the railway company modified the branch and extended it to serve Leasingthorne Colliery (see Bolckow, Vaughan & Co Ltd).

At the height of the "coal famine" in 1871-1872 this extreme southern area of the coalfield was re-examined, and in November 1871 the lease was taken up by **The Chilton Coal Company**, a partnership dominated by the Straker family of Northumberland (see Strakers & Love Ltd). The sinking of a new No.1 shaft began on 29th February 1872 and production began in 1875. To serve this new colliery a new branch was built, not from the NER Chilton branch, but a two mile extension of the sidings which had formerly served Little Chilton Colliery (see J. Robson & Partners), just south of Ferryhill Station. This line was steeply graded, and to work about 1¼ miles of it a **stationary engine** was built near the entrance to the colliery yard. It is assumed that the locomotives below shunted either in the colliery yard or at the exchange sidings with the NER.

In 1875 the colliery was acquired by **The South Durham Coal Co Ltd**, which already owned **SOUTH DURHAM COLLIERY** at Eldon to the west. Faced with the severe economic depression that developed after 1876, the company closed Chilton colliery in 1884, though the equipment was not auctioned, by A.T. Crow, until 6th June 1888. The company continued to operate South Durham Colliery, and was acquired by Pease & Partners Ltd (which see) in 1903.

Meanwhile Chilton Colliery, although closed, was maintained by the Earl of Eldon, the local landowner, and in 1900 the lease was taken on by **Henry Stobart & Co Ltd** (which see).

Reference: *Chilton, Windlestone & Rushyford*, B. Turner, Chilton Press, 1999.

Gauge : 4ft 8½in

CHILTON	0-6-0ST	IC	MW	324	1871	(a)	s/s by /1888
-	0-6-0ST	IC	MW	693	1878	(b)	(1)

(a) ex Eckersley & Bayliss, contractors, Settle & Carlisle railway contract for Midland Railway (c/1875?).

(b) ex Logan & Hemingway, contractors, Tilton – Market Harborough contract, Leicestershire (Great Northern Railway/London & North Western Railway Joint line).

(1) a 15in cylinder MW loco, presumed to be this one, was offered for auction, 6/6/1888; it was acquired by the Earl of Eldon (see above); subsequently sold to South Leicestershire Colliery Co Ltd, Coalville, Leicestershire.

SOUTH DURHAM IRON CO
ALBERT HILL IRONWORKS, Albert Hill, Darlington   **S475**
                                                                                                           NZ 296159

This company was set up by Joseph and Henry Pease (see Pease & Partners) to produce pig iron. The first hearth was tapped on 13th December 1854, and between 1854 and 1857 three blast furnaces were built. The works was served by a link between the Stockton & Darlington Railway, ½ mile south east of Darlington (North Road) Station, and the NER line between Durham and Darlington (Bank Top) Station, though the latter was severed when William Barningham built his ironworks immediately to the east of the South Durham works (see Darlington Steel & Iron Co Ltd). Immediately to the south west of the South Durham works were the premises of the Darlington Forge Co (which see), which were established soon afterwards.

One source records that "incoming material was hauled by a depot engine up an incline". This **stationary engine** was situated in the middle of the works and hauled northwards, though nothing further is known. The works may have had more locomotives than those listed below.

The company went into voluntary liquidation in June 1877. In 1886 (one source gives December 1888) the site was acquired by The Darlington Forge Co Ltd for an extension of its premises.

Gauge : 4ft 8½in

HARRY TURNER	0-4-0ST	OC	MW	447	1873	New	(1)
BLACK PRINCE	0-6-0ST	OC	BH	354	1875	(a)	(2)
BLACK DWARF	?	?	?	?	?	(b)	s/s

(a) ex Mainsforth Coal Co, Mainsforth Colliery, Ferryhill Station.
(b) origin, identity and date of arrival unknown.

(1) to Bowesfield Iron Co Ltd, Stockton-on-Tees (see Cleveland & North Yorkshire Handbook).
(2) to The Darlington Forge Co Ltd, with the site, /1886.

THE SOUTH GARESFIELD COLLIERY CO LTD

This company began operations in 1887 by re-opening **SOUTH GARESFIELD COLLIERY** at Low Friarside (NZ 158573) **(G476)**, a short-lived landsale colliery opened by Thomas Richardson & Co in 1875 and closed in 1879-1880. The colliery, now sea-sale, was served by a NER "branch" alongside the NER Consett/Blackhill branch, linked to Lintz Green Station a mile away. So far as is known, the colliery was always shunted by the NER.

About a mile to the south-east of South Garesfield Colliery lay Lintz Colliery, formerly owned by **The Lintz Colliery Company** (which see). It was previously thought that this colliery was re-opened about 1889 by a man called John Shield, and that The South Garesfield Colliery Co Ltd took it over about 1895. The Mines Inspectorate's Annual Reports show this to be untrue. Instead it would seem that it was The South Garesfield Colliery Co Ltd which took over the Lintz royalty about 1889, and by 1891 they had opened the **LINTZ COLLIERY (ESTHER PIT)** (NZ 159549) **(G477)**. This was sunk on the site of the former West Lintz Drift of The Lintz Colliery Company. It was served by a standard gauge line from the former **LINTZ COLLIERY (ANNA PIT)** (NZ 164561) **(G478)**, which in turn was linked to the southern end of the NER "branch" serving South Garesfield Colliery by a **self-acting incline** about one mile long. The Anna Pit was re-opened about 1900, and the 3rd edition O.S. map also suggests that the line to

139. One of a series of photographs of JF 2849/1876 at Lintz Colliery about 1912 from the extensive collection held by Beamish, The North of England Open Air Museum's Regional Resource Centre.

the Esther Pit was extended for a further ¼ mile to the **STRAIGHTNECK PIT** (NZ 155554) **(G479)**, though whether this extension was standard or narrow gauge is not known; it had closed by 1919.

The company also opened up new drifts as part of South Garesfield Colliery near the NER, working the **Tilley Drift** (NZ 155568) by narrow gauge top endless rope haulage and bringing in coal from the **Hutton Drift** (NZ 164568) by aerial ropeway. The two Lintz Colliery pits, together with the incline, closed in March 1929, but South Garesfield Colliery survived to be vested in NCB Northern Division No. 6 Area on 1st January 1947.

140. Another Beamish photograph, showing the top endless rope haulage taking tubs from the Tilley Drift to South Garesfield Colliery in the distance, about 1902. The tubs were clipped to the constantly-moving rope. Note the under-manager in the centre of the picture.

Gauge : 4ft 8½in

-	0-4-0ST	OC	JF	2849	1876	(a)	(1)
LINTZ No.1	0-4-0ST	OC	see below		1865	(b)	(2)
LINTZ No.2	0-4-0ST	OC	CF	1193	1900	New	(3)

(a) this loco had been offered for sale in *Colliery Guardian*, 8/7/1887; photographs of it at Beamish, The North of England Open Air Museum, County Durham, suggest that it must not have been sold and have passed to this company with the colliery.

(b) ex NER, 1778, 3/8/1893 (another source gives 14/7/1893); construction of loco begun by West Hartlepool Harbour & Railway at Stockton as WHH&R 49 VICTORIA and completed by NER after its takeover of WHH&R, becoming NER 619 VICTORIA; re-numbered 1778 in 4/1889.

(1) one source says that "the first loco", presumably this JF loco, was scrapped in /1912.

(2) to CF for repairs, /1899, but found impractical, and a new loco ordered; the cylinders and motion were returned to be incorporated into an underground haulage engine, the remainder being scrapped by CF.

(3) advertised for sale in *Machinery Market* by Watts, Hardy & Co (1920) Ltd, dealers, Newcastle upon Tyne, 19/4/1929; to South Derwent Coal Co Ltd, West Stanley Colliery, Stanley, /1929.

THE SOUTH HETTON COAL CO LTD
re-registered on 7/5/1898; latterly the company was a subsidiary of **The South Durham Steel & Iron Co Ltd** of West Hartlepool
The South Hetton Coal Co until 18/7/1874

One of those from whom the Marquis of Londonderry had sought financial backing for the construction of the Seaham Harbour was Colonel R.G. Braddyll, who proved more interested in developing the coalfield to the south-west of the harbour. On 1st March 1831, just under four months before the harbour was opened, Braddyll and his partners began the sinking of **SOUTH HETTON COLLIERY** (NZ 383453) and the construction of a railway four miles long to link it to the new dock. The colliery was opened, with the railway, on 5th August 1833, subsequently trading under the title of The South Hetton Coal Company. Meanwhile some partners next formed The Haswell Coal Company and began the sinking of **Haswell Colliery** (NZ 374423), extending the railway to serve it. This colliery was opened, with the extension, on 2nd July 1835. Four months later the Hartlepool Dock and Railway Co opened its line to Haswell, shipping the first South Hetton coal on 23rd November. The Durham & Sunderland Railway was also keen to tap the South Hetton and Haswell traffic and carry it to Sunderland for shipment, and opened its link to Haswell in August 1836. At the beginning the line between South Hetton and Haswell was worked by the Haswell stationary engine; but this became so devoted to Haswell traffic that in May 1837 the South Hetton company installed its own engine, and in 1845 it replaced the route altogether by a new 1¾ mile branch, the Pespool branch, which ran due south to link to the HD&R.

As with other major coal owners whose collieries were served by the HD&R, the South Hetton company ran its own trains using its own locomotives. Two were named BRADDYLL and NELSON, as they are named in the HD&R minutes. South Hetton ended this traffic in June 1846. However, it is possible, but completely unconfirmed, that two of the locomotives were sold to other coal owners. For details about this period see the entry for the Hartlepool Dock & Railway Co.

The Haswell partnership next went on to develop Shotton Colliery, while the South Hetton partnership turned to the sinking of **MURTON COLLIERY** (NZ 399473). This proved an immense undertaking. Two shafts were begun on 19th February 1838, but in June 1839 a thick layer of waterlogged quicksand was reached. A third shaft was started in July 1840 and more and more pumps were brought in, until over 1600 h.p. was employed. The sinking was not completed until 22nd April 1843, after expenditure estimated to approach £400,000, a huge sum by Victorian standards. The new colliery was linked to the original railway by two ¾ mile branches, one, called the **Murton branch**, joining towards South Hetton and the other, called the **Murton Dene branch**, joining towards Seaham.

For the operation of the line between Haswell and South Hetton, see the entry for **The Haswell, Shotton & Easington Coal & Coke Co Ltd**. Haswell Colliery closed in September 1896, although the line had ceased to have any major use long before this. Alongside the stub of this line were situated the South Hetton loco sheds. Traffic for Seaham Harbour was worked as far as Hesleden bank foot, half way between the junctions of the two branches from Murton Colliery, from which the **Cold Hesleden Engine** (NZ 415469) hauled the waggons up to the summit of the line. From here **two self-acting inclines**, the **Stony Cut** and **Swine Lodge Inclines** took the line down into the North, and later the South Dock at Seaham Harbour. At the interchange point between the two inclines there was a link to the Londonderry Railway (see The Marquis of Londonderry).

About 1860 the Marchioness of Londonderry constructed a link from the Pespool branch to serve **Shotton Colliery** (NZ 398412), which was also owned by the Haswell company. A Londonderry Railway locomotive worked this section, presumably working from the South Hetton sheds. How long this went on for is not known; Shotton Colliery closed in 1877. Shortly after this the **Tuthill Limestone Co** opened **Tuthill Quarry** (NZ 388426) and linked it to the Pespool branch by a ½ mile line which re-used part of the route to Shotton. By January 1895 the South Hetton company was undertaking repairs to the quarry company's two locomotives, though again how long this continued is not known. In 1899 the quarry was taken over by **Pease & Partners Ltd**, eventually being closed in 1921. Two years later it was taken over by **The Northern Sabulite Explosives Co Ltd**, later part of **Imperial Chemical Industries Ltd**, for the manufacture of explosives, with the link to the Pespool branch being maintained. ICI Ltd was still continuing to use the quarry for this work when the coal industry was nationalised in 1947.

The two Murton branches were rebuilt as one line, still with junctions in both directions, apparently during the 1890s, and it was probably about the same time that the Cold Hesleden Engine and its bank were replaced by locomotive working. At Murton Colliery itself a small number of beehive coke ovens were operated until shortly after the First World War, and a **brickworks**, making only common bricks, was started in 1910. Apart from this the system remained unchanged until it, with South Hetton and

Murton Collieries and Murton brickworks, were vested into NCB Northern Division No. 2 Area on 1st January 1947.

The railway system was never given an official name, though on Ordnance Survey maps the title **SOUTH HETTON COLLIERY RAILWAY** is found. **J480**

Reference : *The Private Railways of County Durham*, by Colin E. Mountford, Industrial Railway Society, 2004.

A great deal of uncertainty surrounds some of the South Hetton locomotives, and despite extensive discussion over the years, it is unlikely now that an absolutely definitive identity will ever be given for these. As noted above, the company's first locomotives were used to work the company's own trains over the Hartlepool Dock & Railway Co's line to Hartlepool. The first four are all said to have been built by Timothy Hackworth at his Soho Works at Shildon:

Gauge : 4ft 8½in

BRADDYLL	0-6-0	OC	Hackworth	?	New	(1)		
KELLOE	0-6-0	OC	Hackworth	?	New	Scr		
NELSON	0-6-0	OC	Hackworth	?	New	Scr		
WELLINGTON	0-6-0	OC	Hackworth	?	New	Scr		

(1) said to have worked until about 1875, when it was converted into a snowplough. In this form its remains passed to NCB No. 2 Area, with the railway, on 1/1/1947.

Even the names are disputed. One source gives BRADDYLL as BUDDLE, but given the reference above, this would seem to be an error. KELLOE is sometimes given as KELLOR; but Kelloe is the name of a Durham village, which in dialect would sound like Kellor, which is otherwise unknown. NELSON is also recorded as PRINCE ALBERT, but the former name is also mentioned in the Hartlepool Dock & Railway's minutes. The remains of BRADDYLL were examined by an expert team in 1995, who concluded that the loco was probably built in the late 1830s. It does not, of course, follow that the other three were built at the same time, or were even identical.

	DILIGENCE		0-4-0	VC	RS		1826		
		reb	0-6-0	OC	Hackworth		1834	(a)	s/s
No.1			0-6-0	IC	TR	265	1856	New	
		reb	0-6-0ST?	IC	?		?		s/s
1			0-6-0ST	OC	BH	355	1875	New	(1)
2	HAVERHILL		0-6-0T	OC	SS	2358	1873	(b)	(1)
3			?	?	?		?	(c)	(2)
3	(HALSTEAD?)		0-4-2ST	IC	BP	190	1860	(c)	Scr /1902
(No.3)	GLAMORGAN		0-6-0T	IC	HE	396	1886		
		reb			Baker		1907	(d)	
		reb			South Hetton		1926	(e)	(1)
No.4			0-6-0ST	IC	MW	697	1878	New	
		reb			South Hetton		1913		(1)
No.5			0-6-0ST	IC	MW	758	1881	New	
		reb			South Hetton		1910		(1)
6			0-6-0	IC	RS	1913	1869	(f)	Scr /1903
	-		4-4-0T	OC	BP	425	1864	(g)	(3)
No.6			4-4-0T	OC	BP	417	1864	(h)	
		reb	0-6-0T	OC	South Hetton		1909		(1)
No.7			0-6-0ST	IC	Joicey	305	1883	New	
		reb			South Hetton		1908		
		reb			M.Coulson		1935		(1)
No.8			0-6-0	IC	RS	625	1848		
		reb	0-6-0ST	IC	?		1866	(j)	
		reb	0-6-0T	IC	South Hetton		1923		(1)
No.9	SIR GEORGE		0-6-0ST	IC	RS	624	1848		
		reb	0-6-0ST	IC	?		1865	(j)	
		reb			South Hetton		1911		(1)
10	WHITFIELD		0-6-0	IC	SS		1857		
		reb	0-6-0ST	IC	?		1870	(k)	(1)
	JAMES WATT		0-6-0ST	IC	SS	1768	1866	(m)	(4)
11			0-6-0ST	IC	VF	5308	1945	(n)	(1)
12			0-6-0ST	IC	VF	5309	1945	(p)	(1)

(a) ex Stockton & Darlington Railway, No. 4, DILIGENCE, 2/1841; this was the fourth locomotive built by RS, though whether it had a works number and if so what it was is disputed.

(b) ex Colne Valley & Halstead Railway, /1889; previously Cornwall Minerals Railway, 10.

(c) identity, origin and date of arrival unknown; here by 6/1/1895.

(d) ex Colne Valley & Halstead Railway, /1894; previously North London Railway, 42. The BP works number has been questioned, depending on which NLR loco was sold to the Colne Valley & Halstead Railway. The loco would appear to have arrived at South Hetton in 1/1895; the notebook of the South Hetton boilersmith at that time states that work overhauling it began on 25/1/1895, and then that 'loco entered work, 27/3/1895'.

(e) ex P. Baker & Co Ltd, dealers, Cardiff, Glamorgan, c/1907; previously T.A. Walker, contractor, and used on his Barry Railway contracts in Glamorgan.

(f) rebuilt with new frames, new boiler and firebox, new cylinders and new valve gear (= effectively a new locomotive).

(g) ex Metropolitan Railway, 14, /1905.

(h) ex Metropolitan Railway, 6, /1908 (?), via Robert Frazer & Sons Ltd, dealers, Hebburn; there was an oral tradition at South Hetton that this loco came from the Pelaw Main Railway of The Birtley Iron Co, but there was no parallel tradition there, and it seems likely the tradition has become confused with the Metropolitan Railway engines which the Pelaw Main Railway acquired in 1927. It was subsequently fitted with the wheels from JAMES WATT, with the crankpins turned elliptical, and later still fitted with new wheels.

(j) considerable dispute surrounds the identity of these locomotives. The traditional view was that both were ex Alexandra (Newport & South Wales) Docks & Railway in 1898, where RS 624 had been No.1 and RS 625 No.2. RS 624 had been built for the London & North Western Railway (Southern Division) as No.216, renumbered 816 in 1862, 1156 in 8/1864 and 1805 in 11/1871, being also named SIR GEORGE ELLIOT. It was sold by the LNWR to the ADR in 4/1875, who numbered it 1. RS 625 was built for the LNWR (S.D.) as 220, renumbered 820 in 1862, 1199 in 8/1864 and 1807 in 11/1871, being also named LORD TREDEGAR. It too was sold to the ADR in 4/1875, who numbered 2. In the RCTS *Locomotives of the Great Western Railway* Part 10, p.K10, it is suggested that the minutes of the ADR indicate that a new No.23 was built in 1900 using parts of Nos. 1 and 2, while ADR No. 4 (RWH 709/1849) and 7 (a SS loco, possibly 1012/1857), both also ex LNWR, were sold in 1900 to the dealer C.D. Phillips, and so might have been, although this is not stated, the locomotives acquired by The South Hetton Coal Co Ltd. However, after the RCTS book was published, E.R. Mountford, the noted historian of South Wales railways, came to the view that ADR No.23 was actually built from parts of ADR Nos. 3 and 4, and that Nos. 1 and 2 were sold to South Hetton as previously believed. There are certainly similarities in both appearance and dimensions between ADR Nos. 1 and 2 and the South Hetton engines, as well as the otherwise curious co-incidence of the name of RS 624.

According to South Hetton tradition, No.9 was fitted about 1915 with the tank and cab from JAMES WATT (see below).

(k) considerable uncertainty has surrounded this locomotive too. Almost certainly it was one of seven locomotives built for the London & North Western Railway (Northern Division) by Sharp Stewart in 1857, this one numbered 308 and named BOOTH. If this batch was delivered in works number sequence and then given sequential numbers by the LNWR in order of delivery, then 308 would be SS 1014; however, this seems not to have been the case, and instead 308 was given to the second SS loco to arrive, making it probably SS 1010. 308 was re-numbered 1814 in 9/1870 and 1154 in 12/1871. In 7/1874 it was put on the Duplicate List as 1933, but in the same month it was sold to The Chatterley Whitfield Collieries Ltd in Staffordshire, who named it WHITFIELD. Towards the end of 1877 it was sold to the Ebbw Vale Steel, Iron & Coal Co Ltd, Monmouthshire. It was one of the locomotives earmarked for disposal by EVSIC in a list of 7/1903 and offered for sale in the *Colliery Guardian* on 6/5/1904. Despite this, the South Hetton loco carried a plate saying rebuilt by EVSIC in 1904, the loco being said to have come to South Hetton in 1907, almost certainly via C.D. Phillips Jnr, dealer, Cardiff, Glamorgan.

(m) this locomotive is said to have arrived at South Hetton in 1911. Almost certainly it was the JAMES WATT also owned by the Ebbw Vale Steel, Iron & Coal Ltd, Monmouthshire. This loco was purchased new by the firm as one of a batch of three built in 1865-1866, and research now shows that it was almost certainly SS 1768/1866. It was also on EVSIC's disposal list dated 7/1903, and offered for sale in the *Colliery Guardian* advertisement of 6/5/1904. Where it was between 1904 and 1911 is not clear. It too may well have been acquired through a dealer. However, another source recorded that it was scrapped in 1904; presumably this is inaccurate. Its cab, tank and wheels were subsequently re-used on other locomotives.

(n)　ex War Department, 75318, Longmoor Military Railway, Hampshire, 4/1946; loco had to be fitted with dumb buffers in order to shunt the company's chaldron waggons.
(p)　ex War Department, 75319, Longmoor Military Railway, Hampshire, 4/1946; loco had to be fitted with dumb buffers in order to shunt the company's chaldron waggons.

(1)　to NCB No. 2 Area, with the railway, 1/1/1947.
(2)　boiler repairs to this loco ceased on 15/2/1895; s/s.
(3)　it is believed that parts of this loco were used in the rebuilding of 6, BP 417/1864, and the remainder was scrapped c/1909.
(4)　the tank and cab of this loco are said to have been fitted to No.9 about 1915, with the remainder being scrapped.

At nationalisation the company handed over the most elderly and worn out fleet of locomotives in the North East; apart from two War Department locomotives acquired eight months earlier, the newest loco was built in 1886, five were not in working order and two of those that were capable of working had frames that were 99 years old. A large number of 4½ ton chaldron waggons were also still in use.

In March 1871 the company offered for sale a six-coupled saddletank with 12in x 20in cylinders and 3ft0 in diameter wheels; its origin, identity and disposal are unknown.

Once source records that WINGATE 0-4-0ST OC AB 675/1891, from Wingate Quarry, Wingate and owned by The Wingate Limestone Co Ltd, was repaired at South Hetton. The first record of this loco at Wingate is dated 6/1906 and she had left there by 6/1921, so the repair at South Hetton must have fallen between these dates.

THE SOUTH MEDOMSLEY COLLIERY CO LTD
The Owners of South Medomsley Colliery Ltd until 1/5/1931
The Owners of South Medomsley Colliery until 17/2/1898; see also below
SOUTH MEDOMSLEY COLLIERY, near Dipton　　　　　　　　　　　　　　　　　**G481**
　　　　　　　　　　　　　　　　　　　　　　　　　　　　　　　NZ 144531 (Annie Pit)

This colliery had been opened in 1861 as **PONTOP HALL COLLIERY** by D. Baker, but its name was changed to **South Medomsley Colliery** in 1864 with the opening of the **Ann** (later **Annie**) **Pit**. It was served by a NER branch one mile long from the NER Pontop & South Shields branch, one mile east of Carr House N.E. Junction at Leadgate. The colliery was subsequently acquired by the iron company, **Thomas Vaughan & Co** (which see), but it had closed with that company's failure and its plant had been auctioned in 1878. It was re-opened by these new owners in 1884.

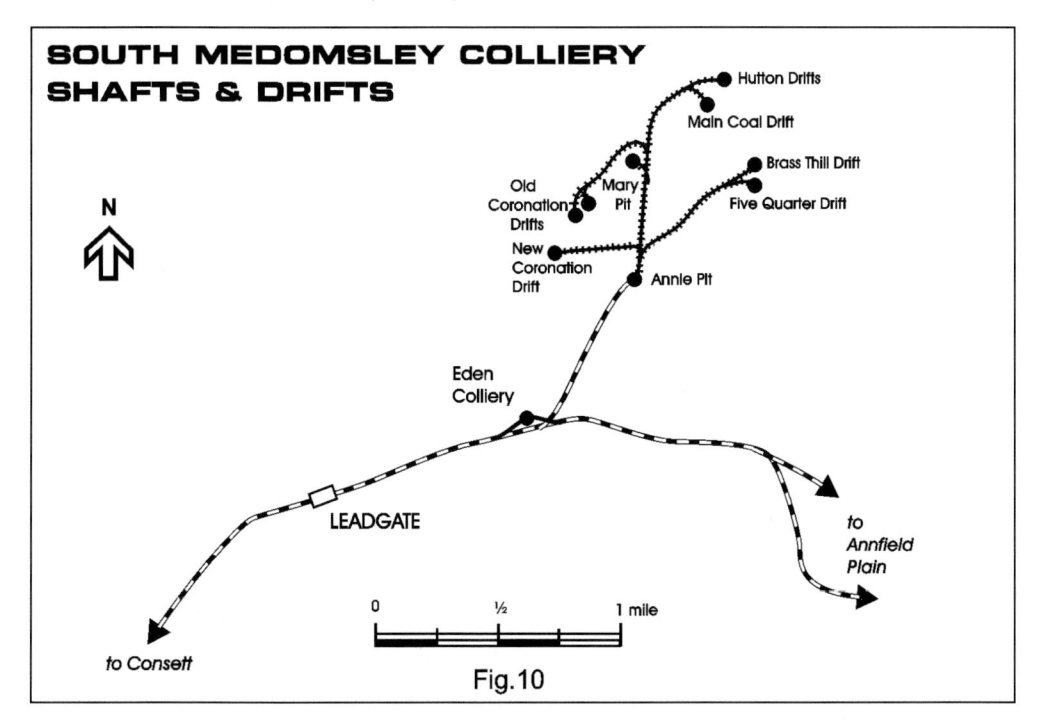

Fig.10

The colliery gradually widened its operations. The **Mary Pit** (NZ 143537), opened in 1867, was sunk 700 yards north of the Annie Pit, to which it was linked by a narrow gauge rope-worked tramway, which also served several drifts en route. This system was subsequently developed considerably, the "main line" being extended to the **Main Coal** (NZ147542) and **Hutton Drifts** (NZ 148543), 1¼ miles from the Annie Pit, with a ½ mile branch from north of the Mary Pit to the **Old Coronation Drifts** (NZ 140535 and 139534), a ½ mile branch from ¼ mile north of the Annie Pit to the **Brass Thill** and **Five Quarter Drifts** (NZ 150536), and then a ½ mile branch from the same point to serve the **New Coronation Drift** (NZ 138532) (see Fig.10). The colliery was vested in NCB Northern Division No. 6 Area on 1st January 1947.

Gauge : 4ft 8½in

SOUTH MEDOMSLEY	0-4-0ST	OC	B	315	1884	New	(1)
-	0-4-0ST	OC	AB	1811	1923	New	(2)
BASIL	0-4-0ST	OC	I'Anson		1875	(a)	(3)
BESSIE	0-4-0ST	OC	HE	205	1878	(b)	(4)

(a) ex Ridley, Shaw & Co Ltd, dealers, Middlesbrough, Yorkshire (NR), /1943.
(b) ex Warner & Co Ltd, Cargo Fleet, Middlesbrough, Yorkshire (NR), loan, /1944.

(1) s/s after /1918; one source gives it to Dorman, Long & Co Ltd, Acklam Works, Middlesbrough, Yorkshire (NR), /1930, another gives it to J.W. Ellis & Co Ltd, Swalwell; neither of these is supported by additional evidence.
(2) to NCB No. 6 Area, with the colliery, 1/1/1947.
(3) to Steel & Co Ltd, Crown Works, Pallion, Sunderland, c/1945.
(4) returned to Warner & Co Ltd, Cargo Fleet, Middlesbrough, Yorkshire (NR), /1944.

SOUTH SHIELDS RURAL DISTRICT COUNCIL
Boldon Colliery   D482

About 1921-1922 this council undertook the construction of a new road from a right-angled junction with Boldon Lane (NZ 363620) to Hubert Street at the southern end of Boldon Colliery village (NZ 347617), a distance of about a mile. Oral tradition recalled at least one steam locomotive, presumably narrow gauge, being used on the work. Today the road, still called New Road, is part of the B1298.

WILLIAM SPENCER & CO LTD
BISHOPLEY CRAG NORTH QUARRY, near Frosterley   U483
started by **Bishopley Crag Mining Co**   NZ 021360 approx.
BISHOPLEY CRAG SOUTH QUARRY, near Frosterley   U484
  NZ 025357 approx.

Bishopley Crag Quarry (later Bishopley Crag North Quarry) was begun about 1871, and was acquired by this company about the mid-1870s. Subsequently the company developed a second quarry to the south called Bishopley Crag South Quarry. Both quarries were served by ½ mile private extensions from the end of the NER Bishopley Branch at White Kirkley, 1¼ miles from Bishopley Junction on the Wear Valley branch, 1¼ miles east of Frosterley Station. It is known that the company used a locomotive here, but details are not known. These quarries were exhausted by about 1902, and the firm then opened:

BOLLYHOPE QUARRY, near Frosterley   U485
  NZ 010349 approx.

This quarry was opened about 1902. It was served by a one mile extension of the line to Bishopley Crag South Quarry. Again, it is known that the company used a locomotive here, but details are unknown. The branch also served the adjacent **Fine Burn Quarry** (NZ 020354 approx.) of **Jacob Walton & Co Ltd** (which see).

In 1907 the company re-opened
BISHOPLEY QUARRY, near Frosterley   U486
  NZ 025361 approx.

This old quarry had been started in 1847 but had been abandoned in the 1870s (see Ord & Maddison Ltd). Which of the former rail links which had once served the quarry (see Map U) was re-laid to serve it is not known, nor is whether the company used a locomotive here.

The company went into liquidation in 1909, and both Bollyhope and Bishopley Quarries were closed.

SPRINGWELL STONE AND BRICK CO LTD (registered in 1908)
OLD SPRINGWELL QUARRY, Springwell, near Gateshead
Northern Stone Firms Ltd until 1908; see also below

C487
NZ 285587

West of Springwell village lay a thick seam of very hard sandstone suitable for grindstones. The quarrying of this began in the mid-nineteenth century, in an area between the Pontop & Jarrow Railway of John Bowes, Esq., & Partners and the village, just west of Springwell level crossing. The quarry was linked to the Blackham's Hill East Incline of the P&JR, and presumably the Blackham's Hill Engine worked traffic in and out of the quarry. Within the quarry shunting was initially done by horses, but eventually a **stationary engine** was built.

In 1904 the quarry was owned by the **Northern Stone Dressing Co Ltd** (registered in 1903), but in the following year it had passed to **Northern Stone Firms Ltd** (registered in 1905), who replaced the stationary engine with the locomotive below. According to oral tradition it was not a success, and horses were re-introduced. In 1907 the quarry was not worked, and in the following year it passed to the **Springwell Stone & Brick Co Ltd**, almost certainly the firm which also owned Springwell Brickworks, ½ mile away at Springwell Bank Head on the P&JR. However, this company did not work the quarry at all, and went into liquidation in 1910, its plant, including the locomotive below, being auctioned by A.T. & E.A. Crow on 23rd November 1910.

The quarry lay unused until 1913, when it was taken over by **J.J. Coxon & Co Ltd**. In 1914 this firm too went into liquidation and the quarry was closed and filled in. During all this a new quarry (NZ 283586), taking the name Springwell Quarry, had been started on the opposite side of the line. This became a much larger quarry and was worked for many years by **Richard Kell & Co** (Richard Kell & Co Ltd from 12/1922), who also owned the Windy Nook Quarries served by the Pelaw Main Railway. It then passed to **James H. Harrison** (which see).

Gauge : 4ft 8½in

| No.9 | DORIS | 0-4-0ST | OC | MW | 1150 | 1891 | (a) | (1) |

(a)   ex Joseph Perrin & Son Ltd, Birkenhead, Cheshire, c10/1906.

(1)   offered for auction, 23/11/1910; s/s.

STANLEY URBAN DISTRICT COUNCIL
HUSTLEDOWN SEWAGE WORKS, near The Middles

G488
NZ 213516

It is believed that this locomotive was used on a ¾ mile narrow gauge line connecting this works (NZ 214517) with the Durham Road (now the B6313) at The Middles, near Craghead, left in place when the contractors building the works (see the Contractors' Section in Part 2) handed the site over to the council. It is not known how long the railway continued in use.

Gauge : 2ft 0in

| - | 4wPM | MR | 903 | 1918 | (a) | s/s after 7/1931 |

(a)   probably came from contractors who built Hustledown Sewage Works at The Middles, near Craghead; it was owned by the council by 10/7/1931, when spares were supplied by MR.

STEETLEY DOLOMITE (QUARRIES) LTD
subsidiary of **The Steetley Co Ltd** from 3/1951
The Steetley Co Ltd until 3/1951
The Steetley Lime & Basic Co Ltd until 22/3/1944, by /1938
formerly **The Steetley Lime Co Ltd**

COXHOE BANK QUARRY, Quarrington Hill
for previous ownership see text below
COXHOE LIMEWORKS, Coxhoe

COXHOE QUARRY, Quarrington Hill

N489
NZ 327363 approx.
N490
NZ 321364
N491
NZ 330367 approx.

Quarrying for limestone in this area was begun by **J. Foster** at **COXHOE BANK QUARRY**, which was nearer to Quarrington Hill than Coxhoe. It was connected by a short link to the private branch to South Kelloe Colliery (see William Hedley & Sons, and Fig.6) from the end of what was originally the Coxhoe branch of the Clarence Railway. Almost certainly the connection joined near the bank head of a rope

141. No.15, AB 688/1891, on 2nd May 1957, awaiting scrapping. Note the aerial ropeway in the background.

incline that took the line down to Coxhoe village. So far as is known locomotives were not used, shunting presumably being done by horses. By the early 1870s Coxhoe Colliery, on the opposite side of the line, was owned by **The Rosedale & Ferryhill Iron Co Ltd** (which see), and it seems likely that by this time this firm also owned the quarry. This company failed in 1879, and the quarry also appears to have closed about this date.

142. To replace its steam locomotives, the company purchased a Yorkshire Engine Company diesel, a make rarely seen in the North-East. YE 2779/1960 is seen on 7th April 1960, shortly after delivery.

On 28th March 1901 **The Coxhoe Limestone Co Ltd** was registered for the purpose of re-opening Coxhoe Bank Quarry, but it would appear from the Annual Reports of the Mines & Quarries Inspectorate that no production took place. In July 1906 the quarry was acquired by **The Steetley Lime Co Ltd**, whose headquarters were in Nottinghamshire, with production beginning in 1907. To serve the quarry a ¾ mile branch was laid from the end of the NER Coxhoe Branch. For most of its length this used the formation of the line abandoned in 1879, but this time the connection to the NER was north-facing because the land where the south-facing link once was had been built over. Most of the branch was a single line **rope incline**, operated by a **stationary engine** near the quarry. Near the bank foot, to the north-east of Coxhoe village, the company developed **COXHOE LIMEWORKS**, building cupolas for the production of doloma on the site of former beehive coke ovens.

By 1919 the firm had begun quarrying on the opposite side of the line, to the east of the former Coxhoe Colliery (whose shafts were used for ventilation for East Hetton Colliery). This was called **COXHOE QUARRY**, and it was also linked to sidings at the incline's bank head. Besides limestone, sand was also quarried. Production at Coxhoe Bank Quarry appears to have ended in the second half of the 1930s. One of the firm's major customers for many years was The Consett Iron Co Ltd.

In 1943 the Ministry of Supply (which see) contracted with The British Periclase Co Ltd, a subsidiary of The Steetley Lime & Basic Co Ltd, to erect a plant known as the Harrington Shore Works near Workington in Cumberland for the manufacture of calcined magnesium. Dolomite limestone was required for this process, and Coxhoe Quarry was chosen to supply it. This operation went under the curious official title of 'Harrington Shore Works, Coxhoe Branch'. It would seem almost certain that a separately-defined part of the quarry was used for this, very probably operated by the Steetley company, but with the Ministry supplying the plant, including locomotives. This work ceased in 1945 and the Ministry disposed of the plant.

With limestone reserves exhausted, the quarry, its branch and most of the works closed in 1967, with the exception of one rotary kiln, which was supplied by rail from the company's new Thrislington Quarry at West Cornforth. The kiln closed in 1979 and the buildings were then demolished, the chimney coming down on 12th September 1981.

Locomotives were used both at the two quarries and at the works. There may well have been more locomotives here than are listed below.

Reference : *A History of Coxhoe*, R. Walton, Coronation Press, 1986.

Gauge : 4ft 8½in

	COXHOE No.1	0-4-0ST	OC	BH	1038	1893		
		reb	CF			1901	(a)	
		reb	YE			1923		
		reb	YE			1933		(1)
No.2		0-4-0ST	OC	AB	233	1881	(b)	(2)
	B No 15	0-4-0ST	OC	RS	2724	1890	(c)	s/s by 5/1949
	B No 13	0-4-0ST	OC	HL	2176	1890	(d)	(3)
No.15		0-4-0ST	OC	AB	688	1891	(e)	
		reb	AB			1946		(3)
	(SENTINEL 1)	4wVBT	VCG	S	7669	1928	New	Scr by 8/1950
	KITCHENER	0-4-0ST	OC	AB	277	1884		
		reb	AB		9410	1915	(f)	Scr c/1942
	COXHOE No.2	0-4-0ST	OC	RSHN	6939	1937	New	(4)
	COXHOE No.3	0-4-0ST	OC	RSHN	6963	1939	New	(4)
No.10		0-4-0ST	OC	?	?	?		
		reb	AB			?	(g)	(5)
(No.4)		0-4-0ST	OC	HC	1734	1942	New	(6)
	-	0-4-0ST	OC	P	2032	1942	(h)	(7)
	-	0-4-0ST	OC	HC	1733	1943	(j)	(8)
	COXHOE No.4	0-4-0ST	OC	HC	1735	1942	(k)	(9)
	COXHOE No.1	0-4-0ST	OC	RSHN	7819	1954	New	Scr 5/1967
	-	0-4-0DE		YE	2779	1960	New	(10)

(a) ex CF, /1901; previously Clifton & Kersley Coal Co Ltd, Clifton, Lancashire.
(b) ex Lonsdale Hematite Iron Co Ltd, Whitehaven, Cumberland.
(c) ex The Consett Iron Co Ltd, Consett, /1922.
(d) ex The Consett Iron Co Ltd, Consett, 6/1923.
(e) ex Sir W.G. Armstrong, Whitworth & Co Ltd, Elswick Works, Newcastle upon Tyne, after 7/1924.

(f) ex Pelaw Main Collieries Ltd, Pelaw Main Railway, c/1930.
(g) origin unknown, via the Ministry of Supply between 1939 and 1945; said to have originally been a GWR engine at Swansea, but nothing can be identified.
(h) ex Ocean Salts (Products) Ltd *, Harrington Shore Works, near Workington, Cumberland, /1945.
(j) ex Ocean Salts (Products) Ltd *, Harrington Shore Works, near Workington, Cumberland, /1948 (by 21/11/1948).
(k) ex The British Periclase Co Ltd *, West Hartlepool, /1954 (here by 18/4/1954).

(1) to Steetley Works, Worksop, Nottinghamshire, /1917; returned by /1938; to William Cory & Son Ltd, Gallions Jetty, Essex, c/1943.
(2) sold for scrap.
(3) to Thos W. Ward Ltd, Haswell, for scrap, 12/1959.
(4) to Thos W. Ward Ltd, Haswell, for scrap, 5/1960.
(5) sold for scrap to M. Coulson & Co Ltd, Spennymoor.
(6) may have been loaned to The British Periclase Co Ltd *, Palliser Works, Hartlepool, sometime between 1943 and 1945; to Birmingham Corporation Gas Department, Swan Village Gas Works, West Bromwich, Staffordshire, 1/1947.
(7) to Birmingham Corporation Gas Department, Swan Village Gas Works, West Bromwich, Staffordshire, 1/1947.
(8) to Magnesium Electron Ltd, Hapton, Lancashire; arrived 28/4/1955.
(9) partially cut up for scrap by 9/9/1966; work completed c2/1967.
(10) to Steetley Ground Limestone Co Ltd, Rainbow Bridge Quarry, Conisborough, Yorkshire (WR), 11/1967.

   * these companies were also subsidiaries of The Steetley Co Ltd

Gauge : 4ft 8½in

The loco below, called a "tram" by the men, had a skip at each end of it which were filled with limestone and coke and which were then lifted up to the top of a set of kilns. The kilns were demolished in 1962.

| - | 0-2-4DM | ? | ? | (a) | s/s c/1967 |

(a) origin unknown – it may have been built at Coxhoe; it had an engine HE 84868/1959.

Gauge : 3ft 5½in

This loco hauled large hopper wagons underneath the kilns to collect the dolomite remaining after the limestone had been burned. The loco had three booms to three overhead wires.

| - | 4wWE | GB | 2319 | 1950 | New | s/s c/1967 |

Gauge : 2ft 0in

These locomotives are believed to have been used within Coxhoe Quarry for a time, bringing tubs from the quarry face.

| - | 4wDM | ? | ? | ? | (a) | Scr |
| - | 4wDM | RH | 211683 | 1941 | (b) | Scr |

(a) origin, identity and date of arrival unknown.
(b) ex Rainbow Bridge Quarry, Conisborough, Yorkshire (WR).

Gauge : 2ft 0in

This system was used to convey tubs of coke and limestone to another set of kilns, where the tubs were emptied into skips for lifting up to the top of the kilns for charging.

-	4wBE	GB	2130	1948	New	s/s c/1967
-	4wBE	WR	5115	1953	New	(1)
-	4wBE	WR	5316	1955	New	(1)

(1) to ?, c/1967, possibly R.E. Trem & Co, Finningley, Yorkshire (ER); sold by R.E. Trem & Co to Tickhill Plant Ltd, Tickhill, near Doncaster, Yorkshire (ER), 11/1968.

G-R STEIN REFRACTORIES LTD
BATTS WORKS, Wolsingham   **T492**
                                                                                            NZ 077369

The locomotive below was moved here from another of the company's sites but was never used.

Gauge : 2ft 0in							
	-	4wDM	RH	170373	1934	(a)	Scr by 8/1974

(a) ex New Bridge Pits, Pickering, Yorkshire (NR), /1963.

THE STELLA COAL CO LTD
The Stella Coal Co until 2/2/1903

Although this royalty was leased by various different owners over the years, and their names are sometimes given in official documents, they all seem to have traded under the style The Stella Coal Company.

The large royalty from Ryton south-westwards down to Chopwell was known as the Stella Grand Lease, the name taken from Stella Hall, the large country house near Ryton. Coal was being worked here in the eighteenth century and carried by wooden waggonway to the River Tyne. In the later years of the eighteenth century the lease was held by George Silvertop, but in 1800 he sub-let it. By 1820 the lease was held by Peregrine Edward Towneley. George Dunn & Sons worked the coal until 1831, after which the royalty lay idle for some years. The southern part eventually passed to the **Marquis of Bute** (which see), while the northern part was taken up in the late 1830s by a partnership including John Buddle (see The Marquis of Londonderry), under the style of **The Stella Coal Company**.

Up to about 1880 the collieries were known as the **Towneley** (sometimes mis-spelt as Townley) and **Stella Collieries**; in the twentieth century the names of the pits became the names of the collieries. As the owners sank new collieries a railway system was developed to serve them, called the

TOWNELEY COLLIERY WAGGONWAY   A493
This name also died out in the early twentieth century.

The first of the nineteenth century collieries to be sunk was what became the **TOWNELEY COLLIERY, STARGATE PIT**, later **STARGATE COLLIERY** (NZ 162635), south-east of Ryton. Sinking began in 1800 and was completed on 16th June 1803. It was served by a waggonway just under 1¼ miles long, running eastwards and then north-east to the **STELLA STAITH** (NZ 176638) on the River Tyne between Ryton and Blaydon. The waggonway was almost certainly wooden at first and used horses. Stella Staith, latterly two shipping points, was the coal staith furthest upstream on the Tyne, some 15 miles from the mouth of the river. Next to be sunk was the **STELLA FREEHOLD** or **BOG PIT** (NZ 166634). This lay alongside the waggonway about ¼ mile east of Stargate. Its sinking was completed on 28th April 1835, but it seems to have ceased production by the 1860s.

The owners next turned their attention to the western area of the royalty, and opened what was at first called **TOWNELEY MAIN COLLIERY**, then **TOWNELEY COLLIERY, EMMA PIT** and latterly **EMMA COLLIERY** (NZ 145640), at Bar Moor near Ryton. Although a sinking was made here in 1823, the sinking of the modern colliery began on 17th March 1845 and was completed on 20th April 1847. As it was sunk deeper it was found that there were nine workable coal seams, and a **brickworks** was also developed here. The colliery was linked to Stargate by a 1¼ mile extension of what was now a railway, though its route followed that of a former waggonway. There is a description of the line in the *Colliery Guardian* of 2nd February 1894. This records a **stationary engine** about ½ mile west of Stargate (NZ 156634), which may well mean that it was built when the line was extended to Emma about 1845. This engine, called in other records the **Stargate Engine**, hauled "fulls from Emma to Stargate. These are then run by gravity down to Stella Staiths and the empties are brought up from the Staiths to Stargate, then run down to Emma." Both the 1st and 2nd edition O.S. maps show these sections to be single lines with no meetings, while the *Colliery Guardian* describes the engine has having two drums, 6ft and 9ft in diameter. All this must mean that the engine worked two inclines, the **western**, or **Emma**, **Incline** about ½ mile up to the engine house and the **eastern**, or **Stella**, **Incline** about 1¼ miles down to the staiths. The April 1871 section of DCOA Return 102 lists only the stationary engine, but soon afterwards locomotive shunting began, for a locomotive was included in the section for November 1876. Presumably it was possible for the loco to propel wagons on to the east bank head for them to be run down to the staiths. Emma itself was worked by a 'jack engine', a small stationary engine at the western end of the colliery yard. The waggons were hauled up to the engine, and then run back by gravity through the screens as required.

The next colliery was sunk to work the coal in the north-eastern sector of the royalty. This was **STELLA COLLIERY, ADDISON PIT**, later **ADDISON COLLIERY** (NZ 167634), at Hedgefield. Sinking work began on 1st February 1864 and the first coal was drawn on 15th December 1865. It lay alongside the Newcastle - Carlisle line of the NER, to which it was connected, one mile east of Ryton Station. Beehive coke ovens were built at all three collieries, but instead of taking the colliery name, they were known collectively as the **HEDGEFIELD COKE OVENS**. The ovens at Addison Colliery, which had begun production on 15th December 1866, were put out in the quarter ending 31st March 1922. A **brickworks** was also started at Stargate Colliery.

The company clearly decided that a link with the NER was better here than where the lines to their staiths now crossed the NER on bridges, and so a link was built from Stargate down to the sidings at Addison just east of the colliery. Known initially as the **Stella & Hedgefield Incline**, and later as the **Stargate Incline**, it was unusual in having a drumhouse at the bank head and also running through a tunnel below the meetings. Various annual costs for the railway are deposited in the Durham Record Office. Those for 1874 make no mention of either this incline or a locomotive, but those for 1877 include both, and since the first locomotive dates from 1875, it is likely the incline began working then too. Later the incline was rebuilt, with a longer meetings and a new tunnel on a larger radius curve. In 1902 the Stargate Engine was replaced by a new engine further west (NZ 156634) and the inclines modified accordingly.

The final colliery to be linked to the system was **GREENSIDE COLLIERY**, Greenside (NZ 139619). A colliery had been worked here in the eighteenth century, but its shafts were ½ mile from the new pit, which served the southern area of the royalty. To serve it a two mile branch was built from the branch to Addison Colliery, just north of the Stargate loco shed. Much of this branch was in cutting to allow easier loco working. A loco shed was built at the south end of the colliery to house the locomotive kept here. Two small sand quarries were also developed alongside the branch, presaging much more extensive quarrying by other companies from the 1940s onwards.

With being so far upstream, shipments through Stella Staiths were latterly limited to wherries, and ceased altogether about 1941-42. After this all coal was sent down the incline to be despatched via the LNER at Addison Colliery. However, the line to the staiths remained open to handle workmen's and landsale coal. Whether this traffic was handled by the Stargate Engine or by locomotive is not known. At the end of the Second World War major changes were made. The rope working to Emma Colliery and the hauler shunting there were replaced by locomotive haulage. Part of the old route near the colliery was replaced by a new line with easier gradients. At the same time the old route to the former staiths was closed. The engine house was converted into premises for the Stargate Colliery fitters.

On 1st January 1947 Stargate, Emma, Addison and Greenside Collieries, together with the railway, were vested into NCB Northern Division No. 6 Area. As at South Hetton, the system was still using chaldron waggons for internal traffic.

Reference : *The Private Railways of County Durham*, Colin E. Mountford, Industrial Railway Society, 2004.

Besides the above, the company also owned two other collieries:

BLAYDON MAIN COLLIERY, Blaydon B494
G.H. Ramsey & Co until c/1883
G H. Ramsey & Son until /1880; formerly **G.H. Ramsey**

The first Blaydon Main Colliery (NZ 182630), opened on 3rd October 1837, was sunk about halfway up Blaydon Bank and was linked by a waggonway 1¾ mile long running eastwards to **Blaydon Main Staith** (NZ 202634) at Derwenthaugh, crossing the then Newcastle & Carlisle Railway there on the level. The 1st edition O.S. map shows the first ½ mile from the colliery to have been a **self-acting incline**, and about ½ mile further on the map an "**Engine Shed**" (NZ 193632) is shown. At the staith a branch curved away eastwards under the line to Garesfield Staith (see Marquis of Bute) to serve a **firebrick** and **bone manure works** (NZ 204633) and then **coke ovens** (NZ 205633) on the western mouth of the River Derwent, all also owned by George Ramsey.

About 1855 this pit was abandoned in favour of a new Blaydon Main Colliery ½ mile away near Axwell Park. Two pits, the **Speculation** and the **Hazard**, were sunk about 50 yards from each other, but with the same surface. As at the original colliery, fireclay was mined below the Hutton seam. They were served by a ½ mile link from the original waggonway, forking to serve the two pits. The engine shed above was demolished and replaced by a new shed at the Hazard Pit, the line to the old colliery being lifted.

Ramsey died in 1880, and the colliery had been acquired by The Stella Coal Co by 1883. By the 1890s coke ovens had been built at the staith and the older ovens sold to Joseph Cowen (see The Priestman Collieries Ltd), while the link to the firebrick works had been abandoned. In the early twentieth century both sites were re-developed by The Consett Iron Co Ltd for its new Derwenthaugh Staiths. The colliery was sold to **The Priestman Collieries Ltd** in 1908 (which see).

CLARA VALE COLLIERY, Clara Vale

A495
NZ 132651

This colliery, close to the River Tyne, was sunk to work the north-western sector of the royalty. Sinking began on 22nd April 1890 and the first coal was drawn on 23rd May 1893. It lay alongside the NER Newcastle – Carlisle line, one mile east of Wylam Station, but with the shafts on the southern side of the railway linked by tramways on bridges to the screens on the northern side. In its early years the colliery's name was spelled as one word.

Addison, Clara Vale, Emma, Greenside and Stargate Collieries passed to NCB Northern Division No. 6 Area on 1st January 1947.

The locomotives were regarded as a pool and were moved round quite frequently as required. There may well have been more locomotives in the early period besides those listed below. The locomotive allocations on 1st January 1947 are given.

Gauge : 4ft 8½in

STELLA	0-4-0ST	OC	AB	70	1868	New	(1)
TOWNELEY	0-4-0ST	OC	RWH	1726	1875	New	s/s
HEDGEFIELD	0-4-0ST	OC	RWH	1817	1880	New	s/s
TOWNELEY	0-4-0ST	OC	HL	2199	1891	New	s/s
MARLEY	0-4-0ST	OC	?	?	?	(a)	Scr
CLARA	0-4-0ST	OC	HL	2281	1894	New	(2)
CLAUDE	0-4-0ST	OC	HL	2349	1896	New	(3)
GERALD	0-4-0ST	OC	HL	2426	1899	New	(3)
STELLA	0-4-0ST	OC	HL	2583	1904	New	
	reb	HL			1931		(4)
JOAN	0-4-0ST	OC	HL	2617	1905	New	(2)
MURIEL	0-4-0ST	OC	HL	2694	1907	New	(2)
ADDISON	0-4-0ST	OC	HL	2702	1907	New	(2)
EMMA	0-4-0ST	OC	HL	2740	1909	New	(2)
VICTORY	0-4-0ST	OC	HL	3438	1920	New	(2)

On 1st January 1947 CLARA was at Clara Vale Colliery, STELLA and EMMA at Addison Colliery, VICTORY at Greenside Colliery and JOAN and MURIEL at Stargate Colliery. The location of ADDISON is not known.

(a) ex "a local colliery".
(1) to "Winter of Winlaton" (identity unknown; no one of this name is listed at Winlaton in any Durham directory between 1890 and 1938), then re-sold to a steelworks in Scotland.
(2) to NCB No. 6 Area, 1/1/1947.
(3) to The Priestman Collieries Ltd, with Blaydon Main Colliery, /1908.
(4) hired to Synthetic Ammonia & Nitrates Ltd, Billingham, /1928 (see Cleveland Handbook); this firm was absorbed by Imperial Chemical Industries Ltd in 1931 and the works became operated by I.C.I. (Fertilizers & Synthetic Products) Ltd; loco returned, /1934; to NCB No. 6 Area, 1/1/1947.

ROBERT STEPHENSON & HAWTHORNS LTD

subsidiary of **Vulcan Foundry Ltd**, Newton-le-Willows, Lancashire, from /1945; see also below
Robert Stephenson & Co Ltd until 6/1937
Robert Stephenson & Co (1914) Ltd until 5/1919
Robert Stephenson & Co Ltd until 13/3/1914
Robert Stephenson & Co until /1886

SPRINGFIELD, Darlington

S496
NZ 300166

The locomotive-building company of Robert Stephenson & Co had been established in Forth Banks, Newcastle upon Tyne, in June 1823, with the Pease family and other Quaker industrialists at Darlington becoming major shareholders. In 1899 it was decided to move the business to new premises built on a green field site at Springfield on the northern outskirts of Darlington. Construction of the new works began on 31st March 1900 and was completed in March 1902; the first locomotive left the works on 4th November 1902. It was served by sidings east of the NER Durham – Darlington line, 1¾ miles north of Darlington (Bank Top) Station. The former Newcastle upon Tyne premises were acquired by R. & W. Hawthorn, Leslie & Co Ltd, whose works lay immediately to the west of the Robert Stephenson site.

The firm was also involved with ship-building, but sold its Hebburn Yard to Palmers Shipbuilding & Iron Co Ltd (which see) in 1912. In June 1937 the company amalgamated with the locomotive-building business of R. & W. Hawthorn, Leslie & Co Ltd in Newcastle upon Tyne, the new company being named Robert Stephenson & Hawthorns Ltd. In general, the former RS works in Darlington concentrated on locomotives for public railways, while the former HL works in Newcastle handled orders for industrial locomotives. In 1944 R. & W. Hawthorn, Leslie & Co Ltd disposed of the whole of its shareholding in RSH to Vulcan Foundry Ltd and North British Locomotive Co Ltd, the former acquiring the latter's interest in 1945. Subsequently Vulcan Foundry Ltd became a subsidiary of English Electric Co Ltd, which took over full control of this works in Darlington on 1st January 1963 (which see). The firm itself continued to exist until 2005, when it was finally wound up.

Gauge : 4ft 8½in

	PHOENIX	0-4-0ST	OC	RS		c1905	(a)	Scr c/1938
(614)	EGBERT	0-4-0CT	OC	HL	2173	1890	(b)	
		reb		RS		1918		
	reb	0-4-0ST	OC	?		?		Scr 11/1952
(615)	WINSTON CHURCHILL	0-6-0ST	IC	MW	2025	1923	(c)	(1)
No.755		0-4-0ST	OC	RSHN	7675	1951	(d)	(2)
		0-4-0DM		RSHN	7869	1956	(e)	(3)
D0227	"THE BLACK PIG"	0-6-0DH		EE	2346	1956		
				VF	D227	1956	(f)	(4)

(a) loco constructed by RS from an unidentified 0-4-0ST acquired from U.A. Ritson & Sons Ltd, Burnhope Colliery, in /1904 and a Joicey locomotive built in 1880, acquired from an unknown company in Northumberland.
(b) ex Hebburn Shipyard, Hebburn (between /1901 and /1912).
(c) ex Cadbury Brothers Ltd, Blackpole Works, Worcester, by /1946.
(d) built new at Forth Banks Works, Newcastle upon Tyne, but painted at Darlington Works.
(e) hire loco previously at Northern Gas Board, Redheugh Works, Gateshead, until /1958; probably hired to Coats Patons Ltd, Darlington (which see).
(f) supplied to British Railways for evaluation trials with D0226, which had diesel electric transmission; withdrawn by BR in 9/1959 and returned to English Electric Co Ltd; sent here c/1960.

(1) to Guy Pitt & Co Ltd, Pensnett Railway, Shutt End, Staffordshire, 29/5/1946.
(2) to NCB East Midlands Division No. 3 Area, Ollerton Colliery, Ollerton, Nottinghamshire, 6/1961.
(3) to Northern Gas Board, Carlisle Works, Carlisle, Cumberland, /1960 (fitted with new plates dated 1960).
(4) to English Electric Co Ltd, with works, 1/1/1963.

HENRY STOBART & CO LTD
Henry Stobart & Co until 24/11/1893

This company was set up in the 1830s and initially developed collieries at Etherley, south-west of Bishop Auckland. During the next 100 years it acquired collieries in the south of the county from Cockfield in the west to Fishburn in the east. On 1st January 1920 its share capital was acquired by Pease & Partners Ltd (which see). In the financial difficulties which beset Pease & Partners from 1929 Henry Stobart & Co Ltd remained profitable, and the transfer of Thrislington Colliery from The North Bitchburn Coal Co Ltd to Henry Stobart & Co Ltd, for which the latter then paid Pease & Partners in lieu of North Bitchburn's debt to its parent company, was a major factor in the parent company's survival.

Not all of the company's collieries are known to have had locomotives, and information about those that did is fairly sketchy. The company had no major locomotive repair facility, and in later years major overhauls were carried out at the Tees Iron Works of Pease & Partners Ltd at Cargo Fleet in Yorkshire (NR). This works often fixed a "rebuild" plate to the locomotive, but this usually meant only an overhaul, perhaps including a new boiler, not a change in the locomotive's appearance or dimensions.

The numbers carrried by locomotives from about 1944 onwards were issued as part of a general re-numbering scheme initiated by Pease & Partners Ltd (which see).

CHILTON COLLIERY & COKE OVENS, near Chilton Buildings   **N497**
NZ 278308

The sinking of this colliery had begun in 1835, but no coal had been produced. It was taken over by **The Chilton Coal Co** in 1871 and passed to **The South Durham Coal Co Ltd** (which see for its early history) in 1876. It was closed in 1884 and its equipment auctioned in 1888. The shafts were maintained by the Earl of Eldon, the royalty owner, until the lease was taken on by Henry Stobart & Co Ltd in 1900.

The colliery's former branch, with its rope incline, to Ferryhill Station was not relaid, and instead the company built a new ¾ mile branch to join the NER Chilton Branch ½ mile west of Chilton Buildings on the alignment of the unfinished Clarence Railway branch of 1835. This had been opened by May 1901. It took three years to clear the colliery of water, though production was begun in October 1902.

A battery of 50 Simplex waste heat by-product coke ovens began production about August 1908. They were closed on 28th March 1921, and replaced by 52 new Simplex waste heat ovens, which began production in the quarter ending 31st December 1923. In 1924 Pease & Partners Ltd (which see) transferred the colliery to its direct ownership in order to combine its workings with those of their Windlestone Colliery to the south, which was then closed.

Gauge : 4ft 8½in

CHILTON No.1	0-6-0ST	OC	AB	961	1902	New	(1)
CHILTON No.2	0-6-0ST	OC	AB	1097	1907	New	(1)
CHILTON No.3	0-6-0ST	IC	P	1219	1910	New	(1)
WINDLESTONE	0-4-0ST	OC	AB	1085	1907	(a)	(1)

(a) ex Pease & Partners Ltd, Eldon Colliery, Eldon.

(1) to Pease & Partners Ltd, with colliery, /1924.

ETHERLEY COLLIERY   Q498

This colliery comprised two pits, some distance apart. The **GEORGE PIT** (NZ 186299), near Escomb, was served by sidings north of the NER line between Bishop Auckland and Crook, about ¾ mile east of Etherley & Witton Park Station. The **JANE PIT** (NZ 175301) was situated at Witton Park, ¼ mile north of the station, opposite Witton Park Iron Works (see Bolckow, Vaughan & Co Ltd). Having initially been called Etherley Colliery, it became **Old Etherley Colliery** until about 1865, when the original name was re-adopted, though the O.S. maps continued to show "Old". Locomotives certainly shunted at the George Pit, and a loco shed is shown on maps; but there does not appear to have been a shed at the Jane Pit, so how this was shunted is unclear. DCOA Return 102 combines the two pits, and shows one loco in April 1871 and two in November 1876. The Jane Pit was abandoned in October 1894, but interestingly it had been re-opened by 1912, by which time the George Pit site had been re-named the **Rush Pit**. This closed in 1917, though the Jane Pit continued to work for a few more years, abandonment dates of 25th April 1925 and 9th March 1929 being recorded.

One steam rail crane was also used here, though at which pit is not known.

Gauge : 4ft 8½in

OLD ETHERLEY	0-4-0T	OC?	GW	175	1864	New	s/s
ETHERLEY	0-4-0ST	OC	HCR	129	1873	New	s/s
RESOLUTE	0-4-0ST	OC	K	1508	1868	(a)	(1)
NILE	0-4-0ST	OC	K	5115	1914	(b)	(2)

(a) origin and date of arrival unknown; originally Thomas Nelson, contractor.
(b) origin and date of arrival unknown.

(1) to Fishburn Colliery, Fishburn, c/1912.
(2) to Lingford, Gardiner & Co Ltd, dealers, Bishop Auckland.

FISHBURN COLLIERY & COKE WORKS, Fishburn   N499
NZ 361318 (colliery)

This colliery was opened in 1914 and was the last to be sunk in this south-eastern part of the coalfield. It was served by a three mile branch from the NER Ferryhill – Stockton line, ¾ mile north of Sedgefield Station. 50 Simplex regenerative coke ovens (NZ 361316) began production in May 1919, the coke being intended for the Skinningrove Iron Works in North Yorkshire, also owned by a subsidiary of Pease & Partners Ltd. The adjacent by-products plant manufactured tar, sulphate of ammonia, crude naphthalene and crude benzole. Both the ovens and the by-products plant were shunted by the colliery locomotives. However, later in 1919 the colliery was flooded and it was not re-opened until 1922. The colliery and its coking plant passed to NCB Northern Division No. 4 Area on 1st January 1947.

Gauge : 4ft 8½in

3	(previously 41)	0-4-0ST	OC	P	1194	1912	(a)	
		reb		TIW	June	1938		(1)
	RESOLUTE	0-4-0ST	OC	K	1508	1868	(b)	(2)
	-	0-4-0ST	OC	T.D.Ridley		?	(c)	(3)
1	(FISHBURN)	0-4-0ST	OC	P	1423	1916	New	
		reb		TIW	May	1936		(4)
	-	0-4-0ST	OC	BLW	45282	1917	(d)	(5)
25		0-4-0ST	OC	HL	2456	1900	(e)	(6)
1		0-6-0ST	OC	AB	961	1902	(f)	Scr
7		0-6-0ST	OC	AB	1097	1907	(f)	(7)
4		0-6-0T	OC	HL	3104	1915	(g)	
		reb		TIW	Dec	1936		(4)
20		0-4-0ST	OC	HL	2713	1907	(h)	
		reb		F'burn	Apr	1945		(4)

It is possible that a second Baldwin loco worked here, but there is no evidence for this.

(a) New here; said at the colliery to have been used in the sinking.
(b) ex Etherley Colliery, George Pit, Etherley, c/1912 (perhaps also used in the sinking of the pit).
(c) ex Newton Cap Colliery and Brickworks, Toronto.
(d) at least one Baldwin loco is said to have worked here, and the available evidence would suggest that it was Baldwin 45282, ex J.F. Wake, dealer, Darlington, /1921; previously War Department, Railway Operating Division, 78.
(e) ex Pease & Partners Ltd, Chilton Quarry, near Ferryhill Station, /1925, apparently on loan.
(f) ex Pease & Partners Ltd, Chilton Colliery, Chilton Buildings, c/1930.
(g) ex Pease & Partners Ltd, Bowden Close Colliery & Coke Ovens, Helmington Row.
(h) ex Pease & Partners Ltd, Bishop Middleham Quarry, Bishop Middleham.

(1) to Newton Cap Colliery and Brickworks, Toronto; returned by 17/10/1933; to NCB No. 4 Area, with colliery and coking plant, 1/1/1947.
(2) scrapped c/1915-1916.
(3) here only a few months; returned to Newton Cap Colliery and Brickworks, Toronto.
(4) to NCB No. 4 Area, with colliery and coking plant, 1/1/1947.
(5) one source records a loco from here, believed to be the Baldwin loco, was sent "to Stockton for scrap, /1928"; a Baldwin loco, rebuilt by Lingford, Gardiner & Co Ltd, dealers, Bishop Auckland, in 1929-1930 (it carried a LG plate) and "lying in Co.Durham", was offered for sale in the *Colliery Guardian* of 24/1/1930 by George Cohen, Sons & Co Ltd, contractors, London; it may be that these two entries refer to the same loco, and that this was the Baldwin loco that worked at Fishburn.
(6) returned to Pease & Partners Ltd, Chilton Quarry & Limeworks, Ferryhill, by 15/8/1931.
(7) to an unknown firm at Spennymoor, for scrap, c/1934.

NEWTON CAP COLLIERY, COKE WORKS & BRICKWORKS, Toronto Q500
NZ 213308

The NER Durham – Bishop Auckland branch was opened on 1st April 1857, and the sinking of this colliery began on 19th November 1857. It was served by ¼ mile branch 1¼ miles north of Bishop Auckland Station. A brickworks had been developed by 1860. In 1913-1914 a battery of 44 Simplex regenerative by-product coke ovens was built here, starting up on 14th September 1914, although the beehive ovens continued to operate until 30th April 1918. The by-product ovens worked until March 1926, but did not re-open after the miners' strike. The colliery was sold to **Herbert Dunn & Alexander W.Miller** (which see) on 17th October 1933.

Gauge : 4ft 8½in

THE COLONEL	0-4-0ST	OC	HG	251	1867	New	(1)
COMET	0-4-0ST	OC	MW	467	1873		
	reb		LG		1902	(a)	(2)
-	0-4-0ST	OC	T.D.Ridley		?	(b)	(3)
TORONTO	0-4-0ST	OC	P	1194	1912	(c)	(4)

(a) ex Lingford, Gardiner & Co Ltd, dealers, Bishop Auckland, initially on hire but then purchased, /1902; originally NER 959, 959 class; later re-numbered 1833, 1800, 2056 and finally 2252, on the NER Duplicate list; also named AUCKLAND.
(b) ex T. D.Ridley, Middlesbrough, Yorkshire (NR) (initially on hire?).

(c)   ex Fishburn Colliery and Coke Works, Fishburn.

(1)   to The North Bitchburn Coal Co Ltd, Rough Lea Colliery & Brickworks, New Hunwick, after 5/1/1924, and returned; to Dunn & Miller, with site, 17/10/1933.

(2)   to The North Bitchburn Coal Co Ltd, Rough Lea Brickworks, New Hunwick, by 15/8/1931, and returned; to Dunn & Miller, with site, 17/10/1933.

(3)   to Fishburn Colliery & Coke Works, Fishburn; there only for a few months, then returned here; s/s by 17/10/1933.

(4)   to Fishburn Colliery & Coke Ovens, Fishburn, by 17/10/1933.

143. THE COLONEL, HG 251/1867, at Newton Cap Colliery, Toronto, near Bishop Auckland, on 5th January 1924. Note the smokebox wingplates, the long handle for operating the sandbox, the large wheels for such a locomotive, 3ft 11½in diameter – and the small door near the makers' plate.

THRISLINGTON COLLIERY & COKE OVENS, and	**N501**
WEST CORNFORTH CHEMICAL WORKS, West Cornforth	**N502**
The North Bitchburn Coal Co Ltd until c8/1932 – see below	NZ 309338 (colliery)

Thrislington Colliery had been sunk in 1867 and had two previous owners before being acquired by The North Bitchburn Coal Co Ltd in 1910. Like Henry Stobart & Co Ltd, this company also became a subsidiary of Pease & Partners Ltd. With the North Bitchburn company in serious financial difficulties, its Directors agreed on 16th August 1932 to sell Thrislington Colliery, together with its coke ovens, closed in 1930, to Henry Stobart & Co Ltd, who would then pay £119,751 to Pease & Partners Ltd in lieu of the North Bitchburn company's debt to its parent company. This transfer was reported as completed at the North Bitchburn Directors' next meeting on 27th September 1932.

Coke production, from 55 Koppers regenerative ovens, was re-commenced in the 1930s. Adjacent to the coke ovens was the **West Cornforth Chemical Works** of **Dent, Sons & Co Ltd**, which produced various tar products, sulphate of ammonia, crude naphthalene and crude benzole. It would appear that its traffic was shunted by locomotives from the colliery. This company was taken over by Henry Stobart & Co Ltd on 1st January 1936. Also adjacent was the slag works of **William Barker & Co** (which see), which used tar from the chemical works to manufacture tarmacadam. Production here is believed to have ended in 1938.

The complex was served by ½ mile branch from Coxhoe Bridge Station on the NER Spennymoor – Hartlepool line, and by a ¾ mile link to the NER Durham – Darlington line, ¾ mile north of

Spennymoor Station. This was a rope incline, worked by a **stationary engine**, although the full wagons were travelling downhill.

The colliery, the coke works and the chemical works were all vested in NCB Northern Division No. 4 Area on 1st January 1947.

Gauge : 4ft 8½in

ISABELLA	0-4-0ST	OC	BH	976	1889	(a)		Scr /1935
THE SIRDAR	0-4-0ST	OC	CF	1187	1899	(a)		Scr /1935
9 (previously 39)	0-4-0ST	OC	HL	2247	1892	(a)		
	reb		TIW	June	1935			(1)
7	0-4-0ST	OC	P	677	1897	(a)		
	reb		TIW	Apr	1935			(2)
6	0-4-0ST	OC	AB	1085	1907	(a)		
	reb		TIW	Jan	1938			(2)
8 (previously 38)	0-4-0ST	OC	HL	2798	1910	(b)		
	reb		TIW	Sept	1935			(2)
10	0-4-0ST	OC	HL	2559	1903			
	reb		TIW	Apr	1937	(c)		(3)
13	0-4-0ST	OC	HL	2823	1910	(d)		(2)
25	0-4-0ST	OC	HL	2456	1900	(e)		(2)

(a) ex The North Bitchburn Coal Co Ltd, with site, c8/1932.
(b) ex Pease & Partners Ltd, St. Helen's Colliery, St. Helen Auckland, /1933.
(c) ex Pease & Partners Ltd, Bishop Middleham Quarry, Bishop Middleham, /1938.
(d) ex Pease & Partners Ltd, Pease's West, Crook, c3/1945.
(e) ex Pease & Partners Ltd, Chilton Quarry, Ferryhill Station.

(1) to Pease & Partners Ltd, Bishop Middleham Quarry, Bishop Middleham, and returned; to Tees Ironworks, Cargo Fleet, Yorkshire (NR), 12/1946.
(2) to NCB No. 4 Area, with colliery and coking plant, 1/1/1947.
(3) to Pease & Partners Ltd, Thorne Colliery, Moorends, near Doncaster, Yorkshire (ER), /1941.

STRAKERS & LOVE LTD

Strakers & Love until 13/5/1925; **Straker & Love** until /1865

This partnership, between Joseph Straker (1784-1867), a shipbuilder and timber merchant in Northumberland, and Joseph Love (1796-1875), a timber merchant in Durham, was formed in 1838. Straker seems to have provided the capital, with Love being the managing partner. With the opening of the NER Bishop Auckland branch in 1857 the partnership leased an additional 6000-7000 acres, which expanded the business considerably. Straker brought his son John (1815-1885) into the business in 1865, and the latter subsequently took a brief interest in Chilton Colliery (which see). Love was also a partner in Ferens & Love (which see) and Joseph Love & Partners (which see). Besides the collieries below, the partnership also owned **BITCHBURN** (or **BEECHBURN**) **COLLIERY**, near Crook (NZ 162346) **(M503)**; this was sold in 1870. John Straker's interest later passed to his son John Coppin Straker (1847-1937), who also acquired colliery interests in Derbyshire. The firm was the last of the major colliery owners in the Durham coalfield to convert itself into a limited company.

BRANCEPETH COLLIERY & BEEHIVE COKE OVENS, Willington — M504
NZ 205357 (colliery)

Although the West Durham Railway was only a small line between an end-on junction with the Clarence Railway at Byers Green and Billy Row, near Crook, and was worked entirely by rope inclines, its opening throughout on 15th June 1841 provided a major stimulus to coal mining in the area. The partnership began in this area with Willington Colliery (see below), following this immediately with the sinking of this colliery. Coal was reached here in April 1842, and the colliery was connected to the West Durham Railway at Tod Hills Bank Foot by a 1¼ mile branch. This included a **self-acting incline** about one mile long and a bridge over the River Wear. However, when the NER Durham-Bishop Auckland branch was opened on 1st April 1857 the colliery was linked to it instead, ½ mile north of Willington Station, and the link to the West Durham line was lifted except for the first ½ mile, which was retained to serve a coal depot in the village; later the route became the road to the colliery.

By the 1870s this colliery and Oakenshaw Colliery (see below) were regarded as one unit, this colliery being Brancepeth 'A' and 'C' Pits (the latter sunk in 1866) and Oakenshaw also known as Brancepeth 'B' Pit. It is assumed that the colliery was shunted by horses before the arrival of the first locomotive

144. BRANDON, MW 200, in 1946. Built in 1867, she passed to the National Coal Board and was scrapped about 1948.

below; but as well as listing two locomotives, DCOA Return 102 includes two **stationary engines** for both April 1871 and November 1876. What was presumably one of these is mentioned in a description of the colliery in the Supplement to the *Mining Journal* on 5th November 1870: "a wagon engine hauls part of the coal in wagons from the A Pit and drops them back to be discharged into coal hoppers at the foot of the Oakenshaw Incline". How the second stationary engine was used is not known.

145. A rare photograph of a R & W Hawthorn 0-4-0ST, STAGSHAW, RWH 1821/1880. Both STAGSHAWs carried a brass stag on the front of the chimney. Note the sparse protection for the driver.

146. The makers' photograph of HELMINGTON, RWH 1882/1882. Note the cover for the crosshead and its slides, the wooden brake blocks and the rodding to operate both sandboxes. Unlike STAGSHAW on previous page, this R. & W. Hawthorn locomotive was driven from the left hand side.

147. The same locomotive after being rebuilt to a four-coupled engine in 1897. The reversing lever is now on the right hand side, suggesting that she was now being driven from there. The brakes are mounted differently and she has also acquired a cab, albeit without much rear end protection; this was added later.

Soon after the opening of the colliery Love established that the coal had excellent coking qualities, and the coke here became some of the most famous in Durham; there were 870 beehive ovens here by 1903. In October 1910 60 Semet-Solway by-product ovens were started up, with 60 more phased in during 1911. They replaced some of the beehive ovens, although the last of these did not close until early in 1924. Nearby was a **brickworks**, certainly in existence by 1890 but probably dating from much earlier, making common bricks, together with a **sanitaryware works** and a **ganister brickworks**. The firm's main workshops were also developed here, carrying out all locomotive repairs. In 1937 a new coke ovens and by-product plant was opened (see below), and in 1940 Brandon Pit House Colliery (see below) was linked to Brancepeth A Pit by a three mile-long aerial ropeway, no doubt to allow Pit House coal to be used at the Brancepeth Coke Ovens.

Willington Colliery (see below) was linked underground to Brancepeth in 1939 and closed, while the 'B' Pit was closed in March 1940. The sanitaryware works and ganister works had also been closed by 1947, but the colliery, coking plant and the brickworks were vested in NCB Northern Division No. 5 Area on 1st January 1947.

In the lists below the "rebuilding" of locomotives was always carried out at Brancepeth Shops, though the actual dates of transfers for this work did not survive in the records still at the colliery in the early 1960s. In most cases the "rebuilding" comprised simply a new boiler or new firebox, and did not involve a major change of dimensions.

Gauge : 4ft 8½in

No.1	BRANCEPETH	0-6-0ST	IC	MW	104	1864	New	(1)
	BRANDON	0-6-0ST	IC	MW	200	1867	New	
		reb				1881		
		reb				1895		
		reb				1909		
		reb				1925		(2)
	TIGER	0-4-0ST	OC	MW	320	1870	New	
		reb				1881		
		reb				1891		
		reb				1901		(3)
	OAKENSHAW	0-4-0ST	OC	HCR	124	1872	(a)	(4)
	BRANCEPETH	0-6-0ST	IC	MW	775	1882	New	
		reb				1897		
		reb				1911		
		reb				1927		Scr 10/1937
	HELMINGTON	0-6-0ST	OC	RWH	1882	1882	New	
	reb	0-4-0ST	OC			1897		
		reb				1905		
		reb				1923		(5)
	HOWDEN DENE No.1 *	0-6-0T	OC	HL	2880	1911	New	(6)
	JUPITER	0-4-0ST	OC	MW	1880	1915	(b)	(7)
	WILLINGTON #	0-4-0ST	OC	AE	1509	1907		
		reb		TJR		1920	(c)	(8)
	STAGSHAW +	0-4-0ST	OC	RWH	1821	1880		
		reb				1891		
		reb				1906	(d)	Scr 4/1927
	STAGSHAW +	0-6-0ST	OC	HL	3513	1927	New	(2)
	MEADOWFIELD	0-6-0ST	OC	FW	289	1876		
		reb		RWH		1881		
		reb				1893		
		reb				1907	(e)	(9)
	HOWDEN DENE No.2	0-6-0ST	OC	RS	4113	1937	New	(2)

* named HOWDEN DENE until /1937.   # named TEST until /1921.
+ carried a home-made brass stag on the front of the chimney, fixed first to RWH 1821/1880 and then transferred to HL 3513/1927

There was also one steam rail crane here.

(a)   ex Brandon Colliery, Brandon, /1880.
(b)   ex Vickers Ltd, Barrow-in-Furness, Lancashire, 6/1920.
(c)   ex Topham, Jones & Railton, contractors, Crymlyn Burrows Depot, Swansea, Glamorgan, 12/1920.

(d)   ex Brandon Colliery by c/1926.
(e)   ex Brandon Colliery, Brandon, c/1933.
(1)   to Brandon Colliery, Brandon, /1880.
(2)   to Brandon Colliery, Brandon, by 24/3/1940.
(3)   to Willington Colliery, Sunny Brow, by 3/1890.
(4)   to Brandon Colliery, Brandon, possibly soon after delivery new; returned from Brandon Colliery, Brandon, /1881; to Oakenshaw Colliery, Oakenshaw, by 3/1890; returned from Willington Colliery, Sunny Brow, /1932; Scr c/1933.
(5)   to Willington Colliery, Sunny Brow, by /1905, and returned; to NCB No.5 Area, with the colliery and coking plant, 1/1/1947.
(6)   to Brandon Colliery, Brandon, 3/1925; returned, via RSH, Newcastle upon Tyne, by 24/3/1940; to NCB No.5 Area, with the colliery and coking plant, 1/1/1947.
(7)   to Brandon Colliery, Brandon, 4/1923.
(8)   to Brandon Colliery, Brandon, 10/1921.
(9)   compelling evidence was produced that this loco was scrapped at Brancepeth in 3/1934; despite this, it was reported as still at Brancepeth on 13/8/1935; s/s.

According to a report in 1951, a 0-6-0ST numbered 18 and built by HL was hired from the Pelton district in 1936. A possibility for this might be the Lambton Railway's 18, 0-6-0ST BH 32/1867, which was rebuilt by HL in 1935 and might have been working on the Beamish Railway, which ran to Pelton, but there is no evidence to support this.

BRANCEPETH COKE OVENS & BY-PRODUCT PLANT, Willington (1)   **M505**
NZ 208359

This was the first by-product plant at the colliery. It was built alongside the NER Durham-Bishop Auckland line, immediately north east of Brancepeth Colliery Junction. It consisted of two batteries of Semet Solvay ovens, the first 60 being started up in October 1910 and the second 60 phased in during 1911. The works included plants to wash coal and to recover tar and sulphate of ammonia, whilst the steam created was used to drive turbines and generators. As the coke was being discharged from an oven it was sprayed with water and slid down a bench into a coke car. Two coke cars, one for each battery, ran on track in the centre of the plant, at the end of which was a hoist for raising the cars to the top of the coke screens, where the coke was discharged and screened into three sizes and dropped into railway wagons.

The description of the plant in the *Colliery Guardian* for 24th November 1911 continues: "The cars are propelled by electrically-operated locomotives, furnished with long stiff bars and automatic couplings. No operator travels with the locomotives; they are remote-controlled from suitable equipment placed in a "pulpit" in the centre of the plant……... Indicators and signals are provided to locate each car, and in the same "pulpit" under the control of the same operator, are assembled all the gear for operating the hoist………. The electrical equipment was supplied by the Electric Construction Company of Wolverhampton, who also planned the remote-controlled locomotives."

This installation would appear to be unique in the Durham coke industry, and would seem to be the first time that locomotives were used on the bench side of the ovens. Nothing more about the locomotives has come to light to add to the description above. It may well be that the Electric Construction Company at Wolverhampton did not build them, and it can probably safely be assumed that they had four wheels. The locomotives presumably lasted until the closure of the plant in 1937 and its subsequent demolition.

Gauge : unknown

| - | see note above | 1910 | New | (1) |
| - | see note above | c1910 | New | (1) |

(1)   survived until closure of the plant in 1937 and its subsequent demolition?

BRANCEPETH COKE OVENS & BY-PRODUCT PLANT, Willington (2)   **M506**
NZ 210360

This plant was opened in 1937 and lay to the north-east of the Semet-Solway ovens, which it replaced. It was operated by **Brancepeth Gas & Coke (Strakers & Love) Ltd**, a subsidiary company of Strakers & Love Ltd.

The locomotives below were leased to Strakers & Love Ltd and used during the construction of this plant between 1935 and 1937. It is not clear why the locomotives were leased to Strakers & Love Ltd rather than the contractors.

Gauge : 2ft 0in

-	4wDM	RH	175121	1935	(a)	(1)
-	4wDM	RH	175399	1935	(b)	(2)

(a) ex RH, 23/7/1935, leased (new loco).
(b) ex RH, 8/10/1935, leased (new loco).

(1) returned to RH; to George Porter, Bellshill, Lanarkshire, Scotland; later to Guanogen Ltd, Castle Bromwich, Warwickshire, via George Cohen, Sons & Co Ltd, dealers, Kingsbury, Warwickshire.
(2) returned to RH, 30/9/1936; to Rowhedge Sand & Ballast Co Ltd, Birch Brook Pits, Essex, 11/1936.

The new plant comprised 59 Woodall-Duckham Becker ovens, and in addition to manufacturing coke and supplying surplus gas to Spennymoor, Bishop Auckland and Darlington, it produced tar products, concentrated ammonia liquor and refined benzole. It is believed to be the third example in Durham of what became the conventional use of an electric locomotive to propel the coke car under the quenching tower for the coke to be cooled, and to haul it back again for the coke to be discharged on to a bench, whence the coke slid down on to a conveyor to be removed. An electricity generating station was also built. As with the earlier plant, the colliery locomotives shunted the site.

The plant continued in the same private ownership after the creation of the National Coal Board in 1947, perhaps because of its supply of town gas, but it was eventually purchased by the NCB in 1949.

Gauge : 4ft 8½in

-	0-4-0WE	HL	3859	1937	New	(1)

(1) to NCB No. 5 Area, with the plant, /1949.

BRANDON COLLIERY

This comprised the **A & B Pits** at Brandon (NZ 244401) **(K/M507)** and the **C Pit** at Meadowfield (NZ 245397) **(K/M508)**.

The sinking of the **A PIT** was begun on 24th October 1844, but initially this was only a landsale colliery. With the planned opening of the NER Bishop Auckland branch in 1857 the colliery was greatly expanded. The A Pit was sunk deeper in 1856 to begin Brandon Colliery proper, to be followed by the **B PIT** a few yards away and in 1860 by the **C PIT** on the opposite side of the NER, at Meadowfield. The complex was ½ mile north of Brandon Station. As at Brancepeth Colliery, **coke ovens** (at both sites), a **firebrick works** and a **sanitaryware (pipe) works** were developed, the brickworks being certainly open by the 1870s. 649 beehive coke ovens were built here, the last of them closing down at the beginning of the miners' strike in March 1921.

To work the remaining coking coal in the Busty and Brockwell seams to the west of Brandon, where the Five Quarter and Main Coal seams were also untouched, the firm began the sinking in 1923 of

BRANDON PIT HOUSE COLLIERY, near Brandon K/M509
NZ 215404

The new colliery was situated in a remote area some 350 feet higher than Brandon C Pit. To serve it a 2½ mile branch was built from exchange sidings at Meadowfield on the opposite side of the LNER from the C Pit. It would appear that the line was opened in 1922 to serve the boreholes then being sunk, but that work was then suspended for two years, although the first sod for the shaft was cut on 18th April 1923. It took a further two years to complete the sinking and the branch, and the colliery was eventually opened in April 1926, but then immediately closed again because of the 1926 miners' strike. The line was the most steeply-graded locomotive-worked line in Durham, with a maximum gradient of 1 in 18, which in turn restricted the downhill load to nine full wagons. With no new houses provided, the firm was obliged to provide a **passenger service** over the branch for the miners working at the colliery. One vehicle was certainly a four-wheeled Maryport & Carlisle Railway 4-compartment carriage (3rd/3rd/1st/2nd), whose remains survived until at least 1950, though two four-wheeled North British Railway coaches are also quoted; later the miners travelled in two four-wheeled box vans fitted with seats built by the company. Latterly the regular locomotive on the passenger service was LEAZES. In 1931 a second shaft was sunk, making the colliery a separately-ventilated colliery. In 1940 the colliery was linked by a three mile aerial ropeway to Brancepeth Colliery, presumably to allow its coal to be used at the Brancepeth Coke Ovens.

The 2nd edition O.S. map of 1897 appears to show loco sheds at both the A/B and C Pits, but on the 3rd edition in 1919 there is only a shed at the C Pit. On the 4th edition (1939) this has gone, and the old shed at the southern end of the A/B Pits' exchange sidings had been re-opened (NZ 243399), doubtless in order to work traffic to and from Pit House, where no loco shed was provided.

Both collieries, together with Brandon Brickworks, were vested in NCB Northern Division No. 5 Area on 1st January 1947.

Gauge : 4ft 8½in

	TIGER	0-4-0ST	OC	MW	320	1870	New	(1)
	OAKENSHAW	0-4-0ST	OC	HCR	124	1872	New	(2)
No.1	BRANCEPETH	0-6-0ST	IC	MW	104	1864	(a)	(3)
	STAGSHAW +	0-4-0ST	OC	RWH	1821	1880	New	
		reb				1891		
		reb				1906		(4)
	MEADOWFIELD	0-6-0ST	OC	FW	289	1876		
		reb		RWH		1881	(b)	
		reb				1893		
		reb				1907		(5)
	WILLINGTON	0-6-0ST	IC	H&C	45	1865	(c)	s/s c/1920
	WILLINGTON	0-4-0ST	OC	AE	1509	1907		
		reb		TJR		1920	(d)	(6)
	JUPITER	0-4-0ST	OC	MW	1880	1915	(e)	(7)
	HOWDEN DENE No.1 *	0-6-0T	OC	HL	2880	1911	(f)	(8)
	LEAZES	0-6-0ST	OC	HL	3830	1934	New	(6)
	BRANDON	0-6-0ST	IC	MW	200	1867		
		reb				1881		
		reb				1895		
		reb				1909		
		reb				1925	(g)	(6)
75256		0-6-0ST	IC	WB	2779	1945	(h)	(6)

* HOWDEN DENE until /1937. + carried a home-made brass stag on the front of the chimney

(a) ex Brancepeth Colliery, Willington, /1880.
(b) ex RWH, /1881; the loco was new to Worcester, Bromyard & Leominster Railway, who purchased it to hire, with an option to purchase, to their contractor, W. Ridler, for his work on the construction of the Yearsett to Bromyard section of the Railway. On the completion of this section Ridler took over its maintenance, but in 1879 this contract was discontinued, and with Ridler not having taken up the purchase option, the Railway offered the locomotive for sale about the summer of 1879. What happened to it between 1879 and 1881 is not known.
(c) ex Willington Colliery, Willington.
(d) ex Brancepeth Colliery, Willington, 10/1921.
(e) ex Brancepeth Colliery, Willington, 4/1923.
(f) ex Brancepeth Colliery, Willington, 3/1925.
(g) ex Brancepeth Colliery, Willington, loan, by 24/3/1940.
(h) ex War Department, Longmoor Military Railway, Hampshire, 75256, 4/1946.

(1) to Brancepeth Colliery, Willington, /1881.
(2) to Brancepeth Colliery, Willington, /1880; returned, from Oakenshaw Colliery, /1893; to Willington Colliery, Sunny Brow, c/1914.
(3) withdrawn from traffic, /1882; to MW, c/1886; rebuilt as MW 966/1888 and sold to T.A. Walker, contractor, Barry Dock contract, Barry, Glamorgan, 5/1888.
(4) to Brancepeth Colliery, Willington, by c1926.
(5) to Brancepeth Colliery, Willington, c/1933.
(6) to NCB No. 5 Area, 1/1/1947.
(7) loaned to The Witton Park Slag Co Ltd, Etherley Works, Witton Park, /1931, and returned, /1931; to NCB No. 5 Area, 1/1/1947.
(8) loaned to Synthetic Ammonia & Nitrates Ltd, Billingham, /1928 (see Cleveland & North Yorkshire Handbook); returned by 2/1929; withdrawn from traffic, /1937; sent to RSH, Newcastle upon Tyne, for repairs, /1939; to Brancepeth Colliery, Willington, by 24/3/1940.

OAKENSHAW COLLIERY, Oakenshaw   M510
NZ 200376

This colliery was also known as **BRANCEPETH B PIT**. It was won in November 1854, and was linked to Brancepeth Colliery's sidings by a 1¼ mile line, worked by a **stationary engine** (NZ 204366) about ¾ mile from the colliery. The first incline, from the colliery to the engine house, was single line, so that full and empty sets must have been worked alternately, while on the shorter bank there was a meetings, so that sets must have been run simultaneously. At the bank foot the line divided, one fork ending at screens where coal was unloaded for Brancepeth's coke ovens, the other linking round to Brancepeth Colliery. Production ceased in March 1940, though the colliery continued as a pumping station. Locomotives do not appear to have been used here normally, and it is likely that a stationary engine at the back of the pit hauled wagons up to itself, for them to be run through the screens as required; but the following locomotive is recorded as having worked here:

Gauge : 4ft 8½in

OAKENSHAW	0-4-0ST	OC	HCR	124	1872	(a)		(1)

(a)   ex Brancepeth Colliery, Willington, by 3/1890.

(1)   to Brandon Colliery, Brandon, /1893.

WILLINGTON COLLIERY, Sunny Brow   M511

The first colliery here, later known as the **Old Pit** (NZ 199341), was sunk by **The Northern Coal Mining Co**. It was being developed during the construction of the West Durham Railway in 1839, to which its sidings were linked at Tod Hills Bank Foot, and the carrying of its coal marked the opening of the first section of the Railway on 12th June 1840, although another press extract records the first coal being dispatched on 27th October 1840. The colliery was acquired by Strakers & Love after the bankruptcy of its original owners in 1848. By 1857 the **New Pit** (NZ 191346) had been sunk, 300 yards to the north-west and served by an extension from the Old Pit. Probably about the same time it was linked to the NER Bishop Auckland branch, ½ mile south of Willington Station; the link to the West Durham Railway being subsequently abandoned. The colliery, which had 277 beehive coke ovens by 1903 and a **brickworks**, was also known as **SUNNY BROW COLLIERY**, and was later called **BRANCEPETH COLLIERY Z PIT**. By 1870 there was also a drift, worked by main-and-tail rope haulage.

For both April 1871 and November 1876 DCOA Return 102 lists a locomotive here, and in April 1871 a **stationary engine** is also included, although it is not listed in November 1876. Its purpose is not known.

The colliery was latterly linked underground to Brancepeth Colliery and was combined with it from 1939, and its surface buildings closed.

Gauge : 4ft 8½in

WILLINGTON	0-6-0ST	IC	H&C	45	1865	New	(1)
TIGER	0-4-0ST	OC	MW	320	1870		
	reb				1881	(a)	
	reb				1891		
	reb				1901		(2)
HELMINGTON	0-6-0ST	OC	RWH	1882	1882		
reb	0-4-0ST	OC			1897		
	reb				1905	(b)	(3)
OAKENSHAW	0-4-0ST	OC	HCR	124	1872	(c)	(4)

(a)   ex Brancepeth Colliery, Willington, by 3/1890.
(b)   ex Brancepeth Colliery, Willington, by /1905, probably c/1898; the loco was rebuilt to 0-4-0ST specifically to negotiate the tight curves at Sunny Brow Colliery.
(c)   ex Brandon Colliery, Brandon, c/1914.

(1)   to Brandon Colliery, Brandon.
(2)   to Low Laithes Colliery Co Ltd, Wakefield, Yorkshire (WR), after 11/1926?
(3)   to Brancepeth Colliery, Willington.
(4)   to Brancepeth Colliery, Willington, /1932.

RICHARD SUMMERSON & CO LTD
formerly **Richard Summerson & Co**

This firm was another set up by the Summerson family of Darlington (see Summerson's Foundries Ltd below). It began as colliery owners, apparently in 1862, owning Vane's Hartley Colliery at Cockfield. This closed in 1883, by which time the firm owned Millfield Grange Colliery, also near Cockfield. Soon afterwards this firm moved into quarrying, and eventually gave up its colliery interests; however, the family continued them through Summersons Ltd (see The Wigglesworth Colliery Co Ltd).

COCKFIELD & CRAG WOOD QUARRIES, Cockfield   P512

It would appear that quarrying here for whinstone was first carried out south of the NER Haggerleazes branch, and then moved southwards up on to Cockfield Fell, where the firm developed **COCKFIELD QUARRY** (NZ 134247 approx.), north of the village of Cockfield, a narrow quarry working north-westwards. This was followed by **CRAG WOOD QUARRY** (NZ 140244 approx.), near Esperley Lane Ends, an equally narrow quarry working south-eastwards (on O.S. maps the spelling is always **Cragg Wood**). The two quarries were not physically connected except by a narrow gauge tramway running from a discharge point on sidings south of the Haggerleazes branch about ½ mile east of Haggerleazes Goods Station. Near to the entrance to Cockfield Quarry a branch curved away to enter Crag Wood Quarry. The loco shed was alongside the discharge point, near Lands Bridge. By 1939 Crag Wood was closed, and Cockfield had extended right up to the LNER, absorbing the site of North End Colliery en route and developing quarrying in small areas northwards from the main quarry.

In the edition of *Machinery Market* of 27th February 1942 the firm, from its address at Barnard Castle, offered two 2ft 4in gauge locomotives for sale, which must clearly be the two below. Production seems finally to have ended about 1946.

Gauge : 2ft 3½in

W.SUMMERSON	0-4-0ST	OC	HE	567	1892	New	(1)
ROSEBERRY	0-4-0ST	OC	BH	1065	1892	(a)	(2)

(a) ex Gribdale Mining Co Ltd, Great Ayton, Yorkshire (NR), /1926 (had been offered for sale by A.T. & E.A. Crow on behalf of this company, 25/8/1926).

(1) offered for sale, 27/2/1942; s/s.
(2) to Greenfoot Quarry, Stanhope, and returned; whether the offer of the loco for sale on 27/2/1942 was before or after these moves is not known; scrapped here c12/1950.

GREENFOOT QUARRY, Stanhope
V513
NY 983392 approx.

On 21st October 1895 the NER opened the extension of its Wear Valley branch, from Stanhope to Wearhead, and this stimulated industrial activity in this area. This quarry was started by **Ord & Maddison** (which see), certainly by 1897, and was served by sidings which ran right into the quarry, north of the NER line 1¼ miles west of Stanhope Station. Initially it was worked for limestone and whinstone, but production of the former ended in 1907. The narrow gauge system below was used within the quarry, running from the discharge point northwards before dividing to serve the various quarry faces. About 1948 the quarry was taken over by the **Durham & Yorkshire Whinstone Co Ltd** (which see), but it was subsequently re-purchased by Richard Summerson & Co Ltd. In October 1963 the firm's share capital was purchased by Ord & Maddison Ltd, itself by then a subsidiary of Tarmac Roadstone Ltd (which see).

Gauge : 2ft 3½in

ROSEBERRY	0-4-0ST	OC	BH	1065	1892	(a)	(1)
-	4wDM		RH	175420	1936	(b)	(2)

(a) ex Cockfield Quarry, Cockfield.
(b) ex The West Hunwick Silica & Firebrick Co Ltd, Hunwick.

(1) to Cockfield Quarry, Cockfield.
(2) to Durham & Yorkshire Whinstone Co Ltd, with quarry, c/1948; returned, with quarry; to Tarmac Roadstone Ltd, with quarry, 10/1963.

SUMMERSON'S FOUNDRIES LTD
subsidiary of **Summerson's Holdings Ltd**; subsidiary of **Thomas Summerson & Sons Ltd** until 31/3/1953

ALBERT HILL FOUNDRY, Albert Hill, Darlington   **S514**
Thomas Summerson & Sons Ltd until 9/1947   NZ 294155
Thomas Summerson & Sons until 17/5/1900; see also below

Summersons always advertised their business as starting in 1840, but this is misleading. 1840 was the year in which **John Harris** (1812-1869) set up his **Hope Town Foundry** (NZ 274172), situated alongside the Stockton & Darlington Railway, ½ mile north of the first Darlington (North Road) Station. He began by contracting for the supply of railway track, with **Thomas Summerson**, who had helped with the construction of the Stockton & Darlington Railway, which opened in 1825, as his chief foreman. By the 1860s Harris had expanded his business into the repair and construction of tank locomotives and railway wagons.

After Harris' death in 1869 Summerson acquired the business, and in 1872 he was joined by his three sons, forming **Thomas Summerson & Sons**. However, the Hope Town Foundry was absorbed into the Whessoe Foundry of Charles I'Anson & Co Ltd (see Whessoe Products Ltd), which lay immediately to its north, while the Summersons moved the business to the new **Albert Hill Foundry**, ½ mile to the south-east of the (new) Darlington (North Road) Station. The new foundry reverted to concentrating solely on the manufacture of railway trackwork, points and crossings. Sometime after 1906 the former premises of the Darlington Wagon & Engineering Co Ltd were incorporated into the works (see Thomas Summerson & Sons Ltd below), and in 1919 the company purchased from its northern neighbours, The Darlington Forge Ltd, the foundry in the area of the latter's premises once owned by William Barningham (see The Darlington Forge Ltd).

The firm undertook at least one track-laying contract, at the War Department's Cannock Chase Military Railway in Staffordshire about 1914, and it may also have been involved with extensions to private railways in Cheshire in the early 1920s.

In 1946 the company decided to develop a new site at Spennymoor for its trackwork business, leaving the Albert Hill works to concentrate on foundry work and setting up **Summerson's Foundries Ltd** as a subsidiary company to run this business. On 31st March 1953 this company became a subsidiary of Summerson's Holdings Ltd.

The company is known to have had more locomotives than those listed below, including at least one before the first in the list; the details that are known are given below the loco list. In 1960 shunting was taken over by a rail crane.

On 25th October 1967 all three Summerson companies were put into receivership and subsequently ceased trading. The Albert Hill works was eventually purchased from the receiver sometime late in 1968 or early in 1969 by F.H. Lloyd & Co Ltd of Wednesbury in Staffordshire, which set up **Lloyd's (Darlington) Ltd** (which see) to re-open it.

When John Harris died in 1869 two locomotives at his premises were offered for sale; both were six-coupled, the first with 12½in x 20in cylinders and 3ft 6in wheels, the second with 15in x 18in outside cylinders. Whether Thomas Summerson took either of these over with the works is not known.

Gauge : 4ft 8½in

-	0-4-0ST	OC	?	?	?	(a)	s/s
-	0-4-0TG	VC	?	?	?	(a)	s/s
-	0-6-0ST	IC	MW	1513	1901	(b)	(1)
HERALD (formerly HASKIN)	0-4-0CT	OC	HL	2468	1900	(c)	Scr
JEANIE	0-6-0T	IC	HC	694	1904	(d)	(2)
-	4wVBT	VCG	S	6076	1925	(e)	(3)
TOM BARRON	0-4-0ST	OC	AB	1287	1914	(f)	(4)

(a) origin and date of arrival unknown; said to have been built by GW or HG.
(b) ex John Scott, contractor.
(c) ex Northern Wood Haskinising Co Ltd, Walker, Newcastle upon Tyne.
(d) ex T. Wrigley, Prestwich, Lancashire, dealer, hire, 2/1922.
(e) ex Derwent Valley Light Railway, Yorkshire (ER), /1927
(f) ex Darlington Corporation, Darlington Gas Works, Darlington, hire, /1940

(1) to War Department, Cannock Chase Military Camp Railway, Staffordshire.
(2) returned to T.Wrigley, Prestwich, Lancashire, 4/1922.
(3) to Lloyd's (Darlington) Ltd, with site, c/1968 (out of use).
(4) returned to Darlington Corporation, Gas Department, Darlington, /1940.

In *The Engineer* for 14/6/1872 the firm offered for sale a geared tank locomotive. This may well have been a four-coupled locomotive. No other details are known.

In *Contract Journal* for 23/3/1910 and also for 4/1/1911 the firm offered for sale a saddletank loco with 10in x 18in cylinders. Again, nothing further is known.

According to the managing director of T.J.Thomson & Son Ltd, the Stockton scrap merchants, his firm cut up "an old diesel locomotive" at Summerson's about 1960. So far as is known, Summerson's never had a diesel locomotive here, and the report remains unsubstantiated.

THOMAS SUMMERSON & SONS LTD
subsidiary of **Summerson's Holdings Ltd** from 31/3/1953
SPENNYMOOR WORKS, Spennymoor   **N515**
                                                                                          NZ 267337 approx.

In 1946 the company decided to move its manufacture of railway trackwork, points and crossings from its Albert Hill works in Darlington to a 25 acre site on the Trading Estate at Spennymoor, retaining the original company to run the latter but setting up Summerson's Foundries Ltd to run the foundry business continuing at Darlington. This new works, which incorporated part of the former Royal Ordnance Factory here (see Ministry of Supply), was served by sidings south of the BR Spennymoor – Ferryhill line, one mile east of Spennymoor Station. It was probably shunted by BR before the arrival of the locomotive below.

Rail traffic ceased about 1965. Like the Albert Hill works above, this works went into receivership on 25th October 1967, to be acquired late in 1968 or early in 1969 by F.H. Lloyd & Co Ltd of Wednesbury in Staffordshire, which set up Lloyd's (Darlington) Ltd to re-open it. It was finally closed about 1975.

Gauge : 4ft 8½in

-		0-4-0DM	JF	4110008	1950	New	(1)

(1) to Leslie Sanderson Ltd, dealer, Birtley, c/1966; re-sold to Associated Portland Cement Manufacturers Ltd, Penarth Works, Penarth, Cardiff, Glamorgan, after 7/1967.

WAGON WORKS, Darlington   **S516**
                                                                                            NZ 295155

An entry in *Machinery Market* advertised an auction by Willman & Douglas on 31st October 1906 on behalf of Thomas Summerson & Sons Ltd at the "Wagon Works, West Street, Darlington". However, there is no West Street in Darlington, and this is clearly an error for York Street. Summerson's Albert Hill works fronted on to York Street, and on the opposite side of the street was the Wagon Works of the **Darlington Wagon & Engineering Co Ltd** (see Blake Boiler, Wagon & Engineering Co Ltd). This works was served by a siding from a curve from Albert Hill Junction to the NER Darlington – Durham line. It closed down about 1905 and the premises were acquired by Summerson's for an extension to their works.

Despite the works being quite small, the auction included three locomotives, with 12in, 11in and 6in cylinders, manufactured by Black, Hawthorn & Co Ltd, The Hunslet Engine Co Ltd and Hopkins, Gilkes & Co Ltd, presumably all standard gauge. Their identity and fate is unknown.

When the works was incorporated into Summerson's premises the former link from the NER was replaced by a link from Summerson's.

THE SUNDERLAND DOCK COMPANY
SOUTH DOCK, Sunderland   **H517**

As the coal trade grew in the nineteenth century there was a steeply-rising demand for dock accommodation. At Sunderland this was first met by the construction of the **North Dock**, owned by the Wearmouth Dock Company and opened on 1st November 1837. It soon proved very inadequate (see the entry for the Port of Sunderland Authority), and a far larger dock, on the opposite side of the river, was soon proposed. This was put forward by the York & Newcastle Railway through a satellite company called The Sunderland Dock Company, formed in 1845 with the backing of the famous railway entrepreneur, George Hudson (1800-1871). The initial work, excavated from the rocks which lay to the south of the river mouth, comprised what was called at first 'The Great Dock', entered from the river by a tidal harbour, followed by a Half Tide Basin. It was to be served from the south by the Durham & Sunderland branch of the York & Newcastle Railway, which the latter had also acquired in 1845. The grand opening took place on 20th June 1850.

Almost immediately the Dock Company set about expanding the dock, with an extension to the south and a second Half Tidal Basin giving access from the sea. The contractors for this work were Messrs. Pawson & Dyson; but by early in 1852 the Dock Company had formed the opinion that the work was proceeding too slowly, and in May 1852 it purchased the contractors' plant, which included "steam and locomotive engines". However, the Dock Company clearly did not purchase everything, for some plant, including a locomotive, was auctioned at the contractors' offices on 31st May 1852 (see Part 2). The Dock Company certainly acquired one locomotive, for later in 1852 W.G. Armstrong of Newcastle upon Tyne sent his men there to repair it. In December 1852 the York, Newcastle & Berwick Railway opened its Pensher Branch from Washington to the dock, giving rail access from the west. One of the first customers was the Earl of Durham (see The Lambton, Hetton & Joicey Collieries Ltd). On 3rd August 1854 the extension of the Londonderry Railway from Seaham to the dock was opened, providing a third rail link (see the Marquis of Londonderry). The extension was opened on 24th November 1855 and the southern Half Tidal Basin and associated works on 5th March 1856, bringing the area enclosed up to 127 acres.

The 1st edition O.S. map, surveyed in 1855, shows an extensive rail system all round the dock and identifies **four engine houses**, one near the landward end of the South Pier at the mouth of the river, one near an iron foundry near the north-eastern corner of the bigger dock and two more near the southern Tidal Basin. Whether any of these were used for railway operations is not known. However, the Dock Company was certainly using a locomotive in 1858, for in the accounts listing expenditure for the half year ending 21st December 1858 there are entries for wharfage charges for the locomotive engine leading timber, etc., and ballast, and a locomotive engineman's wages.

Despite owning the dock the Dock Company did not control the river, and the **River Wear Commissioners**, established in 1717, were able to levy a charge on every vessel leaving it. This made the dock uneconomic, and on 1st August 1859 it was transferred by Act of Parliament to the Commissioners, who immediately began to operate it themselves. The locomotive owned by the Dock Company almost certainly passed to the Commissioners (see The Port of Sunderland Authority).

Gauge : 4ft 8½in (assumed)

At least one locomotive was acquired from Messrs. Pawson & Dyson in May 1852.
A locomotive was handed over to the River Wear Commissioners on 1st August 1859.
Whether the two locomotives are the same is unknown. The latter may have been built by Fossick & Hackworth (see entry for The Port of Sunderland Authority)

SWAN HUNTER SHIPBUILDERS LTD
part of **British Shipbuilders Ltd** from 1/7/1977 to 21/1/1986
Swan Hunter & Tyne Shipbuilders Ltd until 1/1/1968
HEBBURN YARD, Hebburn　　　　　　　　　　　　　　　　　　　　　　　　　　　　　　　　　　**E518**
NZ 305652 approx.

This yard had been opened by Andrew Leslie in 1854 and was latterly owned by **R. & W. Hawthorn Leslie (Shipbuilders) Ltd** (which see) until that company's absorption into Swan Hunter & Tyne Shipbuilders Ltd on 1st January 1968. It was served by a ¾ mile branch north of the BR Gateshead – South Shields line, ½ mile east of Hebburn Station. Immediately east of this yard was the Palmers Hebburn ship-repairing yard, latterly owned by Vickers Ltd (which see). This was taken over by Swan Hunter Shiprepairers Ltd, a sister company, on 19th August 1972, and latterly one of the locomotives below worked in this yard.

The two yards were put on to a care-and-maintenance only basis in 1984, and were finally closed in 1986. The former Hawthorn Leslie yard became largely derelict, but the former Palmers Hebburn yard was subsequently re-opened for ship-repairing; it finally closed in 2001.

Gauge : 4ft 8½in

TRIUMPH	0-4-0DM	RH	304472	1951	(a)	(1)
APOLLO	0-4-0DM	RH	319288	1953	(a)	s/s c/1988

There was also one diesel rail crane here.

(a)　ex R. & W. Hawthorn Leslie (Shipbuilders) Ltd, with yard, 1/1/1968.
(1)　all equipment was removed during 1981-1982; casing and frame passed to British Shipbuilders (Training, Safety & Education) Co, Hebburn (the training arm of British Shipbuilders Ltd), /1982, for conversion to a "steam outline play unit". This project was never completed, and the remains were s/s c/1988.

TARMAC NORTHERN LTD (registered 22/12/1995)
COXHOE QUARRY, Coxhoe　　　　　　　　　　　　　　　　　　　　　　　　　　　　　　　　　　**N519**
Raisby Quarries Ltd until 1/1/2001　　　　　　　　　　　　　　　　　　　　　　　NZ 343354 approx.

The new owners changed the name of the quarry from Garmondsway Quarry to Coxhoe Quarry, perhaps making its location better known. There had been no rail traffic here since the 1970s, with the locomotive below being abandoned in an increasingly-derelict building.

Gauge : 4ft 8½in

| | M14 | 4wDH | S | 10077 | 1961 | (a) | | (1) |

(a)　ex Raisby Quarries Ltd, with the quarry, 1/1/2001.
(1)　to Weardale Railway Trust, Wolsingham, for preservation, 14/12/2002.

TARMAC ROADSTONE (NORTHERN) LTD (registered 6/4/1960)
CORNFORTH QUARRY & LIMEWORKS, Cornforth　　　　　　　　　　　　　　　**N520**
Cornforth Limestone Co Ltd until 1/12/1976 (registered 30/4/1919)　　NZ 317347 approx.

The Cornforth Limestone Co Ltd was a member of the **Tarmac Group** from 1/9/1973; previously, from its inception, it had been a subsidiary of **F.W. Dobson & Co Ltd** (which see).

This quarry began production on 2nd May 1919, working similar deposits of limestone to those worked by Raisby/Garmondsway Quarry nearby (which see). It was served by sidings south of the NER line from Ferryhill to Hartlepool, ¾ mile east of West Cornforth Station. The limeworks closed sometime after the Second World War and latterly the quarry concentrated on road aggregate. The quarry was acquired by Tarmac Roadstone (Northern) Ltd in 1976. Rail traffic ceased on 19th July 1978. The quarry was bought by Raisby Quarries Ltd (which see) on 1st June 1983 and closed on 18th June 1983. Tarmac Roadstone (Northern) Ltd was dissolved on 18th November 2003.

Gauge : 4ft 8½in

-		4wPM	MR	2027	1920	New	(1)
-		4wPM	MR	2262	1929	(a)	
		reb 4wDM	MR		1934		(2)
-		4wDM	MR	5751	1937	New	(3)
JAMES BLUMER		4wDM	RH	236362	1946	New	(4)
R.F. SPALDING		4wDM	RH	262996	1949	New	(5)
CORNFORTH		4wDM	RH	306087	1949	New	Scr 5/1976
(FRED DOBSON)		4wDM	RH	326071	1954	New	(6)

(a)　built by MR for stock in 1923; to Cornforth Limestone Co Ltd as new loco, 12/9/1929.
(1)　last spares order from MR was delivered on 1/3/1929; s/s, but see footnote below.
(2)　rebuilt to diesel by MR, 3/1934; to G.W. Bungey Ltd, dealer, Hayes, Middlesex, 7/1/1949; to Dunlop Rim & Wheel Co Ltd, Coventry, Warwickshire, by 5/1/1950.
(3)　to Wagon Repairs Ltd, Gloucester.
(4)　to F.W. Dobson & Co Ltd, Chilton Quarry & Limeworks, near Ferryhill Station, by 10/8/1950.
(5)　to Raisby Quarries Ltd, with quarry, 1/6/1983.
(6)　to Hawthorn Limestone Co Ltd, Hawthorn Quarry, near Seaham Harbour, c/1965; returned, c5/1970; to Raisby Quarries Ltd, with quarry, 1/6/1983.

About 1930 MR purchased MR 3880 from the Cornforth Limestone Co Ltd, rebuilt it as MR 4231/1931 and resold it to Berry Wiggins & Co Ltd, Kingsnorth, Kent, 23/4/1931. Whether this locomotive has a connection with MR 2027/1920 above is not known.

In *Machinery Market* for 15th August 1947 and *Contract Journal* for 27th August 1947 the company advertised for sale a 65/85 h.p. 4wDM loco built by MR and overhauled by the makers in 6/1947, almost certainly MR 5751/1937 above.

GREENFOOT QUARRY, near Stanhope　　　　　　　　　　　　　　　　　　**V521**
Richard Summerson & Co Ltd until 10/1963　　　　　　　　　　　　NZ 983392 approx.

This quarry was started about 1897 by **Ord & Maddison Ltd** and was subsequently acquired by **Richard Summerson & Co Ltd** (which see). It was sold to **The Durham & Yorkshire Whinstone Co Ltd** in 1948, but subsequently re-purchased by **Richard Summerson & Co Ltd**. In October 1963 the share capital of this firm was acquired by Ord & Maddison Ltd, by then a subsidiary of Tarmac Roadstone Ltd, which now operated the quarry under its subsidiary, Tarmac Roadstone (Northern) Ltd. It was served by sidings that ran right into the quarry, north of the BR Wearhead branch, 1¼ miles west of Stanhope Station. The narrow gauge system below was used within the quarry, bringing whinstone from the quarry faces to the discharge point. It was replaced by road transport soon after Tarmac Roadstone (Northern) Ltd took over. The quarry ceased production about 1965.

Gauge : 2ft 3½in

-		4wDM	RH	175420	1936	(a)	s/s c/1969

(a)　ex Richard Summerson & Co Ltd, with quarry, 10/1963.

148. 2ft 3½in gauge RH 175420/1936, derelict outside its shed in Greenfoot Quarry on 26th May 1969, six years after rail traffic had ceased and four years after the quarry was closed.

ERNEST TAYLOR
GRAVEL WORKS, Witton Park

Q522
NZ 169309

In *Industrial Locomotive* No. 34 there is a reference to a company entitled 'Taylors (Brook) Ltd', which was allegedly formed by a Mr.T. Taylor in 1934 to operate a sand and gravel business at Witton Park. The Durham directories for 1934 and 1938 do not list this firm, but instead show 'Ernest Taylor, sand and gravel merchant' at Witton Park. It would seem likely that these two are linked.

The 1939 O.S. 25in map shows a 'gravel works' at Witton Park situated on the north bank of the River Wear on the western side of the viaduct carrying the LNER Bishop Auckland-Crook line over the river. Access to the works was from a minor road nearby; there was no rail connection. Since Taylor is the only sand and gravel merchant listed at Witton Park, it would seem reasonable to assume that these were his premises. The *Industrial Locomotive* article states that the firm used a "home-made locomotive", from which one might assume that it was an internal combustion vehicle of some kind. The O.S. map shows two straight lines, narrowly apart, running from the gravel beds to buildings. Whether this was a narrow gauge railway line, as opposed to perhaps a conveyor, is unknown.

How long the works operated, and the origin and disposal of the locomotive described, are all unknown.

THARSIS SULPHUR & COPPER CO LTD
incorporated in 1862 and registered in Edinburgh, 27/10/1866

TYNE WORKS, Hebburn

E523
NZ 302642

This was one of four works in Britain owned by this firm, which also owned copper mines in Spain. It was served by sidings west of the NER Gateshead – South Shields line, ½ mile west of Hebburn Station. The works treated pyrites cinders for the recovery of metal after sulphur had been extracted. The firm had a similar works at Willington, on the north bank of the River Tyne. How the Hebburn works was shunted before the arrival of the locomotive below is not known. It closed about 1939.

Gauge : 4ft 8½in

| | COLIN McANDREW 0-4-0ST | OC | AB | 1223 | 1911 | (a) | (1) |

(a) ex Colin McAndrew Ltd, contractors, Redford Military Barracks contract, Edinburgh, per AB, 6/1915.

(1) to N. Greening & Sons Ltd, Warrington, Lancashire, c/1939.

T.J. THOMSON & SON LTD

This firm was formed in 1871 and became a limited company in 1887. Its chief scrapyard has always been located in Stockton-on-Tees, where the firm continues in business.

TYNE DEPOT, Dunston   B524
NZ 225627

This scrap yard was formerly owned by **Clayton & Davie Ltd**, who had established it as a ship-breaking yard on the site of the former Dunston A Power Station (see Central Electricity Generating Board). It was served by a siding north of the BR Redheugh Branch and was shunted by a crane before the arrival of the locomotives below. Rail traffic ceased early in 1988. The name of the yard was changed to **Dunston Engine Works** from about 1989, but soon after this the yard was closed and cleared.

Gauge : 4ft 8½in

D2		0-6-0DM	RH	395303	1956	(a)	(1)
	(CHURCHILL)	0-4-0DM	RH	281270	1951	(b)	(2)
No.3	PSA No.20	0-4-0DM	RH	327969	1954	(c)	(3)
(No.55)		0-4-0DE	RH	381751	1955	(d)	(4)
No.28		0-4-0DH	NBQ	27874	1958	(e)	Scr c11/1988

(a) ex NCB South Durham Area, Mainsforth Colliery, Ferryhill Station, per D. Sep Bowran Ltd (this firm was a subsidiary of T.J. Thomson & Son Ltd), 10/1969
(b) ex D. Sep. Bowran Ltd, Gateshead, c1/1977.
(c) ex D. Sep. Bowran Ltd, Gateshead, 3/1977.
(d) ex Central Electricity Generating Board, Dunston Power Station, Dunston, 11/1981.
(e) ex Central Electricity Generating Board, Blyth Power Station, Blyth, Northumberland, 2/12/1983.

(1) to Millfield Scrap Works, Stockton-on-Tees, Cleveland, by 3/1976.
(2) last reported here, 21/8/1988, gone by 29/1/1989.
(3) last reported here, 15/8/1982, gone by 2/8/1983.
(4) last reported here, 16/4/1990, gone by 10/5/1990.

TYNE DOCK DEPOT, Tyne Dock   D525
NZ 359653

This scrap yard is situated on part of the large area of Tyne Dock that was infilled after coal shipments ceased. This firm developed this site in 2000-01 after a previous yard at the Dock had to be re-located to allow for the construction of a new Port of Tyne logistics shed. The yard is served by a siding from the Network Rail/Port of Tyne Tyne Dock branch. By 2005 all incoming scrap was handled by road transport, as was most out-going traffic, but from time to time scrap is dispatched by rail, when the loco below is required for shunting empty wagons for weighing and then for weighing again after they have been loaded (see photograph No.1).

Gauge : 4ft 8½in

| (9) | | 4wDH | TH | 287V | 1980 | (a) |

(a) ex Millfield Scrap Works, Stockton-on-Tees, Cleveland, c12/1999.

THE THORNLEY COAL CO

The history of Thornley Colliery and the area of east Durham around it is another complex story.

THE THORNLEY COAL COMPANY

The area was first developed by the partners who formed **The Thornley Coal Company**, which began sinking **THORNLEY COLLIERY** (NZ 365395) (**L526**) on 28th January 1834. This was eventually served by a two mile branch from Thornley Colliery Branch Junction from the Hartlepool Dock & Railway's line from Hartlepool to Haswell. The first mile of this branch was owned by the railway company, the second mile by the colliery company. The first coal along this branch was run on 1st January 1835, but only as far as Castle Eden, as the HD&R company had not completed its line. Thornley provided the first coal to be shipped at the Hartlepool Tide Harbour on 9th July 1835, albeit over a temporary line at Hart; the full opening of the whole route did not take place until 23rd November 1835.

The company's next colliery was **CASSOP COLLIERY** (NZ 341382) **(K527)**, where work began on 16th September 1836. It was served by a 1¼ mile extension of the line to Thornley Colliery, and the colliery shipped its first coal on 2nd June 1840. The line was then extended for another 1½ miles to **CASSOP MOOR COLLIERY** (NZ 320392) **(K528)**, which first shipped coal in February 1841. Between the two lay **CASSOP VALE COLLIERY** (NZ 334390) **(K529)**, but when this was sunk is unknown, and it seems to have had only a short life. One source states that the line between Thornley and Cassop was worked by a stationary engine near the latter; but the area in general is relatively level, and the line could have been worked by locomotives, as could the branch to Cassop Moor. A locomotive named CASSOP is recorded as having worked over the HD&R.

There was also a link about ¾ mile long from Cassop Colliery to **Crowtrees Colliery** (NZ 334379) (see William Hedley & Sons). This link was opened in December 1839, and allowed Crowtrees and other collieries west of it to send coal to Hartlepool via the Hartlepool Dock & Railway Co (which see). Again, with the land being virtually level, this link could have been locomotive-worked.

Meanwhile the company was opening up the area north of Thornley with the sinking of **LUDWORTH COLLIERY** (NZ 363415) **(L530)**. This was to have been served by a branch of the Hartlepool Dock & Railway through Ludworth to Littletown, starting at Crow's House Junction about 200 yards from the end of its Thornley branch. In the event work was abandoned after about a mile had been built, though this section remained railway-owned; the final ¾ mile to the colliery was built by the colliery company. The colliery was almost certainly opened in 1842.

Finally, the company opened **TRIMDON COLLIERY** (NZ 378369) **(L531)**, about 2½ miles south of Thornley and 1¼ miles north-east of the village of Trimdon. This was served by a ½ mile branch from the later Trimdon Station on The Great North of England, Clarence & Hartlepool Junction Railway. The colliery's first coal was shipped on 25th February 1843.

Locomotives
According to the HD&R directors' minutes, from 1st October 1835 all coal owners' trains had to be worked by locomotive, so that presumably from the line's full opening on 23rd November 1835, the Thornley Coal Co had to use its own locomotive. If this did indeed happen, nothing is known about the locomotive. At this date the eastern section of the HD&R was operated by rope haulage, but it seems that some of the inclines were short-lived and that by the end of 1837 locomotives were working through to Hesleden Bank Head, about five miles from Hartlepool.

In July 1834 the HD&R paid for the purchase of a locomotive for their contractors Hawthorn & Robson to use to take away the material from the excavation of the Tide Harbour. In the directors' minutes this locomotive was said to be under construction by the contractors themselves, but it would seem very likely that it was RWH 174, which is recorded in the RWH Engine Book as being sent new to Hartlepool about this time. On 16th February 1837 the directors resolved to sell this engine to the Thornley company for £475. The coal company was certainly using a second engine by the end of that year, for in December 1837 a local newspaper reported that a fireman of 'one of the locomotives owned by the Thornley Coal Co' had fallen from his engine and been killed, while a reference in the Northumberland Record Office's Watson Papers, (NRO 3410/Wat/3/45, doc.83b, dated 27th Nov 1838), states that "two engines lead all the Thornley coals". A coal certificate for the company, in the Bell Collection held by the British Geological Survey (BGS 904, Vol.5, item 163), shows a 'Killingworth' type four-wheeled locomotive, though whether this represents an actual engine or is merely a symbol is unknown. Probably about the end of 1839 the company acquired a third locomotive, built by Nesham & Welsh of Stockton, for on 30th January 1840 the HD&R directors resolved to purchase it for its original price of £800, which would suggest that it must have been almost new; but no evidence survives to confirm that this sale went through. The company continued to use its own locomotives to work Thornley and Cassop traffic to Hartlepool; it may well have done the same for Trimdon Colliery traffic, via the Great North of England, Clarence & Hartlepool Junction Railway, and possibly also for other coal owners whose traffic did not justify the purchase of locomotives. All this came to an end on 12th October 1846, when the York & Newcastle Railway took over the HD&R and the GNofE,C&HJR, ended the practice of the colliery owners running their own trains and acquired the latter's locomotives and rolling stock. As the Thornley company, with Cassop, received over £19,000 in compensation (see entry for HD&R), it would appear that a sizeable number of locomotives may well have been involved, but besides the notes above, nothing else survives. Whether the Thornley company retained any locomotives after 1846 to operate on its own system is unknown.

Cassop Moor Colliery seems to have closed by the early 1850s and about 1854 the company sold the two collieries for 'upwards of £75,000' to R.P. Philipson, a Newcastle upon Tyne solicitor, who was also a partner in The Harton Coal Company; for the future history of the Cassop collieries see the entry for Edward Robson & Co. The link between Cassop and Crowtrees had also been abandoned by the mid-1850s, perhaps in 1846 when the York & Newcastle Railway ended coal owners' trains over the Hartlepool Dock & Railway.

By the late 1850s the three remaining collieries were owned by Messrs. Chaytor, T. Wood, Gully and Burrell (Gully being John Gully, the famous prize-fighter), presumably still trading as The Thornley Coal Company. Around this time Trimdon Colliery seems to have been closed, though it was subsequently re-opened by The Trimdon Coal Co; it was finally closed in 1909.

In January 1862 Ludworth Colliery was laid in, leaving only Thornley Colliery still working. In the following year John Gully died and matters passed into the Court of Chancery, which ordered the sale of the two collieries in Newcastle upon Tyne on 31st October 1865.

THE LONDON STEAM COLLIER & COAL CO LTD

This company was formed in London on 12th May 1865 to acquire collieries and to operate a fleet of steam colliers to bring the coal to London. It was registered on 20th May 1865. On 7th July 1865 it purchased the business of Messrs. Gowland & Walton, and on 31st October 1865 these two men, now directors of The London Steam Collier & Coal Co Ltd, purchased Thornley and Ludworth Collieries on behalf of the company for £105,000, taking possession on 24th January 1866. On 30th April 1866 the directors set up a Northern Committee at Sunderland to report on the northern activities to the London Board.

Gowland and Walton were young businessmen who owned a large and successful coal wharf on the Thames at London, but Gowland came originally from Sunderland, Walton from Weardale. Their plan was to mine coal from the Thornley area, ship it at Hartlepool in their own steam colliers (two being ordered for the purpose) and sell it through their London wharf.

Ludworth Colliery was subsequently re-opened, and in 1867 the company began the sinking of **WHEATLEY HILL COLLIERY** (NZ 385393) **(L532)**. This was served by a colliery company-owned loop from a point ¼ mile from Thornley Colliery Branch Junction via the new colliery to a junction where what was now the NER ownership of the Thornley Branch ended.

Locomotive
The company certainly owned at least one locomotive, for in July 1866 a miner wrote to the *Durham Advertiser* praising the company for providing a locomotive to take men to and from work, apparently between Ludworth and Thornley. This report pre-dates the first of the locomotives listed below. Presumably one or more carriages was also provided.

On 9th April 1868 (although not registered until 17th July 1868) the company changed its name to

THE ORIGINAL HARTLEPOOL COLLIERIES CO LTD

Wheatley Hill Colliery began production in 1869. It would seem that at this time coal from Cassop Colliery was still passing through Thornley. On 12th December 1871 the firm offered Ludworth Colliery for sale, unsuccessfully. On 8th May 1875 much of the surface of Thornley Colliery was destroyed in a fire. On 18th January 1877 the firm went into voluntary liquidation, and in February 1877 both Ludworth and Wheatley Hill Collieries were laid in. The three collieries were offered for sale in May 1877, and offered for auction on 9th August 1877, but there were no buyers. Cassop Colliery also closed in 1877. By January 1878 Thornley was also laid in.

However, by the summer of 1879 the coal trade was improving, and the company's administrator re-opened Ludworth and Wheatley Hill in August 1879, with Thornley following by May 1880. One of the company's major mortgagees was William Ford, and on 16th July 1881 he took possession of Thornley and Wheatley Hill Collieries, with Ludworth following three weeks later. Sources do not agree as to the trading title he adopted, with both The Original Hartlepool Coal & Coke Co and the Thornley Colliery Co being found, but it seems most likely that he adopted the title

THE ORIGINAL HARTLEPOOL COLLIERIES CO

William Ford, from Minchinhampton in Gloucestershire, is listed in the 1881 Census as a retired tailor with income from houses, and was 76 years old, perhaps not the time of life to take on three troubled collieries. Ludworth Colliery was closed in August 1882 to enable new sinking to take place. But it seems that Ford rarely visited the collieries and was also ill-advised, for he did not discover until March 1884 that they had made a loss ever since he had taken them over and the miners had not been paid for some time. He closed all of the collieries on 4th April 1884, and was made bankrupt. Thornley and Wheatley Hill were offered for auction by Davison & Son in Newcastle upon Tyne on 21st June

1884, but no one purchased anything. All three collieries and their plant, including three locomotives, were again offered for auction by A.T. & E.A. Crow of Sunderland between 16th-18th February 1885, again without anything being sold.

However, one of the creditors was **The Weardale Iron & Coal Co Ltd**, and in April 1885 this firm concluded an agreement with Ford's representatives to take over all the property (*Colliery Guardian*, 1st May 1885). For their subsequent history see the entry for The Weardale Steel, Coal & Coke Co Ltd.

Gauge : 4ft 8½in

QUEEN	0-6-0ST	OC	BH	31	1867	New	(1)
PRINCESS	0-4-0ST	OC	MW	466	1873	New	(2)
ALBERT	0-4-0ST	OC	MW	492	1874	New	(2)

It is possible that there may have been one or more other locomotives; see below.

(1)　BH 31 was built with 14in x 20in cylinders. In *Colliery Guardian* of 2/4/1875 The Original Hartlepool Collieries Co Ltd offered for sale a 14in six-coupled tank locomotive; it may have been BH 31 or a different locomotive, but if it was the former it remained unsold. In *Colliery Guardian* of 21/9/1877 a six-wheeled coupled locomotive with 14in x 19in cylinders was offered for sale by Jos. Finney, presumably arising from the bankruptcy of 1877; again, it may have been BH 31 or a different loco, but if the former it remained unsold. In the auction of 16-18/2/1885 a six-coupled locomotive was included, which must have been BH 31, as this loco eventually passed to The Weardale Iron & Coal Co Ltd, with the collieries and the railway, in 4/1885.

(2)　two four-coupled locomotives were included in the auction of 16-18/2/1885, which must have been these two locomotives, as they subsequently passed to The Weardale Coal & Iron Co Ltd, with the collieries and railway, in 4/1885.

Note : After the bankruptcy of 1884 the manager of Thornley Colliery is said to have negotiated the sale of a locomotive to the NER, presumably in payment of debt; but the miners, seeing it as an asset which might pay their wages, tore up the track to prevent it leaving the colliery. As there is no trace of this arrangement in NER records, it was presumably rescinded.

TIMBER OPERATORS & CONTRACTORS LTD
SHULL TIMBER CAMP, Wiserley, near Wolsingham　　　　　　　　　　　　　　　　　　T533
Board of Trade Timber Supply Department until 12/1919　　　　　　　　NZ 083345 approx.

On 1st August 1918 the 133rd Company of the Canadian Forestry Corps arrived here under the auspices of the Board of Trade Timber Supply Department to fell timber in the Wiserley and Wiserley Barn Plantations. After the end of the First World War the Corps continued to work the site, latterly under the control of Timber Operations & Contractors Ltd.

The Corps built a narrow gauge line from a sawmill at NZ 083345 for about 2¼ miles to a siding south of the NER Wear Valley Branch, about ¾ mile east of Wolsingham Station. The line is shown on the O.S. 3rd edition map. Given the map and the contours of the area, the last ¾ mile down to the NER was certainly a rope-worked incline. It may well be that the first section away from the sawmill was also worked by a hauler, which may have lowered the wagons down for a short distance on its opposite (northern) side. How the central section, which included a headshunt not far from the top of the incline down to the NER, was worked is not known.

At an auction here on 3rd/4th April 1920 held by Robert T. Handley (for the Board of Trade Timber Supply Department, according to the advertisement), the three locomotives below were offered for sale, together with about 1½ miles of three-line and two-line railway track. It is not known whether any of the locomotives worked here.

Reference : *The Kerry Tramway and Other Timber Light Railways*, D. Cox & C. Krupa, Plateway Press, 1992.

Gauge : 3ft 0in

-	0-4-0ST	reb	Wake	1512?		(a)	(1)

(a)　ex ?, by 3/4/1920; this loco might be WB 1510/1897.

(1)　to Board of Trade Timber Supply Department, Jane Pit Storeyard, Walker-on-Tyne, Northumberland, where it was advertised for sale on 21/12/1920.

Gauge : 2ft 0in

BROMLEY No.1	0-4-0T	OC	AE	1593	1910	(a)	(1)

(a)   ex War Department, Catterick Camp, Catterick, Yorkshire (NR), by 3/4/1920; previously Old Delabole Slate Co Ltd, Delabole, Cornwall.

(1)   this may be the loco acquired by John Pearson Ltd, contractor, Stainton-in Cleveland, Yorkshire (NR).

Gauge : unknown

| - | ? | ? | KS | ? | ? | (a) | s/s |

(a)   ex ?, by 3/4/1920; all other details are unknown.

TOW LAW PATENT FUEL CO LTD
Tow Law   M534
location unknown

Nothing is known about this firm. It clearly went into liquidation in 1906, for an auction of its plant was to be held by A.T. & E.A. Crow on 9th November 1906. The plant included a standard gauge, four coupled locomotive with 9in x 15in cylinders, but nothing else is known.

TYNE & WEAR FIRE AND RESCUE SERVICE
BARMSTON MERE TRAINING CENTRE, Nissan Way, Washington   H535
NZ 330571

The Centre was opened on 23rd October 1996, and the vehicle below was acquired soon afterwards. Most of it, including the driving end, is located within an artificial tunnel. Although it is used for interior training using generated smoke, its main purpose is to allow a car, with dummies inside, to be rammed into it, to simulate collisions on ungated level crossings and thus practice the rescue techniques such accidents need.

Gauge : 4ft 8½in

| 3721 | | 4w-4wRER | MetCam | 1983 | (a) |

(a)   ex London Underground Ltd, c/1997, driving car; formerly used on the Jubilee Line.

THE TYNE BRICK & TILE CO LTD
PHOENIX BRICK WORKS, near Crawcrook   A536
NZ 135627 approx

There was a brickfield at this location, about half way between the villages of Crawcrook and Greenside, by 1860. By 1914 it was owned by **The Phoenix Brick Co** (Davison gives Pheonix), which in 1920 opened a new yard here. It was purchased by **The Tyne Brick & Tile Co Ltd** in 1937 and considerably rebuilt. Only common bricks were produced.

Between 1939 and 1945 the works was closed, and in 1946 the company was acquired by the Newcastle upon Tyne house-building firm of William Leech & Co Ltd, though it continued to trade under its original title. The monorail vehicle below was a powered wagon, almost certainly used to convey the clay from the quarry to the yard. The works was closed in 1965.

Monorail

| - | 1w1DM | RM | 5124 | 1957 | New | s/s |

THE TYNE PLATE GLASS CO LTD
The Tyne Plate Glass Co until 16/7/1886
TYNE PLATE GLASS WORKS, South Shields   D537
R.W. Swinburne & Co until /1868   NZ 359672
Isaac Cookson & Co until /1845

This glass works was situated immediately downstream of the Mill Dam, the southern landing for one of the main ferries across the River Tyne. It was begun about 1696 by the Cookson family, who began the manufacture of plate glass there about 1770; crown and sheet glass was also made. The firm's quay on the river was known for many years as Cookson's Quay. Many ships coming to the quay arrived in ballast, which in the first half of the nineteenth century was almost always chalk. This had to be off-loaded and disposed of, and about 1832 the firm built a waggonway in order to do this. From the quay the line entered a 216-yard tunnel under the works, New Road, Coronation Street and Station Road and passing St. Hilda Colliery (see The Harton Coal Co Ltd), then swinging round south-east before turning north-east across the open fields south of the town to reach the ever-growing "ballast hill" on the beach at The Bents, a distance of about 1¼ miles. Approximately 800 yards from the quay there was a **stationary engine** (NZ 365666). The 1st edition O.S. 25in map shows the whole length as single line, with no sidings or loops, which may well mean that the stationary engine worked inclines on both sides of it, possibly using the same rope for lowering as hauling. The map also shows a link into the "soda works" (probably manufacturing soap) that was situated immediately to the south of St. Hilda Colliery. By the 1850s water was replacing chalk as ship ballast, but the waggonway continued in use for disposing of sand and waste from the works, which had now grown to a considerable size.

By 1840 one of the directors was Nicholas Wood (see John Bowes & Partners Ltd), who was also a director of the nearby Brandling Junction Railway, though the works never had a connection with a public railway. In 1875 the company was acquired by Charles Mark Palmer, more famous for Palmers Shipbuilding & Iron Co Ltd at Jarrow and also Managing Partner in John Bowes, Esq., & Partners. He introduced locomotives on the waggonway in 1879, with severely cut-down mountings in order to pass through the tunnel. The locomotives' initials were those of Palmer and his son Alfred. However, the works was trading at a loss when Palmer acquired it and his efforts to turn it round failed. The firm went into voluntary liquidation on 3rd October 1891.

The firm's failure was the great opportunity that The Harton Coal Co Ltd, headed by Lindsay Wood, Nicholas Wood's son, had hoped for. In 1892 the coal company purchased the works and the waggonway for subsequent development, the premises eventually being largely cleared to make way for the new Harton Low Staiths, while the waggonway was rebuilt and incorporated into the Harton company's electrified system (see The Harton Coal Co Ltd).

Gauge : 4ft 8½in

A.M.P.	0-4-0ST	OC	BH	515	1879	New	(1)
C.M.P.	0-4-0ST	OC	BH	516	1879	New	(1)

(1) to The Harton Coal Co Ltd, with works and waggonway, /1892.

SAMUEL TYZACK & CO LTD
Samuel Tyzack & Co until 5/12/1889
MONKWEARMOUTH IRON & STEEL WORKS, Monkwearmouth, Sunderland   H538
NZ 398583 approx.

This firm was founded in 1857 and developed an ironworks alongside Fulwell Road. This was served by a link to a NER branch serving various premises in the area which joined the Gateshead - Sunderland line ½ mile north of Monkwearmouth Station. Unusually, the company's line completely circumnavigated the works, re-joining the branch a few yards south of the initial junction. From the late 1850s the company also owned **EDMONDSLEY COLLIERY** near Waldridge for some years (see The Charlaw & Sacriston Collieries Co Ltd). The works, which specialised in marine work, closed in 1936, and was dismantled by Thos W. Ward Ltd of Sheffield, Yorkshire (WR), in the following year.

Gauge : 4ft 8½in

No.1	0-4-0ST	OC	JF	1568	1871	New	(1)
No.2	0-4-0ST	OC	BH	529	1880	New	s/s
No.3	0-4-0ST	OC	BH	1098	1896	New	(2)
No.4	0-4-0ST	OC	HL	2533	1902	New	(3)

(1) advertised for sale by W.H. Campbell & Co, Middlesbrough, Yorkshire (NR), 5/1895; s/s.
(2) offered for sale by Thos W. Ward Ltd, 9/1937; s/s.
(3) to The Rolling Stock Co Ltd, Nestfield Works, Darlington, by 14/3/1931.

H.R.VAUGHAN & CO LTD
SPHINX WORKS, Gateshead   C539
NZ 264638

This company entered business in 1906 by developing a brickworks at Bishop Auckland. In 1913 it moved to Gateshead to expand, taking over buildings in South Shore Road, formerly part of a cement works, to manufacture roofing felt and damp-proof courses. Subsequently the works was expanded eastwards along the bank of the River Tyne. The firm became one of the country's leading manufacturers of roofing felt, with branches in England and Scotland, though the Gateshead works remained its headquarters. It was still in existence in the 1950s.

The works was not rail connected, and the reason for acquiring the locomotive below is not known, though it may well have been used for transporting materials between different sections of the premises.

Gauge : 2ft 0in

| - | 4wPM | L | 8975 | 1937 | New | s/s |

THOMAS VAUGHAN & CO

Thomas Vaughan (c1835-1900) was the son of John Vaughan, who with Henry Bolckow was the founder of the famous iron and steel firm Bolckow, Vaughan & Co. Following the death of his father in 1868, Thomas Vaughan was encouraged to go into the iron & steel industry on his own account. Unfortunately he expanded his business rapidly in the early 1870s during a short-lived boom of the economic cycle, only to find he could not survive the severe economic depression after 1875; the firm failed in 1876 owing £1 million. Trustees kept the company going for a short time, but eventually everything had to be closed and sold.

AUCKLAND IRONWORKS, Bishop Auckland Q540
NZ 213294

This works was opened in 1863 by Joseph Vaughan, a nephew of John Vaughan, who had previously been the manager of Bolckow, Vaughan & Co's Witton Park Ironworks. It was set up to make "edge tools", but was subsequently converted into an ironworks. It was served by a ¼ mile branch north of the NER Bishop Auckland – Darlington line, immediately east of Bishop Auckland Station. It passed to Thomas Vaughan & Co. sometime after 1869.

The works closed in June 1877, and the plant was auctioned by C. Willman on 10th October 1878. It was re-opened by new owners in December 1879, but was again closed and its plant, with no locomotives included, was auctioned in November 1884. Part of the site was later taken over by Robert Wilson & Sons Ltd (which see), while the branch which served the ironworks continued on to serve the Auckland Engine Works of locomotive engineers and dealers Lingford, Gardiner & Co Ltd, which may also have been located in part of Vaughan's premises.

Gauge : 4ft 8½in

AUCKLAND	0-4-0ST	OC	MW	87	1863	New	(1)
AUCKLAND	0-4-0ST	OC	MW	270	1869	(a)	(1)
-	0-4-0ST	IC	JF	1571	1872	(b)	(2)

(a) New to 'Joseph Vaughan, Middlesbrough' [Yorkshire (NR)], according to the MW Engine Book, but believed to have come here.
(b) believed ex Clay Lane Ironworks, Middlesbrough, Yorkshire (NR).
(1) both returned to MW, possibly c8/1870, who then re-sold them to Henry Lee & Sons, contractors, Beverwijk, Holland, 9/1870.
(2) to William Benson, Fourstones Colliery, near Hexham, Northumberland, by 5/1884.

SOUTH MEDOMSLEY COLLIERY, near Dipton G541
NZ 144531 (Annie Pit)

This colliery had been opened in 1861 as **Pontop Hall Colliery**, but its name was changed to **South Medomsley Colliery** in 1864 with the opening of the Ann (later Annie) Pit. It was served by a NER branch one mile long from the NER Pontop & South Shields branch, one mile east of Carr House N.E. Junction at Leadgate (see entry for The Consett Iron Co Ltd). It was acquired by Thomas Vaughan & Co in 1871.

Following closure the colliery and its plant was auctioned between 27th-29th May 1878 by Moses Pye & Sons (some sources give the date of the auction as 25th May). It was eventually reopened in 1884 by **The South Medomsley Colliery Co Ltd** (which see).

Gauge : 4ft 8½in (assumed)

In the auction of plant in May 1878 a tank locomotive with 10in cylinders was included; its origin, identity and disposal are unknown.

WHESSOE IRONWORKS, near Darlington S542
NZ 285181

This ironworks has a sad history. In the mid-1860s a group of local workers, disillusioned by numerous industrial disputes in the various ironworks in Darlington and encouraged by the Liberal M.P., resolved to construct and operate a new ironworks based on the principles of the co-operative movement. A site was chosen in the parish of Drinkfield, about two miles from the centre of Darlington and lying east of the NER Bishop Auckland – Darlington line; a company was set up, the **Drinkfield Cooperative Iron Co Ltd** (the Drinkfield Iron Co Ltd is also found), and construction of the works began in March 1867. Work on the **DRINKFIELD IRONWORKS** was well advanced by the late summer of 1867, when the former owner of the land took the company to court over an issue regarding the company's title to the land. By the time this had been resolved, in the company's favour, all the company's remaining money had been spent. The company then found itself unable to raise the extra money needed to complete the works and commence production, and by 1870 it had been compelled to go into liquidation.

After lying in this incomplete state for some time the works was acquired by Thomas Vaughan & Co. It is not known whether he managed to bring the works into actual production, but if he did it could not have been for long, given his own firm's bankruptcy in 1876. The *Colliery Guardian* of 9th November 1877 included a notice that on 28th November 1877 a sale of plant was to be held that included the locomotive below. A further auction to dispose of the remainder was held on 9th January 1879, and the site was cleared.

Gauge : 4ft 8½in

-	0-6-0T+t	OC	?	?	(a)	(1)

(a) the auction advertised for 9/11/1877 lists a six-wheeled coupled tank locomotive with 12in outside cylinders and a tender, weight 24 tons. This has been linked to HOPETOWN, formerly Stockton & Darlington Railway No. 5, which was a six-coupled locomotive with outside cylinders and two tenders, built by W. & A. Kitching in 1841. However, it is not known to have been rebuilt as a tank locomotive, while HOPETOWN had 15in cylinders, 3in bigger than the loco here.

(1) to Mr. Bradley, general dealer, Middlesbrough, Yorkshire (NR), for £45, presumably on 28/11/1877.

The company also owned
HEDLEY HOPE COLLIERY, near Tow Law (NZ 135394) **(M543)**; this was opened in 1866 and was served by the one mile NER Hedley Hope branch from near the summit of the Sunniside Incline on the NER line between Crook and Tow Law, which included a rope-worked incline;

EAST HEDLEY HOPE COLLIERY, near Tow Law (NZ 158404) **(K/M544)**; this was opened by Thomas Vaughan & Co in 1875 and was served by a one mile branch from the NER Deerness Valley branch.

Both of these collieries, having passed through various owners, were acquired in 1936 by The Bearpark Coal & Coke Co Ltd (which see).

CLAY LANE IRONWORKS, Middlesbrough, Yorkshire (NR) – see the Cleveland & North Yorkshire Handbook.

SOUTH SKELTON IRONSTONE MINE, Yorkshire (NR) – see the Cleveland & North Yorkshire Handbook.

VICKERS LTD
Vickers-Armstrongs (Shipbuilders) Ltd until 1/4/1965
Vickers-Armstrongs Ltd until 1/7/1955
PALMERS HEBBURN YARD, Hebburn   E545
Palmers Hebburn Co Ltd until 1/1/1956 (a subsidiary of **Vickers Armstrongs Ltd**)   NZ 306653

After the collapse of **Palmers Shipbuilding & Iron Co Ltd** (which see), this shipyard was taken over in 1934 by Vickers Armstrongs Ltd under a subsidiary company called Palmers Hebburn Co Ltd. It was served by a ½ mile branch from the LNER Gateshead – South Shields line, ½ mile east of Hebburn Station. This line, which for much of its length ran alongside Lyon Street, also served the adjacent shipyard owned by R. & W. Hawthorn, Leslie (Shipbuilders) Ltd (which see). In the autumn of 1957 the two companies jointly set up the Lyon Street Railway Company with a capital of £200 to acquire the branch from its erstwhile owners, the Carr-Ellison Estates. The ship-repairing facilities were shut down on 25th September 1970, and were purchased by Swan Hunter Shiprepairers Ltd on 16th March 1972, though the marine works continued to be operated by Vickers Ltd. About 1972 road transport was adopted in place of rail.

The premises have since passed through a number of owners, passing to A & P Tyne Ltd in 2004. The yard continues in operation, undertaking ship repair work.

Gauge : 4ft 8½in

PALMERS No.1	0-4-0ST	OC	HL	2666	1906	(a)	Scr c12/1959
PALMERS No.2	0-4-0ST	OC	HL	2667	1906	(a)	Scr c12/1959
PALMERS No.1	4wDM		FH	3890	1958	New	(1)
PALMERS No.2	4wDM		FH	3891	1958	New	(2)

(a) ex Palmers Shipbuilding & Iron Co Ltd, Jarrow, via Thos W.Ward Ltd, /1934.

(1) to TH, Kilnhurst, Yorkshire (WR), 26/6/1968; re-sold to South Wales Warehouses Ltd, Penarth, Glamorgan, 10/1968.

(2) disused after c/1972; parts scrapped c11/1979; remainder, including the frame, scrapped by 12/6/1981.

THE WALLSEND & HEBBURN COAL CO LTD
registered 15/2/1892; took over **The Tyne Coal Co Ltd** (see below) on 22/2/1892
HEBBURN COLLIERY, Hebburn   E546

This colliery was begun in 1792, and latterly comprised three pits, the **'A' PIT** (NZ 311656), the **'B' PIT** (NZ 313654) and the **'C' PIT** (NZ 307653). Both the 'A' and 'B' Pits had short links to staiths on the River Tyne, with a short link joining the two lines, while the 'C' Pit was linked by a ½ mile waggonway to the 'A' Pit. The 'B' Pit was closed in the nineteenth century, while the 'A' and 'C' Pits were flooded in 1859 and were not re-opened until about 1871. The colliery was owned by a number of companies, the predecessors to the firm above being **The Tyne Coal Co Ltd**, which acquired the colliery in 1867. After the opening of the new NER line to South Shields in 1872 the colliery was linked to it by a short branch ½ mile east of Hebburn Station. It would seem that this link led the company to introduce locomotives, presumably to replace horses.

150. A drawing by R. H. Inness dated 15th September 1902 of No. 4, 0-4-0WT EB 30/1891.

As its title suggests, the company also owned **WALLSEND COLLIERY** and its replacement, the **RISING SUN COLLIERY**, at Wallsend (see Northumberland Handbook). An unusual feature of the locomotive stock at Hebburn was the battery locomotives used after 1921. The colliery was closed in March 1931. The area where the three pits once were is now the Viking Business Park, whilst the route of the waggonway between the 'A' and 'C' Pits is now Wagonway Road.

149. Another glorious picture by William Parry of South Shields; the Hebburn 'C' Pit yard in September 1922, with so much fascinating detail – and the first two 4wBE locomotives amongst the chaldron waggons.

Gauge : 4ft 8½in

No.1	0-6-0T	IC	JF	1542	1871	New	(1)
No.2	0-4-0ST	OC	BH	237	1873	New	s/s
No.2	0-4-0ST	OC	BH	762	1883	New	s/s
No.3	0-4-0ST	OC	BH	763	1883	New	s/s
No.4	0-4-0WT	OC	EB	30	1891	(a)	(2)
No.1	4wBE		EEDK	518	1921	New	s/s
No.2	4wBE		EEDK	519	1921	New	s/s
No.3	4wBE		EEDK	570	1923	New	s/s

(a) reputed to have come from a glass works.
(1) sold; purchaser and date unknown; one source gives Scr by /1924.
(2) to Wallsend 'G' Pit, Wallsend, Northumberland, c/1902; returned by /1904; s/s.

JACOB WALTON & CO LTD
formerly **Jacob Walton & Co**
FINE BURN QUARRY, near Frosterley   U547
NZ 033373 approx.

This firm was one of the small companies involved in the lead industry in Weardale, and was operating by the late 1850s. Later it expanded into quarrying and took over this quarry about June 1871, apparently from Bolckow, Vaughan & Co Ltd. It was served by a ¾ mile private extension of the NER Bishopley branch. The quarry was closed about 1907, but it was subsequently re-opened by **Pease & Partners Ltd** (which see).

It is believed that the firm used one or more locomotives within the quarry, but no details survive.

STEPHEN WALTON   **No map reference**

This man was one of the original shareholders in the Clarence Railway when it was set up in 1828. As approved by its Acts of 1828 and 1829, the Railway's main line was to run from a junction with the Stockton & Darlington Railway at Simpasture, east of Shildon, to Haverton Hill on the River Tees estuary, with branches to Durham City (only constructed as far as Coxhoe), Byers Green and Stockton-on-Tees. Initially the sole traffic was coal, but from 11th July 1835 a passenger service was begun between Stockton and the Clarence Tavern at Crow Trees, near Coxhoe, for which Walton was the licensee, the company charging him 4d per coach per mile, shortly afterwards reduced to 3d. It would appear that the company provided the coaches; who provided the locomotives is not known. From 1st January 1836 the rights were taken over by the Clarence Coach Company.

During 1836 Walton acquired an estate at Thrislington, north of Ferryhill, and sank **THRISLINGTON COLLIERY** (NZ 309338), which was connected to the Clarence Railway. The railway's directors claimed that it opened in November 1836 (PRO Rail 117/4), but this may not be accurate. It would seem that Walton ran his own coal trains over the line to Haverton Hill, as William Hedley (see above) and Nicholas Wood (see below) were also to do, and that the first locomotive below may well have been acquired for this work.

In November 1837 the Clarence Coach Company gave up its contract, and there was no passenger service until the summer of 1838, when Walton again took the contract, for three years at ¼d per passenger per mile. Passenger traffic between Stockton and Coxhoe (16¼ miles) recommenced on 20th June 1838. VICTORIA is specifically mentioned as hauling the coal traffic from Thrislington, while the Clarence Railway's Engineer in 1839 felt that NORTON was too heavy (at 11tons 11cwts) to be run at more than 16 mph for passengers and 8mph with coals. In May 1839 the company gave Walton permission to run three trips per day and to carry general merchandise, though not as a monopoly. In June 1841 his franchises were renewed, at ½d per mile for passengers and 1½d per mile for merchandise. A week later his rights to use 'the Thames engine on the same terms as before' were also renewed.

In Wishaw's *Railways of Great Britain and Ireland*, 1842, Walton's service is described as consisting of two carriages. One consisted of a central 1st class compartment flanked by two 2nd class compartments, each holding eight people; it had cost £145 and had run 28,000 miles without repairs. The other comprised three 2nd class compartments, holding a total of 30, which had cost £142. The locomotives NORTON and VICTORIA are also mentioned.

By 1841 the Clarence Railway was in severe financial difficulties and passed into the hands of the Exchequer Loan Commissioners. It was put up for sale in 1842 and only rescued by some very peculiar financial manoeuvres. In the same year Thrislington Colliery was put up for sale by order of the sheriff's office under a writ of execution, which would suggest that Walton had gone bankrupt. The colliery appears to have been laid in between 1842 and 1867, when it was re-opened by **The Rosedale & Ferry Hill Iron Co Ltd** (which see). When Fossick & Hackworth took over the complete operation of both the Clarence and Stockton & Hartlepool Railways in November 1844, THAMES was included in the former Clarence stock (see *Industrial Locomotive* 107, 2003), but the fate of the other two remains unknown.

Gauge : 4ft 8½in

NORTON	0-4-0	OC	RWH	218	1836	New	(1)
VICTORIA *	0-4-0	VC	Wm. Lister		1838	New	(1)
THAMES	2-2-0?	OC?	?		1838	(a)	(2)

* Tomlinson (*The North Eastern Railway*, p. 391), gives VICTORY.

(a) the balance of probability would seem to indicate that Walton hired this locomotive from the Clarence Railway, rather than being the owner of it. It had been built for the London & Greenwich Railway as No. 8 THAMES, allegedly by Twells & Co of Birmingham. However, this is the only locomotive attributed to this firm, and it may well have been that they were only acting as agents, with the locomotive being built elsewhere. It was purchased by the Clarence Railway about 1839.

(1) still owned by Walton in 1841; s/s.
(2) returned to Clarence Railway stock, and taken over by Fossick & Hackworth, 11/1844.

THE WASHINGTON COAL CO LTD

registered 10/2/1895; it was owned by William and Frank Stobart, father and son, who were also involved with The Wearmouth Coal Co Ltd. Latterly it became a member of the "empire" developed by M.H. Kellett.

USWORTH COLLIERY, Usworth   H548
Leversons Wallsend Collieries Ltd (registered in 1918) until /1940   NZ 315584
Jonassohn, Gordon & Co Ltd until c/1918
John Bowes & Partners Ltd until /1897 (**John Bowes, Esq., & Partners** until 21/7/1886)
Elliot & Jonassohn until 20/3/1882
originally **George Elliot** (**Sir George Elliot** from 15/5/1874)

The sinking of this colliery by George Elliot began on 7th April 1845 and was completed on 22nd July 1847. It was served by sidings alongside what became the NER line from Washington to Pelaw, ½ mile north of Usworth Station. It was almost certainly shunted by horses before the arrival of the first locomotive below, and there also appear to have been periods when it was shunted by the NER/LNER. It passed to NCB Northern Division No. 1 Area on 1st January 1947.

Gauge : 4ft 8½in

USWORTH	0-4-0ST	OC	BH	38	1867	New	s/s
-	0-6-0ST	IC	BH	716	1882		
	reb		?		1906	(a)	(1)
-	0-4-0ST	OC	P	1460	1916	(b)	(2)

(a) ex James W. Ellis & Co Ltd, Swalwell.
(b) ex John Rogerson & Co Ltd, Wolsingham, via C.W. Dorking & Co Ltd.

(1) to Ryhope Coal Co Ltd, Ryhope Colliery, Ryhope.
(2) to Mid-Durham Carbonisation Co Ltd, Stella Gill Coke Works, Pelton Fell (also owned by M.H. Kellett), loan, and returned; to NCB No. 1 Area, with the colliery, 1/1/1947.

The company also owned **WASHINGTON F COLLIERY** (NZ 303574) **(H549)** and **WASHINGTON GLEBE COLLIERY** (NZ 308563) **(H550)**, both served by a 1¾ mile NER branch from Washington Station and shunted by NER/LNER. Both passed to NCB Northern Division No. 1 Area on 1st January 1947.

WASHINGTON IRON CO

WASHINGTON IRON WORKS, Washington   H551
  NZ 315552

This iron works was built in 1858 and was served by sidings north of the NER Pontop & South Shields branch, ½ mile west of Washington Station. The works seems to have been singularly unsuccessful, its one blast furnace being in blast for only four months in 1858 and 5½ months in 1859 before being shut down. It was put into blast again in 1864, but was shut down again soon afterwards. The works was apparently demolished about 1870 and the site redeveloped, though the title still appears on the 2nd edition of the O.S. map in the 1890s.

It would seem likely that such a works would use one or more locomotives, but no record remains.

WEAR STEEL CO LTD

CASTLETOWN IRONWORKS, Castletown, Sunderland   H552
Wear Rolling Mills Co Ltd (registered 27/7/1880) until /1888   NZ 366583

The Wear Rolling Mills had been opened in 1870 by **Oswald & Co** (which see), but closed in 1875 with Thomas Walter Oswald's bankruptcy. During this period the works had no rail connection. The mills were re-opened by the **Wear Rolling Mills Co Ltd** in 1881, now served by sidings south of what became the NER Hylton, Southwick & Monkwearmouth branch, 1½ miles from Monkwearmouth, which had been opened in 1876. However, this firm went bankrupt in 1887, and the plant was auctioned by George Barnes on 12th January 1888.

The works, re-modelled as a foundry, without blast furnaces, was re-opened by the **Wear Steel Co Ltd** in 1889. This firm went bankrupt in September 1894. The plant was auctioned by A.T. & E.A. Crow on 26th November 1896, and a second auction – at which it was alleged that the plant had only worked for eighteen months – was held on 31st March 1897 by Wheatley, Kirk & Price. The site and its buildings were acquired by **The Wearmouth Coal Co Ltd** for its new Hylton Colliery, whose sinking began in May 1897.

The auction on 12th January 1888 included "locomotives", so there was presumably at least one more locomotive in addition to HC 221/1881 below.

The auction on 26th November 1896 included "two tank locomotives"; if one was AB 641/1889 below, then there was at least one more. Curiously, the auction on 31st March 1897 also included "locomotives"; whether these were the same as those in the auction four months earlier is not known.

Gauge : 4ft 8½in

-	0-4-0ST	OC	HC	221	1881	New	(1)
-	0-4-0ST	OC	AB	641	1889	New	(2)

(1) offered for auction, 12/1/1888(?); to River Wear Commissioners, South Dock, Sunderland, /1888.
(2) believed to have been offered for auction, 26/11/1896 and 31/3/1897; to contractors working on the Hudson Dock, Sunderland, for the River Wear Commissioners; then to The Wearmouth Coal Co Ltd, Monkwearmouth Colliery, Sunderland.

THE WEARDALE LEAD CO LTD

registered 21/6/1883; went into voluntary liquidation 17/9/1900 and a new company with the same name formed on the same day; subsidiary of **Imperial Chemical Industries Ltd** from 4/1962; interests sold to **Swiss Aluminium Mining (UK) Ltd** on 31/5/1977

Galena (lead ore) occurs in vertical veins, and by the eighteenth century this was often exposed by using a large volume of water to scour out a crevasse (a "hush") down a hillside. This method was replaced by driving levels, used both for mining and drainage, into the hillside, and then shafts had to be sunk. Sometimes a working would bisect a series of veins, sometimes it would work along a single vein. Once at the surface the rock was crushed and then various combinations of gravity and water were used to extract all the ore possible. The ore next had to be carried by horse and cart, often for many miles, to the nearest lead smelt mill. For an excellent description and re-construction of the industry and the life and working conditions of the miners, the Killhope, The North of England Lead Mining Museum in Upper Weardale is highly commended. The leading authority on the mines of the Northern Pennines was Professor Sir Kingsley (K.C.) Durham, who began his research in the 1920s and published extensively (e.g. *Geology of the Northern Pennine Orefield, Vol.1, Tyne to Stainmore*, HMSO, 1990), while there is a major deposit of The Weardale Lead Company records (D/WL) in the Durham County Record Office.

From the seventeenth century the majority of the lead mining industry in the Northern Pennines was effectively divided between two big firms, the Beaumont-Blackett family and the London Lead Company, with the remaining third comprising small independent mines run by small companies. Sir William Blackett had obtained his first lease from the Bishop of Durham in 1698. Later the Beaumont family became dominant in the business, trading as W B Lead. From the 1870s the industry began to

Fig.11 — **Lead mines worked by Weardale Lead Co Ltd**

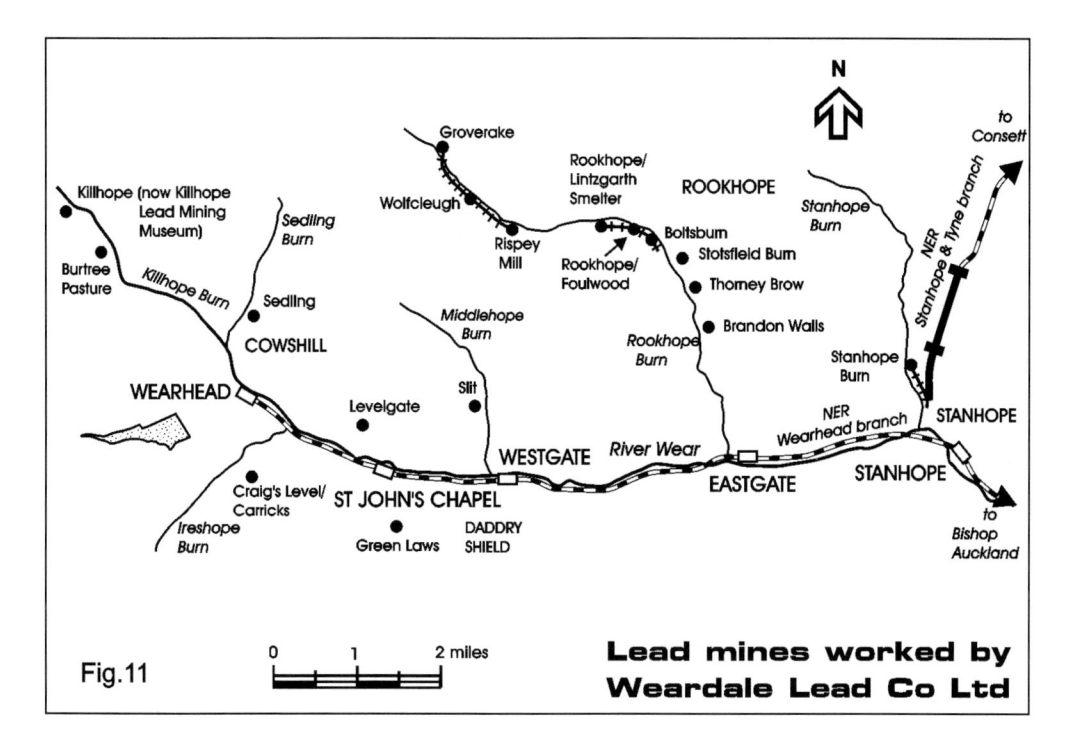

decline, and when the Beaumont family decided to relinquish its interests in Upper Weardale, The Weardale Lead Co Ltd, with its subscribers mainly based in London, was formed to acquire them.

The Mines & Quarries Inspectorate Report for 1883 lists the company as working the 13 mines given below (see also Fig.11). The grid references are those given by Dunham. The dates of "closure" of metalliferous mines are frequently difficult to determine. A mine might "close" (= cease production), but then enter a period of "standing", sometimes for years, in the hope that production might start again; sometimes it did, sometimes it was re-opened for exploratory purposes, only then to close again; sometimes a mine was re-opened by a different company, often to mine a different ore, sometimes it was finally "closed" (= abandoned). For example, Killhope Mine below ceased production in 1910, stood for six years, was re-opened briefly in 1916 but without success and was then just left standing, without any formal abandonment (and is now open again for visitors).

Boltsburn Mine, Rookhope	NY 936428	see below
Brandon Walls Mine, near Rookhope	NY 947410	closed in 1886
Burtree Pasture Mine, Cowshill	NY 830418	closed in 1887
Craig's Level, near Ireshopeburn	NY 861378	closed in 1890s

(re-opened as Carrick's Mine by The Weardale Iron & Coal Co Ltd (which see))

Green Laws Mine, Daddry Shield	NY 889370	closed in 1902

Grove Heads Mine; the location of this mine is uncertain; it is likely to have been part of Craig's Level (see above), where the vein is called Grove Heads Vein.

Killhope Mine, near Lanehead	NY 825430	closed in 1916
Levelgate Mine, near Ireshopeburn	NY 881389	closed in 1890s

Rookhope Mine, Rookhope: this name is only found in the Mines & Quarries' Inspectorate list; it may refer to Foulwood Mine at Rookhope (NY 933429) or part of the Boltsburn complex.

Slit Mine, near Westgate	NY 905392	closed in 1878
Stanhope Burn Mine, Stanhope	NY 986413	see below
Stotsfield Burn, near Rookhope	NY 943423	see below
Thorney Brow Mine, near Rookhope	NY 945414	closed in 1890s

In fact, the company's ownership extended over many more mines during its lifetime, as its 1915 map at the Killhope, The North of England Lead Mining Museum shows. Ore from the various mines and levels was brought by horse and cart to the **ROOKHOPE or LINTZGARTH SMELT MILL** (or Smelter) (NY 926428) **(V553)**, about ¾ mile north-west of Rookhope village. This was opened in 1737 and had been modernised in 1802. With lead prices and production continuing to decline, by 1903 most production was concentrated at Boltsburn Mine and at Wolfcleugh Mine (NY 902432), opened in 1901

2½ miles north-west of Rookhope. The commercial production of **fluorspar**, mined with the lead and previously dumped as waste, was begun in 1897.

Rookhope was served by The Weardale Iron Company's private railway called the Weatherhill & Rookhope Railway, which was subsequently extended to the Slit Mines above Westgate (the Middlehope & Rookhope Railway), while a branch was also built to Groverake Mine (see the entry for The Weardale Steel, Coal & Coke Co Ltd). This handled freight traffic for The Weardale Lead Co Ltd.

The depressed price of lead caused lead mining to end in 1932. For the remainder of its existence the company depended on its production of fluorspar. It survived to become part of the great boom in fluorspar mining that began in the 1960s and its remaining interests were taken over by **Swiss Aluminium Mining (UK) Ltd** on 31st May 1977. This period of its history will be found under the section Fluorspar Mines in Part 2. The entries below list only the locomotives acquired before 1950.

The company's first use of a locomotive was at
GROVERAKE MINE, near Rookhope **V554**
NY 896442 (shaft)

This mine, called in the nineteenth century **GROOVE RAKE**, lay near the head of Rookhope Burn, about three miles from Rookhope village, with the first lead being produced in 1828. From reaching Bolt's Burn about 1848, The Weardale Iron Co determined to extend its railway system up Rookhope Burn, reaching the Rookhope Smelt Mill in 1859 and Groove Rake about 1864. Ironstone production ceased in 1875 and lead mining was suspended the following year.

However, in the 1880s lead production was resumed, and to improve its efficiency, The Weardale Lead Co Ltd built a narrow gauge line on the opposite side of Rookhope Burn from the Weardale Iron Co's line to bring the ore down to **RISPEY MILL** (NY 909428) **(V555)**. This lay about 1¼ miles south-east of Groove Rake Mine, near to the site of the former **Rispey Mine** (NY 911428), formerly worked for ironstone by The Weardale Iron Co Ltd. From here the washed ore was carried by horse and cart the remaining mile to the Smelt Mill. The loco shed was situated just to the south of the bridge over Rookhope Burn ¼ mile south-east of the mine.

151. The firm's first narrow gauge locomotive – 2ft 6in gauge BH 981/1889, near Wolfcleugh.

In 1901 the company re-opened **WOLFCLEUGH MINE** (NY 902432) **(V556)**, about ½ mile north-west of Rispey, which had been worked for lead between 1818 and 1847, and linked it to the line. The opening of Wolfcleugh, mining both lead and fluorspar, led to the suspension of mining at Groverake Mine in 1903, but the line was retained. Wolfcleugh was closed in 1912, but it would seem that locomotive haulage was abandoned three years earlier, when the locomotive was sent to Stanhope Burn Mine, presumably being replaced here by horses. Ganister was being mined at Rispey by 1930, with the tramway still in use, certainly then worked by horses; when the line was finally abandoned and the track lifted is uncertain.

Groverake passed into other hands in the 1930s, while Wolfcleugh Mine was re-opened by the company in 1946; for details see the Fluorspar section in Part 2.

Gauge : 2ft 6in

No.1		0-4-0ST	OC reb	BH HL	981 4109	1889 ?	New (1)	

(1)   to Stanhope Burn Mine, near Stanhope, /1909.

STANHOPE BURN MINE, near Stanhope

V557
NY 986413

This mine had no fewer than ten owners between 1820 and 1982. Lead was discovered in 1820 and from 1846 ironstone was also mined. The main adit was the **Shield Hurst Level** (NY 986413). Until 1864 the mine was worked by **The London Lead Co**, which also owned the **Stanhope Smelt Mill** further up the burn. They were all served by a 1½ mile line from the foot of the Crawleyside Incline on what became the NER Stanhope & Tyne branch, though the iron ore initially went to the Stanhope Burn Iron Works of The Weardale Iron Co (which see) and later to that firm's Tow Law Ironworks. W B Lead took the mine over in 1866, but it was subsequently closed and the railway to it lifted. Still closed, it passed to The Weardale Lead Co Ltd in 1883. The company re-opened it for fluorspar in 1906 and it soon became one of the company's chief sources of supply. To handle the output, the company built a narrow gauge line on the old standard gauge trackbed, running from the mine down to an interchange point within the sidings serving the former Lanehead Quarry (see Ord & Maddison Ltd), where The Consett Iron Co Ltd was still operating the lime kilns. These sidings connected with the NER Stanhope & Tyne line near Crawleyside Bank Foot. At first the line was horse-worked, but eventually the locomotive used between Wolfcleugh and Rispey was transferred here. Production was suspended in December 1930 and the mine was closed in 1933. It was re-opened in 1937 by two local men, **Beaston** and **Elliot**, and then taken over in 1940 by **Fluorspar Ltd** (see Part 2).

Gauge : 2ft 6in

No.1	"LITTLE SALLY"	0-4-0ST	OC reb	BH HL	981 4109	1889 ?	(a)	Scr /1937

(a)   ex Wolfcleugh Mine, near Rookhope, /1909.

BOLTSBURN MINE, Rookhope

V558
NY 936428 (Engine Shaft)

This was an old, rich lead mine in the centre of a village originally called Bolt's Burn but later re-named Rookhope. Initially it gained access to the veins by two Levels. About ¾ mile to the north-west alongside Rookhope Burn lay the **ROOKHOPE SMELT MILL** (see above), while to the south-east lay **BOLTSBURN WASHING MILL** (NY 938425). The three sites were served by a horse-worked tramway, said to have been 2ft 6in gauge. Near the end of 1846 the Weardale Iron Co (which see) began building its Weatherhill & Rookhope Railway, which opened in 1847 or 1848. This line was brought down into Bolt's Burn by the fearsome Bolt's Law Incline. It carried coal and domestic requirements in and took smelted lead out to the junction with what became the NER Stanhope & Tyne branch at Parkhead. Originally shunting around the bank foot was done by horses, but eventually the Weardale Iron Co introduced locomotives, first one and then two sheds being built adjacent to the mine on its south-east side. Rich new lead flats were discovered at Boltsburn in 1892 and by the end of the nineteenth century the levels had been replaced by a shaft (NY 936428), close to the incline bank foot. In 1901 the mine began selling fluorspar in addition to lead.

About 1913 The Weardale Lead Co Ltd rebuilt its line between the Washing Mill, the mine and the Smelter, a distance of about 1¼ miles, and to work it purchased a locomotive, for which a shed was provided on the north-east side of the mine. The smelter, the last smelt mill to operate in the Northern Pennines, closed in May 1919, the lead then being sent to the Bagillt smelter in North Wales. The locomotive thus retired to its shed, where it remained for many years. With the closure of the Weatherhill & Rookhope Railway in 1923, the Lead Company built an aerial flight from Rookhope to the NER's Blakey's Isle siding west of Eastgate. With the fall in lead prices and reserves virtually exhausted, Boltsburn Mine closed in 1931, though the Annual Reports of the Mines & Quarries Inspectorate indicate that the mine was not finally "discontinued" until 1941.

152. 1ft 10in gauge No.2, HL 3029/1913, near its shed at Rookhope, with Bolts Burn Mine behind, served by the Weardale Steel, Coal & Coke Co Ltd's line on the left. This photograph is copied from a book.

The locomotive, possibly uniquely for industrial locomotives in Durham, was fitted with a snowplough when required in the winter.

Gauge : 1ft 10in

| No.2 | "LITTLE NUT" | 0-4-0ST | OC | HL | 3029 | 1913 | New | (1) |

(1)   out of use in shed from /1919; scrapped c/1934.

Underground locomotive

Professor Dunham recorded a underground visit to Boltsburn Mine in 1929 in which he and his colleagues travelled in 'mine cars' (tubs) hauled by a battery locomotive from the main shaft for two miles along the main roadway, called Watts Level, to No. 3 Underground Shaft. The locomotive is believed to have been the one shown below. Whether this service was specially provided on this occasion or was run regularly for the miners is not known.

Gauge : 1ft 10in

| - | 4wBE | BEV | 414 | 1922 | (a) | (1) |

(a)   ex Stotsfield Burn Mine, Rookhope, by /1929.
(1)   to Stotsfield Burn Mine, Rookhope, c/1932 (?).

STOTSFIELD BURN MINE, Rookhope   V559
NY 943423 (Low Shaft)

This mine lay to the south of Bolt's Burn village and was called **STOTSFIELD BURN** until the 1920s. It was begun in 1863 by the **Rookhope Valley Mining Co** and passed through the ownership of a number of small companies. Following a period of closure it was re-opened by The Weardale Lead Co Ltd in 1914 to work fluorspar. It comprised the **Low Shaft** (NY 943423), together with a number of levels, and the **High Shaft** (NY 945423). During a period of suspension the Low Shaft collapsed in 1938. Production was resumed in 1942. For its history after 1950 see the Fluorspar Mines section in Part 2. Locomotive haulage underground was introduced in 1922. There was no main line rail connection.

Gauge : 1ft 10in

| - | 4wBE | BEV | 414 | 1922 | New | (1) |
| - | 4wBE | WR | 4184 | 1947 | New | (2) |

(1) to Boltsburn Mine, Rookhope, by /1929, and returned, c/1932?; for subsequent history see Part 2.
(2) for subsequent history see Part 2.

Another important mine not listed above was **SEDLING MINE** (NY 860411) **(V460)** at Cowshill, which was closed in December 1948.

THE WEARDALE STEEL, COAL & COKE CO LTD
The Weardale Iron & Coal Co Ltd until 28/9/1899
The Weardale Iron Co until 23/7/1863

The original company was formed in 1845 by Charles Attwood (1791-1875), an entrepreneur who had previously operated a soap works at Gateshead (see Imperial Chemical Industries Ltd, Allhusen's Works, Gateshead), and it grew into the second largest iron and coal company in Co. Durham after The Consett Iron Co Ltd. However, by the end of the nineteenth century its iron and steel production had been overtaken by cheaper plants nearer to the coast. At this point it became part of the empire of the rising Teesside entrepreneur Sir Christopher Furness (1852-1912). He had started in shipping, initially with his elder brother Thomas, and in 1891 took over the shipbuilding firm of Edward Withy & Co Ltd to form Furness, Withy & Co Ltd. He soon saw the benefits of expanding into iron and steel production, and in 1898 he acquired the Stockton Malleable Iron Co Ltd and the Moor Iron & Steel Co Ltd, both of Stockton-on-Tees, and the West Hartlepool Steel & Iron Co Ltd and merged them to form the South Durham Steel & Iron Co Ltd. In 1899 he purchased the Weardale Steel & Iron Co Ltd from the Baring family, long its financiers, and re-constituted it. He then used it to take over, on 20th January 1905, the moribund Cargo Fleet Iron Co Ltd of Middlesbrough, though both companies retained their trading identities; subsequently he used the Cargo Fleet company to take over control of The South Durham Steel & Iron Co Ltd. He also owned the engine-making firm of Richardson, Westgarth & Co Ltd of Hartlepool and about 1910 he acquired a controlling interest in Palmers Shipbuilding & Iron Co Ltd of Jarrow. Subsequently other colliery companies were absorbed into this empire, notably The Easington Coal Co Ltd, The South Hetton Coal Co Ltd, The New Brancepeth Colliery Co Ltd, The Trimdon Coal Co Ltd and The Wingate Coal Co Ltd.

Attwood set out to exploit the extensive hematite iron ore deposits in Upper Weardale. He began by taking over an incomplete ironworks near Stanhope, but it was found almost at once that it was more economic to build iron works on the coal field, so the company's main works were built at Tow Law and Tudhoe. Besides the three ironworks, the company had extensive ironstone rights in Upper Weardale, together with a limestone quarry and an extensive railway system, as well as a considerable number of collieries. The firm also developed ironstone mines in North Yorkshire at Spa Wood and Belmont near Guisborough. Unfortunately, because the greatest period of the company's locomotive activity lay in the nineteenth century, and because this was located in remote areas of west Durham and Upper Weardale, not a great deal has survived to inform us which locations used locomotives; so whether all of the locations below actually used locomotives is not known.

Two notes of caution should also be included: some of the ironstone "mines" did not involve shafts or levels, but opencast quarrying, while some of the mines also produced lead, and the working relationship, and sometimes even the ownership, of these in relation to W B Lead and The Weardale Lead Co Ltd (which see) is very unclear. In "joint" situations, like Stanhope Burn and West Slit, the two firms had separate working entrances.

The company began its operations with the
STANHOPE BURN IRON WORKS, near Stanhope **V561**
NY 989399

The construction of this works was begun by Cuthbert Rippon, a local landowner, to be then taken over by Attwood and opened in 1845 or 1846, with one blast furnace. It was served by a ½ mile link to the foot of the Crawleyside Incline at Stanhope on the Stanhope & Tyne Railway. A tramway linked it to deposits of ironstone alongside the Stanhope Burn, the chief of which was **RED VEIN MINE** (NY 988407 – a Level with a surrounding quarry), further up Stanhope Burn. The Stanhope Burn furnace did not work after 1862, but the company continued to use the works as its offices for production in the Stanhope area, and the furnace was not demolished until 1916-1917. Ironstone production at Stanhope Burn Mine ceased in 1878, while Red Vein Mine was officially abandoned on 1st May 1884. However, the company later developed fluorspar mining in the area. The main source of production was the **HOPE LEVEL** (NY 990397), first worked from about 1870, which ceased production in December 1930. This was subsequently re-opened by other owners to mine fluorspar (see Part 2).

153. The first 5, GW 172/1863. Note the smokebox wingplates, the very flared cap to the chimney, the number on the dome and the extraordinarily large cab.

154. The makers' photograph of TUDHOE, BH 57, built in April 1868. She was a large 0-6-0ST for her time, with 16in x 24in cylinders and 4ft 6in wheels.

TOW LAW IRON WORKS, Tow Law   **M562**
NZ 118389 approx.

This works was opened about April 1846 following the opening in the previous May of the Wear & Derwent Junction Railway, a subsidiary of the Stockton & Darlington Railway, between Crook and Waskerley. Tow Law Station was built close to the works, which had six blast furnaces. The first steel rails to be rolled in North-East England were made here in 1862. The works was closed in 1887.

TUDHOE IRON WORKS & COKE OVENS, Spennymoor
N563
NZ 206338 approx.

This works was opened in 1853 and in later years was the company's headquarters. It was served by sidings east of Spennymoor Station on the NER Bishop Auckland – Ferryhill line. The works was the first in the North-East to instal a Bessemer converter for the production of steel in 1861. Eventually four were installed, but they were not very successful and were derelict by the 1890s. The works was closed in 1875, but was subsequently rebuilt and re-opened in 1881. Steel production was stopped abruptly in November 1901. The works' subsequent history is not clear. Dismantling began in 1908 (see below), and more dismantling was done by Thos W. Ward Ltd between 1912 and 1914. However, at least one furnace remained in blast and in October 1917 the NER granted to The South Durham Steel & Iron Co Ltd (the parent company) powers for the latter to run its locos on to NER sidings "in connection with the blast furnaces". It would seem that iron production of some kind continued until at least 1920, possibly until 1926. However, at least part of the ironworks site was derelict from 1911. After this the main output was coke, from batteries of by-product ovens, and there was also a Tar Plant.

155. The Tow Law works clearly had major engineering facilities, used not only to repair locomotives but to build its own. One was the 0-4-0ST 12, originally named STAR.

There were two loco sheds at the iron works, locomotives from them shunting not only the ironworks and coke ovens but also **TUDHOE COLLIERY** and **TUDHOE GRANGE COLLIERY** (for both see below). The latter closed in March 1885, but Tudhoe Colliery, latterly combined with Croxdale Colliery (see below), continued to produce coal until September 1935. The modern coke works began in November 1907 with 60 Koppers regenerative ovens, to which 60 more were added in the summer of 1908. 60 of the ovens had been rebuilt by 1934, the remaining 60 then being closed down. The coke car was a combined car and traction unit, built by James Buchanan of Liverpool in 1931. The by-products plant manufactured various tar products, sulphate of ammonia and crude benzole. Tudhoe Coke Ovens & By-Products Plant passed to NCB Northern Division No. 5 Area on 1st January 1947.

*

Charles Attwood needed a railway system in Weardale to enable him to bring out iron ore to his works at Tow Law, but with the Stockton & Darlington Railway declining to build a line, he was compelled to build his own. With the village of Bolt's Burn (later Rookhope) his first target, construction began in 1846. This first section was called the

WEATHERHILL & ROOKHOPE RAILWAY
V564
This began at a junction with the Wear & Derwent Junction Railway's line from Stanhope to Consett (later the NER Stanhope & Tyne branch) at Parkhead, two miles north of Stanhope and ½ mile north

of the Weatherhill Engine. Although the first 4¼ miles were relatively level, the line reached 1670 feet, the highest railway ever built in Britain, and crossed bleak, treeless, windswept moorland to reach Bolt's Law. Almost certainly this section was worked by a locomotive from the beginning, housed in a shed at Bolt's Law. But from here a fearsome incline was needed to get down into Rookhope, 2000 yards long on a gradient generally at 1 in 12, but with sections as steep as 1 in 6. This was worked by the **Bolt's Law Engine** (NY 949442).

Rookhope was a major lead mining centre, and the new railway was extended to serve **Bolt's Burn Mine** (NY 936428) owned by W B Lead Co (see the entry for The Weardale Lead Co Ltd). A large **DEPOT** (NY 930431) was opened here, and a loco was allocated to Rookhope to shunt the area, certainly by 1854.

Three branches were subsequently built from this section. The first, from a junction ¾ mile east of the Bolt's Law Engine, served gravel pits nearby. About 1862 the Iron Co was persuaded to use the same point from which to build the **Sikehead branch**, to serve the lead mines at **Sikehead** (NY 955465) and **Whiteheaps** (NY 948473), owned by the **Derwent Lead Mining Company**. The first 600 yards to Sikehead were owned by the Iron Co, but who owned the incline from here down to Whiteheaps, worked by a **stationary engine**, is not clear. The mines closed about 1872, probably also closing the branch.

156. Another was 15, built in 1873 and formerly named WOLSINGHAM. The shape of the entrance to the cab was a distinctive Weardale feature.

The final branch ran from Rookhope up the Rookhope Burn to Groove (Grove) Rake, a distance of about three miles. One source suggests that construction of the line began about 1851 and was extended in stages, perhaps not being completed until about 1860. This served **FOULWOOD MINE** (NY 933429), the **Rookhope** or **Lintzgarth Smelt Mill** (NY 926427), owned by W B Lead, **RISPEY MINE** (NY 911428) and **GROOVE RAKE MINE** (NY 896442), and may have been extended to the extensive ironstone deposits at **FRAZER'S HUSH** (NY 884446), which were opencasted.

The company's main ironstone workings lay west beyond Rookhope and to reach them a much longer line was needed, the

ROOKHOPE & MIDDLEHOPE RAILWAY V565

Details of the construction of this line are scant and contradictory, and it may have been opened in three or four sections. The first, short, section was level, serving the **Boltsburn Washing Mill** (NZ 938425), owned by W B Lead, but then the line needed to rise to 1300 feet. Here the **Smailsburn Incline** was built, but although basically a self-acting incline, because full traffic had to pass in both directions, old tenders were used as counter-balances, being filled with water when they were to travel downhill and then being emptied at the bottom. This incline may also have been where JF 1142/1868 (see below) worked – an extraordinary locomotive fitted with a drum, driven by a winch and mounted between the frames, around which the incline rope was wound, with the drum flanked by two boiler barrels fitted to one firebox. Near the meetings of the incline was the **SMAILSBURN MINE** (NY 941415); how its traffic was incorporated into the operation of the incline is not known.

The next 1½ miles, to Heights, was virtually level and was worked by main-and-tail rope haulage controlled by the **Bishop's Seat Engine** (NY 936405). At Heights, **HEIGHTS MINE** (NY 925388) was developed, producing lead and ironstone, but in the 1880s this was abandoned in favour of developing **HEIGHTS QUARRY** to exploit the large limestone deposits there. By 1858 the line had been extended for a further 1¼ miles to run up Middlehope Burn, north of Westgate village, to **SLIT PASTURE MINE** (NY 909384), though the ironstone was actually opencasted. Nearby there were several mines producing both lead and and ironstone, and for these to bring their production by horse and cart to the railway, the company opened **SCUTTERHILL DEPOT** (NY 909390), 300 yards above Westgate. The line was eventually extended again for a further ¾ mile to **MIDDLEHOPE MINE** (NY 904402), from where a reverse ran back down the western side of the burn for ½ mile to serve **WEST SLIT MINE** (NY 903393), giving a total distance from Parkhead of about 11 miles.

With no other railway in the area, the Scutterhill Depot rapidly became the railhead not only for the upper parts of Weardale, but also Teesdale, Tynedale and Allendale. Thousands of tons of food, drink, coal, animal feed and ironmongery came to be handled through the depot, all taken away by horse and cart struggling up the 1 in 5 hill from Westgate. To relieve this, the company decided to build the **Scutterhill Incline**, a ½ mile bank at 1 in 6 from a junction ½ mile east of the depot and worked by the **Scutterhill Engine** (NY 912389) near the bank head.

When locomotive working on the Rookhope & Middlehope Railway was introduced is not known, but presumably it began with the opening of the line westwards from Heights, which may also have seen the demise of the Bishop's Seat Engine. The first locomotive known to have worked here arrived in 1868. Curiously, none of the Ordnance Survey maps for this section show a building which might have been a locomotive shed, though there must have been one.

Despite the opening of the Stockton & Darlington Railway's line to Stanhope in 1862, this was some way from the communities which the Iron Co's line served. So in 1863 the company purchased a second-hand passenger coach, followed by another in 1871. A passenger service between Rookhope and Parkhead operated three times a week, though it would seem unlikely that the coaches ran over the Bolt's Law incline. How long this continued is not known, although it was still being operated about 1898.

The quantity of ironstone obtained from Weardale declined slowly in the last quarter of the nineteenth century, until by the mid 1890s production from sources along the railway had ceased altogether. Although the company was now obtaining iron ore from mines in North Yorkshire, it did continue to operate **CARRICK'S MINE**, near Ireshopeburn (NY 861378) **(V566)**, formerly called Craig's Level and worked by The Weardale Lead Co Ltd (which see). During the 1914-18 War a **rope incline** was installed between the mine and the B6293, part of it through a tunnel; how long this was operated is not known. Production here suffered various periods of suspension, although the mine was not finally abandoned until May 1949. With the abandonment of Grove Heads Mine nearby by the company in May 1950 ironstone production in Weardale finally came to an end.

The opening in 1895 of the NER's line between Stanhope and Wearhead took further traffic away, and the Scutterhill Incline closed in 1897, although the depot remained open. By contrast, the output from Heights Quarry eventually rose beyond the capacity of the railway to handle it, and so in 1915 German prisoners of war were used to build a self-acting incline, known as the **Cambo Keels Incline**, from Heights down to the NER line between Eastgate and Westgate. The Scutterhill Depot was officially closed in 1917, although the line remained open to a point ½ mile east of the depot in order to load timber felled from a nearby plantation.

The Rookhope Smelt Mill closed in May 1919, and the remaining section of the former Grove Rake branch, between Rookhope and Rispey, closed on 20th February 1923. In the previous year the company had constructed an aerial flight between Heights Quarry and the NER, which brought the closure of both the Cambo Keels Incline and the line between Bolt's Burn and Heights, although it was short-lived, for the quarry itself closed in 1923. These closures left only the section between Rookhope and Parkhead, and with the construction by The Weardale Lead Co Ltd of an aerial flight between Rookhope and Eastgate, this last section of the system ceased to be operated on 15th May 1923.

Even despite its closure, much of the system was maintained and the locomotives were repaired in their sheds; the last did not depart until 1943. From 1st June 1923 a horse-drawn truck was operated twice weekly between Parkhead and Bolt's Law to provision the inhabitants of the cottages there. In addition, there was the unique service provided on "The Stanhope, Rookhope & District Railway" by A.E. Bainbridge of Stanhope Castle. He owned the grouse shooting on Stanhope Common, and to assist his guests in their shooting parties between 12th August and December he brought to the line between Parkhead and Rookhope a four-wheeled, battery-operated truck, even issuing (humorous) tickets, including dog tickets, and railway rules. His gamekeepers used the truck during other times of the year. This "service" had almost certainly ceased by 1939. The section itself was eventually lifted in June 1943, although the derelict truck remained at Parkhead until 1958, when it was removed by British Railways.

References : *The Private Railways of County Durham*, Colin E. Mountford, Industrial Railway Society, 2004, and *The Remarkable Rookhope Railway*, L.E. Berry, *Back Track*, Vol.10, Nos 8 & 9 (August and September 1996), Atlantic Publishing.

The company also worked the various collieries below, though whether locomotives were used at all of them is not known:

BLACK PRINCE COLLIERY, Tow Law — M567
NZ 121396

The sinking of this colliery began in 1846, so it was probably opened in 1847 or 1848. It was served by a ½ mile branch from what became the NER Bishop Auckland – Consett branch, ½ mile west of Tow Law Station. Probably about the beginning of the twentieth century the **Victoria Pit** (NZ 128405) was sunk about one mile to the north-east, and a narrow gauge tramway, presumably rope-worked, was laid to serve it. The colliery suffered various closures, including between 1914 and 1923, and was finally shut down in June 1929, its remaining coal being combined with West Thornley Colliery (see below). There is evidence that at least one locomotive was kept here, perhaps two at times, but when the colliery re-opened in 1923 no loco was used here, the colliery presumably being worked by the LNER, A "coke ovens apparatus engine" was used on the beehive coke ovens here; these were closed in April 1914.

CROXDALE COLLIERY, Croxdale — N568
known as **SUNDERLAND BRIDGE COLLIERY** until /1875
NZ 266371

This colliery should not be confused with an earlier Croxdale Colliery (NZ 267373), with a different rail link and not owned by this company. This later colliery, about 1½ miles south of its namesake, was opened in 1875 and was served by a ½ mile branch from the NER Durham – Darlington line at Croxdale Station. There were also beehive ovens here. These closed in January 1915 and the colliery followed in March 1915, but was kept on care-and maintenance. By this time it was combined with Tudhoe Colliery (see below), to the south-west, although it retained a separate surface. The Croxdale area recommenced production in 1918, though with its output wound at Tudhoe for using at Tudhoe Coke Ovens. The combined colliery was finally closed in September 1935.

The loco shed was situated on the north side of the colliery, though few details are known about the locomotives that worked here.

HEDLEY HILL COLLIERY & COKE OVENS, near Waterhouses — K569
NZ 165412

This colliery was opened about 1876 and was served by a ½ mile branch from the former Stanley Bank Foot on the NER Stanley branch. It was closed between 1908-1912, 1914-1915 and 1921-1923, and was finally shut down in November 1929. As elsewhere, there were beehive coke ovens here. When the colliery was working it would seem that one locomotive was kept here.

It would seem that for a short time in the later nineteenth century the company also owned **HEDLEY HOPE COLLIERY** (NZ 135394) **(M570)** and **EAST HEDLEY HOPE COLLIERY** (NZ 158404) **(K/M571)** both near Tow Law (for earlier history see Thomas Vaughan & Co and for later history see both Sir B. Samuelson & Co Ltd and The Bearpark Coal & Coke Co Ltd).

THORNLEY COLLIERY, Thornley — L572
The Original Hartlepool Collieries Co until 4/1885
NZ 365395
LUDWORTH COLLIERY, Ludworth — L573
The Original Hartlepool Collieries Co until 4/1885
NZ 363415
WHEATLEY HILL COLLIERY, Wheatley Hill — L574
The Original Hartlepool Collieries Co until 4/1885
NZ 385393

Thornley Colliery, opened in 1835, Ludworth Colliery, probably opened in 1842 and Wheatley Hill Colliery, opened in 1869, were taken over in April 1885 from the bankrupt William Ford, trading as The Original Hartlepool Collieries Co (which see). Thornley Colliery was served by what had become the two mile NER Thornley Branch from Thornley Junction on what had become the NER Sunderland &

Hartlepool line via Wellfield. Ludworth Colliery was served by a 1¾ mile branch from Crow's House Junction, a mile from Thornley Junction, though the final ¾ mile was owned by the company. Wheatley Hill Colliery was served by a one mile company-owned loop from the Thornley Branch, beginning at Wheatley Hill Junction, ¼ mile from Thornley Junction. The system was worked by locomotives from a shed at Thornley Colliery. All three collieries were badly flooded. Ludworth was re-opened on 31st October 1887, but Thornley and Wheatley Hill were not re-started until about August 1889.

By 1900 a ¼ mile branch had been built from the north-west link between Wheatley Hill Colliery and the Thornley branch to serve Wheatley Hill village, while another ¼ mile NER line from Crow's House Junction served **CROW'S HOUSE BRICK WORKS** (NZ 383399) **(L575)**. Production at Ludworth was suspended between 9th August 1902 and either December 1913 or January 1914, and finally closed with the beginning of the miners' strike in March 1921, while the brickworks also subsequently closed.

There is a reference to a workman's train being operated between Ludworth and Wheatley Hill on 13th August 1909. How long before and after this date this service ran is not known.

Thornley and Wheatley Hill Collieries passed to NCB Northern Division No. 3 Area on 1st January 1947.

TUDHOE COLLIERY, Tudhoe, Spennymoor   **N576**
NZ 257356

The sinking of this colliery began on 7th November 1864 and was completed on 7th July 1866. It was served by a branch just over 1½ miles long from the NER Ferryhill – Bishop Auckland line, ¼ mile east of Spennymoor Station, with a link into Tudhoe Iron Works (see above). Beehive coke ovens were built, and a "coke oven apparatus engine" was used here, as at Black Prince Colliery above. A **brickworks** was developed ¼ mile south of the colliery. The colliery, coke ovens and brickworks were worked by locomotives from the two loco sheds at Tudhoe Iron Works. From 1916, and possibly before this, the colliery was combined with Croxdale Colliery (see above), though retaining separate surfaces. The combined colliery was closed in September 1935. In 1941 the **Tudhoe Park Coal Co Ltd** re-opened Tudhoe as **Tudhoe Park Colliery**, which passed to NCB Northern Division No. 5 Area on 1st January 1947.

TUDHOE GRANGE COLLIERY, Tudhoe, Spennymoor   **N577**
NY 265338

The sinking of this colliery began on 5th May 1869 and was completed on 2nd September 1870. It was also served by the branch serving Tudhoe Colliery, though in fact it was situated adjacent to Tudhoe Iron Works. It was closed in March 1885 and apparently never re-opened; presumably its coal was worked from Tudhoe Colliery.

The company also owned **MIDDRIDGE COLLIERY**, near Shildon (NZ 248253) **(R578)** which was opened in 1871 but had its production "suspended" in 1911, and **WEST THORNLEY COLLIERY**, near Tow Law (NZ 136386) **(M579)** started in 1861. There were beehive coke ovens here, which were put out in April 1914. The colliery was closed from May 1925 to 1929 and production was suspended in 1931. It was re-opened by **The Wingate Coal Co Ltd** (which see), who closed it in 1939, whereupon production was resumed by the Weardale company; it was closed soon afterwards, but passed on 1st January 1947 to NCB Northern Division No. 4 Area, who then re-opened it. No evidence survives of locomotives being used at either location (though one locomotive was named MIDDRIDGE).

The locomotives
With much of the company's activity concentrated in the second half of the nineteenth century, and also located in the more remote areas of west Durham, the surviving information about its locomotives is very limited. The list below is an attempt to summarise what it known, but the authors would be the last to vouch for any cast-iron accuracy. It would seem that at first locomotives carried only names, and that a numbering scheme had been introduced by the mid-1880s, though on what basis is not known. However, a photograph of GW 172/1863 shows this loco with '5' on the dome, and if this is a makers' photograph of the loco as new, then clearly the numbering scheme might even have begun with the first purchased. However, this photograph is the only one known which shows a locomotive carrying both a name and a number together; the vast majority of known photographs show locomotives carrying only a number, which would suggest that the names were removed at an early date.

Such locomotive transfers as are reported are given, but these too are very sketchy. Where a location is given but no date it means that the locomotive concerned is known to have been here but the dates when it arrived and left are unknown. "New" only means new to the Weardale Coal & Iron Co Ltd, not New to a known location unless stated. It would seem that locomotive repairs and rebuildings were at first carried out at Tow Law Iron Works and then at Tudhoe Iron Works. After this works closed, repairs were undertaken at the workshops at Thornley Colliery.

The loco sheds are coded as follows:

BL	Bolt's Law	R	Rookhope
BP	Black Prince Colliery	T	Thornley Colliery
C	Croxdale Colliery	TC	Tudhoe Colliery
HH	Hedley Hill Colliery	TCO	Tudhoe Coke Ovens
HQ	Heights Quarry	TIW	Tudhoe Iron Works
M	Middlehope (exact location unknown)	TL	Tow Law Iron Works

Gauge : 4ft 8½in

No.	Name	Wheel	Cyl	Maker	Works No.	Date	Notes	Disposal
	-	0-4-0	IC	Fairbairn		1842		
	reb	0-4-2	IC	?		1850	(a)	(1)
	at R in 12/1854							
	-	0-4-0	IC	Fairbairn		1842		
	reb	0-4-2	IC	?		1850	(b)	
	at BL, /1856 – R by late 1860s – TIW, probably 12/1879							Scr c /1880
	-	0-6-0	OC	Hackworth		1842	(c)	
	M – R – BL							(2)
	-	0-6-0	OC	Hackworth		1845	(c)	
	M – R – BL							(2)
No.1	JOHN RENNIE	0-4-0T	OC	GW	178	1864	New	Scr
No.1		0-4-0ST	OC	N	2280	1878	(d)	(3)
No.1		0-4-0ST	OC	CF	1196	1900	New	
	C							Scr
2		0-6-0ST	OC	?		?	(e)	
	TIW-T-TIW	reb		Thornley		1920		(4)
3	CRAWLEYSIDE	0-4-0T	OC	GW	128	1861	New	Scr
4	(WEARDALE)	0-4-0ST	OC	Joicey	210	1869	(f)	
	reb	0-6-0ST	OC	?		1876		
	reb			HL	7409	1897		
	BL – M – HQ – R (not necessarily in this order) – BL; repaired at BL, /1924 – T c1930; rebuilt at T, /1931							(5)
5	CHARLES ATTWOOD	0-4-0T	OC	GW	172	1863	New	
	reb	0-4-0ST	OC	Thornley		?		
	T							Scr
5		0-4-0ST	OC	?		?	(g)	
	reb	0-6-0ST	OC	?		?		
	reb			HL		1910		
	reb			Thornley		1925		
	TIW- T, latterly working TCO							(6)
6	TOW LAW	0-6-0	OC	J.Bond,	c1870-1874		New	
	reb	0-6-0ST	OC	Tow Law		?		
	BP – T c/1914							(7)
6		0-4-0ST	OC	KS	4027	1919	New	
	T							(8)
7	SPA WOOD	0-4-0ST	OC	Joicey	?	1872	New	
	reb			Tudhoe		1903		(9)
8	MIDDRIDGE	0-4-0ST	OC	Joicey	?	1870	New	(9)
9	(SEDGEFIELD)	0-4-0ST	OC	Joicey	?	1874	New	
	R; TL; R (when railway closed, /1923)							(10)
10	BELMONT	0-4-0VBT	OC	HW	?	?	(h)	
	TIW							s/s
10	MICKLETON	0-4-0ST	OC	AB	305	1888	(j)	
	TIW							(9)
11	ZEPHYR	0-4-0ST	OC	J.Bond, Tow Law		?	New	s/s

12	(STAR)	0-4-0ST	OC	Tow Law	?	New		
		reb		Tudhoe	1909			s/s
12	later 21	0-4-0ST	OC	KS	4028	1919	New	
	at TIW by 8/1939, working TCO							(6)
13	CROXDALE	0-4-0ST	OC	Joicey	? c1875	New		
	reb	0-6-0ST	OC	?	?			s/s
14	(BLACK PRINCE)	0-4-0ST	OC	Joicey	? c1875	New		
	reb	0-6-0ST	OC	Tudhoe	c1899			
	reb			Thornley	1928			
	TC							(11)
15	(WOLSINGHAM)	0-6-0ST	OC	Tow Law	1873	(k)		
		reb		Tudhoe	1895			
		reb		Tudhoe	1905			
	TC-T?							Scr by 7/1939
16	(TUDHOE)	0-6-0ST	OC	BH	57 1868	New		
		reb		Tudhoe	?			Scr
No.16		0-4-0ST	OC	AB	1066 1906	New		
	T							Scr after 4/1918
17	BISHOP MIDDLEHAM	0-6-0ST	OC	RWH	1554 1872	New		(12)
No.17		0-6-0T	OC	KS	3098 1918	(m)		
		reb		Thornley	1938			
	T							(13)
18	(SUNDERLAND BRIDGE)	0-6-0ST	OC	RWH	1622 1874	New		
	TC (latterly for TCO), by 8/1939 – (11) - T							(14)
19	(formerly No.19)	0-6-0ST	OC	BH	704 1882	New		
		reb		Tudhoe	1909			
	HQ – TIW c12/1899 - HQ c12/1907 – TIW /1908 – HQ /1909							(15)
20		0-6-0ST	OC	Tow Law	?	(n)		
		reb		Tudhoe	1908			
	R – HQ /1899 – R 1908							(16)
21	(QUEEN)	0-6-0ST	OC	BH	31 1867	(p)		
		reb		Tudhoe	1896			
	T – TIW – R – TIW – R – TIW							(17)
No.22	PRINCESS	0-4-0ST	OC	MW	466 1873	(p)		
	T – M – HQ – R – BL (not necessarily in that order) – HH c/1901							(18)
23	(ALBERT)	0-4-0ST	OC	MW	492 1874	(p)		
	T; used normally at HQ or R							(19)
No.24		0-6-0ST	OC	AB	1321 1913	New		
		reb		Thornley	1921			
		reb		Thornley	1929			
	T							(13)
25		0-4-0ST	OC	P	521 1891	(q)		
		reb		Thornley	1922			
	R (see also note r)							(20)
26		0-4-0CT	OC	D	2051 1884	(r)		
	TIW							(9)
27	later No.27	0-6-0T	OC	KS	3100 1918	(m)		
		reb		Thornley	1942			
	T							(12)
-		2-2wTG	IC	JF	1142 1868	(s)		s/s
No.1		0-4-0ST	OC	HL	2412 1899	(t)		
	TIW							(21)
75317		0-6-0ST	IC	VF	5307 1945	(u)		
	T							(13)

There were also two steam rail cranes at Tudhoe Iron Works. At least one was here by the 1890s.

(a) built new for the Manchester & Leeds Railway, No.43, RHINE; became Lancashire & Yorkshire Railway No.143 in /1850; rebuilt to 0-4-2 in 1850; purchased by Stockton & Darlington Railway, 8/1854; on loan to The Weardale Iron Company by 12/1854.

(b) built new for the Manchester & Leeds Railway, No.41, ELBE; became Lancashire & Yorkshire Railway No.141 in /1850; rebuilt to 0-4-2; ex Lancashire & Yorkshire Railway, 1/1855; delivered to Tow Law Iron Works, 5/1855; whether this loco was later allocated or carried a number is not known.

(c) these were both built new for the Stockton & Darlington Railway, as No.8 LEADER (Hackworth 1842) and No.31 REDCAR (Hackworth 1845). In 1863 they passed to the NER, from whom they were purchased in 11/1868; here they were named JENNY LIND and BELLEROPHON, but it is not known which engine carried which name. It would appear likely that these locomotives were allocated/carried numbers, but what they were is not known.

(d) ex Yorkshire & Derbyshire Iron Co, New Carlton Colliery, Yorkshire (WR).

(e) identity and origin unknown; one source suggested it might have been built by BH.

(f) New; the Joicey works number is believed to have been 210, although 240 has been quoted; delivered to Bolt's Law, 16/1/1870.

(g) identity and origin unknown; may have been a BH loco; had '886' stamped on the motion.

(h) probably built between 1870-1875; whether acquired new or second-hand is unknown; one source gave this loco as numbered 11.

(j) ex Joseph Torbock, dealer, Middlesbrough, Yorkshire (NR); previously owned by The Cargo Fleet Iron Co Ltd, Middlesbrough, Yorkshire (NR).

(k) New; another version gives built at Tudhoe Iron Works and rebuilt at Thornley in 1895.

(m) ex Furness Shipbuilding Co Ltd, Haverton Hill, by 1/1929 (see Cleveland & North Yorkshire Handbook).

(n) New; believed built at Tow Law Iron Works in 1880s, possibly with parts supplied by BH.

(p) ex The Thornley Colliery Co, with Thornley, Ludworth & Wheatley Hill Collieries, 4/1885.

(q) ex Broomhill Collieries Ltd, Broomhill, Northumberland; one source says that the loco worked at BL and HQ in 3/1916, and was one of those stored at Rookhope when the Weatherhill & Rookhope Railway closed in 1923; however, P supplied a new boiler to Broomhill in 1922; the loco numbering scheme would suggest that it may well have been here before 1922.

(r) ex The Cargo Fleet Iron Co Ltd, Cargo Fleet Iron Works, Middlesbrough, Yorkshire (NR).

(s) New; see special note on page 411.

(t) ex South Durham Steel & Iron Co Ltd, Hartlepool Works, West Hartlepool, /1917 (see Cleveland & North Yorkshire Handbook).

(u) believed to have come from John Bowes & Partners Ltd, Springwell Bank Foot loco shed, Wardley, 8/1946, but owned by War Department, 75317, where her last known location was the Longmoor Military Railway, Liss, Hampshire.

(1) had left Rookhope by /1859; presumably returned to the Stockton & Darlington Railway.

(2) JENNY LIND (see note c above) was taken out of service in /1900 and sold for scrap (no details are known). BELLEROPHON was taken out of service in /1902; its disposal is unknown.

(3) new boiler supplied by AB, 7/1908; to The Cargo Fleet Iron Co Ltd, Cargo Fleet Iron Works, Middlesbrough, Yorkshire (NR) (a subsidiary of The Weardale Steel, Coal & Coke Co Ltd).

(4) to The Easington Coal Co Ltd (a subsidiary of The Weardale Steel, Coal & Coke Co Ltd), Easington Colliery, Easington.

(5) to The New Brancepeth Colliery Co Ltd (a subsidiary of The Weardale Steel, Coal & Coke Co Ltd), New Brancepeth Colliery & Coke Ovens, New Brancepeth, after 12/1933.

(6) to NCB No. 4 Area, with Tudhoe Coke Ovens, Spennymoor, 1/1/1947.

(7) one source stated that the loco was scrapped about 1915, another that it was scrapped in 1925.

(8) to The Easington Coal Co Ltd (a subsidiary of The Weardale Steel, Coal & Coke Co Ltd), Easington Colliery, Easington, by 5/1925.

(9) one source stated scrapped at Thornley Colliery; another gave to The Cargo Fleet Iron Co Ltd, Cargo Fleet Ironworks, Middlesbrough, Yorkshire (NR).

(10) Scr /1924, but whether at Rookhope or Tudhoe is not known.

(11) to The Easington Coal Co Ltd, Easington Colliery, Easington, and returned by 18/6/1934; to Tudhoe Coke Ovens, Tudhoe, by 8/1939; to Thornley Colliery, /1940; to NCB No. 3 Area, with Thornley Colliery, Thornley, 1/1/1947.

(12) to Broomhill Collieries Ltd, Broomhill, Northumberland, c/1915-20.

(13) to NCB No. 3 Area, with Thornley Colliery, Thornley, 1/1/1947

(14) loaned to The New Brancepeth Colliery Co Ltd, New Brancepeth, by 24/3/1940; said to have worked here only one day; returned to Tudhoe; to Thornley Colliery; to NCB No. 3 Area, with Thornley Colliery, 1/1/1947.

(15) put into store at Heights Quarry, 8/1922; to The Easington Coal Co Ltd (a subsidiary of The Weardale Steel, Coal & Coke Co Ltd), Easington Colliery, Easington, /1943.
(16) put into store at Rookhope, /1923; taken to Thornley Colliery and scrapped c/1939.
(17) sold to an unknown purchaser in Liverpool, /1921.
(18) hired to Wake & Hollis Ltd, contractors; this firm had agreement to remove ballast and other materials from both Tow Law and Tudhoe Ironworks, and from 1903 the firm was also was controlled by Sir Christopher Furness; returned from Wake & Hollis Ltd, and then normally used at Hedley Hill Colliery; said to have been sent to Thornley Colliery for scrapping.
(19) Scr at Tudhoe Iron Works c /1906; another source stated sent to Thornley Colliery for scrapping.
(20) put into store at Rookhope, /1923; scrapped /1941, but whether at Rookhope or Thornley is unknown.
(21) to The Cargo Fleet Iron Co Ltd, Woodland Colliery, Woodland, c/1918.

Notes

In the early days, probably in the 1850s, it is believed that the company hired at least one locomotive from Robert Stephenson & Co of Newcastle upon Tyne.

In the *Colliery Guardian* for 2nd June 1876 there appeared the following: "One new tank locomotive 14³⁄₈in outside cylinders, four coupled, built specially for steep gradients and sharp curves, for sale, Tudhoe Ironworks". Nothing is known about this locomotive.

In *The Engineer* of 29th September 1882 the company advertised for a six-coupled locomotive with 13in x 18in/14in x 18in cylinders and 3ft 6in driving wheels; whether this was acquired is unknown.

The locomotives SPA WOOD and BELMONT were named after the company's ironstone mines near Guisborough in North Yorkshire.

The Tudhoe Iron Works site was cleared during 1967 and replaced with the Bessemer Park housing estate. The edition of *Municipal Engineering* dated 23rd May 1969, in describing the development of homes for 4,000 people on this estate, states that one of the problems found by the contractors responsible for the sub-surface work were "two railway engines", apparently buried. No further details are given.

The locomotives below were used at the Rolling Mill at Tudhoe Iron Works:

Gauge : 3ft 0in

		0-4-0ST	OC	JF	6680	1892	New	(1)
	-	0-4-0ST	OC	JF	6681	1892	New	(1)
	-	0-4-0ST	OC	Miller		?	(a)	(2)

(a) probably new, but not confirmed; date of construction is unknown; one source gave the loco as built by the firm in 1893, works No.932, but on what authority is unknown; it is believed to be the only locomotive constructed by the firm.

(1) to Thos W. Ward Ltd, included in the contract to demolish part of Tudhoe Iron Works; c/1908; offered for sale by Thos W. Ward Ltd in *Machinery Market* on 20/11/1908, 17/9/1909 and 17/3/1911. One source states sold to John Macdonald, Inverness, Invernesshire, Scotland; this man operated the Mills of Airdrie Plantation Railway, near Ferness in Nairnshire, Scotland, which is believed to have been 3ft 0in gauge, and so it is possible that the locomotives actually went there, but there is no other evidence.

(2) to Thos W. Ward Ltd, included in the contract to demolish part of Tudhoe Iron Works; c/1908; offered for sale by Thos W. Ward Ltd in *Machinery Market* on 20/11/1908, 17/9/1909 and 17/3/1911; this locomotive was also offered for sale in the *Colliery Guardian* on 10/10/1910; s/s.

The locomotive below worked at Tow Law Iron Works:

Gauge : 2ft 8in

| LITTLE GRIMSBY | 0-4-0ST | OCG | I.W.Boulton | 1860 | (a) | (1) |

(a) ex I.W. Boulton, dealer, Ashton-under-Lyne, Lancashire, /1865 (rebuilt from 2ft 0in gauge by Boulton in /1864).

(1) one source says that "a narrow gauge loco was left derelict on the slag heaps when the Tow Law furnaces closed" (in 1887), so presumably it was used to convey slag between the furnaces and the slag heaps; another source says the loco was scrapped.

However, another source says that the loco that worked the slag heaps was a "0-4-0 coffee pot loco" (i.e., was a vertical-boilered locomotive), which was stored until the furnaces were dismantled.

WEARDALE STEEL (WOLSINGHAM) LTD (registered 25/3/1983)

WOLSINGHAM STEELWORKS, Wolsingham T580
Wolsingham Steel Ltd until 11/7/1983 NZ 082372
Wolsingham Steel Co Ltd until 1/11/1981
Doxford & Sunderland Ltd until 1/11/1972
Doxford & Sunderland Shipbuilding & Engineering Co Ltd until 23/3/1970

This works was the final entrepreneurial venture by Charles Attwood (see previous entry); for its earlier history see the entry for John Rogerson & Co Ltd. It was served by sidings north of the BR Wear Valley branch, ½ mile east of the former Wolsingham Station. For many years the works' output concentrated on marine foundry work. After the management and workers' buyout in 1983 the firm widened the range of its products, and it continued in production, surprisingly for such an industry and especially given its situation in rural Weardale. However, on 3rd October 2002 Weardale Steel (Wolsingham) Ltd went into administration. The premises were divided and passed to new owners, but were re-united in 2004 under Weardale Castings & Engineering Ltd (registered 27/5/2004), which continues in business.

For the majority of Doxford's ownership shunting was done by steam crane, until a locomotive was re-introduced in 1970. There were five steam rail cranes here: Smith 8371/1913, which was acquired by the North of England Open Air Museum, Beamish, for preservation; Smith 8718, s/s after 23/8/1979; Booth 5902/1952, acquired by the Tanfield Railway Co for preservation but subsequently returned here to the Weardale Railway Preservation Society, and two unidentified cranes, whose chassis were converted into flat wagons. The loco shed became disused by the early 1980s, and the locomotive was then stabled inside the foundry (NZ 081370). Rail traffic ceased in 1984.

Part of the premises has latterly become the home for the locomotives and rolling stock acquired by the Weardale Railway Preservation Society, subsequently the Weardale Railway Trust, whose aim, with considerable public and private backing, is to re-open the former British Rail line between Bishop Auckland and Eastgate (see Part 2).

Gauge : 4ft 8½in

(No.1)	4wDM	RH	432480	1959	(a)	(1)
(No.2)	4wDM	RH	198325	1940	(b)	(2)

(a) ex Golightly (Developments) Ltd, Ferryhill, acting as dealers, 11/1970; previously Northern Gas Board, Commercial Street Gas Works, Middlesbrough, Yorkshire (NR).
(b) ex Golightly (Developments) Ltd, Ferryhill, acting as dealers, 5/1972; previously Northern Gas Board, Hendon Gas Works, Sunderland.

(1) disused from /1984; scrapped on site between 11/1991 and 12/1991.
(2) used as source of spares for RH 432480; remains scrapped on site by 28/7/1979.

THE WEARMOUTH COAL CO LTD
registered 28/4/1896 to take over the company of the same name registered on 1/1/1878
Bell, Stobart & Co Ltd until 1/1/1878
W. Bell & Partners until c/1870 (owners by late 1850s)

WEARMOUTH COLLIERY, Monkwearmouth, Sunderland H581
called **MONKWEARMOUTH COLLIERY** until the 1890s NZ 394578

For many years this was the deepest colliery in the world. The sinking, on the north bank of the River Wear, was begun on 2nd May 1826. It had to pass through the magnesian limestone, as at Hetton, and was also faced with horrendous problems with quicksand and water. The Maudlin seam was reached on 15th February 1834, at a depth of 1563 feet and an expenditure of £100,000, and the first coal was shipped, from the colliery's own staiths, on 13th June 1835. The Hutton seam was finally reached on 4th April 1846 at 287 fathoms, or 1720 feet. The initial partnership was headed by Messrs. Pemberton and Thompson, and for a time the colliery was called **PEMBERTON MAIN COLLIERY**. To ship the coal **two parallel self-acting inclines** 300 yards long were built down to staiths on the River Wear. The first coal was shipped on 13th June 1835. According to the Supplement to the *Mining Journal* dated 6th August 1870, a **stationary engine** had once been used to haul materials from the quay to the pits. By 1855 there was a link between the colliery and the waggonway from Fulwell Quarry (see Sir Hedworth Williamson's Limeworks Ltd), presumably to allow coal to be used at the latter's lime kilns on the river bank.

For many years the colliery had no main line rail connection, but on 1st July 1876 the Hylton, Southwick and Monkwearmouth Railway was opened, and the colliery was linked to it ½ mile west of

Monkwearmouth Station on the NER Gateshead – Sunderland line. The colliery is not known to have had a locomotive before 1876, so presumably horses were used for shunting until then.

The two inclines down to the river probably survived to around the First World War. They were replaced by a long, steep, curve from the western end of the pit yard, ending in a headshunt, with a reverse back on to the staiths. A photograph of the staiths dating about 1910 does not appear to show this reorganisation, which must therefore date from between 1910 and 1919. It was a hard-working job for one of the six-coupled locos to come back up with the empties.

Following the re-organisation of the company in 1896 it began the sinking of
HYLTON COLLIERY, Castletown, Sunderland **H582**
NZ 366583

This was sunk on the site of the former Castletown Ironworks owned by the **Wear Steel Co Ltd** (which see), which had been auctioned in March 1897 and lay 1¼ miles west of Monkwearmouth Colliery. The work began on 2nd May 1897 and was completed on 25th January 1900. As the new colliery also lay alongside the NER Hylton, Southwick & Monkwearmouth branch the colliery company acquired running powers between the two collieries, under which the company ran its traffic on the northern track and the NER used the southern track. This meant the company had to instal signalling at both collieries. In 1926 the LNER closed its branch west of Hylton, leaving the colliery at the end of the line. A four-coupled locomotive, changed as required, was kept at Hylton for shunting, but otherwise all the traffic was handled by locomotives from the sheds at Wearmouth Colliery, where a locomotive workshop was also established.

The two collieries were vested in NCB Northern Division No. 1 Area on 1st January 1947.

It will be noted that the four-coupled locomotives were numbered, while the six-coupled locomotives carried names, none of the latter type being acquired until working to Hylton Colliery began.

157. An Andrew Barclay catalogue from about 1909 featured WILL, AB 999/1904, hauling a long train of full chaldron waggons over the NER line between Hylton and Wearmouth Collieries in Sunderland.

Gauge : 4ft 8½in

-		0-4-0ST	OC	JF	2834	1876	New	s/s
No.2		0-4-0ST	OC	BH	395	1879	New	Scr c/1922
No.3		0-4-0ST	?	?	?	?	(a)	s/s
No.4		0-4-0ST	OC	AB	641	1889	(b)	s/s
	PHYLLIS	0-6-0ST	IC	AB	833	1898	New	
		reb		HL	7866	1915		(1)
	GEORGE	0-6-0ST	IC	AB	909	1901	New	(2)
	ALEXANDRA	0-6-0T	IC	AB	911	1901	New	
	later BUNNY	reb		HL	6429	1913		(1)
	WILL	0-6-0T	IC	AB	999	1904	New	(1)
	JEAN	0-6-0T	IC	HL	2769	1909	New	(1)
No.3		0-4-0ST	OC	HL	2784	1909	New	(3)
No.5		0-4-0ST	OC	HL	2824	1910	New	(1)
No.1		0-4-0ST	OC	HL	3494	1922	New	(1)

No.2		0-4-0ST	OC	HL	3493	1922	New	(1)
No.4		0-4-0ST	OC	RSHN	6945	1938	New	(1)
	DIANA	0-6-0T	IC	RSHN	7304	1946	New	(1)
No.3		0-4-0ST	OC	RSHN	7307	1946	New	(1)

(a) it is assumed that there was a No. 3 before HL 2784/1909 above, but no details survive.
(b) ex contractors working on Hudson Dock, Sunderland, for River Wear Commissioners; previously Wear Steel Co Ltd, Sunderland.
(1) to NCB No. 1 Area, 1/1/1947.
(2) in *Machinery Market* of 16/4/1915 the company advertised for sale a 16in [cylinder] 0-6-0ST recently overhauled by HL, almost certainly this loco; to War Department, perhaps c/1915, though spares not supplied until 27/2/1917, to Pirbright Camp, Surrey.
(3) to RSH, Newcastle upon Tyne, for repairs, /1945, but repairs found to be impractical; the loco was scrapped at RSH, 4/1946, and replaced by RSHN 7307/1946.

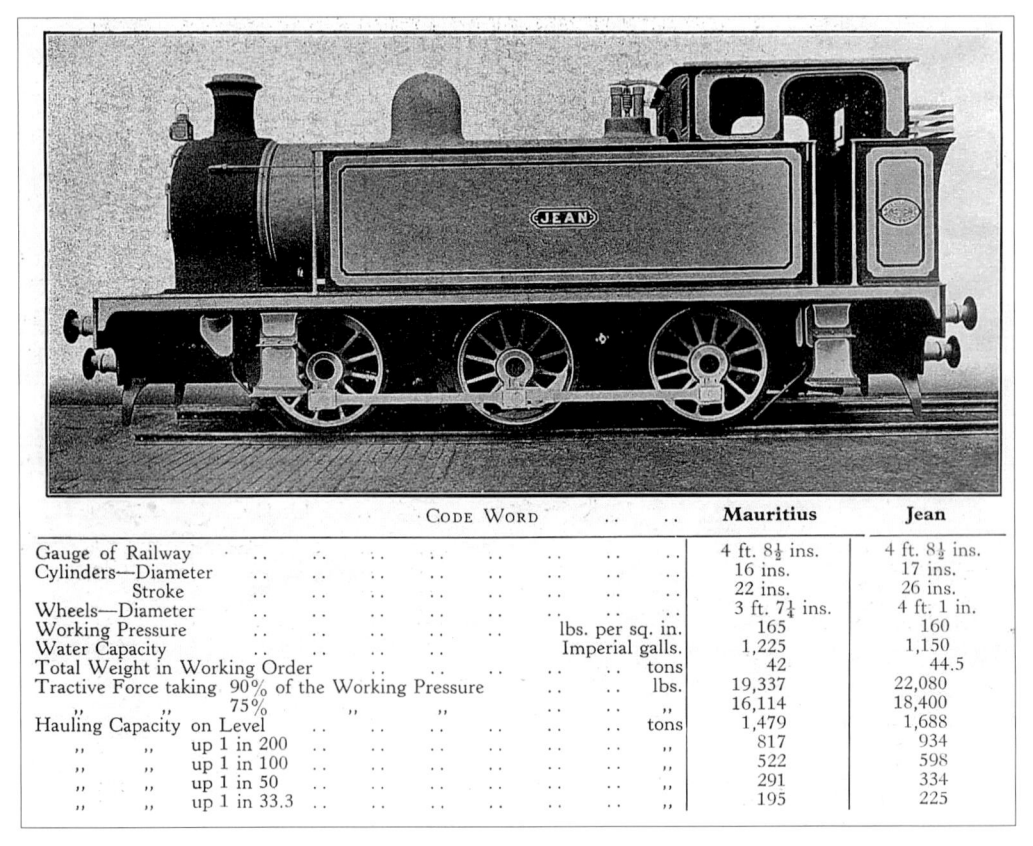

Code Word		Mauritius	Jean
Gauge of Railway		4 ft. 8½ ins.	4 ft. 8½ ins.
Cylinders—Diameter		16 ins.	17 ins.
Stroke		22 ins.	26 ins.
Wheels—Diameter		3 ft. 7¼ ins.	4 ft. 1 in.
Working Pressure	lbs. per sq. in.	165	160
Water Capacity	Imperial galls.	1,225	1,150
Total Weight in Working Order	tons	42	44.5
Tractive Force taking 90% of the Working Pressure	lbs.	19,337	22,080
" " 75% " "	"	16,114	18,400
Hauling Capacity on Level	tons	1,479	1,688
" " up 1 in 200	"	817	934
" " up 1 in 100	"	522	598
" " up 1 in 50	"	291	334
" " up 1 in 33.3	"	195	225

158. A page from a Hawthorn Leslie catalogue of the 1930s, featuring JEAN, HL 2769, presumably as new in 1909, and listing the statistics included to attract potential customers. 'Mauritius' refers to HL 2750, built for Mauritius Railways.

The company and lime making

Although the company only entered lime making for a brief period, it came to acquire two historic waggonways, and so it would seem appropriate not only to include them and their history but also put them into the general context of quarries and lime works at Southwick and Fulwell.

About 1¼ miles north of the River Wear lay a very large area of permian and magnesian limestone, which had been quarried since at least Roman times. As demand for limestone and burnt lime increased in the eighteenth century, quarries were begun at Fulwell (see Sir Hedworth Williamson's Limeworks Ltd) and at Southwick. The first **Southwick Quarry** (NZ 383586 approx.) lay about ½ mile from the river and was begun between 1737 and 1746. The route was probably just a rough track at first, with wooden rails being added later. This is still shown on a map of 1813, but by then it had been superceded by a new and larger **Southwick Quarry** (NZ 381593 approx.) ½ mile further north, which was served by a new waggonway (see Fig.12) down to the river, where there were seven kilns. This was operated by Thomas Brunton of Southwick. Both quarries had been abandoned and their waggonways lifted by the middle of the nineteenth century.

Meanwhile, between the Southwick and Fulwell Quarries two more quarries had been started. These were probably known as **Carley Hill West Quarry** (NZ 387594) and **Carley Hill East Quarry** (NZ 398596), and both were served by waggonways running down to the river. The western system was again operated by Thomas Brunton, while by 1834 the eastern system was owned by Thomas Thompson.

By the time of the 1st edition O.S., surveyed in 1855, both quarries had been superceded by new quarries further north, again probably called **Carley Hill West Quarry** (NZ 379599) and **Carley Hill East Quarry** (NZ 382601), (although only the former, called Carley Hill, is named on the map), with the waggonways extended to serve them, although both systems probably earlier, the last ¼ mile on the Carley Hill West waggonway was operated by a horse-powered windlass, housed in stone buildings straddling the track (see *Railways of Sunderland*, Tyne & Wear County Council Museums, 1985), and it is possible the East waggonway used a similar system over its parallel section. By 1855 both lines are shown with **engine houses** – between the old and new Carley Hill East Quarries (NZ 388598, and named as **Carley Hill Engine**) and at the southern end of Carley Hill West Quarry (NZ 383596), both almost certainly being used to haul waggons out of the quarries. The lines were then single track, with short passing loops for the horse traffic to pass. The "windlass" near the southern end of the West waggonway is also shown. On the river the West Quarry's line ended at the **Wear Lime Works** (NZ 390583), while a few yards to the east the East Quarry's line ended at the **Carley Hill Lime Works** (NZ 391583). The two Carley Hill quarries, their waggonways and limeworks had passed to **E.Burdis & Co** by 1844, and this firm, latterly shown as **E.Burdes & Co**, continued to operate them until at least 1873. By 1877 the West Quarry system was being worked by the **Wear Lime Works Co**.

Information about events in the next fifteen years is confused. From one source it would appear that Carley Hill West Quarry and its waggonway had been taken over by 1881 by **The Wearmouth Coal Co Ltd**, which for reasons unknown had decided to enter the lime burning trade. However, both the minutes of the Southwick Local Board for 1884 and the NER Sidings Map of the same year describe it as owned by Sir Hedworth Williamson, the latter specifically describing the line as "tramway from Fulwell Quarry". The line seems to have closed about 1889. Meanwhile the East Quarry system had passed to **Thompson, Ross & Co** by 1881 and to the **Carley Lime Co** by 1895 – yet in April 1895 the Southwick-on-Wear Urban District Council was discussing repairs to the waggonway with The Wearmouth Coal Co Ltd (Council minutes). Furthermore, from a minute of August 1898 it is clear that the Coal company owned both waggonways. The 2nd edition O.S., surveyed in 1895, shows that the two Carley Hill quarries were connected by this date and called **CARLEY HILL QUARRY**, with an extension of the East waggonway into the former West Quarry, while the West waggonway's link to the quarry had been lifted, leaving only the first ½ mile from the river still in situ. Houses had now been built alongside both lines, and their presence in Cornhill Terrace, later called Southwick Road, was causing problems with the extension of the Sunderland tram system. In the event the Coal company's plan to abandon the East waggonway and to divert its traffic to the West waggonway, which it re-connected to the combined quarry, was approved by the council in March 1901. In September 1902 the company closed Carley Hill Quarry and the West waggonway, ending its involvement in the lime trade. Despite this, the long-standing lime trade from the Quayside to Aberdeen is said to have ended only during the First World War, when the track is also said to have been lifted. The routes of the two lines were subsequently incorporated in the local road system, while the Carley Hill Quarries were absorbed into the new **Southwick Quarry** developed by **Sir Hedworth Williamson's Limeworks Ltd** (which see).

The gauge of the West line was measured from stone blocks examined in the 1970s as 4ft 8½in, the assumption being that the East line was the same. Other than the engine houses described above, the two lines are believed always to have been horse-worked.

WEST HARTLEPOOL HARBOUR & RAILWAY CO/RALPH WARD JACKSON

As this was a public railway this company is strictly outside the scope of this book, but as its involvement with colliery history is so complicated, a brief note was thought to be worthwhile.

During the 1830s and 1840s a considerable expansion of the railway system in the North East took place, due not least to coal owners seeking new shipping outlets for their coal. One of the railways of this period was the Stockton & Hartlepool Railway. Its construction, without an Act of Parliament, began in May 1839, and it was opened for general traffic on 9th February 1841. It shipped its coal at the Hartlepool Tide Harbour (see the entry for the Hartlepool Dock & Railway Co). From 2nd September 1844 the railway leased the Clarence Railway, to which it was connected, and then leased the operation of both railways to Messrs Fossick & Hackworth of Stockton for ten years from 1st November 1844. Fossick & Hackworth took over thirteen Clarence Railway and eight Stockton & Hartlepool Railway locomotives.

The leading promoter and entrepreneur in the Hartlepool area after 1845 was Ralph Ward Jackson (1806-1880). One of his enterprises was the Hartlepool West Harbour & Dock Company, which opened the dock at West Hartlepool on 1st June 1847. He also became a powerful director of the Stockton & Hartlepool Railway. On 30th June 1852 the two enterprises were amalgamated under the title of the West Hartlepool Harbour & Railway Company, which then purchased the Clarence Railway. The enlarged company, of which Jackson was chairman, received its Act of Parliament on 17th May 1853.

Keen to ensure that as much coal traffic as possible came down the railway and used the West Hartlepool dock, Jackson decided to begin acquiring collieries. As it was outside the terms of the company's Act to do this, Jackson did this nominally in his own name. He was helped by the death in 1856 of one John Robson (see entry for J. Robson & Co). By 1855 Robson had built up a major empire of twelve collieries, all in southern Durham. Three of them, Binchester, Little Chilton and Page Bank, had only just begun production, so it may well be that Robson's death came at a financially-inconvenient time. Their output was crucially important to WH&HR, so Jackson acquired them. However, an investigation into the affairs of the Railway led to Jackson having to transfer them to the WH&HR, presumably because he had used the railway company's money. This was done on 23rd February 1860, though *Mineral Statistics* does not record the change until 1862. Jackson was compelled to resign from the railway's board in April 1862, while the Court ruled that it was illegal for the company to own collieries and ordered them to be sold. In the depressed state of the coal trade at the time, this proved to be difficult. In 1865 the WH&HR was taken over by the NER, with five of the collieries still unsold. These were offered for auction on 27th March 1866, and sold as shown.

The twelve collieries are listed below, grouped together according to the royalties on which they were sunk. The second column gives their eventual purchaser and the date of sale, with a footnote to enable their histories to be connected.

Bowburn	James Morrison & Co, 3/1866	(a)
Coxhoe, with West Hetton	James Morrison & Co, 3/1866	(a)
Crow Trees	James Morrison & Co, 3/1866	(a)
Heugh Hall	James Morrison & Co, 3/1866	(a)
South Kelloe	James Morrison & Co, 3/1866	(a)
Binchester	Hunwick & Newfield Coal Co, 1865	(b)
Byers Green	Bolckow, Vaughan & Co Ltd, 1865	(b)
Hunwick	Hunwick & Newfield Coal Co, 1865	(b)
Newfield	Hunwick & Newfield Coal Co, 1865	(b)
Little Chilton	closed, 1865	(c)
Merrington (Spennymoor)	R.S. Johnson & Co	(d)
Page Bank	Bell Brothers	(e)

(a) for previous history see William Hedley & Sons; for James Morrison & Co see The Rosedale & Ferry Hill Iron Co Ltd (in fact, the latter had been registered in June 1864, so presumably the attribution to the former in 1866 is incorrect).

(b) for 1866 *Mineral Statistics* gives The Hunwick Coal Co; for subsequent history of all four collieries see Bolckow, Vaughan & Co Ltd.

(c) see South Durham Coal Co Ltd.

(d) this firm also owned Whitworth Colliery, which shared the same surface with Merrington; the combined site was just east of Spennymoor Station, at the junction with the NER Page Bank branch.

(e) for subsequent history see Dorman, Long & Co Ltd.

WEST HUNWICK REFRACTORIES LTD
Hunwick
M583

The West Hunwick Silica & Firebrick Co Ltd until 6/1963
see also below

NZ 195329 (West Hunwick Colliery)

This complex consisted of **WEST HUNWICK COLLIERY** and a **refractory brickworks**, and lay immediately to the south of Rough Lea Colliery & Brickworks (see The North Bitchburn Coal Co Ltd/The North Bitchburn Fireclay Co Ltd). The colliery was sunk in 1872 by the **Lackenby Iron Co** of Middlesbrough, Yorkshire (NR), whose business was taken over in 1877 by **Lloyd & Co**. On 12th

December 1879 the business was acquired by **Joseph Torbock & Co**, but it had closed down by 1883, presumably due to the depressed economic conditions. Torbock & Co resumed production in the mid-1880s, but the link with the Middlebrough ironworks was subsequently severed, and by 1900 the Hunwick premises were owned by **The Owners of West Hunwick Colliery**, who went out of business in 1910. In 1913 the colliery and the works were re-opened by **The West Witton Ganister Firebrick Co Ltd**. By 1927 this firm had been replaced by **The West Hunwick Silica & Firebrick Co Ltd**, once again based in Middlesbrough. The colliery was abandoned in 1938. The works changed hands again in 1963, as above.

The colliery and its brickworks were served by sidings west of the NER Durham – Bishop Auckland branch, ½ mile north of Hunwick Station. Rail traffic ceased in 1968, and the works was subsequently closed.

159. WEST HUNWICK No. 2, CF 1183/1899, on 24th August 1956. The front end is probably original, but the cab could well be a replacement.

Gauge : 4ft 8½in

-	0-4-0T	OC	FJ	72	1867	(a)	s/s by 7/1939
2	0-4-0ST	OC	CF	1183	1899	(b)	
later WEST HUNWICK No.2		reb	Ridley Shaw		1939		Scr c/1958
FORCETT	0-4-0ST	OC	HL	2465	1900	(c)	(1)
WEST HUNWICK No.3	4wDM		RH	421417	1958	New	(2)

(a) ex Bolckow, Vaughan & Co Ltd, Hunwick Colliery, Hunwick.
(b) ex Pease & Partners Ltd, Tees Bridge Iron Works, Stockton-on-Tees, by 15/8/1931 (see Cleveland & North Yorkshire Handbook).
(c) ex Forcett Limestone Co Ltd, Forcett, Yorkshire (NR), loan, by 13/2/1949, while CF 1183 was away for repairs.

(1) returned to Forcett Limestone Co Ltd, Forcett, Yorkshire (NR).
(2) to Rugby Portland Cement Co Ltd, Strood, Kent, 5/1969.

The locomotive below was ordered by the firm, but there is some doubt as to whether it actually worked at West Hunwick.

Gauge : 2ft 4in

-	4wDM		RH	175420	1936	New	(1)

(1) to Richard Summerson & Co Ltd, Greenfoot Quarry, Stanhope.

WHESSOE PRODUCTS LTD

formed on 25/11/1988 by an amalgamation of **Whessoe Heavy Engineering Ltd** (see below) and **Whessoe Overseas Ltd**; both companies were subsidiaries of **Whessoe plc**

WHESSOE FOUNDRY, Hopetown, Darlington   S584
Whessoe Heavy Engineering Ltd until 25/11/1988
Whessoe Ltd until /1982, when this company was converted to **Whessoe plc** and became a holding company
Whessoe Foundry & Engineering Co Ltd until 8/11/1945
Whessoe Foundry Co Ltd until 9/4/1920
Charles l'Anson & Co Ltd until 30/4/1891
Charles l'Anson & Co until /1862; see also below   NZ 285160 approx.

The origin of this works goes back to William Kitching (1752-1819), who in 1790 opened an ironmonger's shop in Tubwell Row, Darlington, and six years later added a small iron foundry. The business subsequently passed to his two sons, William and Alfred. With the opening of the Stockton & Darlington Railway in 1825 they saw the opportunity to expand their business, and in 1831 they moved out to the Hope Town area of the town and set up north of North Road Station a much larger foundry and engineering works to supply the Stockton & Darlington Railway with locomotives and rolling stock. In 1845 Alfred (1808-1882) bought out his brother. In 1860 the Stockton & Darlington Railway purchased this works for their own use, developing it into the North Road Workshops subsequently owned by the NER, LNER and BR and operated until 1966. Meanwhile in 1859 Alfred, together with his cousin, Charles l'Anson (1809-1884), moved into another works on the opposite side of the railway which they had recently bought from William Lister. This works was named the **Whessoe Foundry** (from Whessoe Grange, a country house nearby), and the business moved into the manufacture of railway structural work, seaside piers, steam cranes, weighbridges and tanks.

Alfred Kitching died in 1882, Charles l'Anson in 1884 and his son James in 1885, and in 1891 Kitching's sons sold the business to Thomas Coates, an energetic and innovative engineer, who set up The Whessoe Foundry Co Ltd. For nearly a century the works was famous for the production of tanks and other pressure vessels, and it also expanded into overseas markets.

Immediately to the south of the works in the 1850s lay the Hope Town Foundry of John Harris (see Summerson's Foundries Ltd), which built a number of locomotives. After Harris' death in 1869 Thomas Summerson acquired this business and moved it to a new site ¾ mile away at Albert Hill and sold the Harris works to Charles l'Anson & Co Ltd for an extension to the Whessoe Foundry. l'Anson also owned the Rise Carr Works (see Darlington & Simpson Rolling Mills Ltd) some ¾ mile to the north of Whessoe, but given that this works manufactured puddled iron, it is almost certain that the five locomotives which Charles l'Anson & Co Ltd is said to have built in the 1870s were made at Whessoe, perhaps in Harris' old premises.

In March 1968 the company took over Ashmore, Benson, Pease & Co Ltd of Stockton-on-Tees. Then in June 1989 the main Whessoe board took the decision to phase out their loss-making manufacturing capacity and instead to develop their design work, sub-contracting manufactured items as required. In June 1990 the majority of the plant and equipment, other than items to be retained for Whessoe Products Ltd, was offered for public auction. The works continues (2006) in business.

There were very probably more locomotives used here than those listed below – none are known in the nineteenth century, for example - but as with other Darlington companies, details are lacking. Rail traffic with BR ceased in 1986, but a locomotive was retained for occasional internal shunting until 1989. The works also used at various times 10 steam rail cranes and 3 diesel rail cranes. The latter were Coles 17595/1958, Coles 18520 and Coles 19302, all last reported here on 16/11/1979; Coles 18520 was subsequently sold to a dockyard in Sunderland, possibly Bartram & Sons Ltd at South Dock.

Reference : *Whessoe*, Dennis Hockin, The Pentland Press, 1994.

Gauge : 4ft 8½in

	HEMINGWAYS	0-4-0ST	OC	?	?	?	(a)	Scr c/1919
	INFLEXIBLE	0-4-0ST	OC	MW	575	1875		
		reb		MW		1901	(b)	(1)
(314)	MONARCH	0-4-0ST	OC	P	1210	1910	(c)	(2)
	CHINA	?		?	?	?	(d)	s/s
	ACKLINGTON	0-6-0ST	OC	?		?		
		reb		T.D.Ridley		1903	(d)	(3)
	EAST LAYTON	0-4-0ST	OC	HL	2871	1911	(e)	(4)

982		0-4-0T	IC	Dar		1923	(f)	(5)
(316)	DERWENT II	0-4-0DE		RH	312988	1952	New	(6)

A locomotive named ASHINGTON is said to have worked here for six months. This <u>could</u> be ASHINGTON 0-4-0ST B 303/1883, owned by The Priestman Collieries Ltd.

(a) ex Hemingways Chilled Rolls Ltd, Haverton Hill, c/1906 (see Cleveland & North Yorkshire Handbook); said by R.H. Inness to have been a 12in cylinder loco built by Thornewill & Wareham for Bass, Ratcliff & Gretton & Co, Burton-on-Trent, Staffordshire; if so, then the locomotive was very probably TW 249/1864, which had been replaced by a new locomotive about 1900.
(b) ex J.F. Wake & Co Ltd, dealer, Darlington, /1919; previously William Rigby & Co Ltd, Bunkers Hill Colliery, Talke, Staffordshire, until /1915.
(c) ex Penderyn Limestone Quarries (Hirwaun) Ltd, Glamorgan, /1926.
(d) ex Ridley, Shaw & Co Ltd, Middlesbrough, Yorkshire (NR), hire.
(e) ex Forcett Limestone Co Ltd, Forcett, Yorkshire (NR), loan.
(f) ex LNER, 982, class Y7, North Road Works, Darlington, /1937 (by 11/1937), loan.

(1) sold for scrap, /1927.
(2) advertised for sale in *Machinery Market*, 5/9/1952, saying that the loco would be available in 11/1952; to Ashmore, Benson, Pease & Co Ltd, Stockton-on-Tees, per Cox & Danks, dealers, by 10/5/1953 (see Cleveland & North Yorkshire Handbook).
(3) returned to Ridley, Shaw & Co Ltd, Middlesbrough, Yorkshire (NR).
(4) returned to Forcett Limestone Co Ltd, Forcett, Yorkshire (NR).
(5) returned to LNER, North Road Works, Darlington, /1937.
(6) to Darlington Railway Preservation Society, for preservation, 11/5/1990, but kept on site; moved to the Society's Hopetown depot, Darlington, by 31/3/1992.

WIGGINS TEAPE LTD
HYLTON PAPER MILL, South Hylton, Sunderland H585
FORD PAPER MILL until /1947 NZ 361574
Ford Paper Mills Ltd until 1/3/1968; this company was transferred to **Alex Pirie & Sons Ltd** in 1947, itself already a subsidiary of Wiggins Teape Ltd; see also below.

This paper mill, built to manufacture white paper from rags, was established by **Vint, Hutton & Co**,

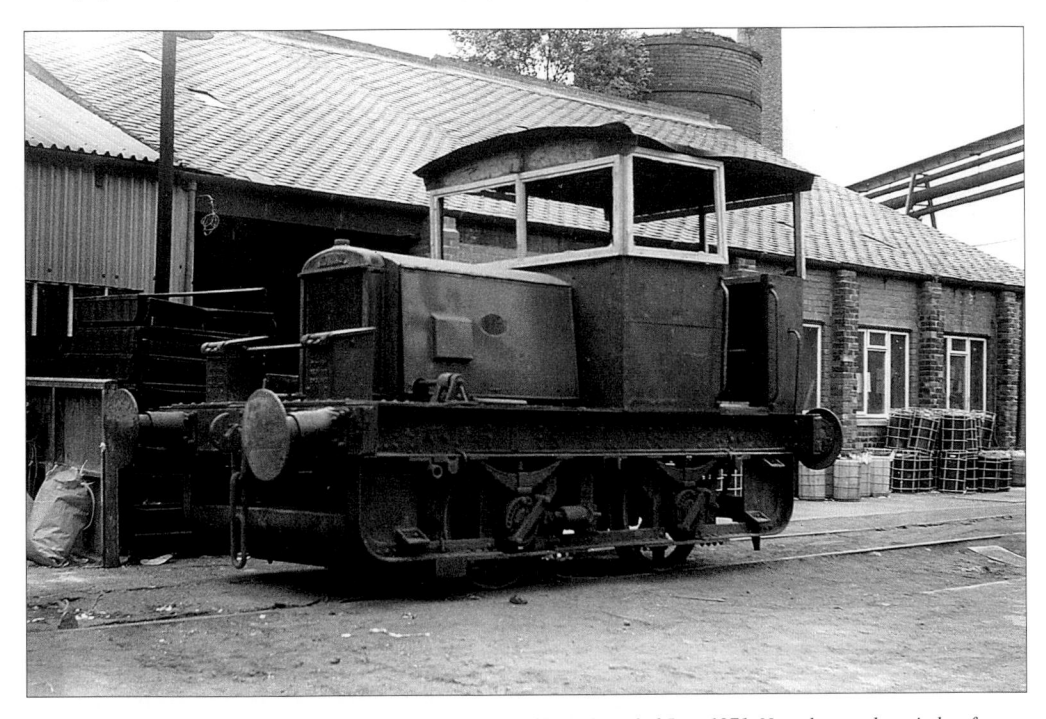

160. A rare photograph of a Howard petrol locomotive: H 965/1930, on 3rd June 1971. Note the wooden window frames and the long footstep below the axleboxes.

who had begun their paper manufacturing business at Deptford, nearer to Sunderland, by converting a saw mill into a paper mill in 1826. Construction of the mill, on the southern bank of the River Wear at South Hylton, began in May 1836, and it was opened on 29th August 1838.

The mill soon went through various changes of ownership. By March 1844 the owners were **Hutton, Fletcher & Co**, which had become **Fletcher, Blackwell & Falconer** by 1850, **J.Blackwell & Co** in 1856 and **The Ford Paper Mill Co** in 1860. In 1864 **The Ford Works Co Ltd** was set up by the new principal shareholder, Thomas Routledge, who perfected the manufacture of paper from esparto grass. He died in 1887 and in 1891 a new company, **The Ford Paper Works Ltd**, was set up (not 'The Ford Works Ltd', set up in 1893, as some sources give). This company was re-constructed in 1923 as **The Ford Paper Works (1923) Ltd**, but it went into liquidation in 1926, and **The Ford Paper Mills Ltd** (see above) was formed in September 1927 to take over the mill and re-commence production. This company was taken over by **Wiggins Teape Ltd** in January 1937. The esparto plant closed in 1959, and the mill went back to the manufacture of paper from rags.

161. The Howard was replaced by HYLTON, FH 3967/1961, also seen on 3rd June 1971. The locomotive was used only around the works and on the river quay, access from British Railways being via a rope incline, so steep that it had to be worked by a stationary engine.

The Pensher branch of the York, Newcastle & Berwick Railway was opened for goods traffic on 20th December 1852 and passed the mill, albeit at a much higher level. The minutes of the Committee of Management of the River Wear Commissioners dated 4th July 1860 approved an application from the works "to extend the line of the railway down to the new shipping place at [the company's] own expense", which would suggest that a link to the Pensher branch had been built by this date. Because of the steepness of the gradient, this was a ½ mile **rope incline** on a large curve, worked by a **stationary engine** alongside the NER. In 1880 a new quay was built alongside the river, with for the first time a locomotive to shunt it. The locomotives were used only to shunt the lines around the works and to the foot of the incline.

The works was closed on 3rd July 1971 and demolished between November 1974 and March 1975. The site was subsequently re-developed for leisure activities.

Gauge : 4ft 8½in

FORD	0-4-0ST	OC	BH	188	1872	(a)	s/s by /1952
FORD	0-4-0ST	OC	MW	1323	1896	(b)	Scr c/1930
-	0-4-0ST	OC	HC	535	1900	(c)	s/s

-	0-4-0ST	OC	HC	1207	1916			
	reb		?		1926	(d)	Scr /1962	
-	4wVBT	VCG	S	6310CH	1926	(e)	(1)	
-	4wPM		H	965	1930	New	(2)	
HYLTON	4wDM		FH	3967	1961	New	(3)	

(a) ex BH, /1880; Manvers Main Colliery Co Ltd, Manvers Main Colliery, Wath-on-Dearne, Yorkshire (WR) until /1879.
(b) ex Sir William Arrol & Co Ltd, Queen Alexandra Bridge contract for NER, Sunderland, c/1909.
(c) ex Thos W. Ward Ltd, Charlton Works, Sheffield, Yorkshire (WR), 1/1920; previously James Pain Ltd, Irchester Quarries, near Wellingborough, Northamptonshire.
(d) ex George Cohen, Sons & Co Ltd, dealers, c/1926; previously Greenwood & Batley Ltd, Leeds, Yorkshire (WR), until /1926.
(e) ex Ipswich Dock Commission, Ipswich, Suffolk, between 2/4/1929 and 10/7/1929; on trial and almost certainly owned by S; loco was a rebuild of 0-4-0ST AB 282/1886.

(1) loco was here about a year, i.e., until mid 1930, and then returned to S; by 2/1938 loco was at Clay Cross Co Ltd, Grin Limestone Quarry, Ladmanlow, near Buxton, Derbyshire.
(2) to Lytham Motive Power Museum, Lytham, Lancashire, /1972, for preservation.
(3) to Ely Works, Cardiff, Glamorgan, c10/1971.

THE WIGGLESWORTH COLLIERY CO LTD (registered 31/5/1922)
HOLLY MOOR COLLIERY, Cockfield P586
formerly **R. Summerson & Co Ltd** (see below) NZ 115247

WIGGLESWORTH COLLIERY (NZ 114245) **(P587)** from which the company took its name, lay about ¾ mile north-west of Cockfield village. About 200 yards to the north-east lay **HOLLY MOOR COLLIERY**, another small mine begun in the nineteenth century and always connected with the Summerson family of Darlington. It was initially worked only for landsale coal and comprised a shaft and a drift a few yards away. R. Summerson & Co Ltd re-opened it in 1911, and Summersons were the chief shareholders in this new firm when it was formed in 1922. From the drift a narrow gauge tramway linked to the shaft and then ran down to screens alongside the LNER line from Bishop Auckland to Barnard Castle, ¼ mile north-east of Cockfield Station. How the ½ mile-long tramway was worked before the arrival of the locomotive below is uncertain; a Sentinel catalogue describing it stated that it replaced an earlier, conventional steam locomotive, but no details of this loco are known.

162. *Taken from a Sentinel catalogue of 1930, this view shows 2ft 0in gauge S 6770CH/1926, hauling a long train of empty tubs from the firm's loading point alongside the LNER near Cockfield Station over the ½ mile across the treeless moorland to Holly Moor Colliery, about 1928. This was an 80 h.p. machine, with outside chain drive from the vertical cylinders mounted at the front end.*

The Sentinel locomotive's stay was brief, for Holly Moor Colliery ceased operations in August 1927 and was abandoned in June 1928, although the company did not go into voluntary liquidation until 27th January 1930.

Gauge : 2ft 0in

-	4wVBT	VCG	S	6770CH*	1926	New	(1)

* this locomotive was allocated works number 6770CH by Sentinel, but it is believed to have carried the makers plates of S 6751CH, which was being built at the same time; see *Industrial Railway Record*, No.126, page 305.

(1) to Northumberland Whinstone Co Ltd, Longhoughton Quarry, Longhoughton, near Alnwick, Northumberland, by 6/1929.

HENRY WILLIAMS LTD
RAILWAY APPLIANCE WORKS, Darlington　　　　　　　　　　　　　　　　　　　　　　　S588
　　　NZ 298159

This firm had commenced business in Glasgow in 1883. In March 1911 it purchased the northern part of the site of Skerne Workworks (see Skerne Ironworks Co Ltd), and built new premises on it to manufacture railway signalling equipment. The works was served by a ¼ mile branch from the NER Durham – Darlington line, 1½ miles north of Darlington (Bank Top) Station.

It is not known whether the firm had a locomotive before 1927. Rail traffic ceased about 1965. The works continues (2006) in production, still owned by this company.

Gauge : 4ft 8½in

-	4wPM	Hardy	908	1925	New	(1)
-	0-4-0DM	HE	2839	1943	New	(2)

(1) to The Rolling Stock Co Ltd, Eryholme Junction, Yorkshire (NR), c/1943.
(2) to Hanratty Bros, Darlington, c/1966.

SIR HEDWORTH WILLIAMSON'S LIMEWORKS LTD
Sir Hedworth Williamson's Fulwell Limeworks Ltd until /1925
Sir Hedworth Williamson until /1912 (and see below)

FULWELL & SOUTHWICK QUARRIES & LIMEWORKS, Fulwell, Sunderland　　　　　　　H589

About 1½ miles north-west of the mouth of the River Wear lies a very large area of permian and magnesian limestone, which was first quarried by the Romans. The largest landowner in the area was the Williamson family, who lived at Whitburn Hall, 1½ miles to the north-east. What might be regarded as modern quarrying began in the 1790s with **FULWELL QUARRY** (NZ 391595 approx.), and it was almost certainly commenced by the then Sir Hedworth Williamson; he was certainly controlling it by 1834. It lay immediately to the west of Fulwell Mill (a windmill, still there, and now a preserved and fully-working ancient monument), and was linked to the River Wear by a waggonway. This was almost certainly built in the 1790s and was about 1¼ miles long, running alongside the Newcastle road for much of its length. Although there were lime kilns at the quarry, the main kilns were situated at Wreath Quay, just upstream of the Wearmouth Bridge, with a quay to off-load the coal needed for the kilns and to dispatch the burnt lime, which was used as an agricultural fertiliser and in building and glass-making; at least some of the output was sent to Scotland. However, the lower section of the line had to be diverted in 1848, because it ran across the area required for the construction of the forecourt of the new Monkwearmouth Station for the York, Newcastle & Berwick Railway, opened on 19th June 1848, and so the line was moved westwards to run alongside the western boundary of the goods yard and into a tunnel about 100 yards long before turning south-west to reach the kilns, now called the **SHEEPFOLD LIME WORKS** (NZ 395575). The 1st edition of the O.S. map for the area, surveyed in 1857, shows that the original quarry had been replaced by a new one, also called **FULWELL QUARRY** (NZ 389600 approx.). The waggonway had been extended to it, but with a **stationary engine** (NZ 391596) in the north-eastern corner of the old quarry, almost certainly used to haul waggons out of the quarry. The remainder of the line southwards is shown as single with passing loops, which would indicate horse working over this section. By this date too a spur had been built from near Monkwearmouth Station into the **Monkwearmouth Colliery** (NZ 394578) of W. Bell & Partners (see The Wearmouth Coal Co Ltd), presumably to obtain coal for the kilns.

Meanwhile the third Sir Hedworth Williamson (1797-1861) was succeeded by his son, also Sir Hedworth Williamson (1827-1900). Precise details for the final forty years of the nineteenth century are unclear. The NER Gateshead - Monkwearmouth line curved round the northern side of the Fulwell workings and it was an obvious development to build a short link of about ¼ mile under the Newcastle road to join it, about one mile north of Fulwell Station. A plaque on the railway side of the parapet of the bridge under the Newcastle road bears the date 1860, so it is reasonable to assume that this was the date when the link was constructed. The waggonway down to the river is believed to have continued in use until the 1870s; it is not shown on the 2nd edition O.S. map, surveyed in 1895.

However, both the minutes of the Southwick Local Board for 1884 and the NER Sidings Map of the same year describe the **CARLEY HILL WEST WAGGONWAY** as owned by Sir Hedworth Williamson, the latter source specifically noting the line as the "tramway from Fulwell Quarry". The history of this line, which ran from the Carley Hill quarries about ½ mile west of Fulwell Quarry down to the river, is described under The Wearmouth Coal Co Ltd, the final company to own it. When and from whom Sir Hedworth Williamson acquired it, with or without Carley Hill Quarries, is not known. The line seems to have been closed about 1889, and by the mid-1890s it had passed to The Wearmouth Coal Co Ltd.

When locomotive working took over from horses is not known. Towards the end of the nineteenth century kilns were built in the quarry, and it may be that the purchase of some of the smallest standard gauge locomotives ever built - BH 1114, 1121 and 1122 had only 6½in x 12in cylinders and 2ft 0½in wheels – were connected with this development.

In 1900 the 5th baronet, also Sir Hedworth Williamson (1867-1943), succeeded his father. At this date approximately 145,000 tons of stone were quarried annually, with 50,000 tons of lime being produced. By the autumn of 1902 The Wearmouth Coal Co Ltd had ceased to work the Carley Hill Quarries, and these were subsequently acquired by Sir Hedworth. Carley Hill East Quarry was absorbed into the Fulwell Quarries, while Carley Hill West Quarry and the old Southwick Quarry were developed as a new **SOUTHWICK QUARRY**, with two large working areas to the north (NZ 378599 approx.) and the east (NZ 382596). The quarry area now covered 140 acres; excavation went as deep as 90 feet, and working faces could be up to a mile long. Meanwhile, the hitherto scattered kilns were replaced by the gradual development of a large **LIMEWORKS** (NZ 386600) on the top level towards the north of the Fulwell area. About 1905 the first Spencer kiln was built here, with six, in two batteries of three, being constructed over the next twenty years. These were between 80ft and 100 feet high and 16ft x 20ft in oval section, and were filled from the top using a locomotive. Nearby, in 1908, the business installed the first hydrating plant in Europe, purchased from Chicago, the limestone being calcinated to produce hydrated lime. A second, much smaller works (NZ 393595) was built in the original Fulwell quarry, just to the north of the Mill. This was later closed and a landsale depot built here.

When the limited company was formed in 1912 Sir Hedworth Williamson became its chairman, and this control was continued by Sir Charles Hedworth Williamson (1903-1946) and his son, Sir Nicholas Hedworth Williamson (b.1937), when he became of age. As the stone in the Fulwell area became exhausted, working was increasingly concentrated at the Southwick Quarry. Foreseeing the end of production at Sunderland, the firm searched for good limestone elsewhere, and in 1925 it began a major development near Kirkby Stephen in Westmorland. In October 1947 the hydrating plant at Fulwell suffered a major fire and was closed, and other production then began to be run down. The quarries and the works closed in February 1957, with the plant, including the Spencer kilns and four locomotives, being auctioned as a single lot by Sanderson, Townend & Gilbert on 11th May 1957. For a considerable time Fulwell Quarry was used for the disposal of household waste by Sunderland Corporation and then by the National Coal Board for tipping colliery waste from Wearmouth Colliery in Sunderland; but in the 1990s a major reclamation scheme saw this area landscaped and development of a golf range and other leisure activities in this area. The quarries on the Southwick side remain largely untouched, apart from reclamation by nature.

By the 1920s the quarries and the works were served by more than ten miles of track. The original **loco shed** (NZ 388601) was situated on the middle level just south of the bridge under the Newcastle road (A184). Later a second **shed** (NZ 385600), with two roads, was added near the limeworks, adjacent to a Fitting Shop and a Joiners Shop, which between them repaired not only the locomotives but also the 70 or so internal wagons used. Many, if not all, of the locomotives were named after Williamson children.

Reference : *British Bulletin of Commerce*, edition for September 1954.

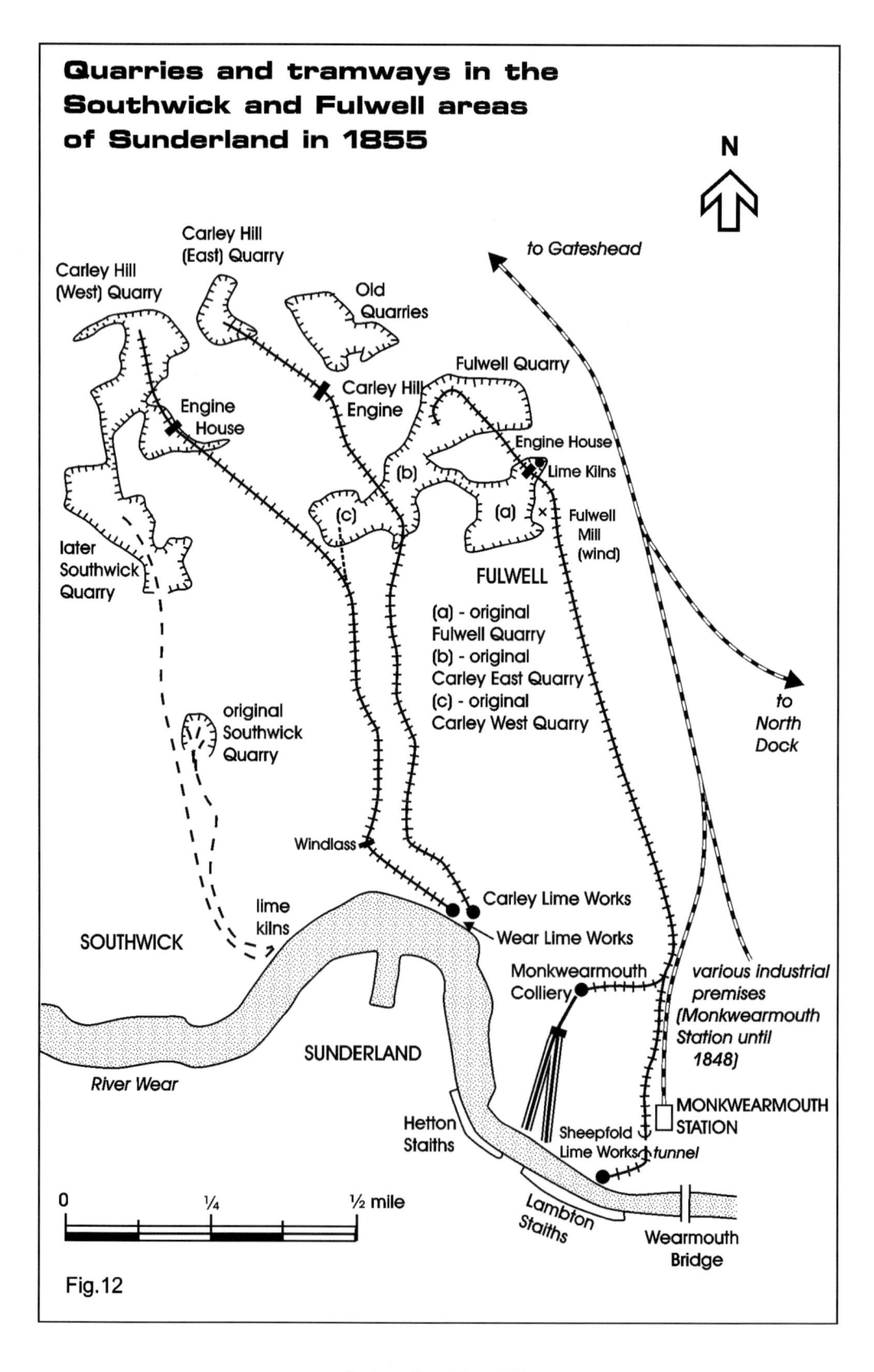

Fig.12 — Quarries and tramways in the Southwick and Fulwell areas of Sunderland in 1855

Fig.13

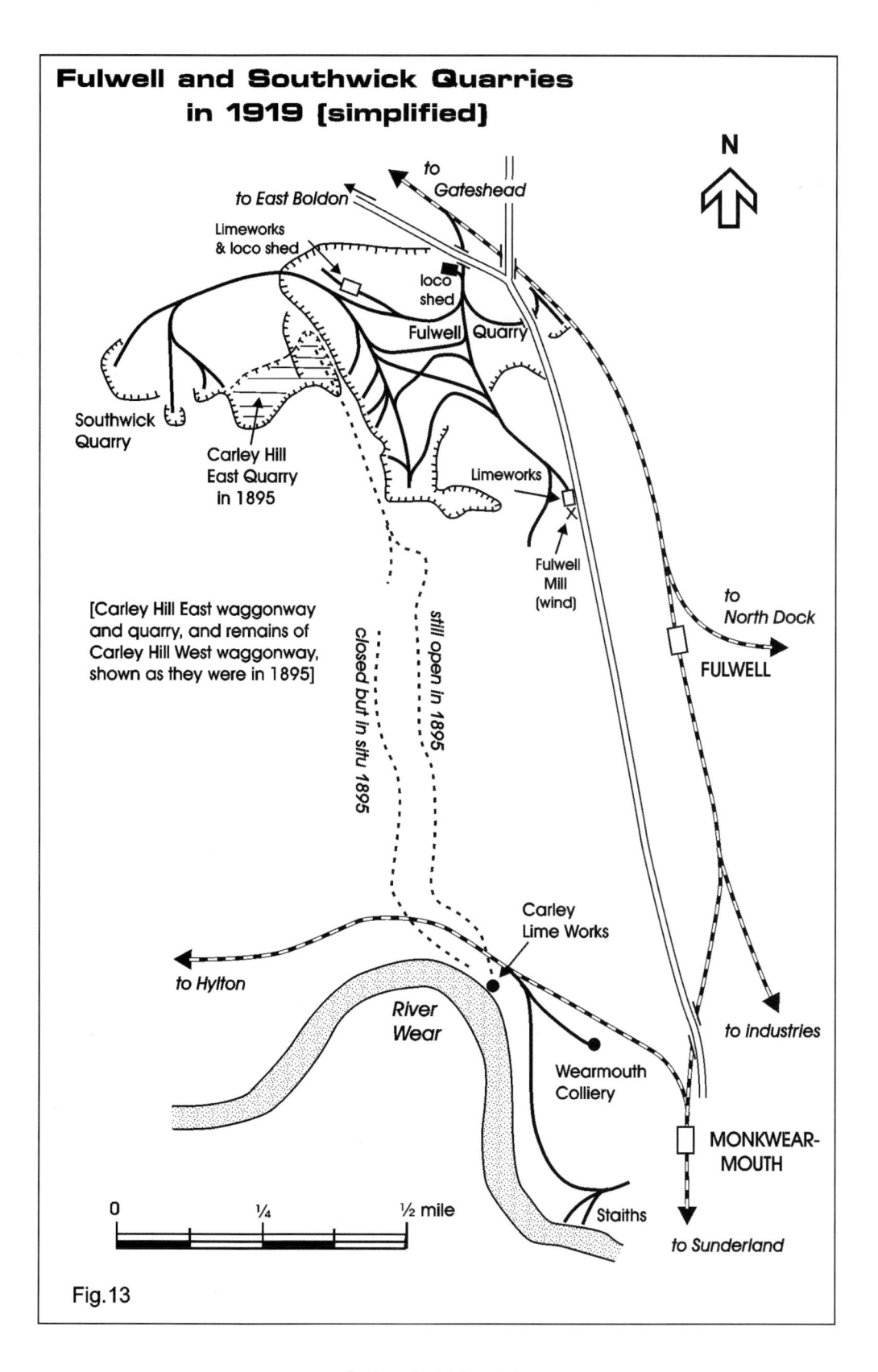

163. The first SYLVIA, LE 238, as built in 1899. Note the plate with the motif on the side of the cab, which several other locomotives also carried. Presumably it had a connection with the Williamson family.

164. The second SYLVIA, very probably carrying the nameplate from the first, in September 1955. HC 1599/1927 stands outside the Low Shed, beyond which the line went under a bridge to sidings alongside the British Railways' line between Pelaw and Sunderland. The recessed curved cab was a feature of all the Hudswell Clarke locomotives here. See the Frontispiece for an example of the work the locomotives here did.

Gauge : 4ft 8½in

-	0-4-0ST	OC	JF	3157	1883	(a)	s/s	
JOE	?	?	?	?	?	(b)	s/s c/1914	
TEDDY (formerly FRITZ)	0-4-0ST	OC	BH	1026	1891	New	Scr	
TEDDY	0-4-0ST	OC	BH	1114	1895	New	s/s after /1918	
DOROTHY	0-4-0ST	OC	BH	1121	1895	New	s/s by /1918	
RACY (RACEY?)	0-4-0ST	OC	BH	1122	1985	New	s/s by /1918	
PHYLLIS	0-4-0ST	OC	LE	230	1897	New		
reb	4wVBT	VCG	S	5988CH	1925		(1)	
SYLVIA	0-4-0ST	OC	LE	238	1899	New	Scr /1927	
CHARLES HEDWORTH	0-4-0ST	OC	HC	683	1903	New		
reb	4wVBT	VCG	S	6218CH	1926		(2)	
JOE	0-4-0ST	OC	HL	2431	1899	(c)	(3)	
BILLY	0-4-0ST	OC	BH	551	1880	(d)	s/s after /1931	
ELIZABETH	0-4-0ST	OC	HC	1484	1922	New	(4)	
FRITZ	0-4-0ST	OC	HC	1493	1923	New	(4)	
SYLVIA	0-4-0ST	OC	HC	1599	1927	New	(4)	
BILLY	0-4-0ST	OC	HL	3806	1934	New	(5)	
TEDDY	0-4-0ST	OC	HL	3887	1936	New	(6)	
NICHOLAS	0-4-0ST	OC	HL	2496	1901			
reb			HL	?	1922	(e)	(4)	

(a) JF records show this loco New to "Sir H. Williamson"; whether the purchaser was Sir Hedworth Williamson has not been confirmed.
(b) origin unknown; said to have been a BH locomotive ordered originally for export to Africa as a wood-burning locomotive.
(c) ex Richard Hill & Co (1899) Ltd, Middlesbrough, Yorkshire (NR), by 12/1914.
(d) ex The Consett Iron Co Ltd, Consett Works, Consett, 10/1919.
(e) ex Steels Engineering Products Ltd, Pallion, Sunderland, /1940.

(1) to Kirkby Stephen Works, Westmorland, /1927.
(2) to Kirkby Stephen Works, Westmorland, c/1928; returned, 5/1948; scrapped c8/1948.
(3) to Kirkby Stephen Works, Westmorland, c/1951 (after 3/1951, by 8/1956).
(4) offered for auction, 11/5/1957; scrapped on site by Thos W. Ward Ltd, /1957.
(5) to Kirkby Stephen Works, Westmorland, by 5/1950.
(6) to Kirkby Stephen Works, Westmorland, c5/1948; returned; to Kirkby Stephen Works, Westmorland.

WILSONS FOUNDRY & ENGINEERING CO LTD (by 3/1959)
Wilsons Forge Ltd from at least 3/1956
formerly **Wilsons Forge (1929) Ltd**
originally **Robert Wilson & Sons Ltd**
RAILWAY FORGE, Bishop Auckland   Q590
NZ 214293

This foundry and engineering works was developed on part of the area formerly occupied by the Auckland Ironworks (see Thomas Vaughan & Co), which had opened in 1863 and had closed in 1877; the site was acquired by Robert Wilson & Sons Ltd in the 1880s, though the business was begun in 1842. It subsequently acquired the adjacent premises of Lingford, Gardiner & Co Ltd, railway engineers and locomotive repairers, which went into liquidation in 1931. The works was served by sidings north of the NER Bishop Auckland – Shildon line, ¼ mile east of Bishop Auckland Station. It is believed to have been shunted by the NER/LNER before the arrival of the first loco below. Rail traffic ceased in the mid 1950s, although the works continued in business until the mid-1990s.

Gauge : 4ft 8½in

1	0-4-0ST	OC	MW	540	1875	(a)	s/s c/1946
-	0-4-0ST	OC	AE	1613	1911	(b)	(1)

(a) ex Lingford, Gardiner & Co Ltd, when its premises were acquired (see above).
(b) ex Geo Cohen, Sons & Co Ltd, dealers, Wood Lane, London, c/1946 (by 4/8/1950); previously Cordes (Dos Works) Ltd, Newport, Monmouthshire.

(1) dismantled by 1956; sold for scrap, 4/1959.

THE WINGATE COAL CO
Both this title and **The Wingate Grange Coal Co** are found in Hartlepool Dock & Railway records.
WINGATE GRANGE COLLIERY, Wingate L591
NZ 398373

This colliery lay near the south-eastern edge of the Durham coalfield and was sunk by Lord and Lady Howden. Coal was reached on 3rd August 1839 and production began in October 1839. It was served by a ¾ mile branch, opened on 18th March 1839, from the Hartlepool Dock & Railway Company, which owned the branch. The owners followed the practice of the South Hetton and Thornley Coal Companies (which see) and operated their own trains over the HD&R between the colliery and Hesleden Bank Head, about five miles from Hartlepool, and perhaps over part of the remaining section. When the York & Newcastle Railway took over the two railways in October 1846 it ended this practice, and in the 1847 compensation payments, the Wingate company was paid over £7300, which is so large that it must have included some locomotives. Whether any of those below were included in the list prepared much later by the NER and reproduced in the Hartlepool Dock & Railway entry (which see) can only be conjecture. Thereafter all traffic, including shunting, was handled by the York & Newcastle Railway and its successors. The company became **The Wingate Coal Co Ltd**, and was latterly a jointly-owned subsidiary of The South Durham Steel & Iron Co Ltd and The Cargo Fleet Iron Co Ltd. Apart from a brief period when it owned West Thornley Colliery (see The Weardale Steel, Coal & Coke Co Ltd), it continued to own only Wingate Grange Colliery, which was vested in NCB Northern Division No. 3 Area on 1st January 1947.

Gauge : 4ft 8½in

-	0-6-0	?	RWH	297	1839	New	s/s
-	0-6-0	?	RWH	308	1840	New	(1)
-	0-6-0	VC	Hackworth		1827	(a)	s/s
-	0-6-0	VC	Hackworth		1829	(b)	s/s

(a) ex Stockton & Darlington Railway, No. 5, ROYAL GEORGE, 12/1840.
(b) ex Stockton & Darlington Railway, No. 8, VICTORY, 2/1841.
(1) to Earl of Durham, Lambton Railway, Philadelphia.

THE WINGATE LIMESTONE CO LTD (registered 31/12/1877)
WINGATE QUARRY, near Trimdon Colliery L592
see below

In 1867 **The Trimdon Coal Company** (not listed in this volume) began the sinking of **Deaf Hill Colliery** (NZ 372368) **(L593)** near Trimdon Colliery village. It was opened in 1873 and was served by a short branch from the NER Ferryhill – Hartlepool line, 1¼ miles east of Wingate Station. About ¾ mile north-west of the colliery, north of the hamlet of Old Wingate, was a large area of limestone, and in 1877 The Wingate Limestone Co Ltd was formed to quarry it. To serve the quarry (NZ 377379 approx.) a branch just under a mile long was built from Deaf Hill Colliery, the quarry company paying wayleave to the Trimdon Coal Co for stone carried over its line.

The 2nd edition O.S. map shows an **engine house** (NZ 378369) near West Moor Farm. The engine was presumably used to haul full wagons from the quarry and lower then down to the colliery, where the NER undertook the shunting. By the end of the nineteenth century a second quarry (NZ 373376 approx.) had been opened to the west of Old Wingate. This was served by a ½ mile link to the original branch. By 1906 the engine house had been replaced by locomotive working, from a **loco shed** (NZ 377374) close to where the lines to the two quarries diverged. The original quarry was eventually closed, probably in the 1920s, and work was concentrated at the western quarry. This closed in May 1930, although the company did not go into voluntary liquidation until 9th September 1932. The dismantling of the quarries was undertaken by W. Blenkinsop & Co Ltd of Middlesbrough, Yorkshire (NR).

The first three locomotives on the list below were said to have been maintained by The South Hetton Coal Co Ltd.

Gauge : 4ft 8½in

WINGATE	0-4-0ST	OC	AB	675	1891	(a)	(1)
BIRKBECK	0-4-0ST	?	?	?	?	(b)	(2)
GREENSIDE	0-4-0ST	?	?	?	?	(b)	(2)
HOLMLEA	0-4-0ST	OC	MW	1887	1915	New	(3)

CARMEL	0-4-0ST	OC	MW	1911	1917	New		(3)
-	0-4-0ST	OC	HL	2177	1890			(4)
		reb	J. Tait	100	1920	(c)		

(a) possibly ex William Beardmore & Co Ltd, Parkhead Forge, Glasgow, Lanarkshire, Scotland; at Wingate by 6/1906.

(b) one of these was 0-4-0ST OC AB 62/1867, which was ex Summerlee Iron Co Ltd, Mossend, Lanarkshire, to AB, who rebuilt it as AB 5979/1908 and then used it as a hire loco; ex AB to Wingate, 9/9/1913; the origin, identity and date of arrival of the other loco are not known.

(c) ex J. Tait & Partners, dealers, Middlesbrough, Yorkshire (NR), /1920; previously The Consett Iron Co Ltd, Consett Iron Works, Consett.

(1) to Northumberland Collieries Ltd, Forestburngate Colliery, near Rothbury, Northumberland, by 6/1921.

(2) in Contract Journal 26/5/1915 and *Machinery Market* for 28/5/1915 the firm advertised for sale a four-coupled locomotive with 12in x 20in cylinders; this would fit AB 62/1867, which was sold to Sir W.G. Armstrong, Whitworth & Co Ltd, Scotswood Works, Newcastle upon Tyne, by 10/1915; the disposal of the other loco is not known.

(3) both of these locomotives passed to W. Blenkinsop & Co Ltd, Middlesbrough, Yorkshire, who received the contract to dismantle the quarries in 1933. One of the locomotives, which is not known, was advertised by the firm in *Contract Journal* on 1/3/1933. Both locomotives subsequently went to Robert Stephenson & Co Ltd, Darlington, for repairs, in 1933. MW 1887/1915 was re-sold to G.&T. Earle Ltd, Hessle, Yorkshire (ER), 3/1934, while MW 1911/1917 was re-sold to S. Taylor, Frith & Co Ltd, Holderness Limeworks, Peak Dale, Derbyshire, per J.C. Oliver Ltd, dealer, Leeds, Yorkshire (WR), /1935.

(5) to W. Blenkinsop & Co Ltd, Middlesbrough, Yorkshire (NR), /1933; re-sold to The Witton Park Slag Co Ltd, Etherley Works, Witton Park.

THE WITTON PARK SLAG CO LTD
The Witton Park Slag Co until 8/12/1928

This firm was set up by Mr.T. Benjamin Maughan, already the owner of William Barker & Co (which see), the West Cornforth firm processing the slag from the former Ferry Hill Ironworks. In July 1912 he purchased from Bolckow, Vaughan & Co Ltd the slag heaps at their former Witton Park Ironworks (which see), possibly initially operating under the title of **Mulholland & Maughan**; the actual date of transfer from Bolckow Vaughan was 6th December 1912. This slag was already being worked by the Auckland Rural District Council (which see), which continued to do so under lease from The Witton Park Slag Co, perhaps as early as 1913 and certainly by 1919. The council gave up its lease on 31st December 1921, leaving The Witton Park Slag Co in sole possession.

In 1926 the company entered on a lease of Broadwood Quarry near Frosterley, formerly worked by Pease & Partners Ltd (which see), but operated it under the trading name of **Broadwood Limestone Company** (which see). From about 1930 the firm entered the road construction business directly, building a new plant at Witton Park to manufacture tarmacadam. The company continues (2006) to trade, though no longer as a manufacturing company.

WITTON PARK SLAG WORKS, Witton Park   Q594
NZ 177305

The slag processing and tarmacadam plant of Auckland Rural District Council lay close to the site of the former Witton Park Ironworks, and when The Witton Park Slag Co decided to develop similar facilities it began work on a site about ¼ mile east of the council's works. The firm was served by a ¼ mile extension of the council's line from Etherley Station on the NER Crook – Bishop Auckland line. A council minute of 7th October 1913 agreed a charge for handling traffic for The Witton Park Slag Co, though how long this was continued is unknown. It would appear that The Witton Park Slag Co acquired its first locomotive about 1919.

The firm completed the construction of its own works here by the end of 1921. It is generally agreed that its production ended in or about 1931. However, the company then opened a new works on the area formerly worked by Auckland Rural District Council (see below).

Gauge : 4ft 8½in

AUBREY LAWRENCE	0-4-0T	OC	KS	2399	1917	(a)	(1)
-	4wPM		MR	1951	1920	New	(2)
JUPITER	0-4-0ST	OC	MW	1880	1915	(b)	(3)

(a)   ex Ministry of Munitions, Sutton Oak, Lancashire.
(b)   ex Strakers & Love Ltd, Brandon Colliery, Brandon, /1931, loan.

(1)   this loco, incorrectly described as having 10in x 18in cylinders, was advertised for sale in *Machinery Market* on 19/8/1927 and in the *Colliery Guardian* on 16/12/1927; it remained unsold, and was still on site, derelict, on 2/8/1935; s/s.
(2)   to Broadwood Limestone Co, Broadwood Quarry, near Frosterley, after 9/1928, by /1930.
(3)   returned to Strakers & Love Ltd, Brandon Colliery, Brandon, ex loan, /1931.

165. AUBREY LAWRENCE, KS 2399/1917, derelict at Witton Park on 5th March 1932. She was still there 3½ years later.

The MR works records list both of the locomotives below as delivered to The Witton Park Slag Co Ltd at Witton Park. However, there is no documentary evidence of a narrow gauge system at Witton Park Slag Works. The company's main commercial activity was road contracting, and so it may well be that the narrow gauge locomotives were actually used on the company's road contracts, though again this lacks confirmation.

Gauge : 2ft 0in

| - | 4wPM | MR | 3833 | 1926 | New | (1) |
| - | 4wPM | MR | 4577 | 1930 | (a) | (2) |

(a)   New to Petrol Loco Hirers, a MR subsidiary company, in 4/1930; sold to The Witton Park Slag Co Ltd on 14/12/1930.

(1)   MR supplied spares for a MR loco, almost certainly this one, on 23/10/1943; almost certainly the loco subsequently went to Broadwood Limestone Co, Broadwood Quarry, near Frosterley.
(2)   to Broadwood Limestone Co, Broadwood Quarry, near Frosterley, by 2/2/1940.

BROADWOOD WORKS, Witton Park   Q595
NZ 175304

When, probably by 1931, the firm had exhausted the slag heaps of the former Witton Park Iron Works, it closed the works completed in 1921 and in its place built a new works for the manufacture of tarmacadam about ¼ mile nearer to the LNER, situated in the area which had been cleared by Auckland Rural District Council (see above). This used limestone brought from Broadwood Quarry near Frosterley (see the entry for Broadwood Limestone Co) and was called (rather confusingly) the **BROADWOOD WORKS**. This works is believed to have been operated until the mid-1950s.

Gauge : 4ft 8½in

		0-4-0ST	OC	HL	2177	1890		
-		reb	J. Tait	100	1920	(a)	s/s by 4/1955

(a) ex W. Blenkinsop, dealer, Middlesbrough, Yorkshire (NR); The Wingate Limestone Co Ltd, Wingate Quarry, near Trimdon Colliery, until /1933.

THE WOLSINGHAM PARK & DINAS FIREBRICK, MINERAL & COAL CO LTD
(registered 17/9/1873), **Salter's Gate, near Waskerley**

This was a major speculation in one of the most remote areas of the county. The mineral deposits in an area of nearly 2500 acres on Wolsingham Park Moor were leased in June 1868, with freestone quarrying added in May 1875. The company built a sizeable works, known as the **WOLSINGHAM PARK BRICK WORKS** (NZ 072431) **(T596)** to manufacture firebricks and freestone flags, and provided workmen's cottages. The works was served by sidings alongside the NER Weardale Extension line between Tow Law and Waskerley, two miles south-east of the latter, and was apparently shunted by the NER. Two lengthy narrow gauge tramways were built to bring raw materials to the works. The first, 2450 yards long, ran north-west to a fireclay quarry known as **No.1 DEPOSIT** (NZ 055438) **(T597)** in the Park Head Plantation, apparently serving en route three quarries known as the **DRYPRY QUARRIES** (NZ 068433) **(T598)** and a lead mine nearby. There were two other quarries near No.1 Deposit, apparently not rail-connected. This line was worked by the locomotive below, its loco shed being at the brickworks. The second line, 1810 yards long, ran south-west and then south, twice crossing Waskerley Beck on bridges, to **TUNSTALL QUARRY** (NZ 068419) **(T599)**, a freestone quarry. This rope-worked line, almost certainly main-and-tail haulage, was worked by a **stationary engine** at the brickworks, the line crossing the standard gauge sidings at a right-angle.

Production lasted only a matter of months. A court order to wind the company up was issued on 20th May 1875. The plant and leases were auctioned at two sales, on 20th September 1877 by Joseph Davison (advertised in the *Glasgow Herald*) and on 22nd October 1879 (advertised in *The Engineer*). The works was demolished and the lines lifted. The company remained under the control of the liquidator until 7th March 1884, when the court issued a winding up order.

Gauge : 2ft 4in

		0-4-0VBT	VC	Chaplin	1651	1874	New	(1)
-								

(1) advertised for auction on 20/9/1877 and on 22/10/1879; s/s

Nearby were two other quarries, **Wolsingham Park Quarry** (NZ 075418) **(T600)** and **Woodburn Quarry** (NZ 084440) **(T601)**. Both were served by tramways, almost certainly narrow gauge, the latter over a mile long, but how they were operated is unknown. Whether they had any connection with the firm above is also unknown.

NICHOLAS WOOD & PARTNERS

It would appear that Nicholas Wood (1795-1865) formed this partnership about 1840, at much the same time that he joined The Marley Hill Coal Co, which became John Bowes, Esq., & Partners, but before he joined The Hetton Coal Co and The Harton Coal Co. The firm's first colliery was **WESTERTON COLLIERY** (NZ 235308) **(N602)**, which was sunk to the south-west of Westerton village and opened in June 1841. A mile to the south-east Leasingthorne Colliery, owned by James Reid of Newcastle upon Tyne, was also being sunk, and this was served by the five mile-long Chilton Branch (called thus, though it never served Chilton Colliery) of the Clarence Railway, with its line to staiths at Stockton and Port Clarence. The privately-owned extension to Westerton Colliery was worked by a **stationary engine** at NZ 244306, which hauled the waggons up from the colliery and then lowered them down to Leasingthorne, joining the original line just east of Leasingthorne. One oral tradition claimed that the section east from the engine house was a self-acting incline, but as the 1st ed. O.S. 25in map shows both sections as single lines this cannot be correct.

With the opening of Westerton Colliery imminent, it would seem clear that Wood decided to work his own traffic over the Clarence Railway, perhaps right through to Port Clarence, and he acquired the first locomotive below in March 1841. However, it was not until September 1841 the Clarence Railway London Committee minutes state that "the Committee had not any objections to Messrs Wood & Co hauling their own coals from the foot of the Chilton Plain [= plane, = incline] provided the Company's Engineer approves the engine employed". The evidence below shows that Wood ran his own trains over the line, and indeed also acquired a second engine.

On 2nd September 1844 the Clarence Railway was leased by the Stockton & Hartlepool Railway, which on 1st November 1844 handed over the operation of the two lines for ten years to Messrs. Fossick & Hackworth of Stockton, although it retained the ownership of the locomotives. The surviving records of the Stockton & Hartlepool Railway are deposited at the National Archives at Kew, and all of the information below is drawn from these records. The Railway's Coal Haulage Account ledgers show that Nicholas Wood & Co was paid for performing coal haulage on the Clarence each month from July 1845 to December 1847, while from March 1848 onwards payments were made to the "owners of Westerton Colliery". Whether this means that Wood's locomotives were hauling other trains besides his own is not known. On 26th February 1849 the Railway agreed to purchase from the "owners of Westerton Colliery" 386 chaldron waggons for £3088 and two locomotives for £540 each, although payment was not made until December 1849, and the Coal Haulage ledgers also show that payments for haulage continued until that month. The locomotives' names are given as NICHOLAS WOOD and ROCKETT, probably a mis-spelling for ROCKET.

On 28th January 1851 Jonathan Backhouse & Co advertised for sake **LEASINGTHORNE COLLIERY** (NZ 252304) **(N603)**, won in July 1842, and **[OLD] BLACK BOY COLLIERY**, near Shildon (NZ 230291) **(Q604)**, opened in 1827. Both were purchased by Nicholas Wood & Partners. This meant that Wood now owned both collieries sending coal down the branch – indeed, on the 1st edition O.S. map (surveyed in 1855) the branch is shown as the "Westerton Colliery Railway", which may well indicate that the ownership of the branch had passed to the firm by this date.

At an unknown date, but certainly after 1855, a line about 1¾ miles was built running south-west from Leasingthorne Colliery, through Coundon village, to connect to the **Gurney Pit** (NZ 237282) of the new **BLACK BOY COLLIERY** (NZ 237282) **(Q605)**, diverting its coal eastwards via Leasingthorne instead of down the Stockton & Darlington Railway as done hitherto. A curve near Black Boy Colliery also linked this line to **Eldon Colliery (John Henry Pit)** (see the entry for Pease & Partners Ltd).

Wood died in 1865, and in 1866 the three collieries passed to The Black Boy Coal Co (see the entry for Bolckow, Vaughan & Co Ltd).

Gauge : 4ft 8½in

| ROCKET | 0-6-0 | OC | RS | 16 | 1829 | (a) | (1) |
| NICHOLAS WOOD | 0-6-0? | IC? | Coulthard | | 1841 | New (b) | (1) |

(a) ex Stockton & Darlington Railway, No.7, ROCKET, c12/1840; however, it would appear that this loco was also offered for sale by Thomas Bowman at New Shildon on 2/3/1841, so it may be that it was not purchased until this date.

(b) a reference to the completion of this locomotive is given in the *Newcastle Courant* of 10th September 1841.

(1) to the Stockton & Hartlepool Railway, 12/1849.

THE WOODIFIELD COAL CO LTD

This firm was registered on 8th August 1922, apparently as an amalgamation of three small coal companies, to which in 1924 was added a fourth. A receiver was appointed on 8th January 1926, and in *Machinery Market* of 1st October 1926 William Pallister, the receiver, advertised an auction of the collieries and their plant, including "locomotives", on 13th October 1926.

The collieries were

COLD KNOT COLLIERY, near Crook **M606**
NZ 148355

The original Cold Knot Colliery (NZ 135359) was situated rather more than ¼ mile souh-west of Craig Lea Colliery (see Finley & Wilkinson/The Harperley Colliery Co). This had been abandoned by the 1890s, and by 1922 a new Cold Knot Colliery (NZ 148355) had been developed much nearer to Crook. It was owned by **The Cold Knot Colliery Co Ltd**. From a drift a narrow gauge tramway ran eastwards for about a mile to screens on the site of Woodifield Colliery (NZ 160354), formerly owned by Bolckow, Vaughan & Co Ltd and closed in 1911. The screens were in turn served by a link to the NER line between Crook and Bishop Auckland, ½ mile south of Crook Station. The Ordnance Survey shows no other building here besides the screens, so at face value it would seem unlikely that a locomotive above worked here.

GLADDOW COLLIERY, near Tow Law **K607**
also known as **CORNSAY FELL COLLIERY** NZ 145437

This was a small drift mine, one of a number worked to the north-west of the village of Cornsay (note : this was a different village from Cornsay Colliery village). It had been worked until the early

1890s and then suffered a lengthy period of closure. In 1922 it was owned by the **Gladdow Coal Co**. It had no rail connection.

WEST BEECHBURN COLLIERY, Howden-le-Wear **M608**
NZ 167331

The colliery was opened in 1914 by **The West Beechburn Coal & Whinstone Co Ltd** (registered in 1913), and was situated alongside a minor road running south-west between Howden-le-Wear and Witton-le-Wear. A branch from Beechburn Station on the NER line between Crook and Bishop Auckland ran westwards for ¼ mile to serve the colliery screens (NZ 170331), to which coal was brought from the colliery by a narrow gauge tramway, almost certainly rope-worked, also ¼ mile long. Unusually, the colliery was also worked for whinstone. Latterly this company had the same Agent as The Harperley Collieries Co Ltd (see Cold Knot Colliery above).

Given the above, it would seem that any "locomotives" which the firm owned would have to have been used on the line between Beechburn Station and the screens of West Beechburn Colliery. No further details are known about them.

THE WOODLAND COLLIERIES CO LTD
registered 2/11/1880; see also below

This company was created by an amalgamation of **The Woodland Colliery Co** and **Fryer & Co**. It operated on the extreme south-western edge of the coalfield. It went into voluntary liquidation on 4th November 1911, but apparently soon afterwards it was acquired by The Cargo Fleet Iron Co Ltd of Middlesbrough, who allowed it to continue trading under the same title until 1914, when the parent company took over direct control (see **The Cargo Fleet Iron Co Ltd**).

Fryer & Co was the older of the two component companies. This firm owned **NEW COPLEY COLLIERY** (NZ 113245) **(P609)**. This was operating by 1868, and lay much nearer the village of Cockfield than Copley. It was served by sidings south of Cockfield Station on the NER Bishop Auckland – Barnard Castle line. The royalty was not linked to the Woodland royalty.

The sinking of **WOODLAND COLLIERY** (NZ 066276) **(P610)** was begun in 1873. The original owners were **The Owners of Woodland Colliery**, becoming **The Woodland Coal Co** in 1876 and then **The Woodland Colliery Co**. Serving this colliery was the Woodland Branch, which left Woodland Junction, at the eastern end of the Lands viaduct on the NER Bishop Auckland – Barnard Castle line 1¼ miles

166. NELSON, K 1786/1871, standing at Woodland Junction, probably just before the First World War. Note that the dome is mounted on the front ring of the boiler.

north-east of Cockfield Station and climbed for five miles on a circuitous route up to 1150 feet to reach the colliery. In 1874 the company opened **CRAKE SCAR COLLIERY** (NZ 082276) **(P611)**, about 1¼ miles north-east of Woodland Colliery.

To work the coal west of Woodland the company developed **WOOLLEY HILLS COLLIERY** (actually a drift) (NZ 045249) **(P612)**. This was opened by 1877 and was served by a narrow gauge main-and-tail rope haulage line 1½ miles long, with the **stationary engine** situated at Woodland Colliery. This drift was standing by 1883, re-opened in the early 1890s but standing again by 1897 and then merged with Woodland Colliery, though it was not officially abandoned until 1911.

With the closure of Woolley Hills, the company decided to work the coal around the Arn Gill, about 1¼ miles south of Woodland. For this **ARN GILL COLLIERY** (NZ 072243) **(P613)** was opened and a new narrow gauge main-and-tail system built, utilising the **stationary engine** that had formerly worked the Woolley Hills line. From Woodland Colliery this line ran underneath Woodland village in a tunnel, then crossed a wooden gantry into another tunnel before crossing a second wooden gantry, from where a short branch curved westwards to serve **COWLEY COLLIERY** (NZ 067255) **(P614)**. From here the "main line" ran through a cut-and-cover tunnel and then crossed Hindon Beck on an embankment before running through a fourth tunnel to reach Arn Gill Colliery. From Arn Gill a second line just over ½ mile long was built westwards to serve **MIDDLE ARN GILL COLLIERY** (NZ 066243) and **ARN GILL WEST COLLIERY** (NZ 062243), given separate names, but collectively part of Arn Gill Colliery This line was worked by a separate narrow gauge main-and-tail system, with its **stationary engine** at Arn Gill.

Meanwhile, the Woodland Branch itself attracted the development of new collieries, by different owners. From Crake Scarr (see above) the next was **Crane Row Colliery** (NZ 093276) **(P615)** and then **Morley Colliery** (NZ 110273) **(P616)**, which was owned by A. Mein and opened by 1883, though it was closed in 1893.

The branch itself, although a private line, was initially operated by the NER @ ten shillings (50p) per loco per hour. However, in 1885 the company took over the operation of the line itself. Initially this was done in an orthodox manner; but in the mid-1890s the driver of AB 694/1891 lost control of his downhill train whilst the colliery engineer was on the footplate. As a result the colliery engineer banned the use of locomotives on downhill trains, ordering that instead the wagons should run the five miles to Woodland Junction by gravity, with the loco fireman acting as a brakesman to pin down or release brakes on the wagons and the locomotive following behind to push when necessary. The usual number of wagons dispatched in a train was supposed to be 24, but oral tradition reported that many more were run on occasion, sometimes very quickly, with 66 wagons the highest known! On arrival at Woodland Junction the locomotive would haul back the empties.

When the company went into liquidation in 1911 the Arn Gill system had been closed down, and on the Woodland royalty it was operating Woodland and Crake Scar (as Crake Scarr had now become), both served by the Woodland Branch. On the Copley royalty there was New Copley Colliery, from which a ½ mile long narrow gauge line, almost certainly rope-worked, ran down to screens alongside the NER, where there were also beehive coke ovens, and **LANGLEYDALE COLLIERY** (NZ 086236) **(P617)**, which was served by sidings alongside the NER Bishop Auckland – Barnard Castle line, 2½ miles south-west of Cockfield Station, on the Copley royalty. So far as is known the NER shunted the traffic at both of these locations. New Copley and Langleydale Collieries were acquired by The New Copley Collieries Ltd, registered on 6th March 1913; it went into liquidation on 29th June 1921.

For developments on the Woodland royalty after 1911 see The Cargo Fleet Iron Co Ltd.

Gauge : 4ft 8½in

NELSON	0-6-0ST	IC	K	1786	1871	(a)	
		reb	LG		?		(1)
-	0-4-0ST	OC	B	208	1873	(b)	(2)
ELEANOR	0-6-0ST	OC	AB	694	1891	New	(1)
-	0-6-0T	IC	Dar		1876		
		reb	Dar		1891	(c)	(1)

(a) ex Thomas Nelson, contractor, Middlesbrough, Yorkshire (NR), /1885.
(b) ex Joseph Torbock, contractor, Middlesbrough, Yorkshire (NR), on hire.
(c) ex NER, 1293, 1037 class, 26/4/1911.

(1) to The Cargo Fleet Iron Co Ltd, with the collieries, /1911.
(2) boiler exploded, 8/11/1886; returned to Joseph Torbock, Middlesbrough, Yorkshire (NR) by 6/1887.

Locomotive works plates of Durham builders on Durham locomotives:
167. E No 2, The Consett Iron Co Ltd. John Roe was The Consett Iron Co Ltd's Chief Engineer, who worked with Black Hawthorn to develop this new design of rail crane.
168. HOLMSIDE No.2, The Holmside & South Moor Collieries Ltd; despite the date here, it was not despatched from the works until 31st October 1901.
169. 16, The Seaham Harbour Dock Company.
170. Lambton Railway 5, The Lambton, Hetton & Joicey Collieries Ltd.

Locomotive works' plates by R. & W. Hawthorn and R. & W. Hawthorn, Leslie & Co Ltd, Newcastle upon Tyne on Durham locomotives:
171. 18, The Weardale Steel, Coal & Coke Co Ltd.
172. BROWNIE, R & W Hawthorn, Leslie & Co Ltd, Hebburn.
173. STAGSHAW, Strakers & Love Ltd.
174. A No 4, The Consett Iron Co Ltd – an unusual plate, because it was put on to a locomotive the year after the company had amalgamated its locomotive business with Robert Stephenson & Co Ltd.

Index 1

Alphabetical list of owners

Where titles <u>are known</u> to have included the definite article, this has been included. Names changed by only a minor detail, for example, a firm becoming a limited company, have not been included. Numbers in **bold** refer to main entries in the text for the firm concerned.

Owner	Pages
The Abbey Wood Coal Co	164
John Abbot & Co Ltd	**31**, 182
Adams Pict Furnace Brick Co Ltd	**31**
C. Allhusen & Son Ltd	182, 203
Allison, English & Co Ltd	**32**, 43
Anglo-Austral Mines Ltd	**32**, 187
Sir William Angus, Sanderson & Co Ltd	**33**, 243
Sir W. G. Armstrong, Whitworth & Co (Engineers) Ltd	33
Sir W. G. Armstrong, Whitworth & Co (Ironfounders) Ltd	33, 34
Sir W. G. Armstrong, Whitworth & Co (Pneumatic Tools) Ltd	33
Sir W. G. Armstrong, Whitworth & Co Ltd	33
Armstrong Whitworth Metal Industries Ltd	33, 34
Armstrong Whitworth Rolls Ltd	**33**, 122
Armstrong Whitworth Securities Co Ltd	**35**
Associated Electrical Industries Ltd	**36**
Associated Portland Cement Manufacturers Ltd	212
Athole G. Allen (Stockton) Ltd	161
Charles Attwood & Partners	203
Auckland Rural District Council	**36**, 435
Aycliffe Lime & Limestone Co Ltd	**37**
Jonathan Backhouse	48, 54, 284
William Barker & Co	**38**, 256
William Barningham	121, 357
J. Bartlett	**39**, 136
The Bauxite Refining Co Ltd	208
The Bearpark Coal & Coke Co Ltd	**39**, 115, 259, 332
Bede Metal & Chemical Co Ltd	42
Bede Metal Co	42
Bell Brothers Ltd	**43**, 121, 125, 126, 127, 131, 135, 137
Bell, Hawks & Co	43
Bell, Stobart & Co Ltd	418
W. Bell & Partners	239, 418
The Birtley Brick Co Ltd	**43**
The Birtley Co Ltd	74, 294, 298
Birtley Iron Co	74, 210, 294
Black Boy Coal Co	46, 47, 48, 51, 284
Blake Boiler, Wagon & Engineering Co Ltd	**44**, 335
Blue Circle Industries plc	212
Blythe & Sons (Birtley) Ltd	**45**
Board of Trade Timber Supply Department	393
Bolckow, Vaughan & Co Ltd	**45**, 125, 126, 127, 128, 129, 130, 135, 136, 289, 321, 326, 400
John Bowes & Partners Ltd	**55**, 144, 156, 274, 401
D. Sep. Bowran Ltd	**61**
Bradley's (Weardale) Ltd	144, 290
The Brancepeth Coal Co Ltd	42, **61**
Brancepeth Gas & Coke (Strakers & Love) Ltd	380
British Crane & Excavator Corporation Ltd	**62**
British Electricity Authority	75
British Gas Corporation	**62**, 261
British Railways Board	**63**, 227
British Ropes Ltd	**64**
British Steel Corporation	**64**
Broadwood Limestone Co	**69**, 286, 435
E. Burdes & Co	420

Company	Pages
J. Burlinson & Co	70
J. H. Burn	356
Burn, Clark, Swainston & Co	356
Marquis of Bute	70
Butterknowle Colliery Co Ltd	250
Robert Candlish & Sons Ltd	347
The Cargo Fleet Iron Co Ltd	**71**, 248, 439
The Carlton Iron Co Ltd	**72**, 125, 130, 256, 336
G. & W. H. Carter	**73**
Castle Eden Coal Co Ltd	**74**
Caterpillar (U.K.) Ltd	**74**
Central Electricity Authority	**75**
Central Electricity Generating Board	**75**
The Charlaw & Sacriston Collieries Co Ltd	**79**, 242
Henry Chaytor	**80**, 292
The Chemical & Insulating Co Ltd	**81**
Chester-le-Street Urban District Council	267
The Chilton Coal Co	357, 372
Clarke Chapman Ltd	**83**
Clarke, Watson & Gurney	**83**
Clarke Chapman-John Thompson Ltd	**83**
The Cleveland Bridge & Engineering Co Ltd	**84**
Cleveland Bridge UK Ltd	**84**
Cleveland Structural Engineering Ltd	**84**
Coal & Alllied Industries Ltd	237
Coast Road Joint Committee	**85**
Coats Patons Ltd	**86**
J. P. Coats, Patons & Baldwins Ltd	**86**
Cochrane & Co Ltd	248
Coldberry Lead Co Ltd	**87**
The Cold Knott Colliery Co Ltd	438
The Consett Iron Co Ltd	64, 68, 70, **87**, 124
Consett Waterworks Co	**114**
Isaac Cookson & Co	394
Cornforth Limestone Co Ltd	388
Joseph Cowen & Co	315, 318, 370
Crossley Sanitary Pipes Ltd	42, **115**
Crouch Mining Ltd	**116**
Crown Coke Co Ltd	**116**, 172
Crowtrees Coal Co	185
W. C. Curteis & Co	272
Darlington Corporation Electricity Department	75
Darlington Corporation Gas Department	261
Darlington Forge Ltd	68, **116**, 357
Darlington Iron Co Ltd	121
Darlington Rolling Mills Co Ltd	119
Darlington & Simpson Rolling Mills Ltd	119
Darlington Steel & Iron Co Ltd	**121**, 357
Darlington Wagon & Engineering Co Ltd	44, 386
Davy Roll Co Ltd	33, 34, **122**
Deanbank Chemical Co Ltd	50
Dent, Sons & Co Ltd	256
Derwent & Consett Iron Co Ltd	87
Derwent Iron Co	87, **123**
Robert Dickinson & Partners	**124**, 354, 355, 356
Dinsdale Smelting Co Ltd	227
F. W. Dobson & Co Ltd	**125**, 388
Dorman, Long & Co Ltd	45, **125**, 273
Dorman Long (Steel) Ltd	**136**, 284
William Doxford & Sons Ltd	140
Doxford & Sunderland Ltd	**140**, 418
Doxford & Sunderland Shipbuilding & Engineering Ltd	**140**, 418
Drinkfield Cooperative Iron Co Ltd	397
Dunn & Miller	**143**, 258, 374

Lord Dunsany & Partners Ltd	272
The Dunston Garesfield Collieries Ltd	**143**
Durham Coke & By-Products Co Ltd	339
Durham & Yorkshire Whinstone Co Ltd	**144**, 384, 388
Durham County Council	**144**, 227
Durham County Water Board	114, **147**
Durham Urban District Council	266
Durham Rural Dustrict Council	266, 267
Durhills Ltd	**151**
The Earl of Durham	213, 214
The Easington Coal Co Ltd	**151**
East Hetton Coal Co Ltd	152
East Hetton Collieries Ltd	**152**
East Howle Coal Co	72
East Tanfield Colliery Co Ltd	354
Egis Shipyard Ltd	165
Eldon Brickworks Ltd	**154**, 287
Sir George Elliot	59, 156
Elliot & Jonassohn	401
James W. Ellis & Co Ltd	**154**, 315
James W. Ellis Engineering Ltd	**154**
English Electric Ltd	**156**
The Executors of the late John Fleming	273
The Felling Coal, Iron & Chemical Co Ltd	**156**, 278
Fence Houses Brickworks Ltd	**157**
Ferens & Love (1937) Ltd	**157**, 239
Finley & Wilkinson	**160**
John Fleming & John Milling	273
Fletcher, Blackwell & Falconer	426
The Ford Paper Mills Co Ltd	426
The Ford Paper Works Co Ltd	426
The Ford Works Co Ltd	426
Fordamin Company (Sales) Ltd	**161**
Forestry Commission	**162**
J. Forster/M. Forster	153
Foster, Blackett & Wilson Ltd	**162**
The Fownes Forge & Engineering Co Ltd	**163**
The Framwellgate Coal & Coke Co Ltd	**163**
The Frankland Coal Co (1934) Ltd	**164**
Frazers, Roberts & Co Ltd	156, 278
Sir Theodore Fry & Co Ltd	119
Fry, Ianson & Co Ltd	119
Fryer & Co	439
Gateshead County Borough Council	**164**
Gladdow Coal Co	439
Golightly Tar & Stone Co	165
The Grange Iron Co Ltd	221, 266
William Gray & Co Ltd	165
Greenside Sand & Gravel Co Ltd	166
Halmshaw & Partners	333
The Hamsterley Colliery Ltd	**166**
Hannington & Co Ltd	154, 315
S. Hanratty & Sons	166
Harehope Gill Mining & Quarrying Co Ltd	**168**, 289
The Harperley Colliery Co	160, 161
The Harperley Collieries Co Ltd	161
James Harrison	**168**, 365
John Harris	169, 384, 424
Hartlepool Dock & Railway Co	**169**, 360, 390, 391, 434
Hartlepool Gas & Water Co	**174**
The Harton Coal Co Ltd	**174**
Haswell Coal Co	74, 180, 232
The Haswell, Shotton & Easington Coal & Coke Co Ltd	74, **180**, 360

Hawks, Crawshay & Sons Ltd	31, **182**
Hawthorn Aggregates Ltd	**183**
Hawthorn Limestone Co Ltd	183
R. & W. Hawthorn, Leslie & Co Ltd	42, 183
R. & W. Hawthorn Leslie (Shipbuilders) Ltd	**183**, 387
William Hedley & Sons	**185**, 192, 365
Thomas Hedley & Bros Ltd	192
Hedley Hope Coal Co Ltd	41
The Hedworth Barium Co Ltd	32, **187**
Hendon Paper Works Co Ltd	353
W. T. Henley's Telegraph Works Co Ltd	36
The Hetton Coal Co Ltd	**187**
Heugh Hall Coal Co	172
The Heworth Coal Co Ltd	295
G. Hodsman & Sons Ltd	**191**
The Holmside & South Moor Collieries Ltd	**192**
Hopper, Radcliffe & Co	**198**
The Horden Collieries Ltd	**199**
The Hutton Henry Coal Co Ltd	**201**
Hunwick Coal Co	48, 53
Hunwick & Newfield Coal Co	48, 53
Hutton, Fletcher & Co	426
The Hulam Coal Co Ltd	**202**
Huwood Ltd	**202**
Huwood-Ellis Ltd	154
Charles I'Anson & Co Ltd	119, 169, 424
Ibbotson & Co	**202**
Imperial Chemical Industries Ltd, General Chemicals Group	**203**
Imperial Chemical Industries Ltd, Nobel Division	**205**
Inkerman Colliery Co Ltd	**207**
The International Aluminium Co Ltd	**208**
Ralph Ward Jackson	421
Jarrow Chemical Co Ltd	203, **208**
Jarrow Metal Industries Ltd	34
Jarrow Tube Works Ltd	34
James Joicey & Co Ltd	**209**, 354
Joseph Johnson (Durham) Ltd	337
Jonassohn, Gordon & Co Ltd	401
Richard Kell & Co Ltd	296, 365
Kvaerner Cleveland Bridge Ltd	84
Lackenby Iron Co	53
Lafarge Cement UK	**212**
Lackenby Iron Co	422
The Lambton Collieries Ltd	209, 213
The Lambton & Hetton Collieries	132, 134, 209, 214
The Lambton, Hetton & Joicey Collieries Ltd	157, 209, **213**
Lanchester & Iveston Coal Co Ltd	**225**
Lanchester Rural District Council	267
John Lee & Co	278
Andrew Leslie & Co	183
Leversons Wallsend Collieries Ltd	401
Linthorpe-Dinsdale Smelting Co Ltd	**226**
The Lintz Colliery Co	**228**, 358
Lloyd & Co	422
Lloyds (Darlington) Ltd	**229**, 385
London & North Eastern Railway	**229**
London Lead Co	**229**, 405
The London Steam Collier & Coal Co Ltd	392
The Marquis of Londonderry	163, **231**, 237, 238, 340, 360
The Londonderry Collieries Ltd	234, **236**, 340
George Love & Son	239
Joseph Love & Partners	157, **239**

Low Beechburn Coal Co Ltd 239
Lucas Bros Ltd 297
Lumley Brickworks Ltd 157
The Lunedale Whinstone Co **240**
Lyon Street Railway Co Ltd 183

The Mainsforth Coal Co 73, **240**
Marley Hill Coal Co 55
Marley Hill Coal & Chemical Co Ltd 55
Christopher Mason 51
Mawson, Clark & Co Ltd **241**
Sir Robert McAlpine & Sons Ltd **241**
McIntyre & Co 277
McLean & Prior 228
T. B. Maughan 38
Middleton Iron Co Ltd 226
The Mid-Durham Carbonisation Co Ltd **241**
Ministry of Defence, Army Department **242**, 243
Ministry of Munitions 33
Ministry of Public Buildings & Works 242
Ministry of Supply 162, 242, **244**
Ministry of Works 242
Modern Fuels Ltd 237
James Morrison & Co 335
William Murphy **246**

National Coal Board No. 1 (North-East Durham) Area 216, 352
National Coal Board No. 3 (South-East Durham) Area 351
National Coal Board No. 4 (South-West Durham) Area 327
Newalls Insulation Co Ltd **246**
The New Brancepeth Colliery Co Ltd **248**
The New Butterknowle Colliery Co Ltd **250**
The Newcastle Chemical Co Ltd 203
Newcastle upon Tyne & Gateshead Gas Co 262
The Newcastle upon Tyne Electricity Supply Co Ltd 75
The Newcastle Shipbuilding Co Ltd **251**
Newlandside Mining Co 53
The New Copley Collieries Ltd 440
The New Jarrow Steel Co Ltd 90, 97
The Newton Cap Brickworks Ltd 258
The North Bitchburn Coal Co Ltd **253**, 256, 327, 340, 372
The North Bitchburn Fireclay Co Ltd 115, 143, 252, **256**
The North Brancepeth Coal Co Ltd 42, 259
The Northern Coal Mining Co 163, 285, 383
The North of England Industrial Iron & Coal Co Ltd 72
North-Eastern Electric Supply Co Ltd 75
North Eastern Paper Mills Co Ltd **259**
North Eastern Railway 110, 229, 267, 268
The North Eastern Iron & Wagon Co Ltd 198
The North Eastern Steel Co Ltd 125
North Eastern Trading Estates Ltd 245, **260**
The North Hetton Coal Co Ltd 231, **264**
Northern Gas Board **261**
Northern Industrial Improvement Trust Ltd **264**
The Northern Sabulite Explosives Co Ltd 205, 360
Northern Sand & Gravel Co Ltd **264**
Northern Stone Firms 365
Northumbrian Water Authority 266

J. M. Ogden 333
Ord & Maddison Ltd 110, **267**, 286, 364, 388
The Original Hartlepool Collieries Co **391**, 412
The Original Hartlepool Collieries Co Ltd 392
Osmondcroft Ltd 271
Oswald & Co **271**, 402
The Owners of Hamsteels Collieries 337

The Owners of Pelton Colliery Ltd	**272**
The Owners of the Priestman Collieries	315
The Owners of Redheugh Colliery Ltd	**273**
The Owners of South Derwent Colliery	354
The Owners of South Medomsley Colliery Ltd	363
The Owners of Victoria Garesfield Colliery	319, 320
The Owners of Waldridge Colliery	320
The Owners of West Hunwick Colliery	423
The Owners of West Stanley Colliery	356
The Owners of Woodland Colliery	439
Page Bank Brick Co Ltd	**273**
Palmer Bros & Co	274
Palmers Hebburn Co Ltd	277
Palmers Shipbuilding & Iron Co Ltd	42, **274**, 396, 397
Patons & Baldwins Ltd	86
H. L. Pattinson & Co	156, 246, **278**
J. W. Pease & Co	124, **279**
Joseph Pease & Partners	279
Pease, Hutchinson & Ledward	352
Pease & Partners Ltd	125, 144, 154, 205, 252, **279**, 357, 360, 372, 373, 400
The Pelaw Main Collieries Ltd	74, 261, **294**
Pelton Brick Co Ltd	**300**
The Phoenix Brick Co	394
Pickford, Holland & Co Ltd	**301**
Alex Pirie & Sons Ltd	425
Port of Sunderland Authority	**301**
Port of Tyne Authority	**309**
The Priestman Collieries Ltd	75, 80, **315**, 370
Priestman & Peile	319
Raine Industries plc	**321**
Rainton Colliery Co Ltd	**324**
Raisby Basic Co Ltd	**324**
Raisby Hill Limestone Co Ltd	**324**
Raisby Quarries Ltd	**324**, 387
G. H. Ramsey & Co	31, 370
G. R. Ramsey	143
Ramshaw Coal Co Ltd	**326**
Randolph Coal Co Ltd	254, **327**
Randolph Coke & Chemical Co Ltd	**327**
Ransome Hoffman Pollard Ltd	**331**
Lord Ravensworth & Partners	192, **328**
Redheugh Iron & Steel Co (1936) Ltd	**329**
James Reid	51
G. Rennoldson	**330**
A. Reyrolle & Co Ltd	**330**
R. H. P. Bearings Ltd	**331**
Richards & Tutt	227
Richardson Brothers	**332**
U. A. Ritson & Sons	332
Ritsons (Burnhope Collieries) Ltd	**332**
River Wear Commission	301, 387
Robinson & Hannon Ltd	**332**
E. Robson & Co	**333**
J. Robson & Co	48, 52, 131, 186, **334**, 422
John Rogerson & Co Ltd	**334**, 385
The Rolling Stock & Engineering Co Ltd	**335**
The Rolling Stock Co Ltd	**335**
The Rosedale & Ferry Hill Iron Co Ltd	186, 256, **335**, 366, 401
The Ryhope Coal Co Ltd	**337**
Sir S. A. Sadler Ltd	239, **337**
Sadler & Co Ltd	**340**
Sir B. Samuelson & Co Ltd	42, 125, 132, 134

J. O. Scott & Co Ltd	296
Walter Scott Ltd	152, **333**
The Seaham Harbour Dock Co	234, **340**
Shell Direct (U.K.) Ltd	**350**
Shell-Mex & B. P. Ltd	350
Short Brothers Ltd	**351**
Shotley Bridge Iron Co	88
Shotton Brick Co Ltd	**351**
Skerne Iron Works Co Ltd	**352**
Slater & Co (Limestone) Ltd	**352**
Howard Smith (Hendon) Ltd	**353**
Joseph Smith	153
Smith, Patterson & Co Ltd	**354**
The South Derwent Coal Co Ltd	124, **354**
The South Durham Coal Co Ltd	286, 287, **357**, 372
The South Durham Iron Co	**357**
The South Garesfield Colliery Co Ltd	228, **358**
The South Hetton Coal Co Ltd	180, 232, 292, **360**
The South Medomsley Colliery Co Ltd	**363**, 396
The South Moor Colliery Co Ltd	192, 194
South Shields Rural District Council	**364**
Spark & Love	239
William Spencer & Co Ltd	**364**
Andrew Spottiswoode	51
Springwell Stone & Brick Co Ltd	**365**
Stanley Urban District Council	**365**
Steel & Co Ltd	62, 351
Steetley Co Ltd	**365**
Steetley Dolomite (Quarries) Ltd	186, **365**
Steetley Lime & Basic Co Ltd	244, 336, **365**
G-R Stein Refractories Ltd	**368**
The Stella Coal Co Ltd	318, **369**
The Stella Gill Coke & Bye-Products Co Ltd	241
Robert Stephenson & Co Ltd	42, 277, **371**
Robert Stephenson & Hawthorns Ltd	156, 261, **371**
Henry Stobart & Co Ltd	143, 252, 286, 326, 357, **372**
Strakers & Love Ltd	239, **376**
Richard Summerson & Co Ltd	144, **383**, 388, 427
Summersons Foundries Ltd	169, **384**
Thomas Summerson & Sons Ltd	384, **386**
Sunderland Corporation Electricity Department	79
Sunderland Corporation Gas Department	262
The Sunderland Dock Co	302, **386**
Swan Hunter Shipbuilders Ltd	387
Swan Hunter & Tyne Shipbuilders Ltd	387
R. W. Swinburne & Co	394
Tarmac Northern Ltd	325, **387**
Tarmac Roadstone (Northern) Ltd	267, 326, **388**
Ernest Taylor	**389**
The Team By-Product Coke Co Ltd	318
Charles Tennant & Partners Ltd	203, 204
Tharsis Copper & Sulphur Co Ltd	**389**
T. J. Thomson & Son Ltd	**390**
The Thornley Coal Co	**390**
The Thrislington Coal Co Ltd	256, 336
Thiedmann & Wallis	320
Timber Operators & Contractors Ltd	**393**
Joseph Torbuck & Co	423
Tow Law Patent Fuel Co Ltd	**394**
The Trimdon Coal Co	392, 434
Trimdon Grange Coal Co Ltd	153
Tuthill Limestone Co Ltd	292, 360
Tyne & Wear Fire and Rescue Service	**394**
Tyne Improvement Commission	309

The Tyne Brick & Tile Co Ltd	**394**
Tyne Coal Co Ltd	398
The Tyne Plate Glass Co Ltd	**394**
Samuel Tyzack & Co Ltd	**395**
The United Alkali Co Ltd	203
H. R. Vaughan & Co Ltd	**395**
Joseph Vaughan	396
Thomas Vaughan & Co	41, 363, **396**, 433
Vickers Ltd	**397**
Vickers-Armstrongs Ltd	42, 397
Victoria Garesfield Colliery Co	319
Waldridge Coal Co Ltd	320
The Wallsend & Hebburn Coal Co Ltd	**398**
Jacob Walton & Co Ltd	289, 364, **400**
Stephen Walton	**400**
War Department	242
The Washington Coal Co Ltd	**401**
The Washington Chemical Co Ltd	246
Washington Iron Co	**401**
Wear Lime Works Co	421
Wear Rolling Mills Co Ltd	402
Wear Steel Co Ltd	**402**
The Weardale Lead Co Ltd	402
Weardale Iron Co	407
Weardale Iron & Coal Co Ltd	393, 407
The Weardale Steel, Coal & Coke Co Ltd	151, 165, 248, **407**
Weardale Steel Co Ltd	418
Weardale Steel (Wolsingham) Ltd	418
The Wearmouth Coal Co Ltd	402, 418
Wearmouth Dock Co	301
The West Beechburn Coal & Whinstone Co Ltd	439
West Durham Wallsend Coal Co Ltd	284
West Hartlepool Harbour & Railway Co	48, 49, 52, 186, **421**
West Hunwick Refractories Ltd	**422**
West Hunwick Silica & Firebrick Co Ltd	422
West Witton Ganister Firebrick Co Ltd	423
Whessoe Foundry & Engineering Co Ltd	**424**
Whessoe Products Ltd	424
Whessoe Ltd	424
The Whitburn Coal Co Ltd	175, 176
Whitwell Coal Co	333
Wiggins Teape Ltd	**425**
The Wigglesworth Colliery Co Ltd	383, **427**
Henry Williams Ltd	352, **428**
Sir Hedworth Williamson's Limeworks Ltd	**428**
Willis & Co	72
Robert Wilson & Sons Ltd	433
Wilsons Forge Ltd	433
Wilsons Foundry & Engineering Co Ltd	**433**
The Wingate Coal Co [Ltd]	413, **434**
The Wingate Limestone Co Ltd	434
The Witton Park Slag Co Ltd	69, 286, **435**
The Wolsingham Park & Dinas Firebrick, Mineral & Coal Co Ltd	437
Nicholas Wood & Partners	47, 49, 51, **437**
The Woodifield Coal Co Ltd	**438**
The Woodland Coal Co	439
The Woodland Collieries Co Ltd	71, **439**
George Wythes & Co	226, 333

Index 2

Locations

Note: When a name is prefixed or suffixed by an addition in round brackets, for example, (New) Herrington, (Monk) Wearmouth, this indicates that this part of the name was subsequently abandoned; where square brackets are used, this part of the name was subsequently added.

(a) brickworks

Axwell Park	315
Bank Foot, Crook	282
Barmston	247
Bath, Washington	44
Batts, Wolsingham	368
Bearpark	39
Birtley	45
Birtley Brick & Tile	295
Birtley Grange Brick & Tile	296
(Birtley) Station	45
Black Boy	49
Blaydon Main	370
Brancepeth	379
Brandon	381
Butterknowle	250
Byers Green	50
Charley, South Moor	196
Cornsay	157
Cowen's, Blaydon	315
Craghead	193
Crow's House, Wheatley Hill	413
Derwenthaugh	31
East Howle	72
Eclipse, Crook	301
Emma, Ryton	369
Eldon	154, 287
Felling	60, 278
Fir Tree	115
Garesfield	112
Hannington, Swalwell	315
Hetton	189
Holmside	196
Jarrow	68
Lambton, Fencehouses	215
Leasingthorne	129
Lilley/Lily, Rowlands Gill	318
Londonderry, Seaham	233, 237
Low Mill and High Yard, Blaydon Burn	315, 316
Lumley	157
Malton	339
Medomsley	94
Murton	360
New Brancepeth	249
Newfield, near Hunwick	53, 136
Newton Cap	143, 258, 373
North Bitchburn	252, 257
Page Bank	131, 273
Pelaw Grange	296
Pelton	300
Phoenix, Crawcrook	394
Randolph, Evenwood	327
Ravensworth, Birtley	296
Rough Lea	253, 258
Ryhope	337
St.Helens	291
Sacriston	79
Shincliffe	239
Shotton	181, 201, 351
Slotburn	294
South Moor	196
South Tanfield	211
Stargate	370
Station, Birtley	45
Sunny Brow	383
Teams	297
Team Valley	32, 43
Templetown	90, 96
Tudhoe	413
Union, Birtley	32, 43
Washington	264
Wellfield	73
West Hunwick	422
Whitwell	333
Willington	384
Wingate	73, 74
Wolsingham Park	437

(b) collieries and drift mines

Abbey Wood Drift	164
Addison	369
Adelaide	284
Adventure Pit, Rainton (old)	231
Adventure, Rainton (new)	232, 233, 324
Alexandrina, Rainton	231, 233
Allerdean/Allerdene	297
Bearpark	39
Beamish Air	210
Beamish Mary	210, 223
Beamish Second [& Park]	210, 223
Belmont	236
Bewicke Main D	296
Billingside Drift	92
Alma, West Pelton	193, 209, 224
Andrews House	55
Arn Gill/Arnghyll	71, 440
Auckland Park	46, 49, 126
Axwell	315
Axwell Garesfield	315
Axwell Park	315
Binchester	47, 334, 422
Bishop Middleham	284
Bitchburn/Beechburn	376
[Old] Black Boy	48
Black Boy	46, 48, 49, 438
Blackburn Fell Drift	56
Blackhall	200

Durham Part 1 Page 450

Blackhill	96
Blackhill Drift	96
Blackhouse H	297
Black Prince	412
Blaydon Burn	315, 316
Blaydon Main	31, 318, 370
Boldon	175, 179
Bowburn (old)	170, 185, 334, 336, 422
Bowburn (new)	126
Bowden Close	285
Boyne	259
Bradley	92
Brancepeth	376
Brancepeth B	376, 382
Brancepeth Z	383
Brandon A	381
Brandon B	381
Brandon C	381
Brandon Pit House	381
Brasside	215
Brass Thill Drift, Lintz	228
Broom's Drift	92
Broompark	259
Broomside	231
Browney	127
Burnhope	42, 192, 332
Burnopfield	55
Butterknowle	250
Byermoor	56
Byers Green	49, 127, 334, 422
Byron	79

Cassop	170, 333, 391
Cassop Moor	170, 333, 391
Cassop Vale	333, 391
Castle Eden	74, 170, 182, 199, 200
Chester South Moor	320
Charlaw	79
Chilton	286, 357, 372
Chop Hill	210
Clara Vale	371
Clay Hole	186
Cocken (George Pit)	214
Cold Knot (original)	438
Cold Knot (later)	161, 438
Copley	250
Cornsay	157
Cornsay Fell	438
Cornforth	170
Cowley	72, 440
Coxhoe	186, 334, 336, 422
Craghead	192
Craig Lea	161
Crake Scar	72, 440
Crane Row	440
Crookbank	55
Crookhall	92
Crookhall [Victory Pit]	92
Crow Trees	170, 185, 333, 334, 391, 422
Croxdale (original)	412
Croxdale (later)	412

Dawdon	237, 344
Deaf Hill	434
Dean & Chapter	50, 128, 136
Deanery	284
Delves	92
(West) Derwent	92
Derwent Crook	296
Diamond Pit, Butterknowle	250
Dipton (Delight)	55
Dorothea, Newbottle	214, 221
Dunston	143
Dunston & Elswick	143
Dunwell Pit, North Hetton	231

(West) Edmonsley	79, 395
Easington	151
East Hedley Hope	41, 397, 412
East Hetton	152, 170
East Howle	72
East Stanley	210, 223
East Tanfield	209, 210, 354
Eden	94
Eighton Moor	295, 297
[Old] Eldon [John Henry]	51, 287
Eldon	154, 287, 438
Eldon Drift	288
Elemore	189
Emma, Roddymoor	281
Emma, Ryton	369
(Old) Etherley	373
Eppleton	189
Esh	288
Evenwood	253

Farnacres	296
Felling	59, 156
Finchale	164
Fir Tree Drift	42, 115
Fishburn	373
Follonsby	56
Framwellgate	163, 232
Frankland	215

Garesfield	70, 88
Garmondsway	170
Gladdow	438
Gordon House (original)	253
Gordon House (second)	254
Gordon House (third)	254, 326
Gordon Pit, Butterknowle	250
Grange	236, 266
Greenwell Wood Drift, near Lanchester	226
Greenside	370

Hamsteels	337
Hamsterley	166
Handen Hold	211, 224
Harperley	161
Harraton (5th Pit)	214
Hartbushes	201
Harton	175
Haswell	170, 180, 232, 360
Hazard Pit, North Hetton	231
Hebburn A	398

Hebburn B	398	Holly Moor	427
Hebburn C	398	Holmside	192
Hedley Hill	412	Horden	199
Hedley Hope	41, 42, 281, 397, 412	Houghall	239
(New) Herrington	215	Houghton	214
Hetton (Lyons)	187, 188	Howden	252
Heugh Hall	170, 185, 334, 336, 422	Humber Hill Drift	92
Heworth	295	Hunwick	52, 334, 422
High Pit, Lintz	228	Hutton Henry	199, 201
Hill Top	339	Hylton	402, 419
Hollingside Drift	39		
Inkerman	157, 207	Iveston	94
Job's Hill	285, 290		
Kelloe	152, 170	Kibblesworth	55, 329
Kepier Grange	236, 266	Kibblesworth Grange Drift	56
Lady Ann, Fencehouses	215, 221	Littletown	133, 214
Lady Park Drift	297	Louisa, Stanley	196
Lambton D	215, 221	Low Beechburn	239
Leasingthorne	51, 129, 438	Ludworth	170, 391, 392, 412
Letch Pit, Rainton	231	Lumley 2nd	214
Lilley/Lily Drift, Rowlands Gill	318	Lumley 3rd	214, 216
Lily, Dipton	124	Lumley 6th	157, 215
Lintz	228	Lumley 7th	214
Lintz (Anna Pit)	358	Lumley 8th	214
Lintz (Esther Pit)	358	Lumley 9th	215
Littleburn	61, 259	Lumley New Winnings	216
Little Chilton	286, 422	Lumley West	214
Mainsforth	73, 130, 240	Mill Drift, Birtley	296
Malton	339	Millfield Grange	383
Margaret, Newbottle	214, 221	Monkwearmouth	428
Marley Hill	55	Morley	440
Marsfield	250	Morrison Busty	196
Marshall Green	294	Morrison North	196
Meadows, Rainton	216, 231, 233	Morton West	214
Medomsley	94	Mount Moor	55, 295, 328
Merrington	55, 334, 422	Mount Pleasant	96
Middridge	413	Murton	232, 360
Nettlesworth	79	Nicholson's, Rainton	216, 233
Newbottle, Dorothea Pit	214	North Biddick	215
Newbottle, Margaret Pit	214	North Bitchburn	252, 257
New Brancepeth	248	North Brancepeth	259
New Copley	439	North Hetton	220, 264
Newfield, near Hunwick	39, 52, 136, 422	Norwood, near Dunston	143, 297
Newfield, Pelton	272	Norwood, near Evenwood	253
Newton Cap	143, 258, 374		
Oakenshaw	376	Ouston A	295
Old Black Boy	48, 437	Ouston B	295
[Old] Croxdale	239	Ouston E	296, 297
Old Durham	236	Oxhill	211
Osmondcroft	271		
Page Bank	131, 334, 422	Pemberton Main	418
Pease's West Brandon	293	Pittington	231, 266
Pelton	272	Pontop Hall	363, 396
Pelton Fell	272, 300		
Quarry Pit, Butterknowle	250		
(Rabbit) Warren Drift	332	Ravensworth Shop	297
Railey Fell	326	Redheugh	273
Ramshaw No.1	326	Riding Drift, Birtley	296, 300
Randolph	254, 327	Roddymoor	280
Ravensworth Ann	297	Rodridge	201
Ravensworth Betty	297	Rough Lea	253, 258
Ravensworth Park Drift	297	Ryhope	337

St. Helen's (Auckland)	290
St. Hilda	175
Sacriston Shield Row Drift	80
Sacriston	79
Salter's Burn Drift	250
Seaham	232, 236, 344
Seaton	232, 236, 265
Sherburn [Lady Durham]	132, 215
Sherburn Hill	132, 134, 215
Sherburn House	132, 215
Sheriff Hill	297
Shildon Lodge	55
Shincliffe	239
Shotton	170, 181, 199, 200, 232, 360
Silksworth	216, 220, 233, 236
Sleetburn	248
South Brancepeth	131
South Derwent	124, 354, 356
South Durham	287, 357
South Garesfield	228, 359
South Hetton	169, 180, 232, 360
South Kelloe	186, 334, 336, 365, 422
South Medomsley	96, 363, 396
Tanfield Lea	209, 211, 225
Tanfield Moor	212, 225
Team(s), Low Fell	296
Tees Hetton	253, 254
Thistleflat	239
Thornley	169, 390, 412
Thrislington	256, 335, 375, 400
Throstle Gill Drift	294
Thrushwood	253
Tindale	290, 292
Urpeth Busty/Urpeth B	295
Urpeth C	295, 297
Vane Tempest	237
Victoria Garesfield	319
Waldridge A	79, 295, 320
Waldridge D	80, 320
Waldridge Shield Row Drift	320
Wardley	56
Washhouses	297
Washington F	401
Washington Glebe	401
Watergate	321
Waterhouses	293
(Monk) Wearmouth	418, 428
West Auckland	54, 326
West Beechburn	439
West Craghead	194
West Hetton	170, 185, 336
West Hunwick	53, 422
West Lintz Drift	228
West Pelton	193, 209
West Shield Row	124, 354, 355
West Stanley	356
West Tees	326
West Thornley	413

South Moor	194
South Pelaw	296
South Pontop	332
South Tanfield	209, 211
South Wingate	170, 199, 201
(High) Spen	70
Springwell	55, 328
Springwell (Vale Pit)	56
Stanley, Crook	281, 291
Stanley, Tanfield	356
Stargate	369
Stella	369
Stella Freehold/Bog Pit	369
Storey Lodge	253
Stormont (Main)	296
Straightneck Pit, Lintz	228, 359
Success, Newbottle	214
Sunderland Bridge	412
Sunniside	291
Sunny Brow	383
Surtees Pit, Dipton	124
Swalwell Garesfield	31, 143
Tin Mill	88, 96
Towneley	369
Trimdon	170, 391, 392
Trimdon Grange	153, 170
Tudhoe	409, 413
Tudhoe Grange	409, 413
Tudhoe Park	413
Tursdale	135
Twizell	209, 212, 224
Twizell Burn Drift	209, 224
Ushaw Moor	292
Usworth	401
Victoria, Howden-le-Wear	253
Westerton (original)	47, 437
Westerton (later)	48, 128, 135
Westoe	175, 178
Westwood	
Wham Drift	250
Wheatley Hill	392, 412
Whitburn	175
White Lee (old)	55, 281
White Lee (new)	281
Whitwell	333
Wigglesworth	427
Willington	379, 383
Windlestone	286, 287, 293, 372
Wingate Grange	170, 434
Witton	79
Woodhouse Close	290
Woodifield	55
Woodland	71, 439
Woodside Winnings	92
Wooley	293
Woolley Hills	440

(c)   by-product coke ovens and beehive ovens using locomotives

Auckland Park	46, 126
Axwell Park, Swalwell (beehive)	315
Bank Foot, Crook (beehive)	281

Bank Foot, Crook	281
Bearpark (beehive)	41
Bearpark	39

Binchester (beehive)	48
Brancepeth (1st set)	380
Brancepeth (2nd set)	380
Bowden Close	285
Bowes	56
Butterknowle (beehive)	251
Byers Green	49, 50, 127
Chilton	128, 286, 372
Chopwell (beehive)	112
Dean & Chapter, Ferryhill	50, 128
Derwenthaugh, Winlaton Mill	112
Eldon	287
Fell, Consett	67, 89, 94, 108
Fishburn	373
Garesfield (at Derwenthaugh)	113
Hamsteels (beehive)	338
Hutton Henry (beehive)	202
Lambton, Fencehouses	216
Langley Park (beehive)	113
Langley Park	113
Leasingthorne	51
Littleburn	259
Malton	339
Marley Hill	56
Marsfield	251
Monkton	56
New Brancepeth	248
Newfield (beehive)	53
Newton Cap	374
Norwood, Dunston	75, 298, 318
Ottovale, Blaydon	316
Randolph, Evenwood	254, 327, 340
St. Helen's, West Auckland	290
Stella Gill, Pelton Fell	241, 300
Templetown, Consett	89, 94
Thrislington	256, 375
Trimdon Grange	154
Tudhoe	409
Ushaw Moor	292
Vane Tempest, Seaham	237
West Auckland	54
Whinfield (Victoria), Rowlands Gill (beehive)	319

(d) iron works, tin works and foundries

Albert Hill, Darlington (Barningham; Darlington Steel & Iron)	121, 357
Albert Hill, Darlington (South Durham)	357
Albert Hill, Darlington (Summersons)	229, 384
Albion, Felling	156
Alliance, Darlington	44
Auckland, Bishop Auckland	396, 433
Birtley	295
Bishopwearmouth	87, 123, 189, 215
Bradley	88, 92
Britannia, Colliery Row	198
Castletown	402
Consett	64, 87, 92, 123
Coxhoe	186
Close, Gateshead	33, 123
Crookhall	87, 88, 92
Darlington Forge	68, 116, 357
Delta, Derwenthaugh	323
Derwent, Berry Edge (Consett)	123
Dinsdale, Middleton St. George	226
Drinkfield, Darlington	396
Felling	60, 156, 278
Ferry Hill, West Cornforth	73, 335
Gateshead (Hawks)	182
Gateshead (Ibbotson)	202
Hannington, Swalwell	154
Hope Town, Darlington	384, 424
Jarrow (Armstrong Whitworth/ Davy Roll)	34, 122
Jarrow (Consett Iron/British Steel)	68, 113
Middleton	226
Monkwearmouth	395
Palmer's, Jarrow	274
Park, Gateshead	31
Pioneer, Blaydon	354
Railway Forge, Bishop Auckland	433
Redheugh	329
Ridsdale/Redesdale, Northumberland	123
Rise Carr, Darlington	119
Shotley Bridge Tin	88, 94
Skerne, Darlington	352, 428
Springfield, Darlington	84, 121
Stanhopeburn, Stanhope	407
Stanners Close, Wolsingham	334, 418
Tow Law	408
Trimdon, Sunderland	174
Tudhoe	165, 409
Victoria, Gateshead	83
Washington	401
Wear, Washington	43
Whessoe, Darlington	396
Whessoe Foundry, Darlington	424
Winlaton (Mill)	70, 322
Witton Park	36, 54
Yarm Road, Darlington	84

(e) barytes, basalt, fluorspar, ganister, ironstone and lead mines

Bolt's Burn (lead & fluorspar)	403, 405, 410
Brandon Walls (lead)	403
Burtree Pasture (lead)	403
Carrick's, Ireshopeburn (ironstone)	403, 411
Closehouse (barytes)	161
Coldberry (lead)	87
Cornish Hush (lead)	229
Cow Green (barytes)	32, 187
Craig's Level, Ireshopeburn (lead)	403
Dubbysike (barytes)	32, 187
Green Laws, Daddry Shield (lead)	403
Groove/Grove Rake (ironstone & lead)	404, 410
Grove Heads (lead)	403
Foulwood, Rookhope (ironstone)	403, 410
Harehope Gill (lead & ganister)	168, 289
Heights, near Eastgate (lead & ironstone)	411
Herrerias, Spain (copper)	42
Isabella, Cow Green (barytes)	187
Killhope (lead)	403
Killingdal, Norway	42
Levelgate (lead)	403
Middlehope, near Westgate (ironstone)	411
Orcanera, Spain (iron ore)	88
Rispey (ironstone & ganister)	404, 410

Rookhope (lead) 403
Sedling (lead) 407
Sikehead (lead) 410
Slit (sometimes Slitt), near Westgate (lead) 403
Slit Pasture, near Westgate (ironstone) 411
Smailsburn (ironstone) 411
Stanhopeburn, Stanhope
(ironstone & lead) 403, 405, 407

Stotsfield Burn (lead) 403, 406
Thornley Brow (lead) 403
West Slit, near Westgate (ironstone) 411
Whiteheaps (lead)..410
Wolfcleugh (lead & fluorspar) 404

(f) quarries

Ashes, Stanhope (limestone) 88, 268
Aycliffe (Aycliffe Lime & Limestone)
(limestone) 38
Aycliffe (Ord & Maddison) (limestone) 269
Bantling, near Annfield Plain (sandstone) 92
Barlow Lane, near Ryton (sand).......................264
Berry Edge Common (sandstone?) 92
Bishopley, near Frosterley (limestone) 268, 364
Bishopley Crag North, near Frosterley
(limestone) 364
Bishopley Crag South, near Frosterley
(limestone) 364
Bishop Middleham, near Ferryhill
(limestone) 284, 335
Bollyhope, near Frosterley (limestone)............364
Brown's House(s), near Frosterley
(limestone) 268, 286
Broadwood, Frosterley (limestone)............69, 285
Burnhills, Ryton Woodside (sand) 166
Burnhope Plantation (whinstone) 148
Butsfield, near Consett (ganister) 90, 111
Carley Hill, near Sunderland (limestone) 421
Carley Hill East [1], near Sunderland
(limestone) 421
Carley Hill East [2], near Sunderland
(limestone) 421, 429
Carley Hill West [1], near Sunderland
(limestone) 421
Carley Hill West [2], near Sunderland
(limestone) 421
Chilton, near Ferryhill (clay, gravel,
limestone & sandstone) 286
Cockfield, near Evenwood (whinstone) 383
Collier Law, near Stanhope (sand) 151
Cornforth, near Coxhoe (limestone) 326, 388
Coxhoe (limestone & sand) 365, 367
Coxhoe, formerly Garmondsway,
(limestone) 324, 387
Coxhoe Bank, Coxhoe (limestone) 186, 365
Crag Wood, near Evenwood
(whinstone) 383, 384
Crookgate (limestone)..56
Crossrigg, near Penshaw (limestone) 215
Crossthwaite (whinstone), near
Middleton-in-Teesdale...269
Drypry, Wolsingham Park (freestone) 437
Eclipse, near Crook (clay) 301
Fine Burn, near Frosterley
(limestone) 289, 364, 400
Folly, Ryton Woodside (sand) 166
Ford, South Hylton (limestone) 248

Fox Cover, Dawdon (limestone).......................344
Frosterley (limestone) 289
Fulwell, near Sunderland (limestone) 428
Garmondsway, near Coxhoe (limestone) 324
Greenfoot, Stanhope (whinstone) 144, 271, 388
Greengates, near
Middleton-in-Teesdale (basalt).......................191
Harehope, near Frosterley
(limestone & ganister) 289
Harthope (ganister) 96
Hawthorn, near Seaham (limestone) 183
Highland, near Ferryhill (limestone) 335
Howe's, Burnhope (limestone & ganister) 148
Kyo, near Annfield Plain (sand) 94
Lanehead, Stanhope (limestone).....................267
Lumley (clay) 157
Lunedale, near Middleton-in-Teesdale
(limestone & whinstone) 191, 240
Marsden Harbour (limestone) 179, 352
Marsden Lighthouse (limestone)..............179, 352
Marsden No.2 (limestone) 179, 352
Marsden (old) (limestone) 175, 179, 352
Middleton, near Middleton-in-Teesdale
(whinstone) 269
Newlandside, Stanhope (limestone) 53, 130
New Woodcroft, Stanhope (limestone)............137
No. 1 Deposit, Salter's Gate (fireclay) 437
North Bishopley, near Frosterley
(limestone) 268, 286
Old Marsden (limestone) 179, 352
Park End, near Middleton-in-Teesdale
(whinstone) 269
Parkhead, near Stanhope (sand) 151
Parson Byers, Stanhope (limestone)................137
Pathhead, near Ryton (sandstone) 315
Raisby Hill (limestone) 324
Redwell Hills, near Annfield Plain
(sandstone)...94
Ridding House, Stanhope (limestone) 137
Rogerley, Frosterley (limestone) 144, 290
Sherburn Hill (limestone) 215
South Bishopley, near Frosterley
(limestone) 268, 286
Southwick [1], Sunderland (limestone)............420
Southwick [2], Sunderland (limestone)....420, 429
[Old] Springwell (sandstone) 365
Springwell (sandstone).....................................365
Stargate (sand) 241
Toad Hole, near Lumley (limestone) 215
Trow Rocks, South Shields (limestone) 309
Tunstalll, Ryhope (limestone)...........................337
Tunstall, Wolsingham Park (freestone) 437

Tuthill, near South Hetton (limestone) 205, 290, 360	Whin Sike (whinstone) 148
Westerton (limestone)..51	Windy Nook, Gateshead (sandstone) 296
	Wingate (limestone) 434

175. An unusual wagon to be built after 1925 – Strakers & Love did not become a limited company until then – and for what purpose?

Index 3

Locomotives

NOTE : The builders below are listed in alphabetical order of their <u>abbreviations</u>, not of their titles.

Information normally relates to the locomotive as built.

Column 1	Works Number (or original company running number for locomotives built in main line workshops without a works number).
Column 2	Date ex-works where known - this may be a later year than the year of building or the year recorded on the works plate. Other dates included as described.
Column 3	Gauge.
Column 4	Wheel arrangement. Where the locomotive was a diesel locomotive, flameproofed for use underground, this is denoted by the suffix F. Otherwise the suffix F indicates that the locomotive involved was of the fireless design.

Steam Locomotives :

Column 5	Cylinder position.
Column 6	Cylinder size (diameter x stroke) in inches.
Column 7	Driving wheel diameter in feet and inches.
Column 8	Miscellaneous information - the manufacturers type designation/crane lifting capacity/public railways' running numbers or names/works numbers of other builders.
Column 9	Page references.

Diesel and Electric Locomotives :

Column 5	Horse power.
Column 6	Engine type.
Column 7	Driving wheel diameter in feet and inches.
Column 8	Weight in working order, as recorded in manufacturers' records.
Column 9	Page references.

Notes on the history of companies in North East England have been included.

ANDREW BARCLAY, SONS & CO LTD, Caledonia Works, Kilmarnock, Ayrshire.　　　　　**AB**
(member of **Hunslet Group** from 21/8/1972)

52	10.1866	4ft 8½in	0-4-0ST	OC	8 x 17	3ft 0in	306
62	7.6.1867	4ft 8½in	0-4-0ST	OC	12 x 20	3ft 6in	434, 435
70	3.2.1868	4ft 8½in	0-4-0ST	OC	9 x 18	3ft 0in	371
233	14.9.1881	4ft 8½in	0-4-0ST	OC	13 x 20	3ft 2in	367
277	24.6.1884	4ft 8½in	0-4-0ST	OC	11 x 18	3ft 0in	298, 367
305	14.1.1889	4ft 8½in	0-4-0ST	OC	14 x 22	3ft 5in	414
641	15.4.1889	4ft 8½in	0-4-0ST	OC	13 x 20	3ft 7in	402, 419
656	7.2.1890	4ft 8½in	0-4-0ST	OC	13 x 20	3ft 2in	54, 130, 139
675	17.2.1891	4ft 8½in	0-4-0ST	OC	12 x 20	3ft 2in	363, 434
688	1.5.1891	4ft 8½in	0-4-0ST	OC	13 x 20	3ft 2in	367
693	2.5.1891	4ft 8½in	0-4-0ST	OC	12 x 18	3ft 0in	31
694	22.4.1891	4ft 8½in	0-6-0ST	OC	14 x 22	3ft 5in	72, 440
698	1.9.1891	4ft 8½in	0-4-0ST	OC	14 x 22	3ft 5in	153, 154
703	21.1.1893	1ft 11⅝in	0-4-0ST	OC	5 x 10	1ft 10in	300
730	18.4.1893	4ft 8½in	0-4-0ST	OC	10 x 18	3ft 0in	241

786	#25.9.1896	4ft 8½in	0-4-0ST	OC	12 x 20	3ft 2in		298
803	26.8.1897	4ft 8½in	0-6-0ST	IC	16 x 24	3ft 9in		321
805	21.10.1897	4ft 8½in	0-4-0ST	OC	14 x 22	3ft 5in		306
833	4.7.1899	4ft 8½in	0-6-0ST	IC	16 x 24	3ft 9in		419
866	20.2.1900	4ft 8½in	0-4-0ST	OC	10 x 18	3ft 0in		251
895	24.4.1901	4ft 8½in	0-4-0ST	OC	13 x 19	3ft 4½in		102
909	17.7.1901	4ft 8½in	0-6-0ST	IC	16 x 24	3ft 9in		419
911	20.3.1901	4ft 8½in	0-6-0T	IC	16 x 24	3ft 9in		419
912	23.7.1901	4ft 8½in	0-6-0ST	OC	12 x 20	3ft 2in		152
961	#1.5.1903	4ft 8½in	0-6-0ST	OC	15 x 22	3ft 5in		286, 373, 374
970	1.4.1903	4ft 8½in	0-6-0ST	OC	14 x 22	3ft 5in		298
973	21.1.1904	4ft 8½in	0-4-0ST	OC	14 x 22	3ft 5in		354, 356
999	1.2.1904	4ft 8½in	0-6-0T	IC	17 x 24	4ft 0in		419
1005	13.4.1904	4ft 8½in	0-6-0ST	IC	16 x 24	3ft 9in		80, 321
1015	8.7.1904	4ft 8½in	0-6-0ST	OC	15 x 22	3ft 5in		200
1066	21.8.1906	4ft 8½in	0-4-0ST	OC	14 x 22	3ft 5in		415
1082	21.11.1907	4ft 8½in	0-4-0ST	OC	16 x 24	3ft 7in		337
1085	30.10.1907	4ft 8½in	0-4-0ST	OC	12 x 20	3ft 2in		256, 286, 288, 373
1097	27.5.1907	4ft 8½in	0-6-0ST	OC	15 x 22	3ft 5in		286, 373, 374
1127	25.11.1907	4ft 8½in	0-4-0ST	OC	14 x 22	3ft 5in		306
1223	14.3.1911	4ft 8½in	0-4-0ST	OC	10 x 18	3ft 0in		390
1256	26.2.1912	4ft 8½in	0-4-0ST	OC	10 x 18	3ft 0in		353
1287	17.6.1914	4ft 8½in	0-4-0ST	OC	10 x 18	3ft 0in		118, 262, 385
1289	26.12.1912	4ft 8½in	0-4-0ST	OC	14 x 22	3ft 5in		325
1305	26.11.1912	4ft 8½in	0-4-0CT	OC	14 x 22	3ft 5in	4T capy.	142
1321	23.7.1913	4ft 8½in	0-6-0ST	OC	14 x 22	3ft 5in		415
1337	22.9.1913	4ft 8½in	0-4-0ST	OC	13 x 20	3ft 2in		323
1453	4.4.1918	2ft 0in	0-4-0WT	OC	6¾ x 10¾	1ft 10in		150
1497	11.12.1916	4ft 8½in	0-6-0ST	OC	14 x 22	3ft 5in		159
1639	29.12.1922	4ft 8½in	0-6-0ST	OC	14 x 22	3ft 5in		177, 180
1659	3.3.1920	4ft 8½in	0-4-0ST	OC	14 x 22	3ft 5in		354
1665	29.5.1920	4ft 8½in	0-4-0CT	OC	14 x 22	3ft 5in	5T capy.	105
1715	13.12.1920	4ft 8½in	0-4-0CT	OC	14 x 22	3ft 5in	5T capy.	105
1724	7.3.1922	4ft 8½in	0-4-0ST	OC	14 x 22	3ft 5in		238
1811	22.12.1923	4ft 8½in	0-4-0ST	OC	14 x 22	3ft 5in		364
1855	30.3.1931	2ft 0in	0-4-0WT	OC	7 x 11	1ft 10in		150
1885	12.1.1926	4ft 8½in	0-4-0ST	OC	14 x 22	3ft 5in		238
1988	31.3.1931	4ft 8½in	0-4-0ST	OC	12 x 20	3ft 2in		150
1991	14.7.1931	2ft 0in	0-4-0WT	OC	7 x 11	1ft 10in		150
1994	8.10.1931	2ft 0in	0-4-0WT	OC	7 x 11	1ft 10in		150
1995	30.11.1931	2ft 0in	0-4-0WT	OC	7 x 11	1ft 10in		150, 326
2029	19.3.1937	4ft 8½in	0-6-0T	OC	16 x 24	3ft 8in		306
2078	21.11.1939	4ft 8½in	0-4-0ST	OC	16 x 24	3ft 8in		101
2091	9.4.1940	4ft 8½in	0-4-0ST	OC	16 x 24	3ft 8in		103
2111	30.6.1941	4ft 8½in	0-4-0CT	OC	14 x 22	3ft 5in	5T capy.	104
2118	23.6.1941	4ft 8½in	0-4-0ST	OC	14 x 22	3ft 5in		323
2119	25.7.1941	4ft 8½in	0-4-0ST	OC	14 x 22	3ft 5in		325
2160	10.3.1943	4ft 8½in	0-4-0ST	OC	14 x 22	3ft 5in		238
2165	11.12.1944	4ft 8½in	0-4-0ST	OC	14 x 22	3ft 5in		238

date of invoice

Unidentified steam locomotives

	KILMARNOCK	4ft 8½in	0-4-0ST	OC	?	?		153
	No. 2	4ft 8½in	0-4-0ST	OC	?	?		120
352	2.4.1941	4ft 8½in	0-4-0DM	80hp	Gardner 6cyl	3ft 5in	14½T	242
380	5.9.1950	4ft 8½in	0-6-0DM	330hp	Crossley 6cyl	4ft 0in	40T	312
381	16.4.1951	4ft 8½in	0-6-0DM	330hp	Crossley 6cyl	4ft 0in	40T	312
382	17.4.1951	4ft 8½in	0-6-0DM	330hp	Crossley 6cyl	4ft 0in	40T	312
384	19.11.1951	4ft 8½in	0-4-0DM	204hp	Gardner 8L3	3ft 4½in	32T	323
423	25.7.1958	4ft 8½in	0-6-0DM	310hp	Natl M4AAU6	3ft 2in	43T	306
548	29.3.1967	4ft 8½in	0-4-0DH	233hp	R-R C6SFL	3ft 9in	32T	306

AVONSIDE ENGINE CO LTD, Bristol.   AE

1054	12.1874	7ft 0in	0-4-0ST	OC	14 x 18	3ft 0in	23T	33, 243
1055	12.1874	7ft 0in	0-4-0ST	OC	14 x 18	3ft 0in	23T	155, 243
1056	1.1875	7ft 0in	0-4-0ST	OC	14 x 18	3ft 0in	23T	33, 243
1387	1898	4ft 8½in	0-4-0ST	OC	14 x 20	3ft 3in	SS #	211, 225
1509	25.9.1907	4ft 8½in	0-4-0ST	OC	12 x 18	2ft 11in	1232	379, 382
1593	8.12.1910	1ft 11in	0-4-0T	OC	7 x 10	1ft 8in	5¾T	393
1613	1911	4ft 8½in	0-4-0ST	OC	8½ x 12	2ft 1in	Std	433
1618	1912	4ft 8½in	0-6-0ST	OC	14 x 20	3ft 3in	B3	312
1769	1917	4ft 8½in	0-4-0ST	OC	12 x 18	2ft 11in	Std	137
1793	2.1918	4ft 8½in	0-4-0ST	OC	14 x 20	3ft 3in	SS3	129
1932	15.7.1924	4ft 8½in	0-4-0ST	OC	12 x 18	2ft 11in	5160	323
1933	1924	4ft 8½in	0-6-0ST	OC	14 x 20	3ft 3in	B3	312
2000	6.8.1930	4ft 8½in	0-6-0ST	OC	14 x 22	3ft 6in	B6	150
2066	7.4.1933	2ft 0in	0-4-0T	OC	7 x 12	2ft 0in	7½T	150
2067	7.4.1933	2ft 0in	0-4-0T	OC	7 x 12	2ft 0in	7½T	150
2071	28.6.1933	2ft 0in	0-4-0T	OC	7½ x 12	2ft 0in	7½T	44, 150
2072	30.6.1933	2ft 0in	0-4-0T	OC	7½ x 12	2ft 0in	7½T	150
2073	6.7.1933	2ft 0in	0-4-0T	OC	7½ x 12	2ft 0in	7½T	150

loco also carried a plate Joicey 457/1898

ALLGEMEINE ELEKTRIZITATS GESELLSCHAFT, Berlin, Germany.   AEG

1565	24.7.1913	4ft 8½in	4w+4wWE	244hp		3ft 1½in	40T	179

ATLAS CAR & MANUFACTURING CO, Cleveland, Ohio, United States of America.   Atlas

2456	1946	2ft 0in	4wBEF	35hp	2ft 0in	7T	238

SIR W. G. ARMSTRONG, WHITWORTH & CO (ENGINEERS) LTD, Scotswood Works, Newcastle upon Tyne.   AW

This long-established company, famous for its armaments work, diversified into locomotive building after the First World War. At first it built only steam locomotives, but after 1930 it experimented with diesel locomotives and railcars with electric transmission. These did not lead to many orders, and this work was abandoned at the outbreak of the Second World War.

D10	12.1932	4ft 8½in	0-4-0DE	95hp	Saurer 6BLD	2ft 9in	15T	75
D21	1933	4ft 8½in	0-4-0DE	85hp	Saurer 6BLD	3ft 0in	15T	75
D22	1933	4ft 8½in	0-4-0DE	85hp	Saurer 6BLD	3ft 0in	15T	331
D24	1933	4ft 8½in	0-4-0DE	85hp	Saurer 6BLD	3ft 0in	15T	43

BARCLAYS & CO, Riverbank Engine Works, Kilmarnock, Ayrshire.   B

208	1873	4ft 8½in	0-4-0ST	OC	12 x ?	?		440
303	1883	4ft 8½in	0-4-0ST	OC	12⅝ x 18	3ft 1in		318, 425
315	1884	4ft 8½in	0-4-0ST	OC	?	?		364

WILLIAM BARNINGHAM, subsequently The Darlington Iron Co.   Barningham

Barningham is said to have built a number of locomotives for use at his own premises, the Pendleton Ironworks in Manchester and the Albert Hill Ironworks at Darlington. Most are said to have been built in Manchester, although one has been attributed to Darlington. The Albert Hill works is said to have used one 0-6-0ST and three 0-4-0STs, one of them rebuilt to a 0-6-0WT. No dates of construction are known. See pages 121-122.

BRUSH BAGNALL TRACTION LTD, Loughborough, Leicestershire.   BBT

The locomotives below were a joint production between W.G. Bagnall Ltd of Stafford (which see), whose works numbers they carried, and Brush Traction Ltd at Loughborough, where they were erected.

3020	22.2.1952	4ft 8½in	0-6-0DE	355hp	Mirrlees TLT6	4ft 0in	55T	65, 107
3021	26.5.1952	4ft 8½in	0-6-0DE	355hp	Mirrlees TLT6	4ft 0in	55T	65, 108

BARR, MORRISON & CO, Britannia Engineering Works, Kilmarnock, Ayrshire.   BarrM

?	1884	1ft 8in	0-4-0ST	OC	5 x 10?	1ft 8in?		308

BRITISH ELECTRIC VEHICLES LTD, Southport, Lancashire.   BEV

414	16.12.1922	1ft 10in	4wBE	?	?	2¾T	406
459	29.9.1923	n.g.	4wBE	?	?	2T	120

E. E. BAGULEY LTD, Burton-on-Trent, Staffordshire.   Bg

3048	5.5.1941	4ft 8½in	0-4-0DM	30hp	Gardner 3L2	2ft 6in	6½T	207
3051	21.3.1942	4ft 8½in	4wPM	20hp	Austin	1ft 8in	?	207
3054 #	4.6.1943	4ft 8½in	4wWE	200hp		2ft 9in	13½T	79

\# sub-contracted from EE, which allocated works number 1214/1943

BLACK, HAWTHORN & CO LTD, Gateshead, County Durham.   BH
Black, Hawthorn & Co until 23/3/1892

This company was formed by William Black and Thomas Hawthorn in 1865 to take over the engine works in Quarry Lane formerly operated by Ralph Coulthard (which see). The firm's main output was for industrial customers, although they also supplied main line railways and customers abroad. Besides locomotives, the firm also built other kinds of steam engine, and these were also allocated numbers in the makers' serial list. Curiously, works numbers ending in zero were very rarely allocated. Although numbers were allocated up to No.1143, the last locomotive built was No.1138. Hawthorn, the Managing Partner, died in 1880. When in 1896 at the age of 73 Black decided to retire, the business was sold to Messrs. Chapman & Furneaux (which see), with the company going into voluntary liquidation on 28th November 1896.

Perhaps more than many firms, Black, Hawthorn frequently built locomotives for stock, and the actual customer's order might come as much as three years later. To try to avoid too much confusion, the date of first ordering is given in the list below, with customer's order dates, if different, shown in a separate list at the end. Very few delivery dates are recorded, but those known are also given. The same problems also give rise to uncertainty about many of the dates carried on the makers' plate; the date given in the main text is that believed to have been carried.

	Date of first ordering							
13	31.10.1866	4ft 8½in	0-4-0ST	OC	10 x 17	3ft 0in		31
17	16.3.1866	4ft 8½in	0-6-0	IC	16 x 24	4ft 6in	%	217
22	7.1867	4ft 8½in	0-4-0ST	OC	10 x 17	3ft 0in		336
31	12.3.1867	4ft 8½in	0-6-0ST	OC	14 x 20	3ft 4in		393, 415
32	12.3.1867	4ft 8½in	0-6-0ST	OC	15 x 20	4ft 0in	@	218, 348
34	+23.5.1867	4ft 8½in	0-6-0	IC	16 x 24	4ft 8in		234
37	26.7.1867	4ft 8½in	0-4-0ST	OC	12 x 20	3ft 3in		291
38	26.7.1867	4ft 8½in	0-4-0ST	OC	12 x 20	3ft 3in		401
48	8.1.1868	4ft 8½in	0-6-0T	OC	14 x 20	3ft 6in		298
52	8.1.1868	4ft 8½in	0-6-0ST	OC	14 x 20	3ft 6in		298
57	6.4.1868	4ft 8½in	0-6-0ST	OC	16 x 24	4ft 6in		415
60	11.6.1868	4ft 8½in	0-6-0T	OC	14 x 20	3ft 6in		118, 298
62	20.6.1868	4ft 8½in	0-4-0ST	OC	9 x 16	2ft 9in		276
95	1.1.1869	4ft 8½in	0-4-0ST	OC	11 x 17	2ft 8½in		336
120	#24.8.1869	4ft 8½in	0-4-0ST	OC	12 x 19	3ft 2in		272
128	1.12.1869	4ft 8½in	0-4-0ST	OC	9 x 16	2ft 9in		276
173	#29.8.1871	4ft 8½in	0-4-0ST	OC	12 x 19	3ft 2in		306
174	17.1.1871	4ft 8½in	0-4-0ST	OC	12 x 19	3ft 2in		272
176	#1.1.1871	4ft 8½in	0-4-0ST	OC	12 x 19	3ft 2in		277
188	#29.1.1871	4ft 8½in	0-4-0ST	OC	10 x 17	3ft 0in		426
191	#24.3.1871	4ft 8½in	0-4-0ST	OC	9 x 16	2ft 10in		104
192	20.4.1871	4ft 8½in	0-4-0ST	OC	9 x 16	2ft 10in		104
202	24.3.1871	4ft 8½in	0-4-0ST	OC	9 x 16	2ft 9in		185
203	#20.4.1871	4ft 8½in	0-4-0ST	OC	9 x 16	2ft 9in		156, 234, 348
204	#20.4.1871	4ft 8½in	0-4-0ST	OC	9 x 16	2ft 9in		204
216	9.12.1871	4ft 8½in	0-4-0ST	OC	14 x 20	3ft 6in		277
223	16.10.1871	4ft 8½in	0-4-0ST	OC	14 x 20	3ft 6in		156
231	#7.5.1872	4ft 8½in	0-4-0ST	OC	10 x 17	2ft 9in		277
237	24.3.1873	4ft 8½in	0-4-0ST	OC	12 x 19	3ft 2in		400
244	11.6.1873	4ft 8½in	0-6-0ST	OC	14¾ x 22	4ft 2⅜in		159
247	#30.7.1872	4ft 8½in	0-4-0ST	OC	9 x 16	2ft 10in		104
248	#30.7.1872	4ft 8½in	0-4-0ST	OC	9 x 16	2ft 10in		104
252	16.11.1873	3ft 0in	0-4-0T	OC	6 x 12	2ft 0in		251
258	+3.2.1873	2ft 0in	0-4-2ST	OC	6 x 12	2ft 0in		229
264	18.4.1873	4ft 8½in	0-4-0ST	OC	12 x 19	3ft 2in		251
267	+12.8.1873	4ft 8½in	0-4-0ST	OC	12 x 19	3ft 2in		153
288	#13.5.1873	4ft 8½in	0-4-0ST	OC	12 x 19	3ft 2in		310
289	12.11.1873	4ft 8½in	0-4-0ST	OC	12 x 19	3ft 2in		100
298	#23.9.1873	4ft 8½in	0-4-0ST	OC	12 x 19	3ft 2in		32, 143

304	9.10.1873	4ft 8½in	0-6-0ST	IC	14 x 20	3ft 6in	57
305	#17.10.1873	4ft 8½in	0-4-0ST	OC	9 x 16	2ft 9in	204, 208
306	#17.10.1873	4ft 8½in	0-4-0ST	OC	9 x 16	2ft 9in	174
315	#1.1.1874	4ft 8½in	0-4-0ST	OC	12 x 19	3ft 2in	277
316	#1.1.1874	4ft 8½in	0-4-0ST	OC	12 x 19	3ft 2in	43
317	#1.1.1874	4ft 8½in	0-4-0ST	OC	12 x 19	3ft 2in	60
318	#1.1.1874	4ft 8½in	0-4-0ST	OC	12 x 19	3ft 2in	153
326	3.2.1874	4ft 8½in	0-4-0ST	OC	12 x 19	3ft 2in	100
327	3.2.1874	4ft 8½in	0-4-0ST	OC	12 x 19	3ft 2in	101
328	25.4.1875	4ft 8½in	0-4-0ST	OC	12 x 19	3ft 2in	101
354	25.2.1875	4ft 8½in	0-6-0ST	OC	14 x 20	3ft 6in	118, 240, 262, 357
355	#25.2.1875	4ft 8½in	0-6-0ST	OC	14 x 20	3ft 6in	361
365	#5.7.1875	4ft 8½in	0-4-0ST	OC	12 x 19	3ft 2in	54, 129, 130, 139
367	#5.7.1875	4ft 8½in	0-4-0ST	OC	12 x 19	3ft 2in	54, 55, 130, 139
368	+5.7.1875	4ft 8½in	0-4-0ST	OC	12 x 19	3ft 2in	55, 183
369	+5.7.1875	4ft 8½in	0-4-0ST	OC	9 x 16	2ft 9in	310
373	#1.9.1875	4ft 8½in	0-6-0ST	OC	14 x 20	3ft 6in	312
387	#1.3.1876	4ft 8½in	0-4-0ST	OC	10 x 17	2ft 10in	118
391	#13.3.1876	4ft 8½in	0-6-0ST	OC	14 x 20	3ft 6in	48
395	#31.5.1876	4ft 8½in	0-4-0ST	OC	13 x 19	3ft 2in	419
402	12.9.1876	3ft 0in	0-4-0ST	OC	5 x 10	1ft 8in	46
403	12.9.1876	3ft 0in	0-4-0ST	OC	5 x 10	1ft 8in	46
418	#12.2.1877	4ft 8½in	0-4-0ST	OC	10 x 17	2ft 10in	185
424	#10.4.1877	4ft 8½in	0-4-0ST	OC	12 x 19	3ft 2in	142
427	5.7.1877	4ft 8½in	0-4-0ST	OC	12 x 19	3ft 2in	46
435	1.9.1877	3ft 0in	0-4-0ST	OC	5 x 10	1ft 8in	46
436	1.9.1877	3ft 0in	0-4-0ST	OC	5 x 10	1ft 8in	46
439	19.9.1877	2ft 10in	0-4-0ST	OC	5 x 10	1ft 8in	338
442	15.10.1877	3ft 0in	0-4-0ST	OC	5 x 10	1ft 8in	46, 48
446	4.1.1878	3ft 0in	0-4-0ST	OC	5 x 10	1ft 8in	46
447	+4.1.1878	3ft 0in	0-4-0ST	OC	5 x 10	1ft 8in	46
459	15.4.1878	3ft 0in	0-4-0ST	OC	5 x 10	1ft 8in	46, 53
476	#3.9.1878	4ft 8½in	0-4-0ST	OC	12 x 19	3ft 2in	277
494	+3.5.1879	3ft 0in	0-4-0ST	OC	5 x 10	1ft 8in	46
495	+3.5.1879	3ft 0in	0-4-0ST	OC	5 x 10	1ft 8in	46
504	6.1879	4ft 8½in	0-6-0ST	OC	14 x 20	3ft 6in	176
515	12.9.1879	4ft 8½in	0-4-0ST	OC	9 x 18	2ft 8in	176, 395
516	12.9.1879	4ft 8½in	0-4-0ST	OC	9 x 18	2ft 8in	176, 395
519	+12.1879	4ft 8½in	0-4-0ST	OC	12 x 19	3ft 2in	55
524	+6.12.1879	4ft 8½in	0-4-0ST	OC	12 x 19	3ft 2in	120
526	+6.12.1879	4ft 8½in	0-4-0ST	OC	12 x 19	3ft 2in	51
529	#31.10.1879	4ft 8½in	0-4-0ST	OC	10 x 17	2ft 10in	395
544	4.1.1881	4ft 8½in	0-4-0ST	OC	12 x19	3ft 2in	48, 54
546	23.3.1879	4ft 8½in	0-4-0ST	OC	12 x 19	3ft 2in	321
551	17.12.1879	4ft 8½in	0-4-0ST	OC	12 x 19	3ft 2in	101, 433
552	17.12.1879	4ft 8½in	0-4-0ST	OC	12 x 19	3ft 2in	101
553	17.12.1879	4ft 8½in	0-4-0ST	OC	12 x 19	3ft 2in	101
557	21.1.1880	3ft 0in	0-4-0ST	OC	5 x 10	1ft 8in	46
558	21.1.1880	3ft 0in	0-4-0ST	OC	5 x 10	1ft 8in	46
559	21.1.1880	3ft 0in	0-4-0ST	OC	5 x 10	1ft 8in	46, 202
561	21.1.1880	3ft 0in	0-4-0ST	OC	5 x 10	1ft 8in	46
567	13.4.1880	2ft 0in	0-4-0ST	OC	6 x 10	1ft 8in	198
569	#16.4.1880	4ft 8½in	0-4-0ST	OC	10 x 17	2ft 10in	185
595	25.11.1880	3ft 0in	0-4-0ST	OC	6 x 10	1ft 10in	46
596	25.11.1880	3ft 0in	0-4-0ST	OC	6 x 10	1ft 10in	46
597	25.11.1880	3ft 0in	0-4-0ST	OC	6 x 10	1ft 10in	46
598	25.11.1880	3ft 0in	0-4-0ST	OC	6 x 10	1ft 10in	46
599	25.11.1880	3ft 0in	0-4-0ST	OC	6 x 10	1ft 10in	46
601	25.11.1880	3ft 0in	0-4-0ST	OC	6 x 10	1ft 10in	46
602	3.8.1880	4ft 8½in	0-6-0ST	OC	17 x 26	4ft 6in	298
606	20.12.1880	4ft 8½in	0-4-0ST	OC	14 x 19	3ft 2in	50, 62, 120, 165
607	20.12.1880	4ft 8½in	0-4-0ST	OC	14 x 19	3ft 2in	48, 51, 126, 135
613	#4.1.1881	4ft 8½in	0-4-0ST	OC	12 x 19	3ft 2in	74, 202
629	26.2.1881	2ft 1in	0-4-0T	OC	5½ x 10	1ft 8½in	192, 240

The gauge of 2ft 1in is believed to be an error for 2ft 6in

645	8.7.1881	4ft 8½in	0-6-0ST	OC	14 x 20	3ft 6in	312
666	18.11.1881	4ft 8½in	0-6-0ST	OC	14 x 20	3ft 6in	312
682	#26.1.1882	4ft 8½in	0-4-0ST	OC	12 x 19	3ft 2in	31, 183
688	9.3.1882	4ft 8½in	0-4-0ST	OC	15 x 20	3ft 8in	218
692	17.4.1882	4ft 8½in	0-6-0ST	IC	14 x 20	3ft 6in	57
698	22.6.1882	4ft 8½in	0-4-0ST	OC	12 x 19	3ft 2in	101
704	#24.8.1882	4ft 8½in	0-6-0ST	OC	14 x 20	3ft 6in	152, 415
716	15.9.1882	4ft 8½in	0-6-0ST	IC	15 x 22	3ft 9½in	155, 176, 337, 401
762	#27.8.1883	4ft 8½in	0-4-0ST	OC	12 x 19	3ft 2in	400
763	19.1.1884	4ft 8½in	0-4-0ST	OC	12 x 19	3ft 2in	400
826	27.5.1884	4ft 8½in	0-6-0ST	IC	15 x 22	3ft 9½in	176
831	15.10.1884	4ft 8½in	0-4-0VBCr	OC	8 x 12	2ft 9in	2T capy. 104
832	20.11.1884	4ft 8½in	0-4-0ST	OC	15 x 22	3ft 8in	218
848	#2.5.1885	4ft 8½in	0-4-0ST	OC	10 x 17	2ft 10in	168
852	#17.5.1885	4ft 8½in	0-4-0ST	OC	12 x 19	3ft 2in	323
854	#17.5.1885	4ft 8½in	0-4-0ST	OC	12 x 19	3ft 2in	70, 102
856	4.8.1885	2ft 10in	0-4-0ST	OC	6 x 10	1ft 10in	338
881	9.11.1886	4ft 8½in	0-4-0ST	OC	10 x 17	2ft 11in	204, 208
888	9.11.1886	4ft 8½in	0-6-0ST	OC	14 x 20	3ft 7in	193
891	9.11.1886	4ft 8½in	0-6-0ST	OC	12 x 19	3ft 3in	57
897	6.1.1887	4ft 8½in	2-4-0VBCr	OC	13½ x 21	3ft 0in	12T capy. 65, 106
898	6.1.1887	4ft 8½in	2-4-0VBCr	OC	13½ x 21	3ft 0in	12T capy. 106
931	27.1.1888	4ft 8½in	0-4-0VBCr	OC	12 x 21½	3ft 0in	7T capy. 106
935	#1888	4ft 8½in	0-4-0ST	OC	12 x 19	3ft 2in	204
937	7.3.1888	4ft 8½in	0-6-2ST	IC	16 x 24	4ft 6⅞in	57
938	7.3.1888	4ft 8½in	0-6-2ST	IC	16 x 24	4ft 6⅞in	57
971	#(?)1889	4ft 8½in	0-6-0ST	OC	14 x 20	3ft 7in	193, 197
976	12.7.1889	4ft 8½in	0-4-0ST	OC	12 x 19	3ft 3in	256
981	3.8.1889	2ft 6in	0-4-0ST	OC	5 x 10	1ft 8½in	405
985	13.11.1889	4ft 8½in	0-4-0ST	OC	14 x 19	3ft 2in	128, 130, 134
991	12.11.1889	4ft 8½in	0-4-0ST	OC	12 x 19	3ft 3in	253
992	12.11.1889	4ft 8½in	0-4-0ST	OC	12 x 19	3ft 3in	135, 139
998	28.11.1889	4ft 8½in	0-4-0ST	OC	12 x 19	3ft 3in	282, 292
1019	1890	New frameplates only, for BH 387/1876					118
1025	3.6.1891	4ft 8½in	0-4-0ST	OC	12 x 19	3ft 3in	262
1026	8.10.1890	4ft 8½in	0-4-0ST	OC	12 x 19	3ft 3in	433
1032	1891	4ft 8½in	0-4-0ST	OC	10 x 17	2ft 11in	356
1034	22.8.1892	4ft 8½in	0-6-0ST	OC	14 x 20	3ft 7in	197, 339
1037	9.6.1891	4ft 8½in	0-4-0ST	OC	12 x 19	3ft 3in	80
1038	9.6.1891	4ft 8½in	0-4-0ST	OC	12 x 19	3ft 3in	367
1048	18.8.1891	4ft 8½in	0-4-0VBCr	OC	12 x 21½	3ft 0in	7T capy. 106
1049	+18.8.1891	4ft 8½in	0-4-0VBCr	OC	12 x 21½	3ft 0in	7T capy. 65, 106
1051	+18.8.1891	4ft 8½in	0-4-0VBCr	OC	12 x 21½	3ft 0in	7T capy. 106
1065	7.4.1892	2ft 3½in	0-4-0ST	OC	6 x 10	1ft 9in	384
1071	21.7.1892	4ft 8½in	0-6-2ST	IC	16 x 24	4ft 6⅞in	58
1095	#20.6.1894	4ft 8½in	0-4-0ST	OC	14 x 19	3ft 2in	51, 128, 130, 134
1096	#20.6.1894	4ft 8½in	0-4-0ST	OC	12 x 19	3ft 3in	42, 61, 259
1098	21.7.1894	4ft 8½in	0-4-0ST	OC	10 x 17	2ft 11in	329
1113	5.4.1895	4ft 8½in	0-4-0ST	OC	12 x 19	3ft 4in	102, 118
1114	24.4.1895	4ft 8½in	0-4-0ST	OC	6½ x 12	2ft 0½in	433
1121	8.8.1895	4ft 8½in	0-4-0ST	OC	6½ x 12	2ft 0½in	433
1122	8.8.1895	4ft 8½in	0-4-0ST	OC	6½ x 12	2ft 0½in	433

% built from material supplied by the Earl of Durham.
@ "reconstruction of loco engine".
+ indicates additional information regarding delivery dates (see below).
# date ordered for stock; for customer's order date see below.

Unidentified steam locomotives

[BUSTY]	4ft 8½in	0-4-0ST	OC	?	?	298
[No.3]	4ft 8½in	0-4-0ST	OC	?	?	120
-	4ft 8½in	0-4-0ST	OC	?	?	200
-	4ft 8½in	0-4-0ST	OC	?	?	243

	-	4ft 8½in	0-4-0ST	OC	?	?		270
	-	4ft 8½in	0-4-0ST	OC	?	?		273
	-	4ft 8½in	0-4-0ST	OC	?	?		314

Number	Date of customer's order	Delivery date	Number	Date of customer's order	Delivery date
34	23.5.1867	5.3.1868	418	7.3.1877	
120	16.1.1871		447	4.1.1878	21.1.1880*
166	17.1.1871		476	8.11.1879	
173	11.12.1872		494		19.7.1879
176	17.9.1872		495		22.7.1879
188	23.7.1872		519	12.1879	19.12.1879
191	20.4.1871		524		7.3.1880
203	10.10.1871		526		30.4.1880
204	30.3.1872		529	6.2.1880	
231	17.2.1874	1.5.1874	546	23.3.1881	
247	30.9.1873	11.11.1873	551		24.4.1880
248	30.9.1873	29.12.1873	552		24.4.1880
258		1874	553		24.4.1880
267		7.11.1873	569	26.1.1881	
288	3.1874		613	21.12.1881	
298	26.4.1875		682	13.11.1883	
305	27.10.1874		704	18.10.1882	
306	19.2.1875		762	12.1883	
315	19.3.1874		848	23.12.1886	
316	29.7.1874		852	22.7.1886	
317	21.6.1875		854	17.5.1885	
318	1.1.1874		935	1888	
355	31.8.1875	5.10.1875	938		26.10.1888
365	31.10.1876		971	7.9.1890	
367	30.3.1877		992	25.4.1890	
368		25.5.1877	1049		11.1892
369		17.3.1876	1051		6.1892
373	8.10.1875		1095	20.6.1894	28.2.1896
387	18.5.1876		1096	20.6.1894	
391	17.1.1877		1098	20.10.1896	
395	19.10.1879	19.10.1879	1113	4.4.1895	

* after display at the Paris Exhibition, 1879

F. BLACKLOCK, Sheriff Hill, Gateshead, Co. Durham.   Blacklock

Frank Blacklock owned The Frankland Coal Co (1934) Ltd and the neighbouring Abbey Wood Coal Co. He also owned a scrapyard at Sheriff Hill in Gateshead, where about 1952-53 he built two locomotives for use on a narrow gauge system serving the collieries. He was also a fine model engineer.

	-	c1952	2ft 0in	4wDM	?	Petter	?	164, 226
	-	c1953	2ft 0in	4wDM	?	Petter	?	164

BLAIR & CO LTD, Norton Road, Stockton-on-Tees, Co. Durham.   Blair

George Blair had been the works manager for the firm of Fossick & Hackworth (see below), and when the latter retired in 1865, Blair joined George Fossick and the firm traded briefly as Fossick & Blair. When Fossick died in 1866, Blair took over the business. So far as is known, no locomotives were built after 1868, with the firm concentrating on marine work. It closed in 1920.

[12]	c1865	4ft 8½in	0-6-0	IC			217
[15]	1868	4ft 8½in	0-6-0	IC	16¾ x 24	4ft 9in	234

BALDWIN LOCOMOTIVE WORKS, Philadelphia, Pennsylvania, United States of America.   BLW

45282	3.1917	4ft 8½in	0-4-0ST	OC	14 x 22	3ft 6in	374
45285	3.1917	4ft 8½in	0-4-0ST	OC	14 x 22	3ft 6in	84

BLYTH & TYNE RAILWAY, Percy Main Workshops, North Shields, Northumberland.   Blyth & Tyne
This railway obtained its Act in 1847. Its main line ran from Blyth to docks on the River Tyne, with branches to Newcastle-upon-Tyne, Morpeth and Newbiggin. It was taken over by the North Eastern Railway in 1874.

[NER 2255]	4.1862	4ft 8½in	0-6-0	IC	16 x 24	4ft 6in	177
[NER 1712]	9.1862	4ft 8½in	0-6-0	IC	16 x 24	4ft 6in	177
[NER 1719]	6.1867	4ft 8½in	0-6-0	IC	16 x 24	4ft 6in	348

J. BOND, Castle Foundry, Tow Law, Co. Durham.   J. Bond
This foundry is reputed to have built two locomotives for the Weardale Iron & Coal Co Ltd in the 1870s, when it was owned by Joseph Bond. The foundry continues (2006) in business.

[6 TOW LAW]	1870-74	4ft 8½in	0-6-0	IC	?	?	414
[11 ZEPHYR]	?	4ft 8½in	0-4-0ST	OC	?	?	414

J. BOOTH & BROTHERS LTD, Leeds.   Booth

164	?	3ft 0in	0-4-0BE	?	?	284

BEYER, PEACOCK & CO LTD, Gorton, Manchester.   BP

190	3.8.1860	4ft 8½in	0-4-2ST	OC	16 x 24	5ft 0in	361
417	29.6.1864	4ft 8½in	4-4-0T	OC	17 x 24	5ft 9in	361
425	5.8.1864	4ft 8½in	4-4-0T	OC	17 x 24	5ft 9in	361
550	8.6.1865	4ft 8½in	0-6-0	IC	17 x 24	4ft 6in	217
770	20.7.1868	4ft 8½in	4-4-0T	OC	17 x 24	5ft 9in	298
772	31.7.1868	4ft 8½in	4-4-0T	OC	17 x 24	5ft 9in	298
868	15.3.1869	4ft 8½in	4-4-0T	OC	17 x 24	5ft 9in	298
7873/BT 443	4.9.1962	4ft 8½in	0-4-0DE	230hp	McLaren LES6	3ft 6in	323
7946/BT 339	13.9.1961	4ft 8½in	0-4-0DE	230hp	McLaren LES6	3ft 6in	323
7947/BT 340	3.10.1961	4ft 8½in	0-4-0DE	230hp	McLaren LES6	3ft 6in	323

THOMAS BROADBENT & SONS LTD, Central Ironworks, Huddersfield, Yorkshire (WR).   Broadbent

-	1924	4ft 8½in	2-2-0PM	?	?	1ft 3in	207

BRUSH ELECTRICAL MACHINES LTD, Falcon Works, Loughborough, Leicestershire.   BT

339/BP 7946	13.9.1961	4ft 8½in	0-4-0DE	230hp	McLaren LES6	3ft 6in	323
340/BP 7947	3.10.1961	4ft 8½in	0-4-0DE	230hp	McLaren LES6	3ft 6in	323
443/BP 7873	4.9.1962	4ft 8½in	0-4-0DE	230hp	McLaren LES6	3ft 6in	323

BRITISH THOMSON-HOUSTON CO LTD, Rugby, Warwickshire.   BTH

1780	1901	4ft 8½in	4wWE	?		?	79

EDWARD BURY & CO, Liverpool, Lancashire.   Bury

[VICTORIA]	1838	4ft 8½in	0-4-0	IC	12 x 18	5ft 0in	294

BUTTERLEY CO LTD, Butterley, Derbyshire.   Butterley

[B12C]	?	4ft 8½in	0-4-0	OC	?	?	339

WEST YARD LOCOMOTIVE WORKS, Cardiff, Taff Vale Railway.   Cdf

302	11.1894	4ft 8½in	0-6-2T	IC	17½ x 26	4ft 6½in	218
311	10.1897	4ft 8½in	0-6-2T	IC	17½ x 26	4ft 6½in	218

CHAPMAN & FURNEAUX, Quarry Lane, Gateshead, Co. Durham.   CF
When William Black decided to retire from and liquidate Black, Hawthorn & Co Ltd (see above), the business was acquired by three directors of the near-neighbours Clarke, Chapman & Co Ltd (see main text), Abel Chapman, Henry Chapman and John Furneaux. They decided not to continue the marine work of their predecessors, nor did they take over work in progress, all this being auctioned off in January 1897. The BH works number list was continued, with numbers ending in zero now included. By 1902 both Abel Chapman and Furneaux were in poor health, with the business already running down. The last locomotive to be built at Gateshead was No.1212, though Nos. 1213-1215 were built by Hudswell, Clarke & Co Ltd of Leeds. A dismantling sale took place in May 1902; the goodwill, drawings and patterns were acquired by R. & W. Hawthorn, Leslie & Co Ltd of Newcastle upon Tyne in November 1902, and sixteen months later the premises were sold to the electrical engineers, Ernest Scott & Mountain Ltd, who transferred their business from Newcastle upon Tyne and re-named the premises the Close Works. For their later history see the entry for Armstrong Whitworth Rolls Ltd in the main text.

1155	+21.12.1897	4ft 8½in	0-6-0ST	OC	14 x 20	3ft 7in	40, 42
1158	+4.3.1898	4ft 8½in	0-6-2T	IC	17 x 24	4ft 3in	58, 76
1163	+16.5.1898	4ft 8½in	0-4-0ST	OC	13 x 19	3ft 4½in	102
1183	#25.4.1899	4ft 8½in	0-4-0ST	OC	12 x 19	3ft 3in	423
1187	#1.7.1899	4ft 8½in	0-4-0ST	OC	14 x 19	3ft 3in	256
1189	2.8.1899	4ft 8½in	0-4-0ST	OC	14 x 19	3ft 2in	38
1190	#9.8.1899	4ft 8½in	0-4-0ST	OC	14 x 19	3ft 2in	316
1193	#21.11.1899	4ft 8½in	0-4-0ST	OC	12 x 19	3ft 3in	356, 359
1196	#15.12.1899	4ft 8½in	0-4-0ST	OC	14 x 19	3ft 2in	414
1198	#3.4.1900	4ft 8½in	0-4-0ST	OC	12 x 19	3ft 3in	316
1202	10.7.1900	4ft 8½in	0-4-0ST	OC	14 x 19	3ft 2in	187
1203	16.10.1900	4ft 8½in	0-4-0ST	OC	15 x 22	3ft 8in	57, 61
1204	31.10.1900	4ft 8½in	0-6-0ST	OC	16 x 24	3ft 8in	193, 197
1205	+12.11.1900	4ft 8½in	0-4-0ST	OC	13 x 19	3ft 4½in	101
1206	12.11.1900	4ft 8½in	0-4-0VBCr	OC	12 x 21½	3ft 0in	7T capy. 65, 106
1210	#18.4.1901	4ft 8½in	0-4-0ST	OC	14 x 19	3ft 2in	80
1211	18.4.1901	4ft 8½in	0-4-0ST	OC	14 x 19	3ft 2in	50, 51, 53, 137

\# date of order for stock; for customer's order date see below.
\+ indicates additional information regarding delivery dates (see below).

Number	Date of customer's order	Date of delivery	Number	Date of customer's order	Date of delivery
1155		2.1898	1193	30.4.1900	
1158		29.12.1898	1196	11.6.1900	
1163		16.12.1898	1198	29.10.1901	
1183	30.6.1899		1205		30.7.1901
1187	30.11.1899		1210	15.1.1902	
1190	15.11.1899				

ALEXANDER CHAPLIN & CO LTD, Cranstonhill Works, Glasgow.   **Chaplin**

708	6.8.1866	4ft 8½in	0-4-0VBT	VCG	5¼ x 11	?	708, 272
1651	6.1.1874	2ft 4in	0-4-0VBT	VCG	4½ x 9	?	437

COCHRANE & CO LTD, Ormesby Ironworks, Middlesbrough, Yorkshire (NR).   **Cochrane**

-	?	4ft 8½in	0-4-0VBT	VC	?	?	227

STEELS ENGINEERING PRODUCTS LTD, Crown Works, Pallion, Sunderland, Co.Durham.   **Coles**

For the history of this company see the entry for British Crane & Excavator Corporation Ltd in the main text. It manufactured cranes under the trading name of Coles Cranes. It built the locomotive below for its own use.

16640	1956	4ft 8½in	4wDE	60hp	?	?	25T	62

CONSETT IRON CO LTD, Consett, Co.Durham.   **Consett**

For the history of this company see its entry in the main text. The firm had a well-equipped Locomotive Department at Consett. The locomotives below were built following the arrival of George Cowell from The Hunslet Engine Co Ltd in Leeds as Locomotive Superintendent.

[B42]	1954	4ft 8½in	0-4-0ST	OC	16 x 24	3ft 8in		103
[B No 14]	9.1955	4ft 8½in	0-4-0F	OC	16½ x 22	3ft 0in		101
[9]	1956	4ft 8½in	0-6-0DM	300hp	Mirrlees J4	3ft 7in	35T	65
[10]	1958	4ft 8½in	0-6-0DM	300hp	Mirrlees J4	3ft 7in	35T	65

JOHN (later RALPH) COULTHARD & CO, Quarry Lane, Gateshead, Co. Durham.   **Coulthard**

In 1840 John Coulthard, formerly one of the senior engineers in the firm of Losh, Wilson & Bell of the Walker Ironworks in Newcastle upon Tyne, set up a general engineering business in Gateshead, to be joined in the following year by his brother Ralph. In 1841 the firm built its first locomotive, and other orders followed, most for public railways. In 1853 John Coulthard died and the title of the firm became Ralph Coulthard & Co. The business went into decline in the early 1860s, and in a press announcement dated 12th July 1865 Ralph Coulthard announced that he had retired from the business and sold the premises to Messrs. Black, Hawthorn & Co (see above). No reliable estimate of the number of locomotives built can be made, but it is unlikely to exceed thirty.

-	9.1841	4ft 8½in	0-6-0	IC	?	?	438
[6]	c1853?	4ft 8½in	0-6-0	IC	16 x 24	4ft 6in	217
[7]	c1853?	4ft 8½in	0-6-0	IC	16 x 24	4ft 6in	217
[8]	c1853?	4ft 8½in	0-6-0	IC	16 x 24	4ft 6in	217

DUBS & CO, Glasgow Locomotive Works, Polmadie, Glasgow. D

1758	1883	4ft 8½in	0-4-0CT	OC	12 x 22	3ft 6in	7T capy.	104
2051	1884	4ft 8½in	0-4-0CT	OC	12 x 22	3ft 6in	7T capy.	415
2063	1884	4ft 8½in	0-4-0CT	OC	12 x 22	3ft 6in	5T capy.	104
2365	1888	4ft 8½in	0-4-0CT	OC	12 x 22	3ft 6in	5T capy.	105
2366	1888	4ft 8½in	0-4-0CT	OC	12 x 22	3ft 6in	5T capy.	105

DARLINGTON WORKS, Darlington, Co.Durham, North Eastern Railway/London & North Eastern Railway. Dar

This works was opened in 1863 by the Stockton & Darlington Railway, just before its take-over by the NER. It was developed on the site of the former works of W. & A. Kitching (see the entry in the main text for Whessoe Products Ltd).

[NER 1293]	11.1876	4ft 8½in	0-6-0	IC	17 x 26	4ft 8½in	72, 440
[NER 494]	6.1887	4ft 8½in	0-6-0T	IC	16 x 22	4ft 7¼in	200
[NER 1144]	7.1892	4ft 8½in	0-6-0T	IC	16 x 22	4ft 7¼in	337
[NER 1953]	6.1898	4ft 8½in	0-6-0	IC	18½ x 24	4ft 7¼in	176
[LNER 982]	8.1923	4ft 8½in	0-4-0T	IC	14 x 20	3ft 6¼in	425
[LNER 983]	8.1923	4ft 8½in	0-4-0T	IC	14 x 20	3ft 6¼in	35
[LNER 8088]	9.1923	4ft 8½in	0-4-0T	IC	14 x 20	3ft 6¼in	155

DREWRY CAR CO LTD, London. (suppliers only) DC

#2157	1941	4ft 8½in	0-4-0DM	153hp	Gardner 6L3	24½T	242
#2164	1941	4ft 8½in	0-4-0DM	153hp	Gardner 6L3	26½T	351

\# built by EE at its Dick Kerr Works and given EE works numbers 1188 and 1195 respectively.

DICK, KERR & CO LTD, Preston, Lancashire. DK

#[No.3]	1908	4ft 8½in	4wWE	?	?	75
#[No.4]	1910	4ft 8½in	4wWE	?	?	75
[51]	1918	4ft 8½in	4wBE	?	?	218
#[No.5]	1920	4ft 8½in	4wBE	?	?	75

\# it is believed that these locomotives were ordered from The British Westinghouse Electric & Manufacturing Co Ltd of Manchester, who sub-contracted the actual construction to Dick, Kerr & Co Ltd, with Westinghouse supplying the motors. No.3 carried only a Westinghouse makers' plate, and it is possible that the others were similar.

DONCASTER WORKS, Doncaster, Yorkshire (WR), Great Northern Railway/LNER. Don

213	10.1876	4ft 8½in	0-6-0ST	IC	17½ x 26	4ft 8in	298

E. BORROWS & SONS, St.Helens, Lancashire. EB

30	1891	4ft 8½in	0-4-0WT	OC	14½ x 20	3ft 4in	400

EARL OF DURHAM, Lambton Engine Works, Philadelphia, Co.Durham. ED

This works was being developed by 1835, and came to service all of the Earl's industrial undertakings – his collieries and their associated industries of coke and brickmaking, his railway, his ships and their staiths. It passed to the National Coal Board on 1st January 1947. For a more detailed account of the history of the works, see *The Private Railways of County Durham*, Colin E. Mountford, Industrial Railway Society, 2004.

[9]	1877	4ft 8½in	0-6-0	IC	17 x 24	4ft 6in	217
[25]	1890	4ft 8½in	0-6-0	IC	17 x 24	4ft 6in	218
[26]	1.1894	4ft 8½in	0-6-0	IC	17 x 24	4ft 6in	218

ENGLISH ELECTRIC CO LTD, Dick Kerr Works, Preston, Lancashire. EEDK

512	1920	4ft 8½in	4wBE	60hp	?	?	331
518	1921	4ft 8½in	4wBE	60hp	?	?	400
519	1921	4ft 8½in	4wBE	60hp	?	?	400
570	1923	4ft 8½in	4wBE	60hp	?	?	400
#1188	1941	4ft 8½in	0-4-0DM	153hp	Gardner 6L3	24½T [DC 2157]	242
#1195	1941	4ft 8½in	0-4-0DM	153hp	Gardner 6L3	26½T [DC 2164]	351
1214	4.6.1943	4ft 8½in	4wWE	200hp		? [Bg 3054]	79
2346	1956	4ft 8½in	0-6-0DH	500hp	EE6RKT	48T [VF D227]	156, 372

\# built at VF.

ENGLISH ELECTRIC CO LTD, Vulcan Works, Newton-le-Willows, Lancashire. EEV

D924	1966	4ft 8½in	0-6-0DH	305hp	Cummins NHRS6B1	?	116
D1121	1966	4ft 8½in	0-6-0DH	305hp	Cummins NHRS6B1	?	349
D1123	1966	4ft 8½in	0-4-0DH	305hp	Cummins NHRS6B1	34T	247
D1126	1966	4ft 8½in	0-4-0DH	305hp	Cummins NHRS6B1	34T	247
D1191	1967	4ft 8½in	0-6-0DH	305hp	Cummins NHRS6B1	?	349
D1192	1967	4ft 8½in	0-6-0DH	305hp	Cummins NHRS6B1	?	349
D1193	1967	4ft 8½in	0-6-0DH	305hp	Cummins NHRS6B1	?	349
D1194	1967	4ft 8½in	0-6-0DH	305hp	Cummins NHRS6B1	?	349
D1195	1967	4ft 8½in	0-6-0DH	305hp	Cummins NHRS6B1	?	349
D1201	1967	4ft 8½in	0-6-0DH	391hp	Cummins NT400	?	116
D1202	1967	4ft 8½in	0-6-0DH	391hp	Cummins NT400	?	116
D1230	1969	4ft 8½in	0-6-0DH	380hp	Cummins NT400	?	116
3994	1970	4ft 8½in	0-6-0DH	391hp	Cummins NT400	?	116

WM FAIRBAIRN & SONS, Canal Street Works, Manchester. Fairbairn

-	10.1842	4ft 8½in	0-4-0	IC	14 x 20	4ft 6in	414
-	12.1842	4ft 8½in	0-4-0	IC	14 x 20	4ft 6in	414

F. C. HIBBERD & CO LTD, Bedford and (from 1932) Park Royal Works, London. FH

It is believed that most locomotives built before 1932 were assembled by sub-contractors.

1652	1.1930	2ft 0in	4wPM	20SX	Dorman 2JOR	?	2½T	146
1655	1.1930	2ft 0in	4wPM	20SX	Dorman 2JOR	?	2½T	146
1782	23.12.1931	2ft 4in	4wPM	16/25hp	Meadows 4EH	?	?	273
1829	8.1933	4ft 8½in	4wPM	40SXR	Dorman 4JO	?	8T	301
1886	4.1935	2ft 0in	4wPM	20SX	Dorman 2JOR	?	2½T	165
1892	12.11.1934	2ft 8½in	4wPM	24hp	Ford 4B	?	4T	273
2064	28.7.1937	2ft 6in	4wPM	20SX	Paxman 2RQT	?	2½T	330
3374	21.2.1950	4ft 8½in	4wDM	75hp	National DA4	?	11T	159, 333, 354
3583	30.9.1952	4ft 8½in	4wDM	75hp	Dorman 4DL	?	?	168
3865	5.2.1958	4ft 8½in	4wDM	117hp	Dorman 6DL	?	23T	353
3890	1.11.1958	4ft 8½in	4wDM	170hp	Dorman 6KUD	?	24T	398
3891	6.11.1958	4ft 8½in	4wDM	170hp	Dorman 6KUD	?	24T	398
3967	8.1961	4ft 8½in	4wDH	?	Ford 590E	?	16T	427

Unidentified locomotive

?	?	2ft 0in	4wPM	?	?	?	?	79

FOSSICK & HACKWORTH, Stockton-on-Tees, Co. Durham F&H

Thomas Hackworth (brother of Timothy Hackworth) had owned the Soho Engine Works at Shildon, but on being compelled to give this up, he entered into partnership with George Fossick to set up this new business in the spring of 1840. Hackworth retired in 1865, leaving Fossick to be joined by his erstwhile works manager, George Blair. When Fossick died in the following year, Blair took over the business (see Blair & Co Ltd above).

Besides the construction of locomotives, the firm undertook two noteworthy railway operating contracts, for the Clarence Railway in County Durham between 1844 and 1854, and for the Llanelly Railway & Dock Co in South Wales between 1850 and 1853.

-	4.4.1861	4ft 8½in	0-6-0	OC	?	?	234

Unidentified locomotive

SWIFT	?	?	?	?	?	?	334

See also the entry for the Hartlepool Dock & Railway Company in the main text.

FLETCHER, JENNINGS & CO LTD, Lowca Engine Works, Whitehaven, Cumberland. FJ

31	30.6.1864	4ft 8½in	0-4-0ST	OC	10 x 20	4ft 0in	48, 49, 54
47	8.5.1865	4ft 8½in	0-4-0T	OC	10 x 20	?	50
72	1867	4ft 8½in	0-4-0T	OC	12 x 20	?	53, 423
87	25.5.1871	4ft 8½in	0-4-0T	OC	12 x 20	?	98
112	22.3.1873	4ft 8½in	0-4-0ST	OC	12 x 20	3ft 5in	252, 253, 257, 258
125	26.11.1874	4ft 8½in	0-4-0WT	OC	12 x 20	3ft 6in	57, 61, 273
167	1879	4ft 8½in	0-6-0ST	IC	14 x 20	3ft 6in	320
187	c9.1882	4ft 8½in	0-4-0T	OC	14 x 20	3ft 6in	356

FOX, WALKER & CO LTD, Altlas Engine Works, Bristol.   FW

140	1872	4ft 8½in	0-6-0ST	OC	13 x 20	?		306
169	11.12.1872	4ft 8½in	0-6-0ST	OC	13 x 20	3ft 6½in	B	40, 116
170	14.1.1873	4ft 8½in	0-6-0ST	OC	13 x 20	3ft 6½in	B	159
171	6.1.1873	4ft 8½in	0-6-0ST	OC	13 x 20	3ft 6½in	B	40
249	1874	4ft 8½in	0-6-0ST	OC	13 x 20	3ft 6½in	B	46, 52
265	c12.1.1875	4ft 8½in	0-6-0ST	OC	13 x 20	3ft 6in	B	40
289	11.4.1876	4ft 8½in	0-6-0ST	OC	13 x 20	3ft 6in	B1	379, 382
375	1878	4ft 8½in	0-4-0ST	OC	12 x 18	?	W1	247

GREENWOOD & BATLEY LTD, Armley, Leeds.   GB

2130	1.9.1948	2ft 0in	4wBE	6hp	?	3T	368
2319	13.7.1950	3ft 5½in	4wWE	12hp	?	2T	368
2368	25.3.1952	4ft 8½in	0-4-0WE	80hp	3ft 1½in	17T	67, 108
2848	7.11.1957	2ft 0in	4wBE	5hp	?	2T	83
420306	1972	4ft 8½in	4wWE	90hp	3ft 1½in	20T	67

GENERAL ELECTRIC CO, Witton, Birmingham.   GEC

-	1928	4ft 8½in	4wWE	100hp	?	17T	83

GIBB & HOGG, Victoria Engine Works, Airdrie, Lanarkshire, Scotland.   GH

[HARE]	1908	4ft 8½in	0-4-0ST	OC	?	?	46, 54, 126, 130, 139

GATESHEAD WORKS, Gateshead, Co. Durham, North Eastern Railway.   Ghd

This works was begun in 1844 by the Newcastle & Darlington Junction Railway and extended in 1853-54 by the York, Newcastle & Berwick Railway, shortly before its absorption into the North Eastern Railway. From the mid-1880s Works numbers began again at the start of each new year. New construction ceased in 1910, with the Central Drawing Office being moved to York. It continued to handle repairs until August 1932, when it was closed. It was re-opened for repairs during the Second World War, finally closing in 1959.

	3.1875	4ft 8½in	0-6-0	IC	17 x 24	3ft 6in	[NER 125]	348
	8.1881	4ft 8½in	0-6-0	IC	17 x 24	5ft 0in	[NER 396]	177
	11.1882	4ft 8½in	0-6-0	IC	17½ x 24	5ft 0in	[NER 1453]	180
	4.1883	4ft 8½in	0-6-0	IC	17 x 24	5ft 0in	[NER 1333]	180
34	12.1888	4ft 8½in	0-4-0T	IC	13 x 20	3ft 6¼in	[NER 898]	270
35	12.1888	4ft 8½in	0-4-0T	IC	13 x 20	3ft 6¼in	[NER 900]	298
38	12.1888	4ft 8½in	0-4-0T	IC	13 x 20	3ft 6¼in	[NER 24]	180
3	2.1889	4ft 8½in	0-6-0	IC	18 x 24	5ft 1¼in	[NER 869]	177
23	6.1889	4ft 8½in	0-6-0	IC	18 x 24	5ft 1¼in	[NER 1509]	176
43	12.1889	4ft 8½in	0-6-0	IC	18 x 24	5ft 1¼in	[NER 776]	176, 180
32	9.1891	4ft 8½in	0-4-0T	IC	14 x 20	3ft 6¼in	[NER 1302]	270
37	10.1891	4ft 8½in	0-4-0T	IC	14 x 20	3ft 6¼in	[NER 1308]	298
38	10.1891	4ft 8½in	0-4-0T	IC	14 x 20	3ft 6¼in	[NER 1310]	298
38	12.1892	4ft 8½in	0-6-0	IC	18 x 24	5ft 1¼in	[NER 1616]	176
4	3.1897	4ft 8½in	0-4-0T	IC	13 x 20	3ft 6¼in	[NER 1799]	34, 62
7	3.1897	4ft 8½in	0-6-0T	IC	14 x 20	3ft 6¼in	[NER 1787]	57

GOODMAN BROS, Chicago, Illinois, United States of America.   Goodman

3576	1924	4ft 8¼in	4wWE	160hp	2ft 10½in	13T	67, 108

GRANGE IRON CO LTD, Belmont, Durham City, Co. Durham.   Grange

This company was registered on 27th April 1866 and the first shareholders included many of the best-known names in the coal mining industry at the time. It first established an ironworks on the site of the former Grange Colliery, but then developed a general engineering business, although the title 'Grange Iron Works' continued in use colloquially. Only a very small number of steam locomotives were constructed, although the firm was the country's leading manufacturer of compressed air locomotives in the 1870s and 1880s. Six were exported to California in the United States of America in 1879, and it is believed that compressed air locomotives reported in Yorkshire and South Wales were also built by the firm. It went into voluntary liquidation on 29th August 1925.

[CARBON]	1866	4ft 8½in	0-4-0ST	OC	?	?	253, 256, 327
[JUBILEE] #	1887	2ft 9½in	4wCA	OC	4 x 7	?	222

\# exhibited at the Royal Newcastle Exhibition in 1887.

From 1877 onwards the firm built between 35 and 40 four-wheeled compressed air locomotives for use underground at the Earl of Durham's collieries, including one compound machine. What little is known about these locomotives may be found in the main text under the entry for The Lambton, Hetton & Joicey Collieries Ltd.

GRANT, RITCHIE & CO, Townholme Engine Works, Kilmarnock, Ayrshire, Scotland.   GR

769	1920	4ft 8½in	0-4-0ST	OC	16 x 24	3ft 8in	218

GILKES, WILSON & CO, Tees Engine Works, Middlesbrough, Yorkshire (NR).   GW

108	1861	4ft 8½in	0-4-0ST	OC	?	?	37
128	8.1861	4ft 8½in	0-4-0T	OC	12 x 20	3ft 9in	414
172	8.1863	4ft 8½in	0-4-0T	OC	?	?	414
175	4.1864	4ft 8½in	0-4-0T	?	?	?	373
178	10.1864	4ft 8½in	0-4-0T	OC	12 x 20	?	414
231	4.1866	4ft 8½in	0-4-0T	OC	12 x 20	?	40
[GEORGE LEEMAN]	?	4ft 8½in	0-4-0ST	OC	?	?	336

JAMES & FREDK. HOWARD LTD, Britannia Ironworks, Bedford, Bedfordshire.   H

965	20.1.1930	4ft 8½in	4wPM	61hp	Dorman 6JUL	?	12T	427

HERON & WILKINSON, Elswick, Newcastle upon Tyne.   H&W

A locomotive attributed to this firm is included in the list of locomotives handed over to the York & Newcastle Railway in October 1846 by the coal owners who had hitherto been running their own trains over the Hartlepool Dock & Railway (which see). The information in the Y&N list is not reliable, but is quoted below as given there. Nothing else is known about the firm.

-	1842	4ft 8½in	0-6-0	OC	14 x 18	4ft 0in	173

TIMOTHY HACKWORTH, Soho Works, Shildon, Co. Durham.   Hackworth

Timothy Hackworth was appointed to have "the superintendence of the permanent and locomotive engines" on the Stockton & Darlington Railway in 1825. In 1833 he took over this work under contract, and for this purpose he developed an engine works at Shildon, putting in his brother Thomas as manager. In 1840 he relinquished the Stockton & Darlington Railway contract, and with his brother leaving to set up a business in Stockton with George Fossick (see Fossick & Hackworth), Timothy ran the business until 1849, when it passed to the Shildon Works Co. Hackworth himself died in the following year. The buildings were acquired by the Stockton & Darlington Railway about 1860 and passed to its successors; those that remain, together with Hackworth's house nearby, are now incorporated into 'Locomotion, the National Railway Museum at Shildon'.

[ROYAL GEORGE]	1827	4ft 8½in	0-6-0	VC	11¼ x 20	4ft 0in	434
[VICTORY]	1829	4ft 8½in	0-6-0	VC	12 x 22	4ft 0in	434
[WYLAM]	c1835?	4ft 8½in	0-6-0	OC	?	?	186
[BRADDYLL]	c1838?	4ft 8½in	0-6-0	OC	15 x 18	4ft 0in	361
[KELLOE]	?	4ft 8½in	0-6-0	OC	15 x 18?	4ft 0in?	361
[NELSON]	?	4ft 8½in	0-6-0	OC	15 x 18?	4ft 0in?	361
[WELLINGTON]	?	4ft 8½in	0-6-0	OC	15 x 18?	4ft 0in?	361
[LEADER]	4.1842	4ft 8½in	0-6-0	OC	14 x 18	4ft 0in	414
[PRINCE ALBERT]	c1842	4ft 8½in	0-6-0	OC	?	?	217
-	9.1845	4ft 8½in	0-6-0	OC	15 x 24	4ft 0in	414

HARTLEY, ARNOUX & FANNING, Stoke-on-Trent, Staffordshire.   HAF

54	1891	1ft 8in	0-4-0ST	OC	5 x 10?	1ft 8in?	308

This firm was absorbed in Kerr, Stuart & Co Ltd, whose Engine Register gives the gauge as 2ft 5in, believed to be incorrect.

HANNOVERSCHE MASCHINENBAU-AG (vormals Georg Egestorff), Hannover-Linden, Germany.   Hanomag

5968	1910	2ft 2in	0-4-4-0WE	120hp	[Siemens 460]	114

HARDY RAIL MOTORS LTD, Slough, Berkshire.   Hardy

908	1925	4ft 8½in	4wPM	?	?	?	428

JOHN HARRIS, Hopetown Foundry, Hopetown, Darlington, Co.Durham.   Harris

John Harris (1812-1869), having previously been engaged on permanent way work for the Stockton & Darlington Railway, set up his business in 1840, north of Darlington (North Road) Station. At first he concentrated on contracting for railway track, but by the 1860s he had expanded into the repair and construction of tank locomotives and wagons. After his death the business was acquired by Thomas

Summerson, who moved it to Albert Hill (see the entry for Summerson's Foundries Ltd in the main text), while the premises were acquired by the Whessoe Foundry (see Whessoe Products Ltd).

[VICTORY]	1863	4ft 8½in	0-4-0ST	OC	?	?	298
[DERWENT]	1865	4ft 8½in	0-4-0ST	OC	?	?	298
[6]	1867	4ft 8½in	0-4-0ST	OC	?	?	234
[BYRON]	1868	4ft 8½in	0-4-0ST	OC	?	?	298
[JOHN HARRIS]	?	4ft 8½in	0-4-0ST	OC	?	?	159
[PRINCE]	?	4ft 8½in	0-4-0ST	OC	?	?	40, 120, 159, 166, 356
[B No. 4]	?	4ft 8½in	0-4-0ST	OC	?	?	101

HUDSWELL & CLARKE (until 1870)   **H&C**
HUDSWELL, CLARKE & RODGERS (from 1870 to 1880)   **HCR**
HUDSWELL, CLARKE & CO LTD   **HC**
HUDSWELL BADGER LTD (from 1972)   **HB**
all at Railway Foundry, Leeds.

	Date ex-works						
21	3.4.1865	4ft 8½in	0-6-0ST	IC	13 x 18	3ft 6in	217
30	30.9.1864	4ft 8½in	0-6-0	IC	17 x 24	5ft 0in	217
32	10.6.1864	4ft 8½in	0-4-0ST	OC	8½ x 13	2ft 6in	276
45	26.5.1865	4ft 8½in	0-6-0ST	IC	13 x 18	3ft 6in	382, 383
71	9.8.1866	4ft 8½in	0-6-0	IC	17 x 24	4ft 0in	217
72	30.10.1866	4ft 8½in	0-6-0	IC	17 x 24	4ft 0in	133, 217
76	7.5.1866	4ft 8½in	0-6-0ST	IC	13 x 18	3ft 0in	217
78	8.8.1866	4ft 8½in	0-6-0ST	IC	13 x 18	3ft 0in	133, 134, 217
79	7.2.1868	4ft 8½in	0-4-0ST	OC	14 x 20	3ft 6in	133, 217
96	28.6.1870	4ft 8½in	0-4-0ST	OC	15 x 20	3ft 6in	217
98	9.9.1870	4ft 8½in	0-6-0ST	IC	17 x 24	4ft 0in	217
101	14.11.1870	4ft 8½in	0-4-0ST	OC	9 x 15	2ft 9in	272
103	23.2.1871	4ft 8½in	0-4-0ST	OC	10 x 16	2ft 9in	323, 356
107	4.9.1871	4ft 8½in	0-4-0ST	OC	10 x 16	2ft 9in	251
121	17.5.1872	4ft 8½in	0-4-0ST	OC	10 x 16	2ft 9in	310, 314
122	8.8.1872	4ft 8½in	0-4-0ST	OC	10 x 16	2ft 9in	310, 314
124	29.10.1872	4ft 8½in	0-4-0ST	OC	13 x 20	3ft 6in	379, 382, 383
129	30.9.1873	4ft 8½in	0-4-0ST	OC	10 x 16	2ft 9in	373
130	28.6.1873	4ft 8½in	0-4-0ST	OC	15 x 20	3ft 6½in	218
169	26.10.1875	4ft 8½in	0-4-0ST	OC	15 x 20	3ft 6½in	217
176	15.3.1876	4ft 8½in	0-6-0ST	OC	11 x 17	2ft 9in	339
203	20.12.1878	4ft 8½in	0-4-0ST	OC	10 x 16	2ft 9in	163
221	12.1.1882	4ft 8½in	0-4-0ST	OC	13 x 20	3ft 3in	306, 402
230	8.8.1881	4ft 8½in	0-4-0ST	OC	15 x 20	3ft 6½in	218
266	31.12.1883	4ft 8½in	0-6-0ST	OC	14 x 20	3ft 6½in	306
299	15.2.1888	4ft 8½in	0-6-0ST	IC	13 x 20	3ft 3in	75
332	23.9.1889	4ft 8½in	0-6-0T	IC	14 x 20	3ft 3in	180
338	31.7.1889	4ft 8½in	0-4-0ST	OC	10 x 16	2ft 9in	273
439	2.3.1896	4ft 8½in	0-6-0ST	IC	13 x 20	3ft 3in	319
485	18.6.1897	3ft 0in	0-4-0ST	OC	8 x 12	2ft 0in	112
530	31.8.1899	4ft 8½in	0-6-0ST	IC	12 x 18	3ft 0in	152
535	9.4.1900	4ft 8½in	0-4-0ST	OC	10 x 16	2ft 9in	426
673	9.3.1906	4ft 8½in	0-6-0ST	IC	12 x 18	3ft 1in	294
674	14.8.1903	4ft 8½in	0-4-0ST	OC	12 x 18	3ft 1½in	316, 321
683	28.11.1903	4ft 8½in	0-4-0ST	OC	12 x 18	3ft 1½in	433
694	22.3.1904	4ft 8½in	0-6-0T	IC	15½ x 20	3ft 7in	385
702	23.6.1904	4ft 8½in	0-4-0ST	OC	13 x 19	3ft 4½in	102
724	26.4.1905	4ft 8½in	0-4-0ST	OC	14 x 20	3ft 3½in	316
749	31.1.1906	4ft 8½in	0-4-0ST	OC	16 x 24	3ft 8in	316
764	23.4.1906	4ft 8½in	0-4-0ST	OC	16 x 24	3ft 8in	316
809	29.6.1907	4ft 8½in	0-6-0PT	IC	18 x 26	4ft 2⅝in	100
880	15.4.1910	4ft 8½in	0-6-0ST	OC	14 x 20	3ft 7in	40
1039	18.9.1913	4ft 8½in	0-6-0T	OC	16 x 24	4ft 0in	306
1190	13.1.1916	4ft 8½in	0-4-0ST	OC	14 x 20	3ft 3½in	316, 318, 321
1191	25.1.1916	4ft 8½in	0-4-0ST	OC	14 x 20	3ft 3½in	316, 319, 321
1199	24.1.1916	4ft 8½in	0-4-0ST	OC	14 x 20	3ft 3½in	118
1201	25.8.1916	4ft 8½in	0-4-0ST	OC	14 x 20	3ft 3½in	321

1207	29.2.1916	4ft 8½in	0-4-0ST	OC	10 x 16	2ft 9½in		427
1250	29.9.1916	4ft 8½in	0-6-0T	IC	15½ x 20	3ft 4in		103
1251	15.2.1917	4ft 8½in	0-6-0T	IC	15½ x 20	3ft 4in		319, 320
1335	5.6.1918	4ft 8½in	0-6-0T	IC	16 x 24	3ft 9in		128, 133, 134
1402	23.10.1922	4ft 8½in	0-4-0ST	OC	14 x 20	3ft 3½in		130, 135
1412	20.8.1920	4ft 8½in	0-4-0ST	OC	16 x 24	3ft 9in		217
1448	26.5.1921	4ft 8½in	0-6-0PT	IC	18 x 26	4ft 2⅝in		100
1449	26.5.1921	4ft 8½in	0-6-0PT	IC	18 x 26	4ft 2⅝in		100
1484	4.9.1922	4ft 8½in	0-4-0ST	OC	12 x 18	3ft 1½in		433
1493	28.12.1923	4ft 8½in	0-4-0ST	OC	12 x 18	3ft 1½in		433
1507	30.6.1923	4ft 8½in	0-4-0ST	OC	9 x 15	2ft 9½in		246
1514	19.12.1923	4ft 8½in	0-4-0ST	OC	16 x 24	3ft 8in		316
1524	30.6.1924	4ft 8½in	0-6-0T	IC	18 x 24	4ft 0in		80, 321
1599	25.8.1927	4ft 8½in	0-4-0ST	OC	13 x 18	3ft 1½in		433
1609	29.7.1934	4ft 8½in	0-6-0ST	IC	13 x 20	3ft 3½in		76
1674	11.3.1937	4ft 8½in	0-6-0ST	IC	13 x 20	3ft 3½in		75, 78
1688	29.12.1937	4ft 8½in	0-4-0ST	OC	14 x 22	3ft 3½in		118
1722	30.8.1941	4ft 8½in	0-4-0ST	OC	14 x 22	3ft 3½in		245
1733	10.2.1943	4ft 8½in	0-4-0ST	OC	14 x 22	3ft 3½in		367
1734	28.12.1942	4ft 8½in	0-4-0ST	OC	14 x 22	3ft 3½in		245, 367
1735	11.12.1942	4ft 8½in	0-4-0ST	OC	14 x 22	3ft 3½in		367
D607	14.7.1938	4ft 8½in	0-4-0DM	120hp	Mirrlees-Ricardo	2ft 9½in	25T	43
D624	10.1942	4ft 8½in	0-6-0DM	165hp	McLaren 165	3ft 0in	30T	323
D835	2.12.1954	4ft 8½in	0-6-0DM	300hp	Crossley EST5	3ft 9in	46T	61
D1159	30.9.1959	4ft 8½in	0-4-0DM	204hp	Gardner 8L3	3ft 0in	32T	68, 118
D1161	30.10.1959	4ft 8½in	0-4-0DM	204hp	Gardner 8L3	3ft 0in	32T	68, 118

THE HUNSLET ENGINE CO LTD, Hunslet, Leeds. HE

14	31.10.1866	4ft 8½in	0-4-0ST	OC	10 x 15	2ft 9in		306
18	18.10.1867	4ft 8½in	0-4-0ST	OC	10 x 15	2ft 9in		354
205	12.8.1878	4ft 8½in	0-4-0ST	OC	12 x 18	3ft 0in		364
240	9.3.1880	4ft 8½in	0-4-0ST	OC	12 x 18	3ft 1in		325
286	7.5.1883	4ft 8½in	0-4-0ST	OC	10 x 15	2ft 9in		180
396	9.4.1886	4ft 8½in	0-6-0T	IC	15 x 20	3ft 4in		361
484	16.10.1889	4ft 8½in	0-6-0ST	IC	13 x 18	3ft 1in		133, 134
567	21.7.1892	2ft 3½in	0-4-0ST	OC	8 x 10	1ft 9in		384
580	6.3.1893	4ft 8½in	0-6-0ST	OC	11 x 15	2ft 6in		241
628	30.5.1895	4ft 8½in	0-4-0ST	OC	10 x 15	2ft 10in		348
1506	6.6.1930	4ft 8½in	0-6-0T	IC	18 x 26	4ft 0in		58
1737	25.7.1935	4ft 8½in	4wDM	20hp	Lister 18/2	2ft 9in	6½T	154, 260
1835	16.4.1937	2ft 6in	4wDM	20hp	Ailsa Craig CF2	?	3T	243
2839	1.10.1943	4ft 8½in	0-4-0DM	40/44hp	Fowler 4B	2ft 9in	10T	168, 428
2842	13.10.1942	2ft 0in	4wDM	25hp	McLarenLMRW2	1ft 6in	3½T	243
2843	13.10.1942	2ft 0in	4wDM	25hp	McLarenLMRW2	1ft 6in	3½T	243
2844	13.10.1942	2ft 0in	4wDM	25hp	McLarenLMRW2	1ft 6in	3½T	243
2979	12.5.1944	2ft 8½in	0-4-0DMF	50hp	Gardner 4L2	2ft 0in	8½T	201
2980	26.5.1944	2ft 8½in	0-4-0DMF	50hp	Gardner 4L2	2ft 0in	8½T	201
2982	17.5.1943	2ft 0in	4wDM	25hp	McLarenLMRW2	1ft 6in	3½T	248
3098	7.11.1944	2ft 0in	4wDM	25hp	McLarenLMRW2	1ft 6in	3½T	248
3330	2.12.1946	2ft 8½in	0-4-0DMF	50hp	Gardner 4L2	2ft 0in	8½T	201
3496	21.11.1947	2ft 0in	4wDM	16hp	Lister	1ft 4in	2½T	115
3504	18.11.1947	4ft 8½in	0-6-0DM	186/204hp	Gardner 8L3	3ft 4in	30T	65, 107
3580	27.4.1949	4ft 8½in	0-6-0DM	186/204hp	Gardner 8L3	3ft 4in	30T	65, 107
4010	1.12.1950	4ft 8½in	0-4-0DM	300hp	Crossley	4ft 0in	40T	65, 107
4011	30.11.1950	4ft 8½in	0-4-0DM	300hp	Crossley	4ft 0in	40T	65, 107
4400	30.11.1954	2ft 6in	4wDM	35hp	Perkins P4	1ft 6in	5T	351
4431	30.11.1953	4ft 8½in	0-4-0DM	300hp	Mirrlees J4	4ft 0in	40T	65, 108
4432	22.12.1953	4ft 8½in	0-4-0DM	300hp	Mirrlees J4	4ft 0in	40T	65, 108
4569	9.3.1956	2ft 0in	4wDM	21hp	Ailsa Craig CF2	1ft 6in	3T	161
4675	1.7.1954	2ft 0in	0-4-0DM	15hp	Enfield Mk.II	?	2½T	87
4979	17.2.1955	2ft 0in	0-4-0DMF	15hp	Enfield Mk.IV	1ft 5in	2½T	87, 226
4987	1.8.1956	4ft 8½in	0-6-0DM	204hp	Gardner 8L3	3ft 4in	30T	65, 108
4988	24.1.1957	4ft 8½in	0-6-0DM	204hp	Gardner 8L3	3ft 4in	30T	65, 108

4989	6.5.1957	4ft 8½in	0-6-0DM	204hp	Gardner 8L3	3ft 4in	30T	65, 108
4991	26.4.1955	2ft 0in	0-4-0DMF	15hp	Enfield Mk.IV?	1ft 5in	2½T	226
5173	12.8.1957	4ft 8½in	0-6-0DM	204hp	Gardner 8L3	3ft 4in	30T	65, 108
5174	28.10.1957	4ft 8½in	0-6-0DM	204hp	Gardner 8L3	3ft 4in	30T	65, 108
5175	18.11.1957	4ft 8½in	0-6-0DM	204hp	Gardner 8L3	3ft 4in	30T	65, 108
5282	20.12.1957	2ft 6in	4wDM	38hp	Perkins P4	1ft 6in	5T	351
5375	1.4.1958	4ft 8½in	0-6-0DM	204hp	Gardner 8L3	3ft 4in	30T	65, 108
5376	30.4.1958	4ft 8½in	0-6-0DM	204hp	Gardner 8L3	3ft 4in	30T	65, 108
5377	9.5.1958	4ft 8½in	0-6-0DM	204hp	Gardner 8L3	3ft 4in	30T	65, 69, 108
5378	30.7.1958	4ft 8½in	0-6-0DM	204hp	Gardner 8L3	3ft 4in	30T	65, 69, 108
5379	27.8.1958	4ft 8½in	0-6-0DM	204hp	Gardner 8L3	3ft 4in	30T	65, 69, 108
5380	29.8.1958	4ft 8½in	0-6-0DM	204hp	Gardner 8L3	3ft 4in	30T	65, 69, 108
5381	15.9.1958	4ft 8½in	0-6-0DM	204hp	Gardner 8L3	3ft 4in	30T	65, 108
5384	16.2.1959	4ft 8½in	0-4-0DM	204hp	Gardner 8L3	3ft 4in	28T	65, 69, 108
5385	23.2.1959	4ft 8½in	0-4-0DM	204hp	Gardner 8L3	3ft 4in	28T	65, 69, 108
5386	20.3.1959	4ft 8½in	0-4-0DM	204hp	Gardner 8L3	3ft 4in	28T	65, 69, 108
5387	25.3.1959	4ft 8½in	0-4-0DM	204hp	Gardner 8L3	3ft 4in	28T	108, 323
5392	31.3.1959	4ft 8½in	0-6-0DH	204hp	Gardner 8L3	3ft 4in	43½T	65, 108
5393	30.4.1959	4ft 8½in	0-6-0DH	204hp	Gardner 8L3	3ft 4in	43½T	65, 108
5394	27.5.1959	4ft 8½in	0-6-0DH	204hp	Gardner 8L3	3ft 4in	43½T	65, 108
5664	8.3.1961	4ft 8½in	0-6-0DM	204hp	Gardner 8L3	3ft 4in	30T	63, 108
6663	c1967 #	4ft 8½in	0-6-0DH	325hp	R-Royce C6SFL	3ft 9in	55T	65, 108

\# built for stock and sent to Consett Steel Works as demonstrator. Not sold til 18.8.1969.

THE HETTON COAL CO LTD, Hetton Engine Works, Hetton-le-Hole, Co. Durham. **Hetton**
This company built a number of locomotives for its own use, all of them for shunting purposes.

[LYONS]	c1852	4ft 8½in	0-4-0	VC	9 x 24	3ft 9in	190
[LADY BARRINGTON]	c1854	4ft 8½in	0-4-0	VC	9 x 24?	3ft 9in?	190
[EPPLETON]	#c1900	4ft 8½in	4wVBT	VC	6½ x 8	?	190, 220
[LYONS]	#c1900	4ft 8½in	4wVBT	VC	6½ x 8?	?	190, 220

\# one source states that these were built in the 1870s.

HAIGH FOUNDRY CO LTD, Wigan, Lancashire. **HF**

46	5.1841	4ft 8½in	0-4-2	IC	14 x 18	5ft 0in	234

HOPKINS, GILKES & CO, Tees Iron Works, Middlesbrough, Yorkshire (NR). **HG**

251	10.1867	4ft 8½in	0-4-0ST	OC	12 x 20	3ft 11½in	143, 253, 374
276	11.1870	4ft 8½in	0-4-0ST	OC	12 x 20	3ft 11½in	202

Unidentified locomotive

ALLIANCE	?	4ft 8½in	0-4-0T	OC	?	?	44, 118

HENRY HUGHES & CO, Falcon Engine Works, Loughborough, Leicestershire. **HH**

-	1876	4ft 8½in	0-4-0ST	OC	?	?	139
-	1876	4ft 8½in	0-4-0ST	OC	?	?	139
[HOLMSIDE]	?	4ft 8½in	0-6-0ST	?	?	?	193
LIONEL	?	4ft 8½in	0-4-0ST	OC	?	?	270

HAWTHORNS & CO, Leith Engine Works, Leith, Edinburgh, Scotland. **H(L)**

220	1859	4ft 8½in	0-4-0WT	OC	?	?	298
[HAWK]	?	4ft 8½in	0-4-0ST	OC	?	?	139
[BIRTLEY]	?	4ft 8½in	2-4-0WT	OC	?	?	298

R. & W. HAWTHORN, LESLIE & CO LTD, Forth Banks Works, Newcastle upon Tyne. **HL**

This company was formed on 1st July 1885 by the amalgamation of R. & W. Hawthorn and Andrew Leslie & Co Ltd, shipbuilders, of Hebburn, Co.Durham, and became a limited company on 7th April 1886. The numbering scheme of the former was continued, but which locomotive was the first to carry a HL plate is uncertain. Whereas R. & W. Hawthorn had built chiefly main line locomotives, the new firm concentrated more on industrial customers. This was helped by the demise of Black, Hawthorn & Co Ltd, and their successors Chapman & Furneaux, whose goodwill Hawthorn Leslie bought in 1902. In that same year the departure of their next-door neighbours, Robert Stephenson & Co Ltd (which see), to Darlington, allowed Hawthorn Leslie to expand into their premises. Given the firm's involvement in ship-building, it is perhaps not surprising that it was a major builder of crane tank locomotives. Between the two World Wars the company experimented with other forms of traction besides steam, and developed both an electric coke car locomotive and diesel locomotives.

In May 1937 the locomotive business was sold to Robert Stephenson & Co Ltd, which changed its name on 27th September 1937 to Robert Stephenson & Hawthorns Ltd (which see). Despite this, locomotives continued for a time to be turned out with HL plates, the last being No.3953 in March 1938. The firm itself continued with its shipbuilding business at Hebburn and its marine engine works at St.Peter's in Newcastle upon Tyne, which continued its works numbers. It was absorbed into Swan Hunter & Tyne Shipbuilders Ltd on 1st January 1968.

2039	4.12.1886	4ft 8½in	0-4-0ST	OC	12 x 18	3ft 0½in	120, 185
2073	19.4.1887	4ft 8½in	0-4-0ST	OC	12 x 18	3ft 0in	273
2110	4.6.1888	4ft 8½in	0-4-0ST	OC	14 x 20	3ft 6in	73, 130
# 2113	9.10.1888	4ft 8½in	2-2-2CT	IC	14 x 18	?	10T capy. 277
2135	22.5.1889	4ft 8½in	0-4-0ST	OC	12 x 18	3ft 0in	277
2139	25.10.1889	4ft 8½in	0-4-0ST	OC	12 x 18	3ft 0in	257, 258
2152	5.1890	4ft 8½in	0-4-0ST	OC	14 x 20	3ft 6in	43, 60, 319
2169	12.12.1889	4ft 8½in	0-4-0ST	OC	12 x 18	3ft 0½in	277
2173	29.5.1890	4ft 8½in	0-4-0CT	OC	11 x 15	2ft 9in	4T capy. 278, 372
2176	29.7.1890	4ft 8½in	0-4-0ST	OC	12 x 19	3ft 4in	367
2177	29.7.1890	4ft 8½in	0-4-0ST	OC	12 x 19	3ft 4in	101, 435, 437
2185	1890	4ft 8½in	0-4-0ST	OC	14 x 20	3ft 6in	288
2199	6.1891	4ft 8½in	0-4-0ST	OC	12 x 18	3ft 0½in	371
2235	1.1892	4ft 8½in	0-4-0ST	OC	12 x 19	3ft 4in	101
2236	1.1892	4ft 8½in	0-4-0ST	OC	12 x 19	3ft 4in	102
2247	5.9.1892	4ft 8½in	0-4-0ST	OC	12 x 18	3ft 0½in	247, 256, 284
2249	10.1892	4ft 8½in	0-4-0CT	OC	12 x 15	2ft 10in	4T capy. 83
2272	4.8.1893	4ft 8½in	0-4-0CT	OC	12 x 15	2ft 10in	4T capy. 185
2273	19.10.1893	4ft 8½in	0-4-0CT	OC	12 x 15	2ft 10in	4T capy. 185
2279	9.1893	4ft 8½in	0-4-0ST	OC	12 x 18	3ft 0½in	153
2281	12.1894	4ft 8½in	0-4-0ST	OC	12 x 18	3ft 0½in	371
2330	6.1896	4ft 8½in	0-4-0ST	OC	12 x 18	3ft 0½in	315
2349	11.1896	4ft 8½in	0-4-0ST	OC	12 x 18	3ft 0½in	316, 318, 321, 371
2357	3.1897	4ft 8½in	0-4-0ST	OC	14 x 20	3ft 6in	33
2358	3.1897	4ft 8½in	0-4-0ST	OC	14 x 20	3ft 6in	249
2377	1.7.1897	4ft 8½in	0-4-0ST	OC	13 x 19	3ft 4½in	102
2387	1.9.1897	4ft 8½in	0-4-0ST	OC	12 x 18	2ft 10in	262
2404	1.1899	4ft 8½in	0-4-0ST	OC	13 x 19	3ft 4½in	102
2412	3.1899	4ft 8½in	0-4-0ST	OC	14 x 20	3ft 6in	72, 415
2425	9.1899	4ft 8½in	0-4-0ST	OC	14 x 20	3ft 6in	31
2426	9.1899	4ft 8½in	0-4-0ST	OC	14 x 20	3ft 6in	316, 318, 371
2429	10.1899	4ft 8½in	0-6-0ST	IC	15 x 22	3ft 8in	46
2431	8.1899	4ft 8½in	0-4-0ST	OC	12 x 18	3ft 0½in	433
2447	25.5.1900	4ft 8½in	0-4-0CT	OC	12 x 15	2ft 10in	4T capy. 277, 278
2449	1.1900	4ft 8½in	0-4-0ST	OC	14 x 20	3ft 6in	51, 129
2453	2.1900	4ft 8½in	0-4-0ST	OC	14 x 20	3ft 6in	125, 183, 284, 287
2456	3.1900	4ft 8½in	0-4-0ST	OC	14 x 20	3ft 6in	284, 287, 374
2465	30.7.1900	4ft 8½in	0-4-0ST	OC	14 x 20	3ft 6in	423
2468	12.2.1901	4ft 8½in	0-4-0CT	OC	12 x 15	2ft 10in	4T capy. 385
2478	5.2.1901	4ft 8½in	0-4-0ST	OC	12 x 18	3ft 0½in	159, 183
2479	8.2.1901	4ft 8½in	0-4-0ST	OC	12 x 18	3ft 0½in	143
2481	12.2.1900	4ft 8½in	0-4-0ST	OC	14 x 20	3ft 6in	212, 225

2484	6.10.1900	4ft 8½in	0-6-0ST	IC	16 x 24	4ft 1in		153
2489	29.4.1901	4ft 8½in	0-4-0ST	OC	14 x 20	3ft 6in		277
2496	5.11.1901	4ft 8½in	0-4-0ST	OC	12 x 18	3ft 0½in		62, 433
2499	2.9.1901	4ft 8½in	0-4-0CT	OC	12 x 15	2ft 10in	4T capy.	83, 163
2514	2.10.1901	4ft 8½in	0-4-0ST	OC	12 x 18	2ft 10in		262
2515	19.12.1901	4ft 8½in	0-6-0ST	OC	15 x 22	3ft 9in		57
2517	6.5.1902	4ft 8½in	0-4-0CT	OC	12 x 15	2ft 10in	4T capy.	142
2530	22..8.1902	4ft 8½in	0-4-0ST	OC	14 x 20	3ft 6in		218
2533	31.7.1902	4ft 8½in	0-4-0ST	OC	12 x 18	3ft 0½in		335, 395
2535	12.8.1902	4ft 8½in	0-4-0CT	OC	12 x 15	2ft 10in	4T capy.	142
2545	24.12.1902	4ft 8½in	0-6-0ST	IC	17 x 26	4ft 6in		58
2550	31.7.1903	4ft 8½in	0-4-0CT	OC	12 x 15	2ft 10in	4T capy.	142, 185
2551	9.1903	4ft 8½in	0-4-0CT	OC	12 x 15	2ft 10in	4T capy.	142
2559	23.7.1903	4ft 8½in	0-4-0ST	OC	14 x 20	3ft 6in		284, 285
2563	28.11.1903	4ft 8½in	0-4-0ST	OC	12 x 18	3ft 0½in		144, 289, 290
2583	11.5.1904	4ft 8½in	0-4-0ST	OC	14 x 20	3ft 6in		371
2589	15.11.1904	4ft 8½in	0-4-0ST	OC	14 x 20	3ft 6in		306
2594	25.1.1905	4ft 8½in	0-4-0CT	OC	12 x 15	2ft 10in	4T capy.	142
2595	9.9.1904	4ft 8½in	0-6-0ST	OC	16 x 24	3ft 10in		51, 128
2606	27.3.1906	4ft 8½in	0-4-0CT	OC	12 x 15	2ft 10in	4T capy.	155
2607	3.3.1905	4ft 8½in	0-6-0ST	OC	15 x 22	3ft 9in		128, 134
2612	19.4.1905	4ft 8½in	0-6-0ST	OC	14 x 22	3ft 4in		48, 52, 128, 129
2613	24.10.1905	4ft 8½in	0-6-0ST	OC	14 x 22	3ft 4in		46, 50, 52, 126, 128, 129
2617	21.6.1905	4ft 8½in	0-4-0ST	OC	14 x 22	3ft 6in		371
2632	7.7.1906	4ft 8½in	0-4-0CT	OC	12 x 15	2ft 10in	4T capy.	142
2639	20.2.1906	4ft 8½in	0-4-0ST	OC	13 x 19	3ft 4½in		102
2640	26.2.1906	4ft 8½in	0-4-0ST	OC	13 x 19	3ft 4½in		102
2641	17.4.1906	4ft 8½in	0-6-0PT	IC	18 x 26	4ft 2⅝in		100
2645	22.3.1906	4ft 8½in	0-4-0ST	OC	14 x 22	3ft 6in		247
2654	3.4.1906	4ft 8½in	0-6-0ST	OC	14 x 22	3ft 4in		50, 52, 128, 129
2655	22.3.1907	4ft 8½in	0-6-0ST	OC	14 x 22	3ft 4in		46, 126, 128, 134
2666	21.7.1906	4ft 8½in	0-4-0ST	OC	14 x 22	3ft 4in		277, 398
2667	10.8.1906	4ft 8½in	0-4-0ST	OC	14 x 22	3ft 4in		277, 398
2684	11.2.1907	4ft 8½in	0-4-0ST	OC	13 x 19	3ft 4in		183
2685	1.3.1907	4ft 8½in	0-4-0T	OC	15 x 22	3ft 9in		282, 285
2694	17.7.1907	4ft 8½in	0-4-0ST	OC	15 x 22	3ft 9in		371
2701	2.5.1907	4ft 8½in	0-4-0ST	OC	14 x 22	3ft 6in		237
2702	7.5.1907	4ft 8½in	0-4-0ST	OC	14 x 22	3ft 6in		371
2713	31.8.1907	4ft 8½in	0-4-0ST	OC	14 x 22	3ft 6in		284, 374
2719	7.11.1907	4ft 8½in	0-6-0ST	OC	15 x 22	3ft 9in		58
2732	8.9.1909	4ft 8½in	0-4-0ST	OC	14 x 22	3ft 7in		126, 128
2737	4.3.1908	4ft 8½in	0-6-0ST	OC	16 x 24	3ft 8in		155, 200
2740	25.2.1909	4ft 8½in	0-4-0ST	OC	14 x 22	3ft 6in		371
2769	26.2.1909	4ft 8½in	0-6-0T	IC	17 x 26	4ft 1in		419
2774	27.2.1909	4ft 8½in	0-4-0T	OC	15 x 22	3ft 9in		284
2780	18.6.1909	4ft 8½in	0-4-0ST	OC	14 x 22	3ft 6in		247
2784	29.9.1909	4ft 8½in	0-4-0ST	OC	14 x 22	3ft 6in		419
2789	1909	4ft 8½in	0-4-0ST	OC	16 x 24	3ft 10in		217
2798	15.2.1910	4ft 8½in	0-4-0ST	OC	12 x 18	3ft 0½in		290, 291, 293
2799	4.3.1910	4ft 8½in	0-4-0ST	OC	12 x 18	3ft 0½in		285, 289, 290
2823	4.7.1910	4ft 8½in	0-4-0ST	OC	14 x 22	3ft 6in		282, 285
2824	18.7.1910	4ft 8½in	0-4-0ST	OC	14 x 22	3ft 6in		419
2826	10.8.1910	4ft 8½in	0-4-0ST	OC	15 x 22	3ft 8in		218
2827	18.8.1910	4ft 8½in	0-4-0ST	OC	15 x 22	3ft 8in		142, 218
2833	12.8.1910	4ft 8½in	0-6-0ST	OC	16 x 24	3ft 10in		51, 128
2837	31.10.1910	4ft 8½in	0-4-0ST	OC	14 x 22	3ft 6in		319
2839	31.12.1910	4ft 8½in	0-4-0ST	OC	14 x 22	3ft 6in		118
2871	26.5.1911	4ft 8½in	0-4-0ST	OC	14 x 22	3ft 6in		424
2880	8.8.1911	4ft 8½in	0-6-0T	OC	17 x 24	3ft 8in		379, 382
2890	16.8.1911	4ft 8½in	0-4-0ST	OC	14 x 22	3ft 6in		118
2909	31.10.1911	4ft 8½in	0-6-0ST	OC	14 x 22	3ft 4in		50, 53, 137
2916	22.3.1912	4ft 8½in	0-4-0ST	OC	14 x 22	3ft 6in		52

2917	8.8.1912	4ft 8½in	0-4-0ST	OC	14 x 22	3ft 6in	185
2932	31.7.1912	4ft 8½in	0-6-0ST	IC	15 x 22	3ft 9in	218, 238
2941	21.6.1912	4ft 8½in	0-4-0ST	OC	12 x 18	3ft 0½in	139
2954	31.8.1912	4ft 8½in	0-4-0ST	OC	15 x 22	3ft 8in	218
2956	29.11.1912	4ft 8½in	0-6-0ST	OC	16 x 24	3ft 8in	197
2984	19.1.1914	4ft 8½in	0-4-0VBCr	OC	12 x 21½	3ft 0in	106
2986	19.4.1913	4ft 8½in	0-4-0T	OC	14 x 22	3ft 6in	298
2989	31.3.1913	4ft 8½in	0-4-0ST	OC	14 x 22	3ft 6in	142
2993	16.5.1913	4ft 8½in	0-4-0ST	OC	16 x 24	3ft 10in	282, 285
3003	19.9.1913	4ft 8½in	0-4-0ST	OC	13 x 19	3ft 4½in	102, 120
3004	23.9.1913	4ft 8½in	0-4-0ST	OC	13 x 19	3ft 4½in	102
3022	3.11.1913	4ft 8½in	0-4-0ST	OC	13 x 19	3ft 4½in	103
3023	17.11.1913	4ft 8½in	0-4-0ST	OC	13 x 19	3ft 4½in	103
3024	12.12.1913	4ft 8½in	0-4-0ST	OC	15 x 22	3ft 8in	218
3029	23.12.1913	1ft 10in	0-4-0ST	OC	5 x 10	1ft 8½in	406
3053	10.3.1914	4ft 8½in	0-4-0ST	OC	14 x 22	3ft 6in	289, 290
3055	6.4.1914	4ft 8½in	0-4-0ST	OC	16 x 24	3ft 10in	217
3056	22.4.1914	4ft 8½in	0-4-0ST	OC	16 x 24	3ft 10in	217
3080	9.12.1914	4ft 8½in	0-6-0PT	IC	18 x 26	4ft 2⅝in	100
3090	22.1.1915	4ft 8½in	0-4-0ST	OC	14 x 22	3ft 6in	78
3103	9.4.1915	4ft 8½in	0-6-0ST	OC	17 x 24	3ft 10in	58
3104	26.4.1915	4ft 8½in	0-6-0T	OC	17 x 24	3ft 10in	285, 374
3139	20.7.1915	4ft 8½in	0-4-0ST	OC	12 x 18	3ft 0½in	139
3140	30.7.1915	4ft 8½in	0-4-0ST	OC	12 x 18	3ft 0½in	130, 139
3185	7.7.1916	4ft 8½in	0-6-0ST	OC	16 x 24	3ft 10in	249
3237	14.4.1917	4ft 8½in	0-4-0ST	OC	15 x 22	3ft 8in	120, 277
3248	19.6.1917	4ft 8½in	0-4-0ST	OC	16 x 24	3ft 8in	129
3251	2.10.1917	4ft 8½in	0-4-0ST	OC	13 x 19	3ft 4½in	103
3252	13.10.1917	4ft 8½in	0-4-0ST	OC	13 x 19	3ft 4½in	103
3253	29.10.1917	4ft 8½in	0-4-0ST	OC	13 x 19	3ft 4½in	103
3254	7.11.1917	4ft 8½in	0-4-0ST	OC	13 x 19	3ft 4½in	103
3282	19.11.1917	4ft 8½in	0-6-0ST	OC	14 x 22	3ft 6in	319
3300	22.11.1917	4ft 8½in	0-4-0ST	OC	14 x 22	3ft 6in	118
3319	27.4.1918	4ft 8½in	0-4-0ST	OC	14 x 22	3ft 6in	185
3349	26.9.1918	4ft 8½in	0-4-0ST	OC	14 x 22	3ft 6in	247
3352	17.10.1918	4ft 8½in	0-4-0ST	OC	16 x 24	3ft 8in	349
3354	15.11.1918	4ft 8½in	0-4-0ST	OC	16 x 24	3ft 10in	349
3355	28.11.1918	4ft 8½in	0-4-0ST	OC	16 x 24	3ft 10in	349
3384	19.8.1919	4ft 8½in	0-4-0ST	OC	14 x 22	3ft 6in	51, 129, 130
3390	21.10.1919	4ft 8½in	0-4-0ST	OC	14 x 22	3ft 6in	100
3391	24.11.1919	4ft 8½in	0-4-0ST	OC	14 x 22	3ft 6in	101
3426	30.8.1920	4ft 8½in	0-4-0ST	OC	14 x 22	3ft 6in	127, 139
3438	27.10.1920	4ft 8½in	0-4-0ST	OC	15 x 22	3ft 9in	371
3440	18.11.1920	4ft 8½in	0-6-0ST	OC	16 x 24	3ft 10in	200
3467	30.10.1920	4ft 8½in	0-4-0ST	OC	14 x 22	3ft 6in	166
3471	1.6.1921	4ft 8½in	0-4-0ST	OC	16 x 24	3ft 10in	103
3472	8.6.1921	4ft 8½in	0-4-0ST	OC	16 x 24	3ft 10in	103
3473	16.11.1920	4ft 8½in	0-4-0ST	OC	14 x 22	3ft 6in	101
3474	16.11.1920	4ft 8½in	0-4-0ST	OC	14 x 22	3ft 6in	101
3475	14.12.1920	4ft 8½in	0-4-0ST	OC	14 x 22	3ft 6in	101
3476	21.12.1920	4ft 8½in	0-4-0ST	OC	14 x 22	3ft 6in	101, 349
3492	25.2.1921	4ft 8½in	0-4-0ST	OC	14 x 22	3ft 6in	237
3493	19.6.1922	4ft 8½in	0-4-0ST	OC	14 x 22	3ft 6in	419
3494	12.12.1922	4ft 8½in	0-4-0ST	OC	14 x 22	3ft 6in	419
3495	23.12.1920	4ft 8½in	0-4-0T	OC	14 x 22	3ft 6in	101
3496	22.7.1921	4ft 8½in	0-4-0ST	OC	16 x 24	3ft 10in	13, 323, 348
3497	11.8.1921	4ft 8½in	0-4-0ST	OC	16 x 24	3ft 10in	103
3504	18.1.1923	4ft 8½in	0-4-0ST	OC	14 x 22	3ft 6in	242
@ 3513	31.1.1927	4ft 8½in	0-6-0ST	OC	14 x 22	3ft 6in	379
3527	19.6.1922	4ft 8½in	0-6-0ST	OC	16 x 24	3ft 10in	348
3528	17.8.1922	4ft 8½in	0-6-0ST	OC	16 x 24	3ft 8in	193, 197
3543	11.9.1923	4ft 8½in	0-4-0ST	OC	15 x 22	3ft 8in	218
3544	11.9.1923	4ft 8½in	0-4-0ST	OC	15 x 22	3ft 8in	218
3568	30.7.1923	4ft 8½in	0-6-0ST	OC	16 x 24	3ft 10in	200
3569	6.11.1923	4ft 8½in	0-6-0ST	IC	18 x 26	4ft 6in	58

	3572	25.9.1923	4ft 8½in	0-4-0ST	OC	12 x 18	3ft 0½in		139
	3573	25.9.1923	4ft 8½in	0-4-0ST	OC	12 x 18	2ft 10in		262
	3576	29.10.1923	4ft 8½in	0-4-0ST	OC	12 x 18	2ft 10in		262
	3577	21.12.1923	4ft 8½in	0-4-0ST	OC	12 x 18	3ft 0½in		354
%	3584	1924	4ft 8½in	0-4-0BE	100hp		3ft 3in	15½T	247
	3641	7.6.1926	4ft 8½in	0-4-0ST	OC	14 x 22	3ft 6in		75
	3651	19.11.1926	4ft 8½in	0-4-0ST	OC	14 x 22	3ft 6in		76, 78
	3654	5.4.1927	4ft 8½in	0-4-0ST	OC	14 x 22	3ft 6in		43
	3732	14.8.1928	4ft 8½in	0-4-0ST	OC	14 x 22	3ft 6in		76
	3744	29.4.1929	4ft 8½in	0-4-0ST	OC	16 x 24	3ft 10in		102, 348
	3745	12.7.1929	4ft 8½in	0-4-0ST	OC	16 x 24	3ft 10in		102, 323
	3752	21.2.1930	4ft 8½in	0-4-0ST	OC	16 x 24	3ft 10in		102
	3753	21.2.1930	4ft 8½in	0-4-0ST	OC	16 x 24	3ft 10in		102
	3766	13.3.1930	4ft 8½in	0-6-0T	OC	18 x 24	4ft 0in		298
	3772	21.11.1930	4ft 8½in	0-4-0ST	OC	14 x 22	3ft 6in		75
	3806	22.8.1934	4ft 8½in	0-4-0ST	OC	12 x 20	3ft 1in		433
	3830	12.6.1934	4ft 8½in	0-6-0ST	OC	18 x 24	4ft 0in		382
	3834	18.9.1934	4ft 8½in	0-6-2T	IC	18½ x 26	4ft 6in		218
	3859	23.1.1937	4ft 8½in	0-4-0WE	80hp		2ft 9in	14T	381
	3873	27.4.1936	4ft 8½in	0-4-0ST	OC	16 x 24	3ft 8in		102, 349
	3887	4.5.1936	4ft 8½in	0-4-0ST	OC	12 x 20	3ft 1in		433
	3891	14.10.1936	4ft 8½in	0-6-0PT	IC	18 x 26	4ft 2⅝in		99
	3894	16.12.1936	4ft 8½in	0-4-0ST	OC	12 x 20	3ft 1in		155
	3895	10.5.1937	4ft 8½in	0-4-0ST	OC	12 x 20	3ft 1in		35
	3898	30.11.1936	4ft 8½in	0-6-0ST	OC	16 x 24	3ft 8in		348
	3905	30.4.1937	4ft 8½in	0-6-0PT	IC	18 x 26	4ft 2⅝in		100
	3906	26.2.1937	4ft 8½in	0-4-0ST	OC	16 x 24	3ft 8in		101
	3919	19.7.1937	4ft 8½in	0-4-0ST	OC	16 x 24	3ft 8in		349
	3934	14.1.1938	4ft 8½in	0-6-0ST	OC	14 x 22	3ft 6in		261
	3951	21.2.1938	4ft 8½in	0-6-0PT	IC	18 x 26	4ft 2⅝in		99
	3952	11.3.1938	4ft 8½in	0-6-0PT	IC	18 x 26	4ft 2⅝in		99
	3953	8.3.1938	4ft 8½in	0-4-0ST	OC	16 x 24	3ft 8in		65, 101

    #   built to Cross' Patent.
    @   originally built in 1923 as an experimental compressed steam locomotive; rebuilt to a standard steam locomotive in 1927, with a 1927 works plate.
    %   built to Durnall's Patent, and used as works shunter. Sold by RSH in 1939, with new (HL) plate dated 1939; ex works, 23.3.1939.

HOPPER, RADCLIFFE & CO, Britannia Iron Works, Fencehouses, Co. Durham.   **Hopper**
George Hopper & Sons until c1870; originally **George Hopper**.

George Hopper (c1798-1875) began in business in Houghton-le-Spring, probably in the 1820s, firstly as a timber merchant and then as a brick manufacturer. Then in the early 1850s he decided to develop an ironworks and general engineering business, choosing a site at Colliery Row, some two miles west of Houghton; the works was certainly in existence by 1854. It was served by the truncated Chilton Moor branch of the Londonderry Railway (see Marquis of Londonderry). By 1870 the firm had acquired the title above, and not long afterwards the firm acquired the Seaham Foundry at Seaham Harbour. This was offered for sale in November 1874. George Hopper died in September 1875 and in 1876 the works and its effects were auctioned. For further details see the entry for the company in the main text.

The firm probably built no more than six locomotives, all saddle-tanks, one being six-coupled and the others four-coupled. One is mentioned in a sale notice of 1873 and two were offered for auction in 1876. Whether any locomotives were sold to customers in County Durham is not known.

ROBERT HUDSON LTD, Gildersome Foundry, Leeds.   **HU**

?	1924	2ft 0in	4wPM	20hp	Fordson	?	4T	85
38384	23.1.1930	2ft 0in	4wPM	20hp	Fordson	?	4T	146
46851	4.10.1933	2ft 0in	4wPM	20hp	Fordson	?	4T	44

HEAD, WRIGHTSON & CO LTD, Teesdale Iron Works, Thornaby-on-Tees, Yorkshire (NR).   **HW**

21	1870	4ft 8½in	0-4-0VBT	VC	6 x 12	2ft 6in	234, 348
33	1873	4ft 8½in	0-4-0VBT	OC	6 x 12	2ft 4in	234, 348
[10]	?	4ft 8½in	0-4-0VBT	G	?	?	414

CHARLES I'ANSON & CO LTD, Darlington, Co. Durham.   I'Anson

For the detailed history of Charles I'Anson, this company and the two works which it owned, the Whessoe Iron Works and the Rise Carr Iron Works, see the entries for Whessoe Products Ltd and The Darlington & Simpson Rolling Mills Ltd in the main text.

The firm almost certainly built only two, or possibly three, three locomotives. It seems very likely that the first two were built at the Whessoe works, but for use at the Rise Carr works. Both were subsequently sold on. It is possible that the first may also be the third.

-	1875	4ft 8½in	0-4-0ST	OC?	?	?	120, 352
[BASIL]	1875	4ft 8½in	0-4-0ST	OC	?	?	339, 364
[RISE CARR]	1875	4ft 8½in	0-4-0ST	OC	?	?	62, 120
[SHIELD ROW No.1]	?	4ft 8½in	0-4-0ST	OC	?	?	197

ISAAC WATT BOULTON, Ashton-under-Lyne, Cheshire.   IWB

[LITTLE GRIMSBY]	1860	4ft 8½in	0-4-0ST	ICG	5½ x 12	2ft 0in	417

JOHN FOWLER & CO (LEEDS) LTD, Hunslet, Leeds.   JF

# 1142	12.1868	4ft 8½in	4wTG		8 x 12	?	415
1539	25.5.1871	4ft 8½in	0-6-0T	IC	8 x 14	2ft 6in	37, 145, 286
1540	6.1871	4ft 8½in	0-6-0T	IC	8 x 14	2ft 6in	69, 257
1541	17.7.1871	4ft 8½in	0-6-0T	IC	8 x 14	2ft 6in	252, 258, 259
1542	1.8.1871	4ft 8½in	0-6-0T	IC	8 x 14	2ft 6in	400
1568	12.12.1871	4ft 8½in	0-4-0ST	IC	8½ x 14	2ft 8in	395
1571	11.1872	4ft 8½in	0-4-0ST	IC	8½ x 14	?	396
1572	1872	4ft 8½in	0-4-0ST	IC	8½ x 14	2ft 8in	247, 322, 323
2076	12.1873	4ft 8½in	0-4-0ST	OC	12 x 18	?	259
2078	3.2.1874	4ft 8½in	0-4-0ST	OC	12 x 18	?	282
2079	9.3.1874	4ft 8½in	0-4-0ST	OC	12 x 18	?	259
2820	9.9.1876	3ft 0in	0-4-0WTG	OC	5 x 14	?	41
2821	13.12.1876	3ft 0in	0-4-0WTG	OC	5 x 14	?	41
2834	22.6.1876	4ft 8½in	0-4-0ST	OC	12 x 18	?	419
2849	15.7.1876	4ft 8½in	0-4-0ST	OC	8½ x 12	2ft 6in	228, 359
3155	5.1880	4ft 8½in	0-4-0ST	OC	12 x 18	?	202
3157	9.1883	4ft 8½in	0-4-0ST	OC	12 x 18	?	433
5653	31.3.1888	3ft 0in	0-4-0WTG	OC	5 x 8	?	41
5661	16.4.1888	3ft 0in	0-4-0WTG	OC	5 x 8	?	249
5822	30.11.1888	3ft 0in	0-4-0ST	OC	5 x 10	?	283
5883	4.3.1889	3ft 0in	0-4-0WTG	OC	5 x 8	?	249
6680	4.1892	3ft 0in	0-4-0ST	OC	5½ x 10	?	417
6681	4.1892	3ft 0in	0-4-0ST	OC	5½ x 10	?	417
7529	12.1895	2ft 0in	0-4-2WT	OC	4½ x 8	?	332
16991	4.10.1926	2ft 0in	0-6-0T	OC	8 x 12	2ft 0in	150

\# locomotive built to a special patent; see the entry for The Weardale Steel, Coal & Coke Co Ltd.

22061	6.1938	4ft 8½in	0-4-0DM	40hp	Fowler 4B	?	10T	242
22137	29.12.1937	4ft 8½in	0-4-0DM	150hp	Fowler 4C	3ft 3in	29T	242
22900	29.3.1941	4ft 8½in	0-4-0DM	40hp	Fowler 4B	2ft 6in	10T	75
22934	23.1.1941	4ft 8½in	0-4-0DM	150hp	Fowler 4C	3ft 3in	29T	245
22943	30.6.1941	4ft 8½in	0-4-0DM	150hp	Fowler 4C	3ft 3in	29T	245
22947	31.10.1941	4ft 8½in	0-4-0DM	150hp	Fowler 4C	3ft 3in	29T	245
22948	10.11.1941	4ft 8½in	0-4-0DM	150hp	Fowler 4C	3ft 3in	29T	245
22976	11.5.1942	4ft 8½in	0-4-0DM	150hp	Fowler 4C	3ft 3in	29T	242
4000013	29.12.1947	4ft 8½in	0-4-0DM	60hp	Fowler 6B	3ft 0in	15T	258, 259
4110001	11.11.1949	4ft 8½in	0-4-0DM	80hp	McLaren MR3	3ft 0in	19T	257, 259
4110002	31.1.1950	4ft 8½in	0-4-0DM	80hp	McLaren MR3	3ft 0in	19T	258, 259
4110006	14.8.1950	4ft 8½in	0-4-0DM	80hp	McLaren MR3	3ft 0in	19T	34, 35, 84, 123
4110008	28.9.1950	4ft 8½in	0-4-0DM	80hp	McLaren MR3	3ft 0in	19T	386
4240013	4.6.1962	4ft 8½in	0-4-0DH	275hp	Leyland	3ft 6in	?	78, 257
4240015	31.12.1962	4ft 8½in	0-6-0DH	275hp	Leyland	3ft 6in	?	78
4240020	18.3.1964	4ft 8½in	0-6-0DH	275hp	Leyland	3ft 6in	?	78

J. & G. JOICEY & CO LTD, Pottery Lane, Newcastle upon Tyne.   Joicey
J. & G. Joicey & Co until 1900

James Joicey (1806-1863) was a self-made entrepreneur in the coal trade (see the entry for James

Joicey & Co Ltd in the main text). Seeing the demand for engineering products from heavy industry locally, in 1849 he and his second brother George (1813-1856) set up this company, initially to operate a foundry; but from 1854 the firm entered the manufacture of engines for winding, hauling and pumping, together with general colliery and engineering equipment. It also built an unknown number of locomotives, apparently mostly in the 1860s and 1870s, which were numbered in the firm's general works list. For much of the second half of the nineteenth century the firm was controlled by Jacob Joicey (1843-1899), George's eldest son. It was closed down in 1925. Its records are believed not to have survived.

210	1869	4ft 8½in	0-4-0ST	OC	?	?		249, 414
?	1870	4ft 8½in	0-4-0ST	OC	?	?	[8 MIDDRIDGE]	414
?	1870	4ft 8½in	0-4-0ST	OC	?	?	[TOBY]	72
?	1872	4ft 8½in	0-4-0ST	OC	?	?	[7 SPAWOOD]	414
?	1874	4ft 8½in	0-4-0ST	OC	?	?	[9 SEDGEFIELD]	414
?	c1875	4ft 8½in	0-4-0ST	OC	?	?	[13 CROXDALE]	415
?	c1875	4ft 8½in	0-4-0ST	OC	?	?	[14 BLACK PRINCE]	152, 415
305	1883	4ft 8½in	0-6-0ST	IC	?	?		361
377	1885	4ft 8½in	0-4-0ST	OC	?	?		211, 212, 224, 225
429	1894	4ft 8½in	0-4-0ST	OC	?	?		211

\# 0-4-0ST AE 1387/1898 (see above) carried a Joicey 457/1898 plate.

Unidentified locomotives

?	?	4ft 8½in	0-4-0ST	OC	?	?	[ALMA]	210
?	?	4ft 8½in	0-4-0ST	OC	?	?	[LEIGH]	212
?	?	4ft 8½in	0-6-0ST	OC	?	?	[HARPERLEY]	210, 211

KITSON & CO LTD, Airedale Foundry, Leeds.   K

1508	26.2.1868	4ft 8½in	0-4-0ST	OC	10 x 18	3ft 0½in	373, 374
1705	27.5.1871	4ft 8½in	0-4-0ST	OC	12 x 18	3ft 0in	287, 292
1786	31.10.1871	4ft 8½in	0-6-0ST	IC	13 x 20	3ft 6in	72, 440
1790	7.3.1872	4ft 8½in	0-4-0T	OC	12 x 18	3ft 0½in	289
1843	27.9.1872	4ft 8½in	0-6-0ST	IC	16 x 24	3ft 10in	99
1844	18.10.1872	4ft 8½in	0-6-0ST	IC	16 x 24	3ft 10in	99
1845	28.10.1872	4ft 8½in	0-6-0ST	IC	16 x 24	3ft 10in	99
1998	23.1.1875	4ft 8½in	0-6-0ST	IC	16 x 24	4ft 0in	99
2509	28.7.1883	4ft 8½in	0-6-0PT	IC	17½ x 26	4ft 2in	99
2510	24.8.1883	4ft 8½in	0-6-0PT	IC	17½ x 26	4ft 2in	99
3069	2.12.1887	4ft 8½in	0-6-2T	IC	17½ x 26	4ft 6in	218
3580	22.11.1894	4ft 8½in	0-6-2T	IC	17½ x 26	4ft 6in	218
3905	23.10.1899	4ft 8½in	0-6-0PT	IC	18 x 26	4ft 2⅝in	99
3906	26.10.1899	4ft 8½in	0-6-0PT	IC	18 x 26	4ft 2⅝in	99
4051	12.6.1901	4ft 8½in	0-6-0PT	IC	18 x 26	4ft 2⅝in	99
4263	18.3.1904	4ft 8½in	0-6-2T	IC	19 x 26	4ft 6in	218
4294	27.10.1904	4ft 8½in	0-6-0T	IC	17½ x 26	4ft 6in	210, 224
4532	24.10.1907	4ft 8½in	0-6-2T	IC	19 x 26	4ft 6in	218
4533	24.10.1907	4ft 8½in	0-6-2T	IC	19 x 26	4ft 6in	218
5115	8.5.1914	4ft 8½in	0-4-0ST	OC	14 x 21	3ft 2½in	130, 373
5179	13.6.1917	4ft 8½in	0-6-0PT	IC	18 x 26	4ft 2⅝in	100

KENT CONSTRUCTION & ENGINEERING CO LTD, Ashford, Kent   KC

KERR, STUART & CO LTD, California Works, Stoke-on-Trent, Staffordshire.   KS

1047	25.2.1908	2ft 0in	0-4-2ST	OC	7 x 12	2ft 0in		150
1142	7.2.1911	2ft 0in	0-4-2ST	OC	7 x 12	2ft 0in		150
1144	30.9.1911	2ft 0in	0-4-2ST	OC	7 x 12	2ft 0in		150
1145	22.1.1912	2ft 0in	0-4-2ST	OC	7 x 12	2ft 0in		150
1202	18.5.1911	4ft 8½in	BoBoWE	280hp	#	2ft 9½in	40T	179
1203	30.6.1911	4ft 8½in	BoBoWE	280hp	#	2ft 9½in	40T	179
1291	8.2.1915	2ft 0in	0-4-2ST	OC	7 x 12	2ft 0in		150
2395	23.2.1917	2ft 0in	0-4-2ST	OC	7 x 12	2ft 0in		150
2399	19.9.1917	4ft 8½in	0-4-0T	OC	10 x 16	2ft 3in		435
3074	14.9.1917	4ft 8½in	0-6-0T	OC	17 x 24	4ft 0in		218, 348
3097	14.3.1918	4ft 8½in	0-4-0ST	OC	15 x 20	3ft 3in		277
3098	26.3.1918	4ft 8½in	0-6-0T	OC	15 x 20	3ft 6in		415
3100	8.4.1918	4ft 8½in	0-6-0T	OC	15 x 20	3ft 6in		415

3126	22.10.1918	4ft 8½in	0-4-0ST	OC	15 x 20	3ft 3in	165
4001	20.12.1918	2ft 0in	0-4-0ST	OC	6 x 9	1ft 8in	150
4003	12.1918	2ft 0in	0-4-0ST	OC	6 x 9	1ft 8in	85
4004	12.1918	2ft 0in	0-4-0ST	OC	6 x 9	1ft 8in	85
4027	9.5.1919	4ft 8½in	0-4-0ST	OC	15 x 20	3ft 3in	152, 414
4028	9.5.1919	4ft 8½in	0-4-0ST	OC	15 x 20	3ft 3in	415
4029	20.5.1919	4ft 8½in	0-4-0ST	OC	15 x 20	3ft 3in	277
4030	23.5.1919	4ft 8½in	0-4-0ST	OC	15 x 20	3ft 3in	57, 61
4143	8.12.1919	4ft 8½in	0-4-0ST	OC	15 x 20	3ft 3in	40, 42, 259
4246	31.1.1922	2ft 0in	0-4-0ST	OC	6 x 9	1ft 8in	70, 85
4290	7.2.1923	2ft 0in	0-4-0ST	OC	6 x 9	1ft 8in	150
4291	6.4.1923	2ft 0in	0-4-0ST	OC	6 x 9	1ft 8in	150

\# sub-contracted from Siemens Bros Ltd, Stafford, Staffordshire, which supplied the motors and other electrical components and had to approve the overall KS design.

W. & A. KITCHING, Hopetown, Darlington, Co.Durham.   Kitching

The origin of this works goes back to William Kitching (1752-1819), who in 1790 opened an ironmonger's shop in Tubwell Row, Darlington, and six years later added a small iron foundry. The business subsequently passed to his two sons, William and Alfred. With the opening of the Stockton & Darlington Railway in 1825 they saw the opportunity to expand their business, and in 1831 they moved out to the Hope Town area of the town and set up north of North Road Station a much larger foundry and engineering works to supply the Stockton & Darlington Railway with locomotives and rolling stock. In 1845 Alfred (1808-1882) bought out his brother. In 1860 the Stockton & Darlington Railway purchased this works for their own use, developing it into the North Road Workshops subsequently owned by the NER and its successors.

[S&DR 26]	6.1840	4ft 8½in	0-6-0	OC	14 x 20	4ft 0in	114
[S&DR 25]	24.11.1845	4ft 8½in	0-6-0	OC	15 x 24	4ft 0in	114, 283

R. A. LISTER & CO LTD, Dursley, Gloucestershire.   L

8975	21.4.1937	2ft 6in	4wPM	6hp	JAP	1ft 1in	1½T	396
20882	9.1.1943	2ft 0in	4wPM	6hp	JAP	1ft 1in	1½T	79

LOWCA ENGINEERING CO LTD, Whitehaven, Cumberland.   LE

230	1897	4ft 8½in	0-4-0ST	OC	8½ x 16	?	433
238	1899	4ft 8½in	0-4-0ST	OC	8½ x 16	?	433

STEPHEN LEWIN, Dorset Foundry, Poole, Dorset.   Lewin

?	c12.1874	1ft 10in	0-4-0WT	ICG	?	?	[SAMSON] 230
683	1877	4ft 8½in	0-4-0WT	OC	9 x 18	2ft 6in	234, 348

LINGFORD, GARDINER & CO LTD, Auckland Engine Works, Bishop Auckland, Co. Durham.   LG

This works was established in part of what had previously been the Auckland Iron Works, owned by Thomas Vaughan & Co (which see), sometime in the last twenty years of the 19th century. Besides repairing locomotives and other equipment, the firm was a dealer in second-hand locomotives and also operated a hire service. It would seem unlikely that it ever built a locomotive from scratch, those attributed to it being rebuilds of other builders' locomotives.

WILLIAM LISTER, Hopetown, Darlington, Co. Durham.   Wm. Lister

William Lister set up his business on the opposite side of the Stockton & Darlington Railway to that run by William and Alfred Kitching. Like theirs, his work was mainly for the Stockton & Darlington Railway. In 1859 he was bought out by Alfred Kitching and his cousin, Charles I'Anson, who re-named the works the Whessoe Foundry and greatly expanded its activities (see the entry for Whessoe Products Ltd in the main text).

[VICTORIA]	1838	4ft 8½in	0-4-0	VC	13 x 18	4ft 6in	401

LILLESHALL CO LTD, Oakengates, Shropshire.   Lill

-	1863	3ft 6in	tank loco	?	5 x 12	?	209

JOHN BOWES, ESQ., & PARTNERS, Marley Hill, Sunniside, near Gateshead, Co. Durham.   Marley Hill

This locomotive was built by the firm for use on its Pontop & Jarrow Railway, presumably in the loco shed at Marley Hill.

-	1854	4ft 8½in	0-4-0ST	IC		[DANIEL O'ROURKE] 57

METROPOLITAN-CAMMELL LTD, Saltley Works, Saltley, Birmingham. **MetCam**

| [LUL 3721] | 1983 | 4ft 8½in | 4w-4wRER | ? | | ? | | 394 |

MUIR HILL (ENGINEERS) LTD, Trafford Park, Manchester. **MH**

| L116 | 19.9.1939 | 4ft 8½in | 4wPM | 30hp | Fordson | ? | 6½T | 239 |

MILLER & CO LTD, Coatbridge, Lanarkshire, Scotland. **Miller**

| 932? | 1893? | 3ft 0in | 0-4-0ST | OC | 7 x 10 | ? | | 417 |

MOTOR RAIL LTD, Simplex Works, Bedford, Bedfordshire. **MR**
Motor Rail & Tram Car Co Ltd until 1931

288	2.3.1917	600mm	4wPM	20hp	Dorman 2JO	1ft 6in	2½T	144
429	24.8.1917	600mm	4wPM	40hp	Dorman 4JO	1ft 6in	6¾T	85, 300
866	7.5.1918	600mm	4wPM	20hp	Dorman 2JO	1ft 6in	2½T	144
903	31.5.1918	600mm	4wPM	20hp	Dorman 2JO	1ft 6in	2½T	365
914	10.6.1918	600mm	4wPM	20hp	Dorman 2JO	1ft 6in	2½T	164
1016	17.9.1918	600mm	4wPM	20hp	Dorman 2JO	1ft 6in	2½T	144
1216	2.12.1918	600mm	4wPM	20hp	Dorman 2JO	1ft 6in	2½T	144
1728	23.7.1918	600mm	4wPM	20hp	Dorman 2JO	1ft 6in	2½T	144
1792	24.9.1918	600mm	4wPM	20hp	Dorman 2JO	1ft 6in	2½T	144
1951	14.5.1920	4ft 8½in	4wPM	40hp	Dorman 4JO	3ft 1in	8T	69, 435
2027	4.9.1920	4ft 8½in	4wPM	40hp	Dorman 4JO	3ft 1in	8T	388
2077	1923	2ft 0in	4wPM	20hp	Dorman 2JO	1ft 6in	2½T	229
2096	28.6.1922	4ft 8½in	4wPM	40hp	Dorman 4JO	1ft 6in	8T	69, 144, 165
2103	8.4.1921	2ft 0in	4wPM	20hp	Dorman 2JO	1ft 6in	2½T	229
2104	13.4.1921	2ft 0in	4wPM	20hp	Dorman 2JO	1ft 6in	2½T	229
2195	28.12.1922	600mm	4wPM	20hp	Dorman 2JO	1ft 6in	2½T	164
2262	12.9.1929	4ft 8½in	4wPM	65hp	Dorman	1ft 9in	15T	388
3833	15.10.1926	600mm	4wPM	20hp	Dorman 2JO	1ft 6in	2½T	70, 435
3869	2.11.1928	2ft 0in	4wPM	20hp	Dorman 2JO	1ft 6in	2½T	39
4045	15.10.1926	2ft 0in	4wPM	20hp	Dorman 2JO	1ft 6in	2½T	326
4577	16.4.1930	600mm	4wPM	20hp	Dorman 2JO	1ft 6in	2½T	70, 435
5067	7.5.1930	2ft 0in	4wPM	20hp	Dorman 2JO	1ft 6in	4T	146, 150
5232	15.11.1930	2ft 0in	4wPM	20/35hp	Dorman 4MRX	?	2½T	165
5751	26.1.1934	4ft 8½in	4wPM	65/85hp	Dorman 4DL	?	20T	150, 388
7604	18.4.1939	2ft 0in	4wDM	16/24hp	Ailsa Craig	?	2½T	146
7605	17.4.1939	2ft 0in	4wDM	16/24hp	Ailsa Craig	?	2½T	146
7606	17.4.1939	2ft 0in	4wDM	16/24hp	Ailsa Craig	?	2½T	146
8717	21.3.1941	2ft 0in	4wDM	20/28hp	Dorman 2DWD	1ft 6in	2½T	166, 241
8725	4.4.1941	2ft 0in	4wDM	20/28hp	Dorman 2DWD	1ft 6in	2½T	264
8747	14.2.1942	2ft 0in	4wDM	20/28hp	Dorman 2DWD	1ft 6in	2½T	247
8931	9.11.1944	600mm	4wDM	20/28hp	Dorman 2DWD	1ft 6in	2½T	166, 241
8995	18.3.1946	2ft 0in	4wDM	20/28hp	Dorman 2DWD	1ft 6in	2½T	166, 241
9103	11.2.1941	2ft 0in	4wDM	20/26hp	Dorman 2JO	1ft 6in	2½T	162
40S273	16.8.1966	2ft 0in	4wDM	40hp	Dorman 2LB	?	2½T	247

Unidentified locomotive

| ? | ? | 2ft 0in | 4wPM | ? | ? | ? | ? | 326 |

MANNING, WARDLE & CO LTD, Boyne Engine Works, Leeds. **MW**

							Maker's class	
17	30.11.1860	4ft 8½in	0-6-0ST	IC	11 x 17	3ft 1½in	old I	306
23	21.3.1861	4ft 8½in	0-6-0ST	IC	11 x 17	3ft 1½in	old I	181
49	27.5.1862	4ft 8½in	0-6-0ST	IC	12 x 17	3ft 1½in	old I	292
57	5.11.1862	4ft 8½in	0-6-0ST	IC	11 x 17	3ft 1½in	old I	306
60	28.3.1862	4ft 8½in	0-4-0ST	OC	9 x 14	2ft 9in	E	73
87	27.6.1863	4ft 8½in	0-4-0ST	OC	9 x 14	2ft 9in	E	396
104	27.1.1864	4ft 8½in	0-6-0ST	IC	13 x 18	3ft 0in	M	379, 382
112	15.4.1864	4ft 8½in	0-4-0ST	OC	6 x 12	2ft 6in	B	98, 352
126	30.1.1864	4ft 8½in	0-6-0ST	OC	12 x 17	3ft 1½in	K	270
144	27.2.1865	4ft 8½in	0-4-0ST	OC	9½ x 14	2ft 6in	E	291
148	14.3.1865	4ft 8½in	0-6-0ST	IC	13 x 18	3ft 0in	M	159, 239
152	9.5.1865	4ft 8½in	0-6-0ST	IC	13 x 18	3ft 0in	M	217

153	8.8.1865	4ft 8½in	0-6-0T	IC	15 x 22	4ft 0in	WYT	126, 129, 134
171	2.10.1865	4ft 8½in	0-6-0ST	IC	12 x 17	3ft 1½in	K	199
194	13.6.1866	4ft 8½in	0-6-0ST	IC	13 x 18	3ft 0in	M	46, 49
199	7.6.1866	4ft 8½in	0-4-0ST	OC	9½ x 14	2ft 9in	E	336
200	6.2.1867	4ft 8½in	0-6-0ST	IC	13 x 18	3ft 0in	M	379, 382
241	4.1.1868	4ft 8½in	0-6-0ST	IC	11 x 17	3ft 1½in	old l	159
242	31.8.1867	4ft 8½in	0-6-0ST	IC	14 x 18	3ft 0in	M	181
270	1.12.1869	4ft 8½in	0-4-0ST	OC	6 x 12	2ft 0in	A	396
274	2.4.1870	4ft 8½in	0-6-0ST	IC	12 x 17	3ft 1½in	K	337
320	13.9.1870	4ft 8½in	0-4-0ST	OC	12 x 18	3ft 6in	H	379, 382, 383
324	9.2.1871	4ft 8½in	0-6-0ST	IC	13 x 18	3ft 0in	M	357
341	24.5.1871	4ft 8½in	0-6-0ST	IC	13 x 18	3ft 0in	M	340
344	21.4.1871	4ft 8½in	0-4-0ST	OC	12 x 18	3ft 0in	H	218
416	21.1.1872	4ft 8½in	0-4-0ST	OC	12 x 18	3ft 0in	H	55
447	10.7.1873	4ft 8½in	0-4-0ST	OC	12 x 18	3ft 0in	H	358
455	20.8.1874	4ft 8½in	0-4-0ST	OC	10 x 16	2ft 9in	F	80, 323
465	15.8.1873	4ft 8½in	0-4-0ST	OC	12 x 18	3ft 0in	H	334
466	8.9.1873	4ft 8½in	0-4-0ST	OC	12 x 18	3ft 0in	H	393, 415
467	22.9.1873	4ft 8½in	0-4-0ST	OC	12 x 18	3ft 0in	H	143, 253, 258, 259, 374
479	3.2.1874	4ft 8½in	0-6-0ST	IC	15 x 22	3ft 9in	O	181, 182
480	20.2.1874	4ft 8½in	0-6-0ST	IC	15 x 22	3ft 9in	O	338
492	20.4.1874	4ft 8½in	0-4-0ST	OC	12 x 18	3ft 0in	H	393,415
498	18.5.1874	4ft 8½in	0-4-0ST	OC	12 x 18	3ft 0in	H	253, 292
540	29.9.1875	4ft 8½in	0-4-0ST	OC	12 x 18	3ft 0in	H	253, 433
569	28.9.1875	4ft 8½in	0-6-0ST	IC	15 x 22	3ft 9in	O	338, 339
573	25.11.1875	4ft 8½in	0-4-0ST	OC	12 x 18	3ft 0in	H	294
575	9.8.1876	4ft 8½in	0-4-0ST	OC	12 x 18	3ft 0in	H	424
693	2.4.1878	4ft 8½in	0-6-0ST	IC	15 x 22	4ft 0in	O	357
697	7.5.1878	4ft 8½in	0-6-0ST	IC	15 x 22	3ft 9in	O	361
744	25.3.1880	4ft 8½in	0-4-0ST	OC	12 x 18	3ft 0in	H	291
756	5.10.1880	4ft 8½in	0-4-0ST	OC	12 x 18	3ft 0in	H	139, 227
758	7.3.1881	4ft 8½in	0-6-0ST	IC	15 x 22	3ft 9in	O	361
775	16.2.1882	4ft 8½in	0-6-0ST	IC	13 x 18	3ft 0in	M	379
777	4.2.1881	4ft 8½in	0-4-0ST	OC	12 x 18	3ft 6in	H	38
813	22.12.1881	4ft 8½in	0-4-0ST	OC	8 x 14	2ft 8in	D	139
838	10.7.1885	4ft 8½in	0-4-0ST	OC	10 x 18	3ft 1in	10in Spl	348
881	20.3.1884	4ft 8½in	0-4-0ST	OC	9 x 14	2ft 9in	E	204
926	25.10.1886	4ft 8½in	0-6-0ST	IC	13 x 18	3ft 0in	M	288
1037	8.11.1887	4ft 8½in	0-4-0ST	OC	10 x 16	2ft 9in	F	273
1042	25.1.1888	4ft 8½in	0-4-0ST	OC	12 x 18	3ft 0in	H	310
1138	2.8.1889	4ft 8½in	0-6-0ST	IC	15 x 22	3ft 6in	15in Spl	51
1150	16.3.1891	4ft 8½in	0-4-0ST	OC	10½ x 16	2ft 9in	H	365
1323	22.9.1896	4ft 8½in	0-4-0ST	OC	10 x 16	2ft 9in	F	426
1327	19.3.1897	4ft 8½in	0-4-0ST	OC	12 x 18	3ft 0in	H	354
1328	29.3.1898	4ft 8½in	0-4-0ST	OC	12 x 18	3ft 0in	H	127, 130, 139
1422	9.1.1899	4ft 8½in	0-4-0ST	OC	12 x 18	3ft 0in	H	126, 130, 132, 135
1469	15.1.1900	4ft 8½in	0-6-0ST	IC	16 x 22	3ft 6in	16in Spl	52, 130
1513	2.9.1901	4ft 8½in	0-6-0ST	IC	12 x 18	3ft 0in	L	385
1517	19.12.1900	4ft 8½in	0-4-0ST	OC	12 x 18	3ft 0in	H	126, 127, 130, 132, 139
1557	5.9.1902	4ft 8½in	0-4-0ST	OC	12 x 18	3ft 0in	H	166
1566	24.6.1902	4ft 8½in	0-6-0ST	IC	13 x 18	3ft 0in	M	288
1602	30.4.1903	4ft 8½in	0-6-0ST	OC	15 x 22	3ft 6in	15in Spl	249
1658	30.6.1905	4ft 8½in	0-4-0ST	OC	12 x 18	3ft 0in	H	127
1659	17.7.1905	4ft 8½in	0-4-0ST	OC	13½ x 18	3ft 0in	alt l	165
1664	25.8.1905	4ft 8½in	0-6-0ST	IC	13 x 18	3ft 0in	M	165
1669	30.10.1905	4ft 8½in	0-6-0ST	IC	13 x 18	3ft 0in	M	323

1691	10.7.1907	4ft 8½in	0-6-0ST	IC	12 x 18	3ft 0in	L		183
1697	10.9.1906	4ft 8½in	0-4-0ST	OC	12 x 18	3ft 0in	H		126, 127
									132, 135
1813	7.4.1913	4ft 8½in	0-6-0T	IC	18 x 24	4ft 2in	18in Spl		218
1880	13.4.1915	4ft 8½in	0-4-0ST	OC	16 x 24	3ft 8in	16in Spl		379,
									382, 435
1887	22.9.1915	4ft 8½in	0-4-0ST	OC	14 x 18	3ft 1in	14in Spl		434
1903	25.7.1916	4ft 8½in	0-4-0ST	OC	14 x 18	3ft 6in	P alt		38
1911	25.1.1917	4ft 8½in	0-4-0ST	OC	14 x 18	3ft 1in	14in Spl		435
1934	29.9.1917	4ft 8½in	0-6-0ST	OC	17 x 24	3ft 9in	17in Spl		218, 348
2023	9.4.1923	4ft 8½in	0-4-0ST	OC	15 x 22	3ft 9in	15in Spl		218
2025	28.6.1923	4ft 8½in	0-6-0ST	IC	12 x 18	3ft 0½in	12in Spl		372
2035	19.5.1924	4ft 8½in	0-4-0ST	OC	15 x 20	3ft 9in	15in Spl		218
2036	28.5.1924	4ft 8½in	0-4-0ST	OC	15 x 20	3ft 9in	15in Spl		218

Unidentified locomotives

1	?	4ft 8½in	?	IC	?	?	?	176
[11]	?	4ft 8½in	0-4-0ST	OC	?	?	?	177
[HENRIETTA]	?	4ft 8½in	0-4-0ST	OC	?	?	?	126
MOSTYN	?	4ft 8½in	0-4-0ST	OC	?	?	?	40, 42, 256, 327
VICTORIA No.1	?	4ft 8½in	0-4-0T	OC	?	?	?	320

NEILSON & CO, Hyde Park Works, Springburn, Glasgow, Scotland.   **N**

2280	8.10.1877	4ft 8½in	0-4-0ST	OC	14 x 20	3ft 8in	414

Unidentified locomotive

?	?	4ft 8½in	0-4-0ST	OC	?	?	[BEACONSFIELD] 60

NESHAM & WELSH, Portrack Lane Iron Works, Stockton-on-Tees, Co. Durham.   **N&W**

David Nesham and Humphrey Welsh had entered into partnership and established this works by 1837. They seem to have built fewer than half a dozen locomotives and probably went out of business in the mid 1840s.

-	1837	4ft 8½in	0-6-0	OC	?	?	[EXILE] 186
-	1839	4ft 8½in	0-6-0?	OC?	?	?	391

See also the entry for the Hartlepool Dock & Railway Company in the main text.

NORTH BRITISH LOCOMOTIVE CO LTD, Glasgow, Scotland.

A	North British Locomotive Co Ltd, Atlas Works, Glasgow.	**NBA**
Q	North British Locomotive Co Ltd, Queen's Park Works, Glasgow.	**NBQ**

A	16628	2.1905	4ft 8½in	0-6-0ST	IC	18 x 26	4ft 3in		58
Q	21522	1.1917	4ft 8½in	0-4-0CT	OC	12 x 22	3ft 6in	5T capy.	105
Q	27874	by 5.6.1958	4ft 8½in	0-4-0DH	330hp	MAN W6V	3ft 9in	40T	390

NEW LOWCA ENGINEERING CO LTD, Whitehaven, Cumberland.   **NLE**

249	1908	4ft 8½in	0-6-0PT	IC	18 x 26	4ft 2⅝in	100

NEILSON, REID & CO LTD, Hyde Park Works, Springburn, Glasgow, Scotland.   **NR**

5408	5.1899	4ft 8½in	0-6-2T	IC	17½ x 26	4ft 6½in	218

PECKETT & SONS LTD, Atlas Works, Bristol.   **P**

maker's class

452	5.3.1889	4ft 8½in	0-4-0ST	OC	10 x 14	2ft 9in	M3	204
521	28.7.1891	4ft 8½in	0-4-0ST	OC	14 x 20	3ft 2in	W4	415
525	18.10.1892	4ft 8½in	0-6-0ST	OC	14 x 20	3ft 7in	B1	153
560	1.2.1893	4ft 8½in	0-4-0ST	OC	14 x 20	3ft 2in	W4	153
583	13.12.1894	4ft 8½in	0-4-0ST	OC	14 x 20	3ft 2in	W4	288
615	20.5.1896	4ft 8½in	0-4-0ST	OC	14 x 20	3ft 2in	W4	218, 220, 265
629	29.4.1896	4ft 8½in	0-6-0ST	IC	16 x 22	3ft 10in	X	51, 128, 134
634	24.2.1897	4ft 8½in	0-4-0ST	OC	12 x 18	3ft 0½in	R1	227

644	9.11.1896	4ft 8½in	0-4-0ST	OC	10 x 14	2ft 6½in	M4	240	
669	28.7.1897	4ft 8½in	0-4-0ST	OC	14 x 20	3ft 2in	W4	126, 130, 134, 135	
677	24.8.1897	4ft 8½in	0-4-0ST	OC	12 x 18	3ft 0½in	R1	252, 256	
703	25.1.1899	4ft 8½in	0-4-0ST	OC	14 x 20	3ft 2in	W4	142	
774	23.2.1899	4ft 8½in	0-6-0ST	IC	16 x 22	3ft 10in	X	298	
845	13.9.1900	4ft 8½in	0-4-0ST	OC	14 x 20	3ft 2in	W4	227	
880	14.5.1901	4ft 8½in	0-4-0ST	OC	14 x 20	3ft 2in	W4	227	
916	16.10.1901	4ft 8½in	0-4-0ST	OC	12 x 18	3ft 0½in	R1	53, 54, 137, 326	
971	9.9.1903	4ft 8½in	0-4-0ST	OC	14 x 20	3ft 2in	W4	351	
1040	28.12.1905	4ft 8½in	0-6-0ST	IC	16 x 22	3ft 10in	X	292	
1052	4.9.1905	4ft 8½in	0-6-0ST	OC	14 x 20	3ft 7in	B2	348	
1058	30.7.1906	4ft 8½in	0-4-0ST	OC	14 x 20	3ft 2in	W4	38, 227	
1083	12.2.1906	4ft 8½in	0-6-0ST	OC	15 x 21	3ft 7in	F	348	
1092	13.7.1906	4ft 8½in	0-6-0ST	OC	14 x 20	3ft 7in	B2	288, 291	
1099	21.10.1907	4ft 8½in	0-4-0ST	OC	12 x 18	3ft 0½in	R1	310	
1180	# 3.5.1912	4ft 8½in	0-4-0ST	OC	15 x 21	3ft 7in	E	80	
1194	19.2.1912	4ft 8½in	0-4-0ST	OC	14 x 20	3ft 2½in	W5	374	
1210	25.8.1910	4ft 8½in	0-4-0ST	OC	12 x 18	3ft 0½in	R2	424	
1219	26.6.1910	4ft 8½in	0-6-0ST	IC	16 x 22	3ft 10in	X2	286, 373	
1310	6.5.1914	4ft 8½in	0-6-0ST	OC	14 x 20	3ft 7in	B2	200	
1337	27.10.1913	4ft 8½in	0-4-0ST	OC	12 x 18	3ft 0½in	R2	252, 256, 327	
1392	12.4.1915	4ft 8½in	0-4-0ST	OC	15 x 21	3ft 7in	E	277	
1403	28.1.1916	4ft 8½in	0-6-0ST	OC	14 x 20	3ft 7in	B2	337	
1413	2.12.1915	4ft 8½in	0-4-0ST	OC	15 x 21	3ft 7in	E	277	
1423	10.1.1916	4ft 8½in	0-6-0ST	IC	16 x 22	3ft 10in	X2	374	
1430	26.7.1916	4ft 8½in	0-4-0ST	OC	12 x 18	3ft 0½in	R2	84	
1455	22.7.1918	4ft 8½in	0-6-0ST	OC	14 x 20	3ft 7in	B2	337	
1460	15.12.1916	4ft 8½in	0-4-0ST	OC	12 x 18	3ft 0½in	R2	242, 334, 401	
1467	9.7.1917	4ft 8½in	0-4-0ST	OC	14 x 20	3ft 2½in	W5	282	
1468	18.7.1917	4ft 8½in	0-4-0ST	OC	14 x 20	3ft 2½in	W5	204	
1508	4.9.1918	4ft 8½in	0-4-0ST	OC	14 x 20	3ft 2½in	W5	163	
1544	7.10.1919	4ft 8½in	0-4-0ST	OC	12 x 18	3ft 0½in	R2	325	
1589	@ 29.3.1928	4ft 8½in	0-4-0ST	OC	15 x 21	3ft 7in	E	306	
1616	27.4.1923	4ft 8½in	0-6-0ST	OC	14 x 20	3ft 7in	B2	312	
1637	10.9.1923	4ft 8½in	0-4-0ST	OC	12 x 18	3ft 0½in	R2	325	
1648	29.11.1923	4ft 8½in	0-4-0ST	OC	12 x 18	3ft 0½in	R2	325	
1671	20.8.1924	3ft 0in	0-4-0T	OC	8 x 12	2ft 3in	"1287"	112	
1748	7.6.1928	4ft 8½in	0-4-0ST	OC	14 x 22	3ft 2½in	W6	298	
1761	17.4.1929	4ft 8½in	0-4-0ST	OC	14 x 22	3ft 2½in	W6	306	
1952	18.7.1938	4ft 8½in	0-6-0ST	OC	16 x 24	3ft 10in	OX2	312	
1953	7.9.1938	4ft 8½in	0-6-0ST	OC	16 x 24	3ft 10in	OX2	312	
1954	19.10.1938	4ft 8½in	0-6-0ST	OC	16 x 24	3ft 10in	OX2	312	
1955	15.12.1938	4ft 8½in	0-6-0ST	OC	16 x 24	3ft 10in	OX2	312	
2016	15.10.1941	4ft 8½in	0-4-0ST	OC	14 x 22	3ft 2½in	W7	245, 260, 261	
2032	12.11.1942	4ft 8½in	0-4-0ST	OC	14 x 22	3ft 2½in	W7	245, 367	
2042	29.3.1943	4ft 8½in	0-4-0ST	OC	14 x 22	3ft 2½in	W7	245, 260, 261	
2049	8.1944	4ft 8½in	0-4-0ST	OC	12 x 20	3ft 0½in	R4	142	
2093	22.9.1947	4ft 8½in	0-4-0ST	OC	14 x 22	3ft 2½in	W7	299	
2142	4.1953	4ft 8½in	0-4-0ST	OC	14 x 22	3ft 2½in	Spl W7	262	

\#   Peckett records suggest that this locomotive was built in 1909.
@   Peckett records suggest that this locomotive was built in 1922.

G. RENNOLDSON, South Shields, Co. Durham.   Rennoldson

George Rennoldson was originally a millwright. In 1826 he set up a shipyard at Wapping Street, subsequently adding an engineering works, where allegedly he built a number of locomotives, though the works had no railway connection. Subsequently the family concentrated on ship-building, notably of tugs. The company of J.P. Rennoldson & Sons Ltd lasted until 1929.

	11.1837	4ft 8½in?	?	?		?	?	330

RUSTON & HORNSBY LTD, Lincoln, Lincolnshire.   **RH**

					maker's class		weight	
114563	7.1923	2ft 0in	4wPM	21hp	ZRH	?	4T	32
166012	30.11.1932	2ft 0in	4wDM	16hp	16HP	?	3T	151
170373	19.9.1934	2ft 0in	4wDM	16hp	16/21HP	?	3T	369
175121	18.1.1937	2ft 0in	4wDM	21hp	16/21HP	?	3T	381
175399	17.11.1936	2ft 0in	4wDM	21hp	16/21HP	?	3T	381
175420	4.3.1936	2ft 4in	4wDM	21hp	16/21HP	?	3T	144, 384 388, 423
177535	28.3.1936	2ft 0in	4wDM	40hp	36/42HP	?	5T	352
177640	20.7.1936	2ft 0in	4wDM	10hp	16/21HP	?	2½T	151
186322	12.8.1937	2ft 0in	4wDM	20hp	16/20HP	1ft 3in	3T	146
186342	31.8.1937	2ft 0in	4wDM	20hp	16/20HP	1ft 3in	3T	146
187059	25.10.1937	2ft 0in	4wDM	40hp	33/40HP	1ft 8¾in	5T	179, 352
189959	11.3.1938	2ft 0in	4wDM	40hp	33/40HP	1ft 8¾in	5T	179, 352
189963	4.3.1939	2ft 0in	4wDM	40hp	33/40HP	1ft 8¾in	5T	179, 352
195844	6.4.1939	2ft 0in	4wDM	20hp	16/20HP	1ft 3in	3T	146
195849	29.4.1939	2ft 0in	4wDM	20hp	16/20HP	1ft 3in	3T	146
198325	16.12.1940	4ft 8½in	4wDM	88hp	80/88HP	3ft 0in	17T	262, 418
207102	26.3.1941	4ft 8½in	4wDM	48hp	44/48HP	2ft 6in	7½T	301
211641	2.1.1942	2ft 0in	4wDM	20hp	20DL	1ft 4⁵⁄₁₆	3T	243
211683	11.10.1941	2ft 0in	4wDM	30hp	30DL	1ft 4⁵⁄₁₆	3T	368
213836	18.7.1942	2ft 0in	4wDM	20hp	20DL	1ft 4⁵⁄₁₆	3T	264, 301
221642	27.3.1944	4ft 8½in	4wDM	48hp	48DS	2ft 6in	7½T	83
223716	21.1.1944	2ft 0in	4wDM	20hp	20DL	1ft 4⁵⁄₁₆	3T	157
226268	8.6.1944	2ft 0in	4wDM	20hp	20DL	1ft 4⁵⁄₁₆	3T	208
226278	4.7.1944	2ft 0in	4wDN	20hp	20DL	1ft 4⁵⁄₁₆	3T	243
236362	23.4.1946	4ft 8½in	4wDM	88hp	88DS	3ft 0in	20T	125, 388
243081	19.2.1948	4ft 8½in	0-4-0DM	165hp	165DS	3ft 2½in	28T	306
247174	6.1947	2ft 0in	4wDM					115
262996	24.1.1949	4ft 8½in	4wDM	88hp	88DS	3ft 0in	20T	326, 388
265615	13.8.1948	4ft 8½in	4wDM	48hp	48DS	2ft 6in	7½T	36
275881	27.6.1949	4ft 8½in	4wDM	88hp	88DS	3ft 0in	17T	332
280865	3.6.1949	2ft 0in	4wDM	18hp	20DLU	1ft 4⁵⁄₁₆	3T	207
280866	15.6.1949	2ft 0in	4wDM	18hp	20DLU	1ft 4⁵⁄₁₆	3T	207
281270	1.2.1951	4ft 8½in	0-4-0DM	165hp	165DS	3ft 2½in	28T	61, 185, 390
287662	23.6.1950	2ft 0in	4wDM	40hp	40DL	1ft 8¾in	5T	352
294263	30.3.1950	4ft 8½in	4wDM	48hp	48DS	2ft 6in	7½T	310, 312
304471	16.4.1951	4ft 8½in	0-4-0DM	165hp	165DS	3ft 2½in	28T	306
304472	12.4.1951	4ft 8½in	0-4-0DM	165hp	165DS	3ft 2½in	28T	185, 387
305302	13.8.1951	4ft 8½in	4wDM	48hp	48DS	2ft 6in	7½T	332
305303	28.12.1951	4ft 8½in	4wDM	48hp	48DS	2ft 6in	7½T	335
305320	2.2.1951	4ft 8½in	4wDM	88hp	88DS	3ft 0in	17T	262
305323	6.6.1951	4ft 8½in	4wDM	88hp	88DS	3ft 0in	17T	155, 202
306087	1.12.1949	4ft 8½in	4wDM	88hp	88DS	3ft 0in	20T	388
312988	19.9.1952	4ft 8½in	0-4-0DE	155hp	165DE	3ft 2½in	28T	425
319288	3.2.1953	4ft 8½in	0-4-0DM	165hp	165DS	3ft 2½in	28T	185, 387
323600	2.3.1953	4ft 8½in	0-4-0DE	155hp	165DE	3ft 2½in	28T	312
326062	10.11.1952	4ft 8½in	4wDM	88hp	88DS	3ft 0in	20T	183
326071	2.4.1953	4ft 8½in	4wDM	88hp	88DS	3ft 0in	17T	183, 326, 388
327969	17.8.1954	4ft 8½in	0-4-0DM	165hp	165DS	3ft 2½in	28T	61, 83, 306, 390
338424	3.2.1955	4ft 8½in	4wDM	88hp	88DS	3ft 0in	17T	63
349087	18.10.1954	4ft 8½in	0-4-0DE	155hp	165DE	3ft 2½in	28T	312
354013	1.7.1953	1ft 8in	4wDM	20hp	20DL	1ft 4⁵⁄₁₆	3T	83
375360	13.1.1955	1ft 8in	4wDM	20hp	LAT	1ft 4⁵⁄₁₆	3½T	83
375696	12.7.1954	2ft 0in	4wDM	20hp	LBT	1ft 4⁵⁄₁₆	4½T	301
381751	2.2.1955	4ft 8½in	0-4-0DE	155hp	165DE	3ft 2½in	28T	76, 312, 390
381752	11.2.1955	4ft 8½in	0-4-0DE	155hp	165DE	3ft 2½in	28T	312
381753	7.6.1955	4ft 8½in	0-4-0DE	155hp	165DE	3ft 2½in	28T	312

381755	12.8.1955	4ft 8½in	0-4-0DE	155hp	165DE	3ft 2½in	28T		312
395294	3.4.1956	4ft 8½in	0-4-0DE	155hp	165DE	3ft 2½in	28T		306
395303	17.9.1956	4ft 8½in	0-6-0DM	165hp	165DS	3ft 2½in	28T		390
402428	15.2.1956	1ft 8in	4wDM	20hp	LAT	1ft 4$^{5/16}$	3½T		83
412707	28.4.1957	4ft 8½in	0-4-0DE	155hp	165DE	3ft 2½in	28T		76, 78
412714	17.7.1957	4ft 8½in	0-4-0DE	155hp	165DE	3ft 2½in	28T		78
416210	4.10.1957	4ft 8½in	0-4-0DE	155hp	165DE	3ft 2½in	28T		306
418600	10.1.1958	4ft 8½in	0-4-0DE	155hp	165DE	3ft 2½in	28T		312
421417	8.4.1958	4ft 8½in	4wDM	88hp	88DS	3ft 0in	20T		423
425485	4.12.1958	4ft 8½in	4wDM	88hp	88DS	3ft 0in	17T		63
432477	5.1.1959	4ft 8½in	4wDM	88hp	88DS	3ft 0in	17T		63
432480	18.3.1959	4ft 8½in	4wDM	88hp	88DS	3ft 0in	17T		418
443644	17.2.1961	4ft 8½in	4wDM	48hp	48DS	2ft 6in	7½T		301
463152	31.5.1961	4ft 8½in	4wDM	88hp	88DS	3ft 0in	17T		63
476124	9.10.1962	1ft 8in	4wDM	31½hp	LBT	1ft 4$^{5/16}$	3½T		83
476140	12.6.1963	4ft 8½in	4wDM	88hp	88DS	3ft 0in	17T		62, 262

THOMAS D RIDLEY & SONS, North Ormesby, Middlesbrough, Yorkshire (NR). **Ridley**

2	?	4ft 8½in	0-4-0ST	OC	?	?	[MABEL]	270
74	1920	4ft 8½in	0-4-0ST	OC	?	?		288, 291
76	1920	4ft 8½in	0-4-0ST	OC	?	?		291

Unidentified locomotive

?	?	4ft 8½in	0-4-0ST	OC	?	?	374

ROAD MACHINES (Drayton) LTD, West Drayton, Middlesex, and later at Iver, Buckinghamshire. **RM**
All below were monorail powered wagons.

4904	1956	Monorail	1w1PM		BSA 420cc		266, 267
5124	1957	Monorail	1w1PM		Hirth		394
9792	1960	Monorail	2wPH		JAP 5B		45
11451	1963	Monorail	1w1PH		?		266
11836	1963	Monorail	2wPH		JAP 4½hp		266
12330	1964	Monorail	1w1PH		JAP 5B		267
12438	1964	Monorail	1w1PH		JAP 4½hp		266
13629	1965	Monorail	2wPH		JAP 4/44		267
14262	15.12.1965	Monorail	2wGasH		JAP 5E		108
14753	1966	Monorail	2wPH		JAP 4/44		266, 267
14798	19.10.1966	Monorail	2w-H		?		108

ROLLS ROYCE LTD, Sentinel Works, Shrewsbury, Shropshire. **RR**
In 1964 this firm took over its hitherto-subsidiary company Sentinel (Shrewsbury) Ltd, which had built diesel locomotives to Rolls Royce design. The marketing of these until the Sentinel name was then replaced with Rolls Royce, although the works numbers continued in the Sentinel series; see under the Sentinel entry.

ROBERT STEPHENSON & CO LTD, Newcastle upon Tyne and Darlington, Co. Durham. **RS**
This firm was founded in June 1823 with Robert Stephenson, the son of George Stephenson, as managing partner, with some prominent Darlington Quakers, already involved with the Stockton & Darlington Railway, as directors. At first its works was at South Street, Newcastle upon Tyne, but it was soon extended to Forth Banks. The company's reputation was always firmly based as a manufacturer of main line stock, and the majority of its products in industrial service were ordered new by the major colliery railways or obtained second-hand from the North Eastern Railway. In 1902 the firm removed from its cramped and rather obsolete site at Newcastle to a large new works on a 'green field' site at Darlington, and thereafter industrial orders were few. Most of the former premises at Newcastle were absorbed into the works of their neighbours, R. & W. Hawthorn, Leslie & Co Ltd. The firm also operated a ship-yard at Hebburn, Co. Durham, which it sold in 1912.

In May 1937 the company acquired the locomotive-building business of R. & W. Hawthorn, Leslie & Co Ltd, and on 27th September 1937 the combined business was entitled Robert Stephenson & Hawthorns Ltd (which see). The firm had built 4155 locomotives.

Following the closure of the Newcastle works in 1960, some of the Robert Stephenson original works buildings survived, including the former offices in South Street. These have now been taken over by the Robert Stephenson Trust, which is developing them as a monument to both the great engineer and his works, and opens them to the public.

In the list below, all locomotives built after 1901 were constructed at Darlington.

?	12.4.1826	4ft 8½in	0-4-0	VC	10 x 24	4ft 0in	[No.1]	56, 329
?	26.4.1826	4ft 8½in	0-4-0	VC	10 x 24	4ft 0in	[No.2]	57, 329
?	9.1825	4ft 8½in	0-4-0	VC	10 x 24	4ft 0in	[LOCOMOTION]	283
?	5.1826	4ft 8½in	0-4-0	VC	10 x 24	4ft 0in	[DILIGENCE]	361
16	1829	4ft 8½in	0-6-0	OC	9 x 24	4ft 0in		438
491	22.12.1845	4ft 8½in	2-4-0	IC	14 x 22	5ft 7¼in		218
624	24.7.1848	4ft 8½in	0-6-0	IC	18 x 24	5ft 0in		361
625	4.9.1848	4ft 8½in	0-6-0	IC	18 x 24	5ft 0in		361
753	20.9.1849	4ft 8½in	0-4-0	OC	14 x 22	4ft 5in		234
795	4.7.1851	4ft 8½in	0-4-0	OC	14 x 22	4ft 4½in		57
816	5.12.1851	4ft 8½in	0-4-0	OC	14 x 22	5ft 0¾in		57
1073	8.12.1856	4ft 8½in	0-6-0	IC	16¾ x 24	4ft 6in		234
1074	10.9.1856	4ft 8½in	0-6-0	IC	16 x 24	4ft 6in		57
1075	29.9.1856	4ft 8½in	0-6-0	IC	15½ x 22	4ft 6in		234
1085	29.10.1857	4ft 8½in	2-4-0T	OC	?	?		98
1096	31.8.1857	4ft 8½in	0-6-0	IC	15½ x 22	5ft 0in		234
1100	19.8.1857	4ft 8½in	2-4-0	OC	15 x 24	4ft 7½in		190, 220
1206	5.8.1859	4ft 8½in	4-4-0	OC	16 x 22	5ft 0in		234
1217	17.11.1859	4ft 8½in	0-6-0	IC	16¾ x 24	4ft 6in		234
1313	17.8.1860	4ft 8½in	0-6-0	IC	16 x 24	4ft 6in		58
1326	17.10.1860	4ft 8½in	0-6-0	IC	16¾ x 24	4ft 6in		234
1327	5.11.1860	4ft 8½in	0-6-0	IC	16¾ x 24	4ft 6in		234
1416	2.4.1862	4ft 8½in	0-6-0	IC	16¾ x 24	4ft 6in		234
1417	4.4.1862	4ft 8½in	0-6-0	IC	16¾ x 24	4ft 6in		234
1516	13.5.1864	4ft 8½in	0-6-0	IC	16 x 24	4ft 6in		57
1611	1.9.1864	4ft 8½in	0-6-0	IC	16 x 24	4ft 6in		58
1612	10.6.1864	4ft 8½in	0-6-0ST	IC	13 x 18	3ft 6in		58
1619	30.1.1865	4ft 8½in	0-4-0T	OC	9 x 16	2ft 9in		276
1649	7.6.1865	4ft 8½in	0-6-0	IC	13 x 18	3ft 6in		190, 220
1800	21.7.1866	4ft 8½in	0-6-0ST	IC	14 x 20	3ft 6in		58
1801	9.7.1866	4ft 8½in	0-4-0T	OC	9 x 16	2ft 9in		276
1913	6.2.1869	4ft 8½in	0-6-0	IC	17 x 24	4ft 7in		361
1919	21.12.1869	4ft 8½in	0-6-0ST	IC	14 x 20	3ft 3in		190, 218, 220, 225
1973	29.11.1870	4ft 8½in	0-6-0	IC	17 x 24	5ft 0in		177
2013	9.2.1872	4ft 8½in	0-6-0ST	IC	14 x 20	3ft 7in		210, 223
2014	9.2.1872	4ft 8½in	0-6-0ST	IC	14 x 20	3ft 7in		210, 224, 225
2017	20.4.1872	4ft 8½in	0-4-0ST	OC	13 x 18	3ft 6in		204
2056	11.6.1872	4ft 8½in	0-6-0	IC	17 x 24	5ft 0⅜in		176
2124	25.8.1873	4ft 8½in	0-4-0ST	OC	9 x 18	2ft 9in		204
2139	24.10.1873	4ft 8½in	0-6-0ST	IC	14¾ x 22	4ft 0in		298
2160	20.2.1874	4ft 8½in	0-6-0	IC	17 x 24	5ft 0⅜in		176
2238	22.5.1875	4ft 8½in	0-6-0ST	IC	14¾ x 22	4ft 0in		348
2239	17.6.1875	4ft 8½in	0-6-0ST	IC	14¾ x 22	4ft 0in		298
2240	23.6.1875	4ft 8½in	0-6-0ST	IC	14¾ x 22	4ft 0in		298
2241	24.6.1875	4ft 8½in	0-6-0ST	IC	14¾ x 22	4ft 0in		348
2244	24.7.1875	4ft 8½in	0-6-0ST	IC	14¾ x 22	4ft 0in		298
2260	31.7.1876	4ft 8½in	0-6-0	IC	17 x 24	4ft 7½in		218
2308	20.11.1876	4ft 8½in	0-4-0ST	OC	15 x 20	3ft 8in		218
2325	# 3.2.1899	4ft 8½in	0-4-0ST	OC	13 x 18	3ft 6½in		284, 285, 292
2326	16.1.1889	4ft 8½in	0-4-0ST	OC	13 x 18	3ft 6½in		278
2381	3.3.1880	4ft 8½in	0-6-0ST	OC	13 x 18	3ft 6in		204
2554	23.11.1885	4ft 8½in	0-6-0ST	OC	13 x 18	3ft 6in		204
2587	29.12.1884	4ft 8½in	0-6-0	IC	17 x 26	5ft 1in		176
2620	3.8.1887	4ft 8½in	0-6-0ST	IC	14 x 20	3ft 7in		320
2629	15.12.1887	4ft 8½in	0-6-0ST	IC	15 x 22	?		176
2637	24.4.1888	4ft 8½in	0-4-0ST	OC	13 x 18	3ft 6in		204
2654	11.10.1888	4ft 8½in	0-4-0ST	OC	12 x 19	3ft 4in		101
2655	29.11.1888	4ft 8½in	0-4-0ST	OC	12 x 19	3ft 4in		101
2668	30.5.1889	4ft 8½in	0-4-0ST	OC	13 x 18	3ft 6in		204
2724	7.7.1890	4ft 8½in	0-4-0ST	OC	12 x 19	3ft 4in		102, 367
2725	19.7.1890	4ft 8½in	0-4-0ST	OC	12 x 19	3ft 4in		102
2730	17.3.1891	4ft 8½in	0-6-0T	IC	17 x 24	4ft 0in		210, 224
2811	28.12.1893	4ft 8½in	0-4-0ST	OC	rebuild of BH 328			101
2821	1894	4ft 8½in	0-6-0ST	IC	rebuilt of P & J Rly No.9			57
2822	5.1895	4ft 8½in	0-6-0T	IC	17 x 24	4ft 0in		201, 224

2840	1.8.1896	4ft 8½in	0-4-0ST	OC	15 x 20	3ft 4in		316
2847	24.9.1896	4ft 8½in	0-6-0ST	OC	15 x 20	3ft 7in		320
2852	14.12.1896	4ft 8½in	0-4-0ST	OC	12 x 19	3ft 4½in		182
2853	21.8.1897	4ft 8½in	0-4-0VBCr	OC	13½ x 21½	3ft 0in	12T capy.	106
2854	5.2.1888	4ft 8½in	0-4-0VBCr	OC	13½ x 21½	3ft 0in	12T capy. 65,	106
2875	10.12.1896	4ft 8½in	0-4-0ST	OC	12 x 19	3ft 4in		292
2879	27.2.1900	4ft 8½in	0-6-0ST	IC	15 x 20	3ft 7in		320
2902	20.6.1898	4ft 8½in	0-6-0ST	IC	rebuild of RS 1516			57
2915	27.2.1899	4ft 8½in	0-6-0PT	IC	rebuild of K 2510			99
2987	25.7.1900	4ft 8½in	0-4-0ST	OC	rebuild of B 303			318
2993	4.12.1901	4ft 8½in	0-6-0ST	IC	15 x 20	3ft 7in		211, 225
3057	20.5.1904	4ft 8½in	0-4-0ST	OC	14 x 20	3ft 3in		298
3072	13.12.1901	4ft 8½in	0-6-0ST	OC	16 x 24	4ft 0in		312
3075	1901	4ft 8½in	0-4-0ST	OC	12 x 18	3ft 1in	126, 130, 316,	332
†	c1905	4ft 8½in	0-4-0ST	OC	?	?		372
3376	30.6.1909	4ft 8½in	0-4-0ST	OC	15 x 22	3ft 6in		316
3377	5.11.1909	4ft 8½in	0-6-2T	IC	18½ x 26	4ft 6in		217
3378	12.11.1909	4ft 8½in	0-6-2T	IC	18½ x 26	4ft 6in		217
3801	28.2.1921	4ft 8½in	0-6-2T	IC	18½ x 26	4ft 6in		218
4112	23.8.1935	4ft 8½in	0-6-0ST	OC	16 x 24	3ft 10in		238
4113	19.3.1937	4ft 8½in	0-6-0ST	OC	16 x 24	3ft 10in		379

\# carried works plate dated 1894.
† built for own use from parts of other locomotives.

Unidentified locomotives

?	?	4ft 8½in	2-4-0	IC	?	?	[GIBSIDE]	56

ROBERT STEPHENSON & HAWTHORNS LTD — RSH
D   Robert Stephenson & Hawthorns Ltd, Harrowgate Hill, Darlington, Co. Durham.   **RSHD**
N   Robert Stephenson & Hawthorns Ltd, Forth Banks, Newcastle upon Tyne.   **RSHN**

This company was formed on 27th September 1937 following the purchase by Robert Stephenson & Co Ltd of the locomotive-building business of R. & W. Hawthorn, Leslie & Co Ltd. The 4155 locomotives built by the former were added to the 2983 locomotives built by the latter, making the first RSH locomotive Works No. 6939. The Darlington works continued to concentrate on main line and export orders, while the Newcastle works handled industrial work. In general the firm concentrated on a standard range of products based on Hawthorn Leslie designs.

In 1938 the company took over the goodwill of Kitson & Co Ltd of Leeds and, through them, the goodwill of Manning, Wardle & Co Ltd, also of Leeds. In 1944 it became a subsidiary of The Vulcan Foundry Ltd of Newton-le-Willows, Lancashire, itself a subsidiary of The English Electric Co Ltd. In later years Newcastle concentrated on steam locomotive work, while the Darlington works handled mainly diesel and battery electric locomotives sub-contracted from English Electric. The Forth Banks Works was officially closed on 30th March 1960, the last locomotive leaving the works on 2nd July 1960. The buildings were sold for other uses, with some surviving to pass into the hands of the Robert Stephenson Trust, which now opens them to the public. In 1962 the parent company took over full control of the Darlington Works, which closed in March 1964, bringing to an end industrial locomotive building in North East England.

The actual company was not finally wound up until March 2005, Its title, together with the titles of the companies from which it was formed, has now passed into private hands, as have other famous names in industrial locomotive building.

In the following list those locomotives assembled at Darlington are denoted with a D, whilst the remainder were built at the Newcastle Works. In the main text these are denoted by RSHD and RSHN.

	6939	9.12.1937	4ft 8½in	0-4-0ST	OC	12 x 20	3ft 1in		367
	6940	27.1.1938	4ft 8½in	0-4-0ST	OC	12 x 20	3ft 1in		257, 258, 259
	6943	22.4.1938	4ft 8½in	0-6-0ST	OC	16 x 24	3ft 8in		197
	6945	14.10.1938	4ft 8½in	0-4-0ST	OC	14 x 22	3ft 6in		420
	6963	29.8.1939	4ft 8½in	0-4-0ST	OC	12 x 20	3ft 1in		367
	7006	28.6.1940	4ft 8½in	0-4-0CT	OC	12 x 15	2ft 10in	4T capy.	142
	7007	10.7.1940	4ft 8½in	0-4-0CT	OC	12 x 15	2ft 10in	4T capy.	142
D	7011	19.9.1940	4ft 8½in	0-4-0ST	OC	16 x 24	3ft 8in		103
D	7013	1.2.1941	4ft 8½in	0-4-0ST	OC	14 x 22	3ft 6in		68, 118, 262
D	7016	24.12.1940	4ft 8½in	0-4-0ST	OC	16 x 24	3ft 8in		103, 348
D	7022	23.3.1941	4ft 8½in	0-4-0ST	OC	16 x 24	3ft 8in		103
	7027	6.3.1941	4ft 8½in	0-6-0PT	IC	18 x 26	4ft 2⅝in		99

	7028	24.3.1941	4ft 8½in	0-6-0PT	IC	18 x 26	4ft 2⅝in		99
D	7029	20.6.1941	4ft 8½in	0-6-0PT	IC	18 x 26	4ft 2⅝in		100
D	7036	17.7.1941	4ft 8½in	0-4-0ST	OC	16 x 24	3ft 8in		349
	7043	30.1.1942	4ft 8½in	0-4-0ST	OC	12 x 20	3ft 1in		327, 328
	7046	27.10.1941	4ft 8½in	0-4-0ST	OC	14 x 22	3ft 6in		245
	7063	29.6.1942	4ft 8½in	0-4-0ST	OC	14 x 22	3ft 6in		76
	7066	30.9.1942	4ft 8½in	0-4-0ST	OC	12 x 20	3ft 1in		75
	7068	4.2.1943	4ft 8½in	0-4-0ST	OC	16 x 24	3ft 8in		247
	7069	23.10.1942	4ft 8½in	0-4-0CT	OC	12 x 15	2ft 10in	4T capy.	142
	7070	30.10.1942	4ft 8½in	0-4-0CT	OC	12 x 15	2ft 10in	4T capy.	142
	7073	11.2.1943	4ft 8½in	0-4-0ST	OC	16 x 24	3ft 8in		120
	7117	1.3.1944	4ft 8½in	0-4-0ST	OC	16 x 24	3ft 8in		247
	7126	24.12.1943	4ft 8½in	0-4-0CT	OC	12 x 15	2ft 10in	4T capy.	185
	7138	19.5.1944	4ft 8½in	0-6-0ST	IC	18 x 26	4ft 3in		312
	7159	13.2.1945	4ft 8½in	0-4-0ST	OC	12 x 20	3ft 1in		327, 328
	7160	23.8.1945	4ft 8½in	0-4-0ST	OC	16 x 24	3ft 8in		120
	7178	20.12.1944	4ft 8½in	0-6-0ST	IC	18 x 26	4ft 3in		200
	7212	25.5.1945	4ft 8½in	0-6-0ST	OC	16 x 24	3ft 8in		312
	7297	5.12.1945	4ft 8½in	0-4-0ST	OC	14 x 22	3ft 6in		35
	7304	28.6.1946	4ft 8½in	0-6-0T	IC	18 x 26	4ft 6in		420
	7305	22.7.1946	4ft 8½in	0-6-0T	IC	18 x 26	4ft 6in		200
	7307	5.7.1946	4ft 8½in	0-4-0ST	OC	14 x 22	3ft 6in		420
	7308	7.8.1946	4ft 8½in	0-4-0ST	OC	12 x 20	3ft 1in		125, 183
	7340	28.2.1947	4ft 8½in	0-4-0ST	OC	16 x 24	3ft 8in		349
	7342	21.3.1947	4ft 8½in	0-4-0ST	OC	16 x 24	3ft 8in		349
	7345	21.5.1947	4ft 8½in	0-4-0ST	OC	16 x 24	3ft 8in		349
	7346	6.6.1947	4ft 8½in	0-4-0ST	OC	16 x 24	3ft 8in		349
	7347	16.6.1947	4ft 8½in	0-4-0ST	OC	16 x 24	3ft 8in		349
	7409	27.7.1948	4ft 8½in	0-4-0ST	OC	12 x 20	3ft 1in		262
	7660	24.2.1950	4ft 8½in	0-4-0ST	OC	16 x 24	3ft 8in		120
	7674	16.2.1952	4ft 8½in	0-4-0ST	OC	16 x 24	3ft 8in		323
	7675	8.1.1952	4ft 8½in	0-4-0ST	OC	16 x 24	3ft 8in		372
	7679	21.6.1951	4ft 8½in	0-4-0ST	OC	14 x 22	3ft 6in		76
	7743	24.4.1953	4ft 8½in	0-4-0ST	OC	14 x 22	3ft 6in		78
	7744	29.4.1953	4ft 8½in	0-4-0ST	OC	14 x 22	3ft 6in		78
	7796	2.12.1954	4ft 8½in	0-4-0ST	OC	14 x 22	3ft 6in		78
	7819	4.10.1954	4ft 8½in	0-4-0ST	OC	16 x 24	3ft 8in		367
	7869	12.12.1956	4ft 8½in	0-4-0DM 107hp	Gardner 6LW	2ft 7½	14T	86, 262, 372
	7899	24.6.1958	4ft 8½in	0-4-0DM 107hp	Gardner 6LW	2ft 7½	16T	262

R. & W. HAWTHORN, Forth Banks, Newcastle upon Tyne.   RWH

Robert Hawthorn began a steam engine business on the western side of Forth Banks in January 1817. He was joined by his brother William in 1820, and quickly expanded eastwards. The firm built the first steam crane in the same year. The firm's first locomotive was No.123, built in 1831 for the Stockton & Darlington Railway, but the firm continued to build a wide variety of steam engines for many years, and increasingly developed its marine work. Robert Hawthorn died in 1867 and his brother retired in 1870, in which year the marine branch moved to a new works at St. Peter's, about 1½ miles downstream on the River Tyne. In 1885 the firm amalgamated with the ship-building firm of Andrew Leslie & Co Ltd of Hebburn to form R. & W. Hawthorn, Leslie & Co Ltd (which see).

171	c1833	4ft 8½in	?	?	?	?		265
174	1834	4ft 8½in	0-4-0?	OC?	?	?		391
192	1835	4ft 8½in	0-4-0	OC?	?	?		186
218	11.1836	4ft 8½in	0-4-0	OC	14 x 15	?		401
297	1839	4ft 8½in	0-6-0	OC	14 x 18	4ft 0in		434
308	1840	4ft 8½in	0-6-0	OC	14 x 18	4ft 0in		217, 434
476	1846	4ft 8½in	0-6-0	IC	15 x 24 #	4ft 3in		234
479	1846	4ft 8½in	0-6-0	IC	15 x 24 #	4ft 3in		56
1019	9.10.1857	4ft 8½in	0-4-0ST	OC	13 x 18	4ft 0in		249
1422	1867	4ft 8½in	0-6-0ST	IC	15 x 22	3ft 9in	58, 190, 218, 220
1430	1868	4ft 8½in	0-6-0ST	IC	14 x 18	3ft 0in	190, 218, 220, 225
1478	1870	4ft 8½in	0-6-0ST	IC	17 x 24	4ft 6in	190, 218, 220, 225
1554	1872	4ft 8½in	0-6-0ST	OC	16 x 24	4ft 6in		415
1564	14.7.1873	4ft 8½in	0-6-0	IC	17 x 24	5ft 0in		177
1622	12.6.1874	4ft 8½in	0-6-0ST	OC	16 x 24	4ft 6in (?)	249, 415
1635	1.12.1874	4ft 8½in	0-4-0ST	OC	13 x 19	3ft 6in		249

1657	10.1875	4ft 8½in	0-6-0ST	IC	15 x 22	4ft 0in		298
1662	12.1875	4ft 8½in	0-6-0ST	IC	15 x 22	4ft 0in		298
1666	2.1876	4ft 8½in	0-6-0ST	IC	15 x 22	4ft 0in		298
1669	2.1876	4ft 8½in	0-6-0ST	IC	15 x 22	4ft 0in		298
1726	1877	4ft 8½in	0-4-0ST	OC	11 x 18	3ft 0in		371
1789	1879	4ft 8½in	0-4-0ST	OC	10 x 15	2ft 9in		183
1817	26.4.1880	4ft 8½in	0-4-0ST	OC	10 x 15	2ft 9in		371
1821	1.5.1880	4ft 8½in	0-4-0ST	OC	15 x 20	3ft 6in		279, 382
1847	1881	4ft 8½in	0-4-0ST	OC	12 x 18	3ft 0½in		339
1877	15.12.1882	4ft 8½in	0-4-0CT/WT	OC	10 x 15	2ft 5in		277
1882	20.3.1882	4ft 8½in	0-6-0ST	OC	13 x 18	3ft 0in		379, 383
1969	9.2.1884	4ft 8½in	0-6-0ST	IC	16 x 24	4ft 6in		190, 217, 320
1986	19.3.1886	4ft 8½in	0-6-0ST	OC	15 x 20	3ft 6in		57
2026	24.6.1885	4ft 8½in	0-4-0ST	OC	12 x 18	3ft 0in		277
2029	28.7.1886	4ft 8½in	0-4-0ST	OC	10 x 15	2ft 9in		306

\# Another version gives 14in x 21in

SENTINEL (SHREWSBURY) LTD, Battlefield Works, Shrewsbury, Shropshire. S
formerly **Sentinel Wagon Works (1936) Ltd**; originally **Sentinel Wagon Works Ltd**.
The company was taken over by Rolls Royce Ltd in 1956, but the new owners continued to market its locomotives under the Sentinel name.

Note : CH indicates built at the firm's Chester Works, Chester (the letters were on the works plate).

5988CH	1925	4ft 8½in	4wVBT	HCG	6¾ x 9		[rebuild of LE 230]	433
6076	4.1925	4ft 8½in	4wVBT	VCG	6¾ x 9		?	229, 385
6218CH	5.1925	4ft 8½in	4wVBT	VCG	6¾ x 9		[rebuild of HC 683]	433
6310CH	1926	4ft 8½in	4wVBT	VCG	6¾ x 9		[rebuild of AB 282]	427
6770CH	1926	2ft 0in	4wVBT	VCG	6¾ x 9		?	6T	428
6902	1930	2ft 0in	4wVBT	VCG	6¾ x 9	1ft 8in	6T	150
6936	1927	4ft 8½in	4wVBT	VCG	6¾ x 9	2ft 6in	16T	298
7062	9.1927	4ft 8½in	4wVBT	VCG	6¾ x 9	2ft 6in	22T	353
7669	1928	4ft 8½in	4wVBT	VCG	6¾ x 9	2ft 6in	19T	367
7852	3.1929	4ft 8½in	4wVBT	VCG	6¾ x 9	2ft 6in	21T	155
9558	9.1953	4ft 8½in	4wVBT	VCG	6¾ x 9	2ft 6in	25T	34, 35, 76, 348
9561	c11.1953	4ft 8½in	4wVBT	VCG	6¾ x 9	2ft 6in	25T	306, 348, 351
9563	31.5.1954	4ft 8½in	4wVBT	VCG	6¾ x 9	2ft 6in	25T	351
9575	10.1954	4ft 8½in	4wVBT	VCG	6¾ x 9	3ft 2in	34T	348
9597	9.11.1955	4ft 8½in	4wVBT	VCG	6¾ x 9	2ft 6in	25T	76
9618	24.1.1957	4ft 8½in	4wVBT	VCG	6¾ x 9	3ft 2in	34T	348

Works numbers 6770CH and 6902 were 80hp; 6076/7062/7669/7852/9558/9561/9563 and 9597 were 100hp and 9575 and 9618 were 200hp.

Diesel locomotives, built to a Rolls Royce design. Until 1964 they were marketed as Sentinel, but thereafter they were marketed under Rolls Royce, although continuing the same sequence of works numbers.

10003	7.5.1959	4ft 8½in	4wDH	207hp	R-R C6SFL	3ft 2in	34T	348
10031	15.3.1960	4ft 8½in	4wDH	207hp	R-R C6SFL	3ft 2in	34T	325
10050	29.12.1960	4ft 8½in	0-6-0DH	311hp	R-R C8SFL	3ft 6in	48T	66
10066	18.5.1961	4ft 8½in	0-6-0DH	311hp	R-R C8SFL	3ft 6in	40T	66, 69
10067	4.7.1961	4ft 8½in	0-6-0DH	311hp	R-R C8SFL	3ft 6in	40T	66
10069	22.8.1961	4ft 8½in	0-6-0DH	311hp	R-R C8SFL	3ft 6in	40T	66, 69
10071	25.5.1961	4ft 8½in	0-6-0DH	311hp	R-R C8SFL	3ft 6in	46T	66
10077	12.7.1961	4ft 8½in	4wDH	200hp	R-R C6SFL	3ft 2in	34T	325, 388
10079	6.9.1961	4ft 8½in	0-6-0DH	311hp	R-R C8SFL	3ft 6in	46T	66
10080	19.10.1961	4ft 8½in	0-6-0DH	311hp	R-R C8SFL	3ft 6in	46T	66
10081	9.11.1961	4ft 8½in	0-6-0DH	311hp	R-R C8SFL	3ft 6in	46T	66
10082	9.11.1961	4ft 8½in	0-6-0DH	311hp	R-R C8SFL	3ft 6in	46T	66
10084	14.9.1961	4ft 8½in	0-4-0DH	311hp	R-R C8SFL	3ft 6in	40T	66, 69
10088	1.12.1961	4ft 8½in	0-6-0DH	311hp	R-R C8SFL	3ft 6in	46T	66
10097	8.2.1962	4ft 8½in	4wDH	230hp	R-R C6SFL	3ft 2in	34T	247
10153	4.9.1963	4ft 8½in	0-6-0DH	325hp	R-R C8SFL	3ft 6in	50T	66

ROLLS ROYCE LTD, Sentinel Works, Shrewsbury, Shropshire.   RR

10197	29.10.1964	4ft 8½in	4wDH	230hp	R-R C6SFL	3ft 2in	34T	213
10232	27.5.1965	4ft 8½in	4wDH	230hp	R-R C6SFL	3ft 2in	34T	213
10256	27.6.1966	4ft 8½in	0-6-0DH	325hp	R-R C8SFL	3ft 6in	48T	349
10278	14.3.1968	4ft 8½in	0-6-0DH	375hp	R-R C8TFL	3ft 6in	48T	65
10285	4.7.1969	4ft 8½in	0-6-0DH	375hp	R-R C8TFL	3ft 6in	48T	65
10286	16.7.1969	4ft 8½in	0-6-0DH	375hp	R-R C8TFL	3ft 6in	48T	65
10287	24.7.1969	4ft 8½in	0-6-0DH	375hp	R-R C8TFL	3ft 6in	48T	65
10288	31.8.1969	4ft 8½in	0-6-0DH	375hp	R-R C8TFL	3ft 6in	48T	65
10289	7.9.1970	4ft 8½in	0-6-0DH	375hp	R-R C8TFL	3ft 6in	48T	66
10290	23.10.1970	4ft 8½in	0-6-0DH	375hp	R-R C8TFL	3ft 6in	48T	66

MARQUIS OF LONDONDERRY, Londonderry Engine Works, Seaham Harbour, Co. Durham.   Seaham

By the early 1860s the Marchioness of Londonderry had developed further the industrial empire developed by her husband, which included numerous collieries, the Londonderry Railway which served them, her dock at Seaham Harbour and her increasing fleet of colliers. Her Chief Agent, John Daglish, thus decided to develop a central engineering works at Seaham to service all these enterprises, which was almost certainly brought into use during 1865-66. All the locomotives below were built for the Londonderry Railway, the first in the primitive facilities which preceded the establishment of the Engine Works. When the Railway was taken over by the North Eastern Railway on 6th October 1900 the Engine Works was excluded. Re-named the Seaham Harbour Engine Works, it built seven steam (road) wagons for the NER. It was closed in the autumn of 1925.

# [2]	1860	4ft 8½in	0-4-2	IC	?	?	234
[6]	6.1.1883	4ft 8½in	0-6-0ST	IC	?	?	234
[5]	1885	4ft 8½in	0-6-0	IC	17½ x 26	4ft 9½in	234
[2]	1889	4ft 8½in	2-4-0T	IC	15 x 24	4ft 11in	234
[20]	1892	4ft 8½in	0-6-0	IC	17½ x 26	4ft 9in	234
[21]	1895	4ft 8½in	0-4-4T	IC	17 x 24	5ft 4½in	234
@	1902	4ft 8½in	0-6-0T	IC	?	?	234, 348

\# previously regarded as a rebuild of Haigh Foundry 46/1841, but actually a new locomotive, with new frames, boiler motion and tender.

@ this locomotive was under construction at the date of the NER take-over; when completed, it was stored in the Londonderry Wagon Shops, owned by Londonderry Collieries Ltd, until 1906, when it was sold to The Seaham Harbour Dock Co.

SIEMENS SCHUCKERT AG, Berlin, Germany.   Siemens

450	1908	2ft 2in	0-4-4-0WE	66hp	?	?	114
451	1908	4ft 8½in	4wWE	300hp	3ft 3⅜in	26T	179
454	1909	2ft 2in	0-4-4-0WE	66hp	?	?	114
455	1908	4ft 8½in	4wWE	100hp	2ft 7¼in	14½T	179
456	1909	4ft 8½in	4w-4wWE	200hp	2ft 9⅜in	33T	179
457	1909	4ft 8½in	4w-4wWE	200hp	2ft 9⅜in	33T	179
458	1909	4ft 8½in	0-4-0+0-4-0WE	100hp	2ft 9⅜in	17T	179
459	1909	4ft 8½in	0-4-0+0-4-0WE	100hp	2ft 9⅜in	17T	179
460	1910	2ft 2in	0-4-4-0WE	120hp	[Hanomag 5968]		114
862	1913	4ft 8½in	4wWE	100hp	2ft 7¼in	14½T	179

JOSEPH SMITH, Trimdon Ironworks, Sunderland, Co. Durham.   SmithJ

This man was also the owner of Trimdon Grange Colliery in Co. Durham. It would seem very likely from the list of locomotives handed over to the York & Newcastle Railway in October 1846 by the coal owners that had hitherto been operating their trains over the Hartlepool Dock & Railway that the locomotive below was under construction when the take-over took place, and it is possible, although there is no supporting evidence, that the locomotive may well have been built at Smith's Trimdon Ironworks in Sunderland.

-	1847	4ft 8½in	0-6-0	OC?	16 x 24	4ft 0in	173

SHARP, STEWART & CO LTD, Atlas Works, Manchester (till 1888) and Atlas Works, Glasgow (from 1888). SS

#	1857	4ft 8½in	0-6-0	IC	18 x 24	5ft 0in	361
1501	18.4.1864	4ft 8½in	2-2-2WT	IC	15 x 18	5ft 6¼in	177
1768	15.1.1867	4ft 8½in	0-6-0ST	IC	18 x 24	5ft 0in	361
2260	10.1872	4ft 8½in	0-6-0ST	IC	17 x 24	4ft 7in	99
2358	1873	4ft 8½in	0-6-0T	OC	16¼ x 20	3ft 6in	361
4051	12.1894	4ft 8½in	0-6-0ST	IC	17½ x 26	4ft 3in	58, 177, 180
4594	3.1900	4ft 8½in	0-6-0ST	IC	17½ x 26	4ft 3in.	57

\# for the possible identity of this locomotive see The South Hetton Coal Co Ltd in the main text.

THOMAS GREEN & SON LTD, Smithfield Ironworks, Leeds.   TG

| 24 | 1894 | 2ft 6in | 0-4-2T | ? | ? | 246 |

THOMAS HILL (ROTHERHAM) LTD, Vanguard Works, Kilnhurst, Yorkshire (WR).   TH

Locomotives with works numbers suffixed "C" were rebuilds using frames and other parts of Sentinel steam locomotives; the suffix "V" indicated a completely new locomotive.

104C	6.10.1960	4ft 8½in	4wDH	170hp	R-R C8NFL	3ft 2in	36T	349
111C	22.9.1961	4ft 8½in	4wDH	157hp	R-R C6NFL	2ft 6in	26T	84, 85
118C	25.6.1962	4ft 8½in	4wDH	157hp	R-R C6NFL	2ft 6in	26T	120
129V	30.7.1963	4ft 8½in	4wDH	179hp	R-R C6NFL	3ft 2in	26T	120
131V	# 29.11.1963	4ft 8½in	4wDH	179hp	R-R C6NFL	3ft 2in	26T	120
148V	6.5.1965	4ft 8½in	4wDH	175hp	R-R C6NFL	3ft 2in	25T	213
221V	16.6.1970	4ft 8½in	4wDH	275hp	R-R C6NFL	3ft 6in	40T	65
222V	24.7.1970	4ft 8½in	4wDH	275hp	R-R C6NFL	3ft 6in	40T	65
223V	31.8.1970	4ft 8½in	4wDH	275hp	R-R C6NFL	3ft 6in	40T	65
224V	10.9.1970	4ft 8½in	4wDH	275hp	R-R C6NFL	3ft 6in	40T	66
287V	15.1.1980	4ft 8½in	4wDH	270hp	R-R C8NFL	3ft 6in	42T	390

\# delivery date; previously used as a demonstration locomotive by the company.

CHARLES TODD, Sun Foundry, Leeds.   Todd

| - | 1.1847 | 4ft 8½in | 0-6-0 | IC | 15 x 24 | 4ft 9in | 298 |

THOMAS RICHARDSON & SONS, Castle Eden and Hartlepool, Co. Durham.   TR

In 1832 Thomas Richardson (c1795-1850) took a 21-year lease on land at Castle Eden in south Durham on which to set up an iron-working and engineering business. In 1843 he won the contract to operate the locomotives and passenger services of the Hartlepool Dock & Railway, which ran past his works. In 1847 he took over the lease of the Hartlepool Ironworks at Middleton, Hartlepool, although the Castle Eden works continued in operation until April 1853. It had previously been believed that the firm did not begin building locomotives until after the take-over of the Hartlepool works, but the list of locomotives for which the York & Newcastle Railway took over from the coal owners operating over the HD&R in October 1846 would suggest otherwise. Locomotive building was developed under the original owner's son, also Thomas Richardson (1821-1890), its works numbers included within the general list of products. In 1852 the firm built the first-ever narrow gauge steam locomotives, for the Upleatham ironstone mines in North Yorkshire. Locomotive building seems to have ended in 1858, when the firm concentrated on marine work. The firm failed in April 1875, and the engineering side of the business was re-constituted as Thomas Richardson & Sons Ltd. This firm formed one of the constituents of Richardsons, Westgarth & Co Ltd when it was formed in 1900, which in 1977 became part of the nationalised British Shipbuilders. The works closed in 1981.

182	c1851	4ft 8½in	0-6-0T+t #	OC	?	?	153, 234
213	1852	4ft 8½in	0-6-0	IC	?	?	217
236	1853	4ft 8½in	0-6-0	IC	?	?	217
251	10.1854	4ft 8½in	0-6-0	IC	17 x 24	?	217
252	1854	4ft 8½in	0-6-0	IC	15 x 22	4ft 6in	58
254	6.1855	4ft 8½in	0-6-0	IC	15 x 22	?	234, 237, 348
255	1855	4ft 8½in	0-6-0	IC	15 x 22	?	153
256	1855	4ft 8½in	?	?	15 x 22	?	153
265	1854	4ft 8½in	0-6-0	IC	16 x ?	?	361

\# this loco was attached to a tender.

See also the references to Richardson and the Hartlepool Ironworks in the entry in the main text for the Hartlepool Dock & Railway Company.

WEARDALE IRON & COAL CO LTD, Tow Law Ironworks, Tow Law, Co. Durham.   Tow Law

This firm is believed to have built three locomotives for its own use at its Tow Law works, all almost certainly in the early 1870s. The works itself was opened about April 1846 and was closed in 1887.

-	?	4ft 8½in	0-4-0ST	OC	?	? [12 STAR]	415
-	1873	4ft 8½in	0-6-0ST	OC	?	? [15 WOLSINGHAM]	415
-	?	4ft 8½in	0-6-0ST	OC	?	? [20]	415

VULCAN FOUNDRY LTD, Newton-le-Willows, Lancashire.   VF

320	7.1848	4ft 8½in	0-4-2	OC	16 x 18	4ft 7in	234
+	1941	4ft 8½in	0-4-0DM	153hp	Gardner 6L3	? 24½T	242
+	1941	4ft 8½in	0-4-0DM	153hp	Gardner 6L3	? 26½T	351
5288	6.1945	4ft 8½in	0-6-0ST	IC	18 x 26	4ft 3in	58

5298	7.1945	4ft 8½in	0-6-0ST	IC	18 x 26	4ft 3in		58
5299	7.1945	4ft 8½in	0-6-0ST	IC	18 x 26	4ft 3in		218
5300	7.1945	4ft 8½in	0-6-0ST	IC	18 x 26	4ft 3in		218
5305	7.1945	4ft 8½in	0-6-0ST	IC	18 x 26	4ft 3in		152
5307	7.1945	4ft 8½in	0-6-0ST	IC	18 x 26	4ft 3in		58, 415
5308	9.1945	4ft 8½in	0-6-0ST	IC	18 x 26	4ft 3in		361
5309	9.1945	4ft 8½in	0-6-0ST	IC	18 x 26	4ft 3in		361
D227 #	1956	4ft 8½in	0-6-0DH	500hp	EE 12CSVT	3ft 4in	48T	156, 372

+ sub-contracted from EE, which allocated works numbers 1188 and 1195, respectively, and built for DC, which allocated works numbers 2157 and 2164, respectively.
# sub-contracted from EE, which allocated works number 2346.

W. G. BAGNALL LTD, Castle Engine Works, Stafford, Staffordshire.   WB

1002	7.1888	2ft 6in	0-4-0IST	OC	8½ x 12	?		192
1381	11.1891	3ft 0in	0-4-0IST	OC	5½ x 10	1ft 8in		277
1917	2..6.1910	2ft 6in	0-4-0T	OC	8½ x 12	2ft 0½in		192
2058	5.12.1917	3ft 0in	0-4-0ST	OC	7 x 12	1ft 9½in		112
2084	12.3.1919	3ft 0in	0-4-0ST	OC	7 x 12	1ft 9½in		112
2169	23.2.1922	4ft 8½in	0-6-0ST	OC	13 x 18	2ft 9½in		84
2664	20.6.1942	4ft 8½in	0-4-0ST	OC	12 x 18	3ft 0½in		323
2779	26.6.1945	4ft 8½in	0-6-0ST	IC	18 x 26	4ft 3in		382
2898	28.9.1948	4ft 8½in	0-4-0F	OC	18½ x 18	3ft 0½in		86
3160	28.10.1959	4ft 8½in	0-6-0DM	304hp	Dorman 6QAT	3ft 4in	38T	348

WINGROVE & ROGERS LTD, Kirkby, Liverpool.   WR

859	11.4.1934	narrow	4wWE	?	Trolley rail	?	?	64
4146	22.6.1949	1ft 6in	0-4-0BE	4hp	Type W217	?	1½T	32
4147	22.6.1949	1ft 6in	0-4-0BE	4hp	Type W217	?	1½T	32
4148	22.6.1949	1ft 6in	0-4-0BE	4hp	Type W217	?	1½T	32
4149	28.6.1949	1ft 6in	0-4-0BE	4hp	Type W217	?	1½T	161
4184	23.11.1949	1ft 10in	0-4-0BE	5hp	Type W417	?	2½T	406
5115	13.7.1953	2ft 0in	4wBE	12hp	Type W227	?	3T	368
5316	26.1.1955	2ft 0in	4wBE	12hp	Type W227	?	3T	368
5655	20.11.1956	1ft 0in	0-4-0BE	4hp	Type W217	?	1½T	161
6133	22.11.1959	2ft 6in	0-4-0BE	4hp	Type W217	?	1½T	39
6593	9.2.1962	2ft 6in	0-4-0BE	4hp	Type W217	?	1½T	39
6595	25.4.1962	2ft 6in	0-4-0BE	4hp	Type W217	?	1½T	39
D6686	30.4.1964	2ft 0in	4wBE	?	Type W527	?	5T	271
6704	320.8.1962	2ft 6in	0-4-0BE	4hp	Type W217	?	1½T	39
C6710	22.2.1963	2ft 6in	0-4-0BE	4hp	Type W217	?	1½T	39
D6754	28.2.1964	2ft 0in	0-4-0BE	4hp	Type W217	?	1½T	161
P7664	17.9.1975	1ft 8in	0-4-0BE	5hp	Type WR5L	?	1½T	161
Q7731	2.12.1976	1ft 7½in	0-4-0BE	5hp	Type WR5	?	1½T	161

WELLMAN SMITH OWEN ENGINEERING CORPORATION LTD, Darlaston, Staffordshire.   WSO

529	1924	4ft 8½in	4wWE	160hp		2ft 10½in	13T	67, 107
1252	1928	4ft 8½in	4wWE	80hp		2ft 9in	15T	113

THE OWNERS OF WYLAM COLLIERY, Wylam Colliery, Wylam, Northumberland.   Wylam

-	1827-1832	5ft 0in	4w	VCG	9 x 36	3ft 3in	193

YORKSHIRE ENGINE CO LTD, Meadow Hall Works, Sheffield, Yorkshire (WR).   YE

480	1891	4ft 8½in	0-4-0ST	OC	14 x 20	3ft 3in		103
2668	15.5.1958	4ft 8½in	0-6-0DE	400hp	2 x R-R C6SFL	?	48T	348
2766	18.1.1960	4ft 8½in	0-6-0DE	440hp	2 x R-R C6SFL	?	48T	69
2779	4.3.1960	4ft 8½in	0-4-0DE	220hp	R-R C6SFL	?	30T	367
2793	20.2.1961	4ft 8½in	0-6-0DE	440hp	2 x R-R C6SFL	?	48T	69

Other abbreviations used

M. COULSON & CO LTD, Spennymoor, Co. Durham.   Coulson
This company operated a general engineering business, including the repair of locomotives.

CAERPHILLY WORKS, Caerphilly, Glamorgan, Great Western Railway.   Caerphilly
SWINDON WORKS, Swindon, Wiltshire, Great Western Railway.   Swindon